THE SCIENCE OF DEFOAMING

Theory, Experiment and Applications

SURFACTANT SCIENCE SERIES

FOUNDING EDITOR
MARTIN J. SCHICK
1918–1998

SERIES EDITOR
ARTHUR T. HUBBARD
Santa Barbara Science Project
Santa Barbara, California

1. Nonionic Surfactants, *edited by Martin J. Schick* (see also Volumes 19, 23, and 60)
2. Solvent Properties of Surfactant Solutions, *edited by Kozo Shinoda* (see Volume 55)
3. Surfactant Biodegradation, *R. D. Swisher* (see Volume 18)
4. Cationic Surfactants, *edited by Eric Jungermann* (see also Volumes 34, 37, and 53)
5. Detergency: Theory and Test Methods (in three parts), *edited by W. G. Cutler and R. C. Davis* (see also Volume 20)
6. Emulsions and Emulsion Technology (in three parts), *edited by Kenneth J. Lissant*
7. Anionic Surfactants (in two parts), *edited by Warner M. Linfield* (see Volume 56)
8. Anionic Surfactants: Chemical Analysis, *edited by John Cross*
9. Stabilization of Colloidal Dispersions by Polymer Adsorption, *Tatsuo Sato and Richard Ruch*
10. Anionic Surfactants: Biochemistry, Toxicology, Dermatology, *edited by Christian Gloxhuber* (see Volume 43)
11. Anionic Surfactants: Physical Chemistry of Surfactant Action, *edited by E. H. Lucassen-Reynders*
12. Amphoteric Surfactants, *edited by B. R. Bluestein and Clifford L. Hilton* (see Volume 59)
13. Demulsification: Industrial Applications, *Kenneth J. Lissant*
14. Surfactants in Textile Processing, *Arved Datyner*
15. Electrical Phenomena at Interfaces: Fundamentals, Measurements, and Applications, *edited by Ayao Kitahara and Akira Watanabe*
16. Surfactants in Cosmetics, *edited by Martin M. Rieger* (see Volume 68)
17. Interfacial Phenomena: Equilibrium and Dynamic Effects, *Clarence A. Miller and P. Neogi*
18. Surfactant Biodegradation: Second Edition, Revised and Expanded, *R. D. Swisher*
19. Nonionic Surfactants: Chemical Analysis, *edited by John Cross*
20. Detergency: Theory and Technology, *edited by W. Gale Cutler and Erik Kissa*
21. Interfacial Phenomena in Apolar Media, *edited by Hans-Friedrich Eicke and Geoffrey D. Parfitt*
22. Surfactant Solutions: New Methods of Investigation, *edited by Raoul Zana*
23. Nonionic Surfactants: Physical Chemistry, *edited by Martin J. Schick*
24. Microemulsion Systems, *edited by Henri L. Rosano and Marc Clausse*
25. Biosurfactants and Biotechnology, *edited by Naim Kosaric, W. L. Cairns, and Neil C. C. Gray*
26. Surfactants in Emerging Technologies, *edited by Milton J. Rosen*

27. Reagents in Mineral Technology, *edited by P. Somasundaran and Brij M. Moudgil*
28. Surfactants in Chemical/Process Engineering, *edited by Darsh T. Wasan, Martin E. Ginn, and Dinesh O. Shah*
29. Thin Liquid Films, *edited by I. B. Ivanov*
30. Microemulsions and Related Systems: Formulation, Solvency, and Physical Properties, *edited by Maurice Bourrel and Robert S. Schechter*
31. Crystallization and Polymorphism of Fats and Fatty Acids, *edited by Nissim Garti and Kiyotaka Sato*
32. Interfacial Phenomena in Coal Technology, *edited by Gregory D. Botsaris and Yuli M. Glazman*
33. Surfactant-Based Separation Processes, *edited by John F. Scamehorn and Jeffrey H. Harwell*
34. Cationic Surfactants: Organic Chemistry, *edited by James M. Richmond*
35. Alkylene Oxides and Their Polymers, *F. E. Bailey, Jr., and Joseph V. Koleske*
36. Interfacial Phenomena in Petroleum Recovery, *edited by Norman R. Morrow*
37. Cationic Surfactants: Physical Chemistry, *edited by Donn N. Rubingh and Paul M. Holland*
38. Kinetics and Catalysis in Microheterogeneous Systems, *edited by M. Grätzel and K. Kalyanasundaram*
39. Interfacial Phenomena in Biological Systems, *edited by Max Bender*
40. Analysis of Surfactants, *Thomas M. Schmitt* (see Volume 96)
41. Light Scattering by Liquid Surfaces and Complementary Techniques, *edited by Dominique Langevin*
42. Polymeric Surfactants, *Irja Piirma*
43. Anionic Surfactants: Biochemistry, Toxicology, Dermatology, Second Edition, Revised and Expanded, *edited by Christian Gloxhuber and Klaus Künstler*
44. Organized Solutions: Surfactants in Science and Technology, *edited by Stig E. Friberg and Björn Lindman*
45. Defoaming: Theory and Industrial Applications, *edited by P. R. Garrett*
46. Mixed Surfactant Systems, *edited by Keizo Ogino and Masahiko Abe*
47. Coagulation and Flocculation: Theory and Applications, *edited by Bohuslav Dobiás*
48. Biosurfactants: Production Properties Applications, *edited by Naim Kosaric*
49. Wettability, *edited by John C. Berg*
50. Fluorinated Surfactants: Synthesis Properties Applications, *Erik Kissa*
51. Surface and Colloid Chemistry in Advanced Ceramics Processing, *edited by Robert J. Pugh and Lennart Bergström*
52. Technological Applications of Dispersions, *edited by Robert B. McKay*
53. Cationic Surfactants: Analytical and Biological Evaluation, *edited by John Cross and Edward J. Singer*
54. Surfactants in Agrochemicals, *Tharwat F. Tadros*
55. Solubilization in Surfactant Aggregates, *edited by Sherril D. Christian and John F. Scamehorn*
56. Anionic Surfactants: Organic Chemistry, *edited by Helmut W. Stache*
57. Foams: Theory, Measurements, and Applications, *edited by Robert K. Prud'homme and Saad A. Khan*

58. The Preparation of Dispersions in Liquids, *H. N. Stein*
59. Amphoteric Surfactants: Second Edition, *edited by Eric G. Lomax*
60. Nonionic Surfactants: Polyoxyalkylene Block Copolymers, *edited by Vaughn M. Nace*
61. Emulsions and Emulsion Stability, *edited by Johan Sjöblom*
62. Vesicles, *edited by Morton Rosoff*
63. Applied Surface Thermodynamics, *edited by A. W. Neumann and Jan K. Spelt*
64. Surfactants in Solution, *edited by Arun K. Chattopadhyay and K. L. Mittal*
65. Detergents in the Environment, *edited by Milan Johann Schwuger*
66. Industrial Applications of Microemulsions, *edited by Conxita Solans and Hironobu Kunieda*
67. Liquid Detergents, *edited by Kuo-Yann Lai*
68. Surfactants in Cosmetics: Second Edition, Revised and Expanded, *edited by Martin M. Rieger and Linda D. Rhein*
69. Enzymes in Detergency, *edited by Jan H. van Ee, Onno Misset, and Erik J. Baas*
70. Structure-Performance Relationships in Surfactants, *edited by Kunio Esumi and Minoru Ueno*
71. Powdered Detergents, *edited by Michael S. Showell*
72. Nonionic Surfactants: Organic Chemistry, *edited by Nico M. van Os*
73. Anionic Surfactants: Analytical Chemistry, Second Edition, Revised and Expanded, *edited by John Cross*
74. Novel Surfactants: Preparation, Applications, and Biodegradability, *edited by Krister Holmberg*
75. Biopolymers at Interfaces, *edited by Martin Malmsten*
76. Electrical Phenomena at Interfaces: Fundamentals, Measurements, and Applications, Second Edition, Revised and Expanded, *edited by Hiroyuki Ohshima and Kunio Furusawa*
77. Polymer-Surfactant Systems, *edited by Jan C. T. Kwak*
78. Surfaces of Nanoparticles and Porous Materials, *edited by James A. Schwarz and Cristian I. Contescu*
79. Surface Chemistry and Electrochemistry of Membranes, *edited by Torben Smith Sørensen*
80. Interfacial Phenomena in Chromatography, *edited by Emile Pefferkorn*
81. Solid–Liquid Dispersions, *Bohuslav Dobiás, Xueping Qiu, and Wolfgang von Rybinski*
82. Handbook of Detergents, editor in chief: Uri Zoller Part A: Properties, *edited by Guy Broze*
83. Modern Characterization Methods of Surfactant Systems, *edited by Bernard P. Binks*
84. Dispersions: Characterization, Testing, and Measurement, *Erik Kissa*
85. Interfacial Forces and Fields: Theory and Applications, *edited by Jyh-Ping Hsu*
86. *Silicone Surfactants, edited by Randal M. Hill*

87. Surface Characterization Methods: Principles, Techniques, and Applications, *edited by Andrew J. Milling*
88. Interfacial Dynamics, *edited by Nikola Kallay*
89. Computational Methods in Surface and Colloid Science, *edited by Malgorzata Borówko*
90. Adsorption on Silica Surfaces, *edited by Eugène Papirer*
91. Nonionic Surfactants: Alkyl Polyglucosides, *edited by Dieter Balzer and Harald Lüders*
92. Fine Particles: Synthesis, Characterization, and Mechanisms of Growth, *edited by Tadao Sugimoto*
93. Thermal Behavior of Dispersed Systems, *edited by Nissim Garti*
94. Surface Characteristics of Fibers and Textiles, *edited by Christopher M. Pastore and Paul Kiekens*
95. Liquid Interfaces in Chemical, Biological, and Pharmaceutical Applications, *edited by Alexander G. Volkov*
96. Analysis of Surfactants: Second Edition, Revised and Expanded, *Thomas M. Schmitt*
97. Fluorinated Surfactants and Repellents: Second Edition, Revised and Expanded, *Erik Kissa*
98. Detergency of Specialty Surfactants, *edited by Floyd E. Friedli*
99. Physical Chemistry of Polyelectrolytes, *edited by Tsetska Radeva*
100. Reactions and Synthesis in Surfactant Systems, *edited by John Texter*
101. Protein-Based Surfactants: Synthesis, Physicochemical Properties, and Applications, *edited by Ifendu A. Nnanna and Jiding Xia*
102. Chemical Properties of Material Surfaces, *Marek Kosmulski*
103. Oxide Surfaces, *edited by James A. Wingrave*
104. Polymers in Particulate Systems: Properties and Applications, *edited by Vincent A. Hackley, P. Somasundaran, and Jennifer A. Lewis*
105. Colloid and Surface Properties of Clays and Related Minerals, *Rossman F. Giese and Carel J. van Oss*
106. Interfacial Electrokinetics and Electrophoresis, *edited by Ángel V. Delgado*
107. Adsorption: Theory, Modeling, and Analysis, *edited by József Tóth*
108. Interfacial Applications in Environmental Engineering, *edited by Mark A. Keane*
109. Adsorption and Aggregation of Surfactants in Solution, *edited by K. L. Mittal and Dinesh O. Shah*
110. Biopolymers at Interfaces: Second Edition, Revised and Expanded, *edited by Martin Malmsten*
111. Biomolecular Films: Design, Function, and Applications, *edited by James F. Rusling*
112. Structure–Performance Relationships in Surfactants: Second Edition, Revised and Expanded, *edited by Kunio Esumi and Minoru Ueno*
113. Liquid Interfacial Systems: Oscillations and Instability, *Rudolph V. Birikh, Vladimir A. Briskman, Manuel G. Velarde, and Jean-Claude Legros*
114. Novel Surfactants: Preparation, Applications, and Biodegradability: Second Edition, Revised and Expanded, *edited by Krister Holmberg*

115. Colloidal Polymers: Synthesis and Characterization, *edited by Abdelhamid Elaissari*
116. Colloidal Biomolecules, Biomaterials, and Biomedical Applications, *edited by Abdelhamid Elaissari*
117. Gemini Surfactants: Synthesis, Interfacial and Solution-Phase Behavior, and Applications, *edited by Raoul Zana and Jiding Xia*
118. Colloidal Science of Flotation, *Anh V. Nguyen and Hans Joachim Schulze*
119. Surface and Interfacial Tension: Measurement, Theory, and Applications, *edited by Stanley Hartland*
120. Microporous Media: Synthesis, Properties, and Modeling, *Freddy Romm*
121. Handbook of Detergents, editor in chief: Uri Zoller, Part B: Environmental Impact, *edited by Uri Zoller*
122. Luminous Chemical Vapor Deposition and Interface Engineering, *Hirotsugu Yasuda*
123. Handbook of Detergents, editor in chief: Uri Zoller, Part C: Analysis, *edited by Heinrich Waldhoff and Rüdiger Spilker*
124. Mixed Surfactant Systems: Second Edition, Revised and Expanded, *edited by Masahiko Abe and John F. Scamehorn*
125. Dynamics of Surfactant Self-Assemblies: Micelles, Microemulsions, Vesicles and Lyotropic Phases, *edited by Raoul Zana*
126. Coagulation and Flocculation: Second Edition, *edited by Hansjoachim Stechemesser and Bohulav Dobiás*
127. Bicontinuous Liquid Crystals, *edited by Matthew L. Lynch and Patrick T. Spicer*
128. Handbook of Detergents, editor in chief: Uri Zoller, Part D: Formulation, *edited by Michael S. Showell*
129. Liquid Detergents: Second Edition, *edited by Kuo-Yann Lai*
130. Finely Dispersed Particles: Micro-, Nano-, and Atto-Engineering, *edited by Aleksandar M. Spasic and Jyh-Ping Hsu*
131. Colloidal Silica: Fundamentals and Applications, *edited by Horacio E. Bergna and William O. Roberts*
132. Emulsions and Emulsion Stability, Second Edition, *edited by Johan Sjöblom*
133. Micellar Catalysis, *Mohammad Niyaz Khan*
134. Molecular and Colloidal Electro-Optics, *Stoyl P. Stoylov and Maria V. Stoimenova*
135. Surfactants in Personal Care Products and Decorative Cosmetics, Third Edition, *edited by Linda D. Rhein, Mitchell Schlossman, Anthony O'Lenick, and P. Somasundaran*
136. Rheology of Particulate Dispersions and Composites, *Rajinder Pal*
137. Powders and Fibers: Interfacial Science and Applications, *edited by Michel Nardin and Eugène Papirer*
138. Wetting and Spreading Dynamics, *edited by Victor Starov, Manuel G. Velarde, and Clayton Radke*
139. Interfacial Phenomena: Equilibrium and Dynamic Effects, Second Edition, *edited by Clarence A. Miller and P. Neogi*

140. Giant Micelles: Properties and Applications, *edited by Raoul Zana and Eric W. Kaler*
141. Handbook of Detergents, editor in chief: Uri Zoller, Part E: Applications, *edited by Uri Zoller*
142. Handbook of Detergents, editor in chief: Uri Zoller, Part F: Production, *edited by Uri Zoller and co-edited by Paul Sosis*
143. Sugar-Based Surfactants: Fundamentals and Applications, *edited by Cristóbal Carnero Ruiz*
144. Microemulsions: Properties and Applications, *edited by Monzer Fanun*
145. Surface Charging and Points of Zero Charge, *Marek Kosmulski*
146. Structure and Functional Properties of Colloidal Systems, *edited by Roque Hidalgo-Álvarez*
147. Nanoscience: Colloidal and Interfacial Aspects, *edited by Victor M. Starov*
148. Interfacial Chemistry of Rocks and Soils, *Noémi M. Nagy and József Kónya*
149. Electrocatalysis: Computational, Experimental, and Industrial Aspects, *edited by Carlos Fernando Zinola*
150. Colloids in Drug Delivery, *edited by Monzer Fanun*
151. Applied Surface Thermodynamics: Second Edition, *edited by A. W. Neumann, Robert David, and Yi Y. Zuo*
152. Colloids in Biotechnology, *edited by Monzer Fanun*
153. Electrokinetic Particle Transport in Micro/Nano-fluidics: Direct Numerical Simulation Analysis, *Shizhi Qian and Ye Ai*
154. Nuclear Magnetic Resonance Studies of Interfacial Phenomena, *Vladimir M. Gun'ko and Vladimir V. Turov*
155. The Science of Defoaming: Theory, Experiment and Applications, *Peter R. Garrett*

THE SCIENCE OF DEFOAMING

Theory, Experiment and Applications

Peter R. Garrett

CRC Press is an imprint of the
Taylor & Francis Group, an **informa** business

CRC Press
Taylor & Francis Group
6000 Broken Sound Parkway NW, Suite 300
Boca Raton, FL 33487-2742

First issued in paperback 2020

Version Date: 20130607

ISBN 13: 978-0-367-57637-0 (pbk)
ISBN 13: 978-1-4200-6041-6 (hbk)

Library of Congress Cataloging-in-Publication Data

Garrett, Peter R.
The science of defoaming : theory, experiment and applications / Peter R. Garrett.
pages cm
Includes bibliographical references and index.
ISBN 978-1-4200-6041-6
1. Antifoaming agents. I. Title.

TP159.A47G37 2013
660'.293--dc23 2013001660

Visit the Taylor & Francis Web site at
http://www.taylorandfrancis.com

and the CRC Press Web site at
http://www.crcpress.com

Dedication

To my wife Carole for her encouragement and forbearance

Contents

Preface xxiii
Author xxv

Chapter 1 Some General Properties of Foams 1

1.1 Introduction 1
1.2 Structure of Foams 1
1.3 Foam Films 5
1.3.1 Surface Tension Gradients and Foam Film Stability 5
1.3.2 Drainage Processes in Foam Films 9
1.3.3 Disjoining Forces and Foam Film Stability 14
1.4 Processes Accompanying Aging of Foam 18
1.4.1 Capillary Pressure Gradients 18
1.4.2 Foam Drainage 21
1.4.3 Bubble Coarsening (Diffusional Disproportionation) 24
1.5 Summarizing Remarks 28
Acknowledgment 29
References 29

Chapter 2 Experimental Methods for Study of Foam and Antifoam Action 33

2.1 Introduction 33
2.2 Measurement of Foam 33
2.2.1 Bartsch Method: Hand Shaking Measuring Cylinders 33
2.2.2 Automated Shake Tests 35
2.2.3 Ross–Miles Method 35
2.2.4 Tumbling Cylinders 36
2.2.5 Gas Bubbling 37
2.2.6 Direct Air Injection 38
2.2.7 Measurement of Bubble Size Distributions 39
2.3 Observations with Single Foam Films 41
2.3.1 Scheludko Cells 41
2.3.1.1 Films Containing Antifoam Drops 41
2.3.1.2 Measurement of Disjoining Pressure Isotherms of Air–Liquid–Air Foam Films 43
2.3.2 Dippenaar Cell 44
2.3.3 Large Vertical Films 45

2.4 Air–Water–Oil Pseudoemulsion Films....46
2.4.1 Direct Observation of Pseudoemulsion Films....46
2.4.2 Measurement of Disjoining Pressure Isotherms for Air–Water–Oil Pseudoemulsion Films....48
2.4.3 Direct Measurements of Pseudoemulsion Film Rupture Pressures....49
2.5 Spreading Behavior of Oils....52
2.6 Summarizing Remarks....53
References....54

Chapter 3 Oils at Interfaces: Entry Coefficients, Spreading Coefficients, and Thin Film Forces....57

3.1 Introduction....57
3.2 Classic Entry and Spreading Coefficients....58
3.3 Generalized Entry Coefficients, Pseudoemulsion Films, and Thin Film Forces....61
3.3.1 Definitions....61
3.3.2 Case Where Generalized Entry Coefficient, $E_g > 0$...65
3.3.3 Case Where Generalized Entry Coefficient, $E_g < 0$...68
3.3.4 Magnitude of Generalized Entry Coefficients....70
3.4 Mode of Rupture of Pseudoemulsion Films....72
3.4.1 Disjoining Pressures and Stability of Pseudoemulsion Films....72
3.4.2 Surface Tension Gradients and Stability of Pseudoemulsion Films....78
3.5 Generalized Spreading Coefficients and Thin Film Forces....79
3.6 Spreading Behavior of Typical Antifoam Oils on Aqueous Surfaces....85
3.6.1 Hydrocarbon Oils....85
3.6.1.1 Spreading and Wetting Behavior of Hydrocarbons on Surface of Pure Water....85
3.6.1.2 Complete and Pseudo-Partial Wetting Behavior of Hydrocarbons on the Surfaces of Aqueous Surfactant Solutions....87
3.6.1.3 Non-Spreading (Partial Wetting) by Hydrocarbons on the Surfaces of Aqueous Surfactant Solutions....94
3.6.2 Polydimethylsiloxane Oils....96
3.6.2.1 Complete and Pseudo-Partial Wetting Behavior of Polydimethylsiloxanes on the Surfaces of Pure Water and Aqueous Surfactant Solutions....96

3.6.2.2 Rates of Spreading of Polydimethylsiloxanes on the Surfaces of Pure Water and Aqueous Surfactant Solutions 104
3.6.3 Effect of Spread Hydrocarbon and Polydimethylsiloxane Oils on the Stability of Pseudoemulsion Films 106
3.7 Non-Equilibrium Effects due to Surfactant Transport 108
3.8 Summarizing Remarks 108
References 110

Chapter 4 Mode of Action of Antifoams 115
4.1 Introduction 115
4.2 Antifoam Effects due to Solubilized Oils 116
4.3 Effect on Foamability of Mesophase Precipitation in Aqueous Surfactant Solutions 121
4.4 Surface Tension Gradients and Theories of Antifoam Mechanism 128
4.4.1 General Considerations 128
4.4.2 Surface Tension Gradients Induced by Spreading Antifoam 129
4.4.3 Elimination of Surface Tension Gradients 138
4.5 Oil Bridges and Antifoam Mechanism 141
4.5.1 Oil Bridges in Foam Films 141
4.5.2 Line Tensions and Antifoam Behavior of Oil Lenses ... 153
4.5.3 Oil Bridges in Plateau Borders and Stability of Pseudoemulsion Films 157
4.6 Antifoam Behavior of Emulsified Liquids 165
4.6.1 Antifoam Effects of Neat Oils in Aqueous Foaming Systems 165
4.6.1.1 Early Work 165
4.6.1.2 Short-Chain Alcohols 167
4.6.1.3 n-Alkanes 175
4.6.1.4 Neat Polydimethylsiloxane Oils 180
4.6.2 Antifoam Effects of Neat Oils in Non-Aqueous Foaming Systems 183
4.6.3 Partial Miscibility and Antifoam Effects 185
4.6.3.1 Origins of Partial Miscibility in Binary Liquid Mixtures 185
4.6.3.2 Systems with Lower Critical Temperatures 188
4.6.3.3 Systems with Higher Critical Temperatures 198

4.7 Inert Hydrophobic Particles and Capillary Theories of Antifoam Mechanism for Aqueous Systems 201
4.7.1 Early Work 201
4.7.2 Experimental Observations Concerning Contact Angles and Particle Bridging Mechanism 203
4.7.3 Theoretical Considerations Concerning Particle Geometry and Contact Angle Conditions for Antifoam Action by Smooth Particles 216
4.7.3.1 Particles with Curved Surfaces and No Edges 216
4.7.3.2 Particles with Edges 216
4.7.3.3 Models of Foam Film Rupture by Particles Using Surface Energy Minimization 224
4.7.4 Effect of Rugosities on Antifoam Action of Particles 228
4.7.5 Rugosities and Stability of Air–Water–Solid Films 233
4.7.6 Particle Size and Kinetics of Foam Film Rupture ... 239
4.7.7 Antifoam Effects of Calcium Soaps 243
4.7.8 Melting of Hydrophobic Particles and Antifoam Behavior 247
4.8 Mixtures of Hydrophobic Particles and Oils as Antifoams for Aqueous Systems 249
4.8.1 Antifoam Synergy 249
4.8.2 Early Patent Literature 251
4.8.3 Role of Oil in Synergistic Oil–Particle Antifoams....252
4.8.4 Early Hypotheses Concerning Role of Particles in Synergistic Oil–Particle Antifoams 263
4.8.5 Role of Particles in Synergistic Oil–Particle Antifoams 267
4.8.5.1 Experimental Observation 267
4.8.5.2 Spherical Particles, Spread Oil Layers, and Rupture of Pseudoemulsion Films 272
4.8.5.3 Smooth Particles with Edges in Absence of Spread Oil Layers 277
4.8.5.4 Smooth Particles with Edges in Presence of Spread Oil Layers 281
4.8.5.5 Rough Particles with Many Edges 281
4.8.6 Antifoam Dimensions and Kinetics 290
4.9 Summarizing Remarks 292
Acknowledgments 295
Appendix 4.1 295
References 302

Chapter 5 Effect of Antifoam Concentration on Volumes of Foam Generated by Air Entrainment....309

5.1 Phenomenology....309
5.1.1 Introduction....309
5.1.2 Dispersion of Single Antifoam....310
5.1.3 Mixed Dispersions of Two Antifoams....313
5.1.4 Relative Effectiveness of Antifoam Entities and Foam Structure....318
5.2 Statistical Theory of Antifoam Action....320
5.2.1 Assumptions....320
5.2.2 Factors Determining Number of Antifoam Entities in Foam Film....324
5.2.3 Calculation of Volume of Air in Foam in Presence of Antifoam....330
5.2.4 Limitations of Theory....334
5.3 Summarizing Remarks....336
Appendix 5.1 Effect of Excluded Volume on Antifoam Concentration in a Film Exhibiting Reynolds Drainage....338
A5.1.1 Mean Flow Velocity of Antifoam Entities....338
A5.1.2 Effect of Excluded Volume on Antifoam Concentration in a Draining Film....339
References....341

Chapter 6 Deactivation of Mixed Oil–Particle Antifoams during Dispersal and Foam Generation in Aqueous Media....343

6.1 Introduction....343
6.2 Deactivation of Antifoam Effect of Polydimethylsiloxane Oils without Particles....344
6.3 Early Work with Hydrophobed Silica–Polydimethylsiloxane Antifoams....346
6.3.1 Separation of Silica from Oil....346
6.3.2 Equilibration and Deactivation....347
6.3.3 Deactivation, Emulsification, and Drop Sizes....348
6.4 Deactivation of Hydrophobed Silica–Polydimethylsiloxane Antifoam by Disproportionation....351
6.4.1 Experimental Evidence....351
6.4.2 Theoretical Considerations Concerning Deactivation by Disproportionation....357
6.5 Effect of Oil Viscosity on Deactivation of Hydrophobed Silica–Polydimethylsiloxane Antifoams....363
6.6 Deactivation in Other Types of Oil–Particle Antifoams....366
6.7 Theories of Foam Volume Growth in Presence of Deactivating Antifoam....368

6.7.1 Antifoam–Bubble Heterocoalescence–Kinetic Model of Pelton and Goddard for Foam Generation by Sparging 368
6.7.2 Modified Antifoam–Bubble Heterocoalescence-Kinetic Model Using a Statistical Distribution of Antifoam Drops over Bubbles 373
6.7.3 Combination of Kinetic Model of Antifoam Deactivation with Kinetic Model of Antifoam Action 376
6.7.4 Combination of Kinetic Model of Antifoam Deactivation by Disproportionation with Empirical Expression for Antifoam Action 379
6.8 Summarizing Remarks 383
Acknowledgment 386
References 386

Chapter 7 Mechanical Methods for Defoaming 389

7.1 Introduction 389
7.2 Defoaming Using Rotary Devices 389
7.2.1 Designs of Rotary Devices Described in Scientific Literature 389
7.2.2 Commercial Rotary Defoamers 398
7.2.3 Defoaming Mechanisms of Rotary Devices 400
7.2.3.1 Role of Centrifugal Force 400
7.2.3.2 Role of Shear and Impact Forces on Bubbles in Mechanical Defoaming 404
7.2.3.3 Defoaming by Inherent Liquid Spray 407
7.3 Defoaming Using Ultrasound 409
7.3.1 Brief History of Defoaming by Ultrasound 409
7.3.2 Defoaming Mechanism of Ultrasound 415
7.4 Defoaming Using Packed Beds of Appropriate Wettability 421
7.5 Summarizing Remarks 422
Appendix 7.1 423
References 428

Chapter 8 Antifoams for Detergent Products 431

8.1 Introduction 431
8.2 Powders for Machine Washing of Laundry 433
8.2.1 Front-Loading Drum-Type Textile Washing Machines 433
8.2.2 Use of Fatty Acids and Soaps 435
8.2.3 Use of Non-Soap Particulate Antifoams 439
8.2.4 Use of Hydrocarbon–Hydrophobic Particle Mixtures 441

8.2.4.1 Hydrocarbon Mixtures with Alkyl Phosphoric Acid Derivatives 441
8.2.4.2 Hydrocarbon Mixtures with Non-Phosphorous-Containing Organic Compounds 446
8.2.5 Use of Polydimethylsiloxane-Based Antifoams 450
8.2.5.1 General Properties 450
8.2.5.2 Storage Deactivation and Incorporation in Detergent Powders 453
8.2.5.3 Dispensing 456
8.2.5.4 Enhancement of Antifoam Effectiveness 457
8.2.6 Hydrocarbon-Based Simulation of Dimethylsiloxane-Based Antifoams 460
8.3 Liquids for Machine Washing of Laundry 461
8.3.1 General Properties 461
8.3.2 Incorporation of Polyorganosiloxane–Hydrophobic Silica Antifoams in Detergent Liquids 462
8.4 Machine Dishwashing 467
8.5 General Hard-Surface Cleaning Products 469
8.6 Summarizing Remarks 471
Appendix 8.1 471
References 476

Chapter 9 Control of Foam in Waterborne Latex Paints and Varnishes 481
9.1 Introduction 481
9.2 Foam and Antifoam Behavior 485
9.2.1 General Considerations 485
9.2.2 Effect of Stratified Layers of Polymer Latex Particles on Foam and Pseudoemulsion Film Stability 488
9.3 Specific Issues Concerning Oil-Based Antifoams 489
9.3.1 Incorporation in Paints and Varnishes 489
9.3.2 Defect Formation in Drying Paint Films 492
9.3.2.1 General Considerations 492
9.3.2.2 Experimental and Theoretical Studies of Cratering Caused by Marangoni Effect Induced by Spreading Oil Drops 495
9.3.2.3 Putative Craters Caused by Non-Spreading Oil Drops Bridging Paint Films 497
9.4 Summarizing Remarks 499
References 500

Chapter 10 Antifoams for Gas–Oil Separation in Crude Oil Production........... 503

10.1 Introduction 503
10.2 Surface Activity at Gas–Hydrocarbon and Gas–Crude Oil Interfaces 504
10.3 Causes of Foam Formation in Gas–Crude Oil Systems........ 509
10.3.1 Disjoining Pressures.. 509
10.3.2 Origin of Surface Tension Gradients at Gas–Crude Oil Interfaces 510
10.3.3 Experimental Observations of Foam Behavior........ 511
10.4 Use of Antifoams.. 515
10.4.1 General Considerations 515
10.4.2 Polydimethylsiloxanes and Substituted Polydimethylsiloxanes.. 517
10.4.2.1 Effect of Solubility of Antifoam Oils....... 517
10.4.2.2 Mode of Antifoam Action in Crude Oils.. 520
10.4.3 Other Materials 524
10.5 Summarizing Remarks.. 525
References 526

Chapter 11 Medical Applications of Defoaming 529

11.1 Introduction 529
11.2 Use of Simethicone Antifoam in Treatment of Gastrointestinal Gas 530
11.2.1 Therapeutic Application 530
11.2.2 Use of Simethicone in Endoscopy........................... 532
11.3 Defoaming of Blood during Cardiopulmonary Bypass Surgery 533
11.3.1 Gas Bubble Oxygenators and Use of Antifoams...... 533
11.3.2 Mechanism of Polydimethylsiloxane–Hydrophobed Silica–Coated Porous Defoamers...... 535
11.3.3 Polydimethylsiloxane–Hydrophobed Silica Antifoam as Source of Emboli 537
11.3.4 Cardiotomy Defoaming.. 539
11.3.5 Defoaming in Cardiopulmonary Bypass Blood Circuits, Which Include Membrane Oxygenators and Cardiotomy/Venous Reservoirs.. 540
11.3.6 Defoaming Systems Avoiding Use of Polydimethylsiloxane-Based Antifoam.................. 542
11.3.6.1 Potential Replacements for PDMS–Hydrophobed Silica in Cardiotomy Reservoirs 542

11.3.6.2 Potential Use of Defoamer Elements with High Air–Blood Contact Angles 544
11.3.6.3 Removal of Gaseous Microemboli 545
11.4 Summarizing Remarks 549
References 551

Frequently Used Symbols and Abbreviations 555

Index 559

Preface

It is 20 years since the only other book exclusively concerned with defoaming was published [1]. However, during the intervening years, much new work has been published and many new insights revealed. It therefore seems appropriate to revisit the topic with a new book.

The earlier book was concerned not only with the scientific aspects of the mode of action of antifoams but also with a number of different industrial applications. Indeed the latter formed the greater part of that book. By contrast, this book is more concerned with the basic science of defoaming. In this it replicates the emphasis revealed by published literature over the past two decades. The industrial application of antifoams does in fact produce challenging scientific issues specific to particular applications. Where appropriate, such issues are described here in the relevant chapters. However, they are apparently rarely addressed, at least in published works. Much application still appears to rest on simple empiricism augmented by naive and sometimes completely incorrect views of the mode of action of antifoams. It is hoped that this book will help stimulate a more enlightened approach to application, particularly by those antifoam manufacturers who supply products for specific applications to many different customers.

The main part of the book then concerns the science of defoaming. The general properties of foams are presented in an introductory chapter, which serves as a background to much that follows. Another chapter describes the experimental techniques involved in measurement of the foam properties relevant for study of antifoam action and the various techniques used in establishing the mode of action of antifoams. Various aspects of antifoam behavior and relevant theory are considered in later chapters.

Most commercially effective antifoams are oil based. The relevance of the entry and spreading behavior of such oils at the surfaces of foaming fluids to their antifoam function has been the subject of much debate over the past half century. A chapter is therefore specifically devoted to the entry and spreading behavior of oils and the role of thin film forces in determining that behavior.

The role of bridging foam films by particles and oil drops in the mode of action of antifoams was emphasized in the earlier book. Also included was the first evidence that the role of particles in synergistic oil–particle antifoam mixtures for defoaming aqueous foams concerns rupture of air–water–oil pseudoemulsion films. The past two decades have seen a general acceptance and further development of these aspects of the theory of antifoam action. Particularly noteworthy have been the insights provided by Professor Nikki Denkov and his coworkers at the University of Sofia. All of this is reviewed in a chapter specifically concerned with the mode of action of antifoams.

The volume of foam generated in the presence of antifoam is a function of antifoam concentration. Both the relevant phenomenology and a statistical theory can be shown to be consistent with experiment provided the effectiveness of individual antifoam entities remains constant. However, the effect of the concentration of antifoam can be complicated by the decline in effectiveness of individual entities during foam generation due to disproportionation to form inactive material. Two chapters deal with these issues.

In some applications, contamination of a foaming medium with chemical antifoams is unacceptable. Defoaming can then be achieved, in principle, using mechanical means, which include centrifugal or ultrasonic devices. A review of the mode of the defoaming action of such devices is included here. It reveals a subject relatively rich in opportunity for the application of new insight.

The author is particularly grateful to various colleagues for stimulating discussions on various aspects of the topics covered in this book. Those colleagues included Dr. Paul Grassia (University of Manchester), Dr. Stephen Neethling (Imperial College), Professor Nikki Denkov (University of Sofia), Professor Clarence Miller (Rice University), and Professor Gordon Tiddy (University of Manchester).

Peter R. Garrett

REFERENCE

1. Garrett, P.R. (ed.). *Defoaming, Theory and Industrial Applications*, Marcel Dekker, Surfactant Science Series, Vol. 45, 1993.

Author

After working for Unilever Research at Port Sunlight for more than a quarter of a century, Peter Garrett retired, acquiring visiting professorships at the University of Manchester in the United Kingdom and briefly at Rice University, Houston, Texas.

He has worked on various aspects of interfacial science in a career lasting 40 years. These included the equilibrium and dynamic adsorption behavior of surfactant solutions, solubilization in microemulsions, kinetics of surfactant dissolution and mesophase growth, and of course both the basic science of defoaming and its application especially in the detergents industry. He is author or coauthor of more than 30 journal articles and is named as inventor on 14 patents, the majority of which concerned the application of antifoam technology. He is also editor of the book *Defoaming, Theory and Applications* published in 1993. A member of the Royal Society of Chemistry, he received his PhD degree from the University of Leicester in the United Kingdom in 1971.

1 Some General Properties of Foams

1.1 INTRODUCTION

Before considering the behavior of antifoams, we review the relevant properties of foams. Only a brief summary is given here. It is for the most part only concerned with those aspects that may have relevance for the understanding of antifoam action. For more complete accounts, the reader is referred to the many books [1–9] and reviews on the subject [10–27].

This brief review includes definition of the structural features of foams. A summary of the processes occurring in foam films follows with particular emphasis on the factors that determine the stability of those films. Finally, we include an outline of the processes of drainage and diffusion-driven coarsening, which concern the entire body of a foam and not just the constituent parts.

1.2 STRUCTURE OF FOAMS

We first consider the structure of a polydisperse foam. That structure is exemplified by the photograph reproduced in Figure 1.1. This image depicts a foam that has been aged, where both drainage and diffusion of gas from small bubbles to large bubbles, as a result of differences in capillary pressure, have occurred. In the lower part of the foam, bubbles are spherical (so-called **kugelschaum**) and of small size with a relatively low overall gas volume fraction. Collections of spherical bubbles, without the distortions associated with film formation, form at gas volume fractions, Φ_G^{foam}, of ≤~0.74 in the case of monodisperse bubbles and ≤~0.72 in the case of polydisperse bubbles [28]. As the liquid drains out of, for example, a polydisperse foam so that Φ_G^{foam} becomes > 0.72, the bubbles distort to form polyhedra. This polyhedral foam (**polyederschaum**), with a relatively high gas volume fraction, consists of thin foam films joined by Plateau border channels.

In the case of the foam depicted in Figure 1.1, there is clear segregation of bubble sizes, with larger bubbles being present at the top of the foam column. The extent of such vertical segregation in polydisperse foams varies according to the method of generation. It probably depends on the extent of mixing during foam generation, gravity segregation [5], and even the so-called brazil nut effect [29]. Segregation of such a polydisperse foam can in fact apparently be facilitated by rapid continuous wetting of a foam column from above so that high liquid volume fractions prevail. Bubble movement can then occur without requiring the distortion of bubbles so that gravitational segregation is in turn specifically facilitated [5].

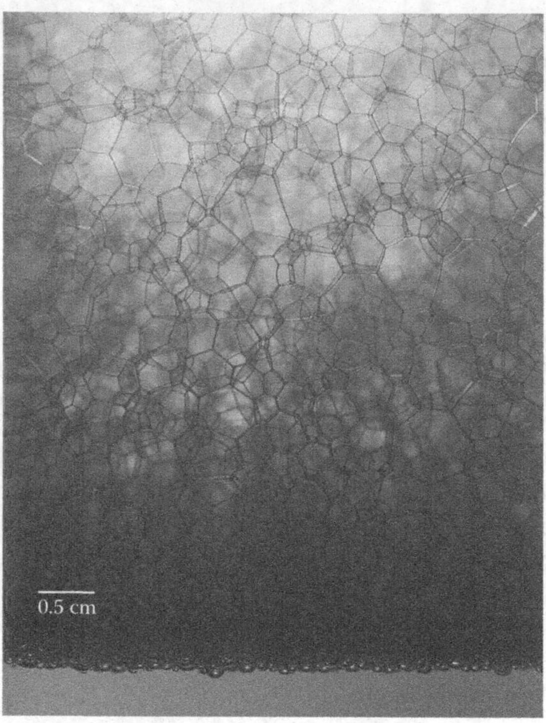

FIGURE 1.1 Aged polydisperse foam with dry polyhedral foam (polyederschaum) at top of column and spherical bubbles (kugelschaum) at bottom of column.

It is worth noting that the Plateau borders between bubbles of different sizes in Figure 1.1 are in fact curved, being concave with respect to the larger bubbles. This implies that the adjacent foam films are also curved so that the capillary pressure is larger inside the smaller bubbles. Such differences drive the process of gas diffusion leading to coarsening of the foam where average bubble sizes increase.

We illustrate the salient structural features of a foam in Figure 1.2, by relating computer-generated simulations reported by Weaire and Hutzler [5] to two actual images of polydisperse foams, each of different gas volume fraction, made by Hartland and Barber [30]. The comparison is, however, intended to be only qualitative, particularly with respect to gas phase volume fractions in the respective experimental images and simulations. The latter concern assemblies of the so-called Kelvin cells, tetrakaidecahedra, with six flat quadrilateral faces and eight curved hexagonal faces. However, polydispersity means that such polyhedra are not at all present in the foams imaged by Hartland and Barber [30].

As shown in Figure 1.2b, Plateau borders join together to form a network of channels, containing almost all the liquid in the foam and through which drainage occurs in the gravity field. The junctions or nodes of the Plateau borders in the interior of a dry foam (where $\Phi_G^{foam} \rightarrow 1$) invariably involve four borders meeting at a regular tetrahedral angle of 109.5°. The Plateau border cross-sections are seen to be concave triangular in shape, with each of the vertices terminating in a foam film where

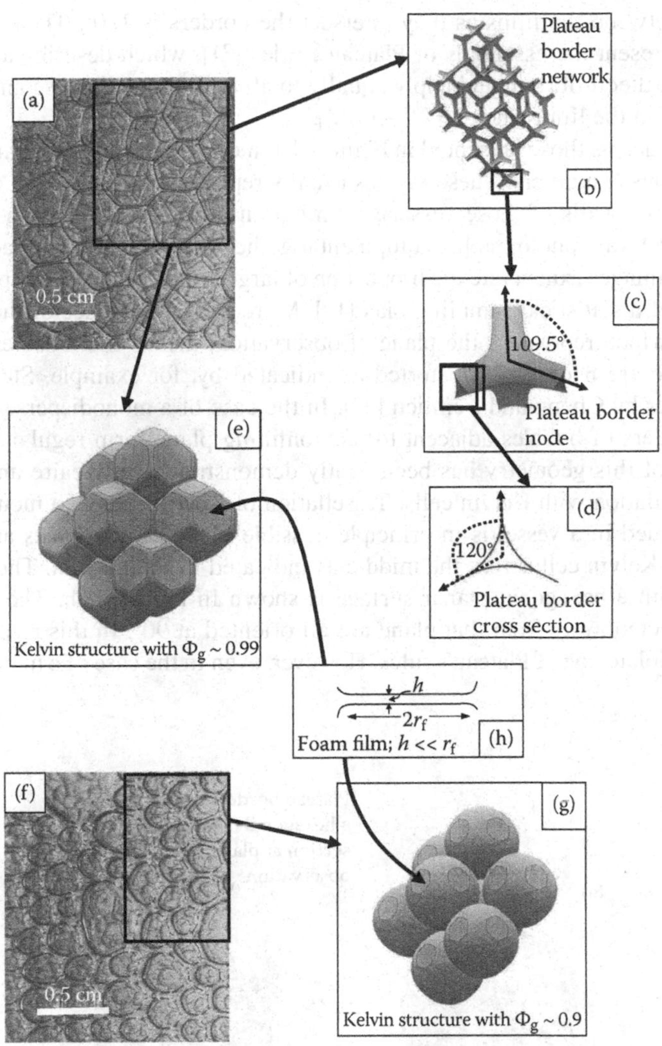

FIGURE 1.2 Structural elements of foam. (a) Image of polyhedral foam against wall of vessel. (b) Simulated three-dimensional Plateau border network for monodisperse Kelvin cell foam. This network contains almost all of the liquid in foam. Liquid drains through network under influence of gravity. (c) Plateau border node where four borders meet at tetrahedral angle. (d) Cross section of plateau border where vertices terminate in foam films. (e) Simulated Kelvin cell "dry" foam with gas volume fraction, Φ_G^{foam}, of ~0.99. (f) Image of "wet" foam or froth with gas volume fraction close to limit of 0.72 (for a polydisperse foam, reference [28]) where Plateau borders and foam films are absent. (g) Simulated Kelvin cell "wet" foam with gas volume fraction, Φ_G^{foam}, of ~0.9. (h) Foam films are present between bubbles in simulated Kelvin cell foams shown in (e) and (g). Such films are characterized by thicknesses several orders of magnitude less than their width. (Images (a) and (f) from Hartland, S., Barber, A.D., *Trans. Inst. Chem. Eng.*, 52, 43, 1974. With permission from Institution of Chemical Engineers. Images (b), (c), (e), and (g) from Weaire, D., Hutzler, S., *The Physics of Foams*, 1999, by permission of Oxford University Press.)

the angle between the films as they intersect the borders is 120°. These structural features represent the essentials of Plateau's rules [31], which describe a condition of mechanical equilibrium and apply equally to all both monodisperse and polydisperse foams in the limit where $\Phi_G^{foam} \rightarrow 1$ [5].

Images, such as those presented in Figures 1.1 and 1.2, are of actual foams, generated in various containers. These images usually represent only the layer of bubbles adjacent to the walls of those vessels, a limitation often imposed by the depth of field of the relevant photographic equipment together with the relative opacity of the foam. Such images exaggerate the proportion of large bubbles in a polydisperse foam as a result of a statistical sampling bias [10]. Moreover, the surface of the containing vessel, which represents the plane of observation, means that bubbles adjacent to that plane are necessarily distorted as indicated by, for example, Steiner et al. [32] and later by Cheng and Lemlich [33]. In the case of a monodisperse foam, the Plateau borders of bubbles adjacent to the confining plane form regular hexagons. The origin of this geometry has been neatly demonstrated by Weaire and Hutzler [5] in a simulation with Kelvin cells. Tessellation of Kelvin cells in a monodisperse foam contained in a vessel is in principle possible if the edge bubbles are formed by slicing a Kelvin cell across the middle as indicated in Figure 1.3a. The resulting structure with a hexagonal planar surface is shown in Figure 1.3b. Those Plateau borders directed away from that plane are all oriented at 90°. In this respect, these structures violate one of Plateau's rules. However, even in the case of a monodisperse

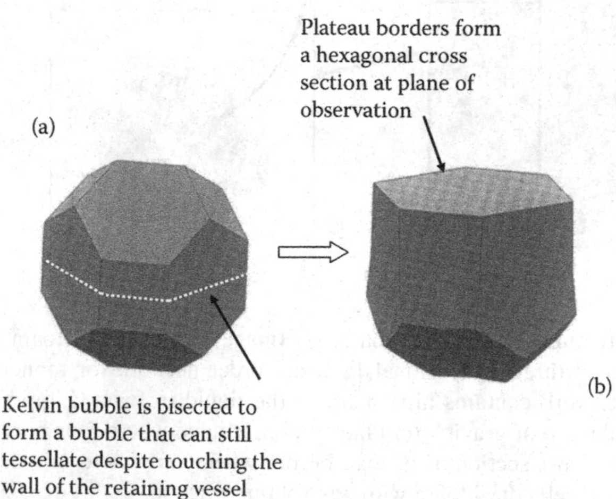
Plateau borders form a hexagonal cross section at plane of observation
(a)
Kelvin bubble is bisected to form a bubble that can still tessellate despite touching the wall of the retaining vessel
(b)

FIGURE 1.3 In a monodisperse Kelvin foam, even bubbles that touch the surface of retaining vessel must tessellate. Simulation suggests that this is permitted if those bubbles have structure formed by bisecting a Kelvin cell as indicated in (a). Bubbles with resulting structure must be extended to restore original volume. As shown in (b), the edge of bubble then presents a hexagonal cross section to plane of observation at wall of vessel. (From Weaire, D., Hutzler, S., *The Physics of Foams*, 1999, by permission of Oxford University Press.)

foam, Kelvin cells only exist one layer into the foam from the surface cells. Beyond that, the structure becomes disordered and Kelvin cells are not apparently found [5].

The dimensions of these structural features of foam are subject to change as a result of the action of various processes. Senescence is facilitated by processes of drainage and diffusional coarsening. However, the ultimate arbitrator of the fate of foams concerns the stability of films, which arguably represents the most important factor in determining both *foamability* (i.e., the tendency to form a foam) and *foam stability*. With thicknesses typically at least three orders of magnitude less than the diameter of the circle circumscribed by the edges of a Plateau border, they represent the most fragile aspect of foam structure. Indeed, if the films break, then the structures described here cannot exist if the continuous phase is liquid and consideration of their putative nature becomes purely academic. Design of effective antifoams tends therefore to be focused on destruction of foam films.

1.3 FOAM FILMS

1.3.1 Surface Tension Gradients and Foam Film Stability

Films formed by adjacent bubbles in a pure liquid are extremely unstable. Pure liquids therefore do not form foams. This arises in part because of the response of the films to any external force such as gravity. Consider, for example, a vertical plane-parallel film of a pure liquid in a gravity field. There is no reason why any element of that film should move in response to the applied gravitational force with a velocity different from that of any adjacent element. No velocity gradients in a direction perpendicular to the plane of the film surface against the air will therefore exist. There will then be no viscous shear forces opposing the effect of gravity. The film will exhibit plug flow (resisted only by extensional viscous forces) with elements accelerated downward tearing it apart. The process is depicted in Figure 1.4a.

This behavior can be drastically altered if we arrange for a tangential force to act in the plane of the liquid–air surface so that the surface is essentially rigid. In the case of a vertical plane-parallel film of a viscous liquid with such rigid surfaces, subject to gravity, a parabolic velocity profile will develop as shown in Figure 1.4b. This means that velocity gradients will exist in a direction perpendicular to the film surfaces. A viscous stress will therefore be exerted at the air–liquid surface. This stress must be balanced by the tangential force acting in the plane of the surface. That force can only be a gradient of surface tension. This balance of viscous forces and surface tension gradients at the liquid–air surface can be written as

$$\frac{\mathrm{d}\sigma_{\mathrm{AL}}}{\mathrm{d}y} = \left(\eta_{\mathrm{L}} \frac{\mathrm{d}u_y}{\mathrm{d}x} \right)_{x=0} \tag{1.1}$$

where σ_{AL} is the air–liquid surface tension of the foaming liquid, η_{L} is the viscosity, u_y is the velocity of flow in the y direction, y is the vertical distance, and x is the horizontal distance in the film.

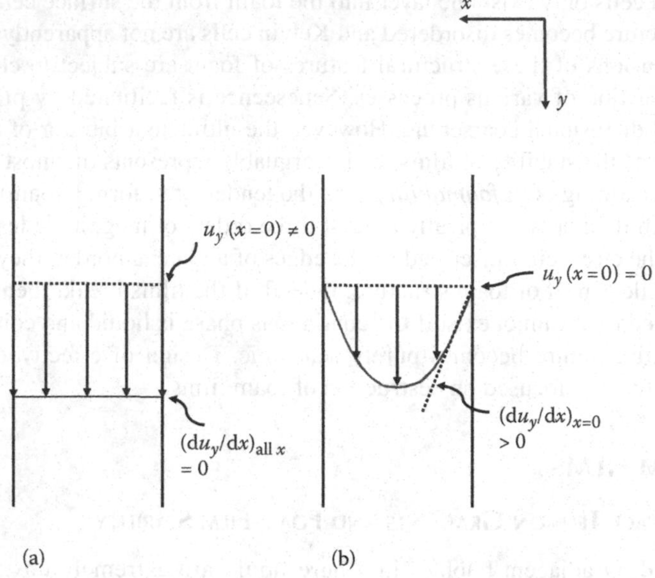

FIGURE 1.4 Limiting range of velocity profiles in draining foam film. (a) Plug flow. (b) Flow with parabolic velocity profile when film surfaces are immobile. Arrows represent magnitude of fluid velocities u_y in the y direction.

Thus, we find that if the force of gravity is to be resisted by the film, then a surface tension gradient must exist at the air–liquid surface. Combining the relevant Navier–Stokes equation with Equation 1.1 as a boundary condition, Lucassen [15] shows that in the case of a vertical film in the gravity field the gradient is

$$\frac{d\sigma_{AL}}{dy} = \frac{\rho_L g h}{2} \tag{1.2}$$

where h is the film thickness, ρ_L is the liquid density, and g is the acceleration due to gravity. This gradient can only exist where differences of surface composition can occur. We therefore require the presence of more than one component in the film. Indeed, it is possible to speculate that in the case of, say, aqueous foams, diffusion of water through the gas phase may rapidly remove any differences in concentration between different parts of a foam film if only one solute is present. In this case, at least two solutes (or three components) would be required.

Surface tension gradients due to differences in the surface excess of soluble surface-active components may exist only when either the surface is not in equilibrium with the bulk composition or there are concomitant differences in bulk composition parallel to the surface. In the case of the former, the magnitudes of the gradients are of course determined by the rate of transport of surfactant to the relevant surfaces. With concentrated surfactant solutions, transport rates by diffusion will be rapid and surface tension gradients will be diminished so that shear rates are only balanced at the surface of films if the air–liquid surfaces become mobile (i.e., so that the velocity $u_y(x = 0) \gg 0$). This would result in enhanced rates of drainage from films. In the extreme, the surface tension gradients will tend to be eliminated altogether. Thus, it has occasionally been reported that foamabilities decline at extremely high concentrations of surfactant in aqueous solution. Conversely, however, if foam films are denuded of surfactant because of extremely slow transport rates, then the maximum surface tension gradients that can be achieved will be small. Such films will therefore be susceptible to rupture when exposed to external stress. However, the complex problem of assessing both the effect of rate of transport on the surface tension gradients in foam films and the overall resultant impact on foam film stability, when subject to an external stress, has not, apparently, been fully addressed.

Differences in bulk composition are possible in a thin foam film as a result of stretching of the film. If the film is sufficiently thin, then any stretching causes a depletion of the bulk phase surfactant solution between the air–liquid surfaces of the foam film as more surfactant adsorbs on those surfaces. Distances perpendicular to the film are small so that, provided the stretching occurs reasonably slowly, the equilibrium inside the film element may be always maintained. Depletion of bulk phase the surfactant concentration will therefore necessarily mean an increase of the surface tension of the film as it is stretched. This will, however, only occur if reduction of the surfactant concentration causes a concomitant increase in surface tension. In the case of a pure surfactant at concentrations above the critical micelle concentration (CMC), this may not always happen.

We find then that it is possible to generate a surface tension gradient in a foam film by stretching various elements of the film to different extents. The increase in surface tension due to stretching imparts an elasticity to the film. This property of foam films was first recognized by Gibbs [34] and is usually referred to as the Gibbs elasticity ε_G. It is defined as

$$\varepsilon_G = \frac{2d\sigma_{AL}}{d\ln A} = -\frac{2d\sigma_{AL}}{d\ln h} \tag{1.3}$$

where A is the film area and the factor 2 arises because of the two surfaces.

A plot of ε_G against concentration for a submicellar aqueous solution of sodium dodecyl sulfate (SDS) is shown in Figure 1.5 by way of example. Here, we see that, except at very low concentrations, decreases in film thickness at constant concentration produce increases in Gibbs elasticity so that $(d\varepsilon_G/dh)c \le 0$. Thus, as the film becomes thinner, stretching will cause a relatively greater depletion of surfactant in the intralamellar liquid and the surface tension will increase to a greater extent.

The plot of Gibbs elasticity against concentration shown in Figure 1.5 clearly reveals a maximum at concentration c_{max}. At extremely low concentrations of surfactant, we find that upon stretching of the film, there is essentially no contribution from the intralamellar liquid, and the surfactant behaves as an insoluble monolayer. Here, with an increase in surfactant concentration, both the surface excess and the elasticity of the monolayer increase. However, further increases in the surfactant concentration will eventually mean that it significantly exceeds that required to compensate for stretching of the air–liquid surface, so $\varepsilon_G \to 0$. These two opposing consequences

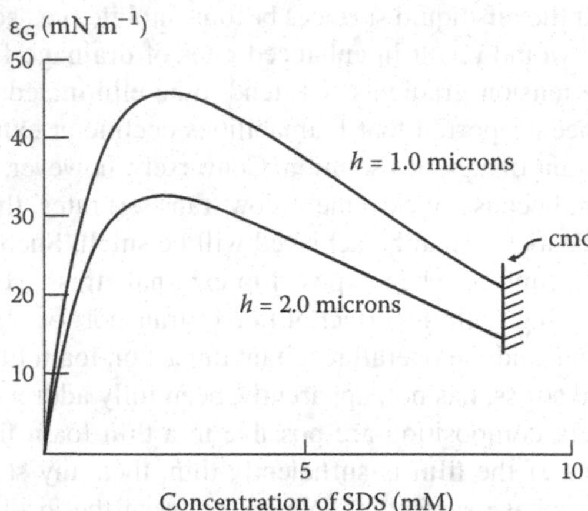

FIGURE 1.5 Gibbs elasticities of submicellar SDS solutions at two different film thicknesses. (After Lucassen, J. Dynamic properties of free liquid films and foams, in *Anionic Surfactants, Physical Chemistry of Surfactant Action* (Lucassen-Reynders, E.H., ed.), Marcel Dekker, New York, Surfactant Sci. Series, Vol. 11, Chapter 6, p. 217, 1981.)

of increasing concentration conspire to produce the maximum in a plot of Gibbs elasticity.

Lucassen [15] considers the effect of Gibbs elasticity on the development of stabilizing surface tension gradients in a foam film in the gravity field. He points out that if ε_G decreases as the film is stretched, it will tend to be dynamically unstable. Under these circumstances, any stretching force will tend to increase the area of the thinnest part of the film. Such situations will tend to prevail for films formed at concentrations on the low side of c_{max}. Thus, we can write

$$\frac{d\varepsilon_G}{dh} = \left(\frac{\partial \varepsilon_G}{\partial h}\right)_c + \left(\frac{\partial \varepsilon_G}{\partial c}\right)_h \frac{dc}{dh} \tag{1.4}$$

where c is the concentration. We can therefore have $d\varepsilon_G/dh > 0$ if $c < c_{max}$ because then $(\partial\varepsilon_G/\partial c)_h > 0$, $dc/dh > 0$, and $(\partial\varepsilon_G/\partial h)_c \rightarrow 0$ as $c \rightarrow 0$.

For vertical films prepared from concentrated solutions, the requirement that the surface tension gradient satisfy Equation 1.2 implies that rapid stretching will occur. In the case of micellar solutions of certain pure surfactants, this may require achievement of submicellar concentrations in order that $\varepsilon_G > 0$ and therefore $d\sigma_{AL}/dx > 0$. For vertical films prepared from extremely dilute surfactant solutions where $c < c_{max}$, Lucassen [15] shows that the magnitude of ε_G for thin elements at the top of the film may be less than that of thinner elements lower down. Any force acting on the film, such as an increase in weight as it grows, could mean catastrophic extension of the thinnest elements because of their lower Gibbs elasticities.

In summary, then, we find that surface tension gradients are necessary if freshly formed foam films are to survive. These gradients may occur if surface tensions depart from equilibrium values. This will happen when foam film air–liquid surfaces are expanded at rates that are fast so that equilibrium with the bulk surfactant concentration cannot be maintained. They may also occur when films are thin so that stretching may deplete intralamellar bulk phase to give rise to a Gibbs elasticity. Unfortunately, there are few experimental observations that clearly reveal the importance of surface tension gradients in determining foam behavior. Perhaps the best examples are reported by Malysa et al. [35] and Prins [36].

1.3.2 Drainage Processes in Foam Films

Any freshly generated foam film that survives will now be subject to a capillary pressure exerted by the curved surfaces of the adjacent Plateau borders. That pressure will tend to suck liquid out of the foam film. The resultant process of film drainage is surprisingly complex.

The simplest description of foam film drainage is obtained if the film is supposed cylindrical with immobile plane-parallel surfaces. Such behavior is represented by the Reynolds equation [37]

$$-\frac{\mathrm{d}h}{\mathrm{d}t} = \frac{2h^3 \Delta P}{3\eta_{\mathrm{L}} r_{\mathrm{f}}^2} \tag{1.5}$$

where h is the film thickness, r_{f} is the film radius, and ΔP is equal to the capillary pressure jump, $p_{\mathrm{c}}^{\mathrm{PB}}$, at the air–liquid surface of the Plateau border. $p_{\mathrm{c}}^{\mathrm{PB}}$ is in turn equal to $\sigma_{\mathrm{AL}}|\kappa_{\mathrm{PB}}|$, where $|\kappa_{\mathrm{PB}}|$ is the modulus of the air–liquid curvature of the Plateau border. If the film is sufficiently thin (<~100 nm), then the applied capillary pressure will be modified by the air–liquid–air (ALA) disjoining pressure, $\Pi_{\mathrm{ALA}}(h)$, so that the total applied pressure becomes $p_{\mathrm{c}}^{\mathrm{PB}} - \Pi_{\mathrm{ALA}}(h)$. Drainage will therefore cease if $p_{\mathrm{c}}^{\mathrm{PB}} - \Pi_{\mathrm{ALA}}(h)$, where a metastable equilibrium can prevail. We consider the effects of disjoining pressure on film stability in Section 1.3.3.

The behavior of real foam films is rarely represented by the Reynolds equation. Films of radius >~100 microns are, for example, generally not plane parallel [6]. The surface of a draining foam film is characterized by a balance of the surface tension gradient and the relevant viscous stress as indicated by Equation 1.1. In turn, this surface tension gradient implies a gradient in surfactant adsorption that will be subject to relaxation by diffusion from the bulk phase and along the surface. Maintenance of the balance of the surface tension gradient and viscous stress will therefore always require a countervailing flow in the surface in the direction of the bulk phase flow. Complete immobility of the air–liquid surface (so that the velocity $u_y(x = 0) = 0$) is therefore never achieved [6].

In the case of relatively large foam films >~100 microns, where the air–liquid surface is close to a condition of immobility, drainage occurs in an axisymmetric manner but with a non-uniform thickness. A thick region develops in the center of the film, the so-called dimple, while a thinner region surrounds this dimple—the so-called barrier ring [38]. A schematic cross-section of a film with a dimple is depicted in Figure 1.6a. The behavior of such a film, formed inside a cylindrical glass cell (the

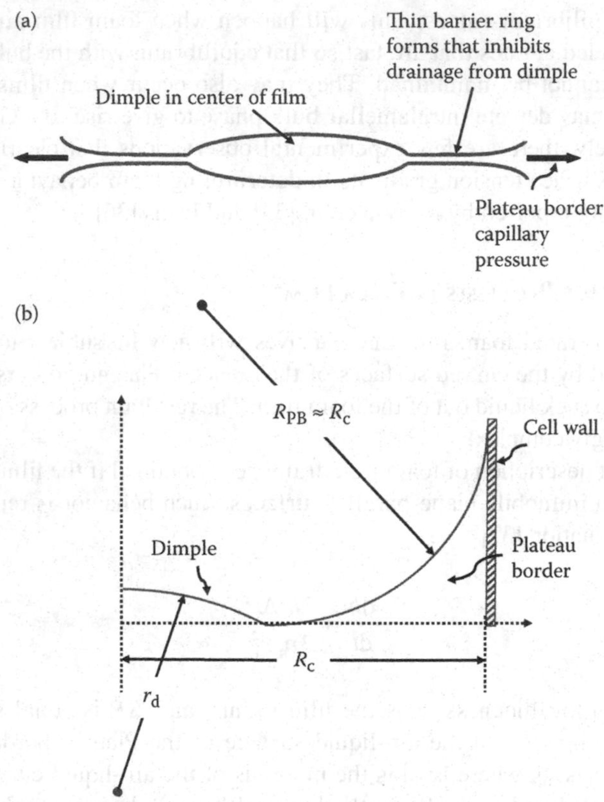

FIGURE 1.6 Axisymmetric drainage of foam film with immobile air–water surfaces. (a) Dimple and barrier ring that form. (b) Defining maximum dimple size where barrier ring is absent and where film is formed in a cylindrical Scheludko cell (references [13, 39]; see Section 2.3.1). Radius of curvature, R_{PB}, of Plateau border air–water surface is approximated by radius, R_c, of cell—r_d is radius of curvature of air–water surface of the dimple. (Reprinted with permission from Joye, J. et al., *Langmuir*, 8, 3083. Copyright 1992 American Chemical Society.)

Scheludko cell [13, 39]; see Chapter 2, Section 2.3.1), has been analyzed numerically by Joye et al. [38]. They deduce that the formation of a dimple requires that the capillary pressure inside the maximum putative dimple that can be formed be less than the capillary pressure in the Plateau border meniscus. This means that $\left(r_d^{-1}/R_{PB}^{-1}\right) < 1$, where r_d is the radius of curvature of the maximum dimple and R_{PB} the radius of curvature of the Plateau border as depicted in Figure 1.6b. Suction from the latter will not then be matched by sufficient pressure from the dimple so that liquid is selectively removed from the region adjacent to the meniscus, forming a thin barrier ring. If, on the other hand, the film is small and the capillary pressure in the maximum dimple is greater than the capillary pressure in the Plateau border, then liquid flow from any putative dimple readily matches that sucked into the border and neither a barrier ring nor a dimple are formed. If a dimple is formed, then the concomitant restricted liquid flow in the barrier ring contributes to a diminished rate of drainage relative to that predicted by the Reynolds equation. As the barrier ring becomes thinner, then eventually repulsive disjoining forces are predicted to become important in determining the final rates of film thinning. At high electrolyte concentrations, the barrier will become extremely thin so that the dimple drains extremely slowly [38].

In practice, *axisymmetric* foam film drainage is not usually observed in the case of solutions of simple surfactants. Although dimples are usually formed with such systems, the resulting films exhibit a hydrodynamic instability where dimples disgorge directly into one side of the Plateau borders [40]. This *asymmetric* drainage is always significantly more rapid than the axisymmetric drainage. The latter appears to be observed only when asymmetric drainage is suppressed by extreme surface rheological behavior of the adsorbed surfactant. We illustrate this in Figure 1.7 where the drainage rates of horizontal foam films formed in a Scheludko cell [13, 39], and prepared from solutions of SDS and SDS–dodecanol mixtures, are compared with the surface shear viscosities of those solutions. Rapid asymmetric drainage is associated with the low values of the surface shear viscosity typical of aqueous solutions of simple soluble surfactants.

Joye et al. [40] offer an explanation, based on the earlier work of Stein [42, 43], for the hydrodynamic instability that produces asymmetric drainage. We illustrate this in Figure 1.8. Here it is supposed that thickness fluctuations occur in the barrier ring (linearized in the figure). In the thicker regions the flow of liquid from the dimple to the Plateau border meniscus will increase relative to the unperturbed flow rate. This will in turn produce an increased surface tension gradient in the direction of the flow. Conversely in the case of the thinner regions, the flow rate in the direction of the meniscus will be slow relative to the unperturbed flow rate. In turn this will produce a decreased surface tension gradient. As a consequence of these two opposing trends in surface tension gradient in the direction of flow from the dimple to the meniscus, another surface tension gradient orthogonal to the unperturbed gradient

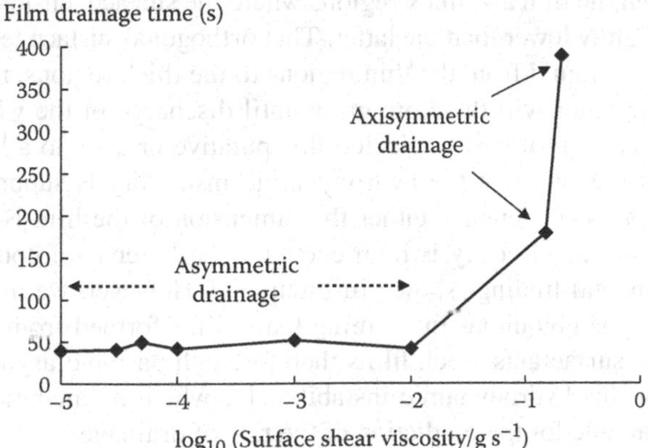

FIGURE 1.7 Film drainage time as function of surface shear viscosity for SDS and SDS + dodecanol (100:1) solutions using a Scheludko cell and film radius of 100 microns. (Reprinted with permission from Joye, J. et al., *Langmuir*, 8, 3083. Copyright 1994 American Chemical Society, using data from Djabbarah, N.F. et al., *Colloid Polym. Sci.*, 256, 1002, 1978 [41].)

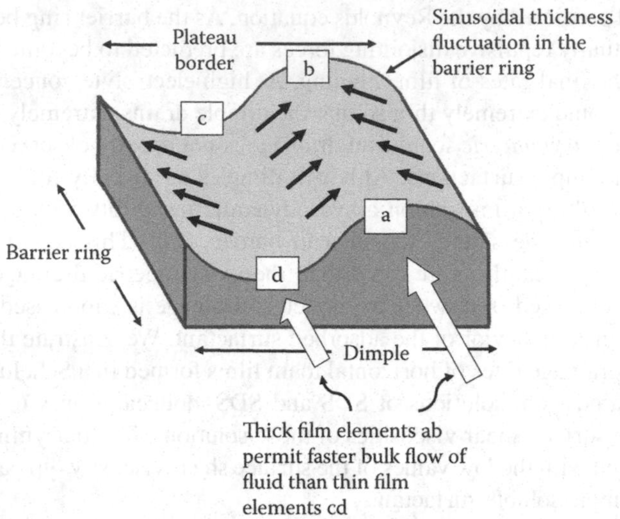

FIGURE 1.8 Origin of hydrodynamic instability driving asymmetric drainage in horizontal foam films. Thickness fluctuation in barrier ring means that region ab is thicker than contiguous region cd. Fluid flow rate from dimple to Plateau border therefore increases relative to cd in ab. Both relative shear stress and corresponding surface tension gradient must also increase in ab according to Equation 1.1. This means that surface tension in ab will be generally greater than cd, which will mean additional surface tension gradient, orthogonal to that between dimple and Plateau border, driving fluid from the thin to the thick parts of barrier ring. This will reinforce instability. However, establishment of differences in surface tension gradients ab and cd between dimple and Plateau border will be resisted by high values of surface shear viscosity leading to suppression of instability. (Adapted with permission from Joye, J. et al., *Langmuir*, 8, 3083. Copyright 1994 American Chemical Society.)

will exist between the thin and thick regions where the surface tension of the former region will be slightly lower than the latter. That orthogonal surface tension gradient will therefore drive liquid from the thin regions to the thick regions, reinforcing the perturbation. The latter will therefore grow until discharge of the whole dimple is facilitated. Joye et al. [40] have subjected this putative process to a linear stability analysis and have shown that the hydrodynamic instability is suppressed, so that axisymmetric drainage occurs, if either the dimension of the film is small enough or if the surface shear viscosity is high enough. The latter is of course consistent with the experimental findings shown in Figure 1.7. However, we must emphasize that the instability is ubiquitous in draining foam films formed from aqueous solutions of common surfactants. Such films therefore exhibit rapid asymmetric drainage by means of this hydrodynamic instability for which no analytical theoretical expression is available for the prediction of the rate of drainage.

A related phenomenon to asymmetric drainage in horizontal cylindrical foam films is that of the so-called marginal regeneration. This process occurs in vertical foam films subject to both capillary suction from the Plateau borders and gravity. It is manifest as an apparent turbulent motion on the margin of foam films where thin elements of film are drawn out of the Plateau border and thick elements are sucked into the border causing continuous regeneration of the margin of the film. The thin elements of film then rise in the gravity field until they reach a height where the film has the same thickness. The phenomenon is readily observable in reflected white light where interference colors distinguish film elements of different thickness. Marginal regeneration can easily be seen in the large vertical foam films (of size >~1 cm) held in metal or glass frames, which are often used for study of aspects of antifoam action (see, e.g., references [44–46]). It is also often visible in the large bubbles formed at the top of foam columns generated by air entrainment (e.g., during hand dishwashing). A detailed description of the phenomenon is to be found in the monograph of Mysels et al. [3]. Images of films exhibiting marginal regeneration are given in Figure 1.9.

Joye et al. [40] argue that marginal regeneration is driven by the same hydrodynamic instability as that which causes asymmetric drainage in horizontal cylindrical foam films. The main difference concerns the role of gravity in marginal regeneration, which causes thin film elements resulting from the growing perturbations to rise up the film. The net effect of the process is relatively rapid drainage of the film. As with asymmetric cylindrical film drainage, no analytical expression is available for the prediction of the rate of film drainage. The hydrodynamic instability is suppressed if the surface dilatational modulus is sufficiently high [15, 47]. It is probable that high surface shear viscosities may also lead to suppression of the instability as is found with small horizontal films. We note, however, that high surface shear viscosities often apparently accompany high surface dilatational moduli, which renders selection of the key property difficult. The resultant "rigid" films drain by a Poiseuille-like flow for which Mysels et al. [3] present an analytical expression. Vertical films, which exhibit marginal regeneration, typically drain at rates two orders of magnitude faster than rigid films in which the instability is suppressed [3].

It should be clear then that asymmetric drainage of horizontal cylindrical films and marginal regeneration in vertical foam films, although often apparent, are not always exhibited. Liquid flow by capillary suction from the Plateau border is,

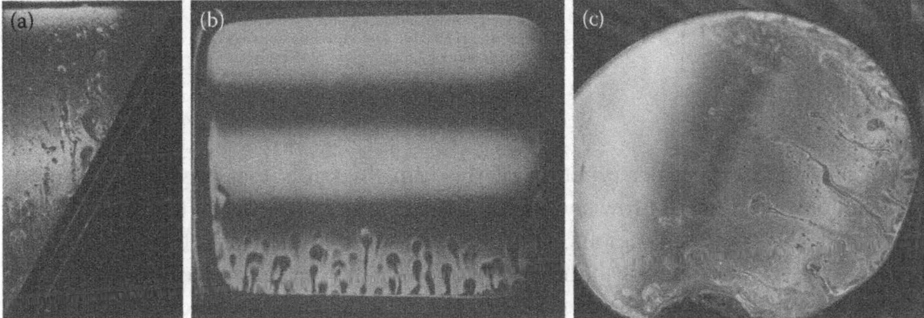

FIGURE 1.9 Marginal regeneration. (a) Film in inverted triangular glass frame. (b) Film in rectangular glass frame illustrating marginal regeneration at bottom of film (it is absent from top of film). (c) Large film (~5.5 cm radius) in cylindrical vessel illustrating marginal regeneration at angle of only ~10° to horizontal. In all cases, rising film elements move up to film regions of the same thickness with which they coalesce. (Image (a) after Mysels, K.J. et al., *Soap Films, Studies of Their Thinning*, Pergamon, London, 1959. Image (b) after Nierstrasz, V.A., Marginal regeneration, PhD Thesis, Delft University of Technology, 1996 [48].)

however, always present. It is therefore misleading for Weaire and Hutzler [5] to state that "the traditional but somewhat obscure phrase for this (liquid flow by capillary suction) process is marginal regeneration." For the relevant hydrodynamic instabilities to occur and cause marginal regeneration, the air–liquid surfaces should additionally have the required rheological characteristics (i.e., low surface viscosity, etc.).

1.3.3 Disjoining Forces and Foam Film Stability

By means of these processes, foam films drain until they either form metastable films or rupture. Which fate awaits a given film is decided largely by the nature of the forces that exist across a foam film as it thins. These forces are usually termed "disjoining" forces and are positive if they resist the thinning of the film and negative if they assist thinning of the film. They are, however, only significant in comparatively thin films of ≤100 nm. The metastability of the film requires that resulting disjoining pressure balances the capillary pressure exerted by the Plateau borders. The disjoining pressure is a function of the film thickness—plots of disjoining pressure against film thickness (disjoining pressure isotherms) largely determine the properties of those metastable films.

The disjoining pressure, $\Pi_{ALA}(h)$ in air–liquid–air foam films, stabilized by adsorbed surfactant, is made up of contributions from several components. These include electrostatic forces (from overlapping electrostatic double layers), van der Waals forces, structural forces (arising from the presence of close-packed layers of micelles or nanoparticles in films [49]), steric forces (derived from the overlapping head group—head group interactions of surfactants or chain–chain interactions of polymers adsorbed on opposite sides of the film), and oscillatory forces, etc. These various contributions conspire to produce disjoining pressure isotherms of various forms.

Arguably, the simplest isotherm is one dominated entirely by the van der Waals forces, which are invariably present. Such an isotherm is shown schematically in Figure 1.10, where the disjoining pressure is always negative and where it becomes more negative as the film thins so that

$$\frac{\mathrm{d}\Pi_{ALA}(h)}{\mathrm{d}h} > 0. \tag{1.6}$$

In consequence, any perturbation of the film thickness that produces thin and thick regions will tend to grow spontaneously because molecules in the thin regions will transfer to the thick regions. However, any perturbation of a film to produce thick and thin regions must also inevitably increase the surface area of the film. This will increase the number of molecules in the relatively weak attractive force field close to the air–liquid surface. Such an increase in surface area will therefore be resisted by an opposing force—the surface tension [51, 52].

Thickness perturbations, of a thermal or mechanical nature, may be considered to be of a wavelike nature. A symmetrical sinusoidal perturbation is shown in Figure 1.11. Here the thin part of the film is subject to two opposing forces. Thus, a capillary pressure due to the surface tension tends to suck liquid back into the thin part of the film, and a disjoining force tends to push liquid away.

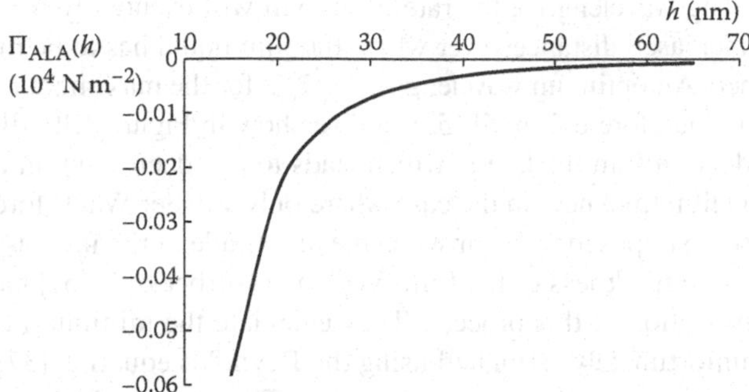

FIGURE 1.10 Plot of disjoining pressure isotherm for a plane-parallel air–water–air film with only van der Waals interactions. Here disjoining pressure, Π_{ALA}, is given by $\Pi_{ALA}(h) = A_H/6\pi h^3$ where A_H is Hamaker constant of 3.7×10^{-20} J. (From Isrealachvilli, J.N. *Intermolecular and Surface Forces with Applications to Colloidal and Biological Systems*, Academic Press, London, 1985 [50].)

The magnitude of the capillary pressure is determined by the curvature of the film surface. For a given amplitude of the perturbation, the curvature is determined by the wavelength. Thus, the shorter the wavelength, the more marked the curvature and the stronger the capillary pressure. Vrij and Overbeek [51, 52] deduce a critical wavelength, λ_{crit}, above which disjoining forces will dominate over the capillary pressure and the perturbation will spontaneously grow. The critical wavelength is

$$\lambda_{crit} = \left[\frac{2\pi^2\sigma_{AL}}{d\Pi_{ALA}(h)/dh}\right]^{\frac{1}{2}} \tag{1.7}$$

The rate of growth of the perturbation will increase with increasing wavelength λ for $\lambda > \lambda_{crit}$ because the damping effect of the capillary pressure will decrease. However,

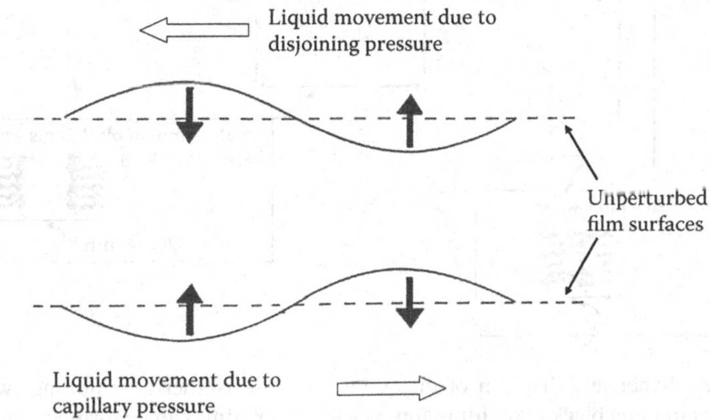

FIGURE 1.11 Sinusoidal thickness perturbations in thin liquid foam film.

for sufficiently long wavelengths, the rate of growth will eventually begin to decline because of the increased distances over which the film liquid has to be moved against viscous resistance. An optimum wavelength of $\sqrt{2_{crit}}$ for the maximum rate of growth of a perturbation therefore exists [51, 52]. As we show in Figure 1.10, $d\Pi_{ALA}(h)/dh$ is strongly dependent on film thickness, which leads to λ_{crit} decreasing in proportion to the square of the film thickness in the case where only van der Waals forces prevail.

Film collapse is supposed to occur when the amplitude of the fastest-growing perturbation equals the thickness of the film. Vrij and Overbeek [51, 52] have produced the simplest description of this process. They calculate the minimum total time for film drainage (unfortunately estimated using the Reynolds equation [37]) and subsequent growth of the fastest-growing perturbation. The average thickness of the film at the moment of rupture is the critical thickness h_{crit}. Experimental measurements of h_{crit} for microscopic foam films are in the region of a few tens of nanometers (see, e.g., reference [53]).

Films that are completely dominated by van der Waals forces and exhibit the type of instability described by Vrij and Overbeek [51, 52] are likely to be either free of surfactant or have low levels of surfactant adsorption. The presence of high levels of adsorbed surfactant means that positive contributions to the disjoining pressure become significant. In particular, long-range electrostatic and van der Waals forces, together with extremely short range steric forces, can produce isotherms with two minima. Such isotherms are often exhibited by aqueous surfactant solutions [6, 54]. This type of isotherm is illustrated schematically in Figure 1.12. It is characterized by two regions where $d\Pi_{ALA}(h)/dh > 0$ and two regions where $d\Pi_{ALA}(h)/dh < 0$.

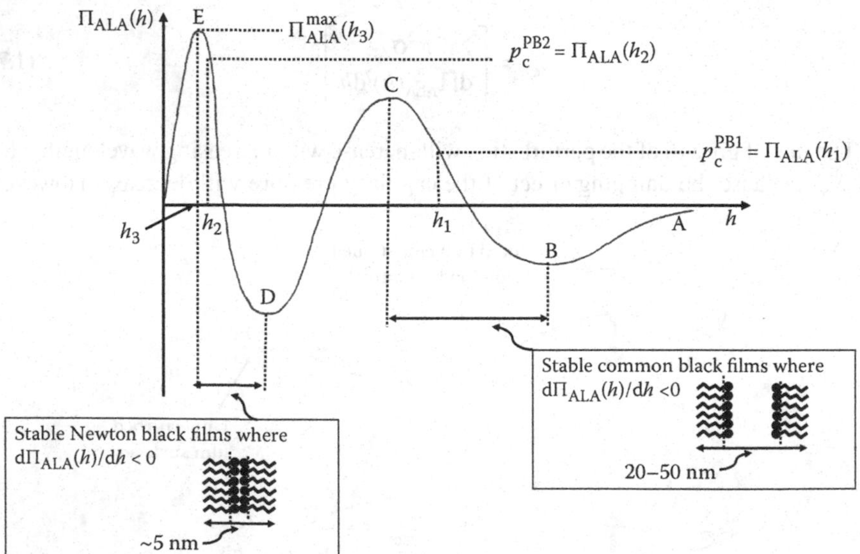

FIGURE 1.12 Schematic diagram of disjoining pressure isotherm exhibiting two stable regions where common black foam films and Newton black films can be formed, respectively (see text for full explanation). Π_{ALA} is air–liquid–air (ALA) foam film disjoining pressure and h is film thickness.

Bergeron [55] has argued that adsorbed surfactant present at the surfaces of the film can also produce other complications in the analysis of the effect of thickness perturbations. Dilatational effects, for example, due to adsorbed surfactant, tend to dampen the fluctuations rendering rupture less probable. Moreover, fluctuations due to lateral adsorption density fluctuations could lead to lateral fluctuations in disjoining pressure to further complicate the picture. These issues are far from fully resolved.

Despite these complications, disjoining pressure isotherms such as that shown in Figure 1.12 offer an explanation for much of the behavior of the thin liquid films, formed by aqueous solutions of simple surfactants, subject to an applied capillary pressure. Consider, for example, the regions in those isotherms where $d\Pi_{ALA}(h)/dh < 0$. This means that the disjoining pressure increases as the film thins so that any perturbation in film thickness will be resisted. Such regions are therefore stable. Conversely in regions where $d\Pi_{ALA}(h)/dh > 0$, any perturbation in film thickness will be enhanced leading to instability. For the isotherm shown in Figure 1.12, there are therefore thickness regions where stable films can be formed (regions BC and DE) and thickness regions where unstable films are formed (regions AB and CD). In consequence, regions AB and CD are not accessible to experimental measurement. In this type of isotherm, the region AD is largely determined by electrostatic and van der Waals contributions and can be readily accounted for by the Derjaguin–Landau–Verwey–Overbeek (DLVO) theory (see, e.g., reference [6]). However, the region DE, if it is present, is usually largely due to short range steric interactions between molecules adsorbed on opposite sides of the film and cannot be accounted for by that theory.

Consider now the likely evolution of a film with a disjoining pressure isotherm similar to that shown in Figure 1.12, which is subject to a capillary suction, p_c^{PB1}, from the adjacent Plateau border. Such a film will drain in the unstable region AB until the critical thickness is reached whereupon perturbations in thickness will not grow to produce holes and film rupture but will rather produce stable regions, or "spots," which continue to expand and cover the whole film until it reaches an equilibrium thickness h_1 where the capillary pressure equals the disjoining pressure (i.e., $p_c^{PB1} = \Pi_{ALA}(h_1)$). The film thicknesses involved at this stage are in the region of 20–50 nm. Such films are so thin that destructive interference of light occurs so that they (and indeed the "spots") appear black in reflected light. They are usually termed "common black films."

If now the capillary pressure increases to p_c^{PB2} so that it exceeds the maximum at C in the isotherm then the film will drain to another unstable region and will jump from C to form another stable film at D. Equilibrium will again be established when the film thins to thickness h_2 where $p_c^{PB1} = \Pi_{ALA}(h_2)$. Such films are extremely thin, being little more than bilayer leaflets of thickness in the region of 5 nm. They are usually termed "Newton black films." Further increases in capillary pressure to $p_c^{PB3} > \Pi_{ALA}^{max}(h_3)$ should simply lead to film rupture. However, thickness fluctuations cannot exist in such thin films. It has therefore been argued by Kashchiev and Exerowa [56] that rupture occurs by nucleation of holes caused by thermal fluctuations in these films.

The transition from a disjoining pressure isotherm dominated by van der Waals forces, such as that shown in Figure 1.10, to one with at least a region where $d\Pi_{ALA}(h)/dh < 0$ is present can be observed at a certain surfactant concentration. That is of course represented by a transition from rupture to black spot formation. The concentration at which this transition occurs is designated c_{black}. It occurs because of the effects of changes in the adsorption layer of surfactant on the thickness dependence of $\Pi_{ALA}(h)$ and $d\Pi_{ALA}(h)/dh$. Foam film stability tends to increase markedly at concentrations above c_{black}.

Values for c_{black} in microscopic films have been measured for a number of surfactants [57]. In the case of aqueous solutions of surfactants, c_{black} is generally significantly lower than the CMC. Thus, for example, for SDS, $c_{black} = 1.6 \times 10^{-6}$ M [57] and the CMC = 8.4 mM [58], and for dodecyl hexaethyleneglycol ($C_{12}EO_6$), $c_{black} = 4.9 \times 10^{-6}$ M [57] and the CMC = 8.7×10^{-6} M [59] at 25°C.

We have seen then that film rupture may occur because surface tension gradients are not sufficiently high to enable the film to withstand stress, because $d\Pi_{ALA}/dh$ is always positive so that rupture is inevitable at a certain critical thickness, or because the Plateau border capillary pressure exceeds any maximum in the relevant disjoining pressure isotherm. However these phenomena are associated with low concentrations of surfactant (at least if we consider films formed slowly so that equilibrium between the air–liquid surface and the intralamellar liquid is maintained). Thus for example, we have $c_{max} \ll$ CMC for $d\varepsilon_G/dh < 0$ and $c_{black} \ll$ CMC. Elimination of both causes of rupture should therefore be readily achieved at sufficiently high concentrations of surfactant. The poor discrimination in foamability often found with relatively concentrated aqueous micellar solutions of surfactants may well be attributable to that cause. Interesting differences in foamability are, however, often revealed when films are either formed rapidly so that equilibrium adsorption is not obtained and conditions for stability are thereby violated or if antifoam is added to the solution.

1.4 PROCESSES ACCOMPANYING AGING OF FOAM

1.4.1 Capillary Pressure Gradients

Foam film stability is, as we have seen, determined in part by the lack of balance between the disjoining pressure and the capillary pressure applied to the films by the Plateau borders. The capillary pressure also drives the process of film thinning, which precedes film rupture. This in turn influences the frequency of foam film rupture. The relative magnitudes of the capillary pressure and the hydrostatic head in the foam also determine the bulk drainage behavior of the foam. If the capillary pressure at the top of the foam balances the hydrostatic head, then bulk drainage will not occur. As we show in later chapters, the stability of the films between antifoam entities and the gas liquid surface—the so-called pseudoemulsion films [60]—may also be determined by the lack of balance between the disjoining pressure in the pseudoemulsion film and the Plateau border capillary pressure. It is therefore important to clearly define the nature of the pressure distribution in the continuous phase of a foam as represented by the system of Plateau border channels. In this, we follow closely the arguments of Princen [61].

We first consider a polydisperse foam under mechanical equilibrium where drainage is absent. The upper region of such a foam is shown schematically in Figure 1.13. The continuous gas–liquid surface at $y = 0$ covers the dome-shaped tops of bubbles and the upper Plateau borders. The Laplace pressure jump at the gas–liquid surface of the Plateau borders is the difference between the atmospheric pressure, P_{atm}, and

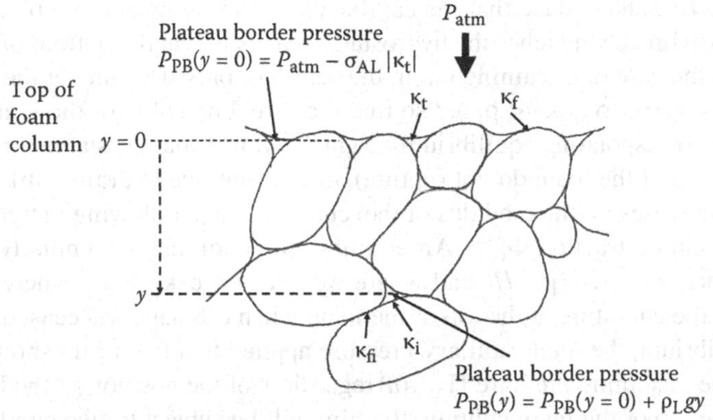

FIGURE 1.13 Schematic illustration of factors contributing to Plateau border pressure in foam. (Adapted with permission from Princen, H.M. *Langmuir*, 4, 164. Copyright 1988 American Chemical Society.)

the pressure $P_{PB}(y = 0)$ in the Plateau borders at $y = 0$. We therefore define the Plateau border capillary pressure, $p_c^{PB}(y = 0)$, by

$$p_c^{PB}(y = 0) = P_{PB}(y = 0) - P_{atm} = -\sigma_{AL}\left|\kappa_t\right| \tag{1.8}$$

where κ_t is the curvature of the air–liquid Plateau border surface at the top of the foam column where $y = 0$ and where we note that κ_t is negative with respect to the fluid in the border. As stated by Princen [61], κ_t is a constant everywhere at the top surface of the foam and is independent of the detailed shape of that surface.

For the total Plateau border pressure $P_{PB}(y)$ at some distance y into the bulk of the foam, we must allow for the hydrostatic head and must therefore write

$$P_{PB}(y > 0) = P_{atm} + \rho_L g y + p_c^{PB}(y = 0) = P_{atm} + \rho_L g y - \sigma_{AL}\left|\kappa_t\right| \tag{1.9}$$

All this is illustrated in Figure 1.13.

At equilibrium the pressure in the liquid phase at the bottom of a foam column must equal the atmospheric pressure plus that due to the weight of the foam. That pressure must also equal the Plateau border pressure at the bottom of the foam. Therefore, from Equation 1.9, we can write

$$P_{PB}(y = H_0^e) = P_{atm} + \rho_L g H_0^e - \sigma_{AL}\left|\kappa_t\right| = P_{atm} + \rho_L g \int_0^{H_0^e} (1 - \Phi_G^{foam})dy \tag{1.10}$$

where H_0^e is the height of the equilibrium foam. The integral accounts for the weight of the foam column where Φ_G^{foam} is the gas phase volume fraction (and where the density of the air is neglected). In the case of an equilibrium foam in the dry limit, the average gas volume fraction $\Phi_G^{foam} \rightarrow 1$. We can therefore often neglect the integral in Equation 1.10 and deduce that the capillary pressure jump at the top of the foam $\sigma_{AL}\,|\kappa_t|$ approximately matches the hydrostatic head $\rho_L g H_0$ at the bottom of the foam. However in the case of a draining foam, the capillary pressure jump at the top of the foam, $\left|p_c^{PB}(y=0)\right| = \sigma_{AL}\left|\kappa_t\right| < \rho_L g H$ so that the actual height H of the foam column exceeds the corresponding equilibrium height H_0. If this foam is stable (in that foam films at the top of the foam do not rupture), it will continue to drain until $H = H_0^e$ as a result of increases in (the modulus of the) curvature, $|\kappa_t|$, following increases in the gas phase volume fraction Φ_G^{foam}. An equilibrium condition is eventually achieved where $\sigma_{AL}\left|\tilde{\kappa}_t\right| = \rho_L g H = \rho_L g H_0^e$ and where we must have $\left|\tilde{\kappa}_t\right| > \left|\kappa_t\right|$ where $\left|\tilde{\kappa}_t\right|$ is the modulus of the curvature at the top of the foam when drainage has ceased.

At equilibrium, the total capillary pressure applied to a foam film should be balanced by the disjoining pressure $\Pi_{ALA}(h)$ regardless of the position of the film in the foam. At the top of the foam column, the film will be subject to two capillary pressures—that due to the curvature of the film and that due to the Plateau border, both of which will conspire to force liquid out of the film, which will be resisted by a positive disjoining pressure. Therefore, at equilibrium, the disjoining pressure is given by

$$\Pi_{ALA}(h) = -\left(p_c^{PB}(y=0) + p_c^{film}\right) = \sigma_{AL}\left(\left|\kappa_t\right| + \left|\kappa_f\right|\right) \qquad (1.11)$$

where κ_f is the curvature of the dome-shaped surface of the bubble films at the top of the foam column and p_c^{film} is the capillary pressure jump in those films. In a dry foam where $\Phi_G^{foam} \to 1$, then we have $|\kappa_t| \gg |\kappa_f|$.

The capillary pressure in the interior of the foam where $y > 0$ is given by

$$p_c^{PB}(y>0) = -\sigma_{AL}\left|\kappa_i\right| \qquad (1.12)$$

where κ_i is the curvature of the relevant Plateau borders. At equilibrium, the disjoining pressure in the films in the interior of the foam must therefore be given

$$\Pi_{ALA}(h) = \sigma_{AL}\left(|\kappa_i| \pm |\kappa_{fi}|\right) \qquad (1.13)$$

where κ_{fi} is the curvature of the film (the sign of which may be either positive or negative), and where, according to Princen [61], κ_{fi} is zero in the case of a monodisperse Kelvin foam.

We can use these arguments to describe the evolution of a foam in a gravity field. Immediately after generation, the foam is wet and Plateau borders are thick with relatively low curvatures. Drainage of the foam can therefore occur if Equation 1.10 is not satisfied. That process continues until the capillary pressure at the top of the foam column equals the hydrostatic head whereupon drainage ceases. During this process, foam films also drain until the disjoining pressure equals the capillary pressure everywhere in the foam and a condition of mechanical equilibrium is attained. However, this condition represents an unstable equilibrium because gas diffusion between bubbles will occur in response to differences in capillary pressure. This will in turn reduce the number of bubbles and therefore cause the Plateau borders to begin to swell so that drainage again commences. Moreover, the attainment of the equilibrium requires that the disjoining pressure isotherm exhibits a maximum sufficiently high to match the total capillary pressure at the top of the foam column so that Equation 1.10 is satisfied. If that equation cannot be satisfied, foam collapse will commence as the capillary pressure exceeds that maximum in the disjoining pressure isotherm. This process of foam collapse will start at the top of the foam column where, as indicated by Equations 1.11 through 1.13, the capillary pressure is highest.

1.4.2 Foam Drainage

The drainage of the continuous phase liquid out of a foam is a surprisingly complex phenomenon. It is driven by a combination of gravitational, capillary, and viscous forces. The rate at which it occurs in the case of so-called "free drainage" is easily observed by monitoring the rate of accumulation of liquid at the bottom of a foam column. If the foam is generated from a gas of low liquid solubility, then, as we shall discuss in Section 1.4.3, changes in the structure of the foam due to gas diffusion are likely to be slow so that they may be neglected. General treatments must, however, take account of the coupling of gas diffusion with foam drainage (see, e.g., the treatment of Hilgenfeldt et al. [62]).

The observed drainage behavior of foams has often been successfully described [63] by the classic foam drainage equation of Goldfarb et al. [64]. Derivation of this equation therefore provides a useful introduction to the nature of that process. This approach neglects any contribution to the process from structural changes due to gas diffusion or film rupture and makes the assumption that the liquid content of the foam is entirely contained within the Plateau borders. Drainage is therefore assumed to occur by flow down the interconnected system of Plateau borders (illustrated by a simulation for the case of a foam made up of Kelvin cells in Figure 1.2b). Any contribution to drainage from foam films is neglected. The gas–liquid surfaces are assumed to be immobile. Essentially, the approach assumes that Poiseuille flow occurs in the Plateau borders of a monodisperse foam and that the liquid volume fraction of that foam, (y,t) (where $\Phi_L^{foam}(y,t) = 1 - \Phi_G^{foam}(y,t)$), at any position y and time t, is simply

$$\Phi_L(y,t) = A_{PB}(y,t)L_{PB} \tag{1.14}$$

where y is measured from the top of the foam column and $A_{PB}(y,t)$ is the cross-sectional area of a Plateau border. L_{PB} is the total length of Plateau borders per unit foam volume. The absence of gas diffusion and film rupture means that L_{PB} is not a function of y or time t. The task of the theory then is to determine $A_{PB}(y,t)$, and therefore $\Phi_L(y,t)$, as a function of y and t.

Following Weaire and coworkers [5, 23, 63], we initially suppose the border to be vertically orientated. The cross-sectional area of the border, $A_{PB}(y,t)$, will change with time as it drains subject to gravitational, capillary, and viscous forces. Consider then a segment of Plateau border of length Δy, which is shown schematically in Figure 1.14. The volume of the segment is approximated by $[A_{PB}(y,t) + (\partial A_{PB}(y,t)/\partial y)_t \Delta y/2]\,\Delta y$, which is the limit as $\Delta y \to 0$ becomes simply $A_{PB}(y,t)\Delta y$. The total gravitational force driving drainage in the segment is $\rho g\, A_{PB}(y,t)\Delta y$ or simply ρg per unit volume. The gravitational force is resisted in part by a capillary pressure gradient,

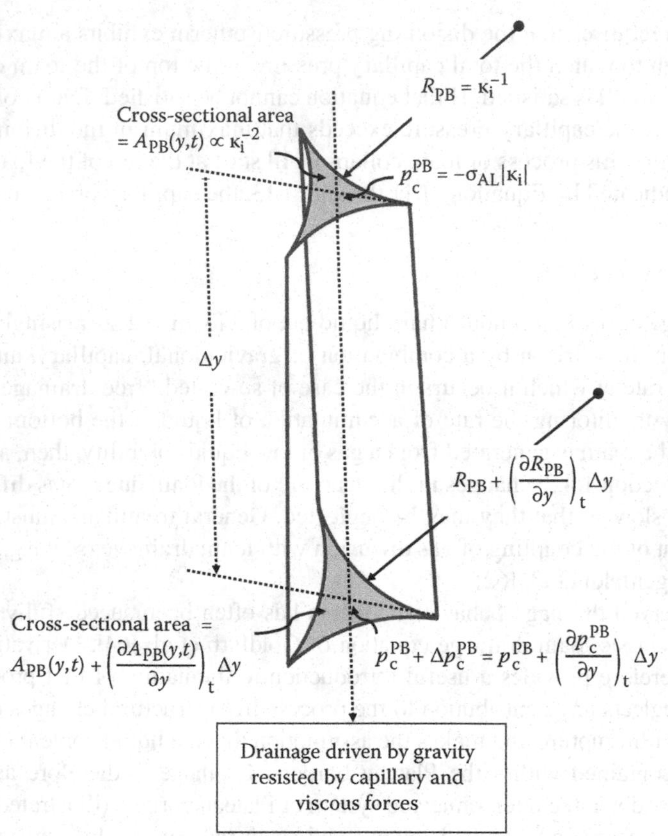

FIGURE 1.14 Plateau border segment subject to increase in cross-sectional area and decrease in capillary pressure along length Δy and where R_{PB} and A_{PB} are radius of curvature and cross-sectional area of Plateau border, respectively.

$(\partial p_c^{PB}/\partial y)_t$ which varies as a result of narrowing of the Plateau border with increasing y. The segment is therefore subject to a net capillary force of $(\partial p_c^{PB}/\partial y)_t \cdot A_{PB}(y,t)\Delta y$ or simply $(\partial p_c^{PB}/\partial y)_t$ per unit volume. The gravitational force is also resisted in part by viscous friction. By analogy with Poiseuille flow in a cylinder (or even dimensional analysis), the relevant shear force per unit volume is $k_{geom}u_{PB}^{m}(y,t)\eta_L/A_{PB}(y,t)$, where k_{geom} is a dimensionless geometrical constant dependent on the shape of the Plateau border (where $k_{geom} \sim 50$), $u_{PB}^{m}(y,t)$ is the mean liquid velocity over the cross-section of the Plateau border [5, 23, 63]. We can therefore write for the net force/unit volume of the segment of Plateau border that

$$\rho_L g - (\partial p_c^{PB}/\partial y)_t - k_{geom}u_{PB}^{m}(y,t)\eta_L/A_{PB}(y,t) = 0 \qquad (1.15)$$

where the capillary pressure is given by

$$p_c^{PB} = -\sigma_{AL}|\kappa_i| = -\sigma_{AL}k_{geom}/\sqrt{A_{PB}(y,t)} \qquad (1.16)$$

and where $\tilde{k}_{geom}$ is another dimensional geometrical constant ($\tilde{k}_{geom} \sim 0.4$ [5, 23, 63]). Equation 1.15 is in fact analogous to Darcy's law for flow in solid porous media where k_{geom} is the effective reciprocal of the "permeability." However, we should note that, unlike the channels in porous solids, the geometry of the Plateau border channels changes as liquid flows through them.

Drainage from the Plateau border segment must also satisfy the conservation of matter. The resulting continuity equation is

$$\left(\frac{\partial A_{PB}(y,t)}{\partial t}\right)_y = -\left(\frac{\partial A_{PB}(y,t).u^{m}_{PB}(y,t)}{\partial y}\right)_t \quad (1.17)$$

If we now combine Equations 1.14 through 1.17 and average over all possible orientations of a Plateau border, we find for the liquid volume fraction $\Phi_L^{foam}(y,t)$ at y for a given time t

$$\left(\frac{\partial \Phi_L^{foam}(y,t)}{\partial t}\right)_y + \tilde{\alpha}\left(\frac{\partial}{\partial y}\left[\tilde{\beta}\Phi_L^{foam}(y,t)^2 - \tilde{\gamma}\sqrt{\Phi_L^{foam}(y,t)}\left(\frac{\partial \Phi_L^{foam}(y,t)}{\partial y}\right)_t\right]\right)_t = 0 \quad (1.18)$$

where $\tilde{\alpha} = (3k_{geom}\eta_L L_{PB})^{-1}$, $\tilde{\beta} = \rho g$, and $\tilde{\gamma} = 0.5\,\tilde{k}_{geom}\sigma_{AL}\sqrt{L_{PB}}$.

Equation 1.18 is a version of the foam drainage equation of Goldfarb et al. [64] and Verbist et al. [63]. In general, it is not amenable to analytical solution. However, a number of analytical solutions are available if the equation is simplified. The simplest concerns the equilibrium liquid volume fraction profile, $\Phi_L^{foam}(y,t)$, which is obtained by setting $\partial\Phi_L^{foam}(y,t)/\partial t = 0$ in Equation 1.18 and integrating with respect to y. Another, due to Kraynik [65], concerns neglect of the capillary pressure gradient and is obtained by integrating Equation 1.18 with $\tilde{\gamma} = 0$. The solution to Equation 1.18 then becomes

$$\Phi_L^{foam}(y,t) = \frac{y}{2\tilde{\alpha}\tilde{\beta}t} \quad \text{if} \quad \frac{y}{2\tilde{\alpha}\tilde{\beta}t} \leq \Phi_L^{foam}(y,t=0) \quad (1.19)$$

Consider now the free drainage of a foam of a constant initial volume fraction where $(\partial\Phi_L^{foam}(y,t=0)/\partial y) = 0$. Equation 1.19 implies that in this case, $\Phi_L^{foam}(y,t)$ increases linearly with distance y from the top of the foam until it reaches that initial volume fraction at a certain distance $\tilde{y}(t) = (2\tilde{\alpha}\tilde{\beta}t\Phi(y,t=0))$ from the top of the foam, whereupon it remains constant. However, that distance, $\tilde{y}(t)$, from the top of the foam is also predicted to increase linearly with time as the foam drains. Despite these limitations, it appears that this equation describes reasonably well the initial stages of free foam drainage.

A number of other solutions to the foam drainage equation are available, not only for free drainage but also for steady drainage and the so-called solitary waves formed by injecting liquid into the top of a foam column [23]. These solutions appear to agree well with experimental observation [23].

The foam drainage equation has also been extended by Neethling et al. [66] to include foam growth and film collapse as bubbles are injected into a liquid during sparging. Film rupture is introduced as an assumed link between the critical Plateau border capillary pressure and the maximum in the relevant disjoining pressure isotherm. This approach may even be applicable to the effect of antifoam on foam growth and stability in the case where antifoam effect is confined to foam films and where low concentrations of antifoam are involved so that total elimination of foam is avoided. However, in the case of weak antifoams, which accumulate in Plateau borders (see Section 4.5.3), an additional complication arises concerning the effect of that material on drainage through the borders.

An obvious limitation of the foam drainage equation concerns the assumption of monodispersity of the foam. Since foam generation by air entrainment, shaking of vessels, etc., necessarily produces polydisperse foams, the present theory would appear to have direct application only to foams generated by slow bubbling through fine capillaries where foams of low polydispersity can be readily prepared (see Chapter 2, Section 2.2). Another limitation of the foam drainage equation is that the gas–liquid surface is assumed to be immobile. This assumption will not of course be generally relevant for all draining foams where more complex surface rheology may often be found. More details about this issue are to be found in references [26, 67].

1.4.3 Bubble Coarsening (Diffusional Disproportionation)

Foams are subject to diffusional disproportionation or coarsening. As we have seen, this phenomenon arises from differences in capillary pressure between bubbles of different sizes and between bubbles in the foam and the ambient gas phase if present. Thus, gas diffuses from both small bubbles to large bubbles and from bubbles to the ambient gas phase, all of which leads to coarsening of the foam. The resulting structural changes in the foam have a potential impact upon foam drainage. The foam drainage equation clearly neglects this factor, which is of course only justified if this process of coarsening is slow compared with drainage.

The essentials of the process may be illustrated by recourse to the simple treatment made by de Vries [10] more than half a century ago. de Vries [10] considered the rate at which small spherical bubbles would shrink by diffusion of gas to their larger neighbors in the case of a polydisperse wet foam. The model is illustrated schematically in Figure 1.15. The capillary pressure difference, Δp_c, between two spherical bubbles of different radii, r_b and $\tilde{r}_b$, is of course

$$\Delta p_c = 2\sigma_{AL}\left(\frac{1}{r_b} - \frac{1}{\tilde{r}_b}\right) \tag{1.20}$$

which reduces to

$$\Delta p_c = 2\sigma_{AL}/r_b \tag{1.21}$$

if $r_b \ll \tilde{r}_b$. According to Fick's first law the molar rate of transport of gas, dn/dt, from the small bubble to the surrounding larger bubbles should be given by

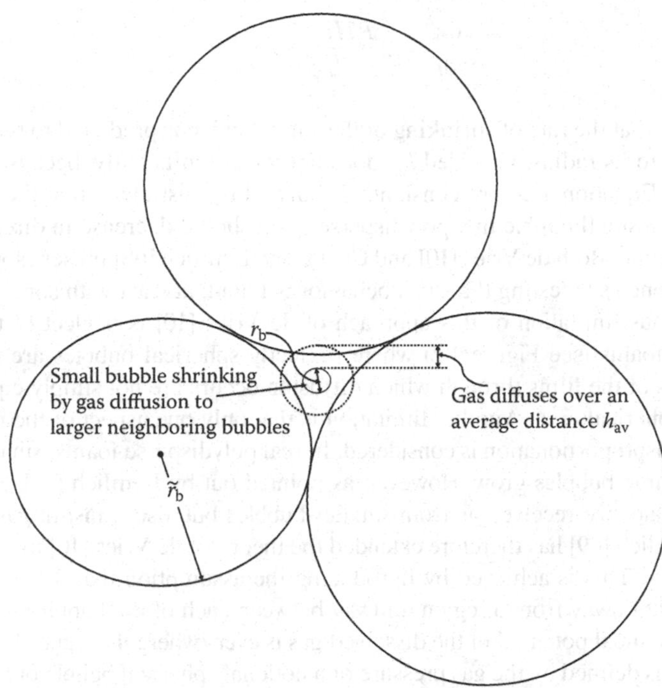

FIGURE 1.15 de Vries model [10] of gas diffusion from small bubbles of radius r_b in a foam to adjacent large bubbles of radius $\tilde{r}_b$, where $r_b \ll \tilde{r}_b$.

$$-\frac{dn}{dt} = DA_b \frac{dc}{dx} \approx DA_b \frac{\Delta c}{h_{av}} \approx DA_b \frac{H_E \Delta p_c}{h_{av}} \tag{1.22}$$

where D is the diffusion coefficient of the gas in water, A_b is the area of the bubble, c is the concentration of dissolved gas in the aqueous phase, and x is the distance between bubbles. The concentration gradient of dissolved gas in the aqueous phase between the small bubble and the surrounding larger bubbles, dc/dx, is approximated rather crudely by $\Delta c/h_{av}$, where h_{av} is a supposed average thickness of the liquid gap between the bubbles. Use of Henry's law means that we can substitute $\Delta c = H_E \Delta p_c$, where H_E is the Henry's law constant for the gas in water.

The number of moles of gas in the small bubble is given by Boyle's law if it is ideal, which means that the rate of transport of gas from that bubble is given by [10]

$$\frac{dn}{dt} = \frac{4\pi r_b^2 P_{atm}}{RT} \frac{dr_b}{dt} \tag{1.23}$$

where the contribution of the capillary pressure to the total pressure in the bubble is small compared with the atmospheric pressure and is therefore neglected. Combining Equations 1.21 through 1.23 together with $A_b = 4\pi r_b^2$ means we can write

$$-\frac{\mathrm{d}r_{\mathrm{b}}}{\mathrm{d}t}=2\frac{RTH_{\mathrm{E}}D}{P_{\mathrm{atm}}}\cdot\frac{\sigma_{\mathrm{AL}}}{h_{\mathrm{av}}r_{\mathrm{b}}} \quad (1.24)$$

which means that the rate of shrinking of the small bubble is predicted to be inversely proportional to its radius, provided h_{av} does not vary significantly, because all other quantities in Equation 1.24 are constants. In turn, this must mean that the square of the radius of a small bubble in a polydisperse foam should decrease in direct proportional to the time. Both de Vries [10] and Cheng and Lemlich [68] present some experimental evidence suggesting that such behavior is found, at least with some systems.

One obvious limitation of this approach of de Vries [10] is neglect of the actual structure of foams (see Figure 1.2) where perfectly spherical bubbles are not found and the areas of the films through which diffusion occurs are not simply equal to the equivalent spherical area. Another limitation is that only one aspect of the process of diffusional disproportionation is considered. In real polydisperse foams, small bubbles shrink and large bubbles grow. However, as pointed out by Lemlich [69], bubbles of an intermediate size receive gas from smaller bubbles but also transmit gas to larger bubbles. Lemlich [69] has therefore extended the theory of de Vries [10] to account for this generality. This is achieved by introducing the assumption that the gas diffuses either toward or away from a region midway between each of the bubbles in the foam where the chemical potential of the dissolved gas is everywhere the same. That chemical potential is defined by the gas pressure of a notional spherical bubble of radius r_{mid}. Following a similar argument to that of de Vries [10], Lemlich [69] shows that the rate of change of any bubble of radius r_b, regardless of relative size, is given by

$$\frac{\mathrm{d}r_{\mathrm{b}}}{dt}=2\frac{RTH_{\mathrm{E}}D}{P_{\mathrm{atm}}}\cdot\frac{\sigma_{\mathrm{AL}}}{h_{\mathrm{av}}}\left(\frac{1}{r_{\mathrm{mid}}}-\frac{1}{r_{\mathrm{b}}}\right) \quad (1.25)$$

where r_{mid} is given by the mean bubble radius by second and first moments and is of course time dependent. Bubbles for which $r_b < r_{mid}$ shrink with time and bubbles for which $r_b > r_{mid}$ grow. Equation 1.25 can be solved by method of finite difference to produce simulations of foam coarsening for a wet polydisperse foam. A simulated plot of the distribution of dimensionless bubble radii as a function of dimensionless time is shown in Figure 1.16 [68, 69]. This simulation used an initial bubble size distribution, $P(\bar{r}_b, t=0)$ at zero time, of the form [68].

$$P(\bar{r}_{\mathrm{b}},t=0)=\frac{\pi}{2}\bar{r}_{\mathrm{b}}\exp\left(-\pi\bar{r}_{\mathrm{b}}^{2}/4\right) \quad (1.26)$$

where the dimensionless bubble radius, $\bar{r}_b$, is equal to r_b divided by the initial arithmetic mean radius. The progressive move to larger bubble sizes with increasing time as a consequence of diffusion, as found in real foams, is clearly illustrated by the simulation.

It is obvious from both Equations 1.24 and 1.25 that the rate of diffusional disproportionation is dependent on the solubility of the gas phase through the magnitude of the Henry's law constant H_E. The magnitude of H_E for carbon dioxide in water is,

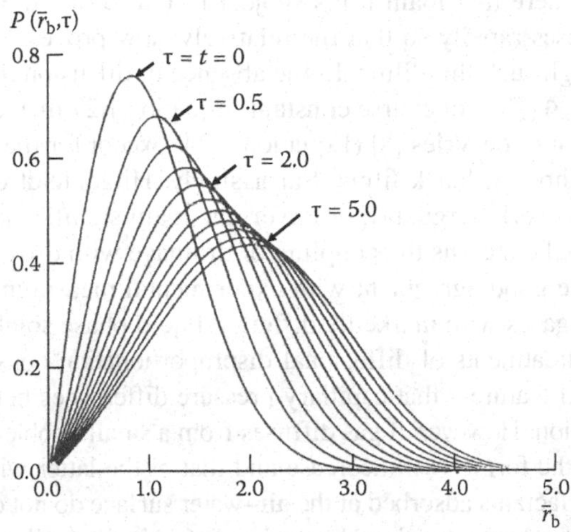

FIGURE 1.16 Simulation of effect of diffusional disproportionation on initial bubble size distribution, $P(\bar{r}_b, \tau = 0)$, where $\bar{r}_b$ is the dimensionless bubble radius and τ is the dimensionless time. (Adapted with permission from Cheng, H.C. and Lemlich, R., *Ind. Eng. Fundam.*, 24, 44. Copyright 1985 American Chemical Society.)

for example, about 50 times greater than for nitrogen. This means that the process is much slower for foams prepared from nitrogen than for those prepared from carbon dioxide. It is also clear from Equations 1.24 and 1.25 that the process will be coupled with foam drainage through the role of film thickness on diffusion and the effect of changes in bubble size distribution on the flow of liquid out of the foam.

An obvious limitation of these approaches concerns the assumption of spherical geometry for the bubbles. Even in the case of "wet" polydisperse foams, of gas volume fraction $0.72 < \Phi_G^{foam} < 0.95$, bubbles are likely to be better considered as deformed polyhedra with Plateau borders and distinct films (see Figure 1.2). Recent work by, for example, Hilgenfeldt et al. [62] has sought to address this issue in the case of polyhedral foam with a relatively low polydispersity. In such a foam, the films separating adjacent bubbles can be curved, which necessarily implies a capillary pressure difference equal to $4\sigma_{AL}\kappa_{mean}$, where κ_{mean} is the mean curvature of the films [62]. That pressure difference will of course drive diffusion of gas from bubble to bubble through the films. The mean curvature then becomes $\kappa_{mean} = 1/\beta r_b$, where r_b is the radius of an equivalent spherical bubble and β is an empirical constant that presumably depends on Φ_G^{foam}. If we suppose that r_b is proportional to the average polyhedral bubble edge length, then the expression of Hilgenfeldt et al. [62] for diffusion-driven coarsening can be rewritten as

$$\frac{dr_b}{dt} = Z(\Phi_G^{foam}).\frac{RTH_E D}{P_{atm}}.\frac{\sigma_{AL}}{h_{bf} r_b} \tag{1.27}$$

where $Z(\Phi_G^{foam})$ is a dimensionless empirical function dependent on the foam geometry (and therefore the gas phase volume fraction) and h_{bf} is the thickness of a black film. It is assumed here that foam films subject to Plateau border suction drain to black film thicknesses rapidly so that the relatively slow process of diffusion must always occur through such thin films. In the absence of diffusion to the ambient gas phase and drainage, Φ_G^{foam} is of course constant. Equation 1.27 then closely resembles the original equation of de Vries [8] (Equation 1.24) except for the assumption concerning diffusion through black films. Surprisingly, Hilgenfeldt et al. [62] do not even cite the earlier work. Arguably, however, the main significance of the work of Hilgenfeldt et al. [62] concerns the coupling of drainage with diffusional disproportionation to produce good agreement with experimental measurements of drainage rates in the case of gases with markedly different liquid phase solubilities.

The theoretical treatments of diffusional disproportionation considered here have one common central feature—that capillary pressure differences between contiguous bubbles drive diffusion. However, if gas diffuses from a small bubble to a large bubble, the surface area of the former will decrease and that of the latter will increase. If the foam-stabilizing surfactants adsorbed at the air–water surface do not desorb in the case where the surface area decreases, then the surface tension will decline and the capillary pressure driving diffusion will be diminished. Similarly in the case where more surfactant does not adsorb as the surface area is increased, the surface tension will increase together with the capillary pressure, which will diminish the tendency of gas to diffuse into the relevant bubble. The tendency for the surface tension to change as the surface area changes is determined by the surface dilatational elasticity defined in the case of aqueous surfaces as $\varepsilon = d\sigma_{AL}/d\ln A$. Unlike the Gibbs elasticity defined in Section 1.2.1, this quantity is general and does not simply concern depletion in foam films. Gibbs [34] has shown that provided $\varepsilon > \sigma_{AL}/2$, diffusional disproportionation is suppressed. In general, for simple surfactants in aqueous solutions, this condition is not usually satisfied. However, for slowly adsorbing/desorbing substances such as proteins, it does suggest a limit to the rate at which diffusional disproportionation can occur. Indeed, Ronteltap and Prins [70] have shown that changes in surface dilatational properties can have a marked effect on gas diffusion from bubbles.

1.5 SUMMARIZING REMARKS

It is obvious that foam is a complex phenomenon. One aspect of that complexity concerns the coupling of various processes that are often studied experimentally in isolation. We have, for example, seen that surface rheology can influence both diffusional disproportionation and foam drainage. Indeed diffusional disproportionation is necessarily coupled to foam drainage. It would seem possible that polydispersity enhances the rate of diffusional disproportionation, which in turn could mean significant coupling with foam drainage at earlier foam ages than would be the case for the corresponding monodisperse foam. Relatively slow transport of surfactant to gas–liquid surfaces may in principle influence not only surface rheology and surface tension gradients but could also produce non-equilibrium disjoining pressures, which lead to diminished foam film stability.

Incomplete understanding of these coupled phenomena must represent a challenge for future research. One particular challenge concerns the relation between the mode of foam generation and the state of the relevant air–liquid surfaces. Different methods of foam generation in general mean differences in the state of air–liquid surfaces as foam films are formed, which in turn can mean differences in the stability of those films. Patist et al. [71], for example, show that solubilized dodecanol enhances the foamability of aqueous micellar solutions of SDS in the case of foam generation by sparging (use of gas bubblers) but diminishes it in the case of foam generation by shaking. These differences seem to concern coupling of the adverse effect of dodecanol on the rate of micelle breakdown [71] and the relative hydrodynamic conditions and rates of formation of air–water surfaces with these two methods. We return to this particular example in more detail in Section 4.2. This issue is also of direct relevance for the performance of antifoams because their effectiveness is in some cases coupled with the rate of transport of surfactant to gas–liquid surfaces and therefore the state of those surfaces (see, e.g., references [72–75]).

ACKNOWLEDGMENT

The author is particularly grateful to Dr. Paul Grassia (of the University of Manchester) for helpful comments concerning Sections 1.4.1 and 1.4.2.

REFERENCES

1. Boys, C.V. *Soap Bubbles, Their Colours and Forces Which Mold Them*, Dover Publications, New York, 1959 (original edition 1890).
2. Bikerman, J.J. *Foams*, Springer-Verlag, Berlin, 1973.
3. Mysels, K.J., Shinoda, K., Frankel, S. *Soap Films, Studies of their Thinning*, Pergamon, London, 1959.
4. Isenberg, C. *The Science of Soap Films and Soap Bubbles*, Tieto Ltd., Clevedon, 1978.
5. Weaire, D., Hutzler, S. *The Physics of Foams*, Clarendon Press, Oxford, 1999.
6. Exerowa, D., Kruglyakov, P.M. *Foams and Foam Films*, Elsevier, Amsterdam, 1998.
7. Cantat, I., Cohen-Addad, S., Elias, F., Graner, F., Höhler, R., Pitois, O., Rouyer, F., Saint-Jalmes, A. *Les Mousses, Structure et Dynamique*, Belin, Paris, 2010.
8. Prud'homme, R.K., Khan, S.A. (eds.), *Foams, Theory, Measurements, and Applications*, Marcel Dekker, New York, Surfactant Sci. Series, 1996, Vol 57.
9. Stevenson, P. (ed.), *Foam Engineering, Fundamentals and Applications*, John Wiley & Sons, Chichester, UK, 2012.
10. de Vries, A.J. Foam stability, *Rec. Trav. Chim.*, 77, 81, 209, 283, 383, and 441, 1958.
11. Kitchener, J.A., Cooper, C.F., *Quart. Rev.*, 13, 71, 1959.
12. Kitchener, J.A. Foams and free liquid films, in *Recent Progress in Surface Science* (Danielli, J., Pankhurst, K.G.A., Riddiford, A.C., eds.), 1964, Vol 1, Chpt 2, p 51.
13. Scheludko, A. *Adv. Colloid Interface Sci.*, 1, 391, 1967.
14. Hansen, R.S., Derderian, E.J. Problems in foam origin, drainage and rupture, in *Foams* (Akers, R.J., ed.), Academic Press, London, 1977, p 1.
15. Lucassen, J. Dynamic properties of free liquid films and foams, in *Anionic Surfactants, Physical Chemistry of Surfactant Action* (Lucassen-Reynders, E.H., ed.), Marcel Dekker, New York, Surfactant Sci. Series, 1981, Vol 11, Chpt 6, p 217.
16. de Feijter, J.A. Thermodynamics of thin liquid films, in *Thin Liquid Films* (Ivanov, I.B., ed.), Marcel Dekker, New York, Surfactant Sci. Series, 1988, Vol 29, Chpt 1, p 1.
17. Hartland, S. Coalescence in dense-packed dispersions, in *Thin Liquid Films* (Ivanov, I.B., ed.), Marcel Dekker, New York, Surfactant Sci. Series, 1988, Vol 29, Chpt 10, p 663.

18. Wasan, D.T., Nikolov, A.D., Lobo, L.A., Koczo, K., Edwards, D.A. *Prog. Surf. Sci.*, 39, 119, 1992.
19. Malysa, K. *Adv. Colloid Interface Sci.*, 40, 37, 1992.
20. Kralchevsky, P.A., Dimitrov, K., Ivanov, I.B. Thin liquid film physics, in *Foams; Theory, Measurements and Applications* (Prud'homme, R.K., Khan, S.A., eds.), Marcel Dekker, New York, Surfactant Sci. Series, 1996, Vol 57, Chpt 1, p 1.
21. Pugh, R.J. *Adv. Colloid Interface Sci.*, 64, 67, 1996.
22. Verbist, G., Weaire, D., Kraynik, A.M. *J. Phys. Condens. Matter*, 8, 3715, 1996.
23. Weaire, D., Hutzler, S., Verbist, G., Peters, E. *Adv. Chem. Phys.*, 102, 315, 1997.
24. Bergeron, V. *J. Phys. Condens. Matter*, 11, R215, 1999.
25. Bhakta, A., Ruckenstein, E. Drainage and collapse in standing foams, in *Surface and Interfacial Tension; Measurement, Theory, and Applications* (Hartland, S., ed.), Marcel Dekker, New York, Surfactant Sci. Series, Vol 119, p 1.
26. Saint-Jalmes, A. *Soft Matter*, 2, 836, 2006.
27. Malysa, K., Lunkenheimer, K. *Curr. Opin. Colloid Interface Sci.*, 13, 150, 2008.
28. Princen, H.M., *Langmuir*, 2(4), 519, 1986.
29. Rosato, A., Strandburg, K., Prinz, F., Swendsen, R., *Phys. Rev. Lett.*, 58(10), 1, 1987.
30. Hartland, S., Barber, A.D. *Trans. Inst. Chem. Engrs.*, 52, 43, 1974.
31. Plateau, J.A.F. *Statique Expérimentale et Théorique des Liquides Soumis aux Forces Moléculaires*, Gauthier-Villars, Paris, 1873, 2 Vols.
32. Steiner, L., Hunkeler, R., Hartland, S. *Trans. Inst. Chem. Engrs.*, 55, 153, 1977.
33. Cheng, H.C., Lemlich, R. *Ind. Eng. Chem. Fundam.*, 22, 105, 1983.
34. Gibbs, J.W. In *The Scientific Papers*, Dover, New York, Vol 1, 1961; or Gibbs, J.W. *The Collected Works of J. Willard Gibbs*, Longmans Green, New York, Vol 1, 1928.
35. Malysa, K., Miller, R., Lunkenheimer, K. *Colloids Surf.*, 53, 47, 1991.
36. Prins, A., Dynamic surface properties and foaming behaviour of aqueous surfactant solutions, in *Foams* (Akers, R.J., ed.), Academic Press, London, 1977, p 45.
37. Reynolds, O. *Phil. Trans. Roy. Soc., London*, A177, 157, 1886.
38. Joye, J., Miller, C.A., Hirasaki, G. *Langmuir*, 8(12), 3083, 1992.
39. Scheludko, A., Exerowa, D. *Comm. Dep. Chem., Bulg. Acad. Sci.*, 7, 123, 1959.
40. Joye, J., Hirasaki, G., Miller, C.A. *Langmuir*, 10(9), 3174, 1994.
41. Djabbarah, N.F., Shah, D.O., Wasan, D.T. *Colloid Polym. Sci.*, 256, 1002, 1978.
42. Stein, H. *Adv. Colloid Interface Sci.*, 34, 175, 1991.
43. Stein, H. *Colloids Surfaces A*, 79, 71, 1993.
44. Garrett, P.R., Davis, J., Rendall, H. *Colloids Surf. A*, 85, 159, 1994.
45. Denkov, N.D., Cooper, P., Martin, J. *Langmuir*, 15(24), 8514, 1999.
46. Racz, G., Koczo, K., Wasan, D.T. *J. Colloid Interface Sci.*, 181, 124, 1996.
47. Nierstrasz, V.A. Marginal regeneration, PhD Thesis, Delft University of Technology, 1996.
48. Prins, A., van Voorst Vader, F. *Proc. 4th Int. Cong. Surf. Activ. Subst., Zürich*, 1972, p 441.
49. Wasan, D., Nikolov, A. *Curr. Opin. Colloid Interface Sci.* 128, 13, 2008.
50. Isrealachvilli, J.N. *Intermolecular and Surface Forces with Applications to Colloidal and Biological Systems*, Academic Press, London, 1985.
51. Vrij, A. *Disc. Faraday Soc.*, 42, 23, 1996.
52. Vrij, A., Overbeek, J.Th.G. *J. Am. Chem. Soc.*, 90, 3074, 1968.
53. Radoev, B.P., Scheludko, A., Manev, E.D. *J. Colloid Interface Sci.*, 95, 254, 1983.
54. Scheludko, A., Radoev, B., Kolarov, T. *Trans. Faraday Soc.*, 64(8), 2213, 1968.
55. Bergeron, V. *J. Phys. Condens. Matter*, 11, R215, 1999.
56. Kashchiev, D., Exerowa, D. *J. Colloid Interface Sci.*, 77(2), 501, 1980.
57. Exerowa, D.R. *Izv. Khim.*, 11(3–4), 739, 1978.
58. Goddard, E.D., Benson, G.C. *Can. J. Chem.*, 35, 986, 1957.

59. Corkill, J.M., Goodman, J.F., Harrold, S.P. *Trans. Faraday Soc.*, 60, 202, 1964.
60. Wasan, D., Nikolov, A.D., Huang, D.D., Edwards, D.A. Foam stability: effects of oil and film stratification, in *Surfactant Based Mobility Control* (Smith, D.H., ed.), A.C.S. Symposium Series 373, Washington D.C., 1988, pp 136.
61. Princen, H.M. *Langmuir*, 4(1), 164, 1988.
62. Hilgenfeldt, S., Koehler, S.A., Stone, H.A. *Phys. Rev. Lett.*, 86(20), 4704, 2001.
63. Verbist, G., Weaire, D., Kraynik, A.M. *J. Phys.: Condens. Matter*, 8, 3715, 1996.
64. Goldfarb, I.I.; Kann, K.B.; Shreiber, I.R. Liquid flows in foams, in *Fluid Dynamics* (Official English Translation of Transactions of USSR Academy of Science series Mechanics of Liquids and Gases), 1988, 23, pp 244.
65. Kraynik, A.M., Sandia report SAND 83-0844, 1983.
66. Neethling, S.J., Lee H.T., Grassia, P. *Colloids Surf. A*, 263, 184, 2005.
67. Durand, M., Langevin, D. *Eur. Phys. J.*, E7, 35, 2002.
68. Cheng, H.C., Lemlich, R. *Ind. Eng. Chem. Fundam.*, 24(1), 44, 1985.
69. Lemlich, R. *Ind. Eng. Chem. Fundam.*, 17(2), 89, 1978.
70. Ronteltap, A.D., Prins, A. *Colloids Surf.*, 47, 285, 1990.
71. Patist, A., Axelberd, T., Shah, D. *J. Colloid Interface Sci.*, 208, 259, 1998.
72. Garrett, P.R. *J. Colloid Interface Sci.*, 69(1), 107, 1979.
73. Garrett, P.R., Moore, P.R. *J. Colloid Interface Sci.*, 159, 214, 1993.
74. Ran, L. Foaming of anionic surfactant solutions in the presence of calcium ions and triglyceride-based antifoams, PhD Thesis, University of Manchester, 2011.
75. Marinova, K.G., Denkov, N.D. *Langmuir*, 17(8), 2426, 2001.

2 Experimental Methods for Study of Foam and Antifoam Action

2.1 INTRODUCTION

Here we briefly review some of the more important experimental techniques that have been used in the study of antifoam action. Many of the techniques are common to the study of both antifoam action and foam in the absence of antifoam. That is hardly surprising because establishing antifoam effectiveness requires knowledge of the latter.

Various methods of foam generation that have been used in this context are first described, together with an indication of their respective advantages, disadvantages, and limitations. Although rarely considered in studies of antifoam action, we briefly include the issue of measurement of bubble size distributions because few methods of foam generation conveniently produce monodisperse foam.

We also describe examples of the various approaches used for the study of the rate of drainage and stability of single foam films. Most of the techniques described have also been used for the study of the behavior of antifoam entities in such films.

As we describe in detail elsewhere in this book, effective antifoams are usually both oil-based and insoluble in the medium to be defoamed. In order to function, it is necessary that the oil exhibit the property of emerging into the air–liquid surface of that medium. In turn this requires that the film of foaming liquid separating the oil from the air be unstable. The stability of these so-called pseudoemulsion films [1] is therefore a key issue in determining antifoam effectiveness. We therefore review the methods for studying such films.

Once antifoam oils emerge into the air–liquid surface of the foaming medium, they must exhibit certain types of spreading behavior if they are to be effective in causing foam film rupture. Finally, then, we review the types of spreading behavior to be expected together with brief descriptions of appropriate experimental methods for distinguishing between them.

2.2 MEASUREMENT OF FOAM

2.2.1 Bartsch Method: Hand Shaking Measuring Cylinders

Shaking a glass vessel, such as a measuring cylinder, vigorously for a fixed time interval (typically for 10–15 s) represents an easy approach to assessing the effectiveness of an antifoam in aqueous surfactant solution. The method has occasionally been attributed [2] to Bartsch [3] who published foam results by using it almost a century ago. It has no doubt been reinvented many times since.

The method typically involves rigorous shaking, for 10–15 s, of a measuring cylinder containing about 25% of its nominal capacity of foaming liquid. Ideally the measuring cylinder should be equilibrated in a thermostat bath before making a measurement. If the stability of the foam is of interest, then the cylinder should be returned to that bath after generation. Typically such shaking occurs with a frequency of about 3 Hz and is characterized by a Reynolds number > 10^5. This method has the advantage that the cylinders may be readily cleaned to avoid contamination by, sometimes persistent, antifoam residue from previous experiments.

Antifoam may be added to the relevant solution in the vessel directly by weighing out the relevant quantity onto a microscope slide coverslip. This method has the disadvantage that dispersal of the antifoam and foam generation occur simultaneously. Preferably, therefore, the antifoam should be predispersed using ultrasonics or a suitable emulsifier/mixer (such as an "Ultraturrax") and the dispersal characterized (by light scattering or optical microscopy). The effect of antifoam on the foam as a function of concentration can then be readily determined by dilution of the dispersion with the surfactant solution.

The effect of the antifoam on the foam is best estimated if the total volume of liquid and gas is measured and the volume of air in the foam is used as a measure of either foamability or stability. The method is surprisingly reproducible even when comparing results obtained by different individuals. This is exemplified by a plot of F (= volume in air in the presence of antifoam/volume of air in the absence of antifoam) against $\log_{10}$(antifoam concentration) shown in Figure 2.1. This plot was prepared by three different individuals using a predispersed antifoam [4].

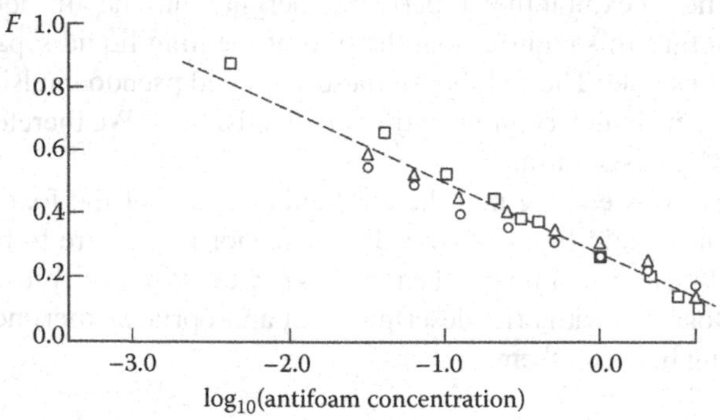

FIGURE 2.1 Plot of F against $\log_{10}$(antifoam concentration) for hand shaking measuring cylinders. Symbols (○, △, □) represent results of three different workers; dotted line represents a linear least square plot of all results (0.5 g dm^{-3} commercial alkylbenzene sulfonate solution containing antifoam of 10 wt.% hydrophobed Aerosil 200 in 90 wt.% liquid paraffin). (Reprinted from *Colloids Surf. A*, 85, Garrett, P.R., Davis, J., Rendall, H., 159. Copyright 1994, with permission from Elsevier.)

2.2.2 Automated Shake Tests

Assessing the durability of antifoam effects by shaking glass vessels can be rather tiresome. Such studies are therefore better conducted using automatic shake tests. A number of workers report the use of such a test for studying polydimethylsiloxane–hydrophobed silica antifoam durability (see, e.g., references [2, 5, 6]). The method utilizes 100 cm^3 surfactant solution in a standard 250 cm^3 glass bottle to which 0.005–0.01 vol. % antifoam is directly added without predispersal. The bottle is shaken at ~4 Hz for 10 s. It is then left quiescent for 60 s. During this quiescent period, the foam typically tends to collapse totally with polydimethylsiloxane–hydrophobed silica antifoams. The time taken for that to occur is noted, and the whole procedure is repeated. Progressive deactivation of the antifoam is indicated by an increase in the time required for total foam collapse during the quiescent period. The procedure is stopped when that time equals 60 s. A disadvantage of this approach is that there would appear to be no easy method for controlling the temperature.

2.2.3 Ross–Miles Method

A description by Ross and Miles of a pour test for measuring foamability was first published 70 years ago [7]. In this technique, the solution is poured through a capillary onto a reservoir of the same solution. The stream of liquid breaks up into drops, which impinge the solution in the reservoir. Air entrainment and foam formation occur as drops of solution first hit the clean air–water surface. However, the drops irrigate the foam as it builds up and air entrainment into the foam continues. This tends to suggest that the gas volume fraction is relatively low, at least where the drops impinge, so that a wet limit is achieved where air entrainment can readily occur. The foam is polydisperse and vertically segregated with small spherical bubbles at the bottom of the column [8]. According to Ross and Miles [7], the volume of foam continues to increase until the flow from the reservoir ceases. Differences in foamability are apparent after a fixed volume of solution has poured onto the reservoir.

This method of foam generation has been turned into a standard method [9, 10]. A diagram of a version of the relevant apparatus suggested by Pauchek and Veber [11] is shown in Figure 2.2. Thermostatting of the apparatus is easily arranged so that it is suitable for study of the effect of temperature on antifoam effects such as those accompanying melting of hydrocarbon particles. Both the reservoir and the pipette typically contain 100–250 cm^3. The capillary diameter is typically 2–4 mm, and the distance from the capillary tip to the surface of the solution in the reservoir is of the order of 1 m. The need for relatively large quantities of solution and therefore antifoam is a potential disadvantage in the case of work with expensive pure materials.

The apparatus can be modified to run continuously if a pump is used to circulate solution from the reservoir to the pipette. In this mode it is suitable for the study of antifoam deactivation. Accumulation of antifoam material on the surface of the flexible tubes used in peristaltic pumps could, however, prove problematic in this context.

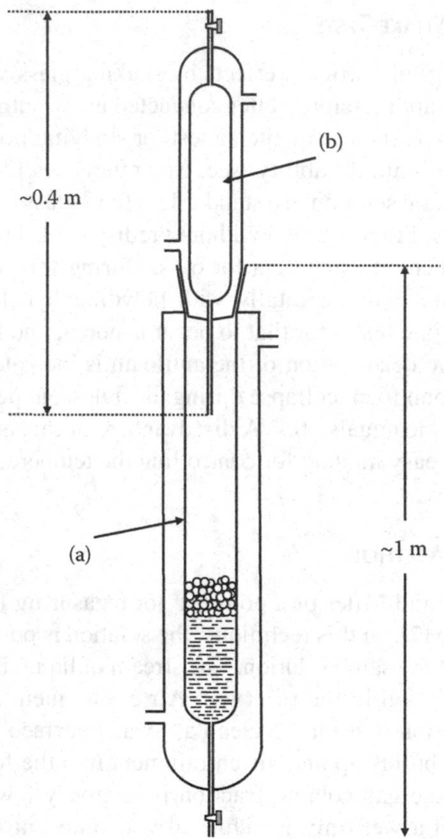

FIGURE 2.2 Ross–Miles foam test using design of Pauchek and Veber (not to scale). (a) Reservoir in which foam is generated. (b) Pipette from which solution is poured. (From Pauchek, M., Veber, V., *Acta Fac. Pharm. Univ. Comenianae*, 27, 27, 1975.)

2.2.4 Tumbling Cylinders

It is possible to generate foam by mechanically tumbling sealed cylinders. Aeration occurs as the liquid falls, affecting the walls of the cylinders. This method is again suitable for monitoring the deactivation of antifoams.

A convenient version of this method involves the use of cylinders of height 30 cm and of total volume ~2000 cm^3, containing 500 cm^3 of liquid, which are rotated continuously at a frequency of ~0.5 Hz [12, 13]. Measurements of the volume of air generated in foam after only 10 rotations have been found to correlate linearly with those obtained with the same solutions using hand shaking of measuring cylinders. In this comparison, solutions of a commercial sodium alkylbenzene sulfonate were used throughout. Variations in foamability were made by partial precipitation of the surfactant with excess calcium and/or by the addition of a triglyceride-based antifoam [13]. Results are shown in Figure 2.3 where the correlation coefficient is 0.95. The existence of such a correlation suggests that similar intensities of agitation prevail

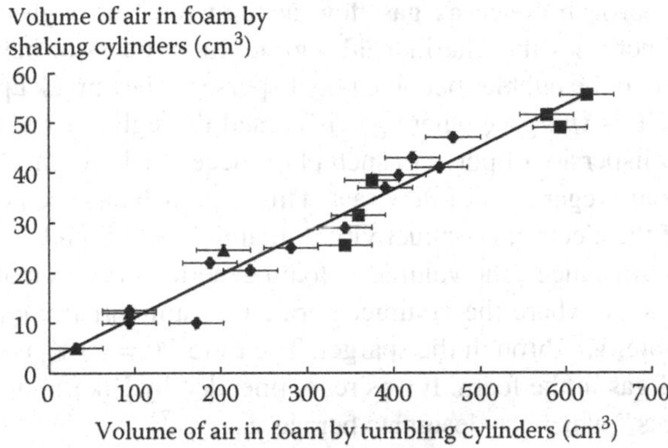

FIGURE 2.3 Comparison of volume of air in foam generated by hand shaking and mechanically tumbling cylinders immediately after foam generation ceased showing a linear correlation with a correlation coefficient of ~0.95. Hand shaking of 100 cm^3 measuring cylinders containing 25 cm^3 of solution for 10 s at ~3 Hz. Mechanically tumbling 2000 cm^3 cylinders containing 500 cm^3 solution for 10 rotations at ~0.5 Hz. Aqueous solution of 2 mM commercial sodium alkylbenzene sulfonate with Ca^{2+} concentration in the range 0–4 and 17 mM NaCl. ■, in absence of antifoam; ◆, in presence of 0.25 g dm^{-3} triglyceride-based antifoam; ▲, in presence of 1.0 g dm^{-3} triglyceride-based antifoam. (After Ran, L., Foaming of anionic surfactant solutions in the presence of calcium ions and triglyceride-based antifoams, PhD Thesis, University of Manchester, 2011.)

with these two rather different methods. However, the correlation deteriorates markedly [13] if comparison is made of the volume of air in the foam after standing. This implies differences in the stability of the foam generated by the two methods, which in turn may be due to the influence of the development of differences in Plateau border capillary pressures on foam film stability as the foam drains.

2.2.5 Gas Bubbling

The method of gas bubbling has often been used to study foam generation in the presence of antifoam. Here we distinguish three types of approaches. The first concerns slow flow through a single narrow orifice in the capillary regime, where bubble sizes are determined by the Laplace equation and secondary bubbles are not usually formed [14]. As suggested by Weaire and Hutzler [15], this regime should produce monodisperse foams. Dynamic surface tensions and the maximum bubble pressure effect [14] mean, however, that formation of such foams also requires that the gas should be supplied at a constant pressure. According to Ruff [16], flow rates of up to 0.1 $cm^3\ s^{-1}$ can produce bubble diameters in the range of 0.2–0.5 cm from orifices of diameters in the range of 0.02–0.4 cm. Generating, for example, 100 cm^3 of monodisperse foam in this way could therefore take more than 15 min. Practicality therefore requires use of several orifices—usually capillary tubes of a small internal diameter (see, e.g., reference [17]).

The second approach concerns gas flow at high rates through single orifices. As the gas flow increases into the inertial regime, the process of bubble formation becomes chaotic and the bubbles become polydisperse as they break up and coalesce. The third approach is *sparging* where gas is passed through porous plugs or perforated plates. Polydispersity of pores in such plugs necessarily implies polydispersity of the resultant foam regardless of flow rate. This approach has been combined with measurements of the electrical conductivity of foam to yield liquid volume fractions [18]. In some circumstances, the volume of foam generated in this manner, $V(t = \infty)$, reaches a steady state where the volumetric rate of foam collapse equals the volumetric gas flow rate, V_G, through the sparger. The ratio $V(t = \infty)/V_G$ is essentially the residence time of gas in the foam. It was recommended by Bikerman [19] as a measure of "foaminess," which we designate here as Σ_{BIK}.

All of these gas bubbling approaches have been used in studying antifoam action. There are, however, some potential limitations in applicability. One concerns the study of the relation between antifoam action and the state of the air–water surfaces during generation of an aqueous foam. If, for example, the solution is micellar where micelle breakdown rates are slow, then this may mean, in the case of methods that involve rapid generation of air–water surfaces, that those surfaces are relatively depleted of surfactant. This could, in turn, render the resulting foam films more susceptible to antifoam action. Indeed, in the extreme, it could represent the only reason why such action is apparent [13]. It is, for example, well known that antifoam action can occur during foam generation but cease after foam generation has ceased, as air–water surfaces assume near-equilibrium properties [20–22]. This would seem to preclude the study of such phenomena using monodisperse foams where generation of surfaces is necessarily slow and near-equilibrium conditions prevail throughout.

In the case of sparging, bubble concentrations can be high on the surface of the porous plug. In these circumstances, foam films may form at that location. It has been shown that the presence of antifoam can cause rupture of such films even though the antifoam is without effect during the free movement of bubbles through the liquid phase, which lies above the plug and below the foam [23]. This may have relevance in formulating any theory concerning the effect of antifoam on bubble size distributions during foam generation.

2.2.6 Direct Air Injection

Some recent studies of antifoam effects have involved the use of a rather strange foam device involving direct injection of air through a valve into a liquid line leading to a circulating pump [24–26]. The apparatus is depicted in Figure 2.4. The relevant volume of surfactant solution is first poured into a clear plexiglass cylinder from which it can be circulated using the pump. Antifoam can be added at this stage, the dispersal being facilitated by circulation of the solution before aeration. Opening the air valve allows subsequent aeration of the circulating solution. The resulting foam accumulates in the cylinder. The rate of formation of foam is indicated by the time taken to fill the cylinder up to a given fiducial mark, a process that typically takes a few minutes. The pump circulation is stopped at that point, and foam stability is monitored as a decline in foam height with time. It is claimed

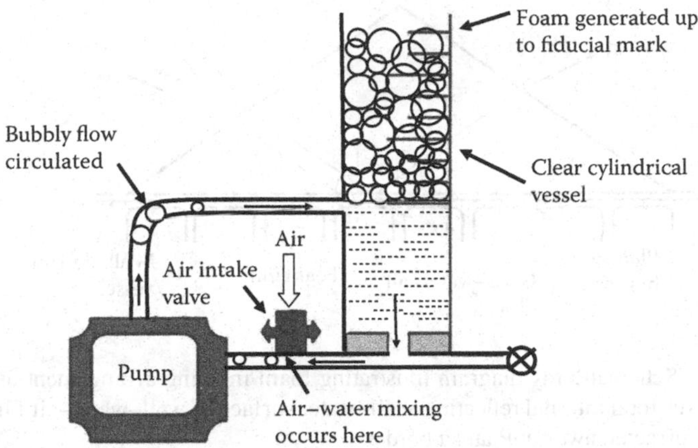

FIGURE 2.4 Foam generation by direct air injection. (Reprinted from *J. Colloid Interface Sci.*, 263, Zhang, H., Miller, C.A., Garrett, P.R., Raney, K., 633. Copyright 2003, with permission from Elsevier.)

that this technique has the twin merits of rapid formation of foam and convenience of use [27].

2.2.7 Measurement of Bubble Size Distributions

Bubble size distributions are of importance in understanding the effect of antifoam concentration on foamabilities (see Chapter 5) and on foam collapse after foam generation has ceased. Observations of the effect of antifoams on bubble size distributions are, however, almost totally absent from the published literature. Possibly this reflects the fact that measurement of the size distribution of bubbles in a polydisperse foam is surprisingly difficult. Intrusive methods using direct probes or sampling methods are likely to be slow, inconvenient, and may produce artifacts. Nonintrusive methods, on the other hand, such as electrical resistance tomography [28], optical tomography [29], and magnetic resonance imaging [30], all have serious disadvantages with respect to routine use for foams with the relevant gas volume fractions, Φ_g, over the range $0.72 < \Phi_g < 0.99$ [31]. Arguably the most convenient method concerns measurement of the size distribution of bubbles at a plane of observation. Even that is fraught with serious difficulty.

If determination of bubble size distributions at a plane of observation is not to be impractically tedious, it must involve computer-aided image analysis. This requires high-contrast images. Such images can be conveniently obtained with the arrangement described by Garrett [31] and depicted in Figure 2.5. Preferably the retaining vessel should have a rectangular cross section. Here observation is made at an angle of incidence, $\hat{e}$, so that total internal reflection occurs at the air–water surface of the wall–water–air film of continuous phase between the bubble and the wall of the vessel. By contrast, light is reflected and refracted away at the vertically orientated Plateau borders, which characterize the bubbles situated at a plane of observation.

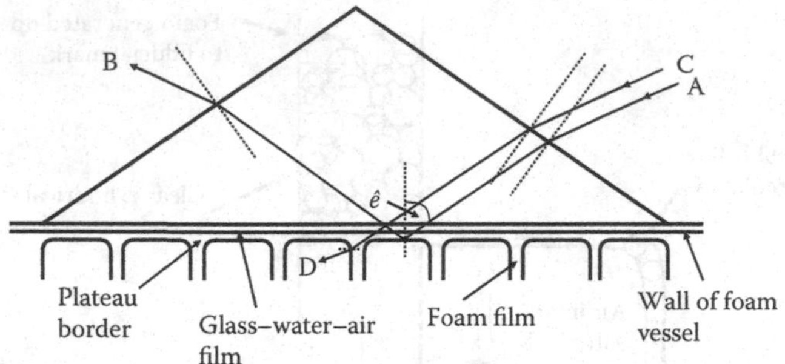

FIGURE 2.5 Schematic ray diagram illustrating foam imaging arrangement using prism. Ray AB achieves total internal reflection at air–water surface of wall–water–air film. Ray CD is by contrast refracted away at Plateau border.

Degradation of the contrast by extraneous reflections can be minimized by use of a prism as indicated in the figure. Typical images obtained in this manner are shown in Figure 2.6. Here we note that the angle of observation causes a distortion of the area of any object (so that circles appear as ellipses). The apparent area must be multiplied by 1/sin (90° – $\hat{e}$) to give the true area. This factor together with the magnification can be determined directly by including a 2D calibration object of known geometry (e.g., an opaque disc) in the overall foam image. This object should preferably be located on the foam side of the plane of observation to avoid errors due to refraction at the film–wall surface.

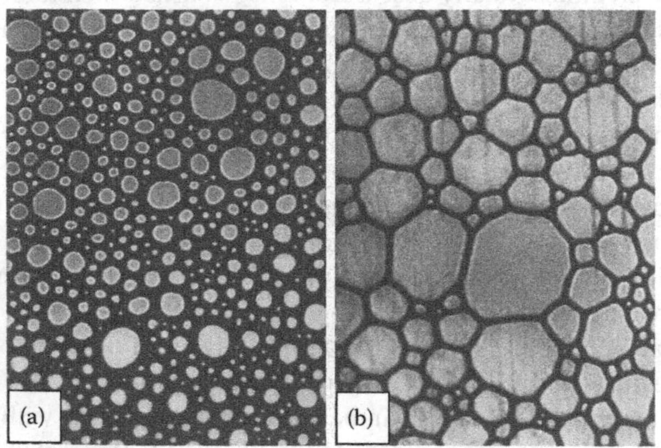

FIGURE 2.6 Images of (a) wet foam and (b) dry polyhedral foam taken using the arrangement shown in Figure 2.5. (From Garrett, P.R., Foams and antifoams, in *Food Colloids, Fundamentals of Formulation*, Dickinson, E., Miller, R., eds., The Royal Society of Chemistry, Special Publication No. 258, p 55, 2001.)

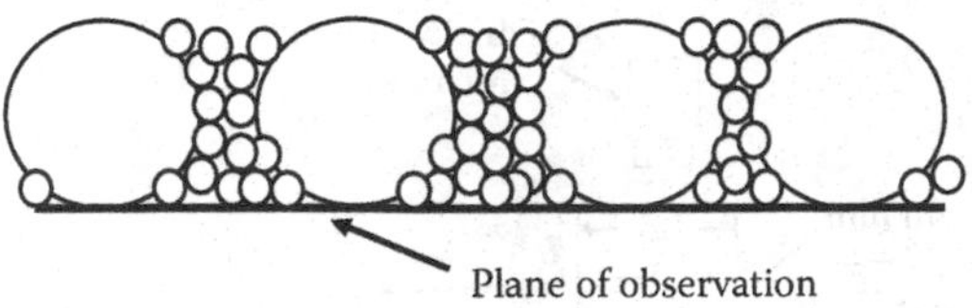

FIGURE 2.7 Schematic illustration of sampling bias in favor of larger bubbles at plane of observation when imaging foams. (From Garrett, P.R., Foams and antifoams, in *Food Colloids, Fundamentals of Formulation*, Dickinson, E., Miller, R., eds., The Royal Society of Chemistry, Special Publication No. 258, p 55, 2001.)

Cheng and Lemlich [32] argue that the radius of an equivalent spherical bubble having the same volume as a polyhedral bubble at the plane of observation is approximated by $\sqrt{(A_{ap}/\pi)}$, where A_{ap} is the area of the polygon formed by the bisector of the Plateau borders at the wall. Even if such an approximation is valid, there remain systematic sampling errors in any distribution calculated from the plane of observation. The first is made obvious if we examine Figure 2.6. Some bubbles are clearly "censored" by the edges of the image. Obviously larger bubbles are more likely to be censored in this manner, which means that the distribution will be skewed by this factor toward smaller sizes. A suitable computational procedure that corrects for this bias has been published [31]. Another sampling bias has been identified by de Vries [32]. It derives from differences in the probabilities of bubbles of different sizes appearing in the plane of observation. This factor is present even if the spatial distribution of bubbles is everywhere similar up to that plane. The problem is illustrated in Figure 2.7 where it is clear that too few small bubbles will be seen at the wall of an enclosing vessel. In contrast to the effect of censoring by the edges of the image, this skews the distribution toward larger bubble sizes. Again, however, a simple computational procedure, first devised by de Vries [33], can be used to correct the bias. All of these corrections can of course be readily applied to the distribution calculated from the raw distribution derived from image analysis. How closely a distribution of equivalent spherical bubble sizes calculated after applying all these corrections approximates to the actual bubble size distribution in the interior of a foam is of course an open question, which could, in principle, be investigated in general by simulation using, for example, the "Surface Evolver" energy minimization software [34] (see Section 1.2).

2.3 OBSERVATIONS WITH SINGLE FOAM FILMS

2.3.1 Scheludko Cells

2.3.1.1 Films Containing Antifoam Drops

The so-called Scheludko cell has found wide application in studies of foam films. This cell was first proposed by Scheludko and Exerowa [35, 36] more than half a century ago. It consists of a (usually) glass cylinder containing a biconcave drop of surfactant solution. A tube inserted into the side of the cylinder permits control of

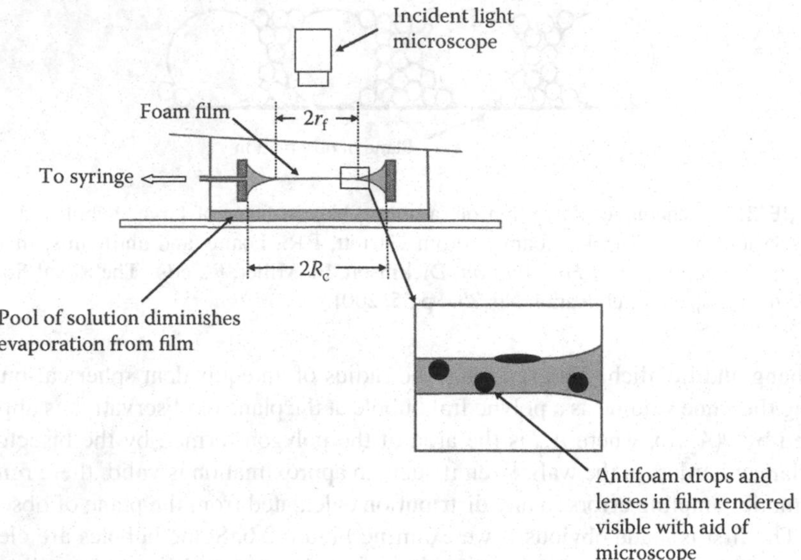

FIGURE 2.8 Typical setup for observation of antifoam drops in foam film using a Scheludko cell. (From Scheludko, A., Exerowa, D., *Commun. Dept. Chem. Bulg. Acad. Sci.*, 7, 123, 1959; Scheludko, A., *Adv. Colloid Interface Sci.*, 1, 391, 1967.)

the drop by using a syringe. Removal of liquid from the drop draws the two surfaces of the drop together until a foam film is formed. The arrangement is shown schematically in Figure 2.8. The foam film can be observed either using a camera with a long focal length lens combined with an external light source or with an incident light microscope as shown in the figure. Dynamic processes occurring in the film can be recorded using a high-speed camera [37]. The cell may be enclosed to minimize evaporation from the film, as shown in Figure 2.8 for example. The presence of a small pool of the relevant solution helps ensure vapor pressure equilibrium and further minimize evaporation. A sloping surface for the cover of the enclosing vessel helps eliminate unwanted light reflections.

This arrangement has been found to be particularly useful in studying the behavior of antifoam drops in draining foam films. It permits observations of both the movement of the drops as they interact with draining films and also the actual events of foam film rupture induced by those drops [24, 37].

Antifoam drop sizes are usually ≥1 micron. Therefore, films containing such drops must be of at least the same order. The thickness profiles of such films can therefore be readily estimated from the interference fringes observed in monochromatic reflected light, which such thickness imply (provided the order of at least one reference fringe is known).

If the syringe controlling the film size is in the same plane as the film, then drainage of that film after the movement of the syringe barrel has ceased is entirely driven by the capillary pressure, p_c^{PB}, in the Plateau border between the film and the walls of the cylinder. That capillary pressure is given as

$$p_c^{PB} = 2\sigma_{AL}\left(\frac{R_c}{R_c^2 - r_f^2}\right) \quad (2.1)$$

provided the inside wall of the cell is perfectly wetted by the liquid. Here σ_{AL} is the air–liquid surface tension, R_c is the internal radius of the cell, and r_f is the radius of the film. R_c is typically in the range of 1–2 mm and r_f is typically in the range of 0.3–0.5 mm.

2.3.1.2 Measurement of Disjoining Pressure Isotherms of Air–Liquid–Air Foam Films

This type of cell can also be used to measure disjoining pressure isotherms in foam films. In this case, however, the films must be much thinner for the disjoining pressure to be non-zero (i.e., <0.1 micron). Near total destructive interference of reflected light will mean that they appear black in reflected light and do not exhibit interference fringes—film thicknesses must therefore be inferred from accurate measurements of reflected light intensities in comparison with the first-order fringe maximum and minimum intensity values in the adjacent meniscus [38, 39].

The disjoining pressures in thin foam films can be as high as 30 kPa (0.3 bar). Such pressures cannot be applied using devices like that shown in Figure 2.8. They can, however, be applied directly to a foam film contained in a cylindrical cell prepared from a porous frit where the menisci in the fine holes in the latter determine the maximum capillary pressure in the Plateau border. In this approach, pressure on the film is directly applied by increasing the surrounding gas pressure. In turn, this produces an enhanced capillary pressure in the liquid trapped in the pores of the frit, presumably as the liquid moves to smaller pore radii. This technique was first used by Mysels and Jones [40] and later refined by Exerowa and Scheludko [41].

Bergeron and Radke [38, 39] have used this approach to measure disjoining pressure isotherms. A schematic diagram of their apparatus is shown in Figure 2.9. Thin films are formed in a hole in a fritted glass disk, which is contained in a hermetically sealed cell. A capillary tube fused to the disc makes connection to an external constant reference pressure. The gas pressure in the cell, P_g, is adjusted by means of a syringe pump and measured using a differential pressure transducer. The film thickness is inferred from measurements of the intensity of reflected light at two different wavelengths. The disjoining pressure $\Pi_{ALA}(h)$ with this arrangement is given by

$$\Pi_{ALA}(h) = P_g - P_L = P_g - P_r + \frac{2\sigma_{AL}}{r} - \rho_L g H_{hyd} \quad (2.2)$$

where P_r is the reference pressure, P_L is the bulk liquid pressure (see Figure 2.9), r is the radius of the capillary tube, and H_{hyd} is the hydrostatic head in the capillary tube. In the case of measurements of high values of $\Pi_{ALA}(h)$ it is possible to set P_r

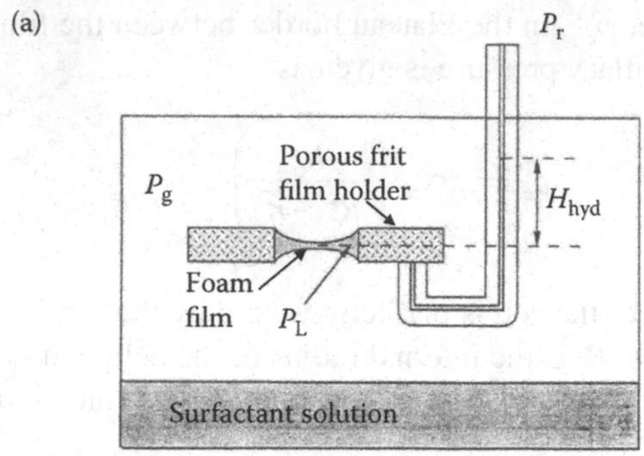

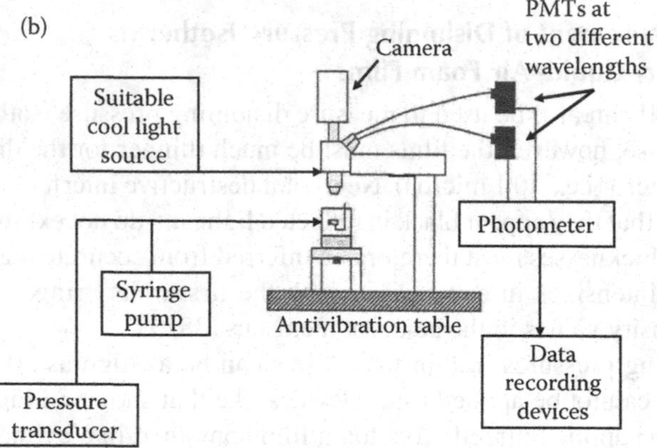

FIGURE 2.9 Measurement of disjoining pressure isotherms of air–liquid–air foam films. (a) Porous frit film holder located inside hermetically sealed cell subject to applied gas pressure, P_g. Bulk liquid pressure is P_L (= $P_r - 2\sigma_{AL}/r + \rho_L g H_{hyd}$; see text for definition of terms) and capillary pressure is $P_g - P_L$. (b) Schematic diagram of setup for measurement of film thickness by monitoring the intensity of light reflected from foam film (where PMT = photomultiplier tube). Applied pressure is controlled by syringe pump and measured using transducer. (Reprinted with permission from Bergeron, V. and Radke, C.J., *Langmuir*, 8, 3020. Copyright 1992 American Chemical Society.)

at the atmospheric pressure. However, for values of $\Pi_{ALA}(h)$ <~100 Pa variability of atmospheric pressure during the course of an experiment produces large errors and some method should be adopted to ensure constant P_r [38].

2.3.2 Dippenaar Cell

Dippenaar used glass cells to observe the effect of the presence of single solid hydrophobic particles on the stability of foam films in transmitted light by using a high-speed cine camera [42]. As we outline in Chapter 4, this technique has produced convincing proof of the bridging mechanism of antifoam action by hydrophobic particles.

A schematic diagram of the cylindrical cell is shown in Figure 2.10. Foam films may be prepared by first introducing a biconcave drop into the cell followed

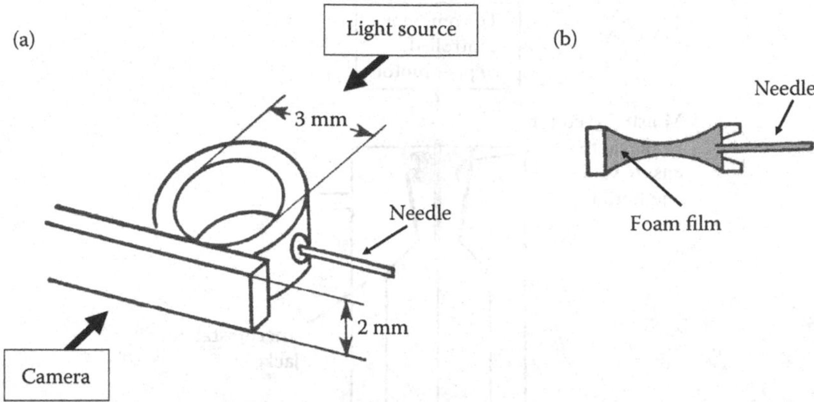

FIGURE 2.10 Dippenaar cell for observing antifoam effects in transmitted light incident parallel with orientation of film. (a) Experimental arrangement, where foam film thickness is controlled by means of screw clamp on a tube connected to needle. Optical glass plate is used to avoid optical aberration caused by use of cylindrical cell. (Reprinted from *Int. J. Miner. Process.*, 9, Dippenaar, A., 1–22, Copyright 1982, with permission from Elsevier.) (b) Cross-section of film in cell.

by sucking liquid out with a syringe in much the same manner as utilized for the Scheludko cell. The arrangement, however, means that observations are confined to relatively large particles (of size ≥ 100 microns). The cell has also been used by Denkov et al. [37] to directly observe the shape of an oil bridge in a foam film. Again, however, the dimensions of the apparatus imply use of large oil drops giving rise to bridges of diameter > 100 microns, which is about two orders of magnitude greater than the size of drops present in antifoam dispersions.

2.3.3 Large Vertical Films

The sizes of films present in actual foams can often be a few orders of magnitude greater than those prepared using the Scheludko cell. Moreover, such films are not invariably horizontal. As we have outlined in Chapter 1, drainage of large vertical foam films may exhibit a different manifestation of the hydrodynamic instability, which characterizes the drainage of all mobile foam films, from that shown by small horizontal films. This is a consequence of the role of gravity in the drainage of vertical films. The manifestation in question is, of course, marginal regeneration (see Section 1.3.2).

The stability and rates of drainage of large vertical foam films can be studied by forming those films using a suitable glass or metal frame. Typically, films of up to 10 cm height can be formed in this manner. The effect of the rate of formation and size of large vertical films on film lifetimes can also be readily varied. In particular the effect of the rate of film formation can, in principle, give information concerning antifoam action in newly formed films where surfaces deviate markedly from equilibrium [13]. The thickness profile of large vertical films can be determined from the interference fringes formed in reflected light. The position of film rupture by any

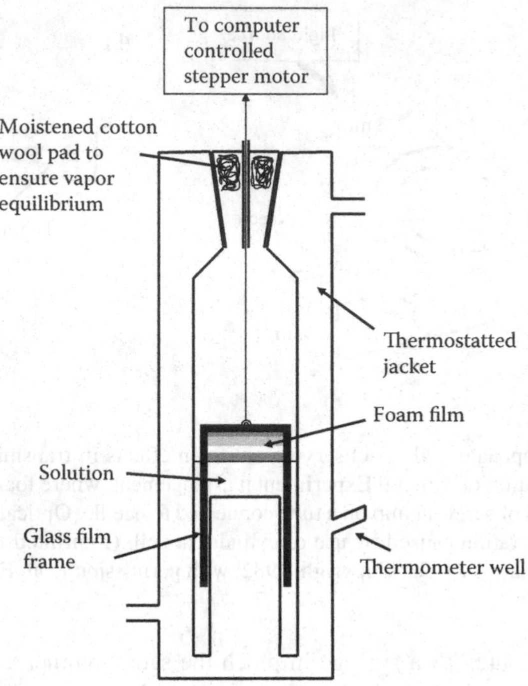

FIGURE 2.11 Schematic of example of a cell for study of stability and rate of formation of vertical foam films. (Reprinted from *Colloid Surf. A*, 307, Garrett, P.R., Wicks, S.P., Fowles, E., 282–283. Copyright 2006, with permission from Elsevier.)

antifoam entities present can also be observed by high-speed video and strobe light illumination [37].

A number of different descriptions of suitable apparatus for forming large vertical foam films exist [4, 13, 37]. One such arrangement is shown schematically in Figure 2.11. A particular feature concerns the use of a moist cotton wool pad at the top of the thermostatted vessel. This helps ensure vapor pressure saturation in such a comparatively large vessel—nevertheless, the usual procedure involves preequilibration for up to 3 h before films are made. The film frame is prepared from a 3 mm glass rod. It is connected to a computer-controlled stepper motor by a polymer monofilament. The motor is programmed to drive the frame upward at various predetermined rates to predetermined heights. Rates can be varied from 0.015 to 45 cm s^{-1}, corresponding to relative rates of surface area increase of 0.003–10.0 s^{-1}. Images can be made with a camera and light source at angles of incidence and reflectance of 45°.

2.4 AIR–WATER–OIL PSEUDOEMULSION FILMS

2.4.1 Direct Observation of Pseudoemulsion Films

As we will discuss at length in Chapters 3 and 4, the stability of the asymmetric air–water–oil films that separate oil antifoam drops from the relevant gas phase is a key factor in determining antifoam activity. After Wasan et al. [1], we use the term pseudoemulsion film to describe such films.

Direct observation of millimeter-size pseudoemulsion films can be made by pressing an oil–water surface against an air–water surface using apparatus similar to that shown in Figure 2.12, the essentials of which were first described by Marinova and Denkov [22] and later by Zhang et al. [24]. A stainless-steel capillary connected to a precision micrometer syringe sits at the center of a glass cell formed by two glass cylinders fused or glued to a glass plate. The capillary is filled with the oil phase and the two cylinders contain the relevant surfactant solution. The inner cylinder is filled as indicated in the figure so that the oil at the tip of the capillary can form a pseudoemulsion film under the control of the syringe. Solution in the outer cylinder ensures vapor pressure equilibrium to avoid evaporative destabilization of films. The apparatus can be mounted on the x–y stage of an incident light microscope equipped with a camera, so that the film can be directly observed, preferably using monochromatic light.

This apparatus permits measurement of pseudoemulsion lifetimes and film thickness profiles as indicated by interference fringes. The effect of hydrophobic particles in the oil phase on film lifetimes can be easily observed. It also permits phenomena such as stratification due to the presence of close-packed hydrophilic particles [43] or micelles to be observed.

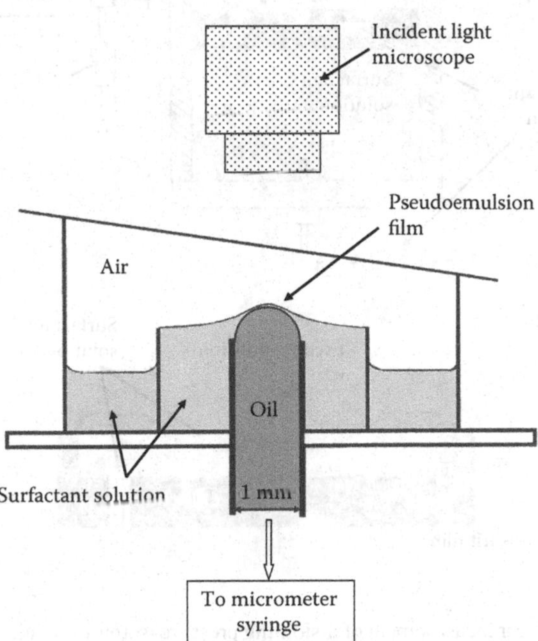

FIGURE 2.12 Direct observation of drainage and stability of pseudoemulsion films using incident light microscope. (Reprinted with permission from Marinova, K.G. and Denkov, N.D., *Langmuir*, 17, 2426. Copyright 2001 American Chemical Society.)

2.4.2 Measurement of Disjoining Pressure Isotherms for Air–Water–Oil Pseudoemulsion Films

The first and apparently only measurements of disjoining pressure isotherms for air–water–oil pseudoemulsion films have been made by Bergeron et al. [39, 44, 45]. These measurements used essentially the same approach as described here in Section 2.3.1.2 and shown in Figure 2.9 for measurements of disjoining pressure isotherms of air–water–air foam films. Some modification of the relevant disjoining pressure cell is, however, necessary. The modified cell is depicted in Figure 2.13a. Again films are prepared in a fritted glass disc; however, for this measurement, the cell containing surfactant solution is placed between an oil layer and the gas phase. In turn the oil layer floats on top of the more aqueous solution.

The fritted glass disc must be designed so that the pseudoemulsion film is plane parallel, which means that the capillary pressure at the air–water surface must equal the capillary pressure at the oil–water surface of the film despite the differences of the air–water and oil–water surface tensions [45]. Design of the fritted glass disc used to achieve this is shown in Figure 2.13b.

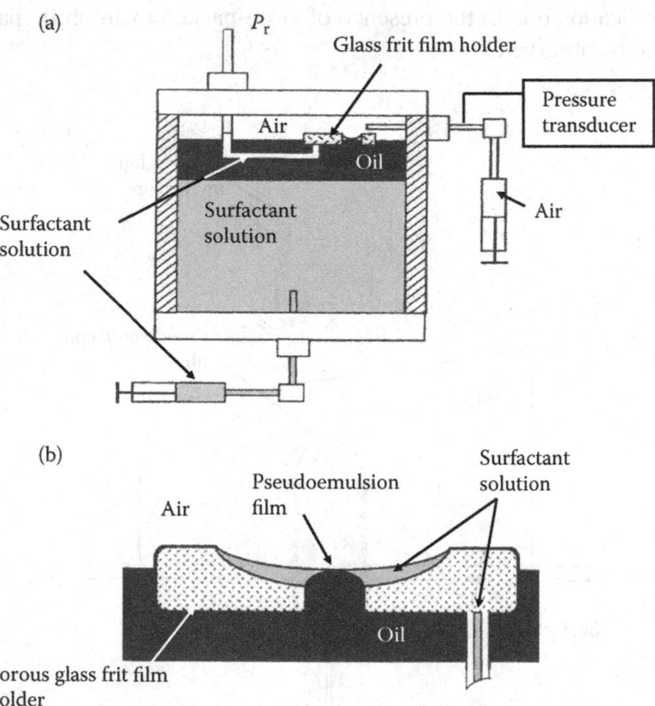

FIGURE 2.13 Cell for measurement of disjoining pressure isotherms of air–water–oil pseudoemulsion films. (a) Arrangement of cell showing control of liquid level and gas pressure. (b) Porous glass frit film holder designed to enable preparation of plane-parallel pseudoemulsion films. (With kind permission from Springer Science+Business Media: *Colloid Polym. Sci.*, 273, 1995, 165, Bergeron, V., Radke, C.J.)

2.4.3 Direct Measurements of Pseudoemulsion Film Rupture Pressures

Bergeron et al. [6] have used an apparatus similar to that depicted schematically in Figure 2.14a to measure the critical pressure required to rupture air–oil–water pseudoemulsion films. Here oil drops extruded from a syringe are pressed against an air–water surface. An interference video microscope is used to observe the drainage and rupture of the film. The disjoining pressure, $\Pi_{ALO}(h)$, in this arrangement is given by Bergeron et al. [6] as

$$\Pi_{ALO}(h) \approx \rho_W g H_{hyd} - \frac{2\sigma_{AW}}{r} + \frac{2\sigma_{OW}}{r_{pef}} \tag{2.3}$$

where σ_{AW} and σ_{OW} are respectively the air–water and oil–water surface tensions, r_{pef} is the radius of curvature of the pseudoemulsion film, H_{hyd} is the head measured by the capillary manometer depicted in the figure, r being the radius of that capillary. The critical disjoining pressure at which rupture occurs, $\Pi^{crit}_{ALO}(h)$, is therefore attained by simply increasing the head until the film is seen to rupture. We should note however that Equation 2.3 neglects a contribution to the pressure in the pseudoemulsion film from the Laplace pressure due to any curvature of the air–water surface.

Koczo et al. [46] have also explored the stability of pseudoemulsion films by measuring the rupture pressure. The film is formed on the surface of an oil drop held in a capillary, as shown schematically in Figure 2.14b. In the case of an unstable pseudoemulsion film, increasing the size of the drop increases the capillary pressure and stretches the film until it ruptures. Stable films survive even as the maximum

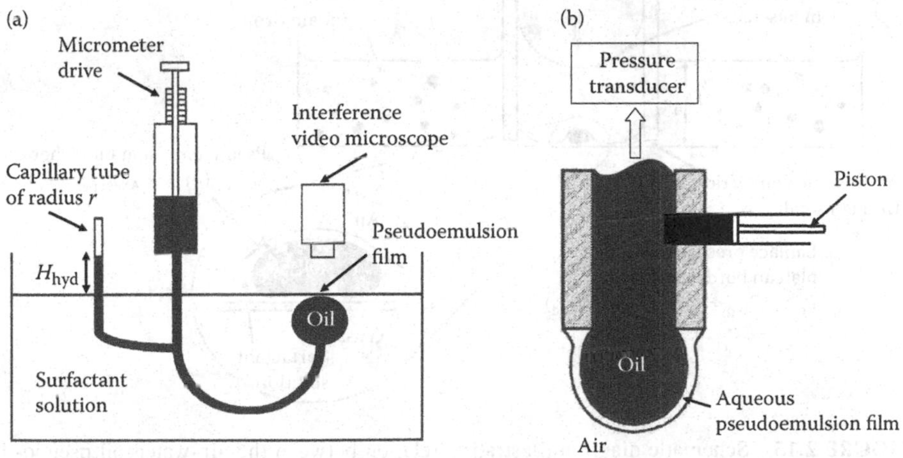

FIGURE 2.14 Measurement of pseudoemulsion film rupture pressure. (a) Apparatus of Bergeron et al. (Reprinted from *Colloids Surf. A*, 122, Bergeron, V., Cooper, P., Fischer, C., Giermanska-Kahn, J., Langevin, D., Pouchelon, A., 103. Copyright 1997, with permission from Elsevier.) (b) Apparatus of Koczo et al. (Reprinted from *J. Colloid Interface Sci.*, 166, Koczo, K., Koczone, J., Wasan, D., 225. Copyright 1994, with permission from Elsevier.)

capillary pressure is achieved as the drop becomes hemispherical. Results of a qualitative nature only have, however, been reported by Koczo et al. [46]—the presence of hydrophobic particles in the film causes rupture; otherwise the film is stable.

Quantitative measurements of the critical capillary pressure required to rupture the pseudoemulsion films preventing emergence of oil drops into air–water surfaces have been made by Denkov and coworkers [2, 22, 47–53] using the so-called film trapping technique (FTT). Such measurements have contributed significantly to our understanding of the mode of action of antifoams. A schematic diagram illustrating the principle of the FTT is given in Figure 2.15, and a diagram of the actual FTT apparatus is given in Figure 2.16. Only a brief summary of this technique is given here, more details are to be found elsewhere [2, 47, 52, 54].

The method of the FTT involves trapping oil drops in an aqueous film covering a hydrophilic surface (e.g., of glass), which is subject to an external pressure. That pressure is increased until the disjoining pressure in the pseudoemulsion film, preventing emergence of the drop into the air–water surface, is overcome. Emergence of the drop can be observed during this process, using a suitable microscope, so that the corresponding applied pressure can be recorded.

The basic principle of the application and measurement of pressures with the FTT is shown schematically in Figure 2.15. A capillary is inserted into a small trough of antifoam oil-in-water emulsion and pressed against the base of the trough so that hydrostatic connection between the liquid trapped in the capillary and that outside

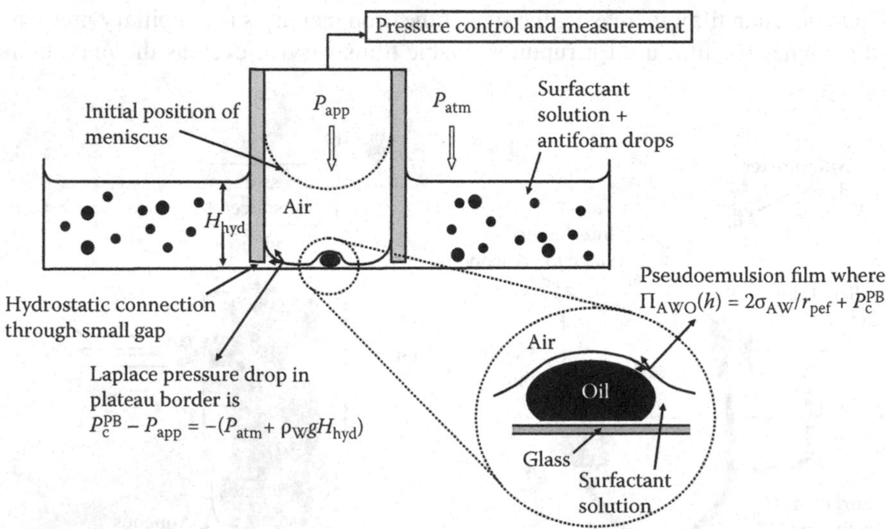

FIGURE 2.15 Schematic diagram illustrating relation between the air–water–oil pseudoemulsion film disjoining pressure, $\Pi_{AWO}(h)$, the applied pressure, P_{app}, and the capillary pressure, p_c^{PB} for an entrapped drop of oil with FTT. (From Denkov, N.D., *Langmuir*, 20, 9463, 2004; Hadjiiski, A. et al., *Langmuir*, 17, 7011, 2001; Hadjiiski, A. et al., Role of entry barriers in foam destruction by oil drops in *Adsorption and Aggregation of Surfactants in Solution* (Mittal, K., Shah, D. eds.), Marcel Dekker, Surfactant Series 109, New York, Chapter 23, p 465, 2003; Hadjiiski, A. et al., *Langmuir*, 12, 6665, 1996.)

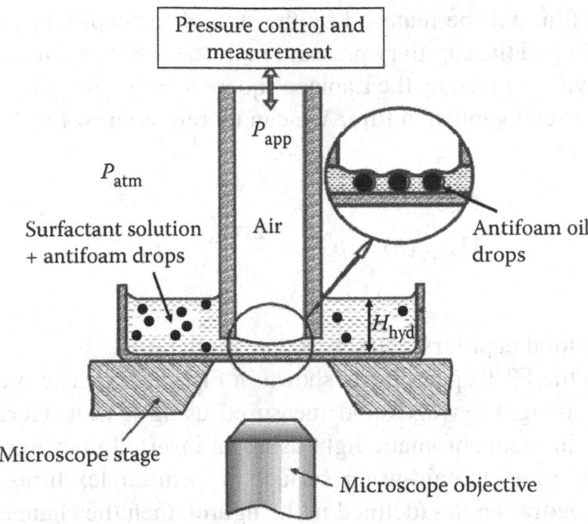

FIGURE 2.16 Diagram illustrating experimental setup for FTT. (Reprinted with permission from Denkov, N.D., *Langmuir*, 20, 9463. Copyright 2004 American Chemical Society.)

is preserved. The pressure inside the capillary, P_{app}, is now slowly increased so that the meniscus moves down to trap oil drops against in the resulting film at the bottom of the trough. As the pressure increases, the air–water surface becomes planar except for a progressively narrowing meniscus at the edge of the capillary as shown in Figure 2.15. This in turn implies the existence of a disjoining pressure across the air–water–glass surface. Here we should note that the gap between the bottom of the capillary and the trough should not be so large that air can leak through at the applied pressures (of order 100 Pa). The Laplace pressure drop across the Plateau border formed at the bottom of the capillary tube is now given by

$$p_c^{PB} - P_{app} = (P_{atm} + \rho_W g H_{hyd}) \tag{2.4}$$

where p_c^{PB} is the capillary pressure in the Plateau border, P_{atm} is the atmospheric pressure, g is the acceleration due to gravity, H_{hyd} is the hydrostatic head in the trough outside the capillary. It is the capillary pressure, p_c^{PB}, that simulates the capillary pressure exerted by an adjacent Plateau border on a drop trapped in a foam film. The critical value of that pressure, p_c^{crit}, at the point of pseudoemulsion film rupture will then give an indication of the stability of the pseudoemulsion film subject to the Plateau border capillary pressures present in a foam. It is therefore the key measurement of the FTT. It should, however, be stressed that this measurement concerns near-equilibrium conditions—the stability of pseudoemulsion films formed where the air–water surface is depleted of surfactant owing to slow transport relative to the rate of air–water surface formation during foam generation is not addressed.

As the pressure increases in the capillary, the trapped drop deforms and the pseudoemulsion film thins. If the build up of the applied pressure, P_{app}, is slow, a condition of mechanical equilibrium will prevail so that the disjoining pressure in the pseudoemulsion film will be matched by the sum of the capillary pressure, p_c^{PB}, in the Plateau border and the capillary pressure in the necessarily curved surface of the film. The latter will be given by the Laplace equation as $2\sigma_{AW}/r_{pef}$, r_{pef} is the radius of curvature of the pseudoemulsion film. We can therefore write for the air–water–oil disjoining pressure

$$\Pi_{AWO}(h) = p_c^{film} = \frac{2\sigma_{AW}}{r_{pef}} + p_c^{PB} \tag{2.5}$$

where p_c^{film} is the total capillary pressure acting on the film.

A diagram of the FTT apparatus is shown in Figure 2.16. The pressure is modified by use of gas-tight syringes and measured using a transducer. The trapped drop is observed in monochromatic light using an inverted microscope. In the case of oils that emerge into the air–water surface to form duplex films or lenses with extremely low dihedral angles (defined in the figure), then the change in shape upon emergence can be dramatic and the rupture of the relevant pseudoemulsion film is easy to observe from the change in interference patterns. Polydimethylsiloxane oils appear to invariably exhibit this type of behavior (see Chapter 3). However, some oils form lenses with high dihedral angles, which can render direct observation of the emergence of a drop into the air–water surface difficult, if not impossible. Ran [13] reports this difficulty in the case of triolein oil. An approach that may be practicable in such cases concerns the greater tendency of drops to coalesce after emergence to form oil lenses. Brewster angle microscopy [55] has been used to detect lens formation at air–water surfaces and may therefore represent another approach but would entail radical redesign of the apparatus. Surprisingly only measurements made by the originators of this FTT technique have been reported thus far.

A limitation of the arrangement of the FTT depicted in Figure 2.16 concerns the minimum capillary pressure, p_c^{PB}, attainable. This is about 20 Pa [2, 52] and is largely determined by the difference $P_{app} - P_{atm}$ in Equation 2.4. That is in turn determined by the dimensions of the maximum radius of curvature of a meniscus in the water-wet glass capillary. Larger radii of curvature are, however, possible if use is made of the so-called gentle FTT cell [2, 52], which permits capillary pressures close to zero at the start of an experiment.

2.5 SPREADING BEHAVIOR OF OILS

The spreading behavior of antifoam oils after emergence onto the relevant liquid substrate represents a key property. Potential contamination of the substrate surface by spreading antifoam is best quantified by simply adding a drop of the oil to the air–liquid surface as the surface tension is monitored with a Wilhelmy plate. Should there be no change in the measured surface tension after such a procedure, then there may be reason to doubt that the oil has emerged onto the air–liquid surface at all. If the oil has not emerged, then reversing the experiment by adding a drop of the substrate liquid to the oil–air surface will produce a reduction in surface tension as the liquid spreads over the oil.

The state of the oil after emergence is important. If the oil emerges without spreading at all, then it will form oil lenses. The geometry of such lenses will determine whether the oil can function as an antifoam. The key geometrical properties are the so-called Neumann angles formed at the relevant three-phase contact line. These angles can be inferred with reasonable accuracy from measurements of the three relevant surface tensions only in cases where the dihedral angle is large. Otherwise resort to interference microscopy, and analysis of the Newton's ring pattern formed by the lens will allow direct determination of at least the relevant three-phase air–water–oil dihedral angle [37].

If the oil spreads over the substrate, then the surface tension of the substrate liquid will necessarily decline. This observation will, however, not of itself distinguish between the two types of spreading behavior that may occur. The simplest involves duplex film formation where the oil completely spreads to form a thick oil film with properties of oil in bulk. The measured surface tension of the film will then equal the sum of the oil–air and substrate–oil surface tensions. Addition of further drops of oil simply produces an increase in thickness of this film. The thickness of the oil film (and even its presence) may be determined by ellipsometry (see, e.g., references [53, 56]).

In the second type of spreading behavior, the oil spreads as a film that may even be duplex but is unstable so that it breaks up to form a thinner film, of lower surface tension than a duplex film, but in equilibrium with oil lenses. Addition of further oil drops simply increases the number or size of oil lenses without increasing the film thickness or changing the surface tension. The dihedral angle of the resulting lenses is often extremely low so that distinguishing between this type of behavior and duplex film formation can be difficult. Measurements of the relevant surface tensions are often not sufficiently accurate to clearly distinguish the nature of the spread oil layer. Again, interference microscopy may be employed to detect the presence of lenses.

The theoretical basis of these generalizations together with actual experimental observations are reviewed in detail in Chapter 3.

2.6 SUMMARIZING REMARKS

There are clearly a number of problems associated with most methods of foam measurement.

Polydispersity is present in most cases and apparently cannot be avoided unless extremely slow methods for forming air–liquid surfaces are employed. This in turn highlights another problem that concerns the actual state of air–liquid surfaces during the generation of foam. It is well known, for example, that the deviation from the equilibrium properties of the surfaces of liquids during foam generation can produce antifoam effects that would not otherwise be apparent (see references [13, 20–22]). Measurement of the dynamic surface tensions of air–liquid surfaces can of course be readily made using such techniques as the versatile maximum bubble method. However, such techniques measure dynamic surface tensions as a function of the surface age. This leaves open the issue of relating a given surface age measured in such a manner to the "effective surface age" of a freshly generated bubble at the top of the foam column in, for example, a Ross–Miles foam test.

Use of the film trapping technique for quantifying the magnitude of the critical capillary pressure required to rupture pseudoemulsion films has produced significant advances in understanding of antifoam action. The existing apparatus design is, however, in some respects difficult to use. Gas leaks, vibrations, and ambient temperature fluctuations can produce frustrating experiences. However, the most serious difficulty concerns detection of the actual emergence of oil drops into the air–water surface in the case of certain types of oils. This important technique would therefore appear to be ready for a significant design upgrade, including possible commercialization.

REFERENCES

1. Wasan, D., Nikolov, A.D., Huang, D.D., Edwards, D.A. Foam stability: effects of oil and film stratification, in *Surfactant-Based Mobility Control, Progress in Miscible Flood Enhanced Oil Recovery* (Smith, D.H., ed.), ACS Symposium Series 373, Washington DC, 1988, Chapter 7, p 136.
2. Denkov, N.D. *Langmuir*, 20(22), 9463, 2004.
3. Bartsch, O. *Kolloidchem. Beihefte*, 20, 1, 1924.
4. Garrett, P.R., Davis, J., Rendall, H. *Colloid Surf. A*, 85, 159, 1994.
5. Pouchelon, A., Araud, C. *J. Disper. Sci. Technol.*, 14(4), 447, 1993.
6. Bergeron, V., Cooper, P., Fischer, C., Giermanska-Kahn, J., Langevin, D., Pouchelon, A. *Colloids Surf. A*, 122, 103, 1997.
7. Ross, J., Miles, G.D. *Oil Soap, May*, 99, 1941.
8. Rosen, M.J., Solash, J. *J. Am. Oil Chem. Soc.*, 46, 399, 1969.
9. DIN 53902, sheet 2, January, 1971.
10. *American Society for Testing and Materials Method*, D1173-53, Philadelphia, 1953.
11. Pauchek, M., Veber, V. *Acta Fac. Pharm. Univ. Comenianae*, 27(5), 27, 1975.
12. Ran, L., Jones, S.A., Embley, B., Tong, M.M., Garrett, P.R., Cox, S.J., Grassia, P., Neethling, S.J. *Colloid Surf. A*, 382, 50, 2011.
13. Ran, L. Foaming of anionic surfactant solutions in the presence of calcium ions and triglyceride-based antifoams, PhD Thesis, University of Manchester, 2011.
14. Garrett, P.R., Ward, D. R. *J. Colloid Interface Sci.*, 132(2), 475, 1989.
15. Weaire, D., Hutzler, S. *The Physics of Foams*, Clarendon Press, Oxford, 1999.
16. Ruff, K. *Chem. Ing. Tech.*, 44(24), 1360, 1972.
17. Denkov, N.D., Marinova, K., Tcholakova, S., Deruelle, M. Mechanism of foam destruction by emulsions of PDMS-silica mixtures, in *Proceedings of 3rd World Congress on Emulsions*, 24–27 Sept., 2002, Lyon, France, paper 1-D-199.
18. Theander, K., Pugh, R.J. *J. Colloid Interface Sci.*, 267, 9, 2003.
19. Bikerman, J.J., *Foams*, Springer, Berlin, 1973, Chapter 3, p 80.
20. Garrett, P.R. *J. Colloid Interface Sci.*, 69(1), 107, 1979.
21. Garrett, P.R., Moore, P.R. *J. Colloid Interface Sci.*, 159, 214, 1993.
22. Marinova, K.G., Denkov, N.D. *Langmuir*, 17(8), 2426, 2001.
23. Hobbs, S.Y., Pratt, C.F. *Am. Inst. Chem. Eng. J.*, 20(1), 178, 1974.
24. Zhang, H., Miller, C.A., Garrett, P.R., Raney, K. *J. Colloid Interface Sci.*, 263, 633, 2003.
25. Zhang, H., Miller, C.A., Garrett, P.R., Raney, K. *J. Colloid Interface Sci.*, 279, 539, 2004.
26. Zhang, H., Miller, C.A., Garrett, P.R., Raney, K. *J. Surfactants Deterg.*, 8(1), 99, 2005.
27. Zhang, H. Effects of oils, soap and hardness on the stability of foams, PhD Thesis, Rice University, Houston, 2003.

28. Wang, M., Cilliers, J.J. *Chem. Eng. Sci.*, 54(5), 707, 1999.
29. Vignes-Adler, M., Monnereau, C. *J. Colloid Interface Sci.*, 202, 45, 1998.
30. Glazier, J.A., Prausse, B., Gonatas, C.P., Leigh, J.S., Yodh, A.M. *Phys. Rev. Lett.*, 75, 573, 1995.
31. Garrett, P.R. Foams and antifoams, in *Food Colloids, Fundamentals of Formulation* (Dickinson, E., Miller, R., eds.), The Royal Society of Chemistry, Cambridge, UK, 2001, Special Publication No. 258, p 55.
32. Cheng, H.C., Lemlich, R. *Ind. Eng. Chem. Fundam.*, 22, 105, 1983.
33. de Vries, A.J. *Rec. Trav. Chim.*, 77, 209, 1958.
34. Brakke, K. *Exp. Math.* 1, 141, 1992; and www.susqu.edu/facstaff/b/brakke/evolver/html/default.htm, 1999.
35. Scheludko, A., Exerowa, D. *Comm. Dept. Chem., Bulg. Acad. Sci.*, 7, 123, 1959.
36. Scheludko, A. *Adv. Colloid Interface Sci.*, 1, 391, 1967.
37. Denkov, N.D., Cooper, P., Martin, J. *Langmuir*, 15(24), 8514, 1999.
38. Bergeron, V., Radke, C.J. *Langmuir*, 8(12), 3020, 1992.
39. Bergeron, V. Forces and structure in surfactant-laden thin liquid films, PhD Thesis, University of California at Berkeley, 1993.
40. Mysels, K.J., Jones, M.N. *Disc. Faraday Soc.*, 42, 42, 1966.
41. Exerowa, D., Scheludko, A. *Chim. Phys.*, 24, 47, 1971.
42. Dippenaar, A. *Int. J. Miner. Process.*, 9, 1–22, 1982.
43. Garrett, P.R., Wicks, S.P., Fowles, E. *Colloid Surf. A*, 282–283, 307, 2006.
44. Bergeron, V., Fagan, M.E., Radke, C.J. *Langmuir*, 9(7), 1704, 1993.
45. Bergeron, V., Radke, C.J. *Colloid Polym. Sci.*, 273, 165, 1995.
46. Koczo, K., Koczone, J., Wasan, D. *J. Colloid Interface Sci.*, 166, 225, 1994.
47. Basheva, E.S., Stoyanov, S., Denkov, N.D., Kasuga, K., Satoh, N., Tsujii, K. *Langmuir*, 17(4), 969, 2001.
48. Hadjiiski, A., Tcholakova, S., Denkov, N.D., Durbut, P., Broze, G., Mehreteab, A. *Langmuir*, 17(22), 7011, 2001.
49. Marinova, K., Denkov, N.D., Branlard, P., Giraud, Y., Deruelle, M. *Langmuir*, 18(9), 3399, 2002.
50. Denkov, N.D., Tcholakova, S., Marinova, K.G., Hadjiiski, A. *Langmuir*, 18(15), 5810, 2002.
51. Marinova, K.G., Denkov, N.D., Tcholakova, S., Deruelle, M. *Langmuir*, 18(23), 8761, 2002.
52. Hadjiiski, A., Denkov, N.D., Tcholakova, S., Ivanov, I.B. Role of entry barriers in foam destruction by oil drops, in *Adsorption and Aggregation of Surfactants in Solution* (Mittal, K., Shah, D. eds.), Surfactant Series 109, Marcel Dekker, NY, 2003, Chapter 23, p 465.
53. Marinova, K.G., Tcholakova, S., Denkov, N.D., Roussev, S., Deruelle, M. *Langmuir*, 19(7), 3084, 2003.
54. Hadjiiski, A., Dimova, R., Denkov, N.D., Ivanov, I.B., Borwankar, R. *Langmuir*, 12(26), 6665, 1996.
55. Bonfilon-Colin, A., Langevin, D. *Langmuir*, 13(4), 599, 1997.
56. Denkov, N.D., Marinova, K.G., Christova, C., Hadjiiski, A., Cooper, P. *Langmuir*, 16(6), 2515, 2000.

3 Oils at Interfaces

Entry Coefficients, Spreading Coefficients, and Thin Film Forces

3.1 INTRODUCTION

Antifoams are usually largely composed of oils that are present as undissolved drops in foaming solutions. The presence of oils in the form of undissolved drops is not, however, a sufficient requirement. Thus, it has often been proposed that such oils should also have the property of emerging into the surface of the foaming liquid, which implies that the films separating that liquid phase from the relevant gas phase should be intrinsically unstable. These asymmetric films are the so-called pseudoemulsion films, first named by Wasan et al. [1]. In addition, it has also been proposed that the emerged oil should spread over the gas–liquid surface of the foaming liquid (see, e.g., references [2–6]). The first proposal that spreading is associated with antifoam behavior was in fact made by Leviton and Leighton [7] more than seven decades ago. Subsequently there have been numerous attempts to correlate the antifoam behavior of undissolved oils with their spreading behavior [4, 8–14]. Underlying much of this work has been the concept that spreading oil drops at gas–liquid surfaces produce shear forces in foam films, leading to the rupture of those films.

Much recent work on the mode of action of antifoams has been concerned with both establishing the importance of the stability of the pseudoemulsion films, which separate antifoam oil drops from gas–liquid surfaces and the necessity that the oil spread at that surface. As we will show in later chapters, the stability or otherwise of those films has been shown to be a key aspect of antifoam mechanism. However, it has also been firmly established that spreading is not a necessary property of the oil [15]. Nevertheless, the presence of spread oil layers at gas–liquid surfaces can have profound effects on the relative effectiveness of antifoams.

To frame a context for describing the mode of action of antifoams in later chapters, we are concerned here with the criteria that determine whether oils will emerge into gas–liquid surfaces and whether they will spread at those surfaces. Those criteria are not confined to consideration of thermodynamic stability alone. In reality metastable states can exist that have a profound effect on whether antifoam action occurs. Complexity also arises because spreading oil molecules can be partially solubilized in the foam-stabilizing surfactant monolayers present at the relevant gas–liquid surfaces. All of these aspects are reviewed here, thermodynamic criteria being represented by the classic entry and spreading coefficients and metastable states by defining generalized entry and spreading coefficients, which include the contributions of colloidal interaction forces in thin films. A review of the mode of rupture of pseudoemulsion films is also included where we highlight some unresolved issues. Finally we review the observed spreading behavior, on the air–water surfaces of surfactant solutions, of the typical antifoam oils represented by hydrocarbons and polydimethylsiloxanes.

3.2 CLASSIC ENTRY AND SPREADING COEFFICIENTS

Here we are concerned with defining the conditions under which a drop of oil can emerge onto the gas–liquid surface of a foaming liquid and subsequently spread on that surface to form a thin film (or layer) of oil. Harkins [16] identified three types of such layers:

- Monolayers: one molecule thick.
- Films thicker than monolayers, but thinner than duplex films.
- Duplex films: the film is sufficiently thick to give complete independence between the energy of the surface and that of the interface where in the present context the surface is the air–oil surface and the interface is the interface between the oil and the foaming liquid.

If the oil is present in the system as a bulk phase and can emerge into the air–foaming liquid surface, then there are in effect three main possible consequences. The simplest concerns formation of an oil lens with no oil film—the so-called partial wetting described by Brochard-Wyatt et al. [17]. At the other extreme we can have duplex film formation where no lens is present—the so-called complete wetting. An intermediate situation can, however, exist where an oil lens is in stable (or even metastable) equilibrium with a film that is thinner than a duplex film and can range in thickness from that of an oil-contaminated surfactant monolayer up to a multimolecular layer of oil superimposed on a surfactant monolayer, of thickness usually ≤100 nm. Brochard-Wyatt et al. [17] refer to this situation as *pseudo-partial wetting*. We will use this terminology throughout.

Here we consider the particular case of oil-based antifoams in aqueous systems. The same arguments can of course be extended to other combinations of fluids as we will, for example, describe in Chapter 10.

If an oil is to emerge into the air–water surface and subsequently spread, then there are certain requirements that must be satisfied concerning the so-called *entry* and *spreading coefficients*. If drops of oil dispersed in an aqueous phase are to emerge spontaneously into the air–liquid surface, the entry coefficient E defined by

$$E = \sigma_{AW} + \sigma_{OW} - \sigma_{AO} \tag{3.1}$$

must be positive. Here σ_{AW} is the air–liquid surface tension of the foaming liquid, σ_{OW} is the antifoam oil–water interfacial tension, and σ_{AO} is the antifoam oil–air surface tension, measured under any conditions of equilibrium or non-equilibrium.

Selection of the appropriate values of the three surface tensions should, however, take account of the extent to which the antifoam and aqueous solution have been pre-equilibrated. Thus, we could follow Ross [4] and define three entry coefficients. First, the initial entry coefficient E^i

$$E^i = \sigma^i_{AW} + \sigma^i_{OW} - \sigma^i_{AO} \tag{3.2}$$

where, σ^i_{AW}, σ^i_{OW}, and σ^i_{AO} are the relevant surface tensions upon initial mixing of the oil with the solution before mutual saturation. Second, we can define a semi-initial entry coefficient

$$E^{si} = \sigma^i_{AW} + \sigma^e_{OW} - \sigma^e_{AO} \tag{3.3}$$

where the air–water surface is uncontaminated with antifoam oil but where the oil and the aqueous phase are mutually saturated so that the oil–water and the air–oil surface tensions are the equilibrium values, σ^e_{OW} and σ^e_{AO}, respectively. Finally we define the equilibrium entry coefficient E^e

$$E^e = \sigma^e_{AW} + \sigma^e_{OW} - \sigma^e_{AO} \tag{3.4}$$

where the tensions are all equilibrium values.

If the oil is also to spread at the surface of the aqueous phase to form a film, then the spreading coefficient S must also be positive. Here we may follow Harkins [18] and define an initial spreading coefficient, S^i

$$S^i = \sigma^i_{AW} - \sigma^i_{OW} - \sigma^i_{AO} \tag{3.5}$$

which determines whether a duplex film will be formed when oil is present at the air–water surface where mutual saturation of the relevant phases has not been achieved. A semi-initial spreading coefficient, S^{si}, is defined

$$S^{si} = \sigma^i_{AW} - \sigma^e_{OW} - \sigma^e_{AO} \tag{3.6}$$

which determines whether a duplex oil film will spread over the air–water surface when the oil and foaming liquid are mutually saturated but only the oil–water and oil–air surface tensions are at equilibrium. A final, or equilibrium, spreading coefficient, S^e

$$S^e = \sigma^e_{AW} - \sigma^e_{OW} - \sigma^e_{AO} \tag{3.7}$$

determines whether a duplex oil film is spread over the air–water surface when all relevant surface tensions are the equilibrium values for the mutually saturated phases.

We should note here that the entry coefficient for an oil to emerge into the air–water surface is equal to minus the spreading pressure for the aqueous phase over the oil and conversely the spreading coefficient for oil over the aqueous phase is equal to minus the entry coefficient for a drop of aqueous phase to emerge into the air–oil surface. However, the entry and spreading coefficients, which are relevant for any putative antifoam action (and which are given in Equations 3.1 through 3.4 and 3.5 through 3.7, respectively), clearly do not possess that equivalence.

In the case of mutually saturated solutions, there are certain limitations on the allowed values of both E^e and S^e [19]. Essentially the limits concern the maximum value of any equilibrium surface tension in such a system. That is determined by the possibility of replacement of the relevant surface by a duplex layer. If the relevant surface has a surface tension higher than the sum of the two surface tensions of the two surfaces of a putative duplex layer, then the latter will spread over that surface. In that case the surface tension of the relevant surface cannot be an equilibrium value. Therefore, we can write, for example, that

$$\sigma_{AW}^{e} \leq \sigma_{OW}^{e} + \sigma_{AO}^{e} \tag{3.8}$$

where the identity

$$\sigma_{AW}^{e} = \sigma_{OW}^{e} + \sigma_{AO}^{e} \tag{3.9}$$

is a definition of Antonow's rule [20]. This rule states that in the case of a system consisting of two liquids and a vapor, the interfacial tension between the two liquids is equal to the difference between their vapor–liquid surface tensions. Systems that obey this rule form thick duplex films where the properties of the liquid in the film, including the two surface tensions, are the same as those of the bulk phase.

We can similarly note that the air–oil surface tension must have a maximum value dictated by the sum of the surface tensions of the two surfaces of any putative duplex film formed by the aqueous phase. This means that we must have

$$\sigma_{AO}^{e} \leq \sigma_{OW}^{e} + \sigma_{AW}^{e} \tag{3.10}$$

Using the definitions of E^e and S^e found in Equations 3.4 and 3.7 together with the conditions 3.8 and 3.10 yields the general conditions

$$0 \leq E^{e} \leq 2\sigma_{OW}^{e} \tag{3.11}$$

and

$$-2\sigma_{OW}^{e} \leq S^{e} \leq 0 \tag{3.12}$$

As pointed out by Rowlinson and Widom [19], reported values of E^e and S^e, which lie outside the limits given by conditions (3.11) and (3.12), are indications of either experimental error in measurement of the relevant surface tensions or failure to reach equilibrium.

If we have $E^e > 0$ and $S^e = 0$, then a drop of oil on the surface of a liquid will in principle be spread over that surface to form a duplex film at equilibrium (obeying Antonow's rule [20]). In contradiction Harkins [18] has claimed that $S^e < 0$ always prevails at equilibrium even if the initial spreading coefficient $S^i > 0$. However, Rowlinson and Widom [19] emphasize convincingly that $S^e = 0$ is allowed so that stable duplex films can often be found between mutually saturated liquids at equilibrium.

If $E^e > 0$ and $S^e < 0$, then an oil drop will form a lens, which may be in equilibrium with the original air–foaming liquid surface (i.e., partial wetting) or that surface after incorporation of material from the oil drop (i.e., pseudo-partial wetting). However, formation of oil lenses on the equilibrium liquid surface does not preclude the possibility that the initial spreading coefficient S^i is positive. If $S^i > 0$ and $S^e < 0$, then pseudo-partial wetting will occur and the oil drop may, for example, spread to form a duplex film, which is, however, unstable. That film will disproportionate to form oil lenses in equilibrium with oil-contaminated air–water surfaces The spreading of a drop of benzene on distilled water to form lenses in equilibrium with a benzene monolayer exemplifies this type of behavior.

3.3 GENERALIZED ENTRY COEFFICIENTS, PSEUDOEMULSION FILMS, AND THIN FILM FORCES

3.3.1 Definitions

These arguments concerning entry and spreading coefficients have dominated the interpretation of antifoam behavior for much of the past century. They are, however, limited in that they ignore the possibility of metastable states, which can mean, for example, that an oil with a positive entry coefficient (as defined by whichever of Equations 3.2 through 3.4 is appropriate) does not emerge into an air–water surface. In this case the metastable state will concern the stability of the oil–water–air pseudoemulsion film. That film must rupture if the oil is to emerge into the air–water surface. Such films will be stabilized by much the same combination of forces as are responsible for metastability in symmetrical aqueous foam films. The very factors that conspire to produce copious amounts of foam will also inhibit the emergence of antifoam oil drops into the relevant air–foaming liquid surfaces. We should stress here, however, that if the equilibrium entry coefficient $E^e = 0$, then the oil cannot emerge into the air–water surface irrespective of these considerations. Unfortunately, unavoidable experimental error in determination of E^e can make it difficult to distinguish between a metastable state and a true equilibrium cause of non-emergence of an oil drop [21].

As we have seen in Chapter 1, the magnitude of the thin film forces in symmetrical foam films is characterized by the disjoining pressure, $\Pi_{AWA}(h)$ (in the case of an aqueous film), which is a function of film thickness h. A similar characterization may be used for the pseudoemulsion films considered here. However, in contrast to foam films, disjoining pressure against film thickness plots have been rarely measured for pseudoemulsion films (for examples of such measurements, see references [22, 23]). We should note here that the optical properties of these asymmetrical pseudoemulsion films are different from those of symmetrical foam films [22, 24, 25]. The phase difference between light reflected from the front and rear surface of pseudoemulsion films is zero. Therefore, constructive interference of light reflected from such surfaces occurs if the films are extremely thin so that the film thickness makes a negligible contribution to the path length of the light. Such films therefore appear white when observed with white light in contrast to the equivalent foam films that appear to be black. Observations suggest that unequivocal analogues of metastable Newton black films [26, 27] and common black films [22] are to be found in pseudoemulsion films, although in those cases perhaps better described as Newton white films and common white films [22, 27]. Indeed it seems most of the phenomenon that occur in foam films are also to be found in pseudoemulsion films [22–25] where, however, the stabilities and magnitude of the disjoining pressures can be markedly different even with the same foaming liquid [22].

As a consequence of these considerations, several workers have introduced the concept of generalized entry coefficients, which allow for metastable states by taking explicit account of the relevant disjoining pressure against thickness isotherms [17, 22, 23, 28]. In adopting this approach, Bergeron et al. [22], considering oil–water systems, define a generalized entry coefficient E_g by replacing the air–oil surface tension in the classic entry coefficients with a thickness-dependent film tension, $\sigma_{AWO}(h)$, which can characterize any film separating the oil from the air. Such films could range from monolayer contamination to extremely thick aqueous duplex pseudoemulsion films where the surfaces have the same properties as the bulk liquid. Bergeron et al. [22] therefore write

$$E_g = \sigma_{AW} + \sigma_{OW} - \sigma_{AWO}(h) \tag{3.13}$$

where σ_{OW} and σ_{OW} are, respectively, the bulk phase air–water and oil–water surface tensions under any (initial, semi-initial, or equilibrium) conditions. The film tension is defined by de Feijter [29] as "the excess tangential force (per unit length of the film) relative to the surrounding phase." Using a representation, attributed to Derjaguin [30] by Scheludko et al. [31], of such a film as a single Gibbs dividing surface associated with volume, Hirasaki [28] writes an augmented Gibbs adsorption equation

$$d\sigma_{AWO}(h) = h d\Pi_{AWO} - \sum_j \left(\Gamma_j^{AW} + \Gamma_j^{OW}\right) d\mu_j \tag{3.14}$$

where $\Pi_{AWO}(h)$ is the disjoining pressure in a pseudoemulsion film separating the oil from the air, μ_j is the chemical potential of species j and where Γ_j^{AW} and Γ_j^{OW} are, respectively, the surface excesses of species j at the air–water and the oil–water surfaces of the film. Integrating Equation 3.14 at constant chemical potential yields for the film tension of a film of thickness h

$$\sigma_{AWO}(h) = \sigma_{AWO}(h = \infty) + \int_{\Pi_{AWO}(h_\infty)=0}^{\Pi_{AWO}(h)} h \, d\Pi_{AWO} \tag{3.15}$$

Here the limit $h = \infty$ represents a duplex pseudoemulsion film where disjoining forces are absent so $\Pi_{AWO}(h_\infty) = 0$ and the surface tensions of the film surfaces are those of the bulk phases so that

$$\sigma_{AWO}(h = \infty) = \sigma_{AW} + \sigma_{OW} \quad (3.16)$$

Therefore, combining Equations 3.15 and 3.16 yields an expression for the film tension of a film of thickness h so that

$$\sigma_{AWO}(h) = \sigma_{AW} + \sigma_{OW} + \int_{\Pi_{AWO}(h_\infty)=0}^{\Pi_{AWO}(h)} h \, d\Pi_{AWO} \quad (3.17)$$

If the integral $\int_{\Pi_{AWO}(h_\infty)=0}^{\Pi_{AWO}(h)} h \, d\Pi_{AWO}$ is negative, then the film tension is less than the sum of the surface tensions of the contiguous bulk phases. In those circumstances the surfaces of those bulk phases form angles of contact at the film. This situation is illustrated in Figure 3.1. Scheludko et al. [31] have shown that the film tension is then given by

$$\sigma_{AWO}(h) = \sigma_{AW} \cos\tilde{\alpha}_{AW} + \sigma_{OW} \cos\tilde{\alpha}_{OL} \quad (3.18)$$

where $\tilde{\alpha}_{AW}$ and $\tilde{\alpha}_{OW}$ are the contact angles made by the air–water and oil–water surfaces, respectively, in an adjacent meniscus with respect to the plane of the film as indicated in Figure 3.1. This implies that Neumann's triangle will still apply despite the presence of a pseudoemulsion film. The angle θ^* between air–water and oil–water

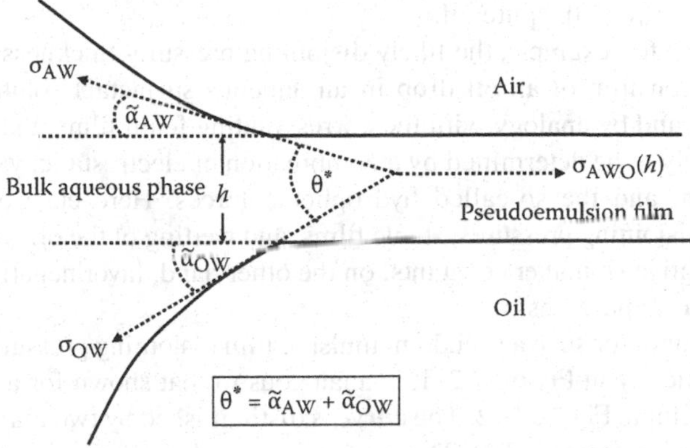

FIGURE 3.1 Contact angles in pseudoemulsion film illustrating Neumann's triangle and definition of angle θ^* in this context.

surfaces is then simply the sum $\tilde{\alpha}_{AL} + \tilde{\alpha}_{OL}$. As we will show in Chapter 4, this angle is of some significance in determining antifoam effects.

Combining Equations 3.13 and 3.17 gives the generalized entry coefficient

$$E_g = -\int_{\Pi_{AWO}(h_\infty)=0}^{\Pi_{AWO}(h)} h\,d\Pi_{AWO} \tag{3.19}$$

If $E_g > 0$, the film tension is less than the sum of the bulk liquid surface tensions and either stable or a metastable films can form with a finite contact angle at any meniscus. If, on the other hand, $E_g < 0$, then the film tension is greater than the sum of the bulk liquid surface tensions and any supposed film to which it refers cannot be realized.

It may be readily shown [28, 31] that the generalized entry coefficient is related to the interaction free energy/unit area, $W(h)$, of the pseudoemulsion film where

$$W(h) = -\int_{h=\infty}^{h} \Pi_{AWO}(h)\,dh \tag{3.20}$$

and where integration by parts reveals that

$$E_g = -h\Pi_{AWO}(h) - W(h) \tag{3.21}$$

If we set $h = 0$ in Equations 3.19 through 3.21 and therefore $\Pi_{AWO}(h) = 0$, then we are considering entry to form an oil–air surface having the properties of the pure oil (i.e., uncontaminated with any film or adsorbed monolayer) whence E_g and $W(h)$ become equivalent. Physically this situation must correspond closely to that represented by the classic initial entry coefficient (or even in some circumstances the classic equilibrium entry coefficient) where the film tension is replaced by the air–oil surface tension, σ_{AO}, of the pure oil.

Consider now for, example, the likely disjoining pressure–thickness isotherm for a pseudoemulsion film of an oil drop in an aqueous surfactant solution. In these circumstances, and by analogy with the corresponding foam films, the shape of the isotherm is likely to be determined by a combination of electrostatic, van der Waals, steric/hydration, and the so-called hydrophobic forces. Here electrostatic forces favor positive disjoining pressures, stable films, and wetting of the oil. van der Waals forces with positive Hamaker constants, on the other hand, favor negative disjoining pressures and unstable films.

A putative curve for such a pseudomemulsion film disjoining pressure isotherm is shown schematically in Figure 3.2. It is analogous to that shown for a symmetrical aqueous foam film in Figure 1.12. The curve is distinguished by two maxima and two minima (see, e.g., reference [31]). The so-called secondary minimum at large thicknesses arises where long range van der Waals attractive forces dominate. However, reduction of the film thickness in the presence of, for example, ionic surfactant leads

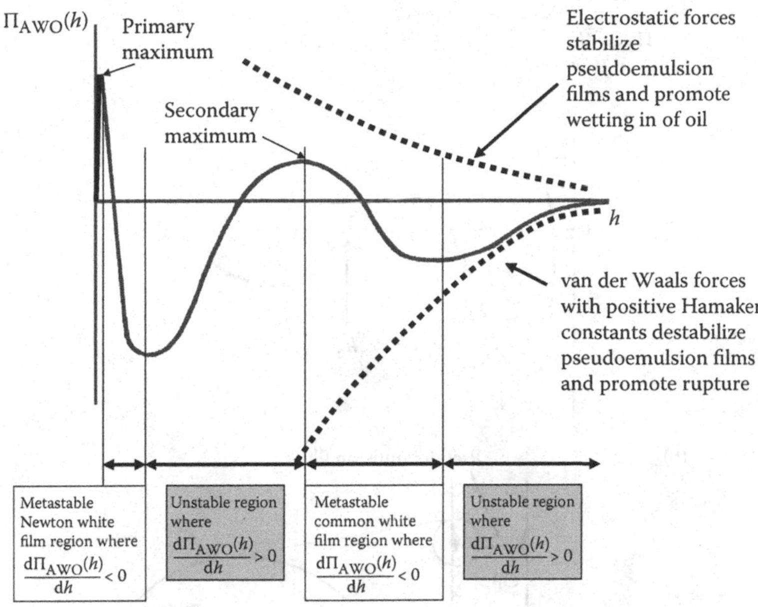

FIGURE 3.2 Schematic illustration of disjoining pressure isotherm in pseudoemulsion film defining metastable and unstable regions and positions of primary and secondary maxima.

to domination by electrostatic repulsion effects and the minimum gives way to a secondary maximum. As the film thins further, again van de Waals forces dominate to produce a primary minimum. Steric/hydration forces can finally contribute to a steep rise in the isotherm at extremely small thicknesses to produce a primary maximum. The two maxima in the disjoining pressure isotherm mean that there are two regions where $d\Pi_{AWO}(h)/dh < 0$, which, as we have described in Chapter 1, is a required condition for the formation of stable films. Here the secondary region, at large thicknesses of order ≤50 nm, permits the formation of common white films and the primary, at lower thicknesses of ≤5 nm, a Newton white film or even just an oil–air surface contaminated with a partial monolayer of water.

3.3.2 Case Where Generalized Entry Coefficient, $E_g > 0$

We now illustrate the arguments of Bergeron et al. [22] for the case where the disjoining pressure isotherm resembles that shown in Figure 3.3a. First we consider the relation between the classic equilibrium entry coefficient and the generalized entry coefficient in the case where both are equal. Bergeron et al. [22] argue that the classic equilibrium entry coefficient concerns equilibrium between the relevant bulk phases and an equilibrium oil surface, which could range from a stable pseudoemulsion film at zero capillary pressure to a clean oil surface where the capillary pressure is absent. In the case of a stable or metastable pseudoemulsion film, the former hypothetical situation would appear to be rather difficult to realize. One possibility involves a vessel containing a bulk oil layer in equilibrium with an underlying bulk aqueous phase.

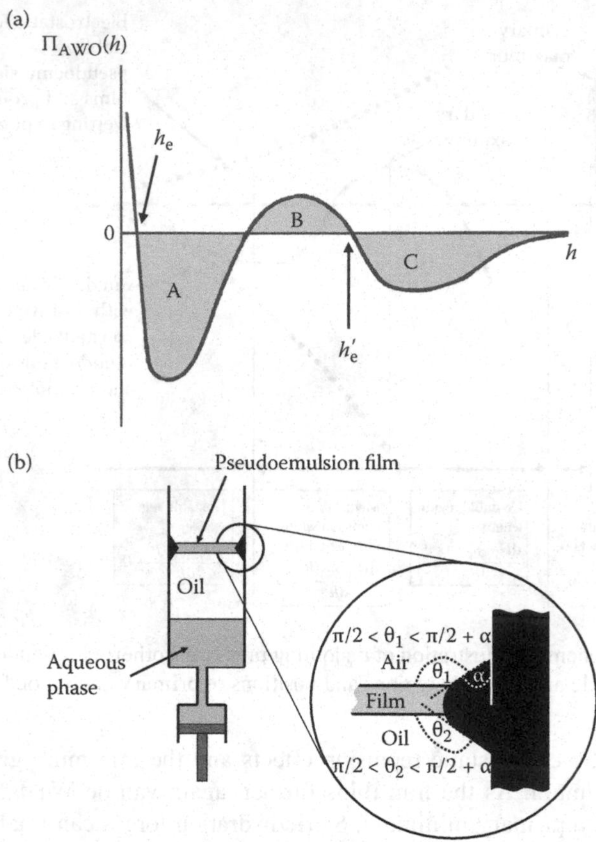

FIGURE 3.3 Formation of stable and metastable equilibrium pseudoemulsion films at zero capillary pressure. (a) Schematic disjoining pressure isotherm plotting disjoining pressure $\Pi_{AWO}(h)$ against film thickness h where stable film exists at h_e and metastable film exists at h_e', both at zero disjoining pressure. (b) Hypothetical arrangement where pseudoemulsion film at zero disjoining pressure is in equilibrium with relevant bulk phases. Cross-section of the rim in vessel containing the film means that film is everywhere planar provided relevant contact angles θ_1 and θ_2 lie within indicated ranges with respect to angle α of rim.

A pseudoemulsion film could be present in the plane of a rim on the inside of the vessel, where the latter is characterized by appropriate geometry and contact angles so that no meniscus exists and the film is everywhere planar. In such an arrangement, the film, at zero capillary pressure, would be in equilibrium with bulk aqueous phase. This putative arrangement is illustrated in Figure 3.3b.

At equilibrium the capillary pressure must equal the disjoining pressure, which must therefore also be zero. Therefore, in the case of the schematic disjoining pressure isotherm for a putative pseudoemulsion film shown in Figure 3.3a, there exist two possible equilibrium thicknesses, h_e and h_e', where for stability $d\Pi_{ALO}/dh < 0$ and where $\Pi_{ALO}(h_e) = 0$ and $\Pi_{ALO}(h_e') = 0$. Now consider the shape of the disjoining pressure isotherm where the hatched areas A, B, and C indicate aspects of the integral of

Equation 3.19. In this example we have $A < 0$, $C < 0$, $B > 0$, $|A| > B$, $|C| > B$, and $|A + C| > B$ so that the integral of Equation 3.19 is negative and the generalized entry coefficient positive for films at both thickness $h = h_e$ and $h = h'_e$. These two situations correspond to stable equilibrium and metastable equilibrium films, respectively, where the former is characterized by the highest entry coefficient and the lowest film tension. The classic equilibrium entry coefficient E^e is concerned with the stable equilibrium so we should have $E^e = E_g(h = h_e)$.

By contrast here we are concerned with oil drops trapped in foam films. In this situation the drop will be subject to a capillary pressure due to both the presence of an adjacent Plateau border and the curvature of the pseudoemulsion film as illustrated in Figure 3.4a. In general, the latter will dominate but we should note that the curvature

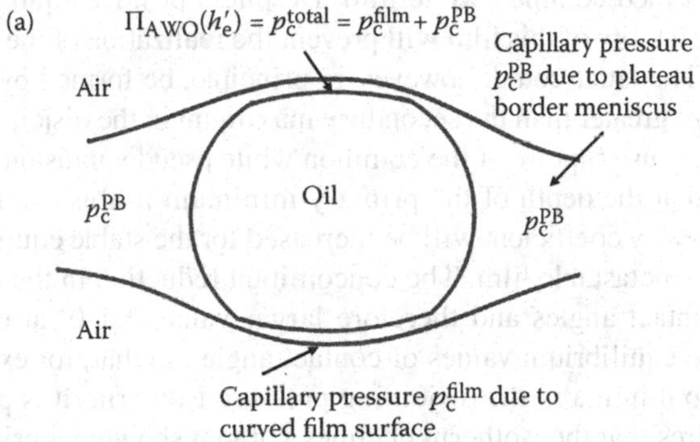

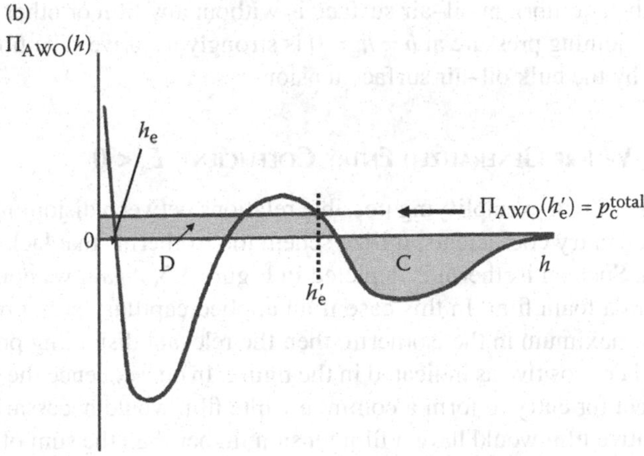

FIGURE 3.4 Formation of metastable pseudoemulsion film in presence of finite imposed capillary pressure. (a) Schematic illustration of origin of capillary pressure in metastable pseudoemulsion films formed by a drop trapped in symmetrical foam film. Capillary pressure is seen to equal the sum of that due to film curvature and that due to an adjacent Plateau border (with the former pressing on film and the latter sucking on film). (b) Schematic disjoining pressure isotherm plotting disjoining pressure $\Pi_{ALO}(h)$ against thickness where disjoining pressure is equal to total capillary pressure p_c^{total}. Metastable pseudoemulsion film is seen to exist at film thickness $h'_e \gg h_e$.

of the film is itself a response to the capillary pressure in the Plateau border as the foam film thins to thicknesses less than the drop diameter. Again equilibrium will require that the disjoining pressure equal the net capillary pressure. This situation is shown schematically in Figure 3.4b where we suppose the disjoining pressure isotherm is identical to that shown in Figure 3.3a. We are therefore considering a situation where the classic equilibrium entry coefficient is positive. If then a positive capillary pressure, p_c^{total}, is applied, which is below the secondary maximum in the isotherm then, as indicated schematically in Figure 3.4b, a meta-stable film of thickness h_e', can in principle be formed where $p_c^{total} = \Pi_{AWO}(h_e')$ because there $d\Pi_{AWO}/dh < 0$. The generalized entry coefficient in this case is determined by the sum of the areas D + C in the figure, which is seen to be negative. The entry coefficient is therefore positive and the metastable film will spontaneously form. Such a film would be expected to have a thickness typical of a so-called common white film. Despite a positive equilibrium entry coefficient, the metastability of this film will prevent the realization of the equilibrium film at $h = h_e \ll h_e'$. The latter could, however, in principle, be formed by application of a capillary pressure greater than the secondary maximum in the disjoining pressure isotherm, which will cause rupture of the common white pseudoemulsion film.

We should note that the depth of the primary minimum in this example is such that the generalized entry coefficient will be increased for the stable equilibrium film relative to that of the metastable film. The concomitant reduction in the film tension will mean larger contact angles and therefore larger values for θ^* at equilibrium. Obviously then large equilibrium values of contact angles so that, for example, $\theta^* > \pi/2$, imply very deep minima in the disjoining pressure isotherm. It is possible that in general this requires that the isotherm declines without showing a primary maximum so that the equilibrium oil–air surface is without any film or other contamination and the disjoining pressure at $h = h_e = 0$ is strongly negative. The film tension is then replaced by the bulk oil–air surface tension.

3.3.3 Case Where Generalized Entry Coefficient, $E_g < 0$

Bergeron et al. [22], in exemplifying possible relations between disjoining isotherms and generalized entry coefficients, utilize schematic isotherms that lack the secondary minimum. Such an isotherm is depicted in Figure 3.5. Again we consider an oil drop trapped in a foam film. In this case if an applied capillary pressure lies below the secondary maximum in the isotherm, then the relevant disjoining pressure integral is everywhere positive as indicated in the figure. In consequence the generalized entry coefficient for entry to form a common white film would necessarily be negative. This putative film would have a film tension higher than the sum of the surface tensions of the contiguous bulk phases, which could not be realized. Moreover if the film cannot be formed in these circumstances, then the capillary pressure cannot be set equal to the disjoining pressure, which is of course a required condition for even metastable equilibrium. Also as pointed out by de Feijter [29] in the context of symmetrical foam films, the contact angles in Equation 3.18 cannot then exist. Indeed, de Feijter [29] states that "this (latter) may give rise to problems for a microscopic description of the contact between the film and the meniscus, because the extrapolated meniscus surface does not meet the film."

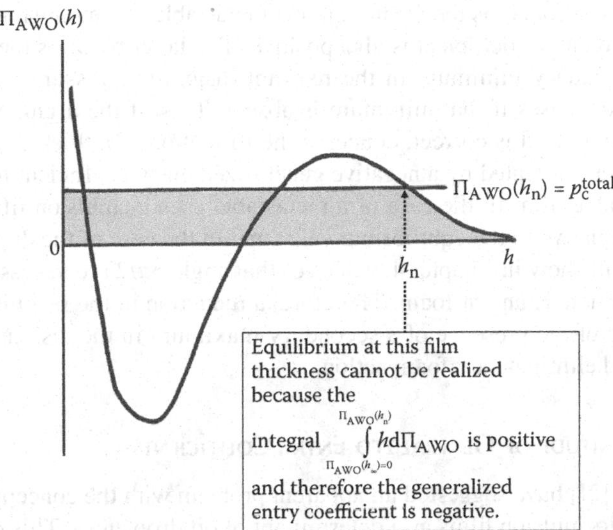

FIGURE 3.5 Schematic disjoining pressure isotherm with negative generalized entry coefficient at imposed capillary pressure p_c^{total} where the latter is lower than the height of secondary maximum in isotherm and secondary minimum is absent (compare with Figures 3.3a and 3.4b). Generalized entry coefficient at thickness h_n is negative so films of this thickness are not realized.

The condition $E_g < 0$ clearly means that the arrangement shown in Figure 3.5 cannot be realized at equilibrium (or at all). As pointed out by Hirasaki [28], it should not therefore be regarded as in conflict with the requirement, given by Equation 3.11, that the classic equilibrium entry coefficient should have a minimum value of zero. Thus the latter derives from Antonow's rule [20], which deals with equilibrium behavior and is only valid for systems in which the equilibrium disjoining pressure is zero.

It is not clear from the literature what arrangement would be realized in the case of a pseudoemulsion film associated with an oil drop in a draining foam film where $E_g < 0$ as indicated in Figure 3.5. Possibly it implies that the curvature of the pseudoemulsion film in the vicinity of the drop would remain negligible as the foam film containing the drop drains. Both the applied capillary pressure and therefore disjoining pressure in the pseudoemulsion film would then be minimal. Pseudoemulsion film thinning would not then occur so that a stable duplex film would exist, with zero contact angles, as suggested by Bergeron et al. [22]. In which case the upper limit of the integral in Equation 3.19 will become $\Pi_{ALO}(h_\infty) = 0$ so that the generalized entry coefficient would also become zero, thereby apparently resolving any conflict with respect to the requirement that the classic entry coefficient should always be ≥ 0!

There would appear then to be two situations where the existence of a secondary maximum in the disjoining pressure isotherm could result in inhibition of entry of an oil drop into the equilibrium configuration at the surface of a foaming solution where the classic equilibrium entry coefficient is positive (and where an oil–air surface is formed). The first concerns the formation of a metastable common white film where the generalized entry coefficient is also positive. The latter requires the presence of a negative secondary minimum in the relevant disjoining pressure isotherm. The second situation arises if that minimum is absent. This, if the argument advanced by Bergeron et al. [22] is correct, concerns the formation of a thick duplex pseudoemulsion film as indicated by a negative generalized entry coefficient for formation of any metastable film. In the case of a metastable pseudoemulsion film, the angle θ^* is reduced relative to the equilibrium value and in the case of the duplex film it is zero. As we will show in Chapter 4, values of that angle $>\pi/2$ are necessary if the oil drop is to function as an antifoam. Therefore, a reduction in the magnitude of θ^* as a consequence of the presence of a secondary maximum in the disjoining pressure isotherm could eliminate antifoam action.

3.3.4 Magnitude of Generalized Entry Coefficients

Aveyard et al. [21] have suggested an apparent problem with the concept of metastability of pseudoemulsion films as a determinant of oil drop entry. This concerns the magnitude of the generalized entry coefficients, which can be calculated from the few reported disjoining pressure isotherms for oil–water–air pseudoemulsion films such as those given Bergeron et al. [22]. Aveyard et al. [21] calculated from the relevant isotherms that E_g is of the order of -10^{-2} mN m^{-1}, which lies within the order of magnitude of the errors in measurements of classic entry coefficients (reported values of $W(h)$ are of a similar order [24]). They then tended to draw the conclusion from this that the stability or otherwise of pseudoemulsion films was determined largely by classic entry coefficients. They implied that instances where positive classic equilibrium entry coefficients did not correlate with ready rupture of pseudoemulsion films should not be attributed to metastable film formation but rather to errors in the measurements of the classic equilibrium entry coefficients. However, in something of a volte-face, Aveyard and Clint [32] later presented evidence that, at least in the case of cyclohexane in a dilute aqueous NaCl solution of AOT, positive classic equilibrium entry coefficients do not necessarily correlate with foam film destabilization. They then concluded that the metastability of pseudoemulsion films can be a key determinant of entry.

We should note here that the low values of the generalized entry coefficients were calculated by Aveyard et al. [21] from the incomplete disjoining pressure isotherms of Bergeron et al. [22]. The method of the latter permits only measurement of the positive parts of the isotherms [22, 33] where films are at least metastable so $d\Pi_{AWO}(h)/dh < 0$. The isotherm for tetralin (1,2,3,4-tetrahydronaphthalene) in sodium dodecyl sulfate (SDS) solution is reproduced in Figure 3.6 by way of example, where the measured part of the isotherm is seen to be everywhere positive. The critical capillary pressure required for rupture of the film associated with the part of the isotherm presented in the figure is reported to be 0.2 kPa. That pressure then presumably represents a maximum in the disjoining pressure isotherm, which is not shown in Figure 3.6. Now again suppose, as suggested by Bergeron et al. [22] and depicted in Figure 3.3a, that the classic equilibrium entry coefficient is given by the integral in Equation 3.19 with the limits $\Pi_{AWO}(h_\infty) = 0$ and $\Pi_{AWO}(h_e) = 0$. The

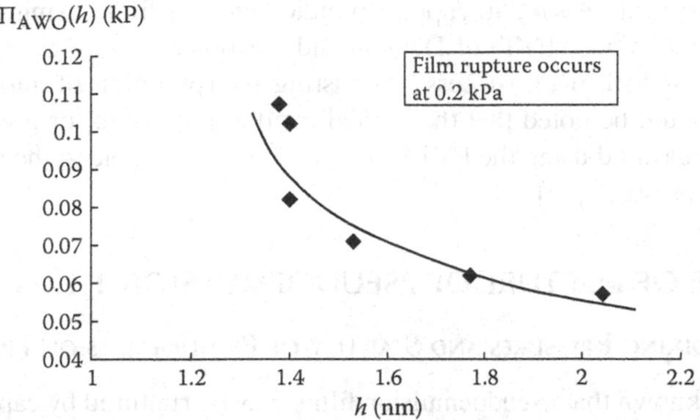

FIGURE 3.6 Experimental partial disjoining pressure isotherm for pseudoemulsion film formed against air by tetralin in ~0.017 M SDS solution. (Reprinted with permission from Bergeron, V. et al., *Langmuir*, 9, 1704. Copyright 1993 American Chemical Society).

requirement that $E^e > 0$ means that there should be at least one minimum in such an isotherm where disjoining pressures are negative. Indeed the quoted surface tensions given by Bergeron et al. [22] for the tetralin–SDS system indicate a positive classic equilibrium entry coefficient of 3.3 mN m^{-1}. The latter, if a true equilibrium value, implies a deep primary minima in the disjoining pressure isotherm, which, in turn, indicates that the quoted segment of the isotherm in Figure 3.6 is unrepresentative of the whole isotherm and cannot contribute significantly to a value of the generalized entry coefficient equivalent to the putative classic equilibrium entry coefficient E^e. The possibility that a secondary minimum also exists cannot be completely rejected. However, we should note that Bergeron et al. [22] find that rupture of pseudoemulsion films during foam flow through porous media correlates equally well with either equilibrium classic entry coefficients or the height of the maximum (measured) positive disjoining pressure. Values of the classic entry coefficients given by these authors are, however, not always consistent with the requirements for equilibrium as indicated by Equation 3.11.*

That direct measurement of the negative parts of disjoining pressure isotherms is not possible using the technique of Bergeron et al. [22, 33] means that this technique cannot therefore produce the complete disjoining pressure isotherm, which as we have seen, is necessary if generalized entry coefficients are to be calculated. The absence of a method of measurement of such isotherms does rather limit the utility of this concept of generalized entry coefficients as a guide to the stability of pseudoemulsion films. Other techniques for studying the stability behavior of pseudoemulsion films are described in Chapter 2. Direct measurement of the applied

* Bergeron et al. [22] also give a partial disjoining pressure isotherm for the SDS–dodecane system where the reported surface tensions also yield an equilibrium classic spreading coefficient > 0, which violates the requirement that $S^e \leq 0$ (Equation 3.11). The latter suggests then that the reported surface tensions are non-equilibrium values (indeed, it is not possible to calculate a value for θ^* from the reported surface tensions, which confirms this conclusion). Therefore, it is not possible to make a similar analysis for dodecane–SDS to that made for tetralin–SDS.

capillary pressure necessary to rupture pseudoemulsion films using the so-called film trapping technique (FTT) of Denkov and coworkers (see, e.g., reference [34]) has in particular had much success in assisting interpretation of antifoam action. However, it should be noted that the critical capillary pressure for pseudoemulsion film rupture measured using the FTT is not directly comparable to the relevant critical disjoining pressure [35].

3.4 MODE OF RUPTURE OF PSEUDOEMULSION FILMS

3.4.1 Disjoining Pressures and Stability of Pseudoemulsion Films

It is now well known that pseudoemulsion films may be ruptured by capillary effects induced by the presence of particles with the appropriate wettability and geometry. We deal with this, by far the most effective mechanism for aqueous systems, in detail in Chapter 4. Here we are concerned only with mechanisms that do not involve the introduction of a fourth phase into the system air–water–oil. In this case rupture of thin liquid films is usually associated with application of a capillary pressure high enough to exceed any relevant maximum in the disjoining pressure isotherm. Rupture of aqueous pseudoemulsion films on hydrocarbon oils by application of total capillary pressures high enough to exceed a primary maximum in the disjoining pressure isotherm has in fact been directly observed by Bergeron et al. [22, 23] in a number of cases. Film thickness transitions from, for example, a common white film to a Newton white film are similarly supposed to involve application of a total capillary pressure higher than the secondary maximum in the relevant disjoining pressure isotherm.

Denkov and coworkers [34, 35] have, however, identified a potential problem with this simple proposition. They have drawn attention to the fact that pseudoemulsion film thicknesses are at least two orders of magnitude less than film diameters or curvatures. The maxima in disjoining pressure isotherms should therefore approximate those found for flat films. In which case the height of the maxima, and therefore the magnitude of the total capillary pressure required for rupture, should be independent of film dimensions and drop size. Denkov and coworkers [34, 35] have attempted to verify this using the FTT (see Chapter 2) using various aqueous surfactant solutions and oils. Here we remember that in the case of the FTT, the relation between the total capillary pressure acting on the film, p_c^{total}, and the disjoining pressure in a pseudoemulsion film is given

$$\Pi_{AWO}(h) = p_c^{total} = \frac{2\sigma_{AW}}{r_{pef}} + p_c^{PB} \tag{3.22}$$

where r_{pef} is the radius of curvature of the pseudoemulsion film, $2\sigma_{AW}/r_{pef}$ is the capillary pressure in the film due to that curvature, and p_c^{PB} is the externally applied capillary pressure. $\Pi_{AWO}(h)$ is the disjoining pressure in an aqueous pseudoemulsion film and σ_{AW} is the air–water surface tension. Here r_{pef} is dependent on both the size of the drop and the magnitude of the applied capillary pressure because the latter induces deformation of the drop. If $\Pi^{crit}_{AWO}(h_{crit})$ is the relevant maximum in the disjoining pressure isotherm, then we can write

$$\Pi^{crit}_{AWO}(h_{crit}) = \frac{2\sigma_{AW}}{r_{pef}} + p^{crit}_{c} \tag{3.23}$$

where p^{crit}_{c} is the quantity measured in the FTT as the critical applied capillary pressure required to rupture the pseudoemulsion film and h_{crit} is the pseudoemulsion film thickness at that point. Since $\Pi^{crit}_{AWO}(h_{crit})$ should be independent of the curvature of the film because the film thickness is $\ll r_{pef}$, then Equation 3.23 would have to imply that p^{crit}_{c} would be inversely proportional to r_{pef}. In turn r_{pef} will be a strong function of both the intrinsic drop size and the effect of deformation of the drop by the applied capillary pressure. We would therefore expect p^{crit}_{c} to both decrease proportionate with r^{-1}_{pef} and to increase with increasing drop size.

In contradiction of this expectation, Denkov and coworkers [34–36] have shown that p^{crit}_{c} for entry of oil drops into the air–water surface of surfactant solutions is usually essentially independent of the equatorial radii of those drops for submicellar and relatively dilute micellar solutions (where concentrations are ≤10 × CMC). Systems included dodecane (and other oils) in aqueous salt solutions of sodium dodecylbenzene sulfonate and polydimethylsiloxane oil in sodium dodecyl polyoxyethylene sulfate solutions. Experimental results [36] for the critical applied capillary pressures, p^{crit}_{c}, as a function of equatorial drop radius for the latter system, are presented in Figure 3.7 to exemplify typical behavior. For relatively low surfactant concentrations, p^{crit}_{c} is seen to be essentially constant. An exception concerns an extremely high surfactant concentration of 200 × CMC where p^{crit}_{c} became strongly dependent

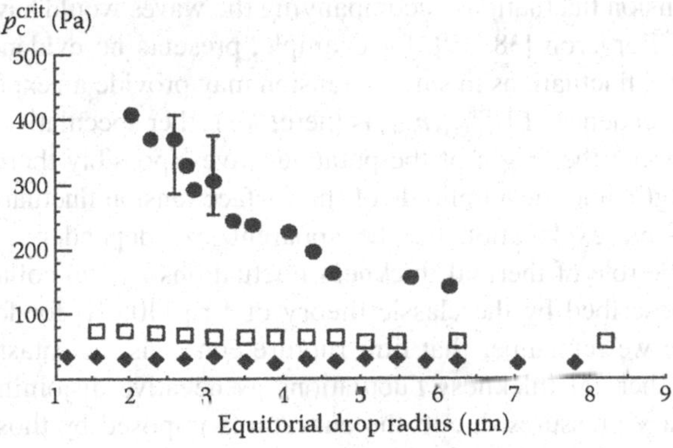

FIGURE 3.7 Plot of critical applied capillary pressure p^{crit}_{c} against equatorial radius of oil drops as measured by FTT. Oil is PDMS (of 5 mPa s, mol.wt. ~700) in sodium dodecyl polyoxyethylene sulfate solutions of different concentrations: ●, 0.1 M (200 × CMC); □, 0.02 M; ◆, 5 × 10^{-4} M. (After Hadjiiski, A. et al., Role of entry barriers in foam destruction by oil drops, in *Adsorption and Aggregation of Surfactants in Solution* (Mittal, K., Shah, D., eds.), Marcel Dekker, New York, Chpt 23, p 465, 2002.)

on drop size, declining by more than a factor of 2 upon doubling the drop radius [36]. Even this latter behavior is inconsistent with the expectation that $\Pi^{crit}_{AWO}(h_{crit})$ is independent of film dimensions because Equation 3.23 implies that p_c^{crit} should then increase with increase in drop size.

If p_c^{crit} is independent of drop size, and therefore r_{pef}, then Equation 3.23 implies that the critical disjoining pressure, $\Pi^{crit}_{AWO}(h_{crit})$, for rupture of a pseudoemulsion film should be inversely proportional to r_{pef}. This implication has been confirmed by calculation of pseudoemulsion film dimensions from experimental knowledge of σ_{AW}, σ_{OW}, p_c^{crit}, and the equatorial drop radius [35] (where σ_{OW} is the oil–water interfacial tension). These calculations reveal that $\Pi^{crit}_{AWO}(h_{crit})$ is also inversely proportional to the effective film radius, $(A_{film}/\pi)^{1/2}$, where A_{film} is the area of the pseudoemulsion film. All of this is therefore clearly not consistent with the usual hypothesis for film rupture where the total film capillary pressure, p_c^{film}, simply exceeds the relevant maximum in the disjoining pressure isotherm, which should be independent of film dimensions. Rather it appears to indicate that the smaller the oil drop the more stable the pseudoemulsion film because the height of that maximum (i.e., $\Pi^{crit}_{AWO}(h_{crit})$) increases with decreasing film size.

Denkov [34] has suggested that the apparent size dependence of $\Pi^{crit}_{AWO}(h_{crit})$ could mean that film rupture may be influenced by lateral fluctuations in surfactant adsorption. Such fluctuations are usually known as longitudinal waves [37]. Bergeron [38, 39] has in fact argued that the presence of such waves may afford explanation for some of his observations of the relative stability of aqueous foam films prepared from solutions of alkyl trimethylammonium bromide homologues. These waves would mean that the height of the relevant maximum disjoining pressure would fluctuate. Capillary pressures too low to exceed the maximum at equilibrium may then in fact do so as a consequence of the fluctuations in surface tension. However, if this is to represent an explanation for the rupture of pseudoemulsion films, then the amplitude of the surface tension fluctuations accompanying the waves would have to be greater for larger films. Bergeron [38, 39], for example, presents no evidence for such an effect. That lateral fluctuations in surface tension may provide an explanation for the apparent size dependence of $\Pi^{crit}_{AWO}(h_{crit})$ is therefore rather speculative in the absence of information about the origin of the putative waves (possibly thermal), their frequency, wavelength, and the amplitude of the surface tension fluctuations.

Another possible explanation for the apparent size dependence of $\Pi^{crit}_{AWO}(h_{crit})$ could concern the role of thermal thickness fluctuations in film collapse. This process has been described by the classic theory of Vrij [40, 41] for foam films (see Chapter 1). Here we remember that film rupture concerns the catastrophic growth of symmetrical thermal thickness fluctuations as negative disjoining forces overcome the capillary pressures due to the curvatures imposed by those fluctuations. For this to happen, the wavelength of the fluctuation should be greater than a critical value λ_{crit}. This in turn means that the size of the film should be greater than λ_{crit}. A similar treatment can be applied to a pseudoemulsion film if we make the simplifying assumption that the oil–water surface tension is significantly lower than the air–water surface tension. This is reasonable in the case of an aqueous surfactant solution at concentrations greater than or equal to the CMC and where the air–water surface tension is then usually at least an order of magnitude greater than that of the oil–water surface. In these circumstances, as pointed out by Kellay et al. [42], the air–water surface will exhibit minimal curvature if it is to match the capillary pressure fluctuations induced by the low surface tension oil–water surface. We therefore follow Kellay et al. [42] and assume a planar air–water surface. This configuration is shown in Figure 3.8. It is a trivial matter to repeat the analysis of Vrij and Overbeek [41] with that configuration, which yields, for sinusoidal fluctuations, a condition for film instability

$$\lambda \geq 2\pi\sqrt{\frac{\sigma_{OW}}{(\mathrm{d}\Pi_{AWO}(h)/\mathrm{d}h)}} = \lambda_{crit} \text{ if } \sigma_{OW} \ll \sigma_{AW} \tag{3.24}$$

where it is obvious that $\mathrm{d}\Pi_{AWO}(h)/\mathrm{d}h$ must be positive if the fluctuations are to grow at wavelengths greater than λ_{crit} until the film ruptures. The value for λ_{crit} given by Equation 3.24 differs by a factor of only $\sqrt{2}$ and the use of σ_{OW} from the symmetrical case obtained by Vrij and Overbeek [41] where both surfaces fluctuate symmetrically (see Chapter 1).

It would, however, appear to be inappropriate for the systems considered by Denkov and coworkers [34, 35] to follow Vrij and Overbeek [41] by only allowing for van der Waals forces in substitution of the relevant expression for $\mathrm{d}\Pi_{AWO}(h)/\mathrm{d}h$ in Equation 3.23. Kellay et al. [42] suggest that in the case where the oil–water interfacial tension is extremely low, there may, for example, be a positive contribution to the disjoining pressure from the sterically frustrated amplitude of the fluctuations. The presence of anionic surfactant implies a repulsive electrostatic contribution and the Hamaker constants for alkane–water–air films are almost all negative [43], which also implies a positive contribution to the disjoining pressure from van der Waals forces. The latter would, however, mean that pure water should spontaneously wet hydrocarbons (in the absence of surfactant), which is clearly not correct! It has been argued therefore that in this case, the so-called short range hydrophobic forces may be responsible for the instability of pseudoemulsion films of pure water on hydrocarbon oils [44].

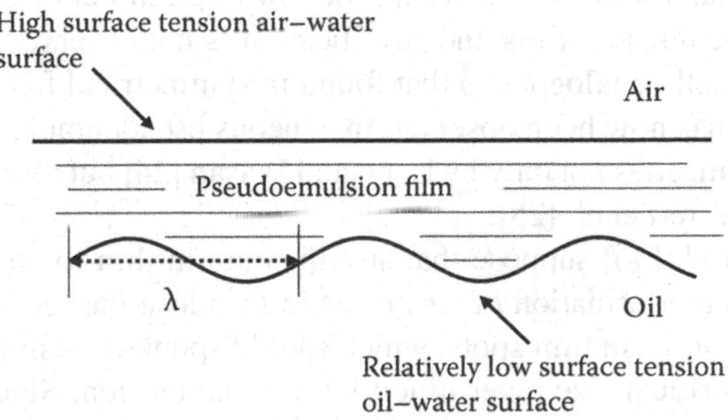

FIGURE 3.8 Schematic illustration of sinusoidal thickness fluctuations, of wavelength λ, in aqueous pseudoemulsion film where $\sigma_{OW} \ll \sigma_{AW}$ so that air–water surface is essentially unperturbed.

Consider now the case of a drop with a pseudoemulsion film drained to a thickness corresponding to a maximum in the disjoining pressure isotherm as a result of the application of an external capillary pressure. If that pressure is sufficiently high, the film will thin further into a region where $d\Pi_{AWO}(h)/dh > 0$ and collapse will occur rapidly if inequality 3.24 is satisfied. However if the drop is small, so that inequality 3.24 cannot be satisfied, then it would not collapse despite the presence of thermally induced thickness fluctuations, but it would continue to drain. It is possible, however, that $d\Pi_{AWO}(h)/dh$ is some inverse function of film thickness (as would be the case with purely van der Waals forces). Inequality 3.24 could then be satisfied at a certain thickness as $d\Pi_{AWO}(h)/dh$ increases whereupon the rapid growth of thickness fluctuations may cause film rupture (or sudden transition to a metastable white film). The overall outcome then clearly depends on the detailed features of the disjoining pressure isotherm—it is even possible that for small drops, inequality 3.24 is not satisfied throughout the region where $d\Pi_{AWO}(h)/dh > 0$. The stabilization of small drops by this type of mechanism is, however, seen to be kinetic, relying on the precise form of the thickness dependence of $d\Pi_{AWO}(h)/dh > 0$. Moreover, it would still be necessary for the applied capillary pressure to exceed the primary maximum, Π_{AWO}^{crit}, in the disjoining pressure isotherm if the film is to rupture. The theory cannot therefore offer any explanation for any supposed decline in that maximum with increasing film size. Another problem concerns the nature of the film at the primary maximum. If that is a Newton white film, then it is little more than a bilayer leaflet and thickness fluctuations are unlikely to occur anyway [38, 45]. Rupture then could perhaps concern spontaneous hole nucleation [45].

As shown in Figure 3.7, Denkov and coworkers [34, 35] find a pronounced increase in p_c^{crit} with decreasing drop size for high surfactant concentrations of 0.1 M and 200 × CMC. Similar observations concerning the enhanced stability of aqueous micellar pseudoemulsion films with decreasing film size have in fact also been reported by Lobo and Wasan [24] for films of high concentrations (~0.026 M and 1300 × CMC) of a non-ionic alcohol ethoxylate surfactant on n-octane. These authors attributed the cause of this phenomenon to the presence of a close-packed micellar structure in the films. That in turn gives rise to a process of stratification where films drain by formation of a succession of thin spots, which spontaneously grow. At lower micellar concentrations, micellar volume fractions are too low to permit close-packed structures to form in pseudoemulsion films and stratification is not observed. This process of stratification is exactly analogous to that found in symmetrical foam films (see, e.g., reference [46]). It has now been observed in aqueous pseudoemulsion films containing close-packed micelles not only by Lobo and Wasan [24] but also by Bergeron and Radke [23] and Garrett et al. [25].

Kralchevsky et al. [46] suppose that stratification in thin foam films containing micelles concerns the formation of vacancies in the close-packed structure that can effectively diffuse to form thin spots, which should spontaneously grow if the spots are larger than a critical size determined by the line tension. Since the number of such vacancies is supposed proportional to the size of the film, then the smaller the film the lower the likelihood that a spot of the critical size will form. Clearly this hypothesis, although concerned with stratification in symmetrical foam films, could give an explanation for the finding that the stability of pseudoemulsion films containing high micellar volume fractions declines with increasing film size and therefore increasing drop size as shown in Figure 3.7 for high surfactant concentrations (of 200 × CMC). However, unfortunately, it makes no reference to the presence of oscillatory disjoining pressures that have been observed in both foam films [33] and pseudoemulsion films [23].

In Figure 3.9 we compare the disjoining pressure isotherms measured by Bergeron and Radke [23] for aqueous micellar pseudoemulsion films of 0.1 M SDS on dodecane with the corresponding aqueous foam films. The positions of the various maxima correlate with each of the thinning steps in a stratifying film and are seen to be essentially identical for both the pseudoemulsion and the foam films. The gaps in the isotherms include the regions where $d\Pi_{AWO}(h)/dh > 0$ and where the films are unstable and the isotherm cannot be measured. Stratification therefore results from the films thinning to this unstable region whereupon a stepwise thickness change occurs until the slope of the disjoining pressure isotherm again becomes negative and stability is restored. The heights of the disjoining pressure maxima are seen to increase with decrease in film thickness. Rupture of both foam and pseudoemulsion films is therefore observed when the applied capillary pressure exceeds the disjoining pressure at the primary maximum. We should note that the primary maximum is two orders of magnitude higher for foam films than for the corresponding pseudoemulsion films. Bergeron and Radke [23] attribute this relatively low stability of the pseudoemulsion film to the so-called hydrophobic force [44] because it cannot be explained by recourse to the DLVO theory. That the rupture pressure at the primary maximum is also significantly higher than the pressures at the transitions accompanying the thinning steps in these stratified films implies that the stratification cannot contribute to enhanced stability, which is determined by the height of the primary maximum.

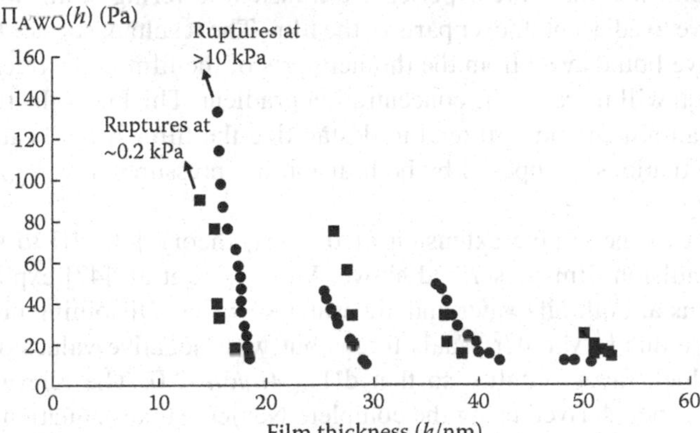

FIGURE 3.9 Disjoining pressure isotherms for stratifying films. ●, 0.1 M SDS foam film; ■, 0.1 M SDS–dodecane pseudoemulsion film (both at 24°C). At each discontinuity in isotherm, film thins by a step of ~10 nm, presumably as a result of loss of layer of micelles. (With kind permission from Springer Science+Business Media: *Colloid Polym. Sci.*, 273, 1995, 165, Bergeron, V., Radke, C.J.)

In summary then we find that there would appear to be no clear unequivocal explanation, based on disjoining pressure behavior, for the findings of Denkov and coworkers [34, 35] that the critical applied capillary pressure, p_c^{crit}, required to cause pseudoemulsion film rupture is independent of film and drop size at relatively low micellar concentrations where micellar stratifying structure is absent. It is also not at all clear how the presence of structure in pseudoemulsion films, due to close-packed micelles, could lead to contrasting behavior where film stabilities decrease with increasing film and drop size. These findings therefore represent a challenge to theory. Clearly more observations of both pseudoemulsion film disjoining pressure isotherms (including regions of negative disjoining pressure) and of p_c^{crit} (using the FTT) as a function of both oil drop dimensions and surfactant concentration using suitable aqueous surfactant solution–oil combinations would also be instructive.

3.4.2 Surface Tension Gradients and Stability of Pseudoemulsion Films

As we have seen (see Chapter 1), gradients in surface tension induce flow in the contiguous bulk phase. Vlakovskaya et al. [47] have presented a theoretical stability analysis of a pseudoemulsion oil–water–air film in which the stability is dominated by the formation of such gradients at the air–water surface. The gradients are assumed to arise when sparingly soluble oil molecules diffuse across the film, form mixed monolayers with the adsorbed surfactant, and therefore lower the surface tension. That some hydrocarbon oils form mixed monolayers with adsorbed surfactants has been amply demonstrated by, for example, Aveyard and coworkers [48, 49]. We consider such effects in detail in Section 3.6.1.

Thermal thickness fluctuations in pseudoemulsion films mean that the amounts of adsorbed oil will vary with film thickness as a consequence of different rates of transport of oil molecules arising from differences in concentration gradient. Thinner parts of the film will therefore experience a transient lowering of air–water surface tension relative to adjacent thicker parts of the film. The resultant surface tension gradient will drive liquid away from the thinner parts of the film to the thicker parts of the film, which will increase the concentration gradient. Thickness fluctuations will therefore be amplified and will tend to destabilize the film. Amplification of such thickness fluctuations is opposed by both disjoining pressures, if $d\Pi_{AWO}(h)/dh < 0$, and capillarity.

In contrast to the simple extension of the Vrij theory [40, 41] to asymmetrical pseudoemulsion films described above, Valkovska et al. [47] explicitly allow for fluctuations at both oil–water and air–water surfaces. Disjoining pressures are assumed to be due to van der Waals forces but with negative values of Hamaker constants, which favor stability so that $d\Pi_{AWO}(h)/dh < 0$. The relevant stability conditions are then derived using the complete Navier–Stokes equation. Generally Valkovska et al. [47] deduce that *decreasing* film thickness, and *increasing* both film radius and surface tension reduction by the oil all conspire to move the film toward instability despite the stabilizing influence of disjoining pressures and capillarity. However, these effects are necessarily transient—as the film approaches equilibrium, the concentration and surface tension gradients, which drive the instability, are eliminated.

It is possible that the presence of surface tension gradients could influence the critical applied capillary pressure p_c^{crit} measured by the FTT. Application of a constant capillary pressure could result in film rupture, due to the action of surface tension gradients irrespective of the nature of the disjoining pressure isotherm. However, stability in these circumstances depends on both film size and thickness. Again then we would expect p_c^{crit} to be some function of drop size. Clearly comparison of measurements of p_c^{crit} as a function of drop size before and after equilibration would be instructive. However, Denkov and coworkers [35] have shown that equilibration actually can diminish the stability of pseudoemulsion films (see Section 3.6.1.3), which is clearly inconsistent with an important role for surface tension gradient–driven effects in this context.

Garrett et al. [25] have shown that surface tension gradient–driven flows can also occur in pseudoemulsion films in the case where the surfactant in the aqueous phase is soluble in the oil. If the surfactant diffuses into the oil, then the surface tension of the air–water surface will increase. The resulting surface tension gradient with respect to the bulk phase in an adjacent meniscus induces a flow increasing the film thickness until the resulting dimple becomes unstable and subsequently discharges into the meniscus. The whole process is then repeated until equilibrium prevails whereupon it ceases. Similar cyclic regeneration in oil–water–oil emulsion films has also been described [50, 51]. If anything, these processes are likely to stabilize pseudoemulsion films.

3.5 GENERALIZED SPREADING COEFFICIENTS AND THIN FILM FORCES

Analogous arguments to those concerning generalized entry coefficients can be applied to spreading (see, e.g., reference [28]). Thus we can define a generalized spreading coefficient

$$S_g = \sigma_{AOW}(h) - \sigma_{OW} - \sigma_{AO} \tag{3.25}$$

where $\sigma_{AOW}(h)$ is a thickness-dependent film tension, which characterizes an oil film separating the aqueous phase from the air. By analogy with Equation 3.17, we can write for the film tension

$$\sigma_{AOW}(h) = \sigma_{OW} + \sigma_{AO} + \int_{\Pi_{AOW}(h_\infty)=0}^{\Pi_{AOW}(h)} h\,d\Pi_{AOW} \tag{3.26}$$

where $\Pi_{AOW}(h)$ is the disjoining pressure in that oil film. The film tension defined in this way describes the state obtained by thinning a duplex film down to a film of thickness h. If $h = 0$, it describes the state obtained by thinning a duplex film down to zero thickness, that is to the surface of the substrate uncontaminated by oil. $\sigma_{AOW}(h = 0)$ is then simply equal to the bulk surface tension, σ_{AW}, of the uncontaminated air–foaming liquid surface.

Combining Equations 3.25 and 3.26 yields for the generalized spreading pressure

$$S_g = \int_{\Pi_{AOW}(h_\infty)=0}^{\Pi_{AOW}(h)} h \, d\Pi_{AOW} \tag{3.27}$$

The classic spreading coefficients can be represented by simply substituting appropriate values for the limits of the integral in Equation 3.27.

Despite mathematical similarity, the generalized spreading coefficient for an oil over water is not of course the obverse of the generalized entry coefficient considered in Section 3.3. Here we are concerned with the spreading or otherwise of an oil film over the aqueous phase not the spreading of the latter over the oil (which is of course the obverse of entry as defined in Section 3.3). There are significant differences deriving from the likely behavior of the relevant disjoining pressure isotherms. Antifoam oils usually possess low dielectric constants, leading to much diminished electrostatic effects deriving from the relative absence of ionic dissociation. Moreover, significant adsorption of surfactants at the oil–air surface is usually absent so that any repulsive contribution to the disjoining pressure arising even from the resulting short-range steric interactions across such a film will be absent (although the presence of polydimethylsiloxanes or perfluoropolymers solutes in, for example, hydrocarbon oils may mean a positive contribution from that source). The attractive so-called hydrophobic force, which destabilizes aqueous films on oils (despite negative Hamaker constants [43]), is a consequence of the hydrogen-bonded nature of the structure of water [44] and is, of course, absent. van der Waals forces in hydrocarbon oil films (for chain lengths > C_8) on water are, however, necessarily present and are attractive as indicated by the relevant positive Hamaker constants [43]. In consequence, such films are unstable and the oils are non-spreading on water. However, as we describe in Section 3.6, both hydrocarbon and polydimethylsiloxane oils can interact strongly with adsorbed surfactant monolayers to produce both complete and pseudo-partial wetting. The nature of the contributions to the disjoining pressure isotherm, which produce that behavior, does not appear to be well understood.

Another difference between oil spreading on the surface of an aqueous phase and antifoam oil drop entry concerns the nature of any capillary pressure present. Consider then the case of an oil drop already emerged onto the surface of a foaming solution. If the drop is small, the air–oil and oil–foaming liquid surfaces will be hemispherical forms a (convex) lens. In this case, the capillary pressure in the drop will oppose film thinning. A stable film could therefore in principle exist in equilibrium with such a drop even if the disjoining pressure in the film is negative.

In the case of pseudo-partial wetting, a condition of stable equilibrium with respect to an oil lens can exist at a film thickness h_e if the disjoining pressure curve is such that we can set $p_c = -\Pi_{AOW}(h_e)$ in the approach to the primary maximum on the disjoining pressure isotherm as shown in Figure 3.10a. Here A, B, C, and E represent the indicated areas in the isotherm. The integral in Equation 3.27 with the upper limit set at $\Pi_{AOW}(h_e)$ is then equal to the generalized equilibrium spreading

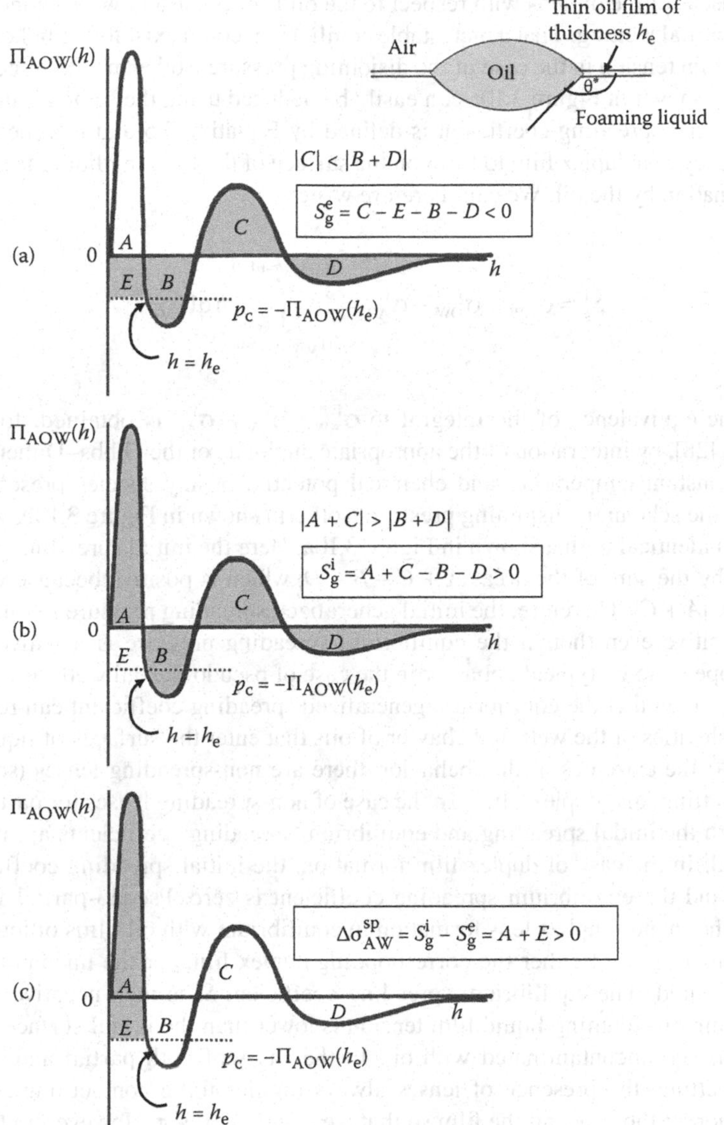

FIGURE 3.10 Pseudo-partial wetting and schematic disjoining pressure isotherm. (a) Generalized equilibrium spreading coefficient, S_g^e. (b) Generalized initial spreading coefficient, S_g^i. (c) Generalized spreading pressure, $\Delta\sigma_{AW}^{sp}$.

coefficient S_g^e at that disjoining pressure. It is given by the sum of the areas $B + C + D + E$ in the figure. Suppose that, in this figure, we have $|B + D| > |C|$, then we must also have $|B + D + E| > |C|$ so the integral is negative. The equilibrium spreading coefficient is therefore negative with the film tension less than the sum of the oil–air and oil foaming liquid surface tensions. This means that a contact angle will exist at the meniscus of the oil lens with respect to the oil film consistent with a condition of pseudo-partial wetting. That a metastable equilibrium could exist for a thicker film at a higher film tension in the case of the disjoining pressure isotherm with a secondary minimum shown in Figure 3.10a can easily be deduced using the same arguments.

The initial spreading coefficient is defined by Equation 3.5 and is a measure of the tendency of a duplex film to form on the surface of the foaming liquid free of any contamination by the oil. We can therefore write

$$S_g^i = \sigma_{AW}^i - \sigma_{OW}^i - \sigma_{AO}^i = \int_{\Pi_{AOW}(h_\infty)=0}^{\Pi_{AOW}(h=0)=0} h \, d\Pi_{AOW} \tag{3.28}$$

where the equivalence of the integral to $\sigma_{AW}^i - \sigma_{OW}^i - \sigma_{AO}^i$ is obtained, following Hirasaki [28], by integration of the appropriate analogue of the Gibbs–Duhem equation at constant temperature and chemical potential of any species present. Now consider the schematic disjoining pressure isotherm shown in Figure 3.10b, which is supposed identical to that shown in Figure 3.10a. Here the initial spreading pressure is given by the sum of the areas $A + C - B - D$, which is positive because we have $|B + D| < |A + C|$. Therefore, the initial generalized spreading pressure in this example is positive even though the equilibrium spreading pressure is negative, all of which appears to be typical behavior in the case of pseudo-partial wetting.

We see then that the equilibrium generalized spreading coefficient can represent the complexities of the wetting behavior of oils that enter the surfaces of liquid substrates. At the extremes of that behavior, there are non-spreading lenses (so-called partial wetting) and duplex films. In the case of non-spreading lenses (or partial wetting), both the initial spreading and equilibrium spreading coefficients are negative and equal. In the case of duplex film formation, the initial spreading coefficient is positive and the equilibrium spreading coefficient is zero. Pseudo-partial wetting, on the other hand, implies lens formation in equilibrium with oil films of lower film tension than those of either the corresponding duplex films or the uncontaminated foaming liquid. The equilibrium spreading coefficient S^e is then negative and the equilibrium air–foaming liquid film tension is lower than the initial surface tension of the surface uncontaminated with oil. In the cases of both partial and pseudo-partial wetting, the presence of lenses always implies finite contact angles at the intersection of the lens and the film so that we have $0 < \theta^* \leq \pi$. However, in the case of a duplex film, $\theta^* = \pi$. All of this equilibrium behavior is summarized schematically in Figure 3.11.

In this context we may define a *spreading pressure* as the reduction in equilibrium surface tension, $\Delta\sigma_{AW}^{sp}$, of the air–water surface caused by the addition of oil to the aqueous surface. In the case of the classic spreading coefficients, it is given by the difference between the semi-initial and the equilibrium spreading coefficients as shown by combining Equations 3.6 and 3.7 since

$$S^{si} - S^e = \sigma_{AW}^i - \sigma_{AW}^e = \Delta\sigma_{AW}^{sp} \tag{3.29}$$

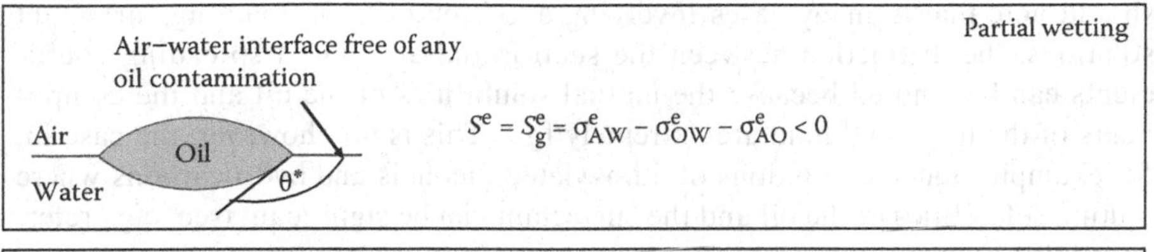

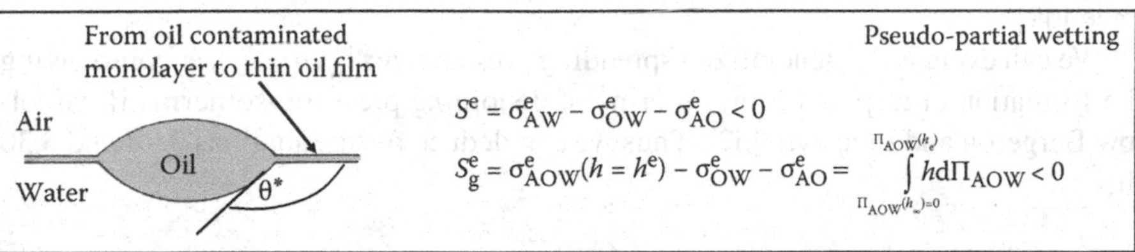

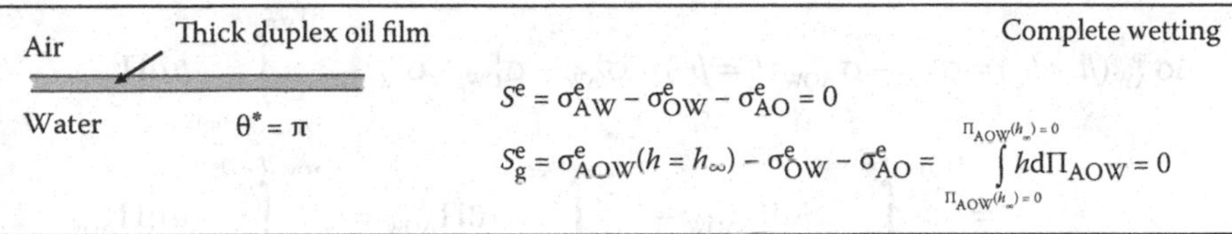

FIGURE 3.11 Schematic illustration of possible equilibrium wetting, and therefore spreading, behaviors of oil that emerges into air–water surface.

where we must have $\Delta\sigma^{sp}_{AW} \geq 0$. Similarly in the case of the equivalent generalized spreading coefficients, we must have by analogy

$$S^{si}_g - S^e_g = \sigma^i_{AW} - \sigma_{AOW}(h = h_e) = \Delta\sigma^{sp}_{AW}(h = h_e) \tag{3.30}$$

where $\Delta\sigma^{sp}_{AW}(h = h_e)$ is a generalized spreading pressure, which is a function of the equilibrium film thickness h_e, which is, in turn, dependent on the applied capillary pressure. In other words, the spreading pressure would be expected to be, in general, dependent on the size of any oil lens in equilibrium with the film because that would determine the capillary pressure.

In the case of duplex film formation at equilibrium, $S^e = 0$ and $\Delta\sigma^{sp}_{AW} = S^{si}$. For partial wetting, $S^{si} < 0$, $S^e < 0$, $S^{si} = S^e$, and $\Delta\sigma^{sp}_{AW} = 0$. However, in the case of pseudo-partial wetting, $S^e < 0$, $\Delta\sigma^{sp}_{AW} > 0$, and we must have in general

$$\Delta\sigma^{sp}_{AW} > S^{si} \tag{3.31}$$

If $S^{si} > 0$, then the oil will initially spread over the air–foaming liquid surface as a duplex film, which subsequently disproportionates to form a film in equilibrium with oil lenses because that film has a lower film tension than either the duplex film or the uncontaminated air–foaming liquid surface. If, however, both $S^{si} < 0$ and $S^e < 0$ where $|S^e| > |S^{si}|$, then we will still have $\Delta\sigma^{sp}_{AW} > 0$ so that the pseudo-partial wetting film forms by direct spreading from an oil drop. Here we should note that in many cases involving antifoam oils and foaming surfactant solutions, the distinction between the semi-initial and initial spreading coefficients can be ignored because the mutual solubilities of the oil and the components of the foaming liquid are extremely low. This is not, however, the case in, for example, aqueous solutions of ethoxylated alcohols and hydrocarbons where mutual solubilities of the oil and the surfactant can be significant (see, e.g., reference [49]).

We can express the generalized spreading pressure, $\Delta\sigma^{sp}_{AW}(h = h_e)$, accompanying the formation of a spread film, in terms of disjoining pressure isotherms if we follow Bergeron and Langevin [52]. Thus we can deduce from Equations 3.28 and 3.30 that

$$\Delta\sigma^{sp}_{AW}(h=h_e)=\sigma^{i}_{AW}-\sigma_{AOW}(h=h_e)=\sigma^{i}_{AW}-\sigma^{i}_{OW}-\sigma^{i}_{AO}-\int_{\Pi_{AOW}(h_\infty)=0}^{\Pi_{AOW}(h_e)} h\,d\Pi_{AOW}$$
$$=\int_{\Pi_{AOW}(h_\infty)=0}^{\Pi_{AOW}(h=0)=0} h\,d\Pi_{AOW}-\int_{\Pi_{AOW}(h_\infty)=0}^{\Pi_{AOW}(h_e)} h\,d\Pi_{AOW}=\int_{\Pi_{AOW}(h_e)}^{\Pi_{AOW}(h=0)=0} h\,d\Pi_{AOW} \tag{3.32}$$

We can illustrate this integral using the same schematic disjoining pressure isotherm shown in Figures 3.10a and 3.10b. Consider then Figure 3.10c. It is clear that in this case, $\Delta\sigma^{sp}_{AW}(h = h_e)$ is positive despite a negative value for the equilibrium generalized spreading coefficient. Such behavior is typical of pseudo-partial wetting. In the case of the schematic examples of Figure 3.10 (all with the same disjoining pressure isotherm), then we find that the relevant areas in the figure reduce to $\Delta\sigma^{sp}_{AW}(h = h_e) = S^{i}_{g} - S^{e}_{g}$ as indicated by Equation 3.29.

During the past decade or so, there has been much investigation of the nature of the transition between the various types of equilibrium spreading (i.e., wetting) behavior exhibited by oil on the surfaces of aqueous solutions [53–58]. Experimental investigations have been facilitated by measurements of the ellipticity of surfaces contaminated with spread films where the ellipticity is proportional to film thickness given the validity of certain assumptions [56]. The transitions can be induced by temperature changes [53, 54]. They can also be induced by changes in the surface excess of adsorbed surfactant in the case of films spread on the surfaces of surfactant solutions [55–58]. The transition from partial wetting to pseudo-partial wetting is discontinuous and is a first-order wetting transition [53]. The transition from pseudo-partial wetting to duplex film formation is continuous and is a so-called critical wetting transition [53]. We consider all this behavior for particular systems in more detail in Section 3.6. These transitions of course concern the equilibrium behavior. However, the nature of the spreading process is uncertain in the case of pseudo-partial wetting. Thus, if an oil drop spreads to form a pseudo-partial wetting film, it is not always clear whether this occurs directly or by formation of non-equilibrium duplex films that subsequently disproportionate to form lenses in equilibrium with pseudo-partial wetting films (after the manner of benzene on water). Perhaps the behavior simply follows the possibilities indicated by Equations 3.30 and 3.31 as outlined here.

3.6 SPREADING BEHAVIOR OF TYPICAL ANTIFOAM OILS ON AQUEOUS SURFACES

3.6.1 Hydrocarbon Oils

3.6.1.1 Spreading and Wetting Behavior of Hydrocarbons on Surface of Pure Water

Hydrocarbon oils often form the main ingredient of antifoams used for the foam control of aqueous systems. As we will describe in Chapter 4, the entry and spreading behavior of such oils is central to both their mode of action and their effectiveness. It is therefore appropriate to review that behavior here.

The initial entry and spreading coefficients of some hydrocarbons on water are shown in Table 3.1. These coefficients are calculated from the relevant surface tensions of the pure oils and of course do not therefore concern mutually equilibrated fluids. However, the low mutual solubilities of water and hydrocarbons mean that there is essentially no difference between initial and semi-initial spreading coefficients. The initial entry coefficients are seen to be positive for all homologues. It is also seen that homologues higher than n-octane exhibit initial spreading coefficients $S^i < 0$ on distilled water at near-ambient temperatures. The initial spreading coefficient for n-octane is too close to zero relative to the likely magnitude of experimental error in the relevant surface tensions to unambiguously state whether $S^i > 0$ or $S^i < 0$. Since for the homologues higher than n-octane we have $E^i > 0$ and $S^i < 0$, this means that lens formation occurs and that a partial wetting regime prevails. In this case the equilibrium and initial spreading coefficients should both be negative and equal. A value of S^e for n-decane reported by Takii and Mori [59] is close to the initial value as shown in Table 3.1. A corollary of this finding is that films of hydrocarbon on water for the higher homologues should be unstable. This in turn means that $d\Pi_{AOW}(h)/dh$ should be positive. The relevant Hamaker constants calculated by Hough and White [43] are also positive, which is consistent with the positive values of $d\Pi_{AOW}(h)/dh$ and a dominance of van der Waals forces in determining the stability of such films.

Hydrocarbons, which are used for antifoams, typically tend to be mixtures of high molecular weight compounds such as mineral oils (or "white" oils, etc.). These, too, do not usually spread on distilled water at ambient temperatures (unless surface-active contaminants are present in sufficiently high concentrations to alter the sign of the relevant spreading coefficient). This is exemplified in the case of the mineral oil quoted in Table 3.1. No change in the surface tension of doubly distilled water upon addition of a drop of the oil to the surface was observed using the Wilhelmy plate method [15].

Hauxwell and Ottewill [62] have measured the spreading pressure, $\Delta\sigma^{sp}_{AW}$, of the lower alkane homologues by diffusion of vapor onto the air–water surface at 15°C. At vapor pressure saturation, all homologues lower than octane had positive initial

TABLE 3.1
Entry and Spreading Coefficients for Some Hydrocarbons on Pure Water

Hydrocarbon	Temp. (°C)	σ_{AO} (mN m^{-1})	σ_{OW} (mN m^{-1})	E^i (mN m^{-1})[a]	S^i (mN m^{-1})[a]	Ref.	S^e (mN m^{-1})	Ref.	Temp. (°C)	$\Delta\sigma_{AW}^{sp}$ (mN m^{-1})	Ref.
n-Pentane	20	16.1	50.2	+107.1	+6.7	[60, 61]			15	+6.8	[62]
n-Hexane	20	18.4	50.8	+105.4	+3.8	[61]	0.0	[59]	15	+3.4	[62]
n-Heptane	20	20.3	51.2	+103.9	+1.5	[60, 61]	–0.2	[63]	15	+1.6	[62]
n-Octane	20	21.7	51.7	+103.0	–0.4	[61]	–1.2	[59, 63]	15	+1.0	[62]
n-Decane	20	23.8	52.3	+101.5	–3.1	[61]	–3.2	[59]			
n-Dodecane	20	25.4	52.8	+100.4	–5.2	[61]					
n-Tetradecane	20	26.5	53.8	+99.8	–6.8	[61]					
n-Hexadecane	20	27.4	53.8	+99.4	–8.2	[61]					
Liquid paraffin	25	29.7	44.9[b]	+87.5	–2.3	[15]					

[a] Taking σ_{AW} = 73.0 mN m^{-1} at 20° C and 72.3 mN m^{-1} at 25°C.
[b] Low, probably due to polar impurities.

spreading coefficients and appeared to form multilayers on the surface of water as indicated by the appearance of interference colors in white light. The spreading pressures were in fact similar to the initial spreading coefficients calculated from the surface tensions of the pure materials shown in Table 3.1 despite a difference in temperature of 5°C. Equation 3.29 would therefore imply that the equilibrium spreading coefficient should be zero for those homologues. Measurements of the equilibrium spreading coefficients reported by Mori and coworkers [59, 63] are in fact close to zero where measured (i.e., for n-hexane and n-heptane) as expected. All of this would tend to support a view that these homologues spread to form duplex films. However, two groups of workers, del Cerro and Jameson [64] and Mori and coworkers [59, 63], both find lens formation and therefore pseudo-partial wetting in these systems. The discrepancy may concern minor amounts of contamination (including the effects of oxidation of the hydrocarbons [59]), differences in temperature between sets of data, or even formation of lenses with extremely small contact angles (between the oil–water and oil–air surfaces) so that the systems are so close to duplex film behavior that the experimental errors in determining spreading coefficients and pressures obscure the distinction.

In the case of n-octane, Hauxwell and Ottewill [62] observed the formation of "microlenses" at saturation. Since the spreading pressure at saturation was positive, this finding tends to suggest a pseudo-partial wetting behavior. However, both the initial and equilibrium spreading coefficients are seen to be negative. The spreading pressure calculated from the difference between the initial and equilibrium spreading coefficients quoted in the table is, in this case, +0.8 mN m^{-1}, which compares with a directly measured value of +1.0 mN m^{-1} reported by Hauxwell and Ottewill [62]. As we speculated in Section 3.5, this could mean that a pseudo-partial wetting film spreads directly from n-octane lenses onto water at these temperatures.

3.6.1.2 Complete and Pseudo-Partial Wetting Behavior of Hydrocarbons on the Surfaces of Aqueous Surfactant Solutions

Those hydrocarbons that do not spread at all on water are exhibiting partial wetting behavior according to the definition of Brochard-Wyart et al. [17]. However, many workers have now shown that some of these hydrocarbons can be induced to undergo a wetting transition if they are added to the surface of surfactant solutions. At a sufficiently high surface excess of surfactant, the hydrocarbon becomes solubilized in the adsorbed surfactant monolayer, producing a reduction in surface tension and therefore a positive initial spreading coefficient. In some cases, this produces pseudo-partial wetting behavior where the mixed film is in equilibrium with excess alkane in the form of lenses. In other cases, however, the resulting films are duplex and they therefore exhibit complete wetting. As we have seen, pseudo-partial wetting implies lenses with finite contact angles and negative equilibrium spreading pressures. Once the chemical potential of the hydrocarbon becomes unity (i.e., the bulk phase is present in the form of lenses), any further addition of hydrocarbon simply increases the size of the lenses without increasing the film thickness. The first-order transition from partial wetting to pseudo-partial wetting is accompanied by a discontinuity in film thickness [57].

These effects have been shown to be apparent for hydrocarbons with anionic, cationic, and non-ionic surfactants [48, 49, 57]. By way of example in Figure 3.12, we show the results of Wilkinson et al. [57] for the reduction of the surface tension of dodecyl trimethylammonium bromide (C_{12}TAB) solutions caused by the addition of a drop of hexadecane to the solution surface. At low surfactant concentrations, the hydrocarbon has no effect on the surface tension so partial wetting prevails (the entry coefficient being presumably positive). However, at a clearly defined surfactant concentration, the presence of the oil produces a reduction in surface tension, film formation, and a transition to pseudo-partial wetting. Wilkinson et al. [57] show that this transition is accompanied by a discontinuity in ellipticity, which in turn means a discontinuity in film thickness, confirming a first-order transition. These mixed surfactant–oil films are of monolayer thickness (i.e., 1–2 nm). The reduction in surface tension (i.e., spreading pressure) increases as the surfactant concentration increases, reaching a maximum well before the CMC. The spreading pressure is constant for

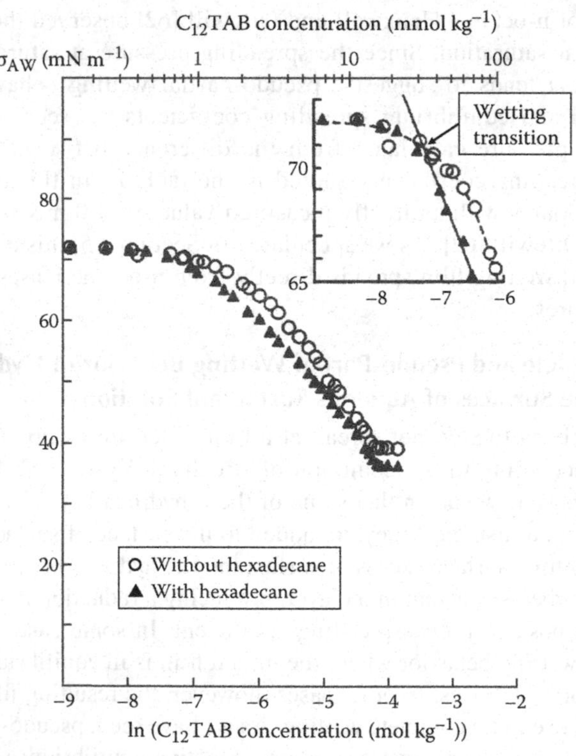

FIGURE 3.12 Reduction of surface tension of aqueous solutions of C_{12}TAB by presence of bulk phase hexadecane on air–water surface at 25°C. Inset shows wetting transition from partial wetting to pseudo-partial wetting induced by increasing surfactant concentration. (From Wilkinson, K., Bain, C.D., Matsubara, H., Aratono, M.: *ChemPhysChem*. 2005. 6. 547. Copyright Wiley-VCH Verlag GmbH & Co. KGaA. Reproduced with permission.)

micellar solutions regardless of surfactant concentration. The so-called dihedral contact angle for the oil lenses was directly measured in this study. Results are presented in Figure 3.13. The angle is seen to decline dramatically to a minimum at the wetting transition after which it increases up to the CMC. The angle is everywhere $<\pi/2$, which means that we must also have $\theta^* > \pi/2$ because $\pi - \theta^*$ is always less than the dihedral angle. Similar results for the same system have been reported elsewhere [56]. That other hydrocarbon surfactant systems show similar contact angle behavior has also been demonstrated for hexadecane on tetramethylammonium dodecyl sulfate solutions [55] and butylcyclohexane on $C_{12}TAB$ solutions [58].

For the lower alkanes this behavior appears to be general. Aveyard et al. [48] have, for example, shown that addition of drops of normal alkanes, from decane to hexadecane, onto the surfaces of aqueous micellar solutions of various ionic surfactants produces significant surface tension reductions and therefore finite spreading pressures, $\Delta\sigma^{sp}_{AW}$, and positive initial spreading coefficients. Results are summarized in Figure 3.14. In all cases, the equilibrium state appears to involve hydrocarbon lenses in equilibrium with a monolayer where, however, the contact angle $\theta^* \rightarrow \pi$ as the surfactant concentration approaches the CMC. This is clearly then a case of pseudo-partial wetting. However, apparently the drops of alkane sometimes first spread as a liquid film, which subsequently disproportionates to form many small lenses as equilibrium is attained. Unfortunately, Aveyard et al. [48] do not distinguish the particular circumstances when this occurs (assuming it is not ubiquitous), although as we have seen (Section 3.5) it could simply imply $S^i > 0$ and $S^e < 0$.

It is clear from Figure 3.14 that the spreading pressures of the alkanes decrease with increasing alkane chain length and increase with increasing surfactant chain length. Aveyard et al. [48], with some reservations, interpret this in terms of a simple

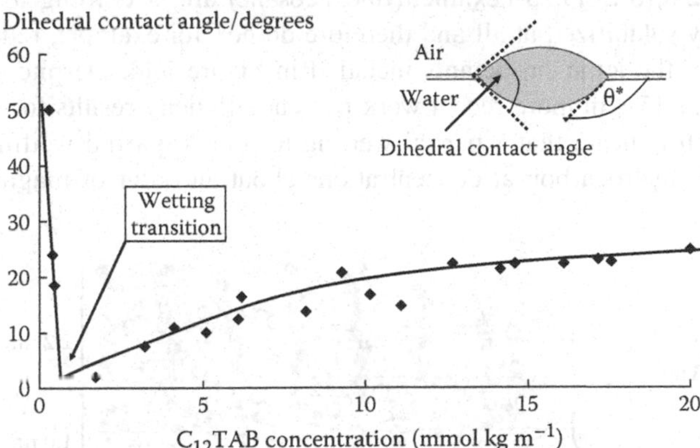

FIGURE 3.13 Dihedral contact angle (defined in figure) of lens of hexadecane on surface of aqueous solution of $C_{12}TAB$ as a function of $C_{12}TAB$ concentration at 25°C showing marked minimum at wetting transition from partial to pseudo-partial wetting. (From Wilkinson, K., Bain, C.D., Matsubara, H., Aratono, M.: *ChemPhysChem.* 2005. 6. 547. Copyright Wiley-VCH Verlag GmbH & Co. KGaA. Reproduced with permission.)

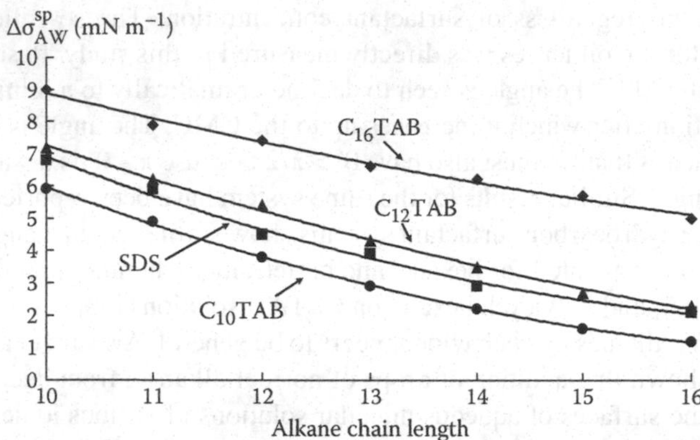

FIGURE 3.14 Spreading pressures $\left(\Delta\sigma_{AW}^{sp}\right)$ of n-alkanes on surfaces of aqueous micellar solutions of various ionic surfactants at 25°C. ◆, hexadecyltrimethylammonium bromide (C_{16}TAB), 1.6 mmol kg^{-1}; ▲, dodecyltrimethylammonium bromide (C_{12}TAB), 20 mmol kg^{-1}; ■, SDS, 10 mmol kg^{-1}; ●, decyltrimethylammonium bromide (C_{10}TAB), 81 mmol kg^{-1}. (After Aveyard, R., Cooper, P., Fletcher, P., *J. Chem. Soc. Faraday Trans.*, 86, 3623, 1990.)

solubilization model in which the surfactant chains become fully extended to accommodate the solubilized alkane. Neutron reflection measurements by Lu et al. [65] have provided evidence that this interdigitation model is reasonable. That is revealed in Figure 3.15, which depicts the structure of a mixed dodecane–tetradecyl-trimethyl-ammonium bromide monolayer in equilibrium with oil lenses as inferred from those measurements. This model suggests that longer-chain hydrocarbons are less readily accommodated, which means that extremely large hydrocarbon molecules such as squalane (2,6,10,15,19,23-hexamethylhexacosane) are, according to Aveyard et al. [48], hardly solubilized at all and therefore do not, for example, reduce the surface tension of the ionic surfactants included in Figure 3.14. Despite this finding, Wilkinson et al. [57] in more recent work present ellipticity results for squalane on C_{12}TAB, which indicate that a partial wetting to pseudo-partial wetting transition occurs for this hydrocarbon at concentrations about an order of magnitude below

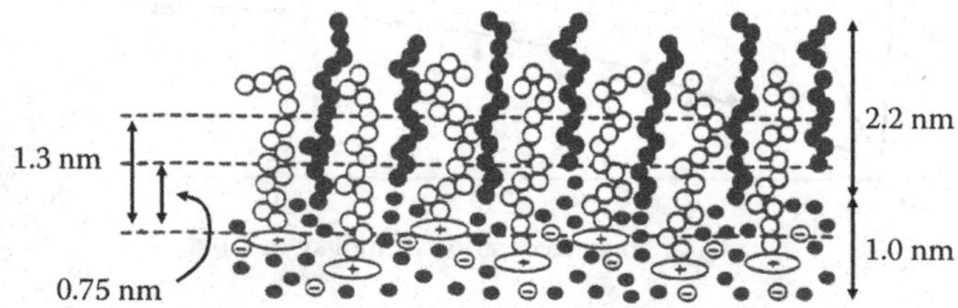

FIGURE 3.15 Schematic illustration of structure of mixed dodecane–C_{14}TAB monolayer showing interdigitation between hydrocarbon and surfactant chains. Structure was deduced from neutron reflection measurements. (Reprinted with permission from Lu, J. et al., *J. Phys. Chem.*, 96, 10971. Copyright 1992 American Chemical Society.)

the CMC. This would imply pseudo-partial wetting, solubilization of squalane in the surfactant monolayer, and necessarily a positive spreading pressure from the surfactant concentration at the wetting transition up to micellar concentrations. The findings of Aveyard et al. [48] and Wilkinson et al. [57] would appear to be irreconcilable in this respect. It is curious that despite this inconsistency with respect to earlier work, Wilkinson et al. [57] present no surface tension results to corroborate their ellipticity findings. It is worth noting also that other combinations of oils and surfactants have been found to produce partial wetting behavior with zero spreading pressure. For example, Aveyard et al. [49] have found that technical grade bis-3,3,5-trimethylhexyl phthalate has essentially zero spreading pressure on the air–water surface of solutions of ethoxylated alcohol surfactants. Another example concerns liquid paraffin, which is mostly a mixture of n-alkanes with a typical chain length distribution from C_{12} to C_{24} [66]. This material has been shown to have zero spreading pressure on aqueous solutions of commercial sodium (C_{10}–C_{14}) alkylbenzene sulfonate (NaLAS) solutions up to and beyond the CMC [15]. Therefore, there is no wetting transition and partial wetting prevails throughout. The alkylbenzene chain in this surfactant is randomly mid-chain substituted by the benzene ring, making this effectively a mixture of two-chain molecules. It is therefore unlikely to be effective in accommodating, by interdigitation, the comparatively long chain n-alkanes present in liquid paraffin.

Solubilization of normal alkanes into monolayers of AOT (sodium bis-diethylhexyl sulfosuccinate) is weak, with surface tension reductions of only 0.4 mN m^{-1} with dodecane [48], and the absence of any wetting transition with hexadecane [57] (possibly because the entry coefficient is negative!). AOT is again a two-chain molecule where each chain is both branched and comparatively short, both features of which minimize the possible configurational changes required to accommodate, by interdigitation, significant amounts of alkane solubilisates. However, Kellay et al. [42, 67] show that multilayer films can be formed by the lower alkanes on the air–water surfaces of micellar AOT solutions at high salinity. In this work the ellipticity of the pseudo-partial wetting film formed by the addition of lower n-alkanes such as decane on the surface of micellar AOT solutions was studied. The thickness of the film was found to vary markedly with addition of NaCl to the surfactant solutions yielding a maximum of about 7 nm at the NaCl concentration where a marked minimum in oil–water interfacial tension is observed. Results for decane are shown in Figure 3.16. The film thickness was independent of the amount of oil added to the surface. Any increase simply resulted in larger lenses. Unfortunately, however, no measurements of the reductions of the air–water surface tension, and therefore of the initial spreading pressures, were made for these alkanes on AOT solutions as a function of salinity.

Kellay et al. [42, 67] attribute the correlation between the maximum film thickness and the minimum oil–water interfacial tension in this decane–aqueous AOT system to the presence of thermally induced fluctuations at the oil–water surface. The maximum film thickness of 7 nm suggests that alkane molecules form a film that separates a probably weakly interdigitated monolayer from the air. Kellay et al. [42, 67] assume this film has a tension equal to that of the bulk oil–air surface. We then have a thin film with one surface having a high surface tension (the oil–air

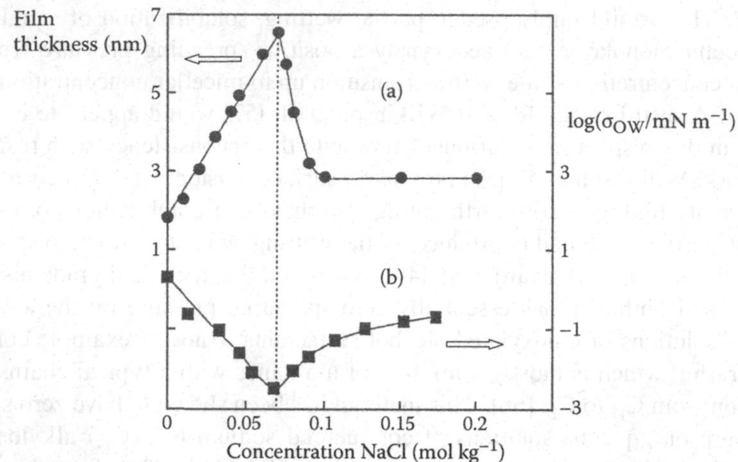

FIGURE 3.16 Effect of increasing salinity on pseudo-partial wetting film thickness of decane on aqueous 3 mM AOT solution illustrating a maximum that correlates with a minimum in oil–water interfacial tension, σ_{OW}. (a) Film thickness, ●; (b) log(interfacial tension), ■. (After Kellay, H. et al., *Phys. Rev. Lett.*, 69, 1220, 1992; and Kellay, H. et al., *Adv. Colloid Interface Sci.*, 49, 85, 1994; from the review of Aveyard, R., Clint, J.H., *J. Chem. Soc. Faraday Trans.*, 91, 2681, 1995. Reproduced by permission of The Royal Society of Chemistry.)

surface) and one surface having a low surface tension (the oil–water surface). This arrangement is of course similar to that described in Section 3.4 in consideration of the effect of thermally induced fluctuations on the stability of pseudoemulsion films. To a first approximation, the high oil–air surface tension film will be planar and the thermal fluctuations will be increasingly frustrated as their amplitude is increased by decreases in oil–water interfacial tension. This steric effect will mean an increase in entropy as the amplitude of the fluctuations increases and will be manifest as a repulsion disjoining force stabilizing the film. If the oil–water interfacial tension is extremely low, then this fluctuation force means that we have $d\Pi_{AOL}(h)/dh < 0$, which becomes more negative as the film thins, resulting in stable, relatively thick, films. However, if the interfacial tension is relatively high, the amplitude of the fluctuations will be low and the fluctuational contribution to the disjoining pressure will therefore be small. Short-range interaction forces then become important as assumed in the classic treatment of film stability by Vrij [40, 41]. The relevant short-range interaction forces in the case of the AOT–decane system are listed by Kellay et al. [42, 67] as a negative contribution to the disjoining pressure from van der Waals forces and a positive contribution due to favorable interactions between surfactant chains and interdigitated hydrocarbon chains. Therefore, at both low and high salt concentrations, where the oil–water interfacial tension is relatively high, short-range forces dominate in determining equilibrium film thicknesses, which means films of only 1–2 nm as depicted in Figure 3.16.

Aveyard et al. [49] have also made a study of the spreading behavior of alkanes on solutions of non-ionic ethoxylated alcohol surfactants. A complication with such systems concerns the solubility of the surfactant in the oil and the solubilization of the oil in the (micellar) surfactant solution to form a Winsor I microemulsion (see reference [49] for details). The reported spreading behavior therefore used oils containing surfactant concentrations in equilibrium with those present in the aqueous phase. In this case semi-initial spreading coefficients are relevant rather than initial spreading coefficients. A plot of the spreading pressures, $\Delta\sigma^{sp}_{AW}$, of various normal alkanes on micellar solutions of typical alkyl ethoxylates is shown in Figure 3.17. Results obtained for two cationic surfactants are included for comparison. Spreading pressures are clearly much larger for the ethoxylated alcohols than for the cationic surfactants of the same hydrocarbon chain length.

A neutron reflectivity study [68] of the films formed by dodecane on both sub-micellar and micellar solutions of $C_{12}EO_5$ indicates little change in the thickness of the surfactant surface excess with increasing amounts of oil, which suggests that the interdigitation model revealed in Figure 3.15 is relevant for this system. Not surprisingly then, in all cases we find that effects decline with increasing alkane chain length. Indeed as for the ionic surfactants discussed above, squalane does not reduce the surface tension of solutions of most of these ethoxylated alcohols—with a spreading pressure on a micellar solution of $C_{12}EO_7$ ($C_{12}H_{25}(OCH_2CH_2)_7OH$) of 2 mN m^{-1} being the only exception. In all cases the oil drops placed on the surfaces of these surfactant solutions initially spread, implying positive initial spreading coefficients. In most cases the "spreading oils retract within about a minute to produce visible stable circular lenses," which implies a pseudo-partial wetting behavior. Equilibrium spreading coefficients, S^e, for such systems are close to zero, which indicates that for these lenses the contact angles $\theta^* \rightarrow \pi$ (as reported earlier by Aveyard et al. [48] for micellar solutions of ionic surfactants). Unfortunately Aveyard et al. [49] do not make clear whether lens formation is a result of break-up of a spread liquid

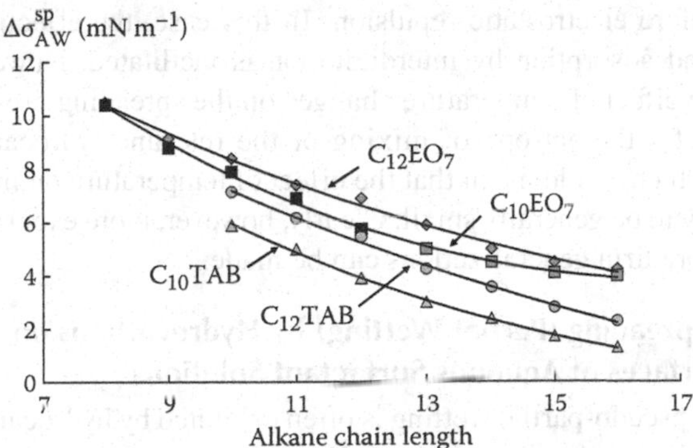

FIGURE 3.17 Comparison of spreading pressures $(\Delta\sigma^{sp}_{AW})$ of n-alkanes on surfaces of aqueous micellar solutions of ethoxylated alcohols and alkyl trimethylammonium bromides of the same hydrocarbon chain length. $C_{12}EO_7$, diamonds; $C_{10}EO_7$, squares; $C_{12}TAB$, circles; decyltrimethylammonium bromide, $C_{10}TAB$, triangles. Black symbols refer to complete wetting (i.e., duplex film formation) and gray symbols to pseudo-partial wetting. (Reprinted with permission from Aveyard, R. et al., *Langmuir*, 11, 2515. Copyright 1995 American Chemical Society.)

film to form many lenses. Exceptionally, however, the oils spread to form stable but irregular duplex films characterized by interference colors when observed in white light. Examples of both types of behavior are indicated in Figure 3.17. In a later study, Binks et al. [68] show that plots of the spreading pressures of a variety of oils (n-alkanes, perfluoro-octane, polydimethylsiloxane, and cyclohexane) on micellar solutions of a wider series of ethoxylated alcohols, CTAB, cetyl pyridinium chloride, and AOT against initial spreading coefficients fall on a straight line with unit gradient and intercept at $S^i = \Delta\sigma^{sp}_{AW} = 0$. As indicated by Equation 3.29, this again suggests that S^e is close to zero, indicating either duplex film formation or pseudo-partial wetting and lenses with low angles so that $\theta^* \rightarrow \pi$.

Almost all these studies of the spreading behavior of hydrocarbons on surfactant solutions are isothermal, with measurements close to, or at, ambient temperatures. However, here we are concerned with the possible effects of spreading from hydrocarbon oils on practical antifoam behavior. Since foam control problems can exist over the whole temperature range at which liquid water can exist, the effect of temperature on spreading pressures and coefficients of hydrocarbons on surfactant monolayers is of some interest. Unfortunately, the effect of temperature has not been extensively studied. Aveyard et al. [69] present a limited study, over the range 10–35°C, of the spreading pressures of dodecane on micellar $C_{12}EO_5$ ($C_{12}H_{25}(OCH_2CH_2)_5OH$) and $C_{12}TAB$ solutions. The effects were rather small, <0.1 mN m^{-1} reduction in the case of $C_{12}EO_5$ and ~0.5 mN m^{-1} increase in the case of the $C_{12}TAB$. It seems likely that the relative effects partially concern the effect of temperature on the effective head group size of the surfactant. Thus, increasing temperature in the case of non-ionic ethoxylated alcohols is known to reduce the head group size because of dehydration of the ethoxy groups (where hydration is an exothermic effect). It is easy to see that this would diminish adsorption of hydrocarbon molecules by interdigitation. Similarly, increase in temperature in the case of ionic surfactants such as $C_{12}TAB$ increases head group counter-ion dissociation (which is an endothermic effect) and therefore electrostatic repulsion. In this case, the effective head group size increases and adsorption by interdigitation is facilitated. However, the small magnitude of the effect of temperature changes on the spreading pressures suggests a dominant role for the entropy of mixing of the relevant hydrocarbon chains in interdigitation, which would mean that the effect of temperature on initial spreading pressures is likely to be generally small. Clearly, however, more experimental studies are required before firm generalizations can be made.

3.6.1.3 Non-Spreading (Partial Wetting) by Hydrocarbons on the Surfaces of Aqueous Surfactant Solutions

We find then that pseudo-partial wetting is often exhibited by hydrocarbon oils on the surfaces of adsorbed surfactant solutions. However, we also note that this behavior is not ubiquitous. Some hydrocarbon–surfactant combinations are known to exhibit non-spreading, the so-called partial wetting, behavior where wetting transitions are absent. The trends revealed by Figures 3.14 and 3.17, concerning the effect of increasing hydrocarbon chain length, suggest that partial wetting will in fact become more prevalent the larger the hydrocarbon molecule.

As we will discuss in Chapter 4, partial wetting systems are rather important in assessing the significance of spreading in the mode of action of antifoams. They have other uses, too. Thus, they have been employed by Aveyard et al. [48, 49] as a means of estimating the surface excess of an oil that is solubilized in a surfactant monolayer in giving rise to pseudo-partial wetting behavior. In this technique, the chemical potential of the latter is varied by mixing with a non-spreading oil such as squalane. The spreading pressure of the mixture can then be measured as a function of composition. Activity coefficients can be obtained by any convenient method if necessary (although mixtures of the relevant hydrocarbons usually only display small deviations from ideality). By utilizing the appropriate form of the Gibbs adsorption equation, it is then possible to infer the surface excess of the oil that is soluble in the surfactant monolayer. A complication concerns the possibility that the presence of the solubilized oil in the surfactant monolayer may induce solubilization of the non-spreading oil. However, Aveyard et al. [49] have measured the surface excesses of dodecane in the monolayer on the surface of a micellar solution of $C_{12}E_5$ in this manner using two non-spreading diluent oils (squalane and bis-3,3,5-trimethylhexyl phthalate) of markedly different chemical structure and have obtained identical results, which suggests that this complication is at least not present for all such systems!

In the case of oils that are non-spreading on water and therefore exhibit lens formation and partial wetting, it is of some interest to establish whether duplex film spreading can be induced by the presence of a surfactant even if the oil is not solubilized in the surfactant monolayer. Such spreading is indicated by positive values of the initial spreading coefficient. Let us assume then that the surfactant is ionic with a single charge, that the mutual solubilities of the surfactant and oil are negligible, and that the surface excess of oil at the air–water surface is always zero. In this case, any distinction between initial and semi-initial entry and spreading coefficients is absent and both oil–water and air–oil surface tensions are equilibrium values for the mutually saturated fluids. The air–water surface tension is then the equilibrium value for the solution in the absence of oil. Then, if we use the Gibbs equation and assume the surfactant solution to be submicellar, we may write for the initial entry coefficient

$$\frac{dE^{si}}{d\ln a_\pm} \approx \frac{dE^{i}}{d\ln a_\pm} = \frac{d(\sigma^{i}_{AW} + \sigma^{e}_{OW} - \sigma^{e}_{OA})}{d\ln a_\pm} = 2RT(-\Gamma_{AW} - \Gamma_{OW}) \tag{3.33}$$

and for the initial spreading coefficient

$$\frac{dS^{si}}{d\ln a_\pm} \approx \frac{dS^{i}}{d\ln a_\pm} = \frac{d(\sigma^{i}_{AW} - \sigma^{e}_{OW} - \sigma^{e}_{OA})}{d\ln a_\pm} = 2RT(-\Gamma_{AW} + \Gamma_{OW}) \tag{3.34}$$

where Γ_{AW} and Γ_{OW} are the equilibrium surface excesses at the air–water and oil–water surfaces of the surfactant solution, and $a_\pm$ is the mean ionic surfactant activity. Contact angle measurements have shown that adsorption at the hydrocarbon solid–water surface is equal to adsorption at the air–water surface for anionic surfactants [70]. This is a consequence of the dominance of the hydrophobic effect in determining adsorption behavior. It seems reasonable to expect that adsorption at the air–water and the liquid hydrocarbon–water surfaces will show the same behavior provided that the chain length of the hydrocarbon is long so that it does not contaminate the air–water surface. Therefore, we may substitute $\Gamma_{AW} = \Gamma_{OW}$ into Equations 3.33 and 3.34 to find

$$\frac{dE^{i}}{d\ln a_{\pm}} = -4RT\Gamma_{AW} \tag{3.35}$$

and

$$\frac{dS^{i}}{d\ln a_{\pm}} = 0 \tag{3.36}$$

Equation 3.34 suggests that addition of surfactant to the aqueous phase may reduce the entry coefficient (and may even "wet-in" the hydrocarbon so that $E^{i} < 0$). However, Equation 3.36 indicates that it will not induce duplex film spreading by a hydrocarbon oil that does not spread on distilled water (except possibly if the assumption $\Gamma_{AW} = \Gamma_{OW}$ is violated). Garrett et al. [15] have in fact observed that the initial spreading coefficient for a liquid paraffin on aqueous solutions of a commercial sodium (C_{10}–C_{14}) alkylbenzene sulfonate surfactant is <0 and is constant within the likely experimental error regardless of concentration so that Equation 3.36 is satisfied and we must have $\Gamma_{AW} \cong \Gamma_{OW}$. Not surprisingly, therefore, $S^{i} = S^{si} = S^{e} < 0$ and the spreading pressure of the liquid paraffin in that system is zero regardless of surfactant concentration. This partial wetting behavior presumably follows because of both the insolubility of the oil in the adsorbed surfactant monolayer and low mutual solubilities of both oil and surfactant in the relevant bulk phases.

3.6.2 Polydimethylsiloxane Oils

3.6.2.1 Complete and Pseudo-Partial Wetting Behavior of Polydimethylsiloxanes on the Surfaces of Pure Water and Aqueous Surfactant Solutions

Polydimethylsiloxane (PDMS) oils are frequently used as antifoam ingredients for control of the foam of both non-aqueous and aqueous liquids. Their molecular structure is depicted in Figure 3.18. In the case of aqueous foaming liquids, PDMS oils are usually mixed with hydrophobically modified silica particles to facilitate rupture of the relevant pseudoemulsion films as we will describe in later chapters. However, as we will also describe in Section 3.6.3, spread films of PDMS oils can influence the rupture of pseudoemulsion films even when such particles are not present.

Polydimethylsiloxane oils are characterized by low air–oil surface tensions of ≤~21 mN m^{-1} at ambient temperatures. Values of the air–oil, σ_{AO}, and the oil–water, σ_{OW}, surface tensions for typical oils in pure water are shown in Table 3.2. The oil–water surface tension against pure water is significantly lower than that of

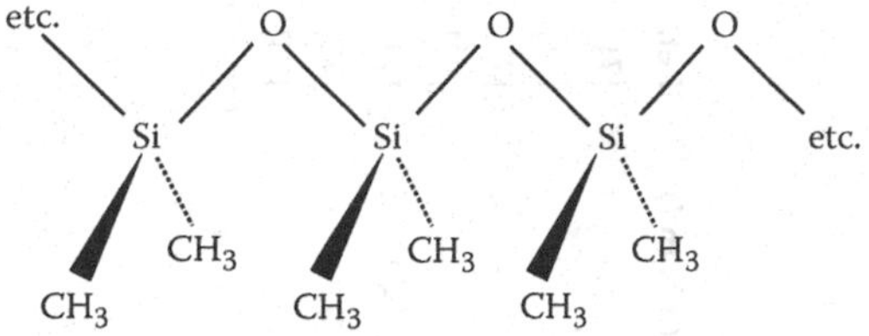

FIGURE 3.18 Molecular structure of polydimethylsiloxanes (PDMSs).

hydrocarbons (compare Tables 3.2 and 3.1) presumably because these polymer molecules can orient parallel to the surface, exposing relatively polar siloxane groups to the water [77]. Entry coefficients are seen to be strongly positive, largely as a consequence of low values of air–oil surface tensions. Low values of both air–oil and oil–water surface tensions conspire to produce strongly positive values of the initial spreading coefficients. The latter tend to be relatively insensitive to increasing molecular weight of the oils (as indicated by increasing viscosities) because the air–oil surface tensions increase and the oil–water surface tensions decrease. Values of the equilibrium spreading coefficients are rarely measured but those few quoted in Table 3.2 are small but positive, which suggests that they are not true equilibrium values. It seems likely, however, that the true values of S^e are either zero or are close to zero, implying either complete wetting (i.e., duplex film formation) or pseudo-partial wetting where oil lenses are characterized by low values of the dihedral angle. Unpublished observations of Garrett and Gratton [74] suggest that, for PDMS oils of viscosity > 1000 mPa s, spreading on water produces films that subsequently break up to produce lenses and pseudo-partial wetting. However, regardless of whether complete or pseudo-partial wetting occurs, it is probable that we must have either $\theta^* = \pi$ or $\theta^* \to \pi$.

The spreading behavior of PDMS oils on the surfaces of surfactant solutions is of course the property of relevance for antifoam behavior. It is only in the past 20 years or so that significant systematic observations of that behavior have been made. Table 3.3 summarizes most of those observations. Similar observations with fully formulated PDMS-based antifoams (see, e.g., reference [12]) are not included but are considered as appropriate in Chapter 4.

The observations listed in Table 3.3 concern micellar solutions of a variety of surfactants together with PDMS oils of a range of viscosities (and therefore molecular weights). In all cases the initial spreading and entry coefficients, together with the spreading pressures, $\Delta\sigma^{sp}_{AW}$, are positive. Equilibrium spreading coefficients, on the other hand, are either zero or close to zero. As with spreading on pure water, this implies either complete wetting (i.e., formation of duplex films) or pseudo-partial wetting where oil lenses are again characterized by low dihedral angles with $\theta^* \to \pi$.

Bergeron, Langevin, and coworkers [52, 75, 76, 83–85] have made an extensive study of the spreading of PDMS oils on the surfaces of micellar surfactant solutions but have mainly confined their observations to two oils with respective molecular weights of ~10^4 and 2.5×10^4. Surfactants included AOT and homologues of alkyl trimethylammonium bromides. It is noteworthy that the spreading pressure of

TABLE 3.2
Entry and Spreading Coefficients for Some Polydimethylsiloxane Oils on Pure Water

Manufacturer	Average Mol. Weight[a]	Viscosity (mPa s)	Temp. (°C)	σ^i_{OA} (mN m^{-1})	σ^i_{OW} (mN m^{-1})	E^i (mN m^{-1})	$\Delta\sigma^{sp}_{AW}$ (mN m^{-1})	S^i (mN m^{-1})	S^e (mN m^{-1})	Ref.
Dow Corning MS 200	2.3×10^2	0.8	24	16.4	42.7	+98.2	–	+12.8	–	[71]
Dow Corning MS 200	6.7×10^2	4.6	24	19.3	42.8	+95.4	–	+9.8	–	[71]
Dow Corning DC 200	$\sim 1.2 \times 10^3$	9.3	20	20.1	39.5	+90.9	–	+11.9	–	[72]
Dow Corning DC 200	4.0×10^3	48.0	20	20.8	34.3	+85.0		+16.4	–	[72]
Dow Corning DC 200	4.0×10^3	48.0	25	21.0	38.0	+88.9	10.5	+12.9	+2.4	[73]
Dow Corning DC 200	$\sim 2.5 \times 10^4$	971.0	20	21.2	36.9	+87.2	–	+13.4	–	[72]
Dow Corning DC 200	$\sim 7.0 \times 10^4$	60,000[b]	20	–	–	–	10.7	–	–	[74]
Rhône-Poulenc 47 v 100	10^4	~200	21 ± 1	20.6	39.1	+91.3	12.2	+13.1	+1.0	[75–77]

[a] Mostly inferred from viscosity using data in reference [77].

[b] Approximate values in mPa s equated with kinematic viscosities in cS, which assumes all densities = 1.0 kg dm^{-3}.

PDMS of molecular weight 10^4 on AOT is very low in comparison with that found for most of the other micellar surfactant solutions listed in Table 3.3. Measurements by Mann and Langevin [77] of the surface excess of AOT at the CMC in the presence of PDMS indicate that the latter has little effect. This could imply little penetration by interdigitation of the surfactant monolayer by the polymer—a situation found also for n-alkanes on the surface of micellar AOT solutions in the absence of added salinity (but see Figure 3.16). Measurements of the ellipticity of the spread PDMS layer on the AOT monolayer reveals the presence of a PDMS layer, which gradually increases in thickness with increasing concentration until a density comparable to that of the bulk oil is realized. This is again apparently consistent with a layer of oil on top of the monolayer with little penetration. By contrast, Mann and Langevin [77], and later Bergeron and Langevin [52] followed by Denkov et al. [80], have indicated that pseudo-partial wetting prevails and that therefore the thickness of that oil film is constant in the presence of bulk phase oil lenses. Addition of more oil simply increases the total volume of lenses without increasing the film thickness. Pseudo-partial wetting implies a negative value for the equilibrium spreading coefficient S^e—the small positive value given in Table 3.3 would then appear to result from experimental error, which is unlikely to be $<\pm 1$ mN m^{-1}. Near zero values of S^e in turn suggest the presence of lenses with extremely small dihedral angles. However, according to Mann et al. [83], the process of spreading, upon addition of a drop of PDMS oil to the surface of a micellar solution of AOT, appears to give rise to rather complex behavior that it is difficult to explain. The oil first spreads, then retracts into a lens and subsequently slowly spreads—the whole process taking several hours. A similar behavior is reported by the same authors [83] for the same PDMS oil on micellar solutions of $C_{10}EO_5$. However, in that case, the whole process takes only a few seconds.

As we have seen (see Section 3.6.1.2), the absence of surfactant monolayer penetration by interdigitation and low values of the spreading pressure has also been reported by Aveyard et al. [48] for dodecane films on micellar AOT solutions (in the absence of added salinity). A comparison of the spreading pressures $\Delta\sigma_{AW}^{sp}$ of PDMS and dodecane on AOT solutions as a function of surfactant concentration is given in Figure 3.19. $\Delta\sigma_{AW}^{sp}$ is seen in both cases to tend to a constant value as the CMC is approached at air–water surface tensions $< \sim 40$ mN m^{-1}. The main distinction concerns the discontinuous *increase* in spreading pressure as the surfactant concentration increases in the case of dodecane and the corresponding *decrease* in spreading pressure in the case of PDMS. The former derives from the first-order wetting transition from partial wetting to pseudo-partial wetting, induced presumably by more favorable interactions between dodecane with the surfactant monolayer than with the pure water surface. Conversely, the absence of a first-order wetting transition with PDMS arises because the oil exhibits complete or pseudo-partial wetting on pure water with which it can presumably more favorably interact than with the hydrophobic surface of the surfactant monolayer. However, the continuous reduction in $\Delta\sigma_{AW}^{sp}$ with decreasing air–water surface tension could imply a second-order wetting transition from complete wetting to pseudo-partial wetting.

In contrast to AOT, the spreading pressure of a PDMS of molecular weight ~2.5×10^4 on a micellar saline solution of another branched chain anionic surfactant, sodium dodecyl 6-phenylbenzene sulfonate, has been shown by Arnaudov et al. [79] to be

TABLE 3.3
Spreading Pressures, Entry, and Spreading Coefficients for Polydimethylsiloxane Oils on Aqueous Micellar Surfactant Solutions

Average Mol Wt of PDMS[a]	PDMS Viscosity (mPa s)	σ^{i}_{OA} (mN m^{-1})	Surfactant[c]	σ^{i}_{OW} (mN m^{-1})	σ^{i}_{AW} (mN m^{-1})	$\Delta\sigma^{sp}_{AW}$ (mN m^{-1})	E^{i} (mN m^{-1})	S^{i} or S^{si} (mN m^{-1})	S^{e} (mN m^{-1})	Observed Equilibrium Behavior[d]	Surfactant Concentration	Temp. (°C)	Ref.
2.3×10^2	0.5	16.0	$C_{12}EO_5$	0.9	30.1	13.3	+15.0	+13.2	−0.1	Duplex	1.5 × CMC (or CAC) in aqueous phase + 1 wt.% in oil phase	25	[73]
–	0.8	17.2		1.3	30.1	12.3	+14.2	+11.6	−0.7	Duplex			
6.7×10^2	4.6	19.5		2.2	30.5	9.1	+13.2	+8.8	−0.3	Duplex			
$\sim 1.2 \times 10^3$	9.3	20.2		2.3	30.5	8.1	+12.6	+8.0	−0.1	Duplex			
$\sim 2.0 \times 10^3$	20[a]	20.5		2.6	30.5	7.4	+12.6	+7.4	0.0	Duplex			
4.0×10^3	48	21.0	C_8EO_5	5.0	32.5	6.3	+16.5	+6.2	−0.1	Duplex			
			$C_{10}EO_5$	3.6	31.1	6.4	+13.7	+6.5	+0.1	Duplex			
			$C_{12}EO_5$	3.0	30.7	6.6	+14.5	+6.7	+0.1	Duplex			
			$C_{14}EO_5$	2.7	30.4	6.6	+12.1	+6.7	+0.1	Duplex			
			$C_{12}EO_2$	2.6	25.9	3.1	+8.0	+2.4	−0.8	Duplex			
			$C_{12}EO_3$	2.3	27.7	4.6	+9.0	+4.4	−0.2	Duplex			
			$C_{12}EO_4$	2.0	28.8	5.8	+9.8	+5.8	0.0	Duplex			
			$C_{12}EO_5$	3.0	30.4	6.4	+12.4	+6.4	0.0	Duplex			
			$C_{12}EO_6$	4.1	32.2	7.3	+15.3	+7.1	−0.2	Duplex			
$\sim 6 \times 10^3$	100[b]	21.0	$C_{12}EO_5$	3.2	30.5	6.5	+12.7	+6.3	−0.2	Duplex			
$\sim 2 \times 10^4$	453[b]	21.2		3.0	30.5	6.3	+12.3	+6.3	0.0	ppw and lenses			
$\sim 2.5 \times 10^4$	971	21.3		3.0	30.4	6.0	+12.1	+6.1	+0.1	ppw and lenses			

10^4	~200	20.6	AOT	4.7	28.0	2.5	+12.1	+2.7	+0.2	ppw and lenses	3 × CMC	21 ± 1	[52, 76, 77]
			$C_{10}EO_5$	3.5	31.5	6.3	+14.4	+7.4	+1.1	ppw and lenses			
			C_9TAB	10.7	41.0	8.8	+31.1	+9.7	+0.9	Duplex			
			$C_{12}TAB$	9.8	38.8	7.5	+28.0	+8.4	+0.9	Duplex			
			$C_{14}TAB$	9.4	37.3	6.6	+26.1	+7.3	+0.7	Duplex			
			$C_{16}TAB$	9.8	37.7	7.1	+26.9	+7.3	+0.2	Duplex			
~2.5 × 10^4	971	21.3	$C_{12}6\phi SO_3Na$ + NaCl			6.0				ppw and lenses	5 mM in 40 mM NaCl	21 ± 1	[78]
~7 × 10^2	5	18.7	$C_{12}\phi SO_3Na$ + NaCl	5.6	30.6	7.1	+17.5	+6.3	–0.8	ppw and lenses	2.6 mM in 12 mM NaCl	25 ± 2	[79]
~2.5 × 10^4	1000	20.6	AOT	4.7	28.5	2.8	+12.1	+2.5	–0.2	ppw and lenses	3.5 × CMC	23 ± 2	[80]
~2.5 × 10^4	1000	20.6	APG	4.6	28.6	4.2	+12.6	+3.4	–0.8	Probably ppw and lenses	0.45M 3 × CMC	23 ± 1	[81]
~7 × 10^2	5	19.8	$C_{12}EO_{2.5}SO_4Na$	6.8	33.2	7.4	+20.2	+6.6	–0.8	ppw and lenses	0.1 M + 66 mM ionic strength	25 ± 1	[82]
			$C_{12}APB$	7.2	31.6	5.3	+19.0	+4.6	–0.7	ppw and lenses	0.1 M + 142 mM ionic strength		
			80/20 mole ratio $C_{12}EO_{2.5}SO_4Na$ + $C_{12}APB$	5.4	30.4	6.4	+16.0	+5.2	–1.2	ppw and lenses	0.1 M + 81 mM ionic strength		

[a] Mostly inferred from viscosity using data in reference [77].

[b] Approximate values in mPa s equated with kinematic viscosities in cS, which assumes all densities = 1.0 kg dm^{-3}.

[c] Surfactant types: $C_nEO_m = C_nH_{2n+1}(OCH_2CH_2)_mOH$; AOT = sodium bis-diethylhexyl sulfosuccinate; $C_nTAB = C_nH_{2n+1}(CH_3)_3N^+Br^-$. $C_{12}6\phi SO_3Na$ = sodium 6-phenyl dodecylbenzene sulfonate; $C_{12}\phi SO_3Na$ = sodium dodecylbenzene sulfonate (branched chain blend); APG = commercial $C_{12/14}$(glucopyranoside)$_{1.2}$; $C_{12}EO_{2.5}SO_4Na$ = commercial sodium dodecyl-polyoxyethylene-2.5-sulfate; $C_{12}APB$ = commercial dodecyl amino-propyl betaine.

[d] Complete (duplex) or pseudo-partial wetting (ppw).

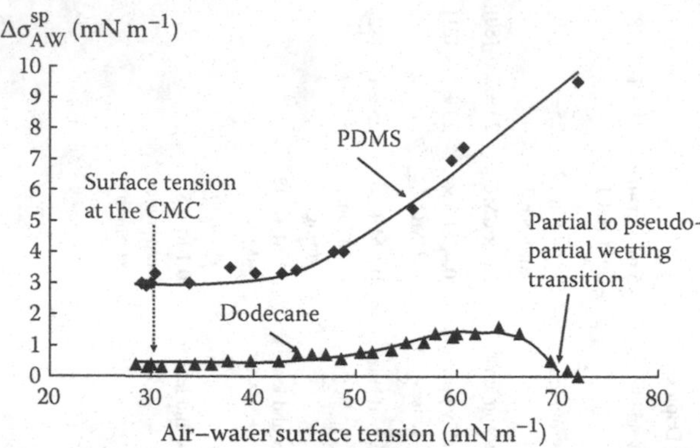

FIGURE 3.19 Comparison of spreading pressure, $\Delta\sigma^{sp}_{AW}$, of dodecane and polydimethylsiloxane oil (of molecular weight ~10^4) on solutions of AOT where variation in surfactant concentration is indicated by corresponding air–water surface tension measured before addition of oils. Both oils exhibit pseudo-partial wetting on AOT solutions. (Dodecane data after Aveyard, R. et al., *J. Chem. Soc. Faraday Trans.*, 86, 3623, 1990; PDMS data after Mann, E., Langevin, D., *Langmuir*, 7, 1112, 1991.)

comparable to that of other micellar solutions listed in Table 3.3. Here the oil initially spreads and the spread layer subsequently disproportionates to form oil lenses and a PDMS contaminated air–water surface—clearly an example of pseudo-partial wetting. Similar conclusions are suggested by the results of Arnaudov et al. [79], albeit with an isomer blend of sodium dodecylbenzene sulfonates.

Bergeron and Langevin [52, 75, 85] have also explored the spreading behavior of a PDMS oil on micellar solutions of homologues of alkyl trimethylammonium bromide. Results are presented in Table 3.3. Although there appears to be some evidence for metastable states in dilute monolayers in the case of chain lengths < C_{14}, it seems all spreading from drops of bulk oil phase produced duplex films [75] for all chain lengths studied. However, a later review by Langevin [85] appears to contradict this generalization, stating that duplex film formation is only observed on micellar solutions of C_{16}TAB, all other lower homologues producing pseudo-partial wetting with lens formation. Unlike the n-alkanes, the spreading pressures of the chosen PDMS oil do not apparently increase significantly with increasing surfactant chain length (compare the relevant results in Table 3.3 with those depicted in Figure 3.14). There do not appear to be any measurements of the relevant Gibbs plots that would establish whether the surface excess of surfactant is significantly changed by the presence of the PDMS oil. The extent of any putative penetration of the alkyl chain region of the alkyl trimethylammonium bromide monolayers by the oil is therefore not known. The absence of significant effects due to the addition of 0.1 M KBr suggests that head group size plays little role in this context [52].

Langevin and coworkers have studied the spreading behavior of PDMS on only one non-ionic surfactant—$C_{10}EO_5$. The effect of PDMS on the relevant Gibbs plot revealed a small reduction in the surfactant surface excess, which does imply some partial penetration of the surfactant monolayer by the polymer [83]. Binks and

Dong [73] report a more detailed study of the spreading behavior of PDMS oils on non-ionic alcohol ethoxylate surfactant solutions. Both the molecular weight of the PDMS (as indicated by the viscosity) and the molecular structure of the surfactants were varied. The study was confined to micellar solutions. A summary of the spreading pressures and coefficients is given in Table 3.3.

As with the corresponding hydrocarbon oils (see Section 3.6.1), a difficulty in this context concerns the solubility of the surfactants in the oil especially in the case of surfactants with long hydrocarbon chains and low CMCs. Reproducible measurements of the spreading pressures of PDMS oils on such solutions could be facilitated, however, by predissolving the relevant surfactant in the oil. In the absence of knowledge of the partition coefficients, the amount dissolved by Binks and Dong [73] was arbitrary. However, the use of micellar concentrations in the aqueous phase at constant surfactant activity meant that spreading pressures (and therefore spreading coefficients) were insensitive to the magnitude of that amount. This procedure therefore presumably produces strictly semi-initial spreading conditions where the oil–water and oil–air surfaces both rapidly attain equilibrium when a PDMS oil drop is added to the air–water surface and maintain that equilibrium as the oil spreads.

Binks and Dong [73] investigated three effects—the effect of increasing molecular weight of PDMS oil for a given micellar surfactant; the effect of increasing the alkyl chain length of the surfactant with a given degree of ethoxylation and a given molecular weight of PDMS; and the effect of increasing the surfactant alkyl chain length for a given degree of ethoxylation with a given molecular weight of PDMS.

The spreading pressures for PDMS oils on micellar $C_{12}EO_5$ solutions decrease with increase in viscosity and therefore molecular weight of the PDMS. Such behavior is qualitatively similar to that found by Aveyard et al. for n-alkanes on both alcohol ethoxylate [49] (see Figure 3.17) and cationic [48] (see Figure 3.14) micellar surfactant solutions. As with the alkanes, this effect is attributed by Binks and Dong [73] to decreasing solubilization by interdigitation in the surfactant monolayer with increase in the molecular weight (and therefore molecular volume) of the PDMS. Direct observation of the spreading behavior suggested duplex film formation except in the case of the higher viscosity oils (of molecular weight 2×10^4–2.5×10^4) where the presence of lenses indicated pseudo-partial wetting.

As shown in Table 3.3, with a PDMS oil of viscosity 48 mPa s (and molecular weight ~4000), increasing the degree of ethoxylation for dodecanol derivatives (i.e., increasing m in $C_{12}EO_m$) produces an increasing spreading pressure, presumably because penetration of the surfactant monolayer is facilitated with increasing size of the head group. However, increasing the alkyl chain length for a given degree of ethoxylation (i.e., increasing n in C_nEO_5) appears to produce little increase in spreading pressure, which does not appear to be consistent with the expected increase in penetration of the surfactant monolayer by interdigitation. In most cases of the examples shown in Table 3.3, direct observation indicated duplex film formation, the only exception being $C_{12}EO_2$ where lenses are seen and which forms a vesicular lamellar phase emulsion rather than a micellar solution.

Somewhat curiously, Binks and Dong [73] report a case of partial wetting with 48 mPa s PDMS on solutions of a non-ionic ethoxylated silicone surfactant of quoted [73] formula

$$(CH_3)_3SiO \times [Si(CH_2)_3(OCH_2CH_2)_{7.5} \times OCH_3] \times OSi(CH_3)_3$$

Even on micellar solutions of this compound, the PDMS oil forms lenses with zero spreading pressure. Similar behavior is seen with submicellar solutions even up to air–water surface tensions of 45 mN m^{-1}. Since the oil spreads on water, this suggests a possible first-order wetting transition between partial wetting and pseudo-partial wetting at even higher air–water surface tensions. Rather surprisingly then, this means that no penetration of the monolayer of this silicone surfactant by the PDMS must occur.

3.6.2.2 Rates of Spreading of Polydimethylsiloxanes on the Surfaces of Pure Water and Aqueous Surfactant Solutions

Early speculations about the mode of action of PDMS-based antifoams assume that duplex film spreading from drops in foam films induces shear in the intralamellar liquid, which leads to foam film rupture. Clearly the rate of spreading would be a key aspect of that mechanism. However, as we describe in Chapter 4, this view of antifoam mechanism is now somewhat discredited. Nevertheless, other aspects of antifoam action, such as the effect of antifoam viscosity on deactivation during prolonged interaction with foam generation (see Chapter 5), could be determined by spreading rates. It is therefore appropriate to briefly review this topic here.

The rate of duplex film spreading of PDMS oils on either water or surfactant solution is given by a simple power law expression provided that, in the case of surfactant solutions, the air–water and oil–water surface tensions are maintained constant by rapid transport of surfactant to the relevant surfaces. Thus, Fay [86], and later Hoult [87], derived by dimensional analysis that the time dependence of the distance y from the origin to the spreading front, in the case of linear spreading as a duplex film, is given by

$$y = a_s \left(\frac{S^i}{\eta_w^{0.5} \rho_w^{0.5}} \right)^{0.5} . t^{3/4} \tag{3.37}$$

where t is time, η_w and ρ_w are the viscosity and density of the aqueous phase substrate, respectively. Hoult [87] has reported a value of the constant a_s of 1.33 after analysis of experimental data. Joos and Pinten [88] deduced a theoretical value of 1.15 by treating the spreading process as a longitudinal disturbance. Bergeron and Langevin [75] have made a theoretical analysis of the analogous problem of radial spreading and derived a similar relation to Equation 3.37 where x becomes the radius of the spreading front and the constant a_s has again the value of 1.15.

The rate of spreading indicated by Equation 3.37 is clearly dependent on the magnitude of the initial spreading coefficient, S^i (and therefore also of the spreading pressure since the equilibrium spreading coefficient in Equation 3.29 is zero for duplex film spreading). However, the rate of spreading is also seen to be independent of the viscosity of the oil. As described by Huh et al. [72], this follows, somewhat paradoxically, from the high viscosity of typical PDMS oils relative to that of the aqueous substrate. Consider then the diagram of a spreading film shown in Figure 3.20. The continuity of shear stress at the oil–aqueous substrate boundary means that we can write

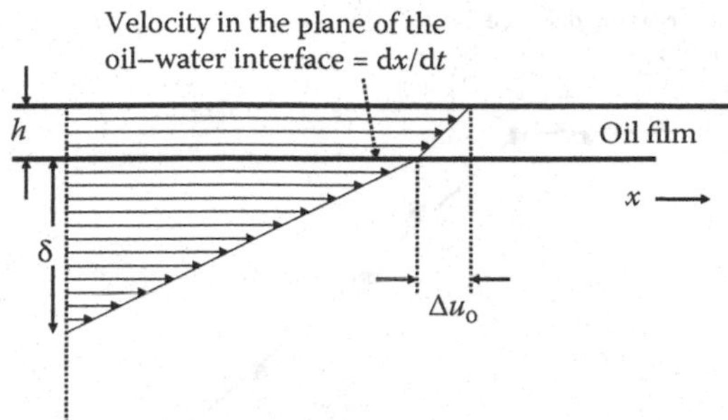

FIGURE 3.20 Schematic illustration of oil film spreading over aqueous surface where the viscosity of oil phase is so high that film moves with almost uniform horizontal velocity. (After Huh, C. et al., *Can. J. Chem. Eng.*, 53, 367, 1975.)

$$\eta_o \dot{\gamma}_o = \eta_w \dot{\gamma}_w \tag{3.38}$$

where $\dot{\gamma}_o$ and $\dot{\gamma}_w$ are the shear rates in the oil film and aqueous substrate, respectively, and η_o is the viscosity of the oil. Substituting for the shear rates in Equation 3.38 means that we can write

$$\eta_o \frac{\Delta u_o}{h} = \eta_w \frac{dy/dt}{\delta} \tag{3.39}$$

where dy/dt is the velocity in the plane of the oil–water interface and Δu_o is the difference between the velocity at that interface and that at the upper surface of the oil film of thickness h. δ is the viscous boundary layer thickness in the aqueous substrate. Therefore, we can write

$$\Delta u_o = \frac{\eta_w}{\eta_o} \cdot \frac{h}{\delta} \cdot dy/dt \approx 0 \tag{3.40}$$

because $\eta_w \ll \eta_o$. The film therefore spreads almost like a rigid solid so that the oil viscosity is irrelevant. Huh et al. [72] show that the linear spreading of a range of PDMS oils of viscosity from 9.3 to 971 mPa s on a pure water substrate follows Equation 3.37 with $a_s = 1.34$ in agreement with estimates of Hoult [87]. This behavior is therefore clearly consistent with duplex film spreading and the irrelevance of the oil viscosity over the range used.

Bergeron et al. [76] have also shown that Equation 3.37 effectively describes radial spreading (if $a_s = 1.15$) over a micellar surfactant solution (i.e., C_{14}TAB at 4 × CMC) in the case of PDMS oils of viscosity from 100 to 1000 mPa s. In this viscosity range, the spreading rate is invariant with viscosity as is shown in Figure 3.21. However, for oils of viscosity > 1000 mPa s, the relation breaks down and the spreading rate declines markedly with increasing viscosity. The relevant experimental results are

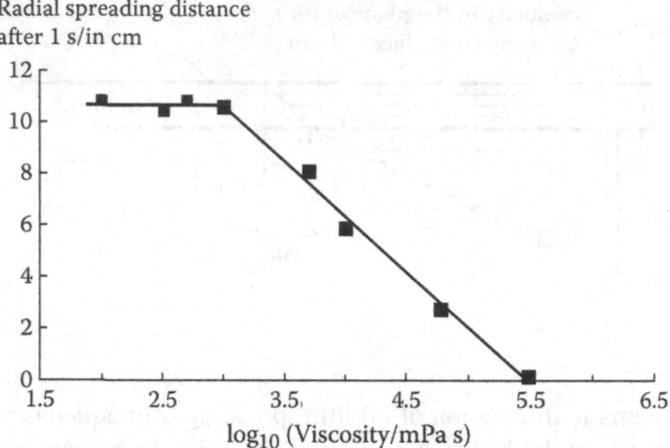

FIGURE 3.21 Radial spreading rate of polydimethylsiloxane oils as function of shear viscosity. Substrate: 14.8 mM aqueous tetradecyltrimethylammonium bromide solution. (Reprinted from *Colloids Surf A*, 122, Bergeron, V., Cooper, P., Fischer, C., Giermanska-Kahn, J., Langevin, D., Pouchelon, A., 103. Copyright 1997, with permission from Elsevier.)

also presented in Figure 3.21. Garrett and Gratton [74] have observed this behavior with PDMS spreading on pure water. Bergeron et al. [76] attribute the lower spreading rates for the high viscosity oils to the dominance of the rate of disentanglement of the PDMS molecules from the bulk phase oil drop.

An experimental study of the rates of radial spreading of a PDMS oil, of molecular weight 10^4 (and viscosity ~200 mPa s), on the surfaces of micellar surfactant solutions has also been reported by Bergeron and Langevin [75]. These solutions were the same as those listed in the relevant part of Table 3.3 (i.e., AOT, $C_{10}EO_5$, C_9–$C_{16}TAB$). Again agreement with Equation 3.37 was found in all cases where, however, we remember that x is then the radius of the spread layer and the constant $a_s = 1.15$. This agreement was found despite there being clear evidence that both complete (i.e., duplex film spreading) and pseudo-partial wetting behavior have been apparently unequivocally found with these systems (e.g., pseudo-partial wetting is unequivocally found with the micellar AOT solutions and complete wetting with all C_nTAB solutions). By contrast, Ye et al. [89], in recent work, have argued that deviation from Equation 3.37 will be apparent if pseudo-partial wetting occurs. These workers have modified Equation 3.37 to account for the forces across the spreading film that give rise to pseudo-partial wetting. They show that in the case of van der Waals forces with negative Hamaker constants, then oil film spreading should slow down markedly at long times relative to that predicted by Equation 3.36. Some experimental evidence for this behavior is presented by Ye et al. [89] in the case of the PDMS films spreading on aqueous SDS solutions.

3.6.3 Effect of Spread Hydrocarbon and Polydimethylsiloxane Oils on the Stability of Pseudoemulsion Films

Wetting by n-alkanes of the adsorbed monolayers of surfactants has been shown by Hadjiiski et al. [35, 36] to have a marked effect on the critical applied capillary pressure, p_c^{crit}, required to rupture the pseudoemulsion films formed by those alkanes. Results are summarized in Table 3.4. It should be emphasized that these results concern only the intrinsic stability of the pseudoemulsion films in the absence of destabilizing hydrophobic particles. A micellar solution (~13 × CMC) of a blend of branched chain sodium dodecylbenzene sulfonates in 12 mM NaCl was used as aqueous phase in these experiments. It is clear from Table 3.4 that with n-decane and n-dodecane, p_c^{crit} is significantly reduced, whereas with hexadecane it is significantly increased. In the case of the lower alkanes, the nature of the spread film is not apparently known. However, in the case of hexadecane, the equilibrium spreading pressure is known to be zero [79], which implies complete wetting as a duplex film despite a spreading pressure of only 0.6 mN m^{-1}.

Denkov et al. [90] have also shown that the presence of a spread layer of a PDMS oil of viscosity 1000 mPa s on micellar AOT solution reduces the applied capillary pressure for pseudoemulsion film rupture from 28 to 19 Pa. As we have seen in this case, pseudo-partial wetting is involved where a thin oil film is in equilibrium with oil lenses.

It is therefore tempting to conclude that complete wetting implies a pseudoemulsion film stability akin to that of an oil–water–oil film in an emulsion that could be relatively stable. On the other hand, pseudo-partial wetting can, as we have seen, imply only solubilization of the oil in the adsorbed surfactant monolayer at the air–water surface. The latter may have an adverse effect on the stability of the pseudoemulsion film through changes in the disjoining pressure isotherm. This is of course more or less pure speculation in the absence of knowledge of the wetting behavior of the lower alkanes in this specific case. However, a systematic experimental study of the effect of spread oils on oil–surfactant combinations with different wetting behavior would appear to be necessary if the effect is to be properly understood.

TABLE 3.4
Effect of Spread n-Alkanes on Stability of Pseudoemulsion Films as Measured by Critical Applied Capillary Pressure Using FTT Technique

n-Alkane	p_c^{crit}/Pa in Absence of Spread Alkane	p_c^{crit}/Pa with Surfactant Monolayer Saturated with Spread Alkane
n-Decane	>70	35 ± 5
n-Dodecane	96 ± 5	48 ± 5
n-Hexadecane	80 ± 5	400 ± 10

Source: Hadjiiski, A. et al., Role of entry barriers in foam destruction by oil drops, in *Adsorption and Aggregation of Surfactants in Solution* (Mittal, K., Shah, D., eds.), Marcel Dekker, New York, Chpt 23, p 465, 2002.

Note: Surfactant solution: 2.6 mM sodium dodecylbenzene sulfonate in 12 mM NaCl (i.e., ~13 × CMC).

3.7 NON-EQUILIBRIUM EFFECTS DUE TO SURFACTANT TRANSPORT

An additional complication associated with both entry and spreading behavior concerns the effect of the finite time required for surfactant to transport to the relevant surfaces during foam generation. Thus, the surface tension of the aqueous solution may be higher than equilibrium even if the solution is saturated with respect to the antifoam. We may then find, for example, that initial spreading coefficients are greater than zero under dynamic conditions simply because $\left(\sigma^{i}_{AW}\right)_{dynamic} > \left(\sigma^{i}_{AW}\right)_{equilibrium}$ despite negative initial spreading pressures on the unperturbed air–water surfaces.

Similar problems arise when considering the rates of spreading of oils on aqueous surfactant solutions. Here the oil–water surface tension of the spreading film may increase if the transport of surfactant is too slow to maintain equilibrium. Similarly the advancing spreading front will tend to compress the air–water surface and decrease the air–water surface tension if transport away from that surface is slow. These effects are considered by, for example, Schokker et al. [91]. Both will tend to slow down the rate of spreading and will mean marked deviations from the behavior predicted by Equation 3.37. That this equation is nevertheless effective in describing the rates of spreading of PDMS oils on the surfaces of the various aqueous micellar solutions described in Section 3.6.2.2 suggests that such effects are absent in those cases.

Hotrum et al. [92] have presented an interesting study of the effect of dynamic factors on the entry and spreading behavior of a triglyceride oil (sunflower oil) on aqueous solutions of both a globular protein (whey protein isolate) and a random coil protein (sodium caseinate). With these systems, the classic initial entry and spreading coefficients were positive at all protein concentrations considered. Spontaneous emergence and spreading to form a surface contaminated with oil lenses, implying pseudo-partial wetting, occurred provided the equilibrium air–water surface tension remained above a critical value. Below that critical value, oil drop emergence and spreading could only be induced if the surface was expanded dynamically using a suitable device. Emergence and spreading required in fact that the air water surface tension be raised to a dynamic value equal to the critical value. It was not clear, however, whether surface tensions less than the critical value represented air–water surface conditions leading to inhibition of the rupture of the relevant pseudoemulsion films, or that they concerned protein adsorptions sufficiently high to cause inhibition of spreading due to monolayer compression.

3.8 SUMMARIZING REMARKS

As we will show in Chapter 4, the entry and spreading behavior of antifoam oils is central to their mode of action. In particular, such oils must enter the air–foaming liquid surface if they are to function. It has been shown that it is not necessary that they spread over that surface [15]. However, it has also been shown [34–36] that the presence of spread films over the air–foaming surface influences the stability of the relevant pseudoemulsion films, the emergence into that surface, and therefore the effectiveness of an antifoam. The relevance of the entry and spreading behavior of oils to antifoam action therefore justifies the detailed review given here.

The conditions under which oils may emerge and spread over the surfaces of foaming liquids can be defined by the classic entry and spreading coefficients, which are, in turn, defined in terms of the relevant surface tensions. These coefficients cannot therefore be directly related to the disjoining forces present in the films to which they refer. Since those forces determine the stability, or otherwise, of the films, this must represent a serious limitation. In particular, they do not account for metastable states. By contrast, the generalized entry and spreading coefficients explicitly incorporate those disjoining forces and accommodate the concept of equilibrium between the disjoining pressure and an applied capillary pressure. Knowledge of the complete disjoining pressure isotherms is of course required if the relevant integrals are to be evaluated. A particular limitation concerns the nature of the primary maximum in the isotherms, which is determined by short-range forces. Unfortunately, experimental measurement of such isotherms is at best difficult and, in the case of the relevant pseudoemulsion and pseudo-partial wetting films, rarely done. The utility of a generalized spreading or entry coefficient is therefore at present apparently limited to that of a conceptual aid. By contrast, classic spreading and entry coefficients are often measured. Unfortunately experimental, so-called equilibrium values, of these coefficients are sometimes presented, which lie outside the bounds set by the thermodynamic necessity implied by Equations 3.11 and 3.12. Such measurements are clearly flawed and can therefore lead to confusion.

It is now recognized that pseudoemulsion film stability is a key factor in determining antifoam action. In the case of commercial antifoams for controlling the foam of aqueous surfactant solutions, hydrophobic particles are usually present to ensure rupture of such films, as we will describe in Chapter 4. In the absence of such particles, positive disjoining forces can lead to metastable films, non-emergence of oil-drops into foam surfaces, and minimal antifoam action despite positive values of classic entry coefficients. It is usually assumed that rupture of those films then requires application of a capillary pressure greater than the relevant maximum disjoining pressure in the disjoining pressure isotherm. However, Denkov and coworkers [34, 35] have presented experimental measurements of the oil drop size dependence of the applied critical capillary needed to rupture pseudoemulsion films, which are difficult to explain using such a concept. Indeed, as shown here, there is no clear explanation at all at present available for those findings. The importance of this issue suggests that a study of pseudoemulsion film stability is required to clarify the matter. Ideally, that could concern a combination of measurement of disjoining pressure isotherms and critical applied capillary pressures using film balance and FTTs, respectively. Some new theoretical insight would appear to be required too!

The spreading behavior of typical antifoam oils on the air–water surfaces of surfactant solutions is clearly complex, often involving solubilization of the oil in the surfactant monolayer to produce finite spreading pressures. In the case of hydrocarbons partial wetting, pseudo-partial wetting and complete duplex film formation are all found. However, for low molecular weight n-alkanes, the predominant equilibrium behavior with all micellar surfactant solutions studied appears to be pseudo-partial wetting but with lenses of extremely low dihedral angle. With high molecular weight hydrocarbons and branched chain surfactants, solubilization by interdigitation is hindered and partial wetting can occur. By contrast, PDMS oils more commonly exhibit complete wetting of the surfaces of micellar surfactant solutions. Pseudo-partial wetting is also found, especially with branched-chain surfactants (AOT and sodium alkylbenzene sulfonate; see Table 3.3). An example of partial wetting has even been reported [73].

Emphasis on equilibrium behavior is a limitation of these studies of the spreading behavior of typical antifoam oils on the air–water surfaces of aqueous surfactant solutions. Information on the process of spreading is often vague and confusing. It is not always clear, for example, whether the equilibrium state is achieved by direct spreading from an oil lens or by initial duplex film spreading followed by disproportionation to form lenses in equilibrium with a spread film. Indeed, more complex behavior than either of these options is sometimes reported (see, e.g., reference [83]). Agreement between experiment and Equation 3.37 for the spreading PDMS oils on the surfaces of some surfactant solutions does suggest initial duplex film spreading in at least those cases. However, Ye et al. [89] point out that if the equilibrium state is pseudo-partial wetting then deviations from that behavior are possible. Similar measurements of the spreading rates of hydrocarbon oils on the air–water surfaces of surfactant solutions appear to be lacking. Improved understanding of the process of spreading of antifoam oils therefore would seem to merit attention not least because of its relevance for understanding the interaction of antifoam with air–water surfaces during foam generation when markedly non-equilibrium conditions prevail.

We should emphasize that the evidence reviewed here suggests that the hydrocarbon and PDMS oils used as antifoams for aqueous surfactant solutions appear to exhibit all types of wetting behavior, from partial wetting with zero spreading pressure to complete wetting. In seeking to understand the basic role of such oils in antifoam action, it would seem preferable that we find a common mechanism that functions irrespective of the nature of the wetting behavior rather than a series of mechanisms, each of which is specific to a particular behavior. That mechanism cannot, however, simply concern the existence of a finite spreading pressure because it has been convincingly shown that antifoam action is possible even with an oil that shows partial wetting behavior [15]. The role of oils in antifoam action is addressed in detail in Chapter 4.

Finally, we should note that antifoam oils such as PDMSs and polyperfluoralkylsiloxanes are used for controlling the foam of non-aqueous liquids. There appears to be a relative absence of studies of the spreading behavior of such oils on the relevant non-aqueous substrates equivalent to those concerning aqueous surfactant solutions.

REFERENCES

1. Wasan, D., Nikolov, A.D., Huang, D.D., Edwards, D.A. Foam stability: Effects of oil and film stratification, in *Surfactant Based Mobility Control* (Smith, D.H., ed.), A.C.S. Symposium Series 373, Washington, DC, 1988, p 136.
2. Kitchener, J.A. In *Recent Progress in Surface Science* (Danielli, I.F., Pankhurst, K.G.A., and Riddiford, A.C., eds), Academic Press, New York, 1964, Vol 1, p 51.
3. Shearer, L.T., Akers, W.W. *J. Phys. Chem.*, 62, 1264 and 1269, 1958.
4. Ross, S. *J. Phys. Colloid Chem.*, 54, 429, 1950.
5. Ewers, W.E., Sutherland, K.L. *Aust. J. Sci. Res.*, 5, 697, 1952.
6. Kitchener, J.A., Cooper, C.F. *Quart. Rev.*, 13, 71, 1959.

7. Leviton, A., Leighton, A. *J. Dairy Sci.*, 18, 105, 1935.
8. Okazaki, S., Hayashi, K., Sasaki, T. *Proceedings of the IV International Congress on SAS, V3*, Brussels, 67, 1964.
9. Ross, S., Young, G.J. *Ind. Eng. Chem.*, 43(11), 2520, 1951.
10. Kulkarni, R.D., Goddard, E.D. *Croatica Chem. Acta*, 50(1–4), 163, 1977.
11. Kroglyakov, M., Taube, P.R. *Zh. Prkl. Khim.*, 44(1), 129, 1971.
12. Jha, B.K., Christiano, P., Shah, D.O. *Langmuir*, 16(26), 9947, 2000.
13. Prins, A. In *Food Emulsions and Foams* (Dickinson, E., ed.), Royal Society of Chemistry Special Publication 58, 1986, p 30.
14. Bisberink, C.G.J. The influence of spreading particles on the stability of thin liquid films, PhD thesis, University of Wageningen, the Netherlands, 1997.
15. Garrett, P.R., Davis, J., Rendall, H.M. *Colloids Surf. A*, 85, 159, 1994.
16. Harkins, W.D. *The Physical Chemistry of Surface Films*, Reinhold Publishing Corp., New York, 1952, p 94.
17. Brochard-Wyatt, F., di Maeglio, J., Quere, D., de Gennes, P. *Langmuir*, 7, 335, 1991.
18. Harkins, W.D. *J. Chem. Phys.*, 9, 522, 1941.
19. Rowlinson, J.S., Widom, B. *Molecular Theory of Capillarity*, Oxford University Press, Oxford, 1982.
20. Antonow G.N. *J. Chim. Phys.*, 5, 372, 1907.
21. Aveyard, R., Binks, B.P., Fletcher, P.D.I., Peck, T.G., Rutherford, C.E. *Adv. Colloid Interface Sci.*, 48, 93, 1994.
22. Bergeron, V., Fagan, M.E., Radke, C.J. *Langmuir*, 9, 1704, 1993.
23. Bergeron, V., Radke, C.J. *Colloid Polymer Sci.*, 273, 165, 1995.
24. Lobo, L., Wasan, D.T. *Langmuir*, 9, 1668, 1993.
25. Garrett, P.R., Wicks, S.P. *Colloids Surf. A*, 282–283, 307, 2006.
26. Kruglyakov, P. DSc Thesis, Novosibirsk, 1978.
27. Exerowa, D., Kruglyakov, P.M. *Foam and Foam Films*, Elesevier, Amsterdam, 1998, p 321.
28. Hirasaki, G. Thermodynamics of thin films and three-phase contact regions, in *Interfacial Phenomena in Petroleum Recovery* (Morrow, N.R., ed.), Marcel Dekker, New York, 1991, p 23.
29. de Feijter, J. Thermodynamics of thin liquid films, in *Thin Liquid Films, Fundamentals and Applications* (Ivanov, I., ed.), Marcel Dekker, New York, 1988, p 11.
30. Martynov, G.A., Dergajuin, B.V. *Colloid J. (USSR)*, 24, 411–417, 1962.
31. Scheludko, A., Radoev, B., Kolarov, T. *Trans. Faraday Soc.*, 64(8), 2213, 1968.
32. Aveyard, R., Clint, J.H. *JCS Faraday Trans.*, 91(7), 2681, 1995.
33. Bergeron, V., Radke, C.J. *Langmuir*, 8, 3020, 1992.
34. Denkov, N. *Langmuir*, 20, 9463, 2004.
35. Hadjiiski, A., Tcholakova, S., Denkov, N., Durbut, P., Broze, G., Mehreteab, A. *Langmuir*, 17, 7011, 2001.
36. Hadjiiski, A., Denkov, N., Tcholakova, S., Ivanov, I. Role of entry barriers in foam destruction by oil drops, in *Adsorption and Aggregation of Surfactants in Solution* (Mittal, K., Shah, D., eds.), Marcel Dekker, New York, 2002, Chpt 23, p 465.
37. Lucassen, J., Barnes, G. *JCS Faraday Trans. I*, 68, 2129, 1972.
38. Bergeron, V. *Langmuir*, 13(13), 3474, 1997.
39. Bergeron, V. *J. Phys. Condens. Matter*, 11, R215–R238, 1999.
40. Vrij, A. *Diss. Faraday Soc.*, 42, 23, 1966.
41. Vrij, A., Overbeek, J.Th.G. *J. Am. Chem. Soc.*, 90(12), 3074. 1966.
42. Kellay, H., Meunier, J., Binks, B. *Phys. Rev. Lett.*, 69(8), 1220, 1992.
43. Hough, D., White, L.R. *Adv. Colloid Interface Sci.*, 14, 3, 1980.
44. Israelachvili, J. *Intermolecular and Surface Forces*, First Edition, Academic Press, London, 1985, p 207.

45. Exerowa, D., Kruglyakov, P. In *Foam and Foam Films, Theory, Experiment, Application*, Elsevier, Amsterdam, 1998, p 238.
46. Kralchevsky, P., Nikolov, A., Wasan, D., Ivanov, I. *Langmuir*, 6, 1180, 1990.
47. Vlakovskaya, D., Kralchevsky, P., Danov, K., Broze, G., Mehreteab, A. *Langmuir*, 16, 8892, 2000.
48. Aveyard, R., Cooper, P., Fletcher, P. *J.C.S. Faraday Trans.*, 86(21), 3623, 1990.
49. Aveyard, R., Binks, B., Fletcher, P., MacNabb, J. *Langmuir*, 11, 2515, 1995.
50. Velev, O., Gurkov, T., Borwankar, R. *J. Colloid Interface Sci.*, 159, 497, 1993.
51. Danov, K., Gurkov, T., Dimitrova, T., Ivanov, I., Smith, D. *J. Colloid Interface Sci.*, 188, 313, 1997.
52. Bergeron, V., Langevin, D. *Macromolecules*, 29(1), 306, 1996.
53. Bonn, D. *Curr. Opin. Colloid Interface Sci.*, 6, 22, 2001.
54. Ragil, K., Meunier, J., Broseta, D., Indekeu, J.O., Bonn, D. *Phys. Rev. Lett.*, 77, 1532, 1996.
55. Aratono, M., Kawagoe, H., Toyomasu, T., Ikeda, N., Takiue, T., Matsubara, H. *Langmuir*, 17, 7344, 2001.
56. Matsubara, H., Ikeda, N., Takiue, T., Aratono, M., Bain, C.D. *Langmuir*, 19, 2249, 2003.
57. Wilkinson, K., Bain, C.D., Matsubara, H., Aratono, M. *ChemPhysChem*, 6, 547, 2005.
58. Matsubara, H., Shigeta, T., Takata, Y., Ikeda, N., Sakamoto, H., Takiue, T., Aratono, M. *Colloids Surf. A*, 301, 141, 2007.
59. Takii, T., Mori, Y. *J. Colloid Interface Sci.*, 161, 31, 1993.
60. Jasper, J., Kerr, E., Gregorich, F. *J. Am. Chem. Soc.*, 75, 5252, 1953.
61. Aveyard, R., Haydon, D. *Trans. Faraday Soc.*, 61, 2255, 1965.
62. Hauxwell, F., Ottewill, R. *J. Colloid Interface Sci.*, 34(4), 473, 1970.
63. Akatsuka, S., Yoshigiwa, H., Mori, Y. *J. Colloid Interface Sci.*, 172, 335, 1995.
64. del Cerro, C., Jameson, G.J. *J. Colloid Interface Sci.*, 78(2), 362, 1980.
65. Lu, J., Thomas, R., Aveyard, R., Binks, B., Cooper, O., Fletcher, P., Sokolowski, A., Penfold, J. *J. Phys. Chem.*, 96, 10971, 1992.
66. Mikhailov, I., Lulova, N., Postnov, V. *Chem. Technol. Fuels Oils*, 11(2), 99–103, 1975.
67. Kellay, H., Binks, B., Hendrikx, Y., Lee, L., Meunier, J. *Adv. Colloid Interface Sci.*, 49, 85, 1994.
68. Binks, B., Crichton, D., Fletcher, P., MacNab, J., Li, Z., Thomas, R., Penfold, J. *Colloids Surf. A*, 146, 299, 1999.
69. Aveyard, R., Binks, B., Fletcher, P., MacNab, J. *Ber. Bunsenges Phys. Chem.*, 100, 224, 1996.
70. van Voorst Vader, F. *Chem. Ing. Tech.*, 49(6), 488, 1977.
71. Kanellopoulos, A., Owen, M. *Trans. Faraday Soc.*, 67, 3127, 1971.
72. Huh, C., Inoue, M., Mason, S. *Can. J. Chem. Eng.*, 53, 367, 1975.
73. Binks, B., Dong, J. *JCS Faraday Trans.*, 94(3), 401, 1998.
74. Garrett, P.R., Gratton, P. unpublished work.
75. Bergeron, V., Langevin, D. *Phys. Rev. Lett.*, 76(17), 3152, 1996.
76. Bergeron, V., Cooper, P., Fischer, C., Giermanska-Kahn, J., Langevin, D., Pouchelon, A. *Colloids Surf. A*, 122, 103, 1997.
77. Mann, E., Langevin, D. *Langmuir*, 7, 1112, 1991.
78. Garrett, P.R. Mode of action of antifoams, in *Defoaming, Theory and Industrial Applications* (Garrett, P., ed.), Marcel Dekker, New York, 1993, Chpt 1, p 83.
79. Arnaudov, L., Denkov, N., Surcheva, I., Durbut, P., Broze, G., Mehreteab, A. *Langmuir*, 17, 6999, 2001.
80. Denkov, N., Cooper, P., Martin, J. *Langmuir*, 15, 8514, 1999.
81. Marinova, K., Denkov, N. *Langmuir*, 17, 2426, 2001.
82. Basheva, E., Ganchev, D., Denkov, N., Kasuga, K., Satoh, N., Tsujii, K. *Langmuir*, 16, 1000, 2000.

83. Mann, E., Lee, L., Hénon, S., Langevin, D., Meunier, J. *Macromolecules*, 26, 7037, 1993.
84. Lee, L., Mann, E., Guiselin, O., Langevin, D., Farnoux, B., Penfold, J. *Macromolecules*, 26, 7046, 1993.
85. Langevin, D. *Rev. I. Fr. Pétrole*, 52(2), 109, 1997.
86. Fay, J.A. In *Oil on the Sea* (Hoult, D.P. ed.), Plenum Press, New York, 1969, p 53.
87. Hoult, D.P. *Ann. Rev. Fluid Mech.*, 4, 341, 1972.
88. Joos, P., Pintens, J. *J. Colloid Interface Sci.*, 60(3), 507, 1977.
89. Ye, X., Cheng, Y., Huang, X., Ma, H. *Chin. Phys. Lett.*, 24(8), 2345, 2007.
90. Denkov, N., Tcholakova, S., Marinova, K., Hadjiiski, A. *Langmuir*, 18, 5810, 2002.
91. Schokker, E., Bos, M., Kuijpers, A., Wijnen, M., Walstra, P. *Colloids Surf. B*, Biointerfaces, 26, 315, 2002.
92. Hotrum, N., van Vliet, T., Cohen-Stuart, M., van Aken, G. *J. Colloid Interface Sci.*, 247, 125, 2002.

4 Mode of Action of Antifoams

4.1 INTRODUCTION

This chapter attempts a complete review of the various mechanisms proposed for the action of antifoams over the past seven or so decades. It is a feature of this subject that some proposed mechanisms, although plausible, have been speculative. Thus, unequivocal experimental evidence has sometimes been lacking. Indeed, the full theoretical implications of proposed mechanisms have also often not been fully developed. In the main, all of this derives from the extreme complexity of the relevant phenomena. As we have seen, foam is itself extremely complex, consisting of (usually) polydisperse gas bubbles separated by draining films. These films exhibit complicated hydrodynamics involving the distinct rheology of air–liquid surfaces and, for thin films, colloidal interaction forces. The nature of the foam film collapse processes that are intrinsic to foam are still imperfectly understood.

Antifoams for aqueous systems are, for example, usually hydrophobic, finely divided, insoluble materials. Their presence therefore further complicates the complexities associated with foam. Indeed commercial antifoams for aqueous solutions usually consist of hydrophobic particles dispersed in hydrophobic oils. The action of such antifoams concerns the effect of a dispersion (of antifoam in foaming liquid) of a dispersion (the antifoam) on yet a third dispersion (the foam).

However, despite this complexity, significant progress has been made over the past three decades concerning the mode of action of antifoams. Much speculation has been replaced with firm theoretical and experimental evidence. In particular, the role of oil lenses in causing foam film collapse has been largely established together with the realization of the importance of the stability of the so-called pseudoemulsion films of foaming liquid, which separate antifoam oil drops from the relevant gas phase. Much progress has also been made in establishing the role of solid particles in antifoam action, particularly their role in breaking aqueous pseudoemulsion films. The latter is central to understanding the reason for the inclusion of hydrophobic particles in commercial oil-based antifoams intended for control of the foam of aqueous solutions. Particularly noteworthy has been the insights made by Denkov and coworkers [1] arising from measurement of the critical applied capillary pressures necessary for rupture of pseudoemulsion films.

Theories of antifoam mechanism appear to fall into two broad categories: those that concern modification of surface tension gradients and those that concern formation of capillary instabilities in foam films. Theories that concern surface tension gradients appear to be rather speculative. Some of these theories attribute antifoam action to the generation of a surface tension gradient that supposedly drives fluid away from the antifoam entity to cause hole formation in foam films. Others attribute antifoam action to the elimination of surface tension gradients. By contrast, theories based on the formation of capillary instabilities appear to be more general in that they explain antifoam action where neither the generation nor the elimination of surface tension gradients necessarily occur.

This chapter is divided into eight sections, ranging over all aspects of antifoam mechanism. It is usually supposed, for example, that effective antifoam action requires the antifoam to be present as a separate phase. Some equivocation about this issue is, however, sometimes expressed in the literature particularly with respect to a possible antifoam effect due to solubilized oils in micellar aqueous solutions. A section is therefore devoted to the issue. Additives that change the state of aggregation in aqueous solution of foam-stabilizing surfactants can cause marked declines in foamability. Such additives are occasionally, and mistakenly, described as antifoams. We address this issue in a separate section. Theories that attribute antifoam action to either generation or elimination of surface tension gradients are described in detail in another section. Oil drops bridging foam films or Plateau borders can form unstable configurations which give rise to film rupture and foam collapse. We summarize the present state of the relevant theories here in Section 4.5. The antifoam effects of neat emulsified liquids, exhibiting partial miscibility through to near complete immiscibility with respect to the foaming liquid are described in yet another section. Such oils occasionally find practical application despite their relative ineffectiveness as antifoams in aqueous solution. However, neat oils are commonly used as effective antifoams for non-aqueous systems, an issue we also address in that section. Also included is a review of the antifoam effects due to cloud point phase separation. We then devote a section to the antifoam mechanism arising from the formation of unstable bridging configurations by otherwise inert hydrophobic particles where we summarize the present understanding of the roles of contact angle and particle geometry. The last section concerns the mode of action of the synergistic mixtures of hydrophobic particles and oils, which form the basis of many of the commercial antifoam concoctions proposed for control of aqueous foams.

4.2 ANTIFOAM EFFECTS DUE TO SOLUBILIZED OILS

One of the earliest generalizations concerning antifoams states that they must be present as undissolved particles (or drops) in the liquid to be defoamed [2–4]. Indeed the presence of antifoam materials at concentrations lower than the solubility limit can even enhance foamability [5, 6]. One well-known example concerns the foam-enhancing effect of dissolved polydimethylsiloxanes (PDMSs) on hydrocarbon lube oils [5]. Arnaudov et al. [7] report a similar, but small, effect for solubilized 2-butyl octanol on the foamability of saline aqueous micellar sodium dodecylbenzene sulfonate solutions where the oil has a significant antifoam effect on the stability of foam when present at concentrations above the solubility limit. Another example concerns the effect of dodecanol on the foam of aqueous micellar anionic surfactant solutions. According to Arnaudov et al. [7] drops of dodecanol in excess of the solubility limit function as weak antifoams—at least in the case of saline micellar solutions of sodium dodecylbenzene sulfonate. By contrast, Patist et al. [8] find that solubilized dodecanol can either enhance or diminish the foam of micellar solutions of aqueous sodium dodecyl sulfate (SDS) solutions depending on the foam generation method used. Thus, hand shaking of measuring cylinders in the presence of dodecanol produces a diminution of foamability and sparging produces an enhancement. Patist et al. [8] present evidence that the dodecanol both increases equilibrium adsorption levels (as revealed by a reduction in the air–water surface tension) and decreases the rate of micelle breakdown. The latter means slower transport of surfactant to air–water surfaces. If the effective age of the surfaces formed during foam generation is relatively low, as appears to be the case with hand shaking, then the adverse effect of dodecanol on transport can dominate so that foamability is lower than in the absence of dodecanol. Conversely if the age of those surfaces is high, as with sparging, then slow transport is less important and foamability is enhanced by the presence of dodecanol as a consequence of higher adsorption levels.

The generalization that antifoams must be present as undissolved entities has, however, occasionally been challenged [6, 9, 10]. A number of authors in fact report experimental results that purport to show antifoam effects due to additives that are solubilized in the foaming solution [11–13]. Thus, Ross and Haak [11], for example, identify two types of antifoam behavior associated with the effect of oils like tributyl phosphate and methyl isobutyl carbinol on the foam behavior of aqueous micellar solutions of surfactants such as sodium dodecylsulfate and sodium oleate. Wherever the oil concentration exceeds the solubility limit, emulsified drops of oil contribute to an effective antifoam action. However, it is claimed [11, 14] that a weak antifoam effect is associated with the presence of such oils even when solubilized in micelles. The consequences of all this behavior are revealed if, for example, tributyl phosphate is added to micellar solutions of sodium oleate [11] at concentrations below the solubilization limit. A marked decrease in foamability is found immediately after dispersing the oil. As the oil becomes slowly solubilized, the foamability increases. However, even after the oil is completely solubilized, the foamability is still apparently less than that intrinsic to the uncontaminated surfactant solution [11]. By contrast, Arnaudov et al. [7] have more recently shown that the significant antifoam effect of n-heptanol on aqueous micellar solutions of sodium dodecylbenzene sulfonate (in the presence of NaCl) is almost completely eliminated after solubilization.

Another approach to studying the effect of solubilized oil has sometimes involved addition of oils to micellar solutions in excess of the solubilization limit, followed by equilibration and subsequent removal of any excess oil. This approach was adopted by Arnaudov et al. [7] in studying the effect of both 2-butyl octanol and isohexyl-neopentanoate on the foam of aqueous saline micellar solutions of sodium dodecylbenzene sulfonate. These oils have little effect on foamability but produce a significant collapse of foam by a stepwise process after several minutes. Removal of the excess oil was attempted by centrifugation (at 5000*g* for 3 h) followed by filtration through a 220-nm membrane filter. Surprisingly, light scattering revealed that this procedure failed to remove all the oil. Foam destabilization was still present but the time scale of the effect was much longer. This weakened antifoam effect was attributed to the continued presence of a small concentration of oil drops rather than to the presence of solubilized oil in the surfactant micelles. A similar approach to studying the possible effect of solubilized oil on foam stability was adopted by

Koczo et al. [15], using a somewhat higher *g*-force and a slightly finer filter. They considered the effect of alkanes such as n-octane and n-decane on the foam behavior of a micellar solution of a blend of branched-chain sodium dodecylbenzene sulfonates at a concentration of ~0.12 M. Verification of the absence of oil drops was attempted using optical microscopy. Koczo et al. [15] found that the foamability of the surfactant solution was unaffected, relative to that in the absence of alkane, after the excess bulk phase oil was removed by this procedure but that the stability was significantly diminished. However, in this case, the presence of drops of emulsified oil in excess of the solubilization limit, rather than destabilizing the foam, had a stabilizing effect. The latter is in turn believed due to a combination of the high stability of the relevant pseudoemulsion films in these systems and the accumulation of oil drops in Plateau borders, which inhibits foam drainage [15], as we will discuss in Sections 4.5 and 4.6. This of course means that, in contrast to the system studied by Arnaudov et al. [7], we cannot find explanation for the apparent adverse effect of solubilized oil on foam stability by recourse to the effect of the presence of residual amounts of bulk phase oil drops.

Binks et al. [16] have reported an interesting study of the effect of n-alkane vapors on the foamability and foam stability of micellar solutions of ethoxylated alcohols. In this study, nitrogen gas, containing known partial pressures of n-alkane, was prepared by bubbling the gas through the neat alkane or through solutions of the alkane in a nonvolatile hydrocarbon (squalane). The gas was then bubbled into aqueous surfactant solutions to generate foam. Surprisingly, the description of the method does not give details of precautions taken to ensure against possible carryover of aerosol drops of oil into the solutions. If we assume that this omission is just a concession to brevity, then this method has some merit in the present context. Thus, if the partial pressure of the alkane is less than that of the vapor pressure of the pure oil, then formation of bulk phase at the air–water surface is precluded. Any ambiguity concerning the presence of bulk phase is thereby avoided. If, on the other hand, the partial pressure is equal to that of the pure oil, then bulk phase may form at the air–water surface by accumulation of adsorbed alkane without the hindrance implied by the stability of pseudoemulsion films (see Chapter 3). A disadvantage, however, concerns partitioning of the oil into the micellar surfactant solution to form oil–water microemulsions [17, 18] in the case of the ethoxylated alcohol–alkane combinations considered by Binks et al. [16] in this study. At sufficiently high micellar surfactant concentrations, for a given total volume of gas, attainment of equilibrium of the initial alkane partial pressure with the surfactant solution will not be realized because the activity of the alkane will be lowered by solubilization.

Plots of foam volume against surfactant concentration for $C_{12}EO_5$ are given in Figure 4.1 where the foam is generated either with nitrogen alone or with nitrogen saturated with decane vapor (i.e., in equilibrium with bulk phase decane). The critical micelle concentration (CMC) of $C_{12}EO_5$ is only 6.4×10^{-5} M [19], which means that the range of surfactant concentrations in the plots mostly concerns micellar solutions. It is seen that the effect of the presence of decane is to move the plot to higher surfactant concentrations. Significant antifoam effects are apparent over much of the surfactant concentration range. However, at a sufficiently high surfactant concentration, both plots reach identical plateau foam volume values. Binks et al. [16] suggest

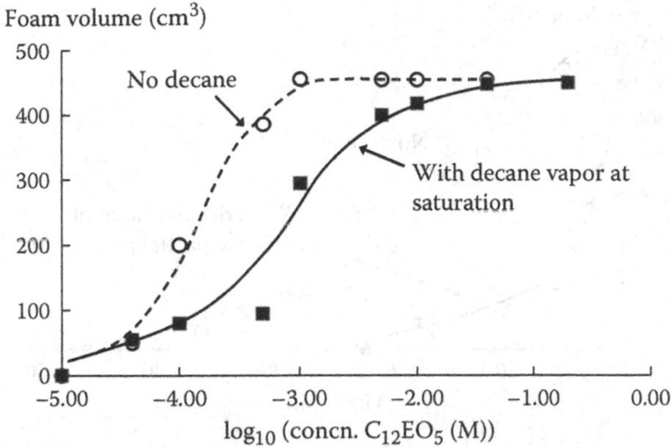

FIGURE 4.1 Effect of saturated decane vapor of foam volume generated by sparging aqueous micellar solutions of $C_{12}EO_5$ for 10 min. (After Binks, B.P. et al., *Colloids Surf. A*, 216, 1, 2003.)

that the effects depicted in Figure 4.1 concern the supposed adverse effect of solubilized decane on transport of $C_{12}EO_5$ to air–water surfaces. Arguably, however, a more convincing explanation for this antifoam behavior concerns formation of bulk phase decane on both of the air–water surfaces of foam films. This must happen at low micellar surfactant concentrations, as equilibrium is attained despite solubilization of part of the oil in the surfactant micelles. For example, taking the vapor pressure of decane as 185 Pa (1.4 mm Hg) at 25°C, we can calculate the molecular ratio of oil to surfactant from the volume of surfactant solution and the total volume of gas bubbled through the solutions. Thus, at a surfactant concentration of 10^{-4} M, where a significant antifoam effect is apparent in the presence of decane, the molecular ratio of decane to $C_{12}EO_5$ is 18, which seems high enough to ensure an excess of bulk phase oil in a so-called Winsor I state even though some of the decane vapor will be carried over in the collapsing foam. It is known that bulk phase decane exhibits pseudo-partial wetting behavior on the surface of micellar $C_{12}EO_5$ [20]—oil lenses will be present in equilibrium with an oil-contaminated monolayer of the surfactant (see Chapter 3). If such lenses bridge foam films, they will lead to foam film collapse, as we will show in Section 4.4. However, this antifoam effect will be eliminated at a sufficiently high surfactant concentration because solubilization will reduce the activity of the oil below unity so that bulk phase will be absent. For example, the onset of a plateau foam volume in the presence of decane shown in Figure 4.1 occurs at a surfactant concentration of about 0.04 M, which corresponds to a molecular ratio of decane to $C_{12}EO_5$ of only 0.45, which is consistent with the absence of bulk phase oil.

These considerations concern the foamability—that is, the amount of foam generated up to the point the gas stream is switched off. However, Binks et al. [16] present results concerning the stability of foam generated from 5×10^{-4} M $C_{12}EO_5$ using decane at a partial vapor pressure significantly less that the vapor pressure of bulk oil. Bulk oil phase cannot then be present during foam formation. Results are shown in Figure 4.2. Even at a relative decane partial pressure of only 0.2, the foam

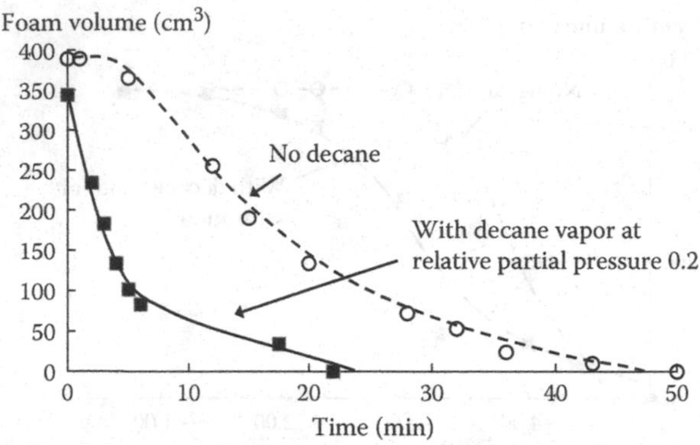

FIGURE 4.2 Effect of relative partial pressure of 0.2 of decane on stability of foam generated by sparging an aqueous solution of 0.5 mM $C_{12}EO_5$ for 10 min. (Reprinted from *Colloids Surf. A.*, 216, Binks, B.P., Fletcher, P.D.I., Haynes, M., 1. Copyright 2003, with permission from Elsevier.)

is seen to be significantly destabilized. These findings are apparently consistent with those of Koczo et al. [15] concerning the effect of solubilized decane on the stability of foam generated from concentrated micellar solutions of sodium dodecylbenzene sulfonate. There would, however, appear to be no clear explanation for this effect—since it concerns foam stability, it clearly cannot concern transport of surfactant to air–water surfaces. It is, of course, known that decane is solubilized in $C_{12}EO_5$ monolayers. However, at such low activities, reductions in surface tension and adsorptions are likely to be rather small (e.g., surface tension reductions <1 mN m^{-1} [20]). Binks et al. [16] suggest that it might concern evaporation of decane from the continuous air–water surface at the top of the foam to produce foam films with different surface tensions on each air–water surface leading to an instability.

There would appear then to be only limited evidence that oils which exhibit antifoam effects, when present as emulsified bulk phase, can also produce antifoam effects when present only as solubilizates in aqueous micellar solutions of surfactants. In many instances, alternative explanations for supposed observations of the latter are possible, which do invoke the presence of the oils as bulk phase. However some of the observations described here are difficult to dismiss. Of particular interest in this context are the findings of Koczo et al. [15], Lobo et al. [21], and Binks et al. [16] concerning the effect of solubilized alkanes on the foam stability of aqueous micellar solutions of various surfactants. Attempts to explain such effects by recourse to dynamic surface tension behavior after the manner of Ross and Haak [11] would appear to be unconvincing (see reference [22]). It is, however, possible that it may concern the effect of the solubilized oil on the relevant disjoining pressure isotherm. Wasan and coworkers [15, 21] have suggested that the phenomenon is a consequence of the effect of solubilization of alkanes on intermicellar interactions. Lobo et al. [21] find that the instability of the foams formed from certain ethoxylated alcohols in the presence of solubilized alkanes depends on the magnitude of the micellar second virial coefficient describing those interactions. Reduction of the magnitude of the second virial coefficient by solubilized alkanes implies reduction of the repulsion forces between micelles. In turn, this is supposed [21] to reduce the stratification in foam films due to the close packing of micelles. In consequence, foam film stability will supposedly be diminished. However, Bergeron and Radke [23] have shown that the critical capillary pressure required to rupture a stratified foam film concerns the maximum in the disjoining pressure isotherm at the last thinning step where micelles are absent anyway. This is illustrated by an example of a relevant disjoining pressure isotherm depicted in Figure 3.9. There appear to be no direct measurements of the disjoining pressure isotherms of foam films, or the critical capillary pressure required to cause film rupture, where solubilized alkane is present in adsorbed surfactant monolayers either with or without the presence of solubilized alkanes in micelles.

We may conclude by noting that, in general, antifoam effects with oils appear to require that the antifoam be undissolved in the foaming medium. Solubilization of the antifoam oil in micelles largely restores the foamability. However, there is some evidence to suggest that the some solubilized antifoam oils may have a weak adverse effect on the foamability or foam stability of some aqueous micellar surfactant solutions. We should note, however, that dissolved antifoam can even enhance foamability [5, 6], at least in the absence of surfactant micelles.

4.3 EFFECT ON FOAMABILITY OF MESOPHASE PRECIPITATION IN AQUEOUS SURFACTANT SOLUTIONS

Addition of soluble inorganic salts can cause significant reductions in the foamability of aqueous solutions of anionic surfactants [24–27]. These effects are usually associated with onset of turbidity and formation of liquid crystalline or even crystalline precipitates. Such precipitation is often simply a consequence of increases in ionic strength upon the phase behavior of the surfactant. Those increases produce compression of electrostatic double layers and a diminution of electrostatic repulsions between charged surfactant head groups. In turn, the latter will mean a reduction in the surfactant head group size and, if significant enough, a change in the so-called molecular packing parameter, $\tilde{p}$. That parameter is defined as $\tilde{v}/\tilde{l}\tilde{a}$, where $\tilde{v}$ is the volume of the surfactant (usually) alkyl chain, $\tilde{l}$ is the extended length of that alkyl chain, and $\tilde{a}$ is the cross-sectional area of the surfactant molecule (see, e.g., reference [28]). The latter is usually determined by the head group size but sometimes by the chain structure in, for example, double-chain surfactants. $\tilde{p}$ determines the nature of the aggregates formed by surfactants in solution—whether, for example, spherical micelles, cylindrical micelles, or (liquid crystalline) mesophases are formed. The concentration at which those aggregates are formed is the CMC in the case of micelles or the critical aggregation concentration (CAC) in the case of macroscopic, liquid crystalline aggregates. If $\tilde{p} \leq 1/3$, spherical micelles are formed; if $1/3 < \tilde{p} < 1/2$, non-spherical or cylindrical micelles; or if $1/2 < \tilde{p} < 1$, liquid crystalline, lamellar phase objects such as vesicles or liposomes [28].

In the case of two-chain surfactant molecules, the alkyl chain length is short for a given alkyl chain volume, $\tilde{v}$. The molecular packing parameter is therefore more readily increased by small changes in the head group size than is the case for single-chain molecules of the same alkyl chain volume. We therefore find that such two-chain molecules are especially susceptible to induced formation of lamellar phase by changes in ionic strength (or indeed other factors such as mixing with cosurfactants). The effect of increases in ionic strength on the foamability of aqueous solutions of such a two-chain anionic surfactant is exemplified by the effect of sodium chloride on the foamability of aqueous micellar solutions of AOT (sodium bis-octyl sulfosuccinate) [24]. Results for 3.8 mM AOT are shown in Figure 4.3 where the foamability was determined by shaking measuring cylinders. The foamability is seen to decline markedly at salt concentrations in excess of ~0.016 M. This decrease in foamability coincides with an increase in the turbidity of the solution. The latter implies the formation of relatively large aggregates, probably lamellar phase emulsion particles of a vesicular structure. The addition of salt has increased the ionic strength, reduced the area $\tilde{a}$, and therefore increased the packing parameter so that lamellar phase rather than micelles are formed. It is therefore tempting to conclude that the low foamability associated with the turbidity concerns slow transport of such entities to air–water surfaces.

Peck [25] reports that increase of AOT concentration above the CAC in 0.05 M NaCl produces only a slight increase in foamability despite markedly increasing turbidity due presumably to the increased concentration of the supposed lamellar phase particles. This implies that transport of surfactant to air–water surfaces by those particles is negligibly slow irrespective of the vesicle concentration—probably as a consequence of extremely low diffusion coefficients of these supposedly large particles and/or slow breakdown kinetics. The diminution of foamability with increasing turbidity shown in Figure 4.3 must therefore largely reflect declining surfactant monomer concentrations with decreasing CACs, augmented by a minor contribution

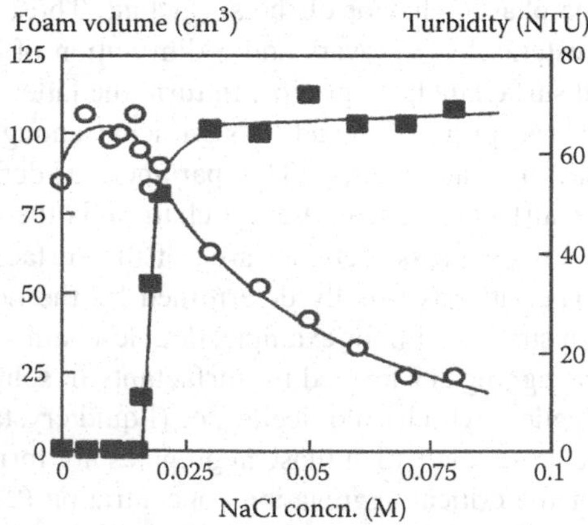

FIGURE 4.3 Foamabilities (○) and turbidities (■) of 3.8 mM AOT solutions in aqueous NaCl solutions. (Reprinted from *Adv. Colloid Interface Sci.*, 48, Aveyard, R., Binks, B.P., Fletcher, P.D.I., Peck, T.G., Rutherford, C.E., 93. Copyright 1994, with permission from Elsevier.)

from transport of vesicular material to air–water surfaces That increasing the concentration of the supposed AOT lamellar phase particles does not cause diminished foamability implies also that the lamellar phase does not exhibit any antifoam effect. As we will discus at length in this chapter, antifoam action involves hydrophobic surfaces. That lamellar phase precipitated in an aqueous environment does not show such effects is hardly surprising because the outer molecular layer of the vesicles is likely to be rich in charged surfactant head groups and will therefore be hydrophilic. This is illustrated in Figure 4.4 where a schematic representation of the molecular arrangement in a vesicle is depicted.

Similar behavior is apparent with other two-chain anionic surfactants. Thus, for example, addition of NaCl at concentrations ≥ 0.025 M to aqueous solutions of sodium 6-phenyltridecanesulfonate produces vesicles of ~500 nm diameter at surfactant concentrations ≤ 0.2×10^{-3} M [29]. The latter compares with the CMC in the absence of salt of 5×10^{-3} M [29]. Addition of increasing amounts of salt at surfactant concentrations higher than the CAC therefore produces a progressively increasing turbidity accompanied by a declining foamability much as observed with AOT [26]. Measurements of the breakdown times of vesicles are in the time range 0.1–10 s, which compares with <1 ms in the case of micelles [29]. This is clearly consistent with a contribution of slow vesicle breakdown rates to slow overall transport of surfactant to air–water surfaces, and therefore to low foamability.

Addition of divalent metal ions, such as Ca^{2+}, to solutions of these two-chain sodium alkylbenzene sulfonates (NaLAS) can also result in the formation of a lamellar phase (Lα) liquid crystalline precipitate of approximate composition $Ca(LAS)_2$. Onset of such precipitation produces a marked reduction in foamability as shown

FIGURE 4.4 Schematic representation of double-walled lamellar surfactant vesicle (not to scale).

by Ran [27]. Results are presented in Figure 4.5 where it is clear that a reduction in foamability is not accompanied by a reduction in the stability of the foam formed. These results concern a scan through the so-called precipitation phase diagram at a constant concentration of sodium dodecyl 4-phenylsulfonate in a dilute salt solution but with increasing concentrations of Ca^{2+}. The phase diagram is reproduced in Figure 4.6a from which it is clear that the scan moves through a micellar region into a micellar–precipitate region where monomer, micelles, and precipitate coexist. The onset of precipitation produced marked turbidity that can be removed by filtration without producing any significant change in the foamability. This implies absence of any antifoam effect due to the precipitate, as expected in the case of a liquid crystalline (probably lamellar) phase. However, measurements of the relevant dynamic air–water surface tensions shown in Figure 4.6b reveal a large increase at the onset of precipitation [27, 30]. Indeed, at a surface age of 0.1 s, the solutions have a surface tension close to that of pure water over much of the micellar–precipitate region [27]. These results strongly suggest that low foamability in that region is attributable to slow transport of surfactant to air–water surfaces. That would mean low surface tension gradients, mobile air–water surfaces, and high capillary pressures in Plateau borders, all of which will conspire to produce rapid foam film drainage in the case of those formed rapidly during foam generation. In turn, low surfactant adsorptions will imply low positive, or even negative, disjoining pressures and unstable films, which readily collapse after draining rapidly. Conversely, any films that survive will remain stable as adsorption levels slowly increase with time—any foam formed would therefore be stable as shown in Figure 4.5.

Precipitation of vesicles or liposomes can also be induced by addition of a strongly interacting cosurfactant to an aqueous micellar solution of a two-chain surfactant. Thus, mixing of *N*,*N*-dimethyldodecylamine butyrate with sodium dodecyl 6-phenylsulfonate at a total surfactant concentration of 5 mM produces a vesicular precipitate

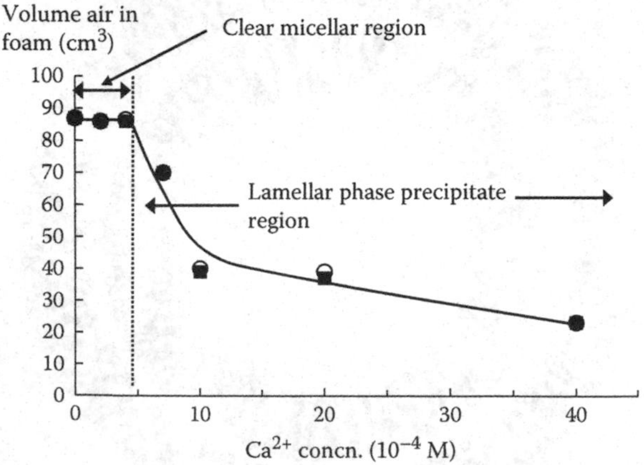

FIGURE 4.5 Effect of Ca^{2+} on foamability (■) and foam stability (O) (after 10 min) of 2 mM sodium dodecyl 4-phenyl sulfonate in aqueous 17 mM NaCl solution at 25°C. (After Ran, L. Foaming of anionic surfactant solutions in the presence of calcium ions and triglyceride-based antifoams, PhD Thesis, University of Manchester, 2011.)

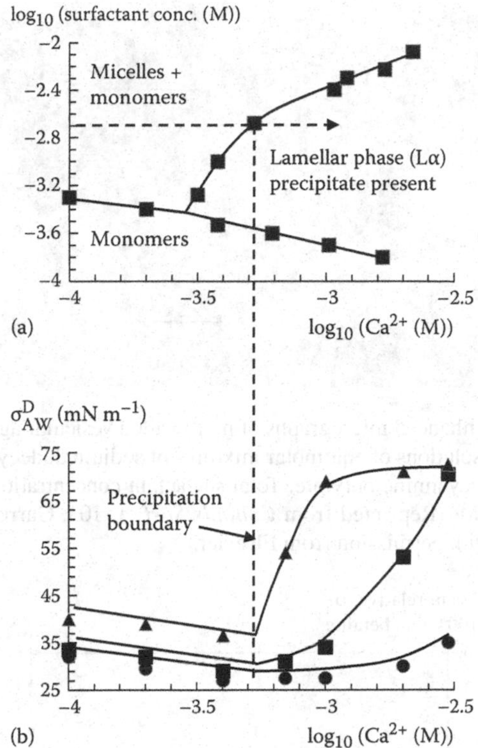

FIGURE 4.6 (a) Precipitation phase diagram for sodium dodecyl 4-phenyl sulfonate/Ca^{2+} system in 17 mM NaCl aqueous solution showing scan at 2 mM surfactant (dotted line) at 25°C. (After Ran, L. Foaming of anionic surfactant solutions in the presence of calcium ions and triglyceride-based antifoams, PhD Thesis, University of Manchester, 2011.) (b) Effect of increasing concentrations of Ca^{2+} on dynamic surface tensions, σ^{D}_{AW}, for 2 mM aqueous solutions of sodium dodecyl 4-phenyl sulfonate in 17 mM NaCl showing large increase with the onset of precipitation. σ^{D}_{AW} at various surface ages: ●, 10 s; ■, 1 s; ▲, 0.1 s. (After Ran, L. Foaming of anionic surfactant solutions in the presence of calcium ions and triglyceride-based antifoams, PhD Thesis, University of Manchester, 2011; Ran, L. et al., *Colloid Surf. A*, 382, 50, 2011.)

of lamellar phase (Lα) over a limited composition range [31]. A video-enhanced micrograph in Figure 4.7 confirms the multiwalled vesicular nature of the precipitate. Foamability as a function of composition is shown in Figure 4.8. Formation of a liposomal precipitate is seen to produce a diminution of foamability that is more marked for cylinder shaking than for Ross–Miles foam generation. Dynamic surface tensions presented in Figure 4.9 again illustrate slow transport of surfactant to air–water surfaces as the likely cause of the low foamability. That the effect is more marked for cylinder shaking than Ross–Miles foam generation presumably then simply reflects faster rates of air–water surface generation with the former and therefore greater deviation from equilibrium adsorption levels. Such differences are

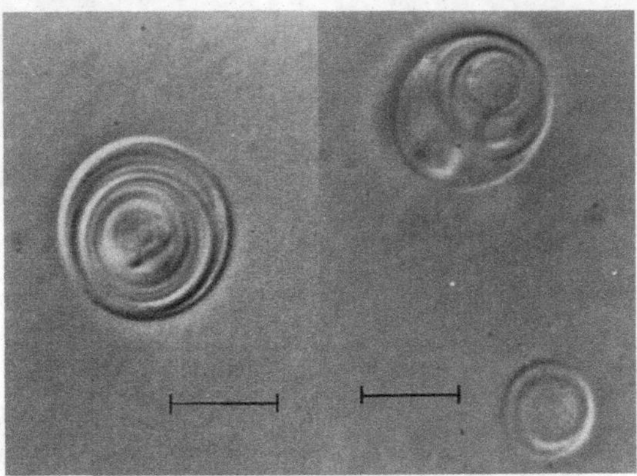

FIGURE 4.7 Video-enhanced micrographs of multiwalled vesicular aggregates of lamellar (Lα) phase in aqueous solutions of equimolar mixtures of sodium dodecyl 6-phenyl sulfonate and *N,N*-dimethyl dodecylamine butyrate. Total surfactant concentration, 5 mM in solution of ionic strength 55 mM. (Reprinted from *Colloids Surf. A*, 103, Garrett, P.R., Gratton, P., 127. Copyright 1995, with permission from Elsevier.)

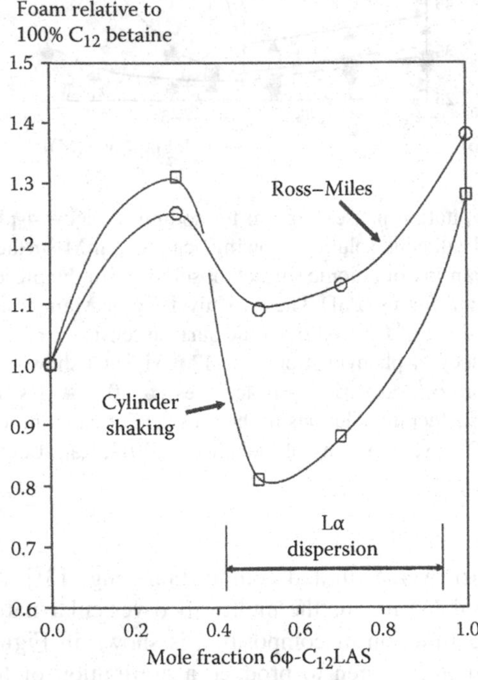

FIGURE 4.8 Plots of foamabilities (relative to those for 100% C_{12} betaine) against mole fraction of sodium dodecyl 6-phenyl sulfonate (6ϕ-C_{12}LAS) for 5 mM mixtures of *N,N*-dimethyl dodecylamine butyrate (C_{12}betaine) + 6ϕ-C_{12}LAS in an aqueous solution of ionic strength 55 mM at 25°C. Foamabilities were measured using both Ross–Miles and hand shaking of measuring cylinder techniques. (Reprinted from *Colloids Surf. A*, 103, Garrett, P.R., Gratton, P., 127. Copyright 1995, with permission from Elsevier.)

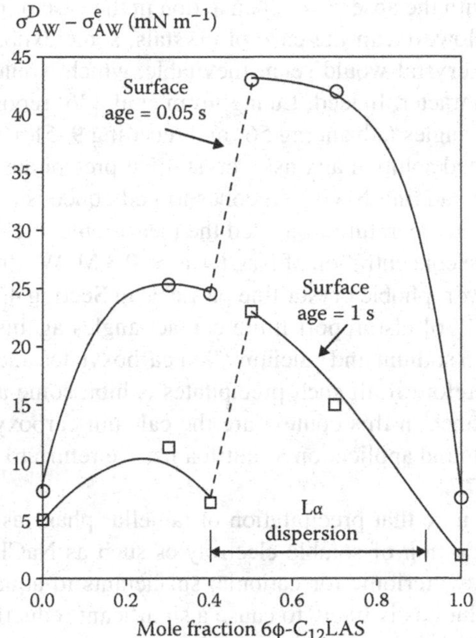

FIGURE 4.9 Plots of difference between dynamic $\left(\sigma^D_{AW}\right)$ and equilibrium $\left(\sigma^e_{AW}\right)$ surface tensions against mole fraction of sodium dodecyl 6-phenyl sulfonate (6ϕ-C_{12}LAS) for 5 mM mixtures of *N*,*N*-dimethyl dodecylamine butyrate + 6ϕ-C_{12}LAS in an aqueous solution of ionic strength of 55 mM at 25°C. (Reprinted from *Colloids Surf. A*, 103, Garrett, P.R., Gratton, P., 127. Copyright 1995, with permission from Elsevier.)

analogous to those reported by Patist et al. [8] concerning the effect of dodecanol on the foamability of micellar SDS solutions.

Addition of soluble inorganic salts can also induce the precipitation from aqueous solutions of crystalline or amorphous crystalline surfactant precipitates. Consider then, for example, the increase in the Krafft temperature of SDS caused by the addition of a common ion, Na^+ [32–34]. This follows from simple mass action because the degree of micelle dissociation < 1 (i.e., the number of bound Na^+ counterions is less than the micelle aggregation number) [35]. Peck [25] has shown, in measurements at 20°C, that the foamability of SDS declines markedly in the presence of added salt at concentrations > 0.3 M. The Krafft temperature of SDS under these conditions is >25°C [34], which means that crystalline SDS particles should be present as indicated by the onset of turbidity. Filtration to remove the turbidity partially restores the foamability [25] to a significant extent, which implies that the crystalline SDS particles exhibit some antifoam behavior. A combination of slow transport of surfactant to air–water surfaces and antifoam action by the crystalline surfactant would account for the almost total loss of foamability in the case of 0.01 M SDS in the presence of >0.3 M NaCl solution [25]. Antifoam action by crystalline particles contrasts markedly with the absence of such action in the case of precipitated lamellar phase particles. However, in the case of crystals, some exposure of the hydrocarbon chains of the crystal would seem inevitable, which would therefore impart some hydrophobic character. Indeed, Luangpirom et al. [36] report finite advancing and receding contact angles (advancing 56° and receding 9–51°) at 30°C for the relevant aqueous saturated solution against a crystalline precipitate of SDS. Here, the latter was prepared by adding NaCl to a concentrated aqueous solution of SDS. To ensure that the Krafft temperature exceeded the measurement temperature of 30°C, it seems likely that the concentration of NaCl was $\gg$ 0.3 M. We discuss the potential antifoam effects of hydrophobic crystalline particles in Section 4.7.3.

Luangpirom et al. [36] also report finite contact angles against precipitates of a variety of single-chain sodium and calcium alkyl carboxylates and sulfates. It would not be surprising therefore if all such precipitates exhibit some antifoam behavior. Of particular importance in this context are the calcium carboxylate soap precipitates that have often found application as antifoams. We return to discussion of such systems in Section 4.7.7.

In conclusion, we note that precipitation of lamellar phase as vesicular or liposomal particles by addition of soluble electrolytes such as NaCl or $CaCl_2$ or even strongly interacting zwitterionic (or cationic) surfactants to aqueous double-chain anionic surfactant solutions is likely to cause a significant reduction in foamability. Antifoam effects involving the precipitate particles are not implicated because of the hydrophilic nature of the precipitates. Removal of the precipitates does not therefore generally restore the foamability. However, such precipitation involves replacement of labile micellar entities with relatively large and stable lamellar entities. In consequence, the transport of surfactant to the rapidly forming air–water surfaces during foam generation is likely to be inhibited. It is this that produces unstable foam films and low foamability. However, any foam films that survive may be stabilized as adsorption slowly increases with time. By contrast, precipitation of crystalline or amorphous crystalline surfactant will not only mean slow transport of surfactant to air–water surfaces but could also mean antifoam effects due to finite contact angles. Therefore, reduction of foamability by precipitation of crystalline particles may be more marked than is the case with precipitation of vesicular, lamellar phase particles.

4.4 SURFACE TENSION GRADIENTS AND THEORIES OF ANTIFOAM MECHANISM

4.4.1 General Considerations

We have seen that in general, dissolved oils and mesophase particles do not give rise to antifoam effects. These generalizations are illustrations of the accepted view that antifoam entities should be present as *undissolved* material having the property of *adhering* to the gas–liquid surfaces of foaming systems. In the literature, the latter property is also often described as "entering in" or "emerging into" those surfaces—we assume equivalence of these terminologies here. In the case of hydrophobic oils, adhesion to gas–liquid surfaces implies partial, pseudo-partial, or complete wetting behavior as described in Chapter 3. In the case of solids, adhesion to such surfaces implies finite contact angles.

The requirement that oils adhere to gas–liquid surfaces implies that entry coefficients be positive and that some agency be present that can destabilize any metastable pseudoemulsion films. In the case of air–water–oil pseudoemulsion films, the latter is usually represented by hydrophobic particles adhering to the oil–water surfaces as we describe in Section 4.8. In the case of solids, surface roughness appears to facilitate emergence into air–water surfaces [37], asperities perhaps helping to nucleate rupture of any metastable film separating the solid from the relevant gas–liquid surface (see Section 4.7.4).

It is well known that oils which satisfy the criteria of low solubility and adhesion to gas–liquid surfaces can also exhibit positive initial spreading coefficients where $S^i > 0$ or $S^{si} > 0$ in Equations 3.5 and 3.6. As we have shown in Chapter 3, this is of necessity a transient state because at equilibrium we have $S^e \leq 0$ [33]. Sparingly soluble solids, which adhere to the gas–liquid surfaces of foaming liquids, can also spread as monolayers on those surfaces. The tendency to spread is in turn measured by the equilibrium spreading pressure on a pure substrate liquid (usually water), which in the case of a surfactant solution is augmented by any tendency to form mixed monolayers with the surfactant adsorbed at the gas–liquid surface (see Chapter 3).

It has often been observed that oils and particles, which exhibit antifoam action, not only have the potential to adhere to the relevant air–liquid surfaces (as represented by positive entry coefficients and finite contact angles) but also exhibit positive initial spreading coefficients. Here, we examine two different approaches to antifoam mechanism that require such behavior. The first concerns the effect of the surface tension gradient, caused by the process of spreading, on the thinning rate of foam films. The second concerns elimination of surface tension gradients as a result of displacement of the stabilizing surfactant monolayer by a spread film of antifoam. Here we should emphasize that these theories concern the spreading property per se. The possible existence of significant barriers to the emergence of putative antifoam entities into the relevant gas–liquid surfaces is ignored. Any agreement with experiment for such approaches is therefore either surprisingly fortuitous or concerns antifoam and surfactant solution combinations for which no such barrier exists.

4.4.2 Surface Tension Gradients Induced by Spreading Antifoam

Here we consider the proposition that antifoams cause destabilizing surface tension gradients in foam films. The proposed mechanism concerns spreading from an antifoam source present on the surface of a foam film. Such spreading is driven by a surface tension gradient between the spreading source and the leading edge of the spreading front. This surface tension gradient will act as a shear force dragging the underlying liquid away from the source to supposedly cause catastrophic thinning and foam film rupture. The process is shown schematically in Figure 4.10. It is clearly a consequence of the well-known Marangoni effect, and we refer to it below as Marangoni spreading.

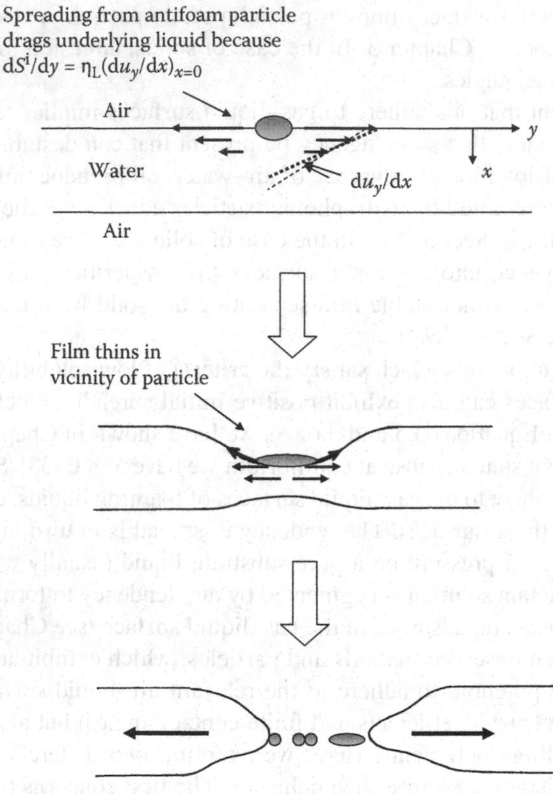

FIGURE 4.10 Foam film rupture caused by spreading from antifoam particle—Marangoni spreading mechanism.

This mechanism was first proposed by Ewers and Sutherland [38]. These authors identified three categories of antifoam substance, which could, in principle, function by this mechanism. These are

- "Solids or liquids containing surface active material other than the substance stabilizing the film"
- "Liquids that contain the foam stabilizer in higher concentration than it is present in the foam"
- "Vapors of surface-active liquids" (including the well-known example of the adverse effect of ether vapor on foam stability)

In the first category, Ewers and Sutherland [38] included all spreading oils that, at least initially, form duplex films for which $S^i > 0$ (or $S^{si} > 0$). In addition, they included monolayer spreading where, for example, a liquid exhibits pseudo-partial wetting. The observations of Binks et al. [16] concerning the antifoam effect observed with saturated alkane vapor clearly fit with the third category. However, as we have shown in Section 4.2 (and in detail Section 4.5), it is possible to explain this effect without recourse to Marangoni spreading.

Here S^i and S^{si} are defined for an aqueous foaming liquid by Equations 3.5 and 3.6 respectively. However these definitions can be readily generalised for non-aqueous systems by appropriate substitutions of the relevant surface tensions in those equations. This is exemplified for the case of natural gas-crude oil-antifoam systems in Section 10.4.1.

The supposed underlying driving force for all of these categories of antifoam substance is the surface tension gradient, or if appropriate, the film tension gradient. If the antifoam is an oil initially spreading as a duplex film, then the film tension adjacent to the drop is equal to the sum of the oil–foaming liquid and the oil–air surface tensions. At the leading edge of the spreading film, it is zero. The gradient in film tension is then $S^i/\Delta x \approx \mathrm{d}S^i/\mathrm{d}x$, where x is the spreading distance. Balancing stresses at the surface of a foam film then means that we can write

$$\frac{\mathrm{d}S^i}{\mathrm{d}y} = \eta_F \left(\frac{\mathrm{d}u_y}{\mathrm{d}x} \right)_{x=0} \tag{4.1}$$

where η_F is the viscosity of the foaming liquid in the foam film, u_y is the velocity of liquid in the y direction, and x is the direction perpendicular to the surface, as shown in Figure 4.10. Equation 4.1 indicates that the flow of liquid in the foam films away from the spreading source is driven by the film tension gradient—an example of the well-known Marangoni effect.

If the oil is solubilized in the surfactant monolayer and spreads directly as a monolayer, then it is appropriate to use the spreading pressure as defined by Equation 3.29. Here we remember that the spreading pressure, $\Delta\sigma^{sp}_{AL}$ (where AL represents a general air–liquid surface) is simply the reduction in the equilibrium surface tension of the monolayer by the oil. The surface tension gradient is therefore $\mathrm{d}\Delta\sigma^{sp}_{AL}/\mathrm{d}y$. It is also appropriate to use the spreading pressure in the case of spreading from a solid where either a mixed monolayer is formed or the spreading material is immiscible with the adsorbed surfactant monolayer so that it totally displaces the monolayer. Since the driving force for Marangoni spreading is the surface (or film) tension gradient, Ewers and Sutherland [38] argued that the greater the magnitude of S^i (or $\Delta\sigma^{sp}_{AL}$), the greater the force and the more effective the antifoam.

This approach of Ewers and Sutherland [38] was advanced with some claim to wide generality. However, as we will show here (see Sections 4.5, 4.7, and 4.8), many substances yield antifoam effects without causing destabilizing surface tension gradients [9, 12, 39, 40–44], which confounds exclusive generality. Moreover, a difficulty with the mechanism concerns the implicit assumption that the antifoam is always to be found in the thinnest and most vulnerable part of the foam. Thus, if the antifoam entity spreads from the Plateau border, this will drag liquid into the adjacent foam films, which will stabilize those films by increasing their thicknesses.

Another limitation of the approach of Ewers and Sutherland [38] is that it is essentially qualitative. However, there have been some attempts to put it on a quantitative basis [5, 45, 46]. The first such attempt was due to Shearer and Akers [5], who were concerned with the antifoam effect of PDMSs on lube oils. Here the effect of antifoam on a monolayer raft of bubbles is modeled theoretically and the results compared with experiment.

In this approach, spreading from drops of antifoam on the foaming liquid surface is supposed to augment gravity drainage from foam films. This gravity drainage is modeled by assuming that the boundaries of the foam films are immobile, in that Poiseuille flow prevails (with a parabolic velocity profile). A schematic illustration of the model where bubbles are supposed to form a raft at the surface of the foaming liquid is shown in Figure 4.11a. Shearer and Akers [5] calculate that the time t_g taken for a thick foam film surrounding a bubble to drain to the thickness h_{crit} at which rupture occurs, is

$$t_g \sim \frac{18\eta_L r_b}{\rho_L g h_{crit}^2} \tag{4.2}$$

Here ρ_L is the density of the foaming liquid and r_b is the radius of the bubble. Equation 4.2 implies that the time t_g is independent of the height of the film above the planar liquid surface at which the bubble is resting. In actuality, the film would become relatively thin at the top of the bubble. Failure to account for this effect is a consequence of incomplete consideration of continuity by Shearer and Akers [5] in the derivation of Equation 4.2.

The treatment of film drainage due to spreading given by Shearer and Akers [5] is summarized here to illustrate the essentials of the argument. Thus, consider one PDMS antifoam oil drop of radius r_{AF} at the center of the foam film as indicated in

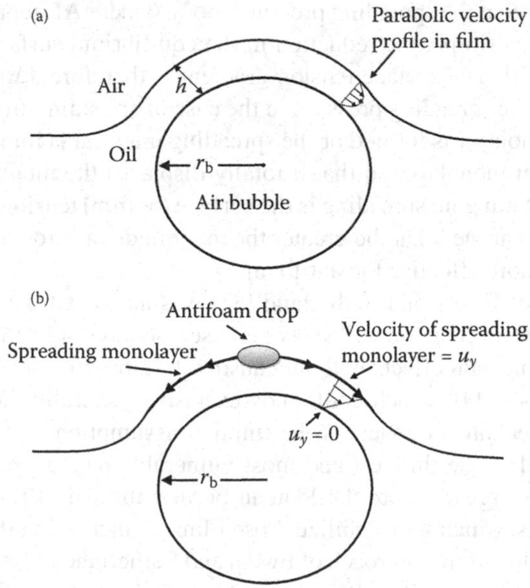

FIGURE 4.11 Model of Shearer and Akers describing effect of gravity drainage and spreading from antifoam drop on film thinning. (a) Foam film drainage under gravity from film separating bubble from air. (b) Foam film drainage induced by spreading from antifoam drop. (From Shearer, L.T., Akers, W.W. *J. Phys. Chem.*, 62, 1264 and 1269, 1958.)

Figure 4.11b. Duplex film spreading is assumed. The total spreading force at the perimeter of the oil drop is assumed to be $2\pi r_{AF}S^i$. It is also assumed that S^i becomes zero at the edge of the bubble in the plane of the liquid surface so that a linear gradient in S^i exists with respect to radial distance from the antifoam drop. The shear stress χ is therefore crudely approximated as

$$\chi = \frac{2r_{AF}S^i}{k_a r_b^2} \tag{4.3}$$

where k_a is a geometrical constant that describes the area of the bubble surface above the plane of the liquid surface. This model of the shear stress obviously ignores one of the essential features of the spreading process, namely that the spreading front will expand over the substrate and the shear stress will decline with time. It will therefore tend to exaggerate the effects of spreading.

Shearer and Akers [5] argue that a linear velocity gradient u_y/h will exist in the film due to the imposed shear stress χ which is therefore equal to $\eta_L u_y/h$. Here h is the film thickness and u_y is the velocity of the spreading surface (which is assumed to be constant throughout drainage). The opposite surface of the film is assumed to be stationary where $u_y = 0$ (see Figure 4.11b). Therefore, it is possible to deduce from Equation 4.3 that

$$u_y = \frac{2r_{AF}S^i h}{k_a \eta_L r_b^2} \tag{4.4}$$

Shearer and Akers [5] use a crude averaging procedure to allow for continuity in the derivation of Equation 4.2. They also use a similar procedure to deduce from Equation 4.4 that the time t_a taken for to cause a thick film to thin down to the critical rupture thickness h_{crit} is given by

$$t_a = \frac{k_b \eta_L r_b^3}{2r_{AF}S^i h_{crit}} \tag{4.5}$$

where k_b is a constant. Using parameters relevant for typical polydimethylsiloxane–lube oil combinations, Shearer and Akers [5] show that $t_a \ll t_g$, so that any contribution of gravity drainage to t_a may be neglected. Indeed, their calculations suggest that t_a/t_g could be of order 10^{-3}. If the crude assumptions upon which they are based are not too misleading, then this would imply that spreading from PDMS drops could significantly increase the rate of foam film rupture in lube oils.

Shearer and Akers [5] allow for the possible presence of several drops of antifoam in a foam film by simply increasing the supposed spreading perimeter to $2\pi N(r_f)r_{AF}$, where $N(r_f)$ is the number of drops per film which will be some function of the film radius r_f. If, however, the drops are evenly dispersed over the surface of foam films, then spreading will occur in opposing directions. Flow of liquid from the film will be less than that predicted if the spreading perimeter is simply increased to $2\pi N(r_f)r_{AF}$.

Equations 4.2 and 4.5 may be combined to give

$$\frac{t_a S^i}{t_g \rho_L} = \frac{k_c r_b^2 h_{crit}}{2N(r_f)_{AF}} \tag{4.6}$$

where k_c is a constant and where allowance is made for the presence of n drops in the film. The group $t_a S^i / t_g \rho_L$ is seen to be independent of the properties of the foaming liquid provided the bubble size can be controlled and h_{crit} is invariant. Indeed, log/log plots of this group against antifoam concentration for three different lube oils with a given PDMS were shown by Shearer and Akers [5] to lie on the same curve. Here t_g and t_a were obtained experimentally by assessing the half-life of bubble rafts. All of this is offered as experimental evidence for the validity of Equation 4.6 and the model from which it is derived. In view of all the assumptions implicit in the approach of Shearer and Akers [5], this experimental accord with theory is remarkable.

Prins [45] has adopted a different approach to the theoretical description of the effect of spreading from heterogeneities on foam film stability. In contrast to Shearer and Akers [5], he recognizes that the spreading process involves a spreading front that advances over the substrate surface. This process has been the subject of a number of independent studies as we have seen in the context of the duplex film spreading of PDMS oils (see Section 3.6.2.2). Thus, Fay [47] and Hoult [48] show from dimensional analysis that an oil (for which $S^i > 0$) will spread as a duplex film over a clean water surface in a linear manner at a rate given by Equation 3.37, which in the interest of clarity we repeat here

$$y = a_s \left(\frac{S^i}{\eta_W^{0.5} \rho_W^{0.5}} \right)^{0.5} t^{3/4} \tag{4.7}$$

where again y is the spreading distance, a_s is a constant which has a value ranging from 1.15 (representing spreading as a longitudinal wave [49]) to 1.33 [48, 50] for linear spreading and 1.15 for radial spreading [51]. η_W and ρ_W are, respectively, the viscosity and density of water. The rate of spreading is seen to be dependent on S^i but is independent of the viscosity of the oil (see Section 3.6.2.2).

As we have seen, the spreading antifoam oil will cause the underlying liquid to move in the same direction. If the depth to which this movement penetrates is δ, we can rewrite Equation 4.1 as

$$\frac{S^i}{y} = \eta_W \frac{dy/dt}{\delta} \tag{4.8}$$

provided the oil spreads as a duplex film (otherwise the surface tension gradient becomes $\Delta\sigma_{AW}^{sp}/y$). Substituting Equation 4.7 in Equation 4.8 using the value $k_s = 1.15$ given by Joos and Pintens [49], yields for the penetration depth

$$\delta = \left(\frac{\eta_W t}{\rho_W} \right)^{1/2} \tag{4.9}$$

Prins [45] supposes that foam film collapse will occur when the penetration depth of a spreading duplex film equals the thickness of the foam film over which it is spreading. Clearly the depth will increase with time (Equation 4.9), and so some estimate of maximum time available is necessary to assess the maximum penetration depth, δ_m. Prins [45] estimates the maximum extent y_m to which the spreading may occur in a radial sense from

$$\frac{4\pi r_{AF}^3}{3} = \pi y_m^2 h_m \tag{4.10}$$

where h_m is the thickness of the spreading film at the maximum extent and r_{AF} is the radius of the oil drop. The time of spreading is then obtained somewhat approximately by substituting y_m into Equation 4.7 using $a_s = 1.33$ for longitudinal wave spreading. Thus, Prins [45] combines Equations 4.7, 4.9, and 4.10 to obtain the maximum penetration depth

$$\delta_m \approx r_{AF}\left(\frac{\eta_W^2}{S^i \rho_W h_m}\right)^{1/3} \tag{4.11}$$

If foam film collapse occurs when the penetration depth equals the thickness of the film, then the larger the value of δ_m the thicker the films that may be ruptured. Prins [45] therefore equates high values of δ_m with high probability of foam film rupture. Equation 4.11 therefore implies that antifoam effectiveness will increase with increasing size of the antifoam drop. At least in this respect, there is agreement with Shearer and Akers [5], although for different reasons (see Equation 4.6). Prins [45] has shown that increasing drop size can increase antifoam effectiveness for small drop sizes. However, this does not continue indefinitely because as the drop size becomes large the probability of an antifoam entity being present in a foam film declines. An optimum particle size for high antifoam effectiveness is therefore obtained. Experimental results for soybean oil antifoam in sodium caseinate solution illustrate this behavior and are reproduced in Figure 4.12.

The crude nature of the model presented by Prins [45] gives rise to an awkward feature in Equation 4.11. Thus, if $S^i = 0$, then we appear to have $\delta_m = \infty$. Equation 4.11 in fact means that antifoam effectiveness is predicted to increase as S^i decreases. Thus, Prins [45] presents a model of antifoam action, based on spreading driven by surface tension forces, which predicts increasing antifoam effectiveness as those forces become weaker. Intuition would suggest an opposite conclusion (which is stated by Ewers and Sutherland [38] and deduced by Shearer and Akers [5]).

We are therefore left without an entirely satisfactory mathematical treatment of the antifoam mechanism proposed by Ewers and Sutherland [38]. Clearly the problem is difficult. However, a solution would permit more serious assessment of the significance of the mechanism particularly with regard to the effect of the magnitude of S^i, or $\Delta\sigma_{AW}^{sp}$. Such a solution should take account of the effect of compression of the surfactant monolayer at the foaming liquid–air surface by the spreading front. This

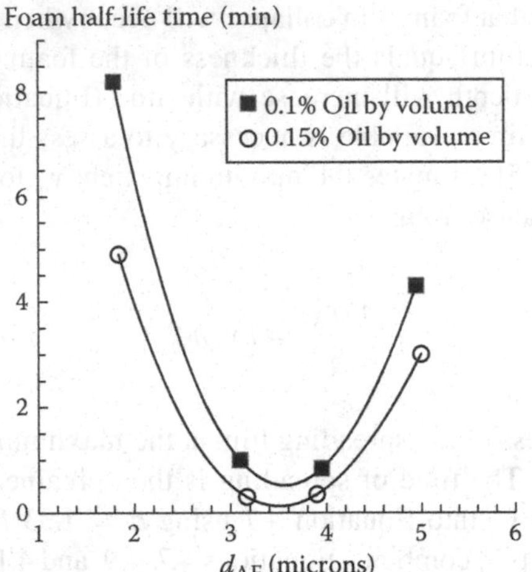

FIGURE 4.12 Foam half-life time of 0.05 wt.% sodium caseinate solutions as function of mean drop diameter d_{AF} of soybean oil. (From Prins, A. Theory and practice of formation and stability of food foams, in *Food Emulsions and Foams* (Dickinson, E., ed.), Royal Society of Chemistry Special Publication 58, 1986, p 30. Reproduced by permission of The Royal Society of Chemistry.)

effect will clearly reduce the rate of spreading and will diminish the threat to foam film stability. Rapid transport of surfactant in solutions of high concentration would, however, be expected to minimize such an effect on spreading rates, as we have discussed in Section 3.7. Prins [46] has, in a later version of the theory described by Equations 4.7 through 4.11, in fact explicitly introduced this factor so that the penetration distance δ_m is dominated by the properties of the longitudinal wave caused by compression of the surfactant monolayer by the spreading layer. Using the expression for the velocity of propagation of such a wave given by Lucassen [52], together with Equation 4.9, Prins [46] deduces that the maximum penetration depth is

$$\delta_m \approx 1.2 r_{AF} \left(\frac{\eta_w^2}{\varepsilon \rho_w h_m} \right)^{1/3} \tag{4.12}$$

which is similar to Equation 4.11 except that the surface dilatational modulus, $\varepsilon = d\sigma_{AW}/d\ln \tilde{A}_{AW}$ where $\tilde{A}_{AW}$ is the area of the air–water surface) is seen to replace S^i. This modulus then replaces the spreading coefficient (or pressure) as the dominant factor and avoids the awkward feature of Equation 4.11 to which we have drawn attention. Equation 4.12 of course introduces another awkward feature in that any measure of the magnitude of the surface tension forces, which drive the putative process, is absent. Moreover, introducing the surface dilatational modulus using the arguments of Lucassen [52] ignores the assumption of the latter concerning the nature of the compressed surface upon which the putative longitudinal wave is propagating.

The treatment of Lucassen [52] assumes that the surface is purely elastic whereas soluble surfactants are viscoelastic. Therefore, Equation 4.12 is not strictly relevant for such solutions. We should emphasize that a requirement for formation of foam is that the relevant surfactants be soluble.

There are other problems with the mechanism. As we have stated, spreading from antifoam entities present in the plateau borders could in principle reinforce foam film stability. Spreading from many different sources in the surface of a foam film could presumably rapidly eliminate surface tension gradients so that rupture induced by surface-tension-driven flow would become unlikely. However, elimination of surface tension gradients can also, in principle, contribute to diminished foam film stability as we show in Section 4.4.3.

Some observations are clearly inconsistent with the mechanism. For example, Kulkarni et al. [53] describe an experiment where a drop of PDMS is added to foam formed from sodium lauryl sulfate in a tray. This caused no foam destruction despite the obvious inference that spreading from the drop would occur, which would in turn cause collapse if the spreading mechanism is effective in this context.

Yet another problem with this mechanism concerns the requirement that the spreading coefficient be positive. As we have discussed at length in Chapter 3, this situation can only occur in non-equilibrated systems because at equilibrium we must have $S^e \leq 0$ (and $\Delta\sigma^{sp}_{AW} = 0$). Of course during foam generation non-equilibrium conditions are expected if for no other reason than that transport of surfactant will not be fast enough to maintain equilibrium air–liquid surfaces. However, after both foam generation and subsequent drainage have ceased, air–water surface tensions should approach equilibrium values and any antifoam action by Marangoni spreading should therefore also cease.

We therefore have a mechanism that relies upon a necessarily transient process. However, materials that have the relevant properties, such as spreading oils, will also be capable of participating in other mechanisms such as the formation of unstable bridging oil configurations in foam films (see Sections 4.5 and 4.8). Unequivocal experimental verification of a role for Marangoni spreading as an antifoam mechanism will therefore represent a formidable challenge. In this regard, Denkov writes [1]:

> Our optical observations, with various systems [7, 54–62] have never revealed a spreading-fluid entrainment type of foam film destabilization (*i.e., Marangoni spreading mechanism*, as drawn in Figure 4.10). In the experiments with silicone oil-based compounds [54], the main observations were in a clear contradiction with this mechanism: we observed rapid rupture of the foam films, when the film surfaces were covered with a prespread layer of oil (hence, spreading and fluid entrainment in the moment of globule entry did not occur). Furthermore, when the film surfaces were free of prespread oil, spreading was indeed observed from the antifoam globules after their entry, but no intensive fluid entrainment was seen and the foam films were very stable.*

* Quoted with permission from Denkov [1]. Copyright 2004 American Chemical Society.

4.4.3 Elimination of Surface Tension Gradients

Kitchener [63] has argued that the role of "a completely effective antifoam agent must be to eliminate surface elasticity, i.e., to produce a surface which has substantially constant tension when subjected to expansion." In effect, this is a corollary of the argument of Lucassen [64] (which we have outlined in Chapter 1) concerning the role of surface tension gradients in stabilizing foam films against external stress. Essentially it means that in the absence of a surface tension gradient, the viscous stress at the gas–liquid surface of, for example, a draining foam film will not be balanced by an opposing force. This would lead to an accelerated plug flow and film collapse analogous to that which occurs in films of pure fluids. A similar argument has been made by Burcik [65]. Ross and Haak [11] have also applied a variation of it to interpretation of putative antifoam behavior in homogeneous systems. Indeed it has been argued that foamability should decline at extremely high surfactant concentrations because of elimination of surface tension gradients due to rapid transport of surfactant to the surfaces of foaming liquids. Relative absence of examples of such behavior suggest, however, that perhaps others factors are also important (see Chapter 1). Here then we examine the application of this argument concerning elimination of surface tension gradients to heterogeneous systems; that is, systems where undissolved antifoam materials are present.

A fundamental difficulty with this proposed antifoam mechanism is that it implies significant change in the properties of the gas–liquid surface of the foaming liquid derived from the spreading of material from dispersed drops or particles. As we have described above, the process of spreading can itself in principle cause foam film collapse as a result of the shear forces exerted by the surface tension gradient driven by the spreading process. Uniquely ascribing any antifoam effect to elimination of those gradients would therefore be difficult in view of this alternative possibility and especially since there are yet other possibilities involving unstable bridging configurations of oil drops or particles in foam films (see Sections 4.5, 4.7, and 4.8).

One suggestion for elimination of surface tension gradients concerns duplex film formation by certain oils. Here we remember that the spread oil layer is relatively thick so the oil–foaming liquid and oil–gas surfaces have the same properties as the surfaces of the respective bulk liquids. Formation of such duplex films is by no means ubiquitous with antifoam oil–surfactant solution combinations. They are, however, for example, common in the case of PDMS oils and aqueous surfactant solutions (see Table 3.3). The oil–air surface of such films will not exhibit surface tension gradients because pure oils are involved. However, the adsorbed surfactant at the oil–water surface of such duplex films will mean that surface tension gradients are sustained at that interface. It would seem unlikely therefore that any antifoam effects [66] attributed to duplex film formation from dispersed oil drops are due to elimination of surface tension gradients but rather to some other cause such as Marangoni spreading or more likely unstable bridge formation, as we will outline below (see Section 3.5.1).

In this context, we note that Ross and Nishioka [67] have examined the stability of air bubbles (of 0.8 cm radius) released under "monolayers or multilayers" of

PDMS oil spread on both distilled water and surfactant solution. With a distilled water substrate, bubbles were more stable than in the absence of PDMS (presumably as a result of reduced film draining rates due to lack of mobility of the PDMS-contaminated surface). With solutions of SDS and $C_{16}TAB$, the presence of a spread layer of PDMS had little effect on bubble stability. Ross and Nishioka [67] therefore deduce that "since a spread layer of polydimethylsiloxane (PDMS) has no destructive effect on bubbles, we conclude that PDMS does not defoam by replacing adsorbed solute at the surface as has been suggested." At this point the authors cite Roberts et al. [68]. Unfortunately, Ross and Nishioka [67] do not clearly characterize these spread layers. However, PDMS is known to form duplex films (i.e., complete wetting) on micellar solutions of $C_{16}TAB$ as we have listed in Table 3.3.

A hypothetical system, which could provide a test of this putative antifoam mechanism, would consist of surface-active solid particles that adhere to the gas–liquid surface of the foaming liquid and spread to form an insoluble monolayer. Suppose then that this monolayer is immiscible with the monolayer of the adsorbed foam stabilizing soluble surfactant. Finally suppose that the spreading pressure of the insoluble monolayer forming surfactant is greater than the surface pressure of the adsorbed surfactant. In this case the insoluble monolayer forming surfactant will spread from particles to displace the soluble surfactant completely. If the gas–liquid surface is now subject to shear, any tendency to set up an opposing surface tension gradient would be in principle eliminated by further spreading from the adhering particles provided that process is sufficiently rapid. This would, however, depend on the rate of spreading from particles compared to the relative rate of gas–liquid surface increase. In the absence of a surface tension gradient the surface would be fully mobile, moving with the same velocity as the underlying fluid. In effect the surface would be expanding at a rate $dA/dt = u_y L$ in the case of a rectangular film where A is the film area, u_y is the surface velocity in the y direction, and L is the width of the film. If m is the total number of insoluble monolayer molecules present on the surface of the film, then we must have $m = \Gamma A$, where Γ is the number of such molecules per unit area. If the surface tension is to be constant, then Γ must also be constant as the surface expands. Therefore, we must have a rate of supply of insoluble monolayer molecules such that

$$\frac{dm}{dt} = \Gamma \frac{dA}{dt} = \Gamma u_y L \tag{4.13}$$

if the surface tension is to remain constant. The supply of such molecules in this model derives from dispersed particles of solid adhering to the gas–liquid surface. The spreading rate, dm_s/dt, from the particles is given by Mansfield [69] as

$$\frac{dm_s}{dt} = K_s S_{per}\left(\Delta\sigma_{AW}^{sp} - \pi_{film}\right) = -K_s S_{per}\left(\sigma_{AW}^{e} - \sigma_{film}\right) \tag{4.14}$$

where K_s is a constant, S_{per} is the perimeter of the solid particles in the film element of area A, $\Delta\sigma^{sp}_{AW}$ is the equilibrium spreading pressure at the air–water surface, and σ^{e}_{AW} is the corresponding surface tension. π_{film} and σ_{film} are the steady-state surface pressure and corresponding surface tension of the foam film, respectively. It is of course obvious that spreading will only occur when $\sigma_{film} > \sigma^{e}_{AW}$. Moreover, as pointed out by Mansfield [69], there will be a gradient in surface tension away from the particles. Supposing, however, that the particles are well distributed over the surface of the film so that such gradients are small and only exist between particles but not along the direction of film drainage. Supply of insoluble monolayer surfactant from the particles will then hold the film surface tension sensibly constant in a steady state where

$$\frac{dm_s}{dt} = \frac{dm}{dt} \tag{4.15}$$

We may combine Equations 4.13 through 4.15 to obtain an expression for the steady-state surface tension of the film where spreading from particles essentially eliminates any surface tension gradient induced by the action of shear accompanying overall film drainage. We can therefore write

$$\sigma_{film} - \sigma^{e}_{AW} = \frac{\Gamma u_y L}{K_s S_{per}} \tag{4.16}$$

where it is clear that the higher the velocity at the surface u_y, the higher the difference between the film surface tension σ_{film} and the equilibrium surface tension of the monolayer σ^{e}_{AW} or the greater the total perimeter S_{per} of the particles for a given $\left(\sigma_{film} - \sigma^{e}_{AW}\right)$. However, we must remember that all this requires that σ_{film} be less than the (dynamic) surface tension determined by the soluble surfactant in the absence of particles of the insoluble monolayer surfactant. Otherwise, the latter would dominate the gas–liquid surface, allowing surface tension gradients to develop, sustained by the adsorption and transport properties of such surfactants.

Unambiguous practical realization of the antifoam situation suggested by Equation 4.16 has not apparently been reported. This is perhaps not surprising because it seems unlikely that the basic assumption of total monolayer immiscibility between the insoluble monolayer surfactant and the soluble foaming surfactant will be realizable. Transport of soluble surfactant to surfaces would mean that mixed monolayer formation could permit surface tension gradients to develop irrespective of the spreading behavior of the particles of insoluble monolayer forming surfactant. It is also worth noting that if spreading from multiple sources in a foam could eliminate surface tension gradients, this would also mean that the mechanism of Marangoni spreading outlined above (Section 4.4.2) could not occur. The latter requires that spreading occurs in foam films, thereby catastrophically enhancing film drainage locally—opposed spreading from sources in the Plateau border, for example, would tend to neutralize such an effect. It is clear then that antifoam action by elimination of surface tension gradients and antifoam action by Marangoni spreading are mutually incompatible mechanisms.

4.5 OIL BRIDGES AND ANTIFOAM MECHANISM

4.5.1 Oil Bridges in Foam Films

It is generally accepted that if oils, in the form of dispersed drops, have the potential to act as antifoams, they must be able to enter (or "emerge into" or "adhere to") the surfaces of foaming liquids. The classic entry coefficients, E^{i}, E^{si}, and E^{e}, should all therefore be >0. However, a positive value for a classic entry coefficient is not a sufficient condition for entry. As we have described in Section 3.3, there always exists the possible formation of metastable pseudoemulsion films, which can prevent oil drops from entering the surfaces of foaming liquids. Formation of such metastable films is a consequence of the presence of positive disjoining pressures, which are explicitly accounted for in the generalized entry coefficient E_g, defined by Equation 3.19. Entry of oil drops into the surface of foaming liquids therefore often requires application of some external pressure to overcome the disjoining pressure in the pseudoemulsion film. In the case of an oil drop in a foam film, that will be the capillary pressure p_c^{PB} exerted by the Plateau border. Entry will then occur if $p_c > \Pi_{ALO}(h_o)$, where h_o is the thickness of a putative pseudoemulsion film corresponding to the highest disjoining pressure $\Pi_{ALO}(h_o)$ in the relevant disjoining pressure isotherm. We have seen, however, that there is evidence that rupture may occur at lower capillary pressures than this consideration would suggest (see Section 3.4). The presence of particles adhering to the interface between the oil and the foaming liquid represents another approach to destabilizing metastable pseudoemulsion films. In fact, formulation of antifoams for aqueous surfactant solutions, which tend to readily produce metastable pseudoemulsion films, normally involves incorporation of such particles. We return to this topic in Section 4.8. Here we are simply concerned with the possible mode of action of antifoam oils in the case where any tendency to form metastable pseudoemulsion films is eliminated by whatever means.

As we have shown in Section 3.6, oils that enter the air–water surfaces of surfactant solutions can adopt three different types of behavior at equilibrium—***complete wetting*** as duplex oil films, ***pseudo-partial wetting*** where the oil contaminates the air–water surface to form mixed films in equilibrium with excess oil as lenses, and ***partial wetting*** where the air–water surface is uncontaminated by the oil which also forms lenses. There is no evidence that antifoam action is confined exclusively to oils that exhibit only one or even two of these types of wetting behavior. There is, however, evidence that the presence of oil films formed as a result of complete or pseudo-partial wetting can give rise to certain enhanced antifoam effects such as destabilization of pseudoemulsion films (see Sections 3.6.3 and 4.8.4) [1]. These effects are of a "second-order" nature since examples exist of oil–surfactant combinations where antifoam action is associated with complete wetting [70], pseudo-partial wetting [1, 54–62] through to partial wetting [43, 71]. Indeed, antifoam action may persist even if the wetting behavior of particular oils is varied by changes in surfactant type and concentration, as is revealed by the claims of manufacturers of antifoams. Ideally, then, a theory of antifoam action by entering oils should be general and not require a specific type of wetting behavior. Here then we explore the possibility that antifoam action concerns formation of unstable oil bridges across foam films. Such bridges could in principle be formed by oils that exhibit any of the possible types of wetting behavior. For the sake of simplicity, we will consider only aqueous foaming liquids but emphasize that the arguments could be equally applied to non-aqueous liquids as we describe in Chapter 10.

We consider first the stability of oil bridges formed from oil lenses where the latter exist if the classic equilibrium spreading coefficient, S^e, is <0 (regardless of whether pseudo-partial or partial wetting is involved). The presence of such lenses in foam films of thickness less than a few microns implies that they must be of dimensions such that capillarity is the dominant force so that the gravitational force can be neglected [1]. The configuration of such an oil lens at the air–water interface is depicted in Figure 4.13. Neglect of the gravitational force means that the air–water surface is planar and the air–oil and oil–water surfaces are both spherical segments. At equilibrium, the capillary pressures across both the oil–water and air–oil surfaces should be equal. The angles formed by the three surfaces at the three-phase contact line are determined by Neumann's triangle of forces [72], as indicated in the figure. Here θ^*_{OA} is the angle made by the air–oil interface with respect to the plane of the air–water interface. Similarly θ^*_{OW} is the angle made by the oil–water interface with respect to that plane and measured through the oil phase. We also define the angle $\theta^* = (180° - \theta^*_{OW})$ which has, as we will see, special significance with respect to antifoam action.

Neumann's triangle and application of the cosine rule means that we can write [72], for example, that

$$\cos\theta^*_{OW} = \frac{1 + \left(\sigma_{OW}/\sigma_{AW}\right)^2 - \left(\sigma_{AO}/\sigma_{AW}\right)^2}{2\sigma_{OW}/\sigma_{AW}} \tag{4.17}$$

where σ_{AO}, σ_{AW}, and σ_{OW} are the oil–air, the air–water, and the oil–water surface tensions prevailing under the given conditions, respectively. Here we must emphasize that the surface tensions should have values so that mechanical equilibrium can occur. Therefore, it cannot be satisfied under initial conditions where, for example, $S^i > 0$. Resolving Neumann's triangle in the perpendicular direction means that we can also write

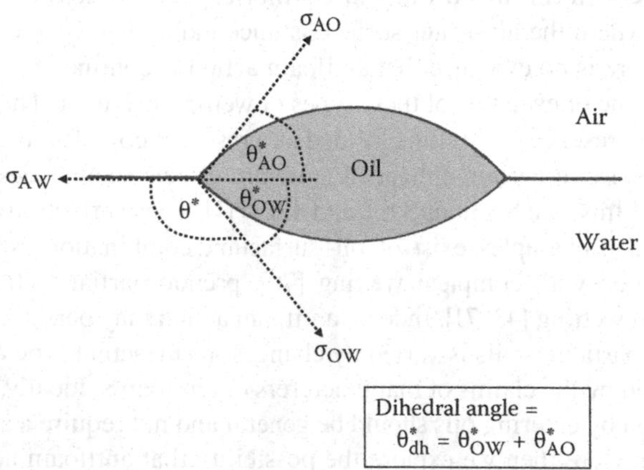

FIGURE 4.13 Neumann's triangle and oil lens at air–water surface. Definition of symbols given in text.

$$\sigma_{OW} \sin\theta^*_{OW} = \sigma_{AO} \sin\theta^*_{AW} \tag{4.18}$$

If such an oil lens is present in a foam film, then there must exist the possibility that it can bridge the foam film to form configurations similar to those depicted in Figure 4.14 where z is the axis of revolution. Figure 4.14a depicts a configuration where $\theta^* > 90°$ and Figure 4.14b depicts a configuration where $\theta^* < 90°$. In both cases, the aqueous film is assumed to be plane parallel. If we are to establish whether those bridging configurations are unstable, all that is necessary is to establish that one of the conditions necessary for equilibrium cannot be satisfied. The first analysis of this problem was made by Garrett [73] more than 30 years ago.

The assumption that the aqueous foam film is plane parallel avoids introducing a pressure gradient in the film because the Laplace pressure jump is therefore everywhere zero. The resulting condition for equilibrium of the bridging drop is then only that the capillary pressure drop across the oil–water surface should equal that across the air–oil surface as is the case for a lens. However, a later treatment by Denkov [55] relaxes this requirement that the film be plane parallel. Here we consider the simpler plane-parallel case first before considering the latter.

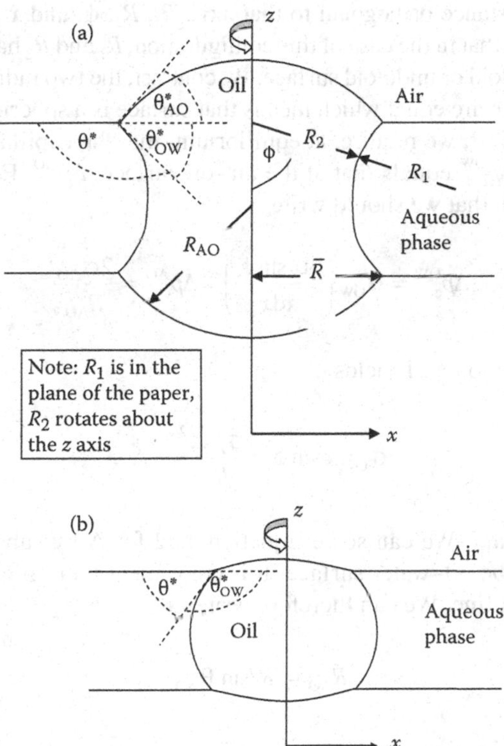

FIGURE 4.14 Bridging oil drops in plane-parallel foam film. (a) Configuration where $\theta^* > 90°$. (b) Configuration where $\theta^* < 90°$.

The capillary pressure Δp_c across a surface of surface tension σ_i is given in a general case by the Laplace equation

$$\Delta p_c = \sigma_i \left(\frac{1}{R_1^g} + \frac{1}{R_2^g} \right) \tag{4.19}$$

where Δp_c is constant at equilibrium over a given surface and where R_1^g and R_2^g are the two orthogonal radii of curvature, which are required to characterize the shape of the surface in the general case. If $R_1^g = R_2^g$, then the surface is a spherical segment. If, however, $R_1^g \neq R_2^g$, the surface has the form of a nodoid or unduloid provided Δp_c is both finite and constant. The latter will be true only in the absence of a perturbing external force such as gravity.

For the specific case of a bridging oil drop where $\theta^* > 90°$, depicted in Figure 4.14a, simple geometrical considerations [72] mean that we can write

$$\frac{1}{R_1} + \frac{1}{R_2} = \frac{\mathrm{d}\sin\phi}{\mathrm{d}x} + \frac{\sin\phi}{x} = \frac{\mathrm{d}x\sin\phi}{x\mathrm{d}x} \tag{4.20}$$

where ϕ is the angle made by the normal to the oil–water surface at the axis of revolution and x is the distance orthogonal to that axis. R_1, R_2, ϕ, and x are defined in the Figure. It is obvious that in the case of this configuration, R_1 and R_2 have different signs, which implies a nodoid or unduloid surface. By contrast, the two radii of curvature R_{AO} of the air–oil surface are equal which means that surface is a spherical segment.

For all values of θ^*, we require, at equilibrium, that the capillary pressure at the oil–water surface Δp_c^{OW} equals that at the air–oil surface, Δp_c^{AO}. Equations 4.19 and 4.20 therefore mean that we should write

$$\Delta p_c^{OW} = \sigma_{OW} \left(\frac{\mathrm{d}x\sin\phi}{x\mathrm{d}x} \right) = \Delta p_c^{AO} = \frac{2\sigma_{AO}}{R_{AO}} \tag{4.21}$$

Integrating Equation 4.21 yields

$$\sigma_{OW}\, x\sin\phi = \frac{\sigma_{AO} x^2}{R_{AO}} + \Lambda \tag{4.22}$$

where Λ is a constant. We can solve Equation 4.22 for Λ by substituting for x and ϕ at any point on the oil–water surface. It is convenient then to take a point on the three-phase contact line. We can therefore write

$$R_{AO} = \bar{R}/\sin\theta_{AO}^* \tag{4.23}$$

$$x = \bar{R} \tag{4.24}$$

$$\phi = \theta_{OW}^* \tag{4.25}$$

where $\bar{R}$ is the radius of the bridging lens in the plane of the air–water surface (see Figure 4.14a). Combining Equation 4.18 with Equations 4.22 through 4.25 yields the result that $\Lambda = 0$ so that we can rewrite Equation 4.22 as

$$\frac{\sin\phi}{x} = \frac{\sin\theta^*_{\mathrm{OW}}}{\bar{R}} \tag{4.26}$$

which is the equation of a sphere [72]. Thus, we find that a state of equilibrium for a bridging drop can only exist if both the oil–air and the oil–water surfaces are spherical segments. Examination of Figure 4.14a, for which the angle $\theta^* > 90°$, clearly illustrates that the oil–water surface cannot adopt such curvature. Indeed a spherical shape can only be found if $\theta^* < 90°$ and then only for a unique combination of drop volume and film thickness [73]. Such a configuration is depicted in Figure 4.14b. We therefore find that unstable bridging configurations are to be found only if

$$90° < \theta^* \le 180° \text{ or } 0 \le \theta^*_{\mathrm{OW}} < 90° \tag{4.27}$$

We have shown in Section 3.3.1 that metastable pseudoemulsion films can exhibit finite contact angles so that $\theta^* > 0$. Therefore, oil bridges are possible where the oil–air surface is replaced by a metastable pseudoemulsion film. However, such angles are likely to be small with $\theta^* \ll 90°$. Lobo and Wasan [74], for example, give estimates of θ^* for an alkyl ethoxylate–octane system of <3° (<0.05 rad) so that condition 4.27 is not satisfied. Therefore, the proposition that formation of metastable pseudoemulsion films can eliminate antifoam action is upheld.

It is also easy to show [73] that in the case of a bridging lens where $\theta^* > 90°$, then we must have at the three-phase contact line

$$\Delta p_{\mathrm{c}}^{\mathrm{OW}} < \Delta p_{\mathrm{c}}^{\mathrm{AO}} \tag{4.28}$$

If the oil–water interfacial area is minimal, then the capillary pressure $\Delta p_{\mathrm{c}}^{\mathrm{OW}}$ should be constant where the interface therefore adopts either a nodoid or unduloid profile. As with the bridging duplex film-forming drop, the pressure imbalance represented by inequality 4.28 would then exist throughout the bridging lens representing a capillary force tending to move the oil–water interface away from the axis of revolution.

Equation 4.17 means that the Condition 4.27 for the realization of unstable configurations can also be expressed as

$$0 < \sigma_{\mathrm{AW}}^2 + \sigma_{\mathrm{OW}}^2 - \sigma_{\mathrm{AO}}^2 \le 2\sigma_{\mathrm{OW}}\sigma_{\mathrm{AW}} \tag{4.29}$$

because $\theta^*_{\mathrm{OW}} = \pi - \theta^*$ and where the upper limit represents duplex film formation. We can also introduce a so-called bridging coefficient B, defined by

$$B = \sigma_{\mathrm{AW}}^2 + \sigma_{\mathrm{OW}}^2 - \sigma_{\mathrm{AO}}^2 \tag{4.30}$$

so that unstable configurations are found if

$$0 < B \leq 2\sigma_{OW}\sigma_{AW} \tag{4.31}$$

where duplex film formation is included in the condition. The upper limit in the inequality corresponds to Antonow's rule where $\sigma_{AW} = \sigma_{OW} + \sigma_{AO}$, where the equilibrium spreading coefficient is zero and a duplex film is present. Any bridging coefficient calculated for a non-equilibrium case where the initial or semi-initial spreading coefficient is relevant and where either S^{i} or $S^{si} > 0$ is therefore meaningless not least because B is based on Neumann's triangle, which requires a condition of mechanical equilibrium between the relevant surface tensions.

As we have shown in Chapter 3 (Table 3.3), duplex film formation is likely to be a relatively common feature of antifoam oil–surfactant combinations where the oil is a PDMS. We therefore consider the implications of this behavior for foam film rupture by bridging in some, albeit qualitative, detail. That duplex oil film-forming drops, which bridge foam films, would also represent an unstable configuration was in fact first proposed by Ross [10] 60 years ago. The probable configurations of a foam film bridged by an oil drop in contact with duplex films on both sides of the film are depicted in Figure 4.15. The film is assumed to be plane parallel except in the vicinity of the drop. The configurations involve only the oil–water and air–oil surfaces, the surface tensions of which having their bulk phase values. Mechanical equilibrium of the bridging drop with plane-parallel duplex films only exists if the oil–water surface of the bridge has a catenoidal profile, which in turn can only occur if both radii of curvature, R_1 and R_2, defining the shape of the oil–water surface of the drop, are of equal size because they must have opposite signs. If, as argued by Denkov [1, 55], the drop forming the bridge is of similar diameter to that of the film, then this condition will not occur—we will have $R_1 > R_2$, where R_2 is the axial radius. The pressure within the bridge will then be greater than the ambient pressure and, as shown in Figure 4.15, the local air oil surfaces will become convex with respect to the plane of the drop equator. This configuration will not, however, represent complete mechanical equilibrium over the whole film—a capillary pressure discontinuity will exist at the edge of the perturbed air–oil surface. This configuration will not, however, prevail as the foam film drains under the influence of the capillary suction at the adjacent Plateau border. Conservation of volume of the drop will mean it will then stretch until $R_1 < R_2$ so that the capillary pressure in the drop becomes negative and the local air–oil surfaces become concave with respect to the drop equator as shown in the figure. The negative capillary pressure may produce a flow of oil from the duplex film into the bridging drop. Further thinning of the foam film increases the negative capillary pressure in the drop and increases the curvature of the local concave air–oil surfaces. Eventually, those two surfaces become close enough for van der Waals forces to cause rupture of the stretched bridging drop and therefore rupture of the foam will increase the curvature of the oil-water surface and therefore the capillary pressure in the aqueous foam film. This will augment stretching of the drop by driving the oil-water surface away from the rotational axis. Increasingly negative capillary pressures in the drop as the film thins will mean that the two local air-oil surfaces become more concave with respect to the drop equator. Eventually, those two surfaces will become close.

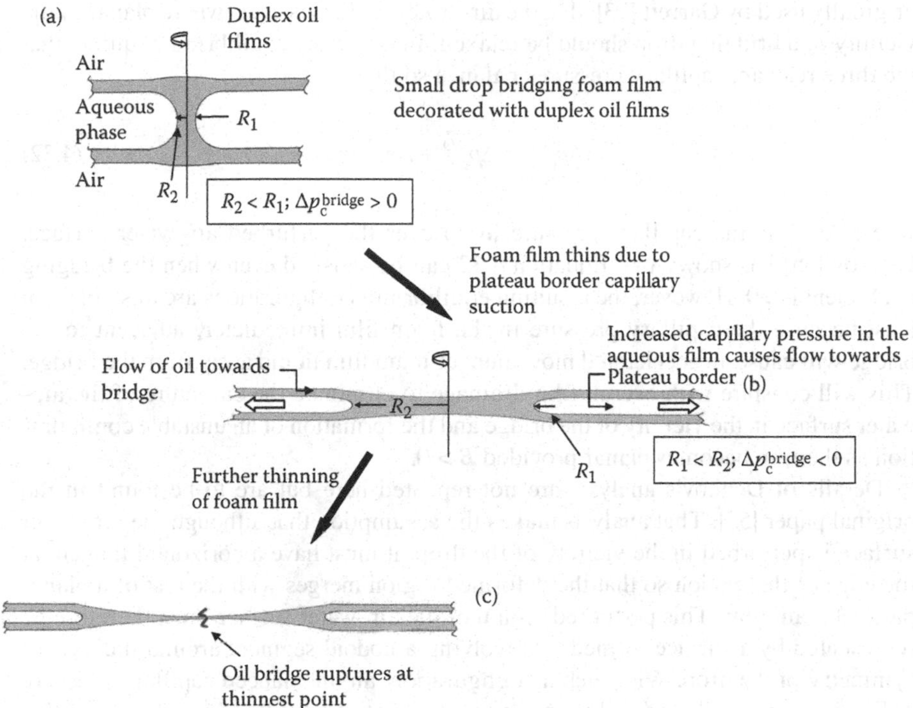

FIGURE 4.15 Schematic illustration (not to scale) of rupture of foam film by bridging by a duplex film-forming oil. (a) Drop forms a bridge where the difference between capillary pressure in bridge and atmospheric pressure, Δp_c^{bridge}, is > 0. Such a drop could potentially disproportionate into two parts, both of which simply merge with the respective duplex oil films covering the foam film. (b) Film thins until Δp_c^{bridge} changes sign—air–oil surfaces must then be concave with respect to equator of bridge. A sufficiently large drop would form such a configuration initially without requiring foam film thinning. Negative capillary pressure in drop will mean oil will be sucked into drop from duplex films as the equatorial radius of the drop expands. The increasing capillary pressure jump at the oil-water surface with increasing curvature as $R_1 \ll R_2$ drives that interface away from the axis of revolution, augmenting stretching of the drp. (c) After further thinning, the two concave air–oil surfaces approach sufficiently closely to cause rupturing.

We are therefore justified in including $B = 2\sigma_{OW}\,\sigma_{AW}$ within the range of unstable bridging configurations even though lens formation on air–water surfaces is not then involved. As suggested by Bergeron et al. [70], such bridging by duplex film-forming drops would be facilitated by high oil viscosities, where drops dewet into both air–water surfaces of a foam film already covered by a duplex film. The presence of hydrophobic particles adhering to oil–water surfaces, which rupture oil–water–oil films, could also facilitate bridging by duplex film-forming drops, as we will show in Section 4.8.

Denkov [55] has further developed the general argument concerning the stability of bridging oil drops in a detailed theoretical analysis. His experimental observations suggested that antifoam oil drops in foam films form a perturbed region of the foam film in the vicinity of a bridging drop [54]. It implies that assumption, originally used by Garrett [73], that the air–water surface is everywhere planar in the vicinity of a bridging drop should be relaxed. In this case, equilibrium requires that the three relevant capillary pressures balance so that

$$\Delta p_c^{OW} = \Delta p_c^{AO} + \Delta p_c^{AW} \tag{4.32}$$

where Δp_c^{AW} is the capillary pressure jump over the perturbed air–water surface. Denkov [55] has shown that Equation 4.32 can be satisfied even when the bridging coefficient is >0. However, the resulting equilibrium configurations are unstable, not least because the capillary pressure in the foam film immediately adjacent to the bridge will cause an accelerated movement of foam film liquid away from the bridge. This will conspire with overall film drainage to eliminate the curvature of the air–water surface in the vicinity of the bridge and the formation of an unstable configuration as the film becomes planar provided $B > 0$.

Details of Denkov's analysis are not repeated here but are to be found in the original paper [55]. That analysis makes the assumption that although the air–water surface is perturbed in the vicinity of the drop, it must have a horizontal tangent at the edge of that region so that the deformed region merges with the rest of a plane-parallel foam film. This perturbed region of the air–water surface must therefore be represented by a surface formed by revolving a nodoid segment around the axis of symmetry of the drop. With such a configuration, an unbalanced capillary pressure will exist between the nodoid segment and the rest of the film. An example of this type of configuration is depicted schematically in Figure 4.16, where R_{pr} is the radius of the perturbed air–water surface about the axis of revolution.

This analysis [55] reveals that formation of unstable bridges, where $B > 0$, is dependent on both θ* and the volume of the drop. The analysis indicates that the presence of very small drops can mean stable bridges especially for high values of θ*. The results of Denkov's calculations [55] for a numerical example are shown in

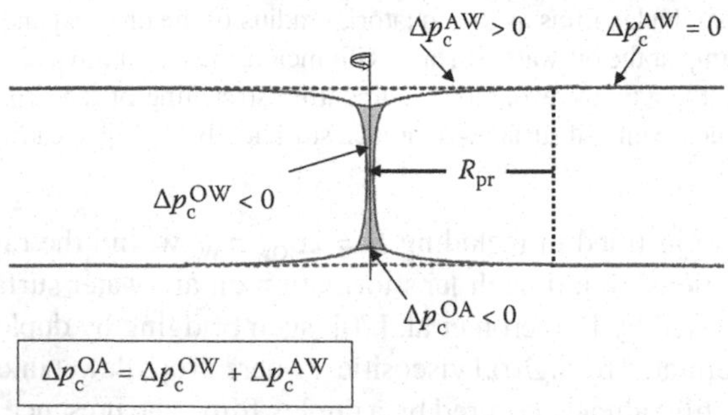

FIGURE 4.16 Example of equilibrium shape of bridging drop with perturbed and curved region of the air–water surface of radius R_{pr} adjacent to drop. Capillary pressure jumps are balanced as required by Equation 4.32 if equilibrium is to prevail. However, beyond the perturbed region, air–water surface is flat and film becomes plane parallel. Therefore, equilibrium between the region defined by radius R_{pr} and the remainder of foam film does not exist. (Reprinted with permission from Denkov, N.D. *Langmuir,* 15, 8530. Copyright 1999 American Chemical Society.)

Figure 4.17a. Here the ratio of the critical volume of a drop, V_B^{crit}, required to produce an unstable bridge, to the volume of a drop with the same diameter as the film thickness V_i is plotted against θ*. In this calculation, values of σ_{OA} and σ_{OW} are selected as characteristic of a PDMS oil in a micellar solution of an anionic surfactant and variations in θ* are made by changing values of σ_{AW}. If a drop of volume > V_B^{crit} and $B > 0$ is present in a film of a given thickness, then the bridging drop is unstable. However, where the drop volume is < V_B^{crit}, then, despite $B > 0$, the bridging drop is predicted to be stable. This is, however, a metastable situation because as the film thins, the ratio V_B^{crit}/V_i will increase until the drop becomes unstable because V_B^{crit} is constant

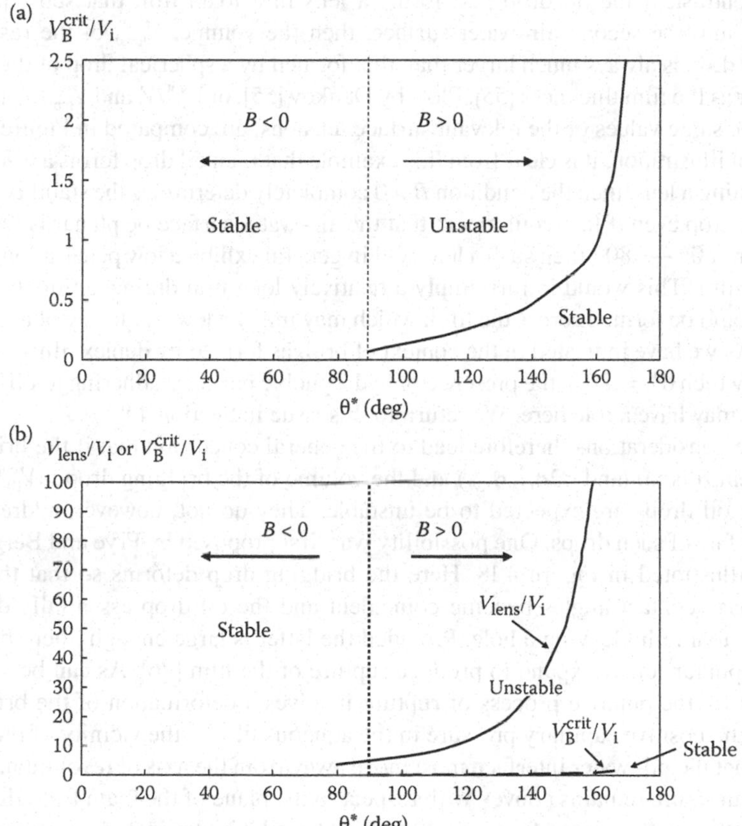

FIGURE 4.17 (a) Calculated regions of stable and unstable bridging drops as determined by ratio of critical bridge volume V_B^{crit} to volume V_i of a drop with the same diameter as the foam film thickness. V_B^{crit}/V_i is presented as a function of angle θ*. Calculations assumed $\sigma_{AO} = 20.6$ mN m^{-1}, $\sigma_{OW} = 4.7$ mN m^{-1}, foam film thickness 2 microns, and $R_{pr} = 100$ microns (where the latter is defined in Figure 4.16). (Reprinted with permission from Denkov, N.D. *Langmuir,* 20, 9463, 2004. Copyright 2004 American Chemical Society.) (b) Comparison of V_B^{crit}/V_i with ratio V_{lens}/V_i where V_{lens} is the volume of an oil lens that penetrates into the foam film by an amount equal to the thickness of that film. (Reprinted with permission from Denkov, N.D. *Langmuir,* 15, 8530. Copyright 1999 American Chemical Society.)

and $V_i = \pi h^3/6$, where h is the film thickness. Clearly the significance of metastable bridges predicted for drops where $V_B^{crit}/V_i < 1$ is purely notional because such drops would be too small to form a bridging configuration anyway. However, high values of θ^* are seen to produce situations where $V_B^{crit}/V_i > 1$ and where in principle a bridging drop with $B > 0$ could be formed in metastable mechanical equilibrium, albeit with a deformed air–water surface. Such drops could, however, become unstable, not only as the aqueous foam film thins, but also as a result of supply of extra oil if a thin oil film is present at the air–water surface. Thus, such a film can not only directly supply extra oil to a bridging drop but can also provide a conduit for transport of oil from any lenses present in the surface [55].

By contrast, if the oil drop first forms a lens in a foam film that subsequently emerges into the second air–water surface, then the volume, V_{lens}, of the resulting bridging drop is always much larger than that formed by a spherical drop of the same diameter as the film thickness [55]. Plots by Denkov [55], of V_B^{crit}/V_i and V_{lens}/V_i against θ^* for the same values of the relevant surface tensions, are compared in Figure 4.17b by way of illustration. It is clear from this example that if an oil drop forms a bridge by first forming a lens, then the condition $B > 0$ completely determines the stability of the bridging drop even if the requirement that the air–water surface be planar is relaxed. However, if $\theta^* \rightarrow 180°$, then such a lens will in general exhibit a low penetration depth into the film. This would in turn imply a relatively long film drainage time before a bridge could be formed across the film, which may imply a low frequency of antifoam action. As we have indicated in the context of bridges formed by duplex film-forming oils (for which $\theta^* = 180°$), the presence of hydrophobic particles adhering to oil–water surfaces may have a role here. We return to this issue in Section 4.8.

These considerations therefore lead to the general conclusion that if the bridging coefficient B is >0 (and $<2\sigma_{OW}\sigma_{AW}$) and the volume of the bridging drop > V_B^{crit}, then bridging oil drops are expected to be unstable. They do not, however, address the ultimate fate of such drops. One possibility was first proposed by Frye and Berg [75]. This is illustrated in Figure 4.18. Here the bridging drop deforms so that the two three-phase contact angles become coincident and the oil drop essentially dewets from the foam film leaving a hole. Provided the latter is large enough, then the hole should spontaneously expand to produce rupture of the film [76]. As can be seen in Figure 4.18, the putative process of rupture involves a deformation of the bridging drop as the positive capillary pressure in the aqueous film in the vicinity of the drop means that the oil–water interface must move away from the axis of revolution. If the air–oil curvature remains convex with respect to the plane of the foam film, then the extent of the oil–water interface must be reduced until it is eliminated as the radius of the bridge increases. There is, however, no theoretical or experimental evidence that this represents a viable process of foam film rupture by an unstable bridging drop.

Yet another possibility is described by Denkov et al. [54] and again ascribed to Frye and Berg [75]. In this mechanism, the oil lens is assumed to be non-deformable so the lens shape is preserved after the lens bridges a foam film. The situation is depicted in Figure 4.19. Again provided $B > 0$ and $\theta^* > 90°$, this configuration would result in foam film rupture as the second air–water surface peels off the lens to produce a hole when the two three-phase contact lines become coincident. Denkov et

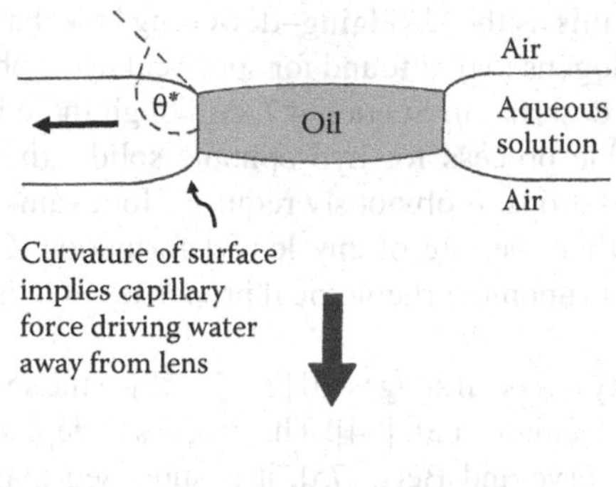

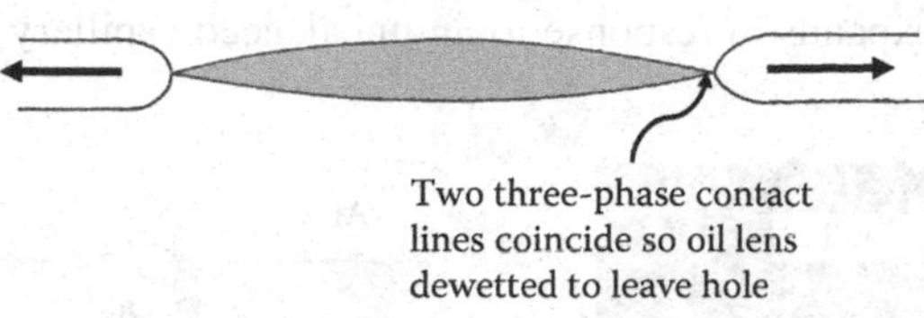

FIGURE 4.18 Foam film rupture mechanism by bridging oil drop proposed by Frye and Berg involving elimination of oil–water surface. (From Frye, G.C., Berg, J.C., *J. Colloid Interface Sci.*, 130, 54, 1989.)

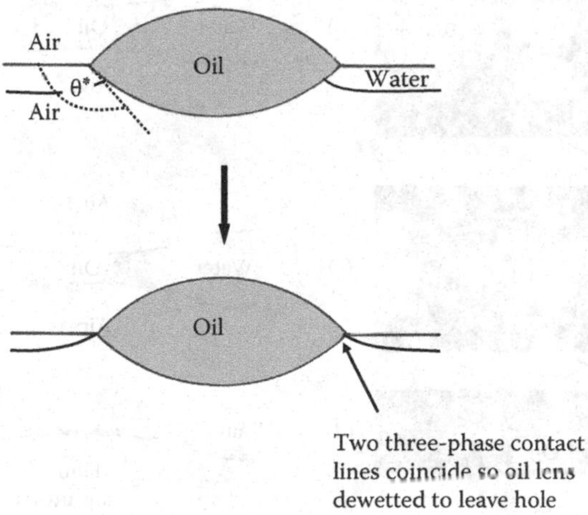

FIGURE 4.19 Bridging–dewetting mechanism of foam film rupture involving bridging of non-deformable oil lens, but again involving elimination of oil–water surface as with the proposed mechanism of Frye and Berg [75] depicted in Figure 4.18. (After Denkov, N.D. et al., *Langmuir*, 15, 8514, 1999; Denkov, N.D., Marinova, K.G. Antifoam effects of solid particles, oil drops and oil-solid compounds in aqueous foams, in *Colloidal Particles at Liquid Interfaces*, Binks, B.P., Horozov, T., eds., Cambridge University Press, p 383, 2006.)

al. [54] therefore describe this as the "bridging–dewetting" mechanism. The mechanism is in fact exactly analogous to that found for spherical hydrophobic solids bridging foam films, which we describe in Section 4.7. Although there is strong evidence that this represents a viable process for hydrophobic solids, there is no evidence that it does for bridging oil drops. It obviously requires, for example, that the rate of dewetting is much faster than the rate of any lens deformation. Clearly this places some significant constraints upon the rheological properties of a drop if it is to function in this manner.

An alternative possibility has been suggested [22, 54, 55]—the so-called "bridging–stretching" mechanism of Denkov et al. [54]. The process is depicted in Figure 4.20. Unlike the mechanism of Frye and Berg [75], it is supposed that the curvature of the air–oil surface becomes concave with respect to the plane of the foam film as the bridging drop expands in response to an unbalanced capillary pressure. Such a

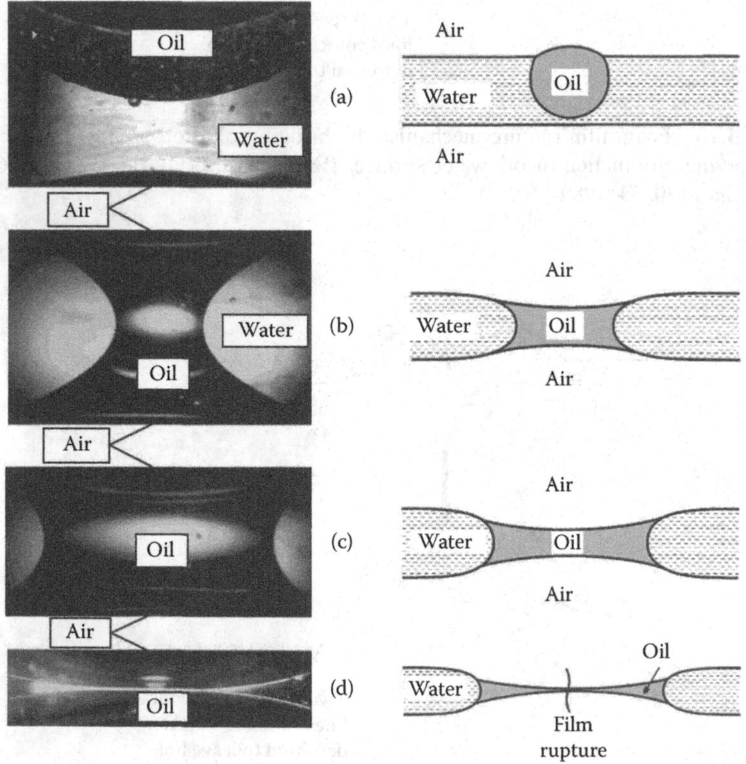

FIGURE 4.20 Bridging–stretching mechanism of foam film rupture—visualization by high-speed video using Dippenaar [77] method (left) and schematic illustration (right). (a, b) Bridging of foam film surfaces by an antifoam drop leads to oil bridge with unbalanced capillary pressures at oil–water and air–water surfaces. (c, d) Bridge stretches until rupture occurs as the two concave air–oil surfaces approach one another. Rupture of bridge in this manner leads to rupture of whole foam film. (Reprinted with permission from Denkov, N.D. *Langmuir*, 20, 9463. Copyright 2004 American Chemical Society.)

configuration would in fact seem to be unavoidable in the case of bridging drops with $\theta^* \rightarrow 180°$. As with the mechanism of Frye and Berg [75], the positive capillary pressure in the deformed film adjacent to the bridging drop means that the oil–water interface moves away from the axis of revolution. However, the concave shape of the resulting air–oil surface means that, as the radius of the drop expands, the two air–oil surfaces will meet and spontaneously rupture at a certain critical thickness. The envisaged process is similar to that first proposed by Ross [10] for the situation of an oil drop bridging a foam film covered with two duplex films which we have already discussed (and which is depicted in Figure 4.15). That this is, in principle, a viable process in the case of a bridging oil lens has been demonstrated by Denkov et al. [54] by observing the rupture of an aqueous film containing a PDMS oil drop in a Dippenaar cell [77]. However, the dimensions of the cell are significantly larger than those prevailing in the usual antifoam action in foam films. Moreover, rupture in the Dippenaar cell of a PDMS bridge was not spontaneous. With this cell foam film drainage was simulated by direct removal of aqueous phase from the aqueous film. The relevant images are reproduced in Figure 4.20. Denkov et al. [54] have in fact reported observations of the interaction of PDMS antifoam drops with foam films using a Scheludko cell [79, 80] in which the film rupture process appeared to copy that of the proposed bridging–stretching mechanism.

We find then that a convincing, but qualitative account, of how the bridging–stretching mechanism gives rise to film rupture in the case of unstable bridging drops has been given by Denkov [55]. However, a quantitative theoretical analysis of the process is lacking. A number of issues therefore remain to be resolved. The role of the magnitude of the bridging coefficient B in determining the process of foam film rupture is, for example, not known. The precise fate of a bridging drop after film rupture represents another issue. The extent of disproportionation could be a significant factor in determining the rate of antifoam deactivation (see Chapter 6).

The complexity of these issues suggests that computer simulation may afford some useful insights. It seems possible therefore that use of the type of energy minimization techniques incorporated in the Surface Evolver software devised by Brakke [81] may have application in this context. However, it is not clear that such an approach could accommodate the effects of disjoining potentials in the supposed process of film rupture. Those potentials would of course be important, for example, in determining the exact circumstance of rupture at the center of an oil drop during the bridging–stretching process. Such analysis would also necessarily ignore the role of viscous dissipation in determining the fate of the relevant processes.

4.5.2 Line Tensions and Antifoam Behavior of Oil Lenses

More than a century ago, Gibbs [82] drew attention to the possibility that the intermolecular interactions near the three-phase contact line made by two fluids at either a solid particle or a liquid lens could give rise to an excess free energy. This excess free energy is manifest as a "line tension," τ, in much the same way as surface free energy is manifest as a surface tension. It arises because the potential energy of molecules in the vicinity of the three-phase edge is different from that close to other surfaces. In the case of an oil lens, this could be interpreted as a slight change in the three relevant surface tensions close to the three-phase edge. This line tension, unlike the surface tension, can in general have either positive or negative values, where positive values indicate a tendency of the line to contract [83].

Existence of a line tension contributes an additional force term to Neumann's triangle of forces in the plane of the air–water surface. That term is the one-dimensional analogue of the Laplace equation, $\tau/\overline{R}$, where $\overline{R}$ is the radius of the line. If then we resolve the horizontal forces in a lens as depicted in Figure 4.21, we can write

$$\sigma_{OW}\cos\theta^*_{OW} + \sigma_{AO}\cos\theta^*_{AO} = \sigma_{AW} - \frac{\tau}{\overline{R}} \tag{4.33}$$

Similarly, we have [84] for θ*

$$\cos\theta^* = -\cos\theta^*_{OW} = -\frac{[\sigma_{AW} - (\tau/\overline{R})]^2 + \sigma^2_{OW} + \sigma^2_{AO}}{2\sigma_{OW}[\sigma_{AW} - (\tau/\overline{R})]} \tag{4.34}$$

which reduces to Equation 4.17 for sufficiently large lenses where the line tension term $\tau/\overline{R} \to 0$. We can also write the corresponding expressions for the classic entry and spreading coefficients. Thus, for example, the equilibrium entry coefficient in the case of a small drop becomes

$$E^e = \sigma^e_{AW} - \tau/\overline{R} + \sigma^e_{OW} - \sigma^e_{AO} \tag{4.35}$$

where we recall that the superscript refers to mutually saturated fluids.

Aveyard and Clint [85] have analyzed the consequences of the presence of a significant line tension on the behavior of lenses for "the role of liquid drops in the rupture of thin liquid films and hence in foam breaking." They note, for example, that Equation 4.35 indicates that a sufficiently high positive value for τ could mean that the equilibrium entry coefficient becomes negative for small drops in the case

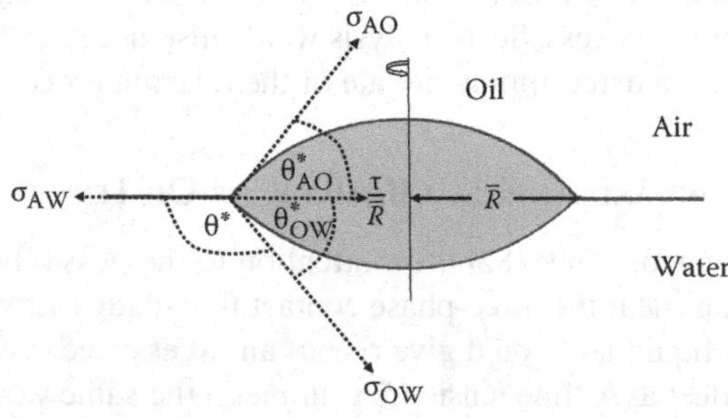

FIGURE 4.21 Relation of line tension, τ, to Neumann's triangle in oil lens where $\overline{R}$ is radius of the line at three-phase contact line.

where the classic equilibrium entry coefficient is positive. This could then in principle result in the non-emergence of antifoam drops into air–water surfaces despite positive values of classic entry coefficients. On the other hand, Garrett et al. [86] have suggested that negative line tensions could mean that entry coefficients could become positive for small enough drops even though classic entry coefficients indicate non-emergence into air–water surfaces. Addition of the line tension term to the expression for the classic spreading coefficient (Equations 3.5 through 3.7) also means that, for small drops and large values of τ, the spreading behavior could be drastically changed. Thus, if τ is negative, a small, otherwise non-spreading drop, could spread until the radius of the drop becomes so large that the sign of the spreading coefficient changes [85]. We also note that Equation 4.34 indicates that positive values of τ imply smaller values of θ^* so that reduction in drop size could in principle mean $B < 0$ and bridging drops become stable.

A problem with these speculations concerns lack of both direct unequivocal experimental observation of the putative effects attributable to line tension and of relevant measurements of line tensions. In illustration of the latter, a list of available line tension measurements for oil lenses at air–water surfaces is presented in Table 4.1. Measurements are seen to be mainly limited to pure hydrocarbon lenses on the surface of pure water. It seems therefore that there is no knowledge of the line tensions of, for example, PDMS lenses at air–water surfaces. Values of τ for hydrocarbons in the main seem to be positive, lying in the range of 10^{-11}–10^{-10} N. This compares with theoretical estimates for τ, lying in the range of 10^{-12}–10^{-10} N [83, 93, 94]. Only Takata et al. [91] report measurements using aqueous solutions of surfactants where a change in sign of line tension is seen to accompany an increase in surfactant concentration. It is worth noting too that the values for dodecane lenses on water shown in Table 4.1 differ not only by five orders of magnitude but are of opposite sign! It is tempting to conclude therefore that the value of τ for this system given by Chen et al. [89] is too high (although values of τ of a similar order of magnitude have also been reported for oil drops on solid surfaces by Neumann and coworkers [95–97]).

In assessing the likely effect of line tension on antifoam behavior, we need only consider the order of magnitude of the term $\tau/\overline{R}$. As we have seen, this force acts in the plane of the air–water surface and therefore has the same effect on lens behavior as changes in the air–water surface tension. The radii of antifoam drops are typically of order ≥1 micron (see, e.g., references [43, 56, 98]). Taking then the range of most reported experimental values of τ for lenses on water of 10^{-10}–10^{-11} N shown in Table 4.1, we can calculate that $\tau/\overline{R}$ for antifoam drops is expected to lie in a range between 10^{-2} and 10^{-1} mN m^{-1}. Since air–water surface tensions of surfactant solutions are ≥20 mN m^{-1}, these values of line tension would be expected to mean negligible effects on entry and spreading coefficients or lens geometry and therefore on antifoam mechanism except in cases where, for example, the magnitude of the spreading coefficient is of the same order as $\tau/\overline{R}$. Thus, if both the classic spreading coefficient and line tension are negative and if $S^e \leq \tau/\overline{R} \sim 10^{-1}$ mN m^{-1}, then a drop will spread over the air–water surface rather than form a lens despite $S^e < 0$ (at least until the inequality is no longer satisfied as $\overline{R}$ increases). Extremely low negative values of the classic spreading coefficient S^e implied by this would of course mean

TABLE 4.1
Experimental Values of Line Tensions for Oil Lenses at Water–Air Surfaces

System	Method	Solute Concentration in Aqueous Phase (M dm^{-3})	Line Tension[a] (N)	Ref.
Water on air–hexadecane surface	Condensation of supersaturated hexadecane vapor onto air–water surface by adiabatic compression	None	$+8 \times 10^{-11}$	[87]
Octane lenses on aqueous NaCl solution	Direct observations of lenses using interference microscopy	3	$+9 \times 10^{-11}$	[88]
		5	$+10^{-9}$	
Dodecane lens on water	Measurement of contact angles and diameter of lens in contact with motor-driven syringe	None	-1.3×10^{-6}	[89]
(Dodecane + 2 mM docecanol) lens on water	Direct observation of lenses using interference microscopy	None	$+1.6 \times 10^{-11}$	[90]
Hexadecane lens on dodecyltrimethylammonium bromide solution	Direct observation of lenses using interference microscopy	0.5×10^{-3}	$+8 \times 10^{-12}$	[91]
		0.91×10^{-3}	-2.5×10^{-11}	
		1.1×10^{-3}	-1.4×10^{-10}	
		0.5×10^{-3}	$+31 \times 10^{-12}$	[92]
		0.91×10^{-3}	-5.5×10^{-11}	
		1.1×10^{-3}	-1.2×10^{-10}	

[a] Quoted line tensions are positive if directed to the oil phase.

values of the lens dihedral angle $\left(\theta^*_{OW} + \theta^*_{AO}\right) \rightarrow 0$ for large drops. It also means relatively large values of $\bar{R}$ for a drop of a given equivalent spherical radius. This factor would decrease the magnitude of $\tau/\bar{R}$ by about a factor of 5 in the case of, for example, a lens of equivalent spherical radius of 1 micron where $\theta^*_{OW} = \theta^*_{AO} = 1°$ (and therefore $\theta^* = 179°$).

Given the magnitude of line tensions, it seems unlikely that the sign of the bridging coefficient could be changed from positive to negative for small drops except in the case where B is extremely small because this requires that $B < \sim 2\sigma_{AW}\tau/\bar{R}$ (see Equation 4.34), remembering that B is a function of σ^2_{AW}. Moreover in the limit approaching duplex film formation where $\theta^* \rightarrow \pi$ and $B \rightarrow 2\sigma_{OW}\sigma_{AW}$, we find that reversal of the sign of B requires that $\sigma_{AO} < \tau/\bar{R}$ (where $\bar{R} \rightarrow \infty$ in those circumstances). It seems extremely unlikely then that such a sign reversal could occur in practice except where ultra low oil–water interfacial tensions prevail which in turn usually correlates with solubilization of significant amounts of the oil in a microemulsion (see, e.g., reference [99]). Such solubilization would of course totally eliminate lens formation irrespective of the magnitude of a line tension unless the oil concentration exceeds the solubilization limit so that Winsor I conditions prevail.

As we have seen, Aveyard and Clint [85] have argued that a positive value for τ could in principle alter the sign of the entry coefficient relative to that indicated by the classic entry coefficients (Equations 3.1 through 3.4) in order to account for the ineffectiveness of oils as antifoams despite positive values for E^e, E', or E. This argument is obviously only applicable if the relevant classic entry coefficient is $E^e < \tau/\overline{R}$ ~10^{-1} mN m^{-1}. Clearly then distinguishing a role for line tension in the inhibition of emergence of oil drops into air–water surfaces is difficult. The evidence that the metastability of pseudoemulsion films is responsible is, however, now overwhelming (see Sections 4.5.3 and 4.8.4). Indeed if the sign and magnitude of line tensions largely determined the emergence of antifoam oil drops into air–water surfaces, then this would leave no obvious role for the particles that are usually mixed with the oil to facilitate rupture of the pseudoemulsion film as described in Section 4.8. The experimental and theoretical evidence that the particles do in fact break metastable pseudoemulsion films to permit ready emergence into air–water surfaces is now extensive. This tends to confirm that non-emergence of an oil drop into the air–water surface, for which $E^e > 0$, concerns a metastable pseudoemulsion film rather than a positive line tension.

4.5.3 Oil Bridges in Plateau Borders and Stability of Pseudoemulsion Films

There is an additional, and obvious, requirement if oil drops are to emerge into air–water surfaces in foam films to form unstable configurations leading to film rupture. Not only must any metastable pseudoemulsion film be ruptured but that must occur before any drop is swept out of the foam film as it drains. That oil drops are likely to be preferentially removed from a foam film exhibiting, for example, Reynolds drainage can in fact be shown to occur as a result of an excluded volume effect (see Section 5.2.2). A number of authors [7, 21, 58, 59, 100–102] have suggested, however, that film rupture may not happen because all drops are swept out of draining films in the case of neat oils where the metastability of the pseudoemulsion film is not undermined by the presence of particles. Indeed, Koczo et al. [98] suggest this may happen even if particles are present in the oils. However, this latter suggestion seems unlikely to be generally true as we will shown in Section 4.8.

Arguably, Wasan and coworkers [21, 100, 101] were the first to suggest that total removal of drops from foam films would lead to their accumulation in Plateau borders where subsequent increases in capillary pressure accompanying foam drainage (see Chapter 1) could in turn lead to rupture of metastable pseudoemulsion films. Emergence of drops into the air–water surface of those Plateau borders then results in rupture of the contiguous foam film. This suggestion arose from observation of the effects of crude oil on the foam of an aqueous anionic surfactant solution using microscopy. Later studies

[15] involving neat lower alkanes and benzene on the foam of aqueous solutions, of both anionic and non-ionic surfactant solutions, also revealed accumulation of oils in Plateau borders. Surprisingly, and in contrast, the presence of oil drops was found to enhance foam stability with little effect on foamability. This study, however, involved extremely high surfactant concentrations (≥100 × CMC) and volume fractions of oil phase up to 0.33. The absence of antifoam effects despite $E > 0$ and $B > 0$ (readily calculated from the data presented in reference [15]) was attributed to the extremely high volume fractions of oil trapped in Plateau borders inhibiting both drainage and the attainment of the sufficiently high capillary pressures required to rupture the relevant pseudoemulsion films. That accumulation of oil drops in Plateau borders, to the exclusion of foam films, can occur is clearly revealed in Figure 4.22. Here octane drops are seen to be trapped in the Plateau borders of a foam prepared from an alkyl ether sulfate solution despite a strongly positive value for the relevant entry coefficient.

Denkov and coworkers [7, 58, 59, 102] have also made a number of studies of the antifoam behavior of different neat oils with a variety of aqueous surfactant solutions. The oils included PDMSs, long-chain alcohols, and alkanes. These studies, in contrast to those of Wasan and coworkers [21, 100, 101], were characterized by use of relatively low oil concentrations of 0.01–0.1 wt.%, which means volume fractions of order 10^{-4}–10^{-3}. With these systems, only antifoam effects were observed and these were largely confined to reductions in the stability of foam. In one particular study, involving a PDMS oil, coarse emulsions containing large drops (of 4–150 microns) proved to be more effective in diminishing foam stability than fine emulsions containing small drops (of 1–15 microns) [58]. Similar findings concerning the effect of drop size were found with other oils. Examination of the draining behavior of single films using a Scheludko cell (see Chapter 2) revealed the rapid ejection of all drops from foam films, which contrasted with effects on foam stability occurring over periods at least an order of magnitude longer [58]. As reported earlier by Koczo et al. [15], direct observation of oil drops in Plateau borders was also reported. This is exemplified by the image of PDMS drops in a foam shown in Figure 4.23. All of this represents strong evidence that foam collapse in the case of neat oils is in fact

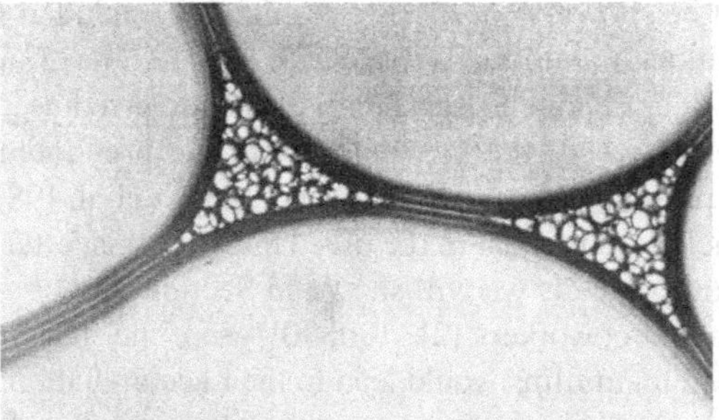

FIGURE 4.22 Image of two-dimensional foam prepared from saline solution of 4 wt.% of commercial ethoxylated alcohol surfactant containing emulsified octane where oil drops have drained out of foam films into Plateau borders. (Reprinted from *J. Colloid Interface Sci.*, 150, Koczo, K., Lobo, L., Wasan, D., 492. Copyright 1992, with permission from Elsevier.)

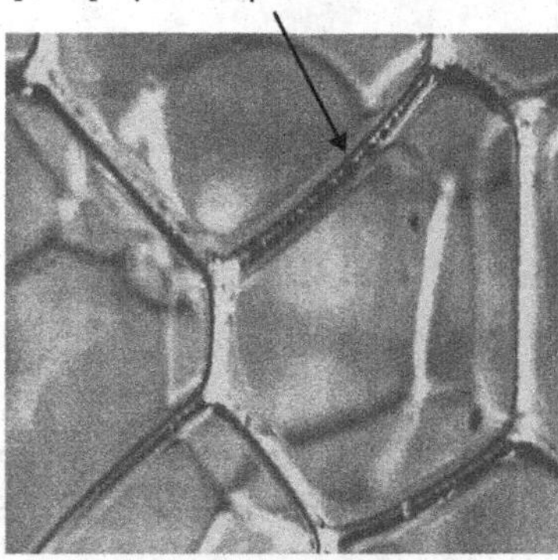

FIGURE 4.23 Image of foam prepared from 0.1 M solution of commercial sample of sodium dodecyl-polyoxyethylene-3-sulfate (contaminated with NaCl and Na_2SO_4) containing 0.1 wt.% emulsified PDMS oil of drop size 4–150 microns. Drops of oil trapped in Plateau borders are clearly visible. (Reprinted with permission from Basheva, E.S. et al. *Langmuir*, 16, 1000. Copyright 2000 American Chemical Society.)

due to the presence of drops in Plateau borders, as suggested earlier by Wasan and coworkers [21, 100, 101].

Convincing evidence that foam film rupture in these systems occurs as a consequence of Plateau border capillary pressures exceeding those necessary to rupture the relevant pseudoemulsion films has been produced by Denkov and coworkers [7, 58, 59, 102]. Most antifoam oils exhibit positive buoyancy in water so that they tend to rise against the drainage flux, which facilitates trapping against borders that narrow with increasing height in the foam. As liquid drains out of the Plateau borders, the trapped drops are supposed compressed against the walls of the borders until the applied capillary pressure exceeds that required to rupture the pseudoemulsion films. Drops then emerge into the air–water surfaces of the Plateau borders, whereupon rupture of adjacent foam films is supposed to occur, possibly as a result of a bridging mechanism. However, if drops are always smaller than the dimensions of the borders, then they will not become trapped and will not therefore be subject to an applied capillary pressure. Such drops will in turn not emerge into the air–water surfaces of the Plateau borders and will therefore not have an adverse effect on foam stability.

The process of foam collapse due to entrapped drops commences from the top of the foam column where the capillary pressure is highest. A significant induction time is usually involved as liquid drains out of the foam and the capillary pressure increases [7, 58, 59, 102]. Once commenced, the foam collapses until the capillary pressures at diminished foam heights are too low to rupture the relevant pseudoemulsion films. No further foam collapse then occurs, leaving a residual foam height. The overall process is shown schematically in Figure 4.24.

Use of the so-called film-trapping technique (FTT, see Chapter 2) by Denkov and coworkers [7, 58, 59, 102] permits direct measurement of the applied critical capillary pressure p_c^{crit} necessary for rupture of the relevant pseudoemulsion films. This

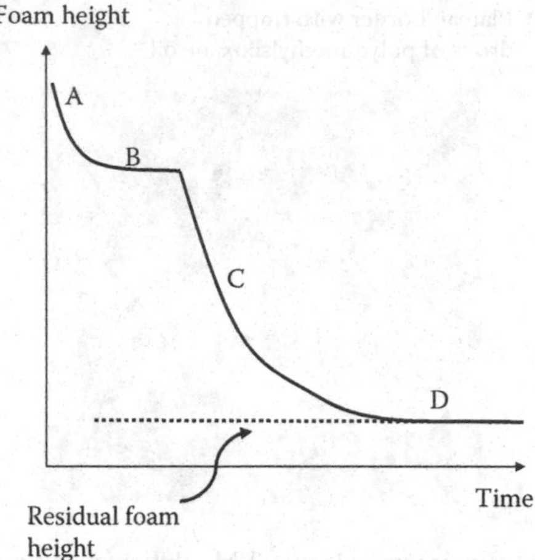

FIGURE 4.24 Schematic of process of foam collapse due to presence of entrapped drops as indicated by the height of foam column measured from level of free liquid. Region A—liquid drains out of foam and capillary pressure increases. Region B—drainage very slow but diffusional disproportionation causes restructuring of foam, oil drops accumulate in Plateau borders. Region C—foam collapse commences as capillary pressure is high enough to rupture pseudoemulsion films of drops trapped in Plateau borders. Region D—foam collapse ceases at residual foam height where capillary pressures are too low to rupture pseudoemulsion films. (After Denkov, N.D., *Langmuir*, 20, 9463, 2004.)

capillary pressure can then be compared with the Plateau border capillary pressure at the top of the foam column p_c^{PB} $(y = 0)$, where the coordinate y measures distance from the top of that column. Following the arguments of Princen [103], outlined in Chapter 1, we can approximate this in the case of an equilibrium aqueous foam where drainage has ceased by

$$p_c^{PB}(y=0) = -\sigma_{AW}\left|\tilde{\kappa}_t\right| \approx \rho_W g H_0 \tag{4.36}$$

where H_0 is the height of the foam column and ρ_W is the density of the aqueous solution. Here we remember that $|\tilde{\kappa}_t|$ is the modulus of the (negative) curvature of the Plateau borders at the top of the foam column. We should note also that if the foam is still draining, then the capillary pressure at the top of the foam will be less than indicated by Equation 4.36.

In making comparison between p_c^{PB} $(y = 0)$ and p_c^{crit}, we note that the former is a suction applied to entrapped drops by the walls of the Plateau borders with negative curvature. The critical capillary pressure p_c^{crit}, measured by the FTT, is, on the other hand, the suction pressure in the aqueous phase surrounding a drop trapped in a film against a solid substrate (see description of the FTT in Chapter 2). In comparing p_c^{crit} with p_c^{PB} $(y = 0)$, we should note that rupture of the relevant pseudoemulsion film is supposed determined by the magnitude of the critical disjoining pressure at the point of film rupture relative to the applied critical capillary pressure as discussed in Section 3.4.1.

Denkov [1] argues that if p_c^{PB} $(y = 0) > p_c^{crit}$, then the pseudoemulsion film will rupture and foam collapse will follow. However, equivalence of the applied capillary pressure measured with the FTT and that prevailing in a foam film due to the Plateau border is not in principle necessarily exact. The relation between the applied capillary pressure and the disjoining pressure in the relevant pseudoemulsion film is given by Equations 3.22 and 3.23 in Section 3.4.1. The applied capillary pressure is seen to be simply the difference between the disjoining pressure and the capillary pressure due to the curvature of the pseudoemulsion film. That the film is curved is of course due to the deformation of the entrapped drop by the applied capillary pressure. However, the magnitude of that applied capillary pressure is small compared with that due to the film curvature. Since the curvature of the film is determined by the radius, and therefore the drop size, we would expect the applied capillary pressure to be also dependent on drop size. Therefore, unless the oil drop used in the FTT experiment has the same size and shape as an oil drop caught in a Plateau border, the equivalence of p_c^{PB} and p_c^{crit} does not necessarily mean equivalence of disjoining pressures and therefore pseudoemulsion film rupture. An additional problem, however, concerns the finding that p_c^{crit} is surprisingly independent of drop size [1, 61, 102]. This finding, which is discussed at length in Section 3.4.1, implies that the critical disjoining pressure is itself a function of drop size [1, 61, 102]. Despite these complications, Basheva et al. [58] argue that nevertheless p_c^{PB} and p_c^{crit} should be of similar order and follow the same trends with respect to drop size so that, to a first approximation, equivalence can be assumed.

We therefore arrive at a depiction of foam destabilization by neat oil drops trapped in Plateau borders when the Plateau border capillary pressure exceeds a certain critical value that is approximated by the value p_c^{crit}, measured by the FTT. Therefore, if

$$\left|p_c^{PB}(y=0)\right| > p_c^{crit} \tag{4.37}$$

then the critical disjoining pressure in the relevant pseudoemulsion film will be exceeded to produce rupture of that film together with the adjacent foam film. Foam collapse would then continue until the foam height H_0 becomes so low that the inequality

$$\left|p_c^{PB}(y=0)\right| = \rho_W g H_0 < p_c^{crit} \tag{4.38}$$

is satisfied whereupon the pseudoemulsion films will not rupture and no further antifoam effect due to the oil will occur.

Simple geometrical considerations enable calculation of the minimum drop radius r_{min} required if it is to be compressed by the walls of the Plateau borders. It is given to good approximation by the radius of a sphere inscribed by those borders. Therefore, if we equate the capillary pressure in the Plateau border to the hydrostatic head H_o, we can write after Arnaudov et al. [7]

$$r_{min} \approx 0.155 |\kappa_t|^{-1} = 0.155 \frac{\sigma_{AW}}{\rho_W g H_0} \tag{4.39}$$

Any drop with a radius less than r_{min} cannot emerge into the air–water surface because no capillary pressure can be exerted to deform the drop and therefore break the pseudoemulsion film. Such a drop will therefore be ineffectual in causing foam film collapse.

From these considerations, Denkov [1] deduces that foam heights in the presence of neat oils will decay until either capillary pressures are too low to cause pseudoemulsion film rupture, as indicated by inequality 4.38, or oil drops are too small with radii less then r_{min}. The resulting residual foam heights are given either from Equation 4.38 by

$$H_{res}^{c} = \frac{p_c^{crit}}{\rho_W g} \tag{4.40}$$

where H_{res}^{c} is the residual height limited by the Plateau border capillary pressure or from Equation 4.39 by

$$H_{res}^{size} = 0.155 \frac{\sigma_{AW}}{\rho_W g . r_{AF}} \tag{4.41}$$

where H_{res}^{size} is the residual height if limited by the size of the oil drops of radius r_{AF}. If $H_{res}^{c} > H_{res}^{size}$, then the residual foam height is determined by the stability of the pseudoemulsion film so that H_{res}^{c} is proportional to p_c^{crit}. If, however, $H_{res}^{c} < H_{res}^{size}$, then the residual foam height is determined by the size of the oil drops relative to the size of the Plateau borders and is independent of the critical capillary pressure required to cause rupture of the relevant pseudoemulsion film.

Denkov and coworkers [1, 102] give experimental evidence that this description of the effect of oils on foam stability is essentially correct. This is reproduced in Figure 4.25 where the residual foam heights, measured using a Ross–Miles apparatus, for a variety of surfactant solutions in the presence of 0.1 wt.% PDMS oil are plotted against the critical capillary pressure, p_c^{crit}. As expected, the plot reveals two regions corresponding to the applicability of Equations 4.40 and 4.41. High values of p_c^{crit} are seen to produce stable foams, whereas low values produce unstable foam that collapses to low residual heights, which are essentially independent of the magnitude of p_c^{crit} and where the drop radii are smaller than the radius of a sphere inscribed by the Plateau borders (i.e., Equation 4.41 is applicable). Unfortunately, the exact specification of the PDMS used is not given in this work.

All of this of course leaves open the question of the precise mechanism for foam film collapse caused by oil drops in Plateau borders upon rupture of the relevant

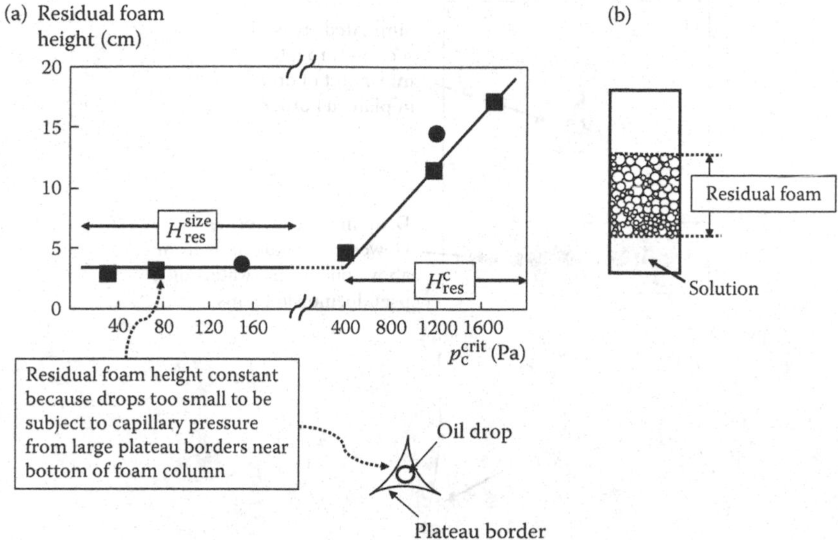

FIGURE 4.25 (a) Experimental observations of residual foam height (by Ross–Miles method) as function of critical applied capillary pressure, p_c^{crit}, measured by FTT for a variety of surfactant solutions containing 0.1 wt.% emulsified neat PDMS oil. Residual foam heights for solutions requiring high values of p_c^{crit} are seen to follow the theoretical estimates of H_{res}^{c} by Equation 4.40. Residual foam heights are seen to be invariant at low values of p_c^{crit} as predicted by Equation 4.41 where drop sizes are constant regardless of solution compositions. ■, Solutions containing sodium dodecyl polyoxyethylene sulfate; ●, solutions containing SDS. (b) Measurement of residual foam height. (After Hadjiiski, A.D. et al., Role of entry barriers in foam destruction by oil drops, in *Adsorption and Aggregation of Surfactants in Solution*, Mittal, K., Shah, D., eds., Surfactant Science Series Vol 109, Marcel Dekker, New York, Chpt 23, p 465, 2003.)

pseudoemulsion films. A simulation study of the configuration of oil drops in Plateau borders has in fact been reported [104]. These simulations were made with the surface energy minimization Surface Evolver software [81]. In this simulation, the pseudoemulsion films were supposed stabilized against Plateau border capillary suction by disjoining forces. However, the film tension was approximated as equal to the sum of the oil–water and air–water surface tensions so that $\theta^* = 0$. The contribution to the film tension due to the disjoining pressure is therefore neglected (both because it is likely to be small and because the Surface Evolver method cannot deal with disjoining potentials). The contribution of disjoining pressure integrals to the pseudoemulsion film tension is discussed in Section 3.3.1.

An example of these simulations is illustrated in Figure 4.26a, which gives two different views of an oil drop trapped in a Plateau border for which the ratio of oil–water to air–water surface tensions is a realistic 0.1 and the ratio of the spherical equivalent drop radius to the reciprocal Plateau border curvature is unity. The equivalent radius of the drop is therefore almost an order of magnitude greater than the minimum radius r_{min}. We see from the figure that the drop is indeed markedly deformed as a result of the high capillary pressure due in turn to the relatively high air–water surface tension. The drop shape in cross-section appears to closely conform to the

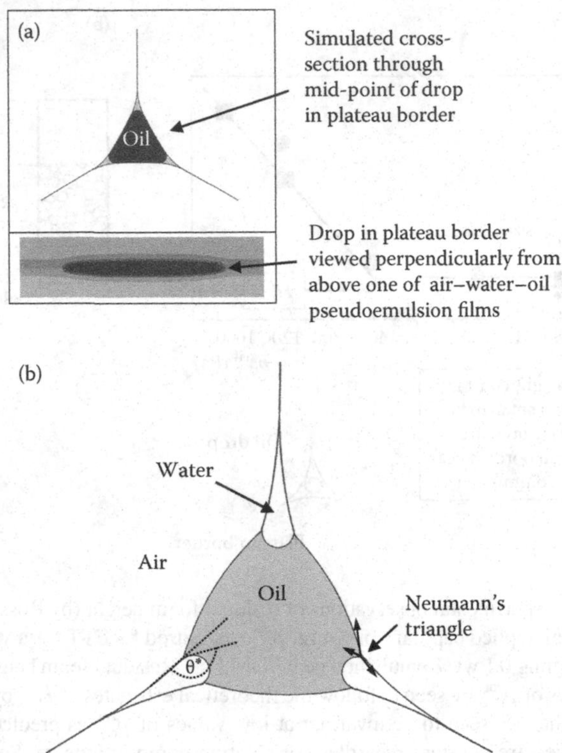

FIGURE 4.26 Oil drops in Plateau borders. (a) Simulation (using Surface Evolver software [81]) of drop in Plateau border with $\sigma_{OW}/\sigma_{AW} = 0.1$ and ratio of equivalent spherical drop radius to reciprocal Plateau border curvature = 1. Perfect wetting of drop by water; pseudoemulsion films are aqueous duplex films. (Reprinted with permission from Neethling, S. et al. *Langmuir*, 27, 9738. Copyright 2011 American Chemical Society.) (b) Schematic illustration of drop bridging a Plateau border after pseudoemulsion films ruptured and air–oil surfaces formed with Neumann's triangle satisfied at three-phase contact lines and $\theta^* > 90°$ (i.e., bridging coefficient $B > 0$).

curved surface of the Plateau border. In consequence, it is also deformed length-wise—squeezed into a shape approximating a long triangular prism. The limited volume of aqueous phase at the three edges of that prism will mean diminished drainage rates through the Plateau borders, as described by Neethling et al. [104]. Ultimately draining will, however, further increase the curvature of the Plateau borders leading to an increase in the capillary pressure. This will in turn destabilize the pseudoemulsion films as the maximum in disjoining pressure isotherm is overcome. Unfortunately, this aspect of the evolution of the configuration of an oil drop in a Plateau border could not be simulated because of the limitations of the Surface Evolver software.

Once the pseudoemulsion films break, then the possible configurations of the bridging drop in a Plateau border are determined by Neumann's triangle and three surface tensions—σ_{AO}, σ_{OW}, and σ_{AW}. There has been no analysis of the stability of the resulting configurations. It is, however, possible to speculate. We show a schematic illustration in Figure 4.26b of a cross section through the middle such a drop for which $\theta^* > 90°$ (i.e., for which $B > 0$, as defined by Equation 4.30). Here we should note that both the air–oil and oil–water surfaces are nondoid or unduloid. If the dihedral angle at the three-phase contact line is sufficiently large, then the air–oil radius of curvature, r_1, in the plane of the page should be everywhere positive with respect to the oil phase. The orthogonal air–oil radius of curvature, r_2, should also be everywhere positive as the drop is of a finite length. The capillary pressure jump Δp_c^{AO} with respect to the atmosphere at the oil–air surface is therefore $\sigma_{AO}(1/r_1 + 1/r_2)$. By contrast, the oil–water radius of curvature, r_3, in the plane of the page should be everywhere negative with respect to the oil drop if $\theta^* > 90°$. However, as with the air–oil surface, the orthogonal oil–water radius of curvature r_4 should be everywhere positive, and of a similar magnitude to r_2. Allowing for the negative radius of curvature, r_5, of the Plateau border, we can write the capillary pressure jump, Δp_c^{OW}, with respect to the atmosphere as $-\{\sigma_{OW}(1/r_3 - 1/r_4) + \sigma_{AW}/r_5\}$. If the configuration of the bridging drop shown in Figure 4.26b is at equilibrium, we must have $\Delta p_c^{AO} - \Delta p_c^{OW} = 0$. All the terms in Δp_c^{AO} are positive and all the terms in Δp_c^{OW} are negative except σ_{OW}/r_4. However, usually $\sigma_{OW} \ll \sigma_{AO}$ and we would expect $r_4 \approx r_2$ so that $\sigma_{AO}/r_2 > \sigma_{OW}/r_4$, which must mean that $\Delta p_c^{AO} \neq \Delta p_c^{OW}$. The schematic illustration shown in Figure 4.26b cannot therefore represent an equilibrium configuration. It is therefore tempting to suggest that such an apparently unstable configuration would lead to rupture of the Plateau border and adjacent film possibly by causing oil to intercept the edge of the foam film. Foam film rupture would then follow a similar bridging–stretching process to that envisaged for bridging drops in foam films. This tentative analysis suggests that the criterion determining this outcome is again simply that $B > 0$. Clearly, however, a rigorous analysis of this issue is desirable.

In the case where the oil can spread over the air–water surface an additional possibility is suggested by Denkov [1]. He has observed that with large (centimeter size) foam films, emergence of a spreading oil in a Plateau border can produce capillary waves in the adjacent film, which seem to result in foam film collapse at large film thicknesses—the process is not, however, observed with the millimeter-size films typical of many foams.

4.6 ANTIFOAM BEHAVIOR OF EMULSIFIED LIQUIDS

4.6.1 Antifoam Effects of Neat Oils in Aqueous Foaming Systems

4.6.1.1 Early Work

Effective antifoams for aqueous solutions usually consist of mixtures of hydrophobic oils and hydrophobic particles. However, we leave consideration of the behavior of such mixtures to Section 4.8. Here we are concerned only with the antifoam effect of emulsified neat oils (i.e., oils without added particulate material). The relative ineffectiveness and ease with which their antifoam performance can be enormously improved by addition of a few percent of finely divided particulate material means that the neat oils are rarely employed as antifoams for aqueous systems. Practical interest in their antifoam effects derives rather from their incidental presence. Such oils may, for example, find application as hair conditioners and skin moisturizers, where adverse effects on the foam behavior of the relevant personal products can be considered undesirable. Knowledge of their mode of action may therefore be expected to assist in mitigation of their adverse effect on foam generation. Understanding the mode of action of these neat oils will of course also add perspective to the task of seeking to understand the enhanced effectiveness associated with the addition of hydrophobic particles.

There appears to be a consensus that necessary properties of antifoam oils for aqueous foams include insolubility in surfactant solution and a tendency to emerge (or "enter") into the air–water surfaces of foam films. Here we review early work concerning the antifoam behavior of neat oils possessing these properties. Of particular interest is the interpretation of the observed behavior using the theories of antifoam mechanism outlined in Sections 4.4 and 4.5.

Robinson and Woods [9] produced perhaps one of the earliest experimental studies of antifoam mechanism. The study concerned the effect of various undissolved oils on the foam behavior of both aqueous and non-aqueous solutions of surfactant. The oils included alkyl phosphates, alcohols (including diols), fatty acid esters, and PDMS. The solutions were of aerosol OT (AOT or sodium diethylhexyl sulfosuccinate) in either ethylene glycol or triethanolamine and sodium alkylbenzene sulfonate in water. Many quoted entry and spreading coefficients, however, violate Equations 3.11 and 3.12, which implies that these coefficients were non-equilibrium (i.e., initial) values where the relevant liquids are not mutually saturated. Robinson and Woods [9] observed that for these systems, wherever $E^i < 0$, no antifoam effect is found. This then represents some evidence that a positive value of the initial entry coefficient is necessary for antifoam action.

Pattle [105] published somewhat singular views about the mode of action of antifoams soon after those of Robinson and Woods [9]. This study included qualitative experimental observations of both the antifoam effect and spreading behavior of 292 combinations of potential antifoam materials and aqueous foaming solutions. The antifoam test employed simply involved allowing a "drop" of the material to contact an existing foam prepared pneumatically. Aqueous solutions of sodium oleate, saponin, egg albumen–isopropanol, and peptone were used. The potential liquid antifoam materials included alcohols, hydrocarbons, esters, various vegetable and animal oils, and some miscellaneous organic compounds such as diethyl ether. Of the combinations of aqueous solutions and potential antifoam materials, some 110 revealed spreading with antifoam effects and 36 revealed spreading without antifoam effects. Included in the list of alcohols were methanol and ethanol, which were considered to have both antifoam effects and spreading behavior. Similar findings with other soluble substances led Pattle [105] to conclude that "a liquid may show antifoam activity even if it is completely miscible with the foaming liquid." Any such effects are, however, likely to be transient and would seem to concern the specific nature of the experiments made by Pattle [105]. Thus, depending on the volume of the "drop" and the volume of liquid in the films and Plateau borders of the foam, it would seem possible that these soluble alcohols significantly lower the surface tension of the aqueous phase at the top of the foam column and cause foam collapse by Marangoni spreading (see Section 4.4.2).

In the same year, Ross [10] reinterpreted the results of Robinson and Woods [9] using a hypothesis concerning the rupture of foam films by oil drops for which $E^i > 0$ and $S^i > 0$. Such oil drops will, as we have seen, emerge into the air–foaming liquid surface and spread as duplex films. If this were to happen with an oil drop simultaneously emerging into both surfaces of a foam film, a duplex film of oil would simultaneously spread over both surfaces of the foam film. This would squeeze out the original liquid in the film to produce a region composed entirely of antifoam oil where rupture would occur. This mechanism would appear to invoke a Marangoni spreading effect, which would cease to function if the foaming liquid and antifoam become mutually saturated so that we have $S^e \leq 0$. However, as we have shown in Section 4.5.1, the presence of duplex oil films on either side of a foam film and joined by a bridge is intrinsically unstable anyway. This instability would exist even with mutually saturated fluids where $S^e = 0$.

Ross [10] noted that of the 40 combinations of oils and solutions reported by Robinson and Wood with $E^i > 0$ that exhibited antifoam behavior, 36 had $S^i > 0$ and 4 had $S^i < 0$. He claimed that this finding represented "definite" agreement with his rupture mechanism, which requires $E^i > 0$ and $S^i > 0$. That conclusion of course ignores the finding that antifoam effects can occur with $S^i < 0$. A rigorous conclusion would therefore be that either spreading (i.e., $S^i > 0$) is not a necessary aspect of the antifoam behavior of these oils or that two different mechanisms may operate corresponding to the two sets of conditions $E^i > 0$, $S^i > 0$ and $E^i > 0$, $S^i < 0$. If, however, we suppose that dispersal of the antifoam in the foaming liquid leads to mutual saturation, then these two sets of conditions condense to $E^e > 0$ and $S^e \leq 0$, which are also consistent with the condition for formation of unstable bridging configurations given by Equation 4.31.

In a later paper, Ross and Young [39] describe other examples of systems for which $E^i > 0$ and $S^i < 0$ where antifoam effects are observed. More recently, Okazaki et al. [12] report that both petroleum ether and olive oil spread rapidly over the surface of 10^{-2} M SDS solution but do not yield antifoam effects with the same solution. By contrast, these authors find that undissolved phenol does not spread on the same solution but does function as an effective antifoam (albeit at the rather high antifoam concentration used in this work). However, despite these obvious ambiguities in experimental evidence concerning the role of spreading, Ross [106] later wrote that "the action of a foam-inhibiting agent is believed to arise from its ability to spread spontaneously over the surface of the foamy liquid." Indeed, in the same paper, he also wrote that spontaneous spreading "is both a necessary and a sufficient condition" for a foam-inhibiting agent. Such statements were not only inconsistent with the literature at the time they were written, but as we show in this chapter, their lack of relevance has been further demonstrated by more recent work.

4.6.1.2 Short-Chain Alcohols

Weakly polar oils such as long-chain alcohols, phenols, and long-chain fatty acid esters have often been shown to exhibit antifoam behavior provided they are not so polar as to show high solubility in the relevant aqueous surfactant solutions (see, e.g., references [9, 10, 39, 105]). Arguably the earliest reported observations concerning the antifoam behavior of an alcohol are those of Sasaki [3] more than 70 years ago. He observed that if n-butanol, at a concentration below the overall solubility limit, was added to an existing and stable aqueous soap foam, complete collapse immediately occurred. Although the alcohol concentration was less than the overall solubility limit, it is possible that the limit was exceeded in the foam where the volume of surfactant solution trapped in Plateau borders and films is relatively low. The foam could, however, subsequently be regenerated, presumably because the alcohol dissolved. However, if n-butanol was added to the soap solution at concentrations above the solubility limit, then no foam could be generated from the solution. Clearly this represents an early observation of the requirement of insolubility for antifoam action. Similar behavior has been reported by Arnaudov et al. [7] for n-heptanol in saline sodium dodecylbenzene sulfonate solutions.

Kruglyakov and coworkers [40, 41, 107] have made a thorough study of the antifoam behavior of the lower normal alcohols. In this study, Kruglyakov and Taube [40] failed to find a correlation between the initial spreading coefficients S^i and antifoam effectiveness. The initial spreading coefficient of, for example, octanol on aqueous solutions of a variety of surfactants was found to be negative despite an optimal antifoam effect with that alcohol under the conditions studied. Indeed, in another paper, Kruglyakov [41] states that among the "numerous systems investigated by Ross and other workers there are many cases where efficient foam breaking was observed at a negative spreading coefficient" (i.e., when $S^i < 0$).

Kruglyakov and Koretskaya [107] have extended this study of normal alcohols to include the effect of surfactant concentration. They find that the optimum alcohol chain length for maximum antifoam effect increases with decreasing concentration of surfactant. This behavior is shown in Figure 4.27 for solutions of a blend of ethoxylated nonyl phenols (in 0.1 M KCl). Here the limiting concentration of antifoam for the foam to exhibit instability is plotted against alcohol chain length for various concentrations of surfactant. Although not entirely clear, it seems instability due to the presence of the alcohols was defined by foam with a lifetime of only 2–3 min where this foam was generated by shaking a volumetric flask. Unfortunately, no information about the effect of these alcohols on the foamability (i.e., the amount of foam generated) is, however, given. Qualitatively similar findings are reported for the antifoam effect of these alcohols on SDS and sodium dodecylbenzene sulfonate solutions.

It is argued by Kruglyakov and Koretskaya [107] that these antifoam effects are due to the formation of "asymmetrical films" where an oil lens is present on one side of a foam film. As we have seen, lenses will form provided $E^i > 0$ and $S^i < 0$. They can, in fact, form even if $S^i > 0$ because the resulting film may be metastable so that pseudo-partial wetting conditions ultimately prevail at equilibrium and we have $E^e > 0$ and $S^e < 0$. These issues are described at length in Chapter 3.

Kruglyakov [41] suggests that these asymmetric films, separating alcohol drops from air–water surfaces, can be metastable in the same sense as symmetrical air–water–air foam films provided entry coefficients are positive. This represents an early appreciation of the significance of what we now know as "pseudoemulsion" films. However, he also argues that the pseudoemulsion film will in general have a different stability from that of the symmetrical foam film with the same surfactant solution. This is supposed to arise in part because of differences in surface excess of

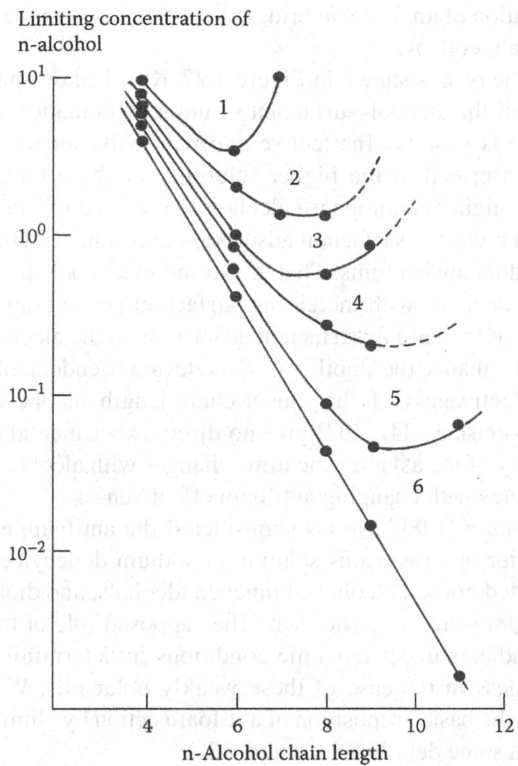

FIGURE 4.27 Limiting alcohol concentrations for foam to exhibit instability at various concentrations of blend of ethoxylated nonyl phenols (in 0.1 M KCl): (1) 1%, (2) 0.1%, (3) 0.5%, (4) 0.025%, (5) 0.005%, and (6) 0.001%. (After Kruglyakov, P.M. Equilibrium properties of free films and stability of foams and emulsions, in *Thin Liquid Films, Fundamentals and Applications*, Ivanov, I.B., ed., Marcel Dekker, New York, p 767, 1988.)

surfactant at the oil–water and air–water surfaces. If the oil is polar, then the adsorption of the surfactant at the oil–water surface will be lower than at the air–water surface. In turn, this will mean diminished stabilizing repulsive disjoining pressures across the film due, for example, to overlapping electrostatic double layers if the surfactant is ionic. The stability of the pseudoemulsion film will therefore be diminished so that the capillary pressure required to break the film is also diminished. Foam film collapse is then supposed to occur when the relatively unstable pseudoemulsion film separating an oil lens from the other air–water surface in the foam film breaks. Kruglyakov [1] does not, however, explain why rupture of the second asymmetric film should lead to foam film rupture. We have, however, seen that an unstable bridge will then form. As we have shown in Section 4.5, the presence of such a bridge will lead to foam film collapse if the bridging coefficient satisfies Equation 4.31. However, there is also the possibility that the oil drops are trapped in Plateau borders as described in Section 4.5.3. Rupture of the relevant pseudoemulsion films could also result in formation of an unstable bridging configuration in the Plateau borders, again leading to foam collapse.

In interpreting the results shown in Figure 4.27, Kruglyakov and Koretskaya [41, 107] note that for all the alcohol–surfactant solution combinations, the equilibrium entry coefficient E^e is positive. Ineffective antifoam behavior for the lower-chain-length alcohols is ascribed to the higher solubility of these materials. Declining effectiveness of the higher-chain-length alcohols is ascribed to a decline in polarity, which leads to an increase in surfactant adsorption and enhanced the stability of the oil–water–air pseudoemulsion films. That the optimum alcohol chain length for antifoam effectiveness declines with increasing surfactant concentration is then attributed to a concomitant increase in surfactant adsorption at the alcohol–water surface. This is supposed to enhance the stability of the relevant pseudoemulsion film, which diminishes the effectiveness of the longer-chain-length alcohols. Unfortunately, Kruglyakov and Koretskaya [41, 107] give no direct experimental evidence to confirm that the stability of the asymmetric films changes with alcohol chain length in a manner that correlates with changing antifoam effectiveness.

Abe and Matsumura [108] have also considered the antifoam effect of alcohols on the foam behavior of an aqueous solution of sodium dodecylbenzene sulfonate. The alcohols included normal alcohols, branched alcohols, and diols. However, Abe and Matsumura [108] were concerned with the supposed role of the elimination of surface tension gradients under dynamic conditions in determining heterogeneous antifoam effectiveness in the case of these weakly polar oils. We have of course already considered the basic proposition of antifoam action by elimination of surface tension gradients in some detail in Section 4.4.3.

Abe and Matsumura [108] followed Ross and Haak [11] and measured the dynamic surface tension of dispersions of alcohols in solutions of sodium dodecyl benzene sulfonate, using the oscillating jet method. They revealed a correlation between the dynamic surface tensions at surface ages of 2×10^{-2} s and foamability in the case of the normal alcohols. Thus, under these conditions, the optimum antifoam is octanol, which produces the most rapid rate of surface tension reduction, as indicated by dynamic surface tension measurements. We should note here though that this correlation concerns measurement of dynamic surface tensions in the presence of 7.68×10^{-3} M alcohol concentration for comparison with foamability in the presence of 3.84×10^{-2} M alcohol (i.e., 5 g dm^{-3} in the case of octanol). However, at 7.68×10^{-3} M, the octanol is essentially ineffective as an antifoam (see Figure 23 in reference [108]) presumably because it is below the solubility limit. Moreover, increase in the octanol concentration above 7.68×10^{-3} M has little effect on the dynamic surface tension at a surface age of 2×10^{-2} s. All of this would tend to suggest that whatever the significance of the correlation, it has limited relevance for the basic proposition that rapid reduction of dynamic surface tension over the time scales measured by the oscillating jet will reduce foamability. Abe and Matsumura [108] in fact demonstrate the absence of any correlation between the rate of surface tension lowering (albeit again with different concentrations between surface tension and foamability measurements) for alkanediols. Here the optimum antifoam effectiveness is found with those compounds that yield the slowest rate of surface tension reduction. More recently, Arnaudov et al. [7] have shown that addition of both primary and secondary alcohols to aqueous solutions of sodium dodecylbenzene sulfonate can produce reductions in dynamic surface tension, which appear to concern the solubilized alcohol (since removal of excess undissolved alcohol does not restore the dynamic surface tension of the surfactant solution). In contradiction of Abe and Matsumura [108], these workers produce some experimental evidence that lowering the dynamic surface tension of surfactant solution by long-chain alcohols actually enhances the foamability. It seems likely then that reduction of the dynamic surface tension implies more rapid transport of surface active material to air–water surfaces and therefore enhanced foam film stability at the low surface ages prevailing during aeration.

Curiously, both Abe and Matsumura [108] and Kruglyakov and Koretskaya [107] have studied the effects of normal alcohols on the foam of aqueous sodium dodecylbenzene sulfonate solutions. Both find octanol to have the optimum chain length for antifoam effect at surfactant concentrations in the region of 1 g dm^{-3} (albeit at different temperatures and ionic strengths). They also find that the antifoam concentration for significant effect under these conditions is extremely high relative to that of the surfactant—in the region of 5–10 g dm^{-3} for octanol. Despite consensus concerning these experimental findings, the two groups of authors have adopted completely different hypotheses describing the mode of antifoam action of these alcohols. Surprisingly, Abe and Matsumura [108] do not even cite the earlier work of Kruglyakov and Koretskaya [107].

Further insight into the potential behavior of long-chain alcohols as antifoams has been provided by the study of Arnaudov et al. [7], which we have already occasionally cited. This study was based on the conclusion of Kruglyakov and coworkers [40, 41, 107] that antifoam effectiveness was determined by two main properties—solubility and polarity. Here high polarity supposedly means low adsorption of surfactant at oil–water surfaces, lower positive disjoining pressures, and therefore low pseudoemulsion stability. In this study, three alcohols were selected, two primary alcohols and a secondary alcohol. The primary alcohols were n-heptanol and n-dodecanol, where the former is distinguished by relatively high solubility and polarity. The secondary alcohol was 2-butyloctanol, which is distinguished by low solubility and low polarity. Despite the latter, Arnaudov et al. [7] argue that the branched chain structure will mean diminished adsorption of surfactant at the oil–water interface and therefore less stable pseudoemulsion films (presumably using an implicit reasoning similar to that quoted in Section 3.6.1.2 concerning interdigitation and monolayer penetration). Antifoam behavior was studied using saline solutions of sodium dodecylbenzene sulfonate (2.6 mM in 12 mM NaCl).

This study involved measurements of equilibrium entry, spreading and bridging coefficients, together with the applied critical capillary pressure, p_c^{crit}, necessary for rupture of the relevant pseudoemulsion films (as measured by the FTT). Results are summarized in Table 4.2. Reference to "initial" bridging coefficients is omitted because they are meaningless with systems where the initial spreading pressure is positive (see Section 4.5.1). Partial miscibility of the primary alcohols with water precluded measurement of initial values of these coefficients. Initial values for 2-butyloctanol are, however, given in the table. In this case, a positive initial spreading coefficient combined with a negative equilibrium spreading coefficient suggests

TABLE 4.2
Applied Critical Capillary Pressures, p_c^{crit}, and Entry, Spreading, and Bridging Coefficients for Short-Chain Alcohols (in Aqueous 2.6 mM Sodium Dodecylbenzene Sulfonate/12 mM NaCl)

Alcohol	E^i (mN m^{-1})	E^e (mN m^{-1})	S^i (mN m^{-1})	S^e (mN m^{-1})	B^e ((mN m^{-1})^2)	p_c^{crit} (Pa)	Temp. (°C)
2-Butyloctanol	+7.7	+4.8	+0.5	−2.4	+78	44	25 ± 2
n-Dodecanol	–	+2.7	–	−9.3	−137	>1500	27 ± 2
n-Heptanol	–	+6.9	–	−2.3	+144	–	25 ± 2

Source: Reprinted with permission from Arnaudov, L. et al. *Langmuir*, 17, 6999. Copyright 2001 American Chemical Society.

pseudo-partial wetting. Absence of a detectable oil film by ellipsometry means that the oil must simply form a mixed monolayer with the adsorbed surfactant in equilibrium with oil lenses.

Observation of the drainage of foam films containing drops of these alcohols using a Scheludko cell revealed that the oil drops were swept out of the films without causing film rupture. Indeed, the drainage behavior and stability of the films were apparently totally unaffected by the presence of the oils despite a reduction in equilibrium surface tension. Arnaudov et al. [7] then concluded that these alcohols must cause foam collapse by rupturing Plateau borders rather than foam films as described in Section 4.5.3.

Foam measurements for emulsified alcohols at a concentration of 0.01 wt.% made with the Ross–Miles technique are given in Figure 4.28. Results at zero time represent foamability measurements. Comparison of foamability results for heptanol and 2-butyloctanol with those in the absence of any alcohol reveals no effect. Since the concentration of heptanol is less than the solubility limit, this is hardly surprising. That there is no effect on foamability with 2-butyloctanol is presumably to be expected because capillary pressures and/or inertial forces during foam generation are too low to permit entry of drops into air–water surfaces. It is tempting to conclude that similar observations of the absence of effects of dispersed n-alcohols on foamability were made by Krugyakov and coworkers [40, 41, 107] for the same reason but were not reported because of their unexciting nature!

By contrast, the presence of dodecanol is seen in Figure 4.28 to have an adverse effect on foamability despite a negative bridging coefficient and an extremely high value of p_c^{crit}. It would seem therefore that the effect of dodecanol does not concern dispersed drops but rather the effect of solubilized dodecanol on the transport of surfactant to air–water surfaces. As we have noted in Section 4.2, Patist et al. [8] have shown that solubilization of dodecanol in micelles of SDS reduces the rate of micelle breakdown and therefore reduces that rate of transport of surfactant to air–water surfaces. In turn, this can reduce the amount of foam generated with certain methods involving high rates of aeration. It seems probable then that this effect could

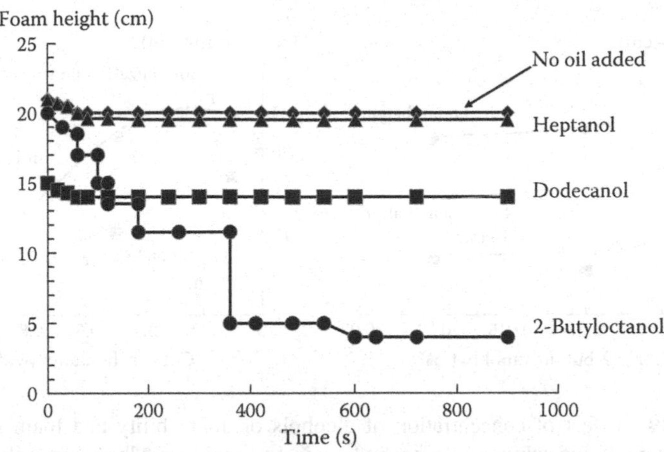

FIGURE 4.28 Effect of addition of 0.01 wt.% of various emulsified alcohols on the foamability and foam stability of 2.6 mM aqueous saline solution of a branched chain sodium dodecylbenzene sulfonate blend (using the dynamic Ross–Miles technique at 25–27°C). (Reprinted with permission from Arnaudov, L. et al. *Langmuir*, 17, 6999. Copyright 2001 American Chemical Society.)

also be responsible for the adverse effect of dodecanol on foamability in the case of sodium dodecylenzene sulfonate solutions. Indeed, Arnaudov et al. [7] show that the gradient of the reciprocal time dependence of the dynamic surface tension of saline solutions of dodecylbenzene sulfonate is increased by the presence of dodecanol, which is consistent with such a view.

Effects concerning the transport of surfactant to air–water surfaces in time scales of <1 s are not of course relevant for foam stability over time scales lasting several minutes. We therefore find that the extremely high values of p_c^{crit} and negative equilibrium bridging coefficients for n-dodecanol listed in Table 4.2 mean that foam generated from dispersions of that alcohol is quite stable. However, relatively low values of p_c^{crit} and positive values of the bridging coefficient for 2-butyloctanol suggest that it should be effective in destabilizing foam. Foam stability measurements given in Figure 4.28 confirm this expectation, where the foam height of 2-butyloctanol is seen to decline markedly by a series of steps. The residual foam height of about 4 cm can, however, be accounted for by Equation 4.41 only if the low probability large drops present in the measured size distribution are considered. The steps have been attributed to an avalanche of collapse events caused by an adjacent Plateau border failure due to the presence of a single drop, which satisfies Equation 4.37 so that the capillary pressure in the Plateau border exceeds p_c^{crit} [7].

The effect of alcohol concentration on both the initial and residual foam heights are shown in Figure 4.29 for both 2-butyloctanol and n-heptanol. The most striking difference concerns the displacement to concentrations almost two orders of magnitude higher in the case of n-heptanol. In part, this reflects the higher solubility of this alcohol. Despite this obvious difference, the plots exhibit some features that are

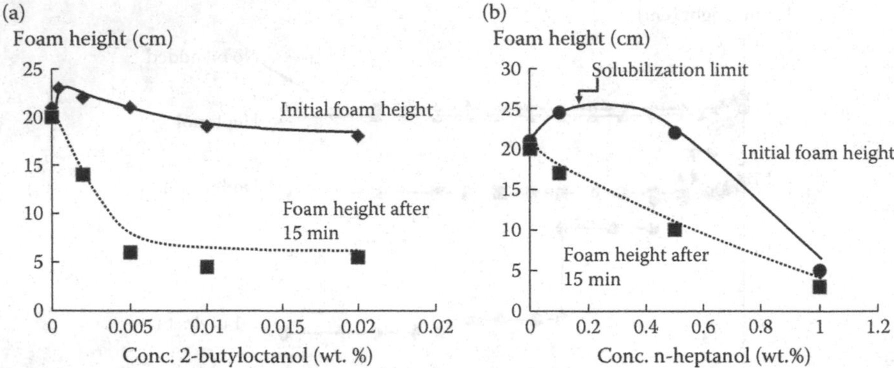

FIGURE 4.29 Effect of concentration of alcohols on foamability and foam stability of 2.6 mM aqueous saline solution of branched chain sodium dodecylbenzene sulfonate blend (using the dynamic Ross–Miles technique at 25–27°C). (a) 2-Butyloctanol. (b) n-Heptanol. (Reprinted with permission from Arnaudov, L. et al. *Langmuir*, 17, 6999. Copyright 2001 American Chemical Society.)

qualitatively similar. At low alcohol concentrations, both show a small increase in foamability (i.e., initial foam height) relative to that of the solution before addition of the alcohols. This may concern stabilization of foam films by the presence of molecularly adsorbed alcohol. Weak antifoam behavior is apparent at higher concentrations. However, in both cases, the foam decays to low residual foam heights, which decrease with increasing alcohol concentration. Consideration of Equation 4.41 suggests that residual heights will asymptotically approach a limit as the tail of the drop size distribution is approached, which appears to have been realized in the case of 2-butyloctanol.

In summary then, we note that Arnaudov et al. [7] have shown that the antifoam effect of emulsified short-chain alcohols probably does not concern formation of unstable bridging configurations in foam films but rather concerns their formation in Plateau borders as described in Section 4.5.3. In the main, this derives from the relatively high stabilities of the relevant pseudoemulsion films. As a consequence, foam collapse is mainly confined to high foam ages where the necessary high capillary pressures are built up in the Plateau borders as the foam drains. Foam collapse is therefore characterized by an induction period that can last several minutes. Direct addition of drops of antifoam to an existing well-drained foam can, however, produce more rapid foam collapse. This is shown by, for example, addition of drops of heptanol or 2-butyloctanol to the drained foam formed by saline sodium dodecylbenzene sulfonate solutions [7]. More rapid foam collapse presumably follows because direct addition of the alcohol obviously diminishes the role of the stability of the pseudoemulsion film. The capillary pressure in the drained foam will also tend to suck the antifoam directly into the plateau borders. However, subsequent regeneration of foam after such collapse also disperses the alcohol so that the stability of the pseudoemulsion film again becomes an issue and the resultant foam collapses more slowly and to a higher residual level. In the case of n-heptanol, the foamability and stability is eventually completely restored to that intrinsic to the surfactant solution as the alcohol is completely solubilized.

Finally, we emphasize that we have been concerned here exclusively with the behavior of short-chain alcohols, which function as liquid oil antifoams. We briefly return to the topic of the antifoam behavior of alcohols when we consider the effect on antifoam action of the melting behavior of various weakly polar and nonpolar compounds, including long-chain alcohols, in Section 4.7.7.

4.6.1.3 n-Alkanes

Aveyard and coworkers [24, 44, 66] have made studies of the effect of n-alkanes on the foam and film behavior of both aqueous hexadecyl ammonium bromide ($C_{16}TAB$) and saline AOT solutions. In the case of the former, the study only concerned dodecane [44]. Addition of drops of this alkane to the air–water surface of aqueous solutions of $C_{16}TAB$ produces pseudo-partial wetting behavior at equilibrium (where both E^i and E^e are positive). The oil first spreads as a duplex film, which retracts to form lenses—behavior that seems to be general for the lower n-alkanes on aqueous micellar solutions of ionic surfactants (see Section 3.6.1.2). We therefore have $S^i > 0$ and $S^e < 0$ where the spreading pressure, $\Delta\sigma^{sp}_{AW}$, is ~7 mN m^{-1} for a micellar solution (see Figure 3.15). Equilibrium bridging coefficients are positive. However, initial bridging coefficients are meaningless, as we have shown in Section 4.5.1, because the initial spreading coefficients are positive.

Foam was generated by cylinder shaking, where the oil was added apparently without pre-emulsification. Addition of 5 vol.% of the oil to a micellar surfactant solution of $C_{16}TAB$ (of 1 mM) produced an initial 40% volume reduction of foam, which slowly approached 90% after about 400 min. The reduction in foamability (i.e., initial foam volume) could be due to bridging drops in foam films where the critical capillary pressures required for pseudoemulsion film rupture are diminished by low surfactant adsorptions due to slow surfactant transport rates for these dilute solutions (all ≤1.6 mM). However, slow collapse of foam over several hours suggests that oil drops not effective in causing initial foam collapse accumulate in Plateau borders. Bridging can then lead to eventual collapse after capillary pressure build up, during drainage, to the critical values needed to overcome positive equilibrium disjoining pressures (as described in Section 4.5.3). Equilibrium or near-equilibrium conditions with pseudo-partial wetting of course mean that negative spreading coefficients apply and Marangoni spreading cannot have a role.

Aveyard et al. [24, 66] have made a more extensive study of the effect of n-alkanes on the foam behavior of a solution of a different surfactant—an aqueous saline solution of AOT (3.8 mM in 0.03 M NaCl). This solution is turbid due to the presence of dispersed lamellar phase as shown in Figure 4.3. The study involved all n-alkanes from hexane to hexadecane together with squalane. Again foam was generated by cylinder shaking with 5 vol.% oil added without pre-emulsification.

The classic semi-initial and equilibrium entry and spreading coefficients, together with equilibrium bridging coefficients and spreading pressures, $\Delta\sigma^{sp}_{AW}$, for this system are listed in Table 4.3. Up to and including undecane, the initial and equilibrium entry coefficients are positive. For higher alkanes, they are seen to be essentially

TABLE 4.3
Classic Entry, Spreading, and Bridging Coefficients for n-Alkanes in 3.8 mM AOT Solutions in 0.03 M NaCl at 25°C

Alkane	E^{si} (mN m^{-1})	E^{e} (mN m^{-1})	S^{si} (mN m^{-1})	S^{e} (mN m^{-1})	$\Delta\sigma^{sp}_{AW}$ (mN m^{-1})[a]	B^{e} (mN^2 m^{-2})
n-Hexane	+7.5	+0.2	+7.4	+0.1	7.3	+3.6[b]
n-Heptane	+5.6	+0.2	+5.4	0.0	5.4	+4.0[b]
n-Octane	+4.1	+0.1	+3.9	0.0	3.9	+0.01[b]
n-Nonane	+3.0	+0.5	+2.8	+0.3	2.5	+18.1[b]
n-Decane	+2.1	+0.5	+1.7	+0.2	1.5	+14.1[b]
n-Undecane	+1.3	+0.3	+0.9	–0.1	1.0	+24.5[b]
n-Dodecane	+0.6	–0.1	+0.2	–0.5	0.7	–15.0[c]
n-Tetradecane	–0.5	–0.1	–0.4	–0.8	–0.4	–20.1[c]
n-Hexadecane	–1.3	–0.2	–0.9	–1.1	0.2	–35.4[c]
Squalane	–1.8	0.0	–2.4	–2.4	0.0	–60.7[c]

Source: Aveyard, R. et al., *JCS Faraday Trans.*, 89, 4313, 1993.

Note: Uncertainties in E^{i}, E^{e}, S^{i}, S^{e}, and $\Delta\sigma^{sp}_{AW}$ are estimated at ±0.2 mN m^{-1}.

[a] $\Delta\sigma^{sp}_{AW}$ is given by Equation 3.29 and must be ≥0.

[b] These positive values of B^{e} are close to the maximum values permitted corresponding to complete wetting by a duplex oil film so that $\theta^* \to \pi$ (according to Equation 4.31). They are therefore in good agreement with the corresponding near-zero values of S^{e}.

[c] These negative values of B^{e} are close to the minimum values permitted corresponding to complete wetting by a pseudoemulsion film so that $\theta^* \to 0$. They are therefore in good agreement with the corresponding near-zero values of E^{e}.

zero if due consideration is given to the likely magnitude of experimental uncertainties. Therefore, the lower alkanes will enter the air–water surface under equilibrium conditions and the higher alkanes will not. The relevant equilibrium bridging coefficients for these higher alkanes are negative and close to the minimum values permitted for complete wetting, implying that $\theta^* \to 0$. These alkanes will not therefore exhibit any antifoam effect under those conditions. The initial spreading coefficients for the lower alkanes, up to and including undecane, are seen to be positive and decline with increasing chain length of the alkane. The equilibrium spreading coefficients are close to zero. The equilibrium bridging coefficients are positive and close to the maximum values for complete wetting by the oil, which implies that $\theta^* \to 180°$. However, as we have seen, Kellay et al. [109, 110] state that addition of decane to the surface of a 3 mM AOT in 0.3 M NaCl forms pseudo-partial wetting films of ~4 nm thickness in equilibrium with lenses (see Section 3.6.1.2). Indeed, they state that pseudo-partial wetting on such solutions is to be expected for all alkanes of chain length up to decane. With uncertainties of at least ±0.2 mN m^{-1}, it is possible then that the equilibrium spreading coefficients listed in Table 4.3 are not sufficiently accurate to distinguish between complete wetting and pseudo-partial wetting where the dihedral angle for the lenses is small. Nevertheless, we find that with these systems, the equilibrium behavior flips at undecane from almost complete oil-on-water wetting for the lower alkanes to almost complete water-on-oil wetting for the higher alkanes.

In all cases, these alkanes reduce the foamability of the AOT solutions—however, the effects are relatively small, involving at most ~30% reduction in foam volume relative to the intrinsic foam volume generated by the surfactant solution alone. Once formed, the resulting foam is stable for a few minutes before further collapse starts. It seems possible then that the phenomenon of reduction in foamability has a different origin from that of any subsequent foam collapse. It may involve large drops of alkane because foam generation and dispersal appear to occur simultaneously with the method used in this work. Indeed, that foamability is reduced by undecane and the higher alkanes, all of which exhibit negative semi-initial and essentially zero equilibrium entry coefficients, presumably must concern low non-equilibrium levels of surfactant adsorption at rapidly forming air–water surfaces during foam generation as a result of slow surfactant transport to those surfaces. None of the reported measurements by Aveyard et al. [66] allow inference about the state of the air–water surfaces under those conditions. It is, however, tempting to suggest that rupture actually occurs in foam films due to the formation of oil bridges as outlined in Section 4.5.1 even if complete wetting prevails, which in any case seems unlikely in view of the findings of Kellay et al. [109, 110].

Several minutes after foam generation has ceased, a process of foam collapse commences with these systems. A plot of the time for onset of that collapse for each alkane is shown in Figure 4.30. The foam is seen to be intrinsically unstable. In the case of undecane and the higher alkanes, the collapse commences at the same time as the foam in the absence of the alkanes as indicated in the figure. In the case of the lower alkanes, foam collapse commences much earlier—after a couple of minutes with hexane through to 6 min for decane, for example. A plot of foam half-life

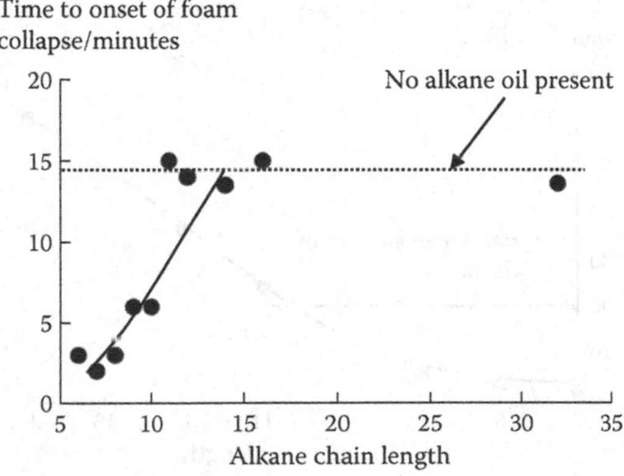

FIGURE 4.30 Time for onset of foam collapse of saline solution of 3.8 mM AOT containing 5 vol.% emulsified alkanes. (Adapted from Aveyard, R., Binks, B.P., Fletcher, P.D.I., Peck, T., Garrett, P.R., *JCS Faraday Trans.*, 89, 4313, 1993. Reproduced by permission of The Royal Society of Chemistry.)

against alkane chain length is shown in Figure 4.31. The foam half-life is seen to be increased by the higher alkanes from undecane. These alkanes actually stabilize the foam. Since most of these oils cannot enter the air–water surface but do reduce foam film drainage rates at film thicknesses significantly less than drop diameters, Aveyard et al. [66] conclude that it is likely that this is a result of their accumulation in Plateau borders as described by Kozo et al. [15]. However, it is therefore difficult to explain that the time for commencement of foam collapse is unaffected by the presence of these higher alkanes.

Foam film collapse by the lower alkanes after an induction time of several minutes does suggest that, as with the alcohols, most oil drops are also swept out of foam films only to accumulate in the Plateau borders despite positive entry coefficients. The induction time then derives from the time required for the capillary pressure at the top of the foam column to increase, as a result of drainage, until the relevant pseudoemulsion films rupture. Absence of information about residual foam heights, drop sizes, and applied critical capillary pressures for pseudoemulsion film rupture means, however, that more detailed interpretation using Equations 4.40 and 4.41 is not possible. Some indication that the stability of the pseudoemulsion film may be significant is revealed by measurements of the half-lifetimes of individual foam films in the presence of the alkanes and pseudoemulsion films. Results are shown in Figure 4.32 where single film half-life times are plotted against alkane chain length for both single foam films and single pseudoemulsion films. That these different measurements correlate well with each other, indicating lifetimes of the order of minutes, appears to be consistent with the view that the rate-determining step in foam collapse concerns rupture of the pseudoemulsion film.

It is, however, difficult to see how the actual foam film rupture events induced by the lower alkanes could concern Marangoni spreading as suggested by Aveyard et al. [66]. The solutions were pre-equilibrated for an hour, the foam was already a few

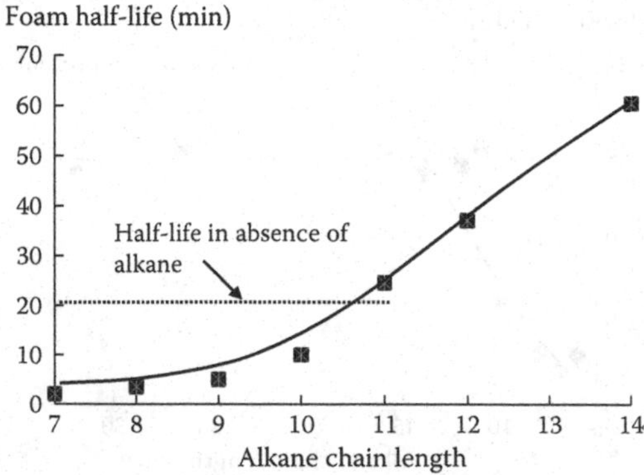

FIGURE 4.31 Effect of 5 vol.% emulsified alkanes on foam half-life of a saline solution of 3.8 mM AOT. (Adapted from Aveyard, R., Binks, B.P., Fletcher, P.D.I., Peck, T., Garrett, P.R., *JCS Faraday Trans.*, 89, 4313, 1993. Reproduced by permission of The Royal Society of Chemistry.)

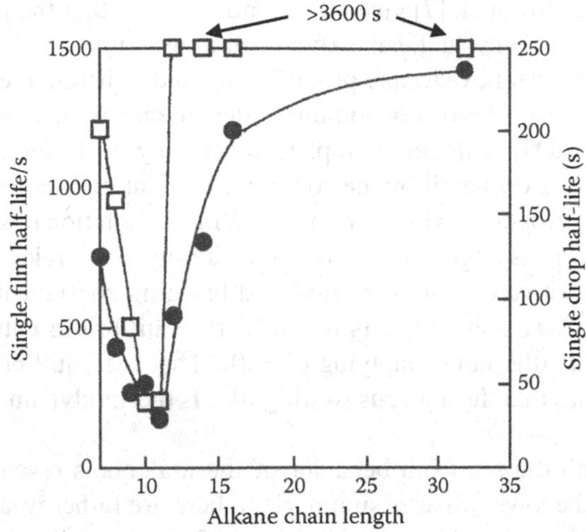

FIGURE 4.32 Effect of chain length of alkanes on half-life of single foam films prepared from 5 vol.% emulsions of the oils in saline solution of 3.8 mM AOT (●) and half-life of single drops of alkane under air–solution surface as an indication of pseudoemulsion film stability (□). (Reprinted from *Adv. Colloid Interface Sci.*, 48, Aveyard, R., Binks, B.P., Fletcher, P.D.I., Peck, T.G., Rutherford, C.E., 93. Copyright 1994, with permission from Elsevier.)

minutes old by the time rupture commenced, and once the top layer of foam is ruptured the air–water surfaces of lower layers will become directly irrigated with oil (where the volume fraction is at least 5 vol.%). Entry of drops into air–water surfaces as a result of rupture of the relevant pseudoemulsion films is therefore unlikely to be accompanied by the rapid spreading indicated by the positive semi-initial spreading coefficients listed in Table 4.3. It is more likely that emerging drops will not spread but will simply form lenses in equilibrium with the already alkane-contaminated surfaces as indicated by Kellay et al. [109, 110]. Foam collapse then concerns the destabilizing effect of an oil lens bridging a Plateau border.

Arnaudov et al. [7] have studied the effect of hexadecane on the foam of a saline micellar solution of sodium dodecylbenzene sulfonate (2.6 mM in 12 mM NaCl). Under these conditions, the equilibrium entry and bridging coefficients are positive and the equilibrium spreading coefficient is zero. The latter implies complete wetting by the oil with duplex film formation. The presence of this oil film apparently increases the applied critical capillary pressure for pseudoemulsion film rupture from ~80 Pa to ~400 Pa (see Section 3.6.3). Observation of the drainage of foam films in a Scheludko cell failed to reveal any film rupture events before the oil was removed as the films thinned. Any destabilization of foam would then have to concern the effect of bridging oil lenses in the Plateau border as the relevant pseudoemulsion films rupture at capillary pressures determined by p_c^{crit}, as suggested by Equation 4.40. All of this implies that hexadecane drops should destabilize the Plateau borders of the foam of this saline sodium dodecylbenzene sulfonate solution provided the foam height exceeds ~4 cm. Using only ~0.011 vol.%, some enhancement of foamability is observed with the Ross–Miles technique with no evidence of instability until after

1 h. However, Arnaudov et al. [7] give no indication of whether the foam height then declines to the expected residual value after even longer times.

We therefore find that hexadecane produces markedly different equilibrium wetting behavior on saline solutions of sodium dodecylbenzene sulfonate from that on saline solutions of AOT, with near-complete wetting by oil in the former case and near-complete wetting of the oil by the solution in the latter case. However, in both cases, little or no antifoam activity is observed. With the solution of sodium dodecylbenzene sulfonate, this is largely due to the metastability of the relevant pseudoemulsion film despite positive equilibrium entry and bridging coefficients [7]. However, in the case of the AOT solution, it is due to zero equilibrium entry and negative bridging coefficients (the latter implying $\theta^* = 0$). That the equilibrium entry coefficient is zero implies that the aqueous wetting film is thermodynamically stable (see Section 3.2).

Comparison with the antifoam behavior of the n-alcohols reveals that the antifoam effects with the lower alkanes summarized here are rather weak. For example, almost total foam collapse of saline solutions of 2.6 mM sodium dodecylbenzene sulfate by ~0.01 wt.% (~0.011 vol.%) 2-butyloctanol (see Figure 4.29) compares with the use of 5 vol.% of alkanes with saline solutions of 3.8 mM AOT in the study described here [66]. Clearly, neat alkane oils do not represent effective antifoams!

4.6.1.4 Neat Polydimethylsiloxane Oils

PDMS oils are well-known ingredients of hydrophobic oil–hydrophobic particle mixed antifoams. They are ubiquitous in their application for aqueous systems. However, the oils alone are relatively ineffective for most aqueous solutions of surfactants with hydrocarbon chains despite extremely low solubilities [111], positive spreading pressures, initial high entry and spreading coefficients, positive equilibrium entry coefficients, and zero or near-zero equilibrium spreading coefficients. The oils therefore enter air–water surfaces where they usually initially spread, exhibiting either complete or pseudo-partial wetting at equilibrium as shown by the summary of typical behavior in Table 3.3. Much of this wetting behavior is attributable to the low air–oil surface tensions of these oils.

Perhaps the main practical interest in the potential antifoam behavior of neat PDMS oils concerns their application as conditioners in high-foam shampoo products. Minimization of any antifoam effect is the main concern in this context. As a consequence of this concern, Denkov and coworkers [1, 7, 58–60, 62] have made a number of studies of the effect of such neat oils on the foam behavior of several surfactant systems, including the role of so-called foam boosters [58, 59]. A summary of the main findings of these studies is given in Table 4.4. Foam was generated by air entrainment by either shaking measuring cylinders or use of the Ross–Miles technique (see Chapter 2). In all cases with these solutions, the generated foam volumes in the absence of PDMS were essentially stable over the relevant periods. Some indication of the effect of PDMS oils on foamability (i.e., initial foam volumes formed immediately after cessation of foam generation) is given by the ratio F. The latter is the ratio of the volume of foam formed in the presence of antifoam to that formed in the absence of antifoam. Despite positive initial spreading coefficients and equilibrium bridging coefficients, F is seen to be close to unity in almost every case,

TABLE 4.4

Comparison of Applied Critical Capillary Pressures, p_c^{crit}, Entry, Spreading, and Bridging Coefficients with Foam Behavior for Neat Polydimethylsiloxane Oils with Various Surfactant Solutions

Surfactant[a]	PDMS Viscosity (mPa s)	Concn. PDMS (wt.%)	E^i (mN m^{-1})	E^e (mN m^{-1})	S^i (mN m^{-1})	S^e (mN m^{-1})	B^e (mN m^{-1})^2	p_c^{crit} (Pa)	Foam Method	F ($t = 0$)[b]	$V(t)/V(0)$ ($t = 300$ s)[c]	$V(t)/V(0)$ ($t = 900$ s)[c]	Ref.
2.6 mM $C_{12}\phi SO_3Na$ + 12 mM NaCl	5	0.01	+17.5	+10.4	+6.3	−0.8	234	>3000	Ross–Miles	1.0	0.90	0.86	[7]
10 mM AOT	1000	0.01	+12.1	+9.6	+2.5	−0.2	246	19	Cylinder shaking	1.0	0.05	0.05	[1, 54, 62]
1 mM Triton X-100	1000	0.01						>200	Cylinder shaking	0.76	0.38	–	[62]
0.45 mM APG	1000	0.005	+12.6	+8.4	+3.4	−0.8	193	>1250	Cylinder shaking	0.92	–	0.95	[1, 60]
100 mM $C_{12}EO_{2.5}SO_4Na$[d]	5	0.1	+20.2	+12.8	+6.6	−0.8	320	180	Ross–Miles	1.2[e]	0.46	0.28	[58]
100 mM $C_{12}APB$[f]	5	0.1	+19.0	+13.7	+4.6	−0.7	350	>7000	Ross–Miles	1.2[e]	0.93	0.91	[58]
80 mM $C_{12}EO_{2.5}SO_4Na$ + 20 mM $C_{12}APB$[g]	5	0.1	+16.0	+9.6	+5.2	−1.2	213	850	Ross–Miles	1.0	0.9	0.90	[58]

Sources: Denkov, N.D., *Langmuir*, 20, 9463, 2004; Denkov, N.D. et al., Mechanism of foam destruction by emulsions of PDMS-silica mixtures, in *Proceedings of 3rd World Congress on Emulsions*, 24–27 September 2002, Lyon, France, paper 1-D-199; Denkov, N.D. et al., *Langmuir*, 15, 8514, 1999; Basheva, E.S. et al., *Langmuir*, 16, 1000, 2000; Marinova, K.G., Denkov, N.D., *Langmuir*, 17, 2426, 2001; Denkov, N.D. et al., *Langmuir*, 18, 5810, 2002.

[a] Surfactant types: $C_{12}\phi SO_3Na$, sodium dodecylbenzene sulfonate (branched-chain blend); AOT, sodium bis-diethylhexyl sulfosuccinate; Triton X-100, octyl phenol decaethylene glycol ether; APG, commercial $C_{12/14}$(glucopyranoside)$_{1,2}$; $C_{12}EO_{2.5}SO_4Na$, commercial sodium dodecyl-polyoxyethylene-3-sulfate; $C_{12}APB$, commercial dodecyl amino-propyl betaine.

[b] F = volume of foam in presence of PDMS/volume of air in absence of PDMS.

[c] $V(t)$ = volume of foam in presence of PDMS at time t; $V(0)$ = volume of foam in presence of PDMS at time $t = 0$.

[d] Also contains 66 mM ionic strength simple electrolyte impurities.

[e] Values of $F > 1$ indicate some apparent enhancement of foamability due to the presence of PDMS.

[f] Also contains 142 mM ionic strength simple electrolyte impurities.

[g] Also contains ~81 mM ionic strength simple electrolyte impurities.

implying little or no antifoam effect at all. Indeed, values of $F > 1$ are even seen to occur!

The effect of the PDMS oil on foam stability was, however, more interesting. In some cases, the ratio of the foam volume after a few minutes to the initial foam volume decreased markedly. Significant foam collapse is seen to correlate with low values of the applied critical capillary pressure rather than any of the coefficients listed in Table 4.4. Examination of the draining behavior of small horizontal films of PDMS emulsions in these surfactant solutions also revealed that the oil drops were removed within <60 s as the films drained without any adverse effect on film stability. Basheva et al. [58, 59] therefore concluded that, as with the neat alcohols and alkanes, the PDMS drops accumulate in Plateau borders and only cause foam collapse when the capillary pressure reaches the critical value required to rupture the pseudoemulsion film, as described in Section 4.5.3. Relatively low volume fractions of PDMS of $\leq 10^{-3}$ are, however, involved. Nevertheless, Basheva et al. [58] produce images verifying the accumulation of continuous strings of PDMS drops in Plateau borders, which implies increased concentrations in the foam as a result of drainage and diffusional disproportionation. Estimates of the critical capillary pressures required for foam collapse calculated from Equation 4.40 using residual foam heights gave values of the same order as directly measured values of p_c^{crit} for sodium dodecyl polyoxyethylene sulfate ($C_{12}E_{2.5}SO_4Na$) and 80/20 mole ratio sodium dodecyl polyoxyethylene sulfate + betaine (C_{12} amino-propyl betaine) mixtures.

Of potential practical importance is the observation that reduction of drop size produces enhanced foam stability with these systems. In part, this derives from the relative dimensions of the Plateau borders and the PDMS drops as revealed by consideration of Equations 4.40 and 4.41. If the drop size is too small, it will simply pass through the Plateau borders without being subject to the applied capillary pressure required to rupture the relevant pseudoemulsion films. However, in the case of 0.1 M solutions of sodium dodecyl polyoxyethylene sulfate, the critical capillary pressure for rupture of the pseudoemulsion film also actually increases with decreasing drop size as shown in Figure 3.7. Enhanced stability of the foam of this solution could also be achieved by addition of a betaine foam booster as shown in Table 4.4. The principle effect of the betaine in this context would appear to be an increase in p_c^{crit}.

All of this would appear to leave open the issue of the precise mechanism by which foam collapse occurs when the pseudoemulsion films in Plateau borders are ruptured. In an attempt to elucidate that mechanism, Basheva et al. [58, 59] have directly observed the rupture of large vertical foam films formed in glass frames. In one experiment, rupture was induced by direct addition of drops of PDMS to the Plateau border of a preformed film. This procedure circumvents any issues associated with the stability of the relevant pseudoemulsion film. In another experiment, films were formed from PDMS emulsions where drops accumulated in Plateau borders. At sufficiently high capillary pressures, the relevant pseudoemulsion films ruptured to release PDMS onto the air–water surface. In both cases, the oil was observed to spread and cause the foam film to *increase* in thickness, as a result of a Marangoni effect before film rupture. It is therefore difficult to see how rupture could be attributed to a Marangoni-spreading mechanism as suggested by Basheva et al. [58]. That would require that the film thickness decrease catastrophically. Basheva et al. [58]

also suggest that emergence of a drop from a Plateau border could mean the release of sufficient mechanical stress to burst the adjacent foam films However, that would not be consistent with the observation of foam film rupture by direct addition of an oil drop to a preformed film. It seems more likely then that the oil spreads to draw out more oil from the Plateau border to produce a bridging–stretching instability as discussed in Section 4.5.3.

We find then that the similarities in antifoam behavior of neat oils represented by alcohols, alkanes, and PDMSs are rather more obvious than the differences. In all cases, it would seem that the dominant factor is the magnitude of the critical capillary pressure required to rupture the relevant pseudoemulsion films. The high stability of that film means that oil drops are readily removed from foam films before causing any rupture. It is the application of increasing capillary pressures upon drops trapped in Plateau borders during foam drainage that causes rupture of the relevant pseudoemulsion films. However, the actual mode of foam film rupture by drops, which then enter the air–water surface, is not well understood.

4.6.2 Antifoam Effects of Neat Oils in Non-Aqueous Foaming Systems

There have apparently been comparatively few scientific studies of antifoam action in non-aqueous media. We therefore have little systematic information concerning the wetting behavior of typical antifoam oils in non-aqueous systems. Moreover, there would appear to be no information concerning the stability of the relevant pseudoemulsion films. Clearly if the antifoams are to be effective they should at least enter the surface of the liquid. The expectation is that the low dielectric constants of non-aqueous liquids such as hydrocarbons will mean little tendency for charge separation to occur and limited role for electrostatic effects in stabilizing both pseudoemulsion films and foam films. In which case the sign of the entry and bridging coefficients may be sufficient to determine antifoam action. Foam rupture by antifoam drops with positive entry and bridging coefficients should therefore be relatively rapid so that accumulation of drops in Plateau borders will not occur. Insolubility of the antifoam oil is also found to be a necessary property [5, 6, 112], although again not without some equivocation (see, e.g., reference [113]). This issue is discussed in some detail in Chapter 10 where we consider the criteria for selection of antifoams for natural gas–crude oil systems.

We have already considered (see Section 4.4.2) the study, made more than 50 years ago, of the antifoam mechanism of PDMS oils on lube oil foam by Shearer and Akers [5]. Here we remember that these authors used a semiquantitative Marangoni spreading argument to account for antifoam action. Complete elimination of foam was observed at concentrations of PDMS of <20 parts per million. Direct experimental observations using high-speed photography confirmed the entry of PDMS drops into the air–oil surface of bubbles where the observations seemed to suggest pseudo-partial wetting. Not unexpectedly, no evidence of electrostatic effects was found. Dissolution of the PDMS oil in the lube oil was found to actually enhance the foamability.

Another rare study of the antifoam mechanism of PDMS oils in a non-aqueous medium was presented 25 years later by Callaghan et al. [114]. In contradiction of

Shearer and Akers [5], Callaghan et al. [114] supposed that the mechanism involves elimination of surface tension gradients (see Section 4.4.3) as indicated by elimination of surface elasticity. These authors studied the effect of PDMSs on the surface elasticity of crude oil. PDMSs are used as antifoams to assist gas–oil separation during crude oil production and are apparently effective at the remarkably low concentration of 1 part per million (which presumably still exceeds the solubility limit). Callaghan et al. [114] find that PDMS diminishes the frequency-dependent dynamic dilational (elastic) modulus $|\varepsilon| = d\sigma_{AO}(t)/d \ln A(t)$ relative to that found for the uncontaminated oil. Here $\sigma_{AO}(t)$ is the time-dependent air–crude oil surface tension, and $A(t)$ is the area of a constrained element of air–crude oil surface subject to time-dependent dilation. The effect is more marked the higher the molecular weight (or viscosity) of the PDMS. This correlates with an enhanced antifoam effectiveness found with increase in molecular weight.

Callaghan et al. [114] argue that all of this represents evidence that the PDMS functions by elimination of surface elasticity and therefore, by implication, surface tension gradients. However, McKendrick et al. [115] find a frequency-independent increase in dynamic dilational modulus of a hydrocarbon oil upon addition of a fluorosilicone antifoam. Moreover, the correlations of Callaghan et al. [114] concern extremely long time scales in the range 10–10^3 s. Whether these correlations are fortuitous or represent clear evidence for the proposition under test requires additional information concerning the relevance of those time scales. The absence of hard evidence concerning the magnitude of the relevant time scales is thrown into high relief if comparison is made between the approaches of Abe and Matsumura [108] (see Section 4.6.1.2) and Callaghan et al. [114]. Both of these two groups of workers have sought evidence for the same basic proposition concerning the elimination of surface tension gradients in foam films by antifoam oils, albeit for PDMS on crude oil in the latter case and short-chain alcohols on aqueous surfactant solutions in the former case. Surprisingly, the time scales considered by these two groups of workers are separated by more than three orders of magnitude despite differences in the bulk phase viscosities of the foaming liquids of only about one order of magnitude! It would also seem unlikely that the use of a pneumatic method of foam generation by Callaghan et al. [114] and the Ross–Miles method by Abe and Matsumura [108] could account for such large differences in time scales.

Effective antifoam oils for non-aqueous systems appear to require low solubility in the foaming liquid together with positive entry and, presumably, positive bridging coefficients and a viscosity significantly higher than that liquid. These requirements are apparently not always readily satisfied by typical PDMS oils, with relatively high solubility occasionally representing a serious limitation [113]. Antifoam manufacturers seek to avoid this limitation by using oils with molecular structures based on a PDMS backbone with grafted polyoxyethylene (EO) or polyoxypropylene (PO) groups, EO–PO copolymeric groups and perfluoro groups (see Section 10.4). Examples of the types of possible structures suitable for use with diesel and "mobile" fuel oils are given by Wu et al. [116].

As we will describe at length in Section 4.8, antifoam oils for aqueous solutions are usually mixed with hydrophobic particles where the main function of the latter concerns rupture of the air–water–oil pseudoemulsion films. Since the stability of those films is often attributable largely to electrostatic effects, it would seem likely that such particles would have a limited role in the case of antifoam oils for non-aqueous systems where such effects are expected to be largely absent. This is consistent with the usual practice of apparently using neat antifoam oils for non-aqueous foaming liquids. However, formation of unstable bridging configurations by the antifoam oil in foam films where either complete wetting or lens formation with small dihedral angles occurs (i.e., where $\theta^* \rightarrow \pi$) would presumably require oils with significantly higher viscosity than that of the foaming liquid [70]. In the case of complete wetting, for example, such an oil drop could in principle dewet into both surfaces of the foam film and cause foam film collapse by the bridging–stretching mechanism described in Section 4.5.1. The absence of modern experimental studies of the antifoam mechanism of neat oils in non-aqueous liquids means of course that such considerations are rather speculative. The practical issues involved in the use of neat oils as antifoams for such liquids are considered in more detail in Chapter 10.

4.6.3 Partial Miscibility and Antifoam Effects

4.6.3.1 Origins of Partial Miscibility in Binary Liquid Mixtures

The free energy of mixing of a pair of completely miscible liquids is everywhere negative and is a simple function of the composition with a single minimum. By contrast, the free energy of mixing $\Delta G(\psi)$ (where ψ is the mole or weight fraction of one component) of a pair of liquids that exhibit the property of partial miscibility has two minima as shown if Figure 4.33. This behavior arises because the contribution of adverse intermolecular interactions between the two liquids opposes the contribution of mixing entropy, which favors mixing. Consider then the compositions lying between the points ψ_1 and ψ_2 on the tangent to $\Delta G(\psi)$ drawn in the figure. All points on the tangent represent the total free energies of various ratios of immiscible liquids of compositions $(\psi_1, 1 - \psi_1)$ and $(\psi_2, 1 - \psi_2)$, which are seen to be lower than the free energies of the corresponding mixed system given by the free energy of mixing

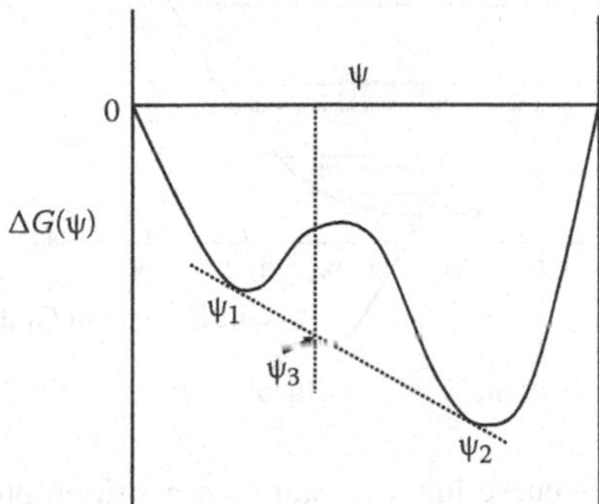

FIGURE 4.33 Schematic showing free energy of mixing $\Delta G(\psi)$ against mole fraction ψ of one component in binary mixture exhibiting partial miscibility. Various proportions of phases having compositions ψ_1 and ψ_2 all have lower free energies of mixing than that of single-phase mixture represented by solid curve.

curve. Therefore, at equilibrium, two separate liquid phases exist over the compositions range $\psi_1 < \psi < \psi_2$ with single phases existing over composition ranges $0 < \psi < \psi_1$ and $\psi_2 < \psi < 1$, giving rise therefore to partial miscibility. The compositions $(\psi_1, 1 - \psi_1)$ and $(\psi_2, 1 - \psi_2)$ are often described as the compositions of conjugate solutions.

In general, the magnitude of the adverse intermolecular interactions that give rise to this behavior are temperature dependent. In some cases, they decrease with increasing temperature. This is often the case with, for example, mixtures of simple organic liquids such as nitrobenzene and hexane. In other cases, the adverse intermolecular interactions increase with increasing temperature. This type of behavior is exemplified by mixtures of water and ethoxylated alcohol surfactants, which are of some special relevance in the present context.

If the temperature is changed in the direction of diminution of the relative significance of adverse intermolecular interactions, then ultimately entropy dominates and the free energy of mixing becomes a simple function of composition with a single minimum—total miscibility is then achieved at the critical (or consolute) point (or temperature). If then we plot the compositions ψ_1 and ψ_2 against temperature, we obtain the coexistence curve in the phase diagram. An example of such a diagram is given in Figure 4.34 for the system $H_2O\text{-}C_{10}H_{21}(OC_2H_4)_4OH$ ($C_{10}EO_4$) [117]. Here immiscibility derives from the effect of increasing temperature on the hydration of

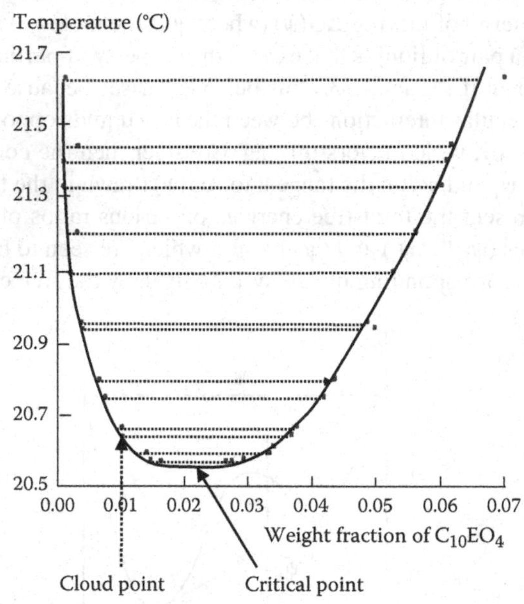

FIGURE 4.34 Coexistence curve in temperature-composition phase diagram for the system $H_2O\text{–}C_{10}EO_4$. Critical point and so-called cloud point are indicated (where the latter is somewhat arbitrarily determined at weight fraction 0.01). At temperatures higher than critical point phase, separation occurs and solution becomes turbid. (Reprinted with permission from Lang, J., Morgan, R., *J. Chem. Phys.*, 73, 5849. Copyright 1980, American Institute of Physics.)

the ethoxy groups of the surfactant so that those groups become more hydrophobic. The two conjugate phases are both micellar—one relatively dilute and the other more concentrated, as indicated by the tie lines. Increasing the temperature at a given composition (usually 1 wt.%) until the coexistence curve is reached results in phase separation whereupon the solution turns "cloudy" with this type of compound. The corresponding temperature is the cloud point.

An issue relevant for potential antifoam behavior for compositions within the coexistence curve concerns the physical state of the two conjugate solutions. Essentially they form an emulsion, the continuous phase of which is in part determined by the relative amounts of each phase. Consider then the tie line shown in Figure 4.33 joining compositions ψ_1 and ψ_2. The relative amounts of each phase $L(\psi_1)$ and $L(\psi_2)$ at the overall composition ψ_3 on the tie line is then given by the lever rule so that

$$L(\psi_1)/L(\psi_2) = (\psi_2 - \psi_3)/(\psi_3 - \psi_1) \tag{4.42}$$

The closer ψ_3 to ψ_1, the greater the ratio of $L(\psi_1)$ to $L(\psi_2)$ and the greater the probability that the phase of composition $(\psi_1, 1 - \psi_1)$ will be the continuous phase. For antifoam behavior within the two-phase region, it is obvious from consideration of the mode of action of antifoams that the dispersed phase should be characterized by both a tendency to emerge into the air–liquid surface of the other phase (so that $E^e > 0$) and a positive value of the bridging coefficient.

A complication in relating the position of coexistence curves to antifoam behavior concerns the possibility of metastable equilibria. Realization of an equilibrium state requires that the system be subject to perturbations to overcome any obstacle to continuous attainment of a state of minimum free energy. Consider then the part of a free energy of mixing function for a system exhibiting partial miscibility shown in Figure 4.35a. A mixture of composition ψ_s within that part of the coexistence region, which is concave upward, can in principle be supersaturated so that its free energy is given as a point on the free energy of mixing curve. Suppose then that such a mixture shows a tendency to separate into two slightly different compositions, $\psi_s + \Delta\psi$ and $\psi_s - \Delta\psi$, as shown in the figure. The line joining those two compositions is seen to lie above the free energy of mixing curve—this therefore implies that any such tendency to spontaneously decompose into two phases of different composition would carry a free energy penalty. It is easy to show then that the metastability implied by this effect requires that $(\partial^2 \Delta G(\psi)/\partial\psi^2)_{\psi_s} > 0$ (see, e.g., references [118, 119]). Where this condition is satisfied, the system will remain as a metastable single phase unless subject to some external perturbation. Conversely, however, if we consider a composition ψ_s' within the part of the coexistence region that is concave downward, the same argument indicates that no such free energy penalty would be implied as shown in Figure 4.35a. In this case the system will spontaneously decompose into two phases. This corresponds to the condition $(\partial^2 \Delta G(\psi)/\partial\psi^2)_{\psi_s} < 0$ [118, 119]. Therefore, in the absence of any external perturbation (shaking, vibration, stirring, etc.), a metastable coexistence curve (or spinodal curve) will be measured. It will lie inside the equilibrium coexistence curve (or bimodal curve) as depicted schematically in Figure 4.35b. The region between the curves represents a region where metastable supersaturated mixtures will exist if significant perturbation is absent.

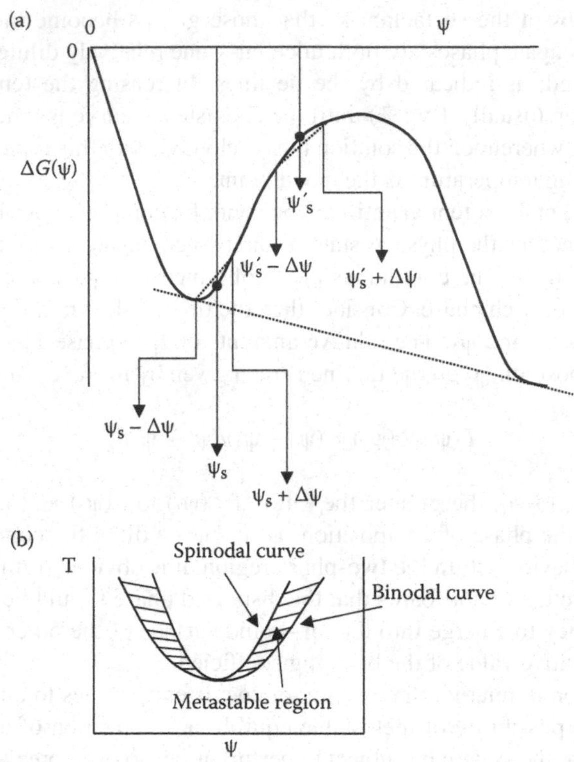

FIGURE 4.35 (a) Free energy of mixing illustrating origins of metastability. Composition ψ_s although unstable will be metastable because any perturbation of composition $\Delta\psi$ accompanying demixing will increase the free energy (i.e., become less negative). Similarly, composition ψ'_s, although also unstable, will readily demix because the perturbation of composition $\Delta\psi$ will decrease the free energy. (b) Schematic showing metastable partial miscibility behavior in a temperature-composition plot with lower critical temperature.

Formation of foam usually involves intense agitation, arguably so intense that metastable supersaturated states are not likely to be present and any phenomena associated with phase separation across the equilibrium coexistence curve will be realized. However, direct independent measurement of points on that curve, such as the so-called cloud point, sometimes appear to involve minimal agitation. In which case it is possible that antifoam effects due to the relevant conjugate phase can be interpreted as evidence for antifoam effects in homogeneous systems. We will keep this possibility in mind when reviewing the antifoam effects associated with partial miscibility.

4.6.3.2 Systems with Lower Critical Temperatures

A. *Phase Separation and Antifoam Effects*

It is well-known that ethoxylated and propoxylated non-ionic surfactants and polymers exhibit high solubility in water at low temperatures and partial miscibility at elevated temperatures. This behavior derives from dehydration of the relevant groups with increasing temperature, which imparts a progressively increasing hydrophobicity and eventual phase separation to form two conjugate micellar solutions—one dilute and the other relatively concentrated [120] Low foam behavior accompanying this phase separation concerns the antifoam effect of the dispersed concentrated micellar conjugate phase upon the foam behavior of the dilute conjugate.

The nomenclature of these ethoxylated and propoxylated compounds is sometimes confusing. We note therefore the equivalence of "polyoxyethylene" group, "polyethylene oxide" group, "polyethylene glycol" group, and "ethoxylated," which refer to compounds containing the molecular group $H(OC_2H_4)_nOH$, often abbreviated to EO_n. Similarly, we have "polyoxypropylene" group, "polypropylene oxide" group, "polypropylene glycol" group, and "propoxylated," which refer to compounds containing the molecular group $H(OCH_2.CHCH_3)_nOH$, often abbreviated to PO_n.

Arguably one of the earliest systematic accounts of the reduction in foamability of ethoxylated non-ionic surfactants at the cloud point is that of Fineman et al. [121]. Here foamability was observed to decrease at temperatures above the cloud point for a series of polydisperse alkyl phenyl ethoxylates. The temperature dependence of foamability for octyl phenyl ethoxylate ($OP.EO_{9.7}$) is shown in Figure 4.36 by way of example. The foamability is seen to decline gradually as the temperature is increased above the cloud point until it is almost eliminated. The authors attribute this behavior to the polydisperse nature of the surfactants, expecting pure surfactants to show steep declines in foamability. No explanation for the decline in foamability is given by these authors. They do, however, note that the marked change in foam behavior at the cloud point is not accompanied by any significant discontinuous change in the air–water surface tension.

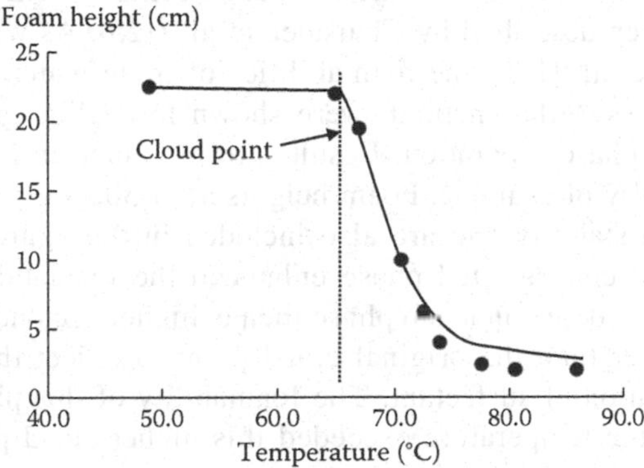

FIGURE 4.36 Foamability as function of temperature of an aqueous solution of octyl phenyl ethoxylate ($OP.EO_{9.7}$) at weight fraction of 0.01 surfactant (by Ross–Miles technique). (Reprinted with permission from Fineman, M. et al. *J. Phys. Chem.*, 56, 963. Copyright 1952 American Chemical Society.)

More than two decades later, Kruglyakov [41] and others [122, 123] argued that the phenomenon of reduction in foamability of ethoxylated non-ionic surfactants at temperatures above the cloud point may concern the formation of oil lenses in foam films by non-ionic-rich cloud phase drops. Clear evidence that such drops contribute to an antifoam effect is presented by Koretskaya [123], who showed that removal of the cloud phase drops by filtration restores the foamability despite the reduction in overall surfactant concentration caused by such a procedure.

Similar experiments have been reported by Bonfillon-Colin and Langevin [124] using aqueous solutions of $C_{10}H_{21}(OC_2H_4)_4OH$ at a concentration of 8 wt.% and at 24°C. Slight extrapolation of the coexistence curve shown in Figure 4.34 confirms that this temperature is significantly higher than the cloud point. Of particular interest is the effect of adding a small drop of the dispersed cloud phase onto the existing foam formed from the dilute conjugate solution. Foam collapse immediately commenced and was complete in about 50 s. Clearly the antifoam effect of the cloud phase is not necessarily confined to an effect on foamability. Measurements of the equilibrium air–water surface tensions of each conjugate phase, together with the interfacial tension between the conjugate phases, were not sufficiently accurate to unambiguously establish the presence of lenses of the cloud phase on the air–water surface of the dilute conjugate phase (the reported positive bridging coefficient unfortunately violates Equation 4.31). However, use of Brewster angle microscopy directly revealed the formation of lenses (of about 5 μm diameter) on the surface of the dilute conjugate solution as the temperature was raised above the cloud point. The relevant images are reproduced as Figure 4.37. Bonfillon-Colin and Langevin [124] concluded that the lenses represent partial wetting so that the actual equilibrium spreading coefficient, S^e, is negative.

The effect of temperature on the foam behavior of a series of polydisperse nonyl phenol ethoxylate surfactant blends (NP.EO$_n$, where n = 8–10) has been described by Chaisalee et al. [125]. As with the studies of Fineman et al. [121], the foamabilities of these materials, measured using the Ross–Miles method, were shown to decline gradually from the onset of phase separation. Results are presented in Figure 4.38 for NP.EO$_9$ by way of example. Foam heights after 300 s give some indication of foam stability and are also included in the figure. Removal of the dispersed concentrated phase enhanced the foamability markedly. The resulting dilute conjugate phase then exhibited a cloud point several degrees higher than the original cloud point to reflect the lower overall concentration of surfactant. The foamability of this phase therefore declined as the temperature exceeded this higher cloud point as phase separation occurred.

Chaisalee et al. [125] again ascribed these effects to bridging of foam films by drops of the dispersed concentrated conjugate phase. However, measurements of the relevant bridging coefficients, although positive, are in violation of Equation 4.31. This presumably reflects experimental error.

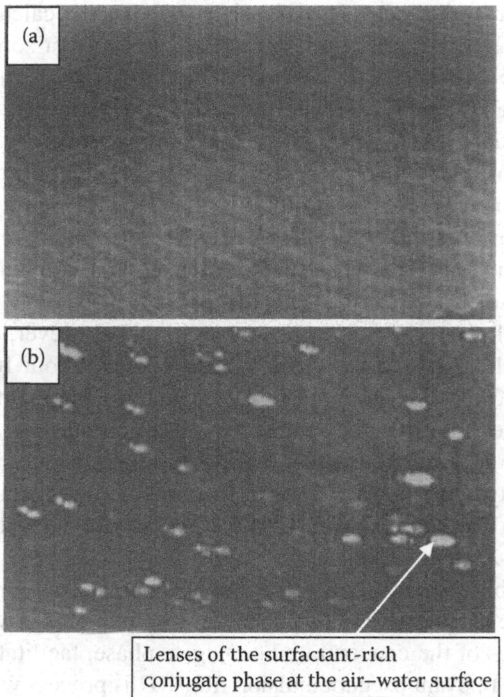

FIGURE 4.37 Images of air–water surface of $C_{10}EO_4$ solution at weight fraction of 0.04 of surfactant by Brewster angle microscopy. (a) At 15°C. (b) At 25°C. Comparison with Figure 4.34 reveals that cloud point at weight fraction 0.04 is only ~20.7°C. Lenses shown are circular, of diameter ~5 microns, but have an elliptical appearance as a result of optical distortion associated with angle of observation. (Reprinted with permission from Bonfillon-Colin, A. and Langevin, D. *Langmuir*, 13, 599. Copyright 1997 American Chemical Society.)

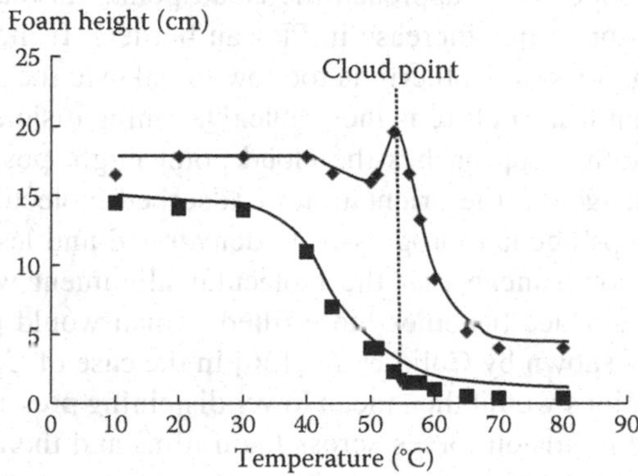

FIGURE 4.38 Foamability (◆) and foam stability (■, after 300 s) by Ross–Miles technique as function of temperature for 0.02 M aqueous solutions of commercial nonyl phenol ethoxylate ($NP.EO_9$). (With kind permission from Springer Science+Business Media: *J. Surfactants Deterg.*, 6, 2003, 345, Chaisalee, R., Soontravanich, S., Yanumet, N., Scamehorn, J., figure 1.)

Moreover, the authors do not make clear whether the calculated bridging coefficients refer to equilibrium surfaces or surfaces at some surface age characteristic of the actual conditions prevailing during measurement of foamability. Presumably, foamability measurements involve air–water surfaces which depart somewhat from equilibrium (especially in the case of commercial blends such as those used in this work). On the other hand, the stability of foam after standing for 300 s, shown in Figure 4.38, is likely to concern surfaces that are close to equilibrium, which are therefore easier to unambiguously relate to surface tension measurements. Significant foam collapse, due presumably to drops of the concentrated conjugate phase, is still observed under these conditions. However, foam stability is shown in the figure to start to decline markedly about 10°C below the cloud point. Chaisalee et al. [125] make no comment about this observation except to state that they do not wish "to pursue the more ambiguous property of foam stability"! Similar findings concerning declining foam stability with increasing temperature below the cloud point are also reported by Koretskaya [123] for an octylphenyl ethoxylate (OP.EO_{10}, Triton X-100).

It is possible then that two processes of foam film rupture are involved with this type of system. The first concerns bridging of relatively thick foam films by drops of the concentrated conjugate phase, facilitated by low stabilities of the relevant pseudoemulsion films. This process will then mainly concern foam films formed both during and immediately after foam generation—it will therefore be apparent in measurements of both foamability and foam stability. The second occurs at temperatures below the cloud point in homogeneous solutions and concerns thin foam films of thickness ≤ 100 nm where disjoining pressures are significant. Such a process would therefore require drainage of the foam and would only be apparent in measurements of foam stability. Suppose then that the disjoining pressures decrease as temperatures approach the cloud point. As the foam drains, then capillary pressures increase in Plateau borders. If the maximum in the disjoining pressure isotherm is too low to balance the capillary pressure, then foam film rupture is inevitable. Declining disjoining pressures as the temperature approaches the cloud point might possibly occur as a result of changes in the orientation of adsorbed molecules as the ethoxy head groups become progressively dehydrated and less hydrophilic. In turn, this could mean that the molecular alignment with respect to the air–water surface becomes more tilted, which would produce lower adsorptions as shown by Colin et al. [126] in the case of $C_{10}H_{21}.EO_5.PO_7$. Lower adsorptions would then mean lower disjoining pressures as a result of diminished repulsion forces across foam films and therefore a greater probability of their rupture.

A study of the foaming behavior of polypropylene glycols (of formulae $OH(CH_2CH(CH_3)O)_nH$ where $3 < n < 33$) has been made by Tan et al. [127]. These materials are surface-active oligomers that show both foaming behavior and cloud points. Tan et al. [127] made measurements of the foamability, by sparging, at increasing concentrations of polypropyl glycols at constant temperature. A marked decrease in foamability was observed as the concentration passed into the two-phase region across the coexistence curve. The effect was attributed to the antifoam behavior of drops of the dispersed phase. The antifoam effect of the dispersed cloud phase on the foamability of the dilute conjugate phase is therefore clearly not confined to alcohol ethoxylates. It is probably a general property of all non-ionic polyoxyethylene and polypropoxylene surfactants and polymers.

The "cloud" phase antifoam effect is clearly large and can concern almost total elimination of foamability (as shown in Figure 4.36) or even rapid elimination of the existing foam formed from the relevant conjugate phase [124]. This is qualitatively different from the antifoam effects of alcohols, alkanes, and PDMSs on the foam behavior of anionic surfactant solutions discussed in Sections 4.6.1.2 through 4.6.1.4. These oils produce at most only slight reductions of foamability in that context. Significant effects on foam behavior are mainly confined to reduced foam stability, which is usually only apparent after elapse of times of up to several minutes or more. It seems likely that this behavior is due to the high stability of the relevant pseudoemulsion films, which means that oil drops are largely removed from foam films before they can cause foam film rupture. Emergence into air–water surfaces of oil drops trapped in Plateau borders as the Plateau border capillary pressure overcomes the disjoining pressure in the pseudoemulsion films initiates foam collapse. This mechanism is described in Section 4.5.3.

By contrast, with these oils, the antifoam effect of cloud phase drops on the foamabilities of the relevant conjugate solutions suggests that they directly induce foam film rupture, probably by ready formation of unstable bridging configurations, as suggested by Bonfillon-Colin and Langevin [124] and described in Section 4.5.1. This must imply that the relevant pseudoemulsion films are of intrinsically low stability. Only low values of an applied capillary pressure are therefore required to cause their rupture, which must occur rapidly before the drops are removed from the draining foam films. Rapid foam film collapse relative to the rate of foam generation is of course necessary if foamability is to be significantly diminished. Direct measurement of the applied critical capillary pressure required for rupture of these pseudoemulsion films would, however, be necessary to provide a firm basis for these considerations.

Other issues need to be placed on a firmer basis. One of which concerns the apparent reduction in foam stability as the cloud point is approached from temperatures below the critical point. Direct measurement of disjoining pressure isotherms as a function of temperature using a film balance would clearly be desirable in that context. Even use of pure surfactants in study of the foam behavior would be useful. With the exception of that of Bonfillon-Colin and Langevin [124], most studies of the foam behavior of solutions of, for example, ethoxylated non-ionic surfactants at temperatures above the cloud point have thus far apparently concerned commercial polydisperse blends.

B. *Cloud Point Antifoams*

Antifoam effects due to the dispersed concentrated conjugate phase can even be apparent when additional surfactants are present. Therefore, nonionic ethoxylated and propoxylated compounds often find application as the so-called cloud-point antifoams (see, e.g., reference [128]). The neat compounds are single-phase liquids with relatively high aqueous phase solubility at temperatures below the critical point. These properties can facilitate distribution and storage of products. A particular application concerns machine dishwashing. However, formation of mixed micelles between the relevant polyethoxylate or polypropxylate compounds and ionic surfactants can significantly increase cloud points and even totally eliminate the relevant phase behavior [129, 130]. Moreover, the effectiveness of cloud point antifoams may require a low tendency of any other surfactant present to adsorb on the surface of cloud phase drops lest that should enhance the stability of the relevant pseudoemulsion films. These considerations may be expected to place significant limitations upon the general use of such antifoams.

Apparently there have been few scientific studies of the antifoam effect of the cloud phase on the foamability of systems containing an additional surfactant, where controlling the foaming propensity of the latter is the objective. However, Nemeth et al. [131] describe such a study. It concerns the effect of copolymeric antifoams on the foam behavior of solutions of an octylphenyl ethoxylate ($OP.EO_{10}$, Triton X-100). The molecular structure of these antifoams consists of PDMS polymer backbones upon which are grafted polyethylene glycol-polypropylene glycol (EO–PO) block copolymers. The presence of the EO–PO groups imparts the familiar partial miscibility with increasing temperature (with a cloud point 40°C lower than that of Triton X-100). This work is particularly interesting because it challenges the usual view that reductions in foamabilities by cloud point antifoam are always associated with the formation of drops of a relatively concentrated conjugate phase which have the property of rupturing foam films under the relevant conditions. We therefore consider it in some detail.

Nemeth et al. [131] observe that both the foamability, and the foam stability, of solutions of Triton X-100 is diminished by the presence of these PDMS–EOPO copolymers. A plot of foam volume immediately after generation (i.e., the foamability) by shaking cylinders is given as a function of temperature in Figure 4.39 for 1% by weight of Triton X-100 solutions containing 0.3 wt.% of copolymer. The specification details of the copolymer are given in the figure legend. A plot of the foam volume after 60 s is also included in the figure. Both the foamability and foam stability plots are seen to decline at a temperature about 5–7°C lower than the cloud point (= 47°C). Also indicated in the figure are the initial foam volume and the foam volume after 60 s for the dilute conjugate solution prepared by centrifuging the dispersion formed above the cloud point at 50°C. Removal of

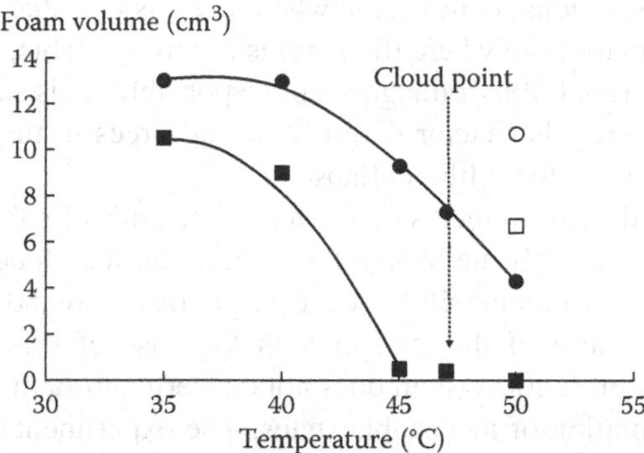

FIGURE 4.39 Foamability (●) and foam stability (■, after 60 s) by hand shaking of cylinders as function of temperature for 1 wt.% of Triton X-100 containing 0.3 wt.% of a PDMS–EOPO copolymer (of mol.wt. 29,000, EO–PO ratio 0.66, PDMS–EOPO ratio 0.26). Open symbols give foam behavior of dilute conjugate solution formed by centrifuging the turbid dispersion at 50°C; foamability (○), foam stability (□, after 60 s). (After Nemeth, Z. et al., *J. Colloid Interface Sci.*, 207, 386, 1998.)

the concentrated dispersed conjugate phase has clearly markedly increased both the foamability and stability, indicating that it functions as a "cloud point" antifoam in this context. However, the onset of antifoam behavior at temperatures several degrees below the cloud point is difficult to explain.

Nemeth et al. [131] follow a similar argument to that given above concerning the instability of foam in homogeneous solutions at temperatures below the cloud point (as exemplified by the results for NP.EO_9 shown in Figure 4.38). Thus, they attribute the antifoam effect of the copolymer, in supposedly homogeneous solutions, to the diminished repulsion between the adsorbed monolayers as they become dehydrated with the approach to the relevant cloud point as the temperature increases. The authors made single vertical film studies that indicated that at temperatures below that at which rapid foam collapse occurs black films were formed, but at higher temperatures film collapse occurred rather than black film formation at the same film thickness profile. The authors then concluded that this indicated that "most probably unstable black spots formed and they immediately ruptured the film."

The main problem with application of this argument in the present context is that foam films must be thin enough for disjoining forces to be significant (i.e., ≤~100 nm). However, examination of Figure 4.39 reveals that the onset of antifoam effects is apparent at temperatures below the cloud point even with initial foam heights; that is, even foamabilities decline at temperatures significantly below the cloud point. Moreover, almost total foam collapse is seen to occur in only 60 s at temperatures below the measured cloud point. Films thin enough for disjoining forces to be apparent are unlikely to be found in those circumstances unless foam drainage is extremely rapid. Nemeth et al. [131] report similar rates of single vertical foam film drainage measured for solutions in the region where foam is unstable (up to the point of film rupture) to those where the foam is relatively stable, which suggests that extremely rapid film drainage is not responsible. This suggests the possibility that some other factor than disjoining forces in thin films may be involved in causing foam film collapse.

All reported measurements of the foam behavior of solutions of ethoxylated compounds indicate a correlation between the cloud point and the onset of declines in foamability (see, e.g., Figures 4.36 and 4.38). Perhaps, therefore, the cause of the exception in the case of this PDMS–EOPO copolymer/Triton X-100 system does not concern antifoam mechanism but rather the formation of metastable states. The experimental technique for measurement of the cloud temperature employed by Nemeth et al. [131] involved heating the relevant solutions without agitation until turbidity increased as indicated by a photodiode. Such a procedure may have failed to detect a metastable state. The interfacial tensions between conjugate phases are usually extremely low—values of as low as 0.1 mN m^{-1} have been reported by Chaisalee et al. [125], for example, and <0.2 mN m^{-1} by Bonfillon-Colin and Langevin [124]. Any metastability does not therefore seem likely to concern nucleation of small drops of the concentrated conjugate phase. However, a metastable state can exist between the equilibrium and spinodal coexistence curves. As we have discussed in Section 4.6.3.1, the intense agitation present during foam generation could decompose that metastable state. Clearly, observation of turbidities of the foaming solution after foam generation at temperatures below the apparent cloud point would be instructive in this context.

Nemeth et al. [132] also report the effect of polyethoxy–polypropoxy block copolymers ($EO_n.PO_m.EO_n$, where $m = 33$ and $2.5 < n < 5.5$) on the foam behavior of a protein—bovine serum albumin (BSA). These block copolymers reduced both the foamability and stability of the foam of BSA solutions at temperatures below the relevant cloud point where the solution was homogeneous. Filtration of the mixed solutions at the temperature of these foam experiments produced essentially no enhancement of foamability or foam stability. However, the filtration was done before foam generation so that the possibility of the decomposition of any putative metastable state during foam generation was not examined. Unlike with the PDMS–EOPO copolymer/Triton X-100 system, the possibility that antifoam effects at temperatures above a measured cloud point for this EO–PO–EO + BSA system, which could be eliminated by removal of the relevant conjugate phase, was not explored.

This polyethoxy–polypropoxy block copolymer is more surface active than the protein and radically alters the vertical foam film drainage behavior of solutions of the latter. The extremely slow rigid film drainage behavior (presumably with suppressed marginal regeneration) of the BSA solutions was shown by Nemeth et al. [132] to be replaced by relatively rapid, apparently mobile, drainage behavior in the presence of copolymer. These vertical films were, however, unstable with lifetimes of the order of a few seconds at concentrations of the polyethoxy–polypropoxy copolymers where foamability is essentially eliminated. In this respect, the stability of the vertical films was similar to that of solutions of the copolymer alone. It is tempting then to conclude that this copolymer displaces BSA from the air–water surface more or less completely. Rapid foam film drainage to yield intrinsically unstable foam films at thicknesses where disjoining pressures are apparent means low foamabilities regardless of the presence or absence of the protein.

Increasing the temperature above the cloud point can result in formation of gel phases rather than the micellar liquids usual in studying effects on foam behavior. Joshi et al. [133] have, for example, described the use of a non-ionic block copolymer derivative of an unknown n-alcohol as an antifoam. This material was shown to exhibit a cloud point of 7°C for a 21 wt.% aqueous solution. A micellar solution exists below the cloud point. However, at the cloud point at that concentration, a solution to gel transition occurs. The gel phase is likely to be an ordered micellar structure, which Joshi et al. [133] show to be viscoelastic with a storage modulus at least an order of magnitude greater than the loss modulus. Preparation of an antifoam dispersion simply involves addition of the neat material to water. Unfortunately, Joshi et al. [133] do not present clear evidence of the phase behavior in this, the relevant concentration region for antifoam effects. Somewhat speculatively, it appears that two conjugate phases form above a supposed critical point—a dilute micellar solution and the gel phase where the latter grows at the interface between the dispersed antifoam drops and the aqueous phase. If this dilute dispersion is then added to a foaming solution of a "recycled paper system" at temperatures above the relevant cloud point, an antifoam effect is immediately apparent—at temperatures below that cloud point, no such effect is apparent. However, the quoted cloud point in this context is that of a concentrated 21 wt.% solution in the absence of the surfactant present in the "recycled paper system."

Joshi et al. [133] present entry, spreading, and bridging coefficients for this system, which are clearly non-equilibrium initial values (which violate Equations 3.11, 3.12, and 4.31, respectively). Direct evidence of coalescence of bubbles by bridging drops of gel phase is also presented. However, it would seem that this happens only if air–water surfaces are at non-equilibrium surface tensions, which in turn could imply negative values of the bridging coefficient under equilibrium conditions. Interestingly, Joshi et al. [133] also show that the antifoam effect diminishes markedly as the temperature is increased well above the cloud point. This effect is attributed to decreases in the bulk phase viscosity of the gel phase, rendering drops more readily deformable and more readily ejected from foam films before rupture of the relevant pseudoemulsion films.

In conclusion, then we note that despite the importance of cloud point antifoams for some applications [128], we have been able to report here only three studies that address the relevant phenomena. However, it should be clear from this brief review that there are many issues that deserve more attention. Examples concern the exact nature of the phase behavior, including the use of pure materials where appropriate, interactions between the foaming surfactant and the antifoam, and the stability of the relevant pseudoemulsion films. The issue of the so-called antifoam effects in homogeneous solutions deserves particular attention.

4.6.3.3 Systems with Higher Critical Temperatures

Mass transport in distillation and fractionation towers can sometimes be adversely affected by the generation of unwelcome, but transient, foam, which is a product of the intrinsic properties of the relevant liquids rather than any inadvertent contaminant. Ross and coworkers have drawn attention to the role played by partial miscibility of those liquids in determining that foam behavior (see, e.g., references [134–137]). Their studies concerned both binary and ternary mixtures of low molecular weight molecules, most of which were non-aqueous. Unlike the aqueous ethoxylated and propoxylated non-ionic surfactant and polymer systems considered in Section 4.6.3.2, these binary systems often exhibit higher critical temperatures so that miscibility occurs with increasing temperature.

The behavior of such systems can be illustrated by considering the foam behavior of a binary mixture of methyl acetate and ethylene glycol reported by Ross and Nishioka more than three decades ago [134]. Foam was generated by sparging until a steady-state volume was achieved. The quantity $\Sigma_{BIK} = V_{st}/V_G$ was used as a measure of "foaminess," where V_{st} is the steady-state volume of air in the foam and V_G is the volumetric gas flow rate. A plot of foaminess against composition of the mixture along a tie line AC at 20°C is shown in Figure 4.40, where an inset sketches the coexistence curve and indicates the critical temperature. Low foaminess is found in the heterogeneous region along the tie line with zero foaminess in the region BC where the methyl acetate–rich conjugate is dispersed in a continuous phase of the ethylene glycol–rich conjugate. The presence of the methyl acetate–rich conjugate is seen to cause a dramatic decline in foaminess at the coexistence curve from a peak in the homogeneous region. The dispersed conjugate would therefore appear to act as an antifoam for the continuous phase conjugate in region BC. In the region AB, the emulsion is inverted so that the methyl acetate–rich conjugate becomes the continuous phase. The ethylene glycol–rich conjugate does not then show significant antifoam behavior, which may indicate that drops of this phase do not emerge into the air–liquid surface of the continuous phase. However, it is surprising that foaminess of the methyl acetate–rich conjugate declines in the region AB as the proportion of ethylene glycol–rich dispersed conjugate phase increases. After all, the continuous phase should have a constant composition in this region. As Ross and Nishioka [134] state, this could be due to partial emulsion phase inversion before complete inversion occurs at an overall proportion of ethylene glycol of 50 wt.%.

The relevant air–liquid surface tensions, σ_{AL}, for the pure liquids and conjugate pair are given in Table 4.5 together with the entry, spreading and bridging coefficients. The interfacial tension between the conjugate phases is seen to be negligible. Since those phases are presumably of an equilibrium composition, it is reasonable to suppose that the surface tensions are also supposed to be equilibrium values. In which case, the entry, spreading, and bridging coefficients should satisfy Equations 3.11,

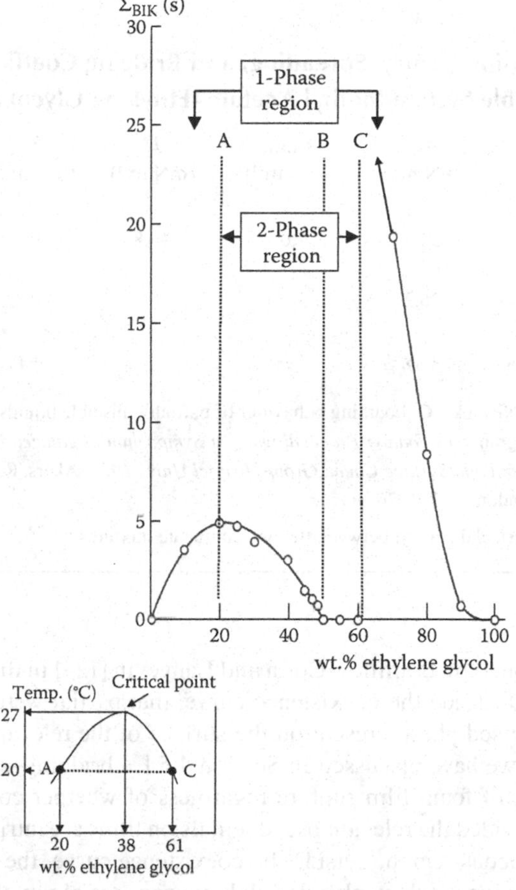

FIGURE 4.40 Variation of foaminess Σ_{BIK} (at 20°C) against composition for the system ethylene glycol + methyl acetate, which exhibits higher critical solution temperature and partial miscibility in the region AC. Schematic showing salient features of relevant coexistence curve is given in inset. (Reprinted from Ross, S., Nishioka, G. Foaming behaviour of partially miscible liquids as related to their phase diagrams, in *Foams; Proceedings of a Symposium Organized by the Soc. Chem. Ind., Colloid and Surface Chem. Group, Brunel Univ., 1975*, Akers, R.J., ed., Academic Press, London, p 17, 1976, with permission from both Elsevier and the Society of Chemical Industry, London.)

3.12, and 4.31, respectively. Unfortunately, none satisfy these equations presumably because of unavoidable experimental error combined with the stringent nature of the requirement imposed by the low value of the interfacial tension between the conjugate phases. It is clearly not therefore possible, for example, to follow Ross and Nishioka [134] to conclude, on the basis of these erroneous spreading coefficients, that spreading of the methyl acetate–rich conjugate over the ethylene glycol–rich conjugate is the cause of the absence of foam in the heterogeneous region BC. It is

TABLE 4.5
Interfacial Tensions, Entry, Spreading, and Bridging Coefficients for the Partially Miscible System Methyl Acetate–Ethylene Glycol at 20°C

	σ_{AL} (mN m^{-1})	$\sigma_{MA/EG}$ (mN m^{-1})[a]	E^e (mN m^{-1})	S^e (mN m^{-1})	B ((mN m^{-1})^2)
Pure methyl acetate	24.1	–	–	–	–
20 wt.% Ethylene glycol conjugate	25.4	<0.1	+0.8	+0.8	+41
61 wt.% Ethylene glycol conjugate	26.2	<0.1	–0.8	–0.8	–41
Pure ethylene glycol	46.9	–	–	–	–

Source: Ross, S., Nishioka, G. Foaming behaviour of partially miscible liquids as related to their phase diagrams, in *Foams; Proceedings of a Symposium Organized by the Soc. Chem. Ind., Colloid and Surface Chem. Group, Brunel Univ., 1975* (Akers, R.J., ed.), Academic Press, London, p 17, 1976.

[a] $\sigma_{MA/EG}$ is the interfacial tension between the two conjugate mixtures.

also possible, as found by Bonfillon-Colin and Langevin [124] in the case of aqueous solutions of $C_{10}EO_4$ inside the coexistence curve, that partial wetting prevails with lenses of the dispersed phase present on the surface of the relevant conjugate solution. However, as we have discussed in Section 4.5.1 a bridging–stretching process is the likely cause of foam film rupture regardless of whether complete or partial wetting occurs provided the relevant pseudoemulsion films are intrinsically unstable.

In the homogeneous region, outside the coexistence curve, the foaminess is significantly higher in the ethylene glycol–rich homogeneous region than in the methyl acetate–rich region. We should, of course, remember that the foam formed by these mixtures is unstable—essentially transient, with a foaminess of only a few seconds. Such foams are sustained only by transient surface tension gradients—rapid relaxation of those gradients because of rapid transport at high concentrations means rapid foam film drainage, possibly leading to plug flow without any thin film stabilization by disjoining forces at any film thickness. In these circumstances then, transient foamability is likely to be dominated by the magnitude of the potential surface tension gradients. Establishment of those surface tension gradients requires that the air–liquid surfaces of foam films experience differential stretching and/or compression before a steady state can be achieved. The greater the difference in surface tension between those regions where the stretching is maximal and those where it is minimal, the greater the surface tension gradient and the greater the resistance to any applied shear force. Andrew [138] has made an analysis of the surface tension increase accompanying expansion of the air–water surface of a binary mixture of two liquids, which yields an estimate of the resulting surface tension increase relative to the equilibrium surface tension. Such a surface tension change should represent a rough indication of the maximum surface tension gradient that can exist with a given rate of film stretching. Andrew [138] deduces that the surface tension change, $\Delta\sigma^{d}_{AL}$, is proportional to the gradient of surface tension with respect to mole fraction of the component with the lowest surface tension so that

$$\Delta\sigma^{d}_{AL} \propto \upsilon.(1-\upsilon)\left(\frac{d\sigma_{AL}}{d\upsilon}\right)^{2} \tag{4.43}$$

where υ is the mole fraction of that component. We can now estimate the likely relative magnitude of the surface tension gradients in the homogeneous regions depicted in Figure 4.40. If we approximate by supposing that in both the methyl acetate–rich and ethylene glycol–rich homogeneous regions the surface tensions vary linearly with mole fraction, then we calculate that $\upsilon(1 - \upsilon)(d\sigma_{AL}/d\upsilon)$ is ~8 (mN $m^{-1})^2$ in the former case and ~149 (mN $m^{-1})^2$ in the latter case. This implies that surface tension gradients should be greater in the case of the ethylene glycol–rich homogeneous mixture, which in turn means slower foam film drainage rates and greater film stability as indicated by higher foaminess.

Further results with similar considerations concerning the effect of partial miscibility on the foam behavior of binary and ternary liquid mixtures may be found in other papers of Ross and coworkers [112, 135–137]. Of particular interest is a study by Ross and Nishioka [112] of the effect of inducing the partial immiscibility of polymers by changing the solvent. Foaminess increased at a constant weight percentage of polymer as the solubilizing quality of the solvent decreased, as indicated by a function derived from solubility parameters (of Hildebrand et al. [139]). However, phase separation eventually occurred with solvents of inadequate solubilizing quality. High foaminess then switched to zero foaminess, much as shown in Figure 4.40, where the coexistence curve is crossed on the high ethylene glycol side. Some apparent indication that foaminess declined with solvents of declining solubilizing quality before phase separation is reasonably attributed by Ross and Nishioka [112] to the limitations of solubility parameter theory rather than to an effect occurring with homogeneous solutions of different solvents.

4.7 INERT HYDROPHOBIC PARTICLES AND CAPILLARY THEORIES OF ANTIFOAM MECHANISM FOR AQUEOUS SYSTEMS

4.7.1 Early Work

Solid particles, which adhere to fluid–fluid surfaces so that part of their surface is exposed to one fluid and part to the other fluid, exhibit a finite contact angle at the relevant surface. For almost three-quarters of a century, it has been known that such particles may have a drastic effect on the behavior of dispersions formed by mixing fluids. Two of the earliest publications are those of Ramsden [140] and Pickering [141] concerning the effect of particles on oil–water emulsion behavior. A few years later, Bartsch [142] described the stabilizing effect of hydrophobed mineral particles on froths formed by aqueous solutions of 3-methylbutanol.

Since then, there have been many publications concerning the effect of particles on the formation and stability of foams, froths, and emulsions. Here we are mainly concerned with the destabilizing effect of particles on foams and froths. Practical examples include the effect of hydrophobed mineral particles and collector precipitates on the stability of mineral flotation froths [77, 143]. Another is the use of wax [144] to control the foam of detergent formulations for automatic washing machines.

Arguably the earliest suggestion that an adverse effect of particles on aqueous froths may be attributable to low wettability (or high contact angle at the air–water surface, measured through water) was made by Mokrushin 60 years ago [145]. This concerned the effect of metal sulfides on froths formed by gelatin solutions. A few years later, Dombrowski and Fraser [146] reported direct observation of the effect of wettability of particles on the disintegration of thin films of water or alcohol formed from spray nozzles. Here, particles that were not wetted by the film liquid caused perforation of the film when the particle size was of the same order as the film thickness. Particles that were wetted by the film liquid had no effect on the manner of disintegration.

A later reference to the role of the wettability (or the contact angle) of dispersed solid material in determining froth stability is due to Livshitz and Dudenkov [147]. Here the destabilizing effect of hydrocarbon particles on froths is attributed to air–water contact angles > 90° by analogy with the effect of particles on oil–water emulsion behavior. Thus, the presence of particles with a contact angle > 90° at the oil–water surface (measured through the water) in an oil–water mixture will produce a water-in-oil emulsion [148]. Livshitz and Dudenkov [147] argue that analogous behavior in the case of an air–water system is tantamount to destruction of the foam. This of course gives no mechanism for either supposedly analogous phenomenon. However, in a later paper, concerned with the effect of hydrophobed metal xanthates and oleate precipitates on flotation froths, Livshitz and Dudenkov [149] suggest that hydrophobic particles may bridge aqueous films. The more hydrophobic the particle, the closer are the wetting perimeters of each air–water surface at the particle. This will supposedly reduce the thickness of the film, which will result in acceleration of film rupture. In the case of a hydrophilic particle, on the other hand, contact angle formation does not take place and there is no film rupture. This mechanism appears to suggest that there should be a continuous improvement in defoaming ability with increasing contact angle.

A difficulty with establishing the role of wettability in determining the effectiveness of particles in breaking foams and froths is measurement of the relevant contact angle. This is particularly so for precipitated particles of sparingly soluble materials. Thus, representative smooth surfaces upon which contact angles can be readily measured may not be easily prepared. In recognition of this, Dudenkov suggested that the solubility products of polyvalent metal salt precipitates of butyl xanthogenates and oleates should correlate with the contact angles [150]. Thus, the lower the solubility product, the more hydrophobic is the precipitate for a given anionic hydrophobe. Whatever the merit of this supposed correlation, Dudenkov [150] produces results that indicate a sharp increase in the volume of air present in a froth for precipitates of oleate or xanthogenate with solubility products greater than a critical value. This effect, however, decreases with increase in frother concentration. Examples of

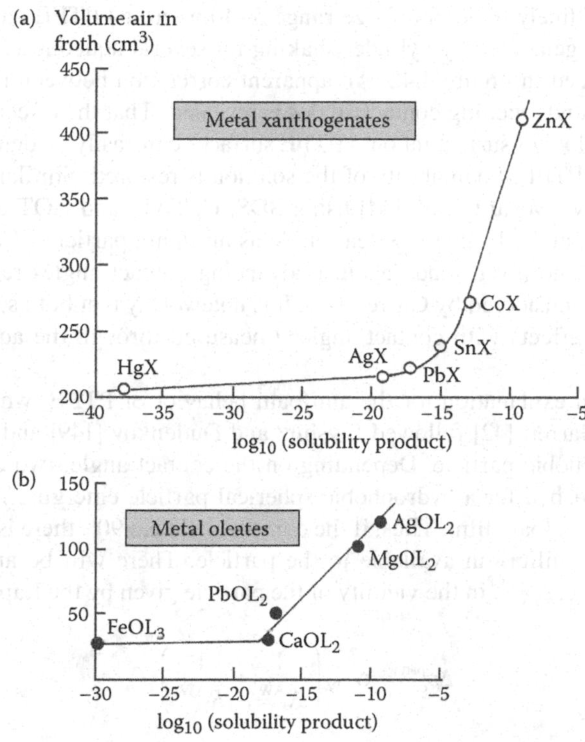

FIGURE 4.41 Relation between volume of air in froth and solubility products of hydrophobic precipitates. (a) Metal butyl xanthogenates at 5×10^{-5} M in 30 mg dm^{-3} commercial frother solution. (b) Metal oleates at 2×10^{-5} M in 10 mg dm^{-3} commercial frother solution. (From Dudenkov, S.V., *Tsvet. Metally*, 40, 18, 1967, reproduced by permission of Primary Sources.)

their results are presented in Figure 4.41. These are interpreted by Dudenkov [150] as indicating the importance of wettability in determining antifoam effectiveness. However, the solubility products given by Dudenkov [150] may be suspect. Thus, the value given for calcium oleate is almost two orders of magnitude lower than that given by Irani and Callis [151].

4.7.2 Experimental Observations Concerning Contact Angles and Particle Bridging Mechanism

Attempts to relate the foam behavior of hydrophobic particles directly to contact angles followed more than a decade after the work of Dudenkov [150]. Of these, the first was a paper by Garrett [42], which concerned finely divided polytetrafluoroethylene (PTFE) as antifoam material. PTFE has the merit of inertness so that any possible contribution from Marangoni spreading to the antifoam effect can be unambiguously eliminated. Moreover, reliable representative contact angles can be measured on polished PTFE plates.

The effect of finely divided (of size range 5–40 microns) PTFE particles on the volume of foam generated by cylinder shaking for several aqueous surfactant solutions is reproduced in Figure 4.42. An apparent correlation between the volume of foam destroyed and receding contact angles is revealed. That the effects do not concern adsorption loss of surfactant onto PTFE surfaces can easily be demonstrated by filtering off the PTFE. Foamability of the solution is restored. Similar results were later obtained by Aveyard et al. [44] using SDS, C_{16}TAB, and AOT as surfactants and PTFE, wax, and ethylene bis-stearamide as antifoam particles. Comparison of foamability was, however, made against advancing contact angles rather than the receding angles considered by Garrett [42]. It is noteworthy that both sets of workers found antifoam effects with contact angles (measured through the aqueous phase) $\theta_{AW} < 90°$.

In seeking an explanation for the antifoam behavior of PTFE, which is related to wettability, Garrett [42] followed Livshitz and Dudenkov [149] and considered a bridging hydrophobic particle. Depending on the contact angle, two consequences may be distinguished for a hydrophobic spherical particle emerging into both air–water surfaces of a foam film. Thus, if the contact angle is >90°, there is no condition of mechanical equilibrium available to the particle. There will be an unbalanced capillary pressure, Δp_c^{part}, in the vicinity of the particle given by the Laplace equation

$$\Delta p_c^{part} = \sigma_{AW}\left(\frac{1}{R_1^{AW}} + \frac{1}{R_2^{AW}}\right) \tag{4.44}$$

where R_1^{AW} and R_2^{AW} are the two radii of curvature describing the curved air–water surface in the vicinity of the particle and σ_{AW} is the air–water surface tension. The

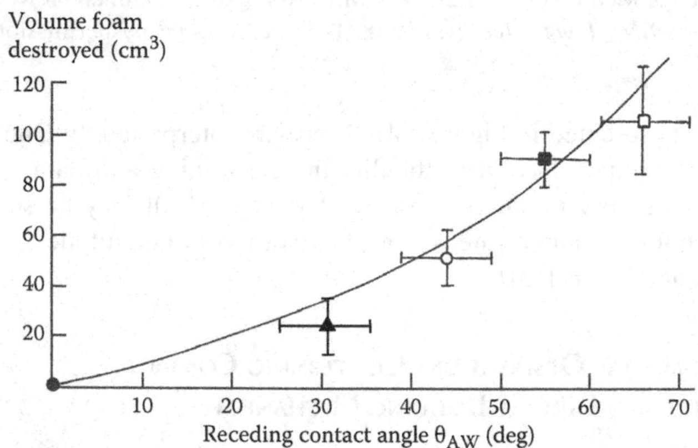

FIGURE 4.42 Initial antifoam effect of polytetrafluoroethylene particles as function of receding contact angle using hand shaking of measuring cylinders for foam generation at ambient temperature (25 ± 2°C). ■, 0.01 M SDS; ○, 0.01 M SDS/0.1 M NaCl; □, 2 mM tetradecylaminopropane sulfonate; ▲, 0.1 mM $C_{14}EO_6$; ●, soap detergent product. (Reprinted from *J. Colloid Interface Sci.*, 69, Garrett, P.R. 107. Copyright 1979, with permission from Elsevier.)

configuration is shown in Figure 4.43 where the particle is assumed to be small so that the effect of gravity upon the shape of the air–water surface may be neglected. The two radii of curvature are of opposite sign so that $R_1^{AW} < 0$, and therefore the direction of the unbalanced capillary pressure depends on the relative magnitudes of R_1^{AW} and R_2^{AW}. The ratio R_1^{AW}/R_2^{AW} is not obvious by inspection, although it has been simply assumed in later work by Frye and Berg [152] that

$$\frac{R_2^{AW}}{R_1^{AW}} \ll 1 \tag{4.45}$$

So that Equation 4.44 becomes

$$\Delta p_c^{part} = \frac{\sigma_{AW}}{R_2^{AW}} \tag{4.46}$$

The capillary force will therefore act in a direction to cause enhanced drainage out of the foam film. The process of drainage will continue until the two three-phase contact lines at the particle surface are coincident, whereupon a hole will form in

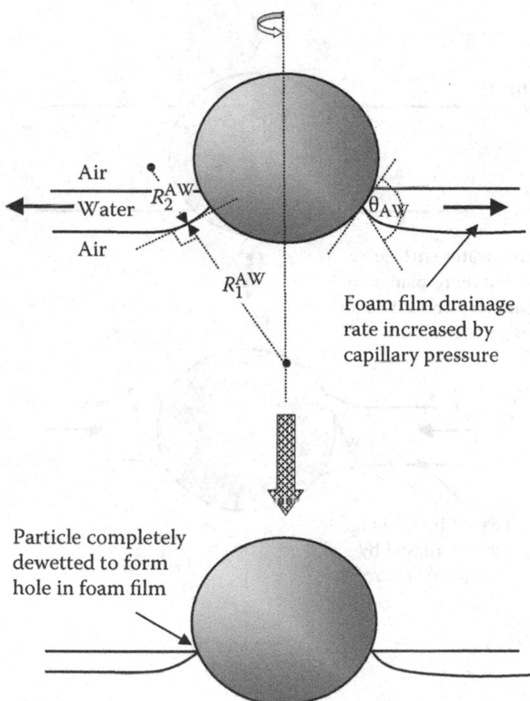

FIGURE 4.43 Foam film rupture caused by spherical particle with contact angle $\theta_{AW} > 90°$.

the foam film. It has been shown by de Vries [76] that if the hole is sufficiently large it will spontaneously increase in size and lead to film collapse. It is trivial matter to deduce that an exactly similar mechanism for foam film collapse will occur for hydrophobic cylindrical rods with a contact angle > 90° where the rotational axis is perpendicular to the foam film. Clearly then, this cannot represent a complete explanation for the results of Garrett [42] and Aveyard et al. [44] because they report antifoam effects with contact angles < 90°. As we shall see, the discrepancy probably largely concerns the non-spherical geometry of the particles used by these workers.

The second consequence of a spherical particle emerging into both air–water surfaces of a foam film follows if the contact angle is <90°. The resulting configurations are depicted in Figure 4.44. Here the initial configuration is similar to that found for particles with contact angles > 90° where, if inequality 4.45 is valid, an unbalanced capillary pressure will enhance drainage out of the foam film. However, as the film drains, it will always attain some thickness where the film is planar and where the capillary pressure becomes zero. Further drainage will then result in a change in sign

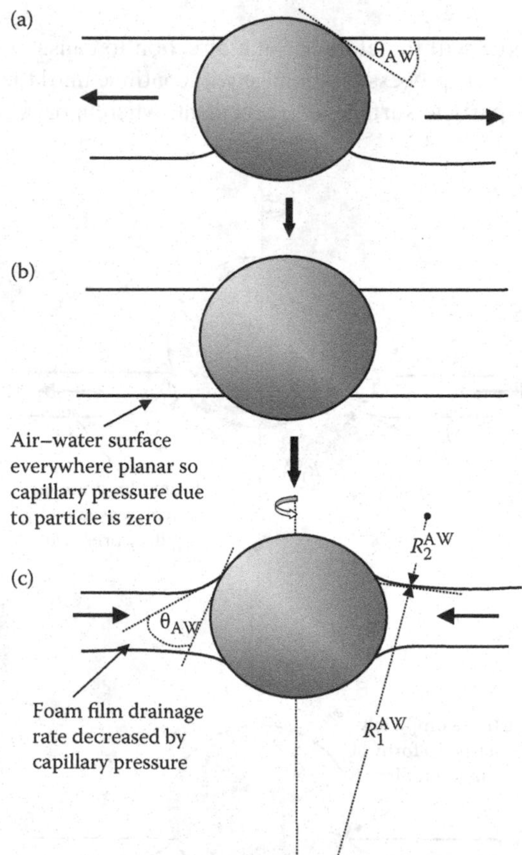

FIGURE 4.44 Effect of presence of spherical particle with contact angle θ_{AW} < 90° on foam film.

of both radii of curvature. If condition 4.45 is assumed, this will mean an unbalanced capillary pressure inhibiting further drainage of the film. It is therefore possible that a particle with this configuration will actually stabilize foam films. That hydrophobic particles may stabilize foams has in fact been frequently reported in the literature (see, e.g., references [77, 147, 153–161].

The essentials of the mechanism of film collapse by spherical particles have been confirmed in a cinematographic study by Dippenaar [77], using hydrophobed glass spheres. Representations of high-speed movie frames are reproduced in Figure 4.45 for a glass bead of 250 μm diameter and a contact angle of 102° in a distilled water film. Here the contact angle formed initially at the lower air–water surface is seen to be apparently less than the equilibrium value. The usual interpretation of such dynamic contact angles is that the equilibrium contact angle always prevails near the three-phase contact line and that the air–water surface exhibits extreme curvature in the same region to give an apparent macroscopic contact angle much lower than the equilibrium value. It is the capillary pressure implied by that curvature that produces the force driving film thinning. Collapse occurs about an order of magnitude faster than would have occurred if the particle had been absent [77]. This is seen to occur as the two three-phase contact lines become coincident. By contrast, a sphere of contact angle 74° had no effect on film rupture times and adopted a configuration similar to that of Figure 4.44c. Clearly, these observations, while being consistent with the mechanisms shown in Figures 4.43 and 4.44, do not, however, quantitatively address

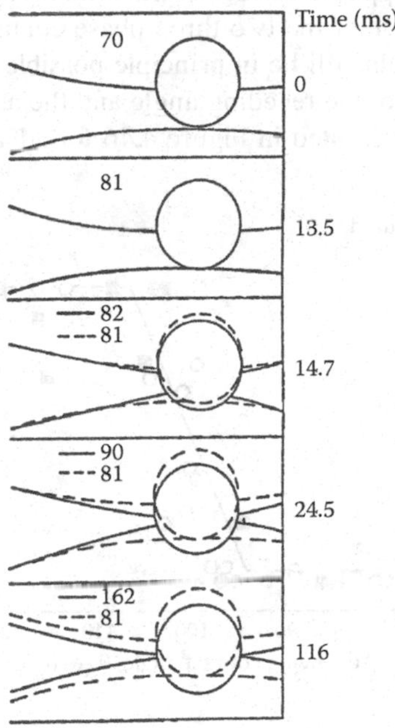

FIGURE 4.45 Reproduction of high-speed cinematographic film frames showing interaction of hydrophobic glass bead (θ_{AW} = 102°) with aqueous foam film. Frame numbers are indicated in the figure (Reprinted from *Int. J. Miner. Process.*, 9, Dippenaar, A., 1. Copyright 1982, with permission from Elsevier.)

the prediction that foam film rupture by spherical particles should critically occur for contact angles precisely >90°.

Some confirmation that contact angles must be precisely >90° for foam film rupture in the case of cylinders has been attempted by Frye and Berg [152]. In this work, cylindrical rods, of ~2 mm diameter, were directly inserted into foam films with the rotational axis of the rods perpendicular to the plane of the film. This procedure eliminates issues concerning the emergence of particles into the air–water surface and minimizes any dynamic effects associated with transport of surfactant to that surface during foam generation. Rods of PTFE and glass hydrophobed with PDMS were used. Contact angles were measured directly on the rods—solutions of different surfactants (cetyltrimethylammonium bromide, Triton X-100, and sodium dodecylbenzene sulfonate) and concentrations permitted variation of those angles. The proportion of films ruptured by inserting the rods is plotted against the advancing contact angle in Figure 4.46. Increasing probability of foam film rupture is seen to occur over a range of advancing contact angles between 95° and 105° with more or less certain rupture at angles > 105° and no rupture at angles < 95°. However, in reality, this experiment will involve both advancing and receding angles where hysteresis will generally be found so that the latter is likely to be lower than the former. Frye and Berg [152] do not present receding contact angles measured on these rods, arguing that the advancing angle is solely relevant because rupture occurred immediately after penetration of the rod into the film and "before the receding angle can be established" at the opposite surface of the film. However, in the case of cylindrical geometry, coincidence of the two three-phase contact lines on the rods, and therefore film rupture, would still be in principle possible if the advancing contact angle is >90° and the sum of the receding angle and the advancing angles is >180°. In the case of the results presented in Figure 4.46 for advancing contact angles of,

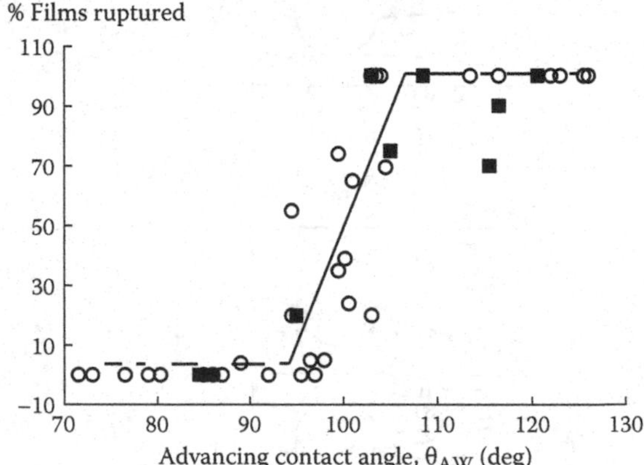

FIGURE 4.46 Film rupture probability caused by PTFE (■) and PDMS-treated glass (○) rods as function of advancing contact angle for films prepared from various aqueous surfactant solutions (hexadecyl trimethylammonium bromide, Triton X-100, and sodium dodecylbenzene sulfonate). (Reprinted from *J. Colloid Interface Sci.*, 127, Frye, G.C.C., Berg, J.C., 222. Copyright 1989, with permission from Elsevier.)

for example, 95°, this relaxed condition would therefore mean that hysteresis of ≤10° would still allow film rupture. However, for advancing contact angles in the range of 90–95° and hysteresis > 10°, then film rupture would not occur when both advancing and receding angles are established on opposite sides of the film. This argument does, of course, ignore the transfer of liquid across the film due to the capillary pressure gradient arising from the different curvatures of the film surfaces adjacent to the rod. If that process is relatively rapid, it would presumably mean that the two contact angles tend to become equal at an average value, which in consequence will not, of course, numerically affect the foregoing argument.

An almost identical study has been made by Aveyard et al. [24, 162] a few years after that of Frye and Berg [152], again using glass rods (but of ~0.2 mm diameter and hydrophobed with octadecyltrichlorosilane under different conditions to produce surfaces with a range of contact angles). However, the contact angles were not directly measured on the rods but were inferred from measurements on glass plates, hydrophobed in the same manner. As with some of the experiments of Frye and Berg [152], foam films were prepared from solutions of cetyltrimethylammonium bromide (C_{16}TAB). A plot of the proportion of films ruptured by inserting these glass rods as a function of the advancing contact angle [24] is shown in Figure 4.47, which reveals a discontinuous jump in the probability of rupture from zero for angles < 90° to near unity for angles > ~92°. This finding is clearly to some extent at variance to that of Frye and Berg [152] about which Aveyard et al. [24] make no comment. One possible explanation for the discrepancy concerns the increase in film rupture time with increase in diameter of rods for a given contact angle. According to Frye and Berg [152], this is to be expected because larger rods mean more liquid has to be removed before a hole can be formed. We will consider this issue in more detail in Section 4.7.5. It is tempting also to conclude that the discrepancy must in some manner also concern measurement of contact angles.

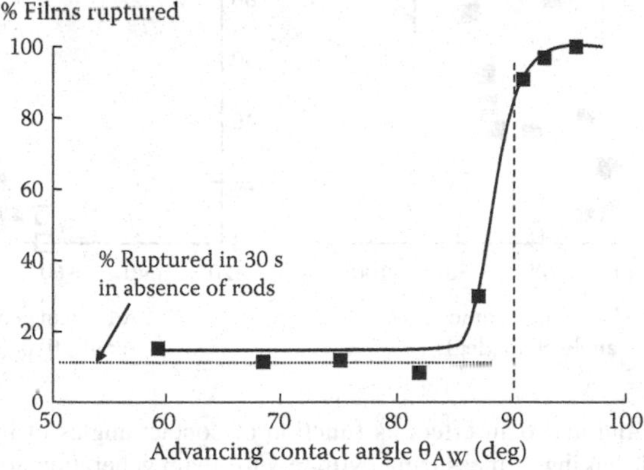

FIGURE 4.47 Film rupture probability caused by octadecyltrichlorosilane-treated glass rods as function of advancing contact angle for films prepared from 0.2 mM aqueous solutions of hexadecyl trimethylammonium bromide. (Reprinted from *Adv. Colloid Interface Sci.*, 48, Aveyard, R., Binks, B.P., Fletcher, P.D.I., Peck, T.G., Rutherford, C.E. 93. Copyright 1994, with permission from Elsevier.)

Attempts have also been made by both Frye and Berg [152] and Aveyard et al. [162] to verify the requirement that contact angles > 90° are necessary for rupture of foam films during actual foam generation using hydrophobed spherical glass particles. The former used particles at a concentration of 1 g dm^{-3} in the size range of 5–50 microns and the latter at a concentration of 2.8 g dm^{-3} in the size range of 35–53 microns. Both sets of workers used a simple hand-shake test for foam generation. Contact angles were in both cases measured on glass plates, hydrophobed in an identical manner to that used for the spherical particles. Here, Frye and Berg [152] followed Garrett [42] in comparing foam behavior with receding contact angles. By contrast, Aveyard et al. [162] used advancing contact angles.

Plots of the proportion of foam initially destroyed by the glass spheres against contact angle are presented in Figure 4.48. Here, contact angles were varied by differences in hydrophobing conditions [162] and changes in surfactant type and concentration [152, 162]. The most striking feature of these plots is the observation that antifoam effects are clearly apparent with contact angles < 90°. Frye and Berg [152] argue that this is due in part to differences between contact angles measured on plates and those actually prevailing in the foam with the spherical particles. Such differences they argue could concern, for example, relatively slow transport of surfactant to air–water surfaces during foam generation, which could lead to higher contact angles than those measured with static air–water surfaces. However, they argue that the main factor concerns contamination of their spherical particles with sharp edged shards, which, as we will see, would involve lower critical contact angles for

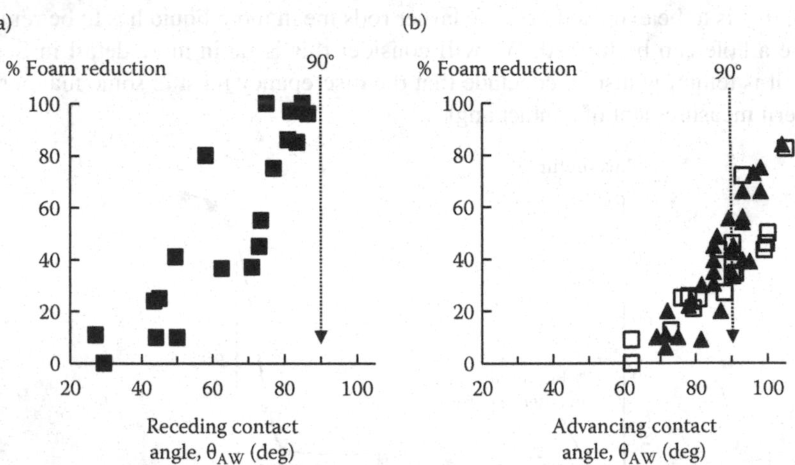

FIGURE 4.48 Initial antifoam effect as function of contact angles of hydrophobed glass spheres using hand shaking of measuring cylinders for foam generation at ambient temperature (25 ± 2°C). (a) Receding contact angles with 1 g dm^{-3} PDMS-treated spheres (5–50 microns) in aqueous solutions of various surfactants ($C_{16}TAB$, sodium dodecylbenzene sulfonate, and Triton X-100). (Reprinted from *J. Colloid Interface Sci.*, 127, Frye, G.C.C., Berg, J.C., 222. Copyright 1989, with permission from Elsevier.) (b) Advancing contact angles with 2.8 g dm^{-3} octadecyltrichlorosilane-treated glass spheres (35–53 microns) in $C_{16}TAB$ (▲) and AOT (□) solutions. (After Aveyard, R. et al., *J. Disper. Sci. Technol.*, 15, 251, 1994.)

foam film rupture. Aveyard et al. [162], on the other hand, simply attribute the apparent occurrence of antifoam effects at advancing contact angles < 90° to differences between measured contact angles on plates and those actually prevailing on particles during foam generation where relatively slow transport of surfactant to surfaces may be expected to occur. Some indirect evidence for this has been obtained if the effect of the particles upon foam stability is considered. Surfactant transport effects should then be minimal.

Aveyard et al. [162] illustrate the effect of hydrophobed spherical glass particles on foam stability by defining a "half-life ratio" τ_{AF} as

$$\tau_{af} = (t_{1/2\ \text{with particles}} - t_{1/2\ \text{without particles}})t_{1/2\ \text{without particles}} \tag{4.47}$$

where $t_{1/2}$ is the half-life of a foam. Obviously $\tau_{AF} > 0$ implies that the particles are stabilizing foam and that $\tau_{AF} < 0$ implies that particles are destabilizing foam. A plot of τ_{AF} against advancing contact angles for various solutions of $C_{16}TAB$ is shown in Figure 4.49. Similar plots are presented by Aveyard et al. [24] for AOT solutions. It is clear from the figure that particles with a contact angle < ~75° have little or no effect, being apparently rapidly flushed out of the foam as it drains. Not surprisingly, particles with contact angles > 95° destabilize the foam. However, particles with contact angles in the range 75–95° actually stabilize the foam. This compares with significant antifoam effects with foam generation by hand shaking over the same contact angle range as shown in Figure 4.48b. The distinction then tends to confirm the hypothesis that antifoam effects can be enhanced by the relatively slow transport of surfactants to the rapidly expanding air–water surfaces formed during foam

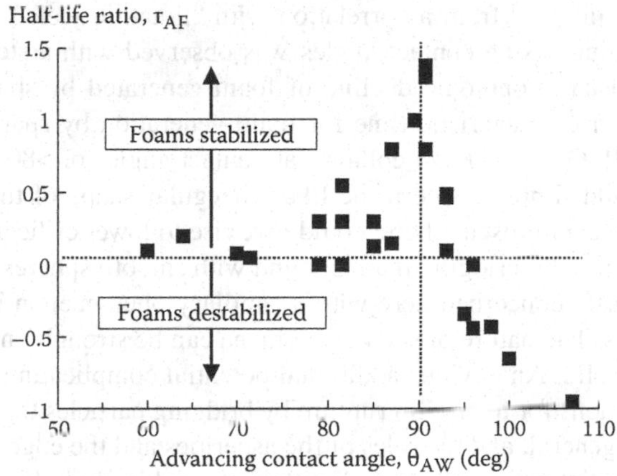

FIGURE 4.49 Half-life ratios (defined by Equation 4.47) indicating the effect of octadecyltrichlorosilane-treated glass spheres, at concentration of 2.8 g dm^{-3} in aqueous solutions of $C_{16}TAB$, on foam stability. Foams were generated by hand shaking of measuring cylinders at ambient temperature (25 ± 2°C). Advancing contact angles were varied by changing both extent of hydrophobing and concentration of surfactant. (After Aveyard, R. et al., *J. Disper. Sci. Technol.*, 15, 251, 1994.)

generation by some methods so that dynamic contact angles significantly higher than static angles prevail.

The quantitative interpretation of the plot given in Figure 4.49 is, however, undermined somewhat by the issue of contact angle hysteresis. It is probable that particles with advancing contact angles > 90° will not be readily separated from air–water surfaces except under conditions where the air is totally eliminated by application of rigorous degassing, which has not been applied in this context. If so, then it is possible that a bridging particle will have at least one advancing contact angle. As we have seen, the situation where one angle is advancing and one angle receding would mean a capillary pressure gradient across the foam film. It seems likely that such a situation could not prevail and perhaps an average intermediate angle would result for both surfaces. Moreover, since one reason for contact angle hysteresis concerns surface roughness (see, e.g., reference [163]), the detailed nature of the surface of the spheres may also be of relevance—it is possible that hysteresis on the particles is less marked than on the equivalent plate surfaces. We are left then with the impression that the role of contact angle hysteresis in potentially determining the behavior of spherical particles with contact angles in the region of 90°, which bridge foam films, is yet to be fully established. This issue is of course coupled to the related difficulty of satisfactorily measuring contact angles on spheres of diameters of <100 microns.

Somewhat similar results to those shown in Figure 4.49 have been obtained by Johansson and Pugh [164] using crushed and ground quartz particles of size range 26–44 microns, hydrophobed with a methylsilane derivative. However, these results concerned steady-state foam generated by sparging rather than some measure of foam stability under quiescent conditions. Aqueous solutions of various "frothers" were used for foam generation. The peak in foamability appeared to occur at a contact angle in the region of 70–80°. Contact angles were, however, not directly determined but rather inferred from a correlation with "flotation yield." That a peak in foamability with increasing contact angles was observed with a steady-state foam rather than the usual monotonic decline of foam generated by shaking may well concern the slower characteristic time for foam generation by sparging, as shown by Patist et al. [8]. Onset of foam collapse at contact angles of >80°, rather than at angles > 90°, could simply concern the likely irregular shape of these quartz particles, which, as we will discuss here, could give rise to lower critical contact angles for foam film rupture by bridging than is found with smooth spheres.

We are obviously concerned here with a capillary phenomenon involving fluids and solid surfaces. The nature of such phenomena can be strongly influenced by the geometry of the solid. An obvious additional potential complicating factor in determining the conditions for foam film rupture by bridging particles therefore concerns particle shape in general, and the roles of the asperities and the edges that characterize crystalline particles in particular. A cinematographic study by Dippenaar [77] of the effect of orthorhombic crystalline particles of xanthanated galena produced the first direct experimental insights concerning the effect of crystalline particle geometry. Galena forms orthorhombic crystals with a square cross-section in the plane orthogonal to the long axis. The xanthanated crystals used by Dippenaar had a contact angle of 80 ± 8° against water. Such particles were observed to adopt two different orientations, with about equal probability, at the air–water surface.

Dippenaar [77] suggested that the two orientations adopted by the orthorhombic galena particles could concern intersection of the air–water surface at different edges. The behavior of air–water surfaces at edges is usually interpreted by supposing that the equilibrium contact angle is always satisfied at the submicroscopic level [165]. The edge is represented as having a submicroscopic circular cross section [165]. The air–water surface may then adopt an apparent single macroscopic angle at the edge for a range of contact angles, any of which may be satisfied at the submicroscopic level. Dippenaar [77] therefore suggested that orthorhombic galena particles could adopt either a horizontal orientation or a diagonal orientation with respect to the air–water surface. These orientations are depicted in Figure 4.50, where it is easy to see that the horizontal orientation should require contact angles in the range $0 < \theta_{AW} < 90°$ and the diagonal orientation should require contact angles in the range $45° < \theta_{AW} < 135°$ so that a particle with a contact angle of 80° could adopt either.

A difficulty with this argument concerns the intersection of the air–water surface at the surface of the orthorhombic particle orthogonal to the edges with which it intersects in the case of the diagonal orientation. As we will discuss in Section 4.7.3, the extra surface energy associated with this aspect of that orientation means that it cannot exist for orthorhombic particles with aspect ratios of unity (i.e., cubic particles). The aspect ratio of the galena particles used by Dippenaar [77] is, however, ~1.4, which permits the existence of the diagonal orientation. Even at such high aspect ratios, the extra surface energy due to the ends of the particle causes an increase in the lower limit of 45° of the contact angle range. Indeed, the lower limit of 45° is only found if the aspect ratio is extremely large so that the contribution to the overall work of emergence from the ends of the particle is negligible.

Dippenaar [77] observed that a hydrophobed galena particle with a horizontal orientation (Figure 4.50a) had little adverse effect on foam film stability. The behavior resembled that of spherical particles of contact angle < 90° shown in Figure 4.44. However, in the case of a particle with a diagonal orientation (Figure 4.50b), two consequences arose when it contacted the second air–water surface of the thinning film. The particle either twisted to a horizontal orientation with two faces in the planes of the upper and lower film surfaces or retained its diagonal orientation. Particles that

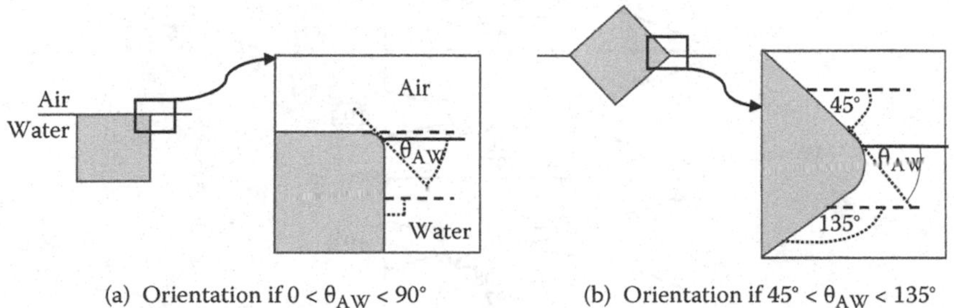

FIGURE 4.50 Supposed orientations of cubic particle at air–water surface. Recent work has, however, indicated that diagonal orientation (b) is only possible for particles with a square cross section if aspect ratio is ≫1 (see Section 4.7.3). (After Dippenaar, A., *Int. J. Miner. Process.*, 9, 1, 1982.)

retained the diagonal orientation caused rapid film collapse. The sequence of events is illustrated in Figure 4.51, where representations of the original high-speed cinematographic frames are reproduced. Here, the air–water surface is seen to move up the inclined left-hand face until the two three-phase contact lines become coincident on the edge, whereupon film collapse occurs. We therefore find that a cubic particle can give rise to film collapse, even though $\theta_{AW} < 90°$, by a mechanism exactly analogous to that shown for spheres in Figures 4.43 and 4.45. Elementary geometry suggests, however, that two planar air–water surfaces can adhere to an edge of circular cross-section simultaneously if $\theta_{AW} < 90°$. Foam film rupture at that edge in this circumstance presumably means that it should be so sharp that its radius of curvature is smaller than the thickness of the maximum in the relevant foam film disjoining pressure isotherm. Rupture is therefore induced by the presence of a sharp edge because the contiguous foam films would be characterized by $d\Pi_{AWA}(h)/dh > 0$, where $\Pi_{AWA}(h)$ is the disjoining pressure and h is the film thickness (see Section 1.2.3).

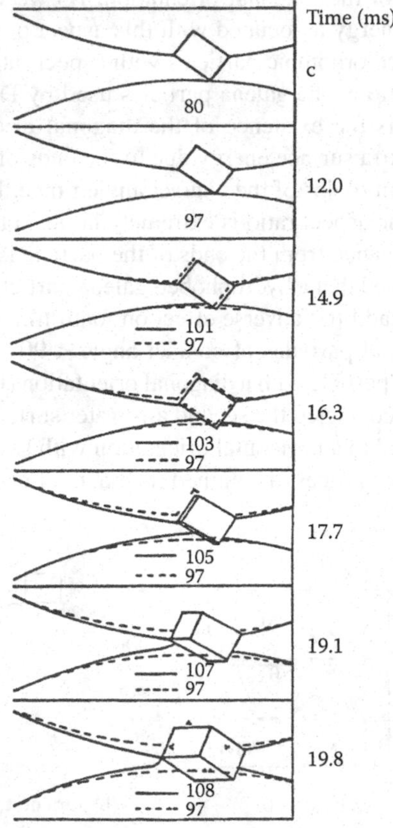

FIGURE 4.51 Reproduction of high-speed cinematographic film frames showing interaction of rhombohedral (of aspect ratio ~1.4) xanthenated galena particle ($\theta_{AW} = 80 \pm 8°$) with aqueous film. Frame numbers are indicated in the figure. (Reprinted from *Int. J. Miner. Process.*, 9, Dippenaar, A., 1. Copyright 1982, with permission from Elsevier.)

The PTFE particles used to obtain the results depicted in Figure 4.42 are not by any means perfect spheres; moreover, the receding contact angles are all < 90°. In attempting to explain this behavior, Garrett [42] dismissed deviations from equilibrium during foam generation on the grounds that if this was significant enough to produce dynamic contact angles > 90°, air–water surfaces would have to be so denuded of surfactant that foam films would not be stable anyway. An alternative explanation for antifoam effects with contact angles < 90° was suggested, which considered the case of disc-shaped particles. Calculation of the Helmholtz free energy of dewetting to bridge and form a hole suggested that this could happen with contact angles < 90° provided the aspect ratio is large enough. However, this argument ignores both viscous dissipation and the possibility of stabilizing configurations analogous to that depicted in Figure 4.44. Examination of the electron micrograph of these PTFE particles given in reference [42] reveals complex highly irregular particles with some obvious sharp edges. This suggests that the explanation for the antifoam effectiveness of this material, despite receding contact angles of <90°, concerns a combination of such edges as suggested by Frye and Berg [152] and possibly dynamic contact angles (where a requirement for angles < 90° does not imply foam film surfaces to be denuded of surfactant).

A particularly convincing demonstration of the effect of sharp edges has in fact been made Frye and Berg [152] who have made a comparison of the antifoam effectiveness of the PDMS hydrophobed smooth spherical particles of Figure 4.48a and ground-glass particles hydrophobed in the same manner. Here, ground glass formed shards with many sharp edges, as shown in Figure 4.52. However, a feature complicating interpretation of this image is the apparent aggregation of particles and/or the apparent presence of rugosities on the shards. As with Figure 4.48a, contact angles were adjusted by using different concentrations of several surfactants. The relative effects of the different types of particle are shown in Figure 4.53. The antifoam effectiveness of ground-glass particles for a given contact angle is clearly much greater than that of smooth spheres. Indeed, Frye and Berg [152] found these

FIGURE 4.52 Scanning electron micrograph of PDMS hydrophobed ground-glass particles used to obtain the results depicted in Figure 4.53. The size bar is ~50 microns. (Reprinted from *J. Colloid Interface Sci.*, 127, Frye, G.C.C., Berg, J.C., 222. Copyright 1989, with permission from Elsevier.)

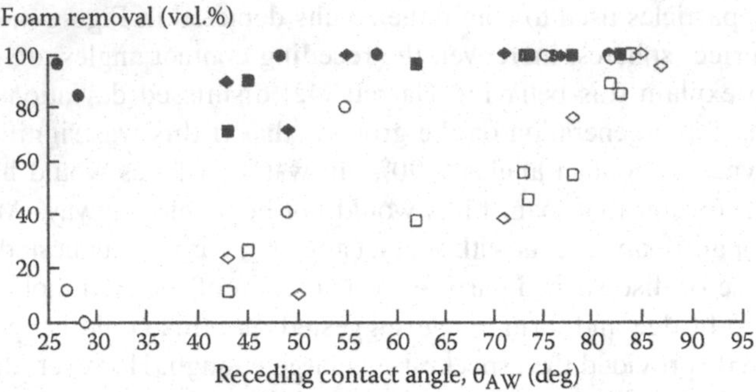

FIGURE 4.53 Antifoam effect of hydrophobed glass particles as function of receding contact angles with various surfactant solutions. Open symbols, smooth spherical particles; filled symbols, ground-glass particles. Foam generation by hand shaking measuring cylinders containing surfactant solution with particle concentration 1 g dm^{-3}. ■□, Sodium dodecylbenzene sulfonate solutions; ●○, Triton X-100 solutions; ◆◇, C_{16}TAB solutions. (Reprinted from *J. Colloid Interface Sci.*, 127, Frye, G.C.C., Berg, J.C., 222. Copyright 1989, with permission from Elsevier.)

ground-glass particles to be more effective than PTFE particles, which they attributed to the greater incidence of sharp edges with the former.

4.7.3 Theoretical Considerations Concerning Particle Geometry and Contact Angle Conditions for Antifoam Action by Smooth Particles

4.7.3.1 Particles with Curved Surfaces and No Edges

We have seen that for particles with smooth curved surfaces and infinite axial symmetry, such as spheres, cylinders, toroidally edged discs, and ellipsoids, the condition for foam film rupture by the process shown in Figures 4.43 and 4.45 is

$$\theta_{AW} > 90° \tag{4.48}$$

where θ_{AW} is the appropriate contact angle at the air–water surface measured through the aqueous phase. However, the results of Dippenaar [77] suggest that it may be possible to devise general rules for more complex particles with lower symmetry and clearly defined edges. The presence of edges and the relative extents of the contiguous surfaces determine the possible configurations that such particles can adopt at air–water surfaces (or in general at air–foaming liquid surfaces) and in principle allow the possibility of foam film rupture at contact angles < 90°.

4.7.3.2 Particles with Edges

Arguably the simplest particles with edges are lens shaped. As with spheres, cylinders, and discs, such particles possess infinite axial symmetry but also one edge. They could, in principle, be prepared by spreading a photo-polymerizable hydrophobic oil on an aqueous substrate to form lenses (i.e., by partial or pseudo-partial wetting), which could then be solidified by UV irradiation (see, e.g., reference [166]). In this case, the edge angle would be the dihedral angle and would be determined by Neumann's triangle. This angle could therefore be varied by changes in the relevant surface tensions using appropriate surfactants. Practicality of preparation of such particles assumes, of course, that polymerization and solidification does not deform the lens. Moreover, it can hardly represent a practical method for preparing large concentrations of antifoam particles.

The orientation of such a solid lens on one surface of a foam film requires that the air–water surface forms the same contact angle at the relevant three-phase contact line at all points on that line. The particle can therefore be orientated either with the three-phase contact line on the edge or at some position on the curved surfaces of the lens. If the particle satisfies the condition $\theta_1 < \theta_{AW} < 180° - \theta_2$, then orientation on the edge is favored where the angles θ_1 and θ_2 are defined in Figure 4.54 so

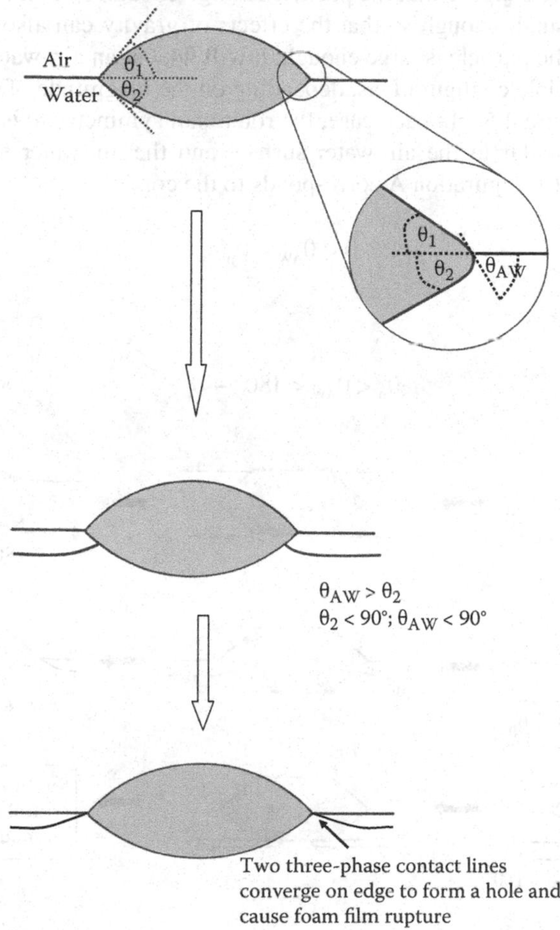

FIGURE 4.54 Role of edge in rupture of foam film by solid lens-shaped particle.

that the dihedral angle of the lens is equal to $\theta_1 + \theta_2$. Provided $\theta_{AW} > \theta_2$, such a lens would then cause film rupture as the relevant three-phase contact lines converge on the edge when the film thins under the action of Plateau border suction and localized capillary pressure in the vicinity of the particle. Since, in general, we would have $\theta_2 < 90°$, film rupture could occur even if $\theta_{AW} > 90°$. This is, of course, analogous to the "bridging–dewetting" mechanism suggested by Denkov et al. [54], as a possibility for the antifoam action of bridging liquid lenses shown in Figure 4.19 (see Section 4.5.1).

We now consider hypothetical particles that possess more than one edge. To simplify, we consider particles with a hexagonal cross-section. First, we examine the putative effect on foam film stability of such a hypothetical particle, which is defined by infinite axial symmetry and three edges but no curved surface that crosses the rotational axis. The geometry of this particle is shown in Figure 4.55, where it is seen that only one angle, θ_p, need to be specified to define all edge angles (where $\theta_p < 90°$). Here we suppose that the particle is large enough so that line tension can be ignored and small enough so that the effects of gravity can also be ignored. If the diameter of the particle is large enough, it will float at an air–water surface with one of three possible configurations, depending on the magnitude of θ_{AW}. These are illustrated in Figure 4.55. In each case, the rotational symmetry axis of the particle will be perpendicular to the air–water surface and the air–water surface will be everywhere flat. Configuration A corresponds to the condition

$$0 < \theta_{AW} < \theta_p \quad (4.49)$$

configuration B to

$$\theta_p < \theta_{AW} < 180° - \theta_p \quad (4.50)$$

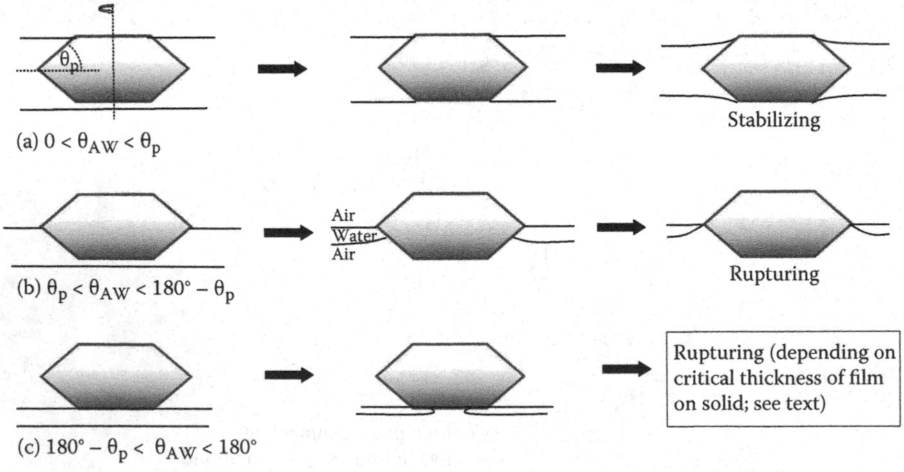

FIGURE 4.55 Effect of axially symmetrical particle with hexagonal cross-section and three edges on foam film stability.

and configuration C to

$$180° - \theta_p < \theta_{AW} < 180° \tag{4.51}$$

Since $\theta_p < 90°$, film rupture will not occur for a particle with the configuration A when bridging across a foam film occurs. This particle will interact with the film to produce a capillary pressure to oppose drainage. On the other hand, particles with configuration B will rupture films according to the process depicted in Figure 4.55. Thus, condition 4.50 implies that particles with the selected geometry may rupture films when $\theta_{AW} < 90°$. Particles with configuration C will only affect film behavior if the thickness for spontaneous rupture of the aqueous film on the solid is greater than that for the aqueous film alone. Such a configuration is only possible if θ_p is in the region of about 70–90° because values of θ_{AW} never usually exceed about 120°.

By contrast, crystalline particles do not usually possess curved surfaces. In particular, even when gravity can be neglected, the absence of an infinite rotational axis orthogonal too the particle edges means that the air–water surface cannot form a plane everywhere around this type of particle unless it intercepts an edge that continues in the same plane right around the particle. This means that if the air–water surface intercepts an edge as a plane, where the edge does not continue around the particle, it must also form a segment of curved surface at some other point on the crystal if the contact angle is to be satisfied. Such a segment of curved surface, in the absence of significant gravitational distortion, must have a catenoid profile if capillary pressure gradients are to be absent under conditions of mechanical equilibrium before a bridging configuration forms. In the case of a bridging crystalline particle, foam film rupture cannot occur at points on the surface where such curvature is present. Despite this complexity, Dippenaar [77] appears to show that a crystal can, in any case, cause foam film rupture on an edge.

Here we are concerned to explore at least some of the salient features of foam film rupture by hypothetical crystal-like particles. Since crystals often possess at least a 2-fold rotational axis together with a mirror plane, we will consider structures with these features. However, we seek to minimize the complexity of real crystals by considering needle-shaped geometries with high aspect ratios. If the aspect ratio of the crystal is large enough and the rotational axis is aligned with the longest dimension, then it seems realistic to suppose that edges should be similarly aligned parallel to that axis. Again we select a hexagonal cross-section to derive a particle with the features illustrated in Figure 4.56a. Here, unlike typical hexagonal crystal morphologies like that of quartz, we truncate the ends orthogonally with respect to the rotational axis.

If the air–water surface is flat and hinges on particular edges of a particle with the geometry shown in Figure 4.56a, then the capillary pressure across that surface is zero. Necessarily, however, the contact angle must be satisfied at both the edges and the vertical face of the truncated ends of the particle. As we have indicated, this implies a curved surface across which the capillary pressure should also be zero. This can be achieved only if the surface is described by two radii of curvature,

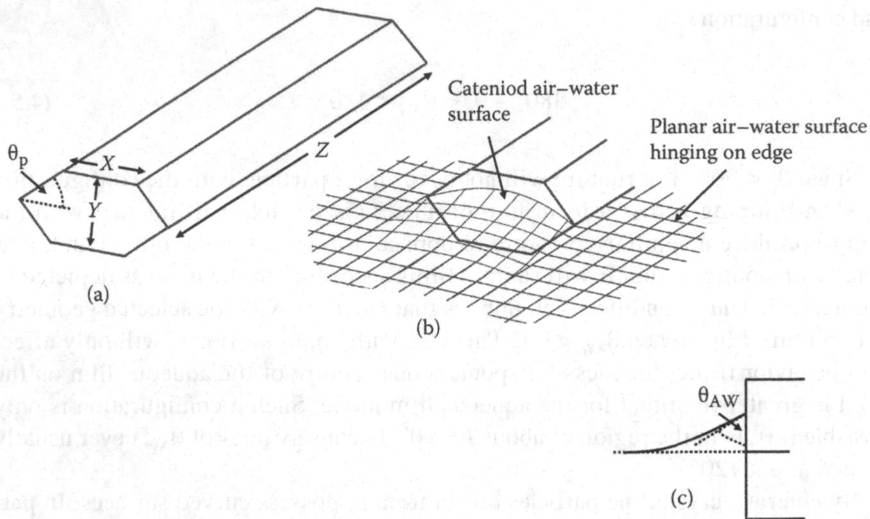

FIGURE 4.56 Hypothetical particle with hexagonal cross-section and high aspect ratio. (a) Definitions of relevant symbols where in general $Y \gg X$. (b) Rough sketch of expected catenoid profile of air–water surface at truncated ends of the particle in case where the particle adopts an orientation with flat air–water surface hinging on "equatorial" edge of particle. (c) Vertical slice through catenoid profile illustrating that air–water contact angle, θ_{AW}, must be everywhere satisfied at truncated end of particle so that the curvature of air–water surface is concave with respect to a plane parallel to rotational axis of particle. In consequence, air–water surface must be convex in the plane orthogonal to the plane perpendicular to rotational axis as shown in sketch (b).

each of opposite sign. We sketch a rough outline of the expected catenoid profile of that surface in Figure 4.56b for the case where the particle lies flat in the air–water surface and where the latter hinges on the "equatorial" edge of the particle. If the contact angle is <90°, the air–water surface adjacent to the end of the particle must be concave with respect to a plane parallel to its rotational axis. It will therefore be convex with respect to a plane orthogonal to the rotational axis of the particle, as shown in the figure if the capillary pressure is to be zero. Unfortunately, these geometrical considerations add appreciable complexity to the task of calculating the work of emergence of such a particle into the air–water surface. However, here we are not concerned to calculate the work of emergence of an actual crystal but rather to illustrate the overall principles involved. We therefore consider a geometry with a sufficiently high aspect ratio so that the contribution to the overall work of emergence, arising from the local configuration at the ends of the particle, can be neglected.

If the aspect ratio is large enough, the configurations in which the particle lies with that projection parallel to the air–water surface are energetically favored. However, by contrast with particles with infinite rotational symmetry, coincidence of both three-phase contact lines on an edge when such particles bridge films cannot cause complete dewetting of the particle. Growth of a hole in the foam film could occur from the edge and will not be axially symmetrical. That this will occur is, of course, confirmed by Dippenaar's observations with galena [77]. The resulting "linear hole" in the foam film must of course spontaneously expand because, unlike an axially symmetrical hole formed at, say, a spherical particle, only one positive curvature is initially involved. However, the hole must also terminate at the ends of the relevant edge so that it will become semicircular as it expands. This would mean that the curvature of the air–water surface at the edge of the expanding hole could then be determined by two orthogonal radii of opposite sign much as described in Section 4.27 in the case of a spherical particle. However, if the length of the edge on the particle is significantly greater than the film thickness, then inequality 4.45 will be satisfied and the capillary pressure will continue to be dominated by the radius of curvature of the air–water surface at the edge of the hole in the plane orthogonal to the plane of the film. The hole will therefore continue to expand.

The number of possible configurations at the air–water surface of a straight-edged particle of a given cross section can be much larger than that of an axially symmetric particle of the same cross section even when we apply the simplifying condition of a high aspect ratio. For example, in the case of straight-edged particles of hexagonal cross-section defined by one angle, θ_p, shown in Figure 4.56a, the number of distinct configurations is 6 when $\theta_p < 90°$, $\theta_{AW} < 90°$, and the aspect ratio is large. These are shown in Table 4.6, where it is clear that not all of these configurations are mutually exclusive for a given contact angle. Most of them are "degenerate" in that the same configuration can be realized by using different edges and surfaces. Only one of the configurations may give rise to film collapse.

We may estimate the works of emergence W_i of the particle from the aqueous phase to form each of the possible configurations i shown in Table 4.6. Here W_i is given by

$$W_i = \sigma_{SA}\,\Delta A_{SA} - \sigma_{SW}\,\Delta A_{SW} - \sigma_{AW}\,\Delta A_{AW} \tag{4.52}$$

where σ_{SW}, σ_{SA}, and σ_{AW} are the solid–water, solid–air, and air–water surface tensions, respectively, and ΔA_{SW}, ΔA_{SA}, and ΔA_{AW} are the changes in solid–water, solid–air, and air–water surface areas accompanying emergence of the particle into the air–water surface, respectively. Since $\Delta A_{SA} = \Delta A_{SW}$ and we have the Young equation

$$\sigma_{SA} = \sigma_{SW} + \sigma_{AW}\cos\theta_{AW} \tag{4.53}$$

then from Equation 4.52 we can deduce

$$W_i = \sigma_{AW}\,(\Delta A_{SA}\cos\theta_{AW} - \Delta A_{AW}) \tag{4.54}$$

It is a matter of elementary geometry to estimate ΔA_{SW} and ΔA_{AW} for the configurations in Table 4.6. By way of example, values of $W_i/\sigma_{AW}X^2$, where X is the cross-sectional dimension defined in Figure 4.56a, for each of these configurations are given in Table 4.6. These are calculated for $\theta_{AW} = 40°$ and particles of length 20× so that the aspect ratio is extremely large. Here we see that the only configuration that can give rise to film collapse has the largest negative work of emergence. However, this is not the only configuration with a negative work of emergence. Clearly, the

TABLE 4.6
Air–Water–Air Foam Film Rupture and Configurations at Air–Water Surface of Particle with Several Edges

Possible Configurations at Air–Water Surface if $\theta_{AW} < 90°$	Conditions for Realization of Configurations	"Degeneracy"	Foam Film Rupture if $\theta_{AW} < 90°$	Work of Emergence W_i for $Z >> X$	$W_i/\sigma_{AW}X^2$ for $Z = 20X$ and $\theta_{AW} = 40°$
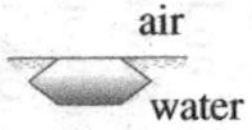 air, water, $i = 1$	$0 < \theta_{AW} < 30°$	2	No	$W_i/\sigma_{AW}ZX = (\cos\theta_{AW} - 1)$	–
 $i = 2$	$0 < \theta_{AW} < 30°$	4	No	$W_i/\sigma_{AW}ZX = (\cos\theta_{AW} - 1)/2\sin\theta_p$	–
 $i = 3$	$60° < \theta_{AW} < 90°$	2	No	$W_i/\sigma_{AW}ZX = (\cos\theta_{AW}/\sin\theta_p - 1)$	–
 $i = 4$	$15° < \theta_{AW} < 45°$	4	No	$W_i/\sigma_{AW}ZX = (1/(2\sin\theta_p) + 1)\cos\theta_{AW} - [(1/(4\sin^2\theta_p)) + 1 - \cos(180 - \theta_p)/\sin\theta_p]^{1/2}$	–8.0
 $i = 5$	$30° < \theta_{AW} < 150°$	2	Yes	$W_i/\sigma_{AW}ZX = (1 + 1/\sin\theta_p)\cos\theta_{AW} - (1 + 1/\tan\theta_p)$	–8.68
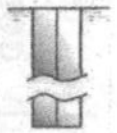 $i = 6$	$0 < \theta_{AW} < 90°$	2	No	$W_i/\sigma_{AW}X^2 = (1 + 1/(2\tan\theta_p))(\cos\theta_{AW} - 1)$	–0.44

Source: Adapted from Garrett, P.R., Mode of action of antifoams, in *Defoaming, Theory and Industrial Applications* (Garrett, P.R., ed.), Marcel Dekker, New York, Chpt 1, p 1, 1993.

Note: Particle geometry as shown in Figure 4.56a with $\theta_p = 30°$ and $Y = X$. Z is the length of the particle as defined in the figure.

effectiveness of the particle in causing film collapse requires consideration of the relative probabilities of achieving all the possible configurations.

There is no obvious rigorous approach that we may employ to calculate these probabilities. The problem does, however, resemble that addressed by classic statistical mechanics where the distribution over energy states of a system subject to thermal motion is considered. Thus, in general, we have various states available to the particles ranging from free dispersion to adhesion at the air–water surface with various configurations. However, during foam generation, agitation will mean that the kinetic energies of the particles will in general far exceed those due to thermal motion alone. We can attempt to allow for this by, according to the particles, a "granular temperature," $\tilde{T}$, in a manner analogous to that described by, for example, Hopkins and Woodcock [167]. Thus, we may write

$$\tilde{T} = \frac{2E_{\mathrm{T}}}{3(n_{\mathrm{T}}-1)\bar{k}} \tag{4.55}$$

where n_{T} is the total number of particles, $\bar{k}$ is the Boltzmann constant, and $3(n_{\mathrm{T}} - 1)$ is the number of degrees of freedom. Here E_{T} is the total kinetic energy of the particles so that

$$E_{\mathrm{T}} = \frac{1}{2}\sum_{\varsigma=1}^{n_{\mathrm{T}}} m_{\varsigma} u_{\varsigma}^{2} \tag{4.56}$$

where m_{ς} and u_{ς} are the velocity and mass of particle ς, respectively. Therefore, $\tilde{T}$ is characteristic of the agitation conditions prevailing during foam generation. If $\tilde{T}$ can be accorded the same significance as temperature in the relevant context, then we can write a Boltzmann-type expression for the number of particles n_i with configuration i

$$n_i \propto \exp(-W_i\beta) \tag{4.57}$$

where $\beta = 1/\bar{k}\tilde{T}$. In the absence of agitation, $\tilde{T} = T$, so $\beta = 1/\bar{k}T$.

The relative probability q_i of any configuration i occurring where the particle adheres to the air–water surface is therefore given by the familiar Boltzmann distribution

$$q_i = \frac{g_i \exp(-W_i\beta)}{\sum_{i=1}^{\Omega} g_i \exp(-W_i\beta)} \tag{4.58}$$

where g_i is the "degeneracy" of configuration i and Ω is the number of possible configurations at the air–water surface. Lower values for $|W_i|$ found for the configurations that do not lead to film collapse shown in Table 4.6 therefore imply low relative probabilities of occurrence.

The significance of β is perhaps easier to see if the relative proportion of configuration $i = 5$ (which will cause film collapse) and configuration $i = 4$, shown in Table 4.6 for $\theta_{AW} = 40°$, are compared. Thus, we have

$$n_4 = g_4 n_5 \exp [(W_5 - W_4)\beta] \tag{4.59}$$

The dimensions of typical antifoam particles are such that X (see Figure 4.56a) is in the region of 1–10 microns, so that $W_5 - W_4$ is in the region of $-(2.3 \times 10^{-12} - 2.3 \times 10^{-14})$ J (for σ_{AW} ~30 mN m^{-1}). In the case of thermal motion alone, $\beta = 1/\bar{k}T \sim 2.5 \times 10^{20}\,\mathrm{J}^{-1}$ at ambient temperatures, and we obtain n_4/n_5 ~0, so that the probability of finding configurations of the particles at the air–water surface that do not cause foam film collapse is essentially zero. However, where the agitation is intense, it is reasonable to suppose that the value of β could be considerably in excess of $1/\bar{k}T$, so that configurations that do not cause foam collapse may become more probable. Increasing n_4/n_5 to, for example, 0.01 would require values of 1/β from ~$(8 \times 10^5 – 8 \times 10^7)\bar{k}T$, depending on the size of the particle.

In principle, it is possible to deduce similar conditions to those depicted in Table 4.6 for any crystal habit. However, consideration of geometries with lower aspect ratios will necessarily introduce significant complexity. In general, it will not then be possible to neglect the inevitable curvature of the air–water surface as the contact angle is satisfied on some parts of the surface of the hypothetical crystal. This could mean that configurations where the crystal rotational axis is tilted with respect to the air–water surface may become energetically favored at certain contact angles. When these complicating factors are combined with the often complex geometry of real crystal habits, it becomes clear that a formidable challenge to theory exists. Use of computer modeling could, however, represent a suitable approach, as we discuss in Section 4.7.3.3.

Comparison of experiment with theory requires assessment of the antifoam potential of crystalline particles of known geometry together with contact angle measurements where the issue of the relative appropriateness of advancing or receding angles is fully resolved. However, apart from the work of Dippenaar [77] concerning galena, no studies of the antifoam behavior of well-defined hydrophobic crystalline particles have been made. As it has been almost 30 years since that work was published, further studies would seem to be timely!

4.7.3.3 Models of Foam Film Rupture by Particles Using Surface Energy Minimization

Morris et al. [168, 169] have used an iterative surface energy minimization technique to establish both the minimum energy configurations of hydrophobic particles of various geometries at air–water surfaces and assess the consequences for foam film stability if such particles bridge foam films. The technique employs the so-called Surface Evolver software developed by Brakke [81]. In this technique, surfaces are represented as a mesh of triangular facets where vertices are constrained to satisfy an equation that describes their overall shapes. A particularly convenient method [170] for representing many different particle shapes as a continuous surface is given by a "superquadratic" equation

$$\left|\frac{x}{s_1\tilde{p}}\right|^{e_1}+\left|\frac{y}{s_2\tilde{p}}\right|^{e_2}+\left|\frac{z}{s_3\tilde{p}}\right|^{e_3}=1 \tag{4.60}$$

where x, y, and z are the coordinates of any point on a particle surface; e_1, e_2, and e_3 determine the shape of the particle; and s_1, s_2, and s_3 determine the relative dimensions of the length, width, and height of the particle, respectively. The overall size of the particle is determined by $\tilde{p}$. If, for example, we set $e_1 = e_2 = e_3 = 2$ and $s_1 = s_2 = s_3 = 1$, then Equation 4.60 becomes the equation of a sphere where $\tilde{p}$ is the square of the radius. In the case of an orthorhombic particle $s_2 = s_3$ and the aspect ratio is simply s_1/s_2, while $e_1 = e_2 = e_3$ where $e_1 \gg 1$—the larger the value of e_1, the sharper the edges of the particle.

The approach has been used by Morris et al. [168, 169] to study both the stable orientations of orthorhombic particles in air–water surfaces and the resultant effect on foam film stability, all as a function of contact angle and aspect ratio. The computational procedure involves first inserting the particle into a planar air–water surface at a given orientation. The air–water surface is characterized by a different surface energy from that of the surface of the particle, which therefore defines the contact angle The energy of the ensemble is then minimized by the iterative procedure included in the Surface Evolver software combined with mesh adjustments. This procedure is then repeated for many orientations with a given contact angle in order to plot an energy profile that reveals stable orientations as minima.

Bridging configurations with these stable orientations of the particles in foam films are now created as another ensemble and subjected to incremental changes in an applied capillary pressure, expressed as an increased curvature of the film surfaces. At each increase in the capillary pressure, the overall surface energy is minimized and the film searched for any points where the opposite sides are touching, indicating film rupture. This procedure therefore makes no allowance for disjoining forces in the film.

The energy minimization approach, using the Surface Evolver software, reveals only two stable orientations in the case of a cubic particle (of aspect ratio unity) at the air–water surface over the contact angle range $0 < \theta_{AW} < 90°$. These are depicted in Figure 4.57 for a contact angle $\theta_{AW} = 45°$. The first concerns a horizontal orientation, shown as a and b in Figure 4.57, which is stable over the contact angle range $0 < \theta_{AW} < 80°$. The second concerns a so-called rotated orientation shown as b and c in Figure 4.57, which is stable in the range $65° < \theta_{AW} < 90°$. Since for the example depicted in the figure, $\theta_{AW} = 45°$, then only the horizontal orientation is stable while the rotational orientation is unstable. The two orientations are, however, both stable over the contact angle range $65° < \theta_{AW} < 80°$. The diagonal orientation suggested by Dippenaar [77] and shown in Figure 4.50b is not stable at all in the case of cubic particles. It seems likely that the reason for this discrepancy concerns the contribution to

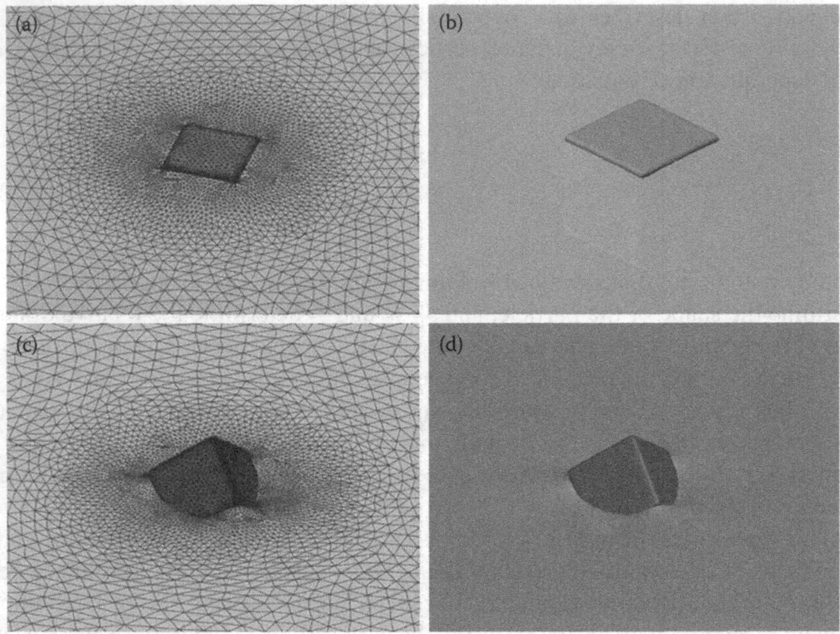

FIGURE 4.57 Orientations of cubic particle with aspect ratio unity at an air–water surface with contact angle $\theta_{AW} = 45°$ by surface energy minimization using Surface Evolver software (with $s_1 = s_2 = s_3 = 1$ and $e_1 = e_2 = e_3 = 20$). (a) Stable horizontal orientation, showing triangular mesh. (b) Stable horizontal orientation omitting mesh. (c) Unstable rotated orientation, showing triangular mesh. (d) Unstable rotated orientation omitting mesh. (Reprinted from *Miner. Eng.*, 23, Morris, G., Neethling, S., Cilliers, J., 979. Copyright 2010, with permission from Elsevier.)

the overall work of emergence of the particle of the curved air–water surface formed at parts of the particle surface not served by an edge as illustrated in Figure 4.56b. The simulation of Morris et al. [168] will of course account for that factor.

Careful examination of the cinematographic film frames of Dippenaar [77] in Figure 4.51 reveals, however, that the orthorhombic galena particle is in fact not a cube—it has an aspect ratio of 1.4. Following the argument given in Section 4.7.3.2, we would expect the relative adverse contribution of the curved air–water surface accompanying the diagonal orientation to be diminished with an increase in the aspect ratio from unity for a cube to 1.4. Morris et al. [168], in considering only the contact angle range 0–90°, show that this increase in aspect ratio renders the diagonal orientation stable provided the contact angle is >60°. This orientation is shown as c and d in Figure 4.58. The geometry of the air–water surface against the vertical surface of the diagonal orientation clearly resembles that of the sketch shown in Figure 4.56b and arises because the capillary pressure should be zero and therefore the shape should be a catenoid segment. At lower contact angles, a horizontal orientation, shown as a and b in Figure 4.58, is stable over a contact angle range $0 < \theta_{AW} < 90°$. Both the diagonal and horizontal orientations for this oblong particle are therefore stable at $\theta_{AW} = 70°$ as shown in the figure and in general over the contact

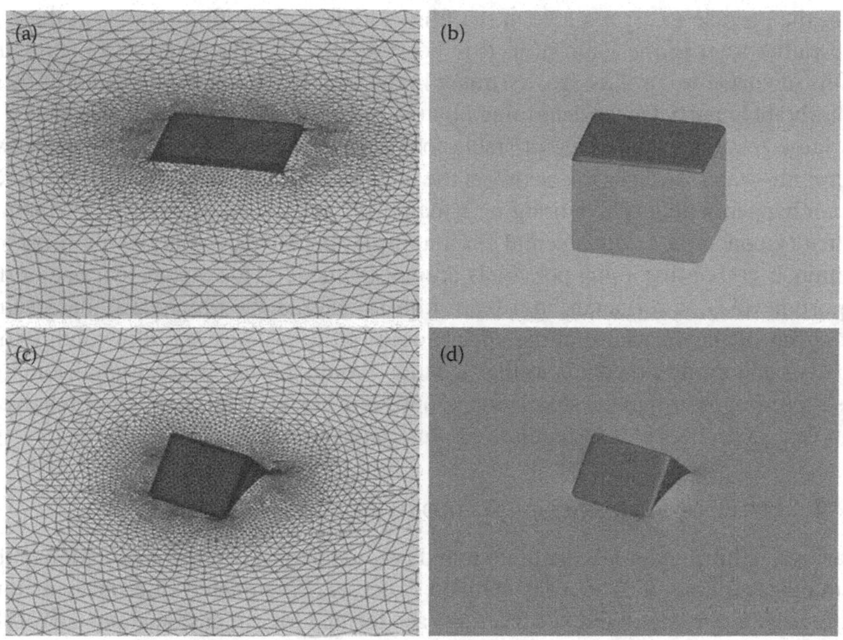

FIGURE 4.58 Orientations of oblong particle with aspect ratio 1.4 at air–water surface with contact angle $\theta_{AW} = 70°$ by surface energy minimization using Surface Evolver software (with $s_1 = 1.4$, $s_2 = s_3 = 1$ and $e_1 = e_2 = e_3 = 20$). (a) Stable horizontal orientation, showing triangular mesh. (b) Stable horizontal orientation omitting mesh. (c) Stable diagonal orientation, showing triangular mesh. (d) Stable diagonal orientation omitting mesh. (Reprinted from *Miner. Eng.*, 23, Morris, G., Neethling, S., Cilliers, J., 979. Copyright 2010, with permission from Elsevier.)

angle range $60° < \theta_{AW} < 90°$. This is consistent with the finding of Dippenaar [77] that galena particles with a contact angle of 80 ± 8° can adopt either orientation. However, the condition for foam film rupture at the edge of the particle is determined by the contact angle condition required to produce a stable diagonal orientation of >60° rather than >45° as suggested by the argument of Dippenaar [77]. The contact angle condition for the diagonal orientation and therefore foam film rupture will, however, presumably become >45° at sufficiently high aspect ratios where the contributions to the work of emergence of the particle due to the surfaces at the ends of the diagonally oriented particle become relatively negligible.

Determination of the critical capillary pressure for foam film rupture in the presence of bridging orthorhombic particles by surface energy minimization appears to produce results at variance with experiment. Regardless of particle orientation and for all contact angles <90°, the simulation tends to find foam film rupture at some point within the film at a higher capillary pressure than would be required without the particle [168]. The particles in the simulation actually appear to invariably stabilize the film. However, the cinematographic film frames of Dippenaar [77] for particles with contact angles < 90°, which are shown in Figure 4.51, clearly reveal antifoam effects where the two relevant three-phase contact lines become coincident upon the particle to form a hole in the film, which leads to film rupture at the particle rather than in the foam film. It is possible that the discrepancy concerns the radius of curvature of the edges assumed in the simulations of the behavior of these orthorhombic particles (as determined by the magnitude of e_1, etc.). If that radius is too large relative to the range of stable foam film thicknesses, it could permit two planar air–water surfaces to coexist on the edge when the contact angle < 90°, much as can happen with a sphere bridging a foam film as shown in Figure 4.44. Another difficulty concerns a fundamental limitation of this surface energy minimization technique in that disjoining potentials are not included. This means that one of the important roles of surfactant in a foam film is ignored. In which case, decreasing the radius of the curvature of the edges could fail to produce the film rupture that negative disjoining potentials at that location would cause. General application of this type of approach to the simulation of antifoam effects with particles of relatively complex geometries clearly requires resolution of these issues.

4.7.4 Effect of Rugosities on Antifoam Action of Particles

Sparingly soluble materials are often found precipitated from aqueous solution in the form of amorphous particles of ill-defined structure. Precipitated silica is an obvious extreme example where fractal structures are formed (of dimension about 2.4) [171]. Such materials may be intrinsically hydrophobic or be rendered hydrophobic by surface treatment. The presence of rugosities on the surfaces of such particles can have a profound effect on their antifoam function.

More than half a century ago, Wenzel [172, 173] showed that the presence of rugosities changes the apparent contact angle of a surface. This effect arises because the actual surface area of a rough surface is increased by those rugosities. The contact angle on a rough surface will therefore be a function of the ratio ϖ of the apparent area to the true area, where the latter accounts for the presence of rugosities. Wenzel [172, 173] has shown (see also references [174–176]) that

$$\cos \overline{\theta}_{AW} = \varpi \cos \theta_{AW} \tag{4.61}$$

where $\overline{\theta}_{AW}$ is the observed contact angle of the rough surface and θ_{AW} is the intrinsic angle. Since the ratio $\varpi < 1$, we must have $\overline{\theta}_{AW} < \theta_{AW}$ if $\theta_{AW} < 90°$ and $\overline{\theta}_{AW} > \theta_{AW}$ if $\theta_{AW} > 90°$. Clearly, then, the effect of surface roughness on the observed contact angle cannot of itself provide any explanation for enhanced antifoam action at least in the case where $\theta_{AW} < 90°$. However, as pointed out by Johnson and Dettre [163], this approach does not take account of metastable configurations—$\overline{\theta}_{AW}$ is simply the angle yielding the lowest free energy configuration.

Random rough amorphous and hydrophobic particles may, in general, adopt many different possible metastable configurations at the air–water surface for a given intrinsic contact angle. Some of these may be associated with film collapse if the requisite contact angle conditions are satisfied. However, the existence of many edges and rugosities will increase the chances of a planar interface hinging at more than one site on a particle. Stable bridging configurations in aqueous foam films may then occur with two planar air–water surfaces. The case of an axially symmetrical rough particle with regular rugosities is shown in Figure 4.59 by way of example. Again we suppose that the size of the particle is such that the effects of line tension and gravity can be ignored.

In this case, only one of the allowed configurations that satisfy the contact angle θ_{AW} condition

$$\theta_r < \theta_{AW} < 180° - \theta_r \tag{4.62}$$

(where $\theta_r < 90°$ and is defined in Figure 4.59) can give rise to film collapse. For that configuration, the air–water surface is depicted in Figure 4.59 as hinging on rugosity $ii = 8$ so that most of the particle is outside the aqueous phase. All other configurations would actually stabilize a foam film against drainage. Moreover, the probability of occurrence of the destabilizing configuration is relatively low if $\theta_{AW} < 90°$. This is revealed by consideration of the relative values of the work W_{ii}^{+}, of emergence of

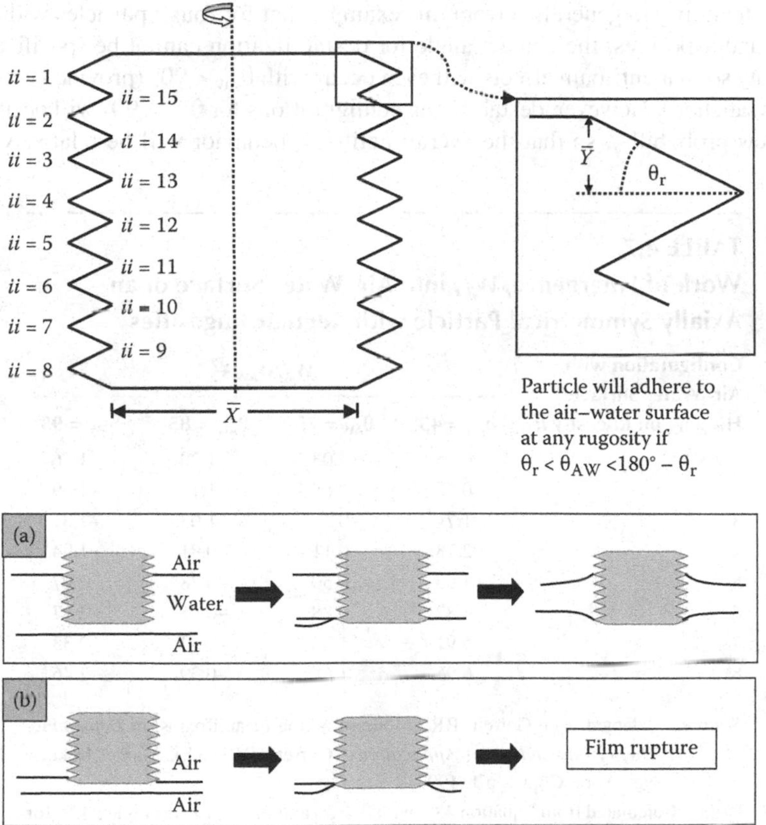

FIGURE 4.59 Axially symmetric particle with many edges. (a) Example of configuration that stabilizes foam film. (b) Configuration with air–water surface hinging on rugosity $ii = 8$ is the only configuration leading to foam film rupture.

each configuration. Here W_{ii}^{+} may be calculated in a similar manner to W_i by using Equation 4.54.

The geometry of the particle shown in Figure 4.59 means that we can write, for W_{ii}^{+}

$$\frac{W_{ii}^{+}}{\sigma_{AW}\bar{X}^2} = \pi\left[p_{ii}\left(\frac{\bar{Y}/\bar{X}}{\sin\theta_r}\right)\left(1+\frac{\bar{Y}/\bar{X}}{\tan\theta_r}\right)+0.25\right]\cos\theta_{AW} - \pi\left[0.5+\frac{\bar{Y}/\bar{X}}{\tan\theta_r}\right]^2 \quad (4.63)$$

where $\bar{X}$ is defined in Figure 4.59 and where p_{ii} is the number of dewetted segments each of height $\bar{Y}$ associated with the air–water surface hinging on rugosity *ii*. Values of $W_{ii}^{+}/\sigma_{AW}\bar{X}^2$, for different configurations of the particle, are presented in Table 4.7, for four different contact angles, by way of example.

Low negative values or even positive values for W_{ii}^{+} found for the destabilizing configuration $ii = 8$ shown in Table 4.7 for $\theta_{AW} < 90°$ imply low relative probabilities of occurrence. Conversely, for $\theta_{AW} > 90°$, the destabilizing configuration has the highest negative value for W_{ii}^{+}, which implies the highest probability of occurrence.

It is tempting to generalize from this example that for rough particles with many edges and asperities, the contact angle for destabilization cannot be specified with certainty so that antifoam effects will even occur with $\theta_{AW} < 90°$ (provided condition 4.62 is satisfied). However, destabilizing configurations for $\theta_{AW} < 90°$ will occur only with low probability, so that the overall antifoam behavior will be relatively weak

TABLE 4.7
Work of Emergence, W_{ii}^{+}, into Air–Water Surface of an Axially Symmetrical Particle with Regular Rugosities

Configuration with Air–Water Surface Hinging on Rugosity *ii*	$W_{ii}^{+}/\sigma_{AW}\bar{X}^2$			
	$\theta_{AW} = 45°$	$\theta_{AW} = 75°$	$\theta_{AW} = 85°$	$\theta_{AW} = 95°$
ii = 1	−0.35	−1.03	−1.29	−1.56
2	0.70	−0.65	−1.16	−1.69
3	1.74	−0.27	−1.03	−1.81
4	2.78	0.11	−0.91	−1.94
5	3.82	0.50	−0.78	−2.07
6	4.87	0.88	−0.65	−2.17
7	5.91	1.26	−0.52	−2.33
8[a]	6.95	1.64	−0.39	−2.46

Source: Adapted from Garrett, P.R., Mode of action of antifoams, in *Defoaming, Theory and Industrial Applications* (Garrett, P.R., ed.), Marcel Dekker, New York, Chpt 1, p 1, 1993.

Note: Calculated from Equation 4.63 with $\theta_r = 30$ and $\bar{Y}/\bar{X} = 0.1$; see Figure 4.59 for definitions of θ_{AW}, $\bar{Y}$, and $\bar{X}$.

[a] *ii* = 8 (see Figure 4.59) is the only configuration that could lead to foam film rupture.

or even negligible in those circumstances. Thus, for film collapse, the particle must adopt a configuration at the air–water surface so that no rugosities exist below that surface at which the second air–water surface of a foam film could hinge. Such configurations of necessity involve removal of most of a particle with many rugosities outside the aqueous phase. For contact angles < 90°, this is energetically unfavorable and will therefore have a relatively low probability of occurrence.

Evidence concerning the role of rugosities in antifoam action by particles is rather limited and not without ambiguity. Aronson [37] has, for example, shown that molten spherical drops of stearic acid, emulsified in surfactant solution, can be cooled to produce particles that are spheroidal assemblies of crystalline plates showing extremely rough surfaces with many edges. Electron micrographs of these entities are reproduced in Figure 4.60. Particle sizes are in the region of 4–6 microns. In 5×10^{-3} M aqueous solutions of sodium tridecylbenzene sulfonate, a concentration of 1 g dm^{-3} of these rough stearic acid particles is sufficient to eliminate about 90% of the foam volume generated by hand shaking cylinders. Similarly, the lifetimes of macroscopic foam films generated under near-equilibrium conditions in the presence of these particles are only a few seconds compared with more than 30 min in their absence. However, this antifoam effect is more or less absent when generating foam by sparging the same solution.

That the antifoam effect of these stearic acid particles is markedly more apparent with cylinder shaking than with sparging in this case suggests that it concerns the relative rates of air–water surface generation by these two techniques. As implied by Patist et al. [8], that rate is higher with hand shaking of cylinders than with sparging. We would therefore expect larger deviations from static contact angles with the former than with the latter technique. In turn, that would produce the observed enhanced antifoam effects. However, the marked effect of the stearic acid particles on the lifetimes of macroscopic foam films formed under near-equilibrium conditions cannot be explained by this factor.

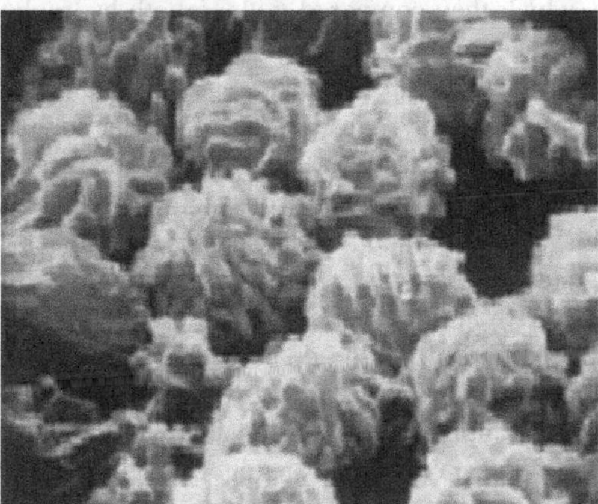

FIGURE 4.60 Scanning electron photomicrograph of stearic acid particles formed by cooling below the melting point ultrasonic dispersions of molten stearic acid in aqueous surfactant solution. Average particle size is 4–6 microns. (Reprinted with permission from Aronson, M. *Langmuir*, 2, 653. Copyright 1986 American Chemical Society.)

This effect could concern the size of the films (of ~2 cm^2), which are likely to be significantly larger than those of foam films generated by, for example, sparging. This could mean the presence of high relative numbers of antifoam entities in the macroscopic films so that even a small probability of one being effective could lead to film rupture.

If we accept the assertion of Aronson [37] that the contact angles of these stearic acid particles are likely to be similar to those measured on hydrophobic quartz with the same surfactant solution, then the advancing contact angle is likely to be ≤64° and the receding angle ≤ 33°. We can therefore make a comparison with the results of Frye and Berg [152] and those of Aveyard et al. [162] shown in Figure 4.48. Both of these authors considered smooth hydrophobed glass spheres (albeit almost an order of magnitude larger than those shown in Figure 4.60) and measured foamability by hand shaking of measuring cylinders. The results of Aveyard et al. [162] indicate little or no elimination of foam at advancing contact angles of smooth glass particles of ≤64°. The results of Frye and Berg [152] indicate little or no reduction in foamability by spherical particles with receding contact angles ≤ 33°. This comparison does tend to suggest that the roughness of the stearic acid particles has contributed to an enhanced antifoam effect. However, as indicated in Table 4.7, antifoam action of a particle with many rugosities will require that the least energetically favored configuration at the air–water surface be adopted in the case of contact angles < 90°. In consequence, the contact angle requirement for antifoam action is similar to that of a smooth sphere—at or close to 90°. However, it is possible that the relatively intense agitation involved in foam generation by cylinder shaking could produce for these rough stearic acid particles both high contact angles close to 90° (but still < 90°) and therefore an enhanced probability of occurrence of foam collapse configurations, which are a consequence of the presence of rugosities seen in Figure 4.60.

Tamura et al. [177–179], in a rather repetitive series of three papers, have also drawn attention to the role of rugosities in enhancing the antifoam effects of hydrophobic particles. In this context, they were concerned with agglomerates of hydrophobic PDMS gel particles, which they compared with smooth hydrophobed silica spheres. The surfaces of the former were characterized by rugosities. Both types of particles were found to have similar air–water contact angles measured through the aqueous phase—in the region of 130° in the presence of a micellar solution of SDS (containing polyvinyl alcohol)! These angles were measured by the sessile drop technique on a close-packed surface, only 70% of which consisted of particles, which accounts for the unreasonably high values. Foamabilities measured by cylinder shaking, in the presence of the agglomerated gel particles, were significantly lower than in the presence of the spheres. This relative antifoam effect was shown to correlate with the ease of rupture of the air–water–solid film—the presence of rugosities in the case of the gel particles apparently facilitating rupture of that film. Indeed, Tamura et al. [178, 179] show that the greater the amplitude of the rugosities, the more marked the antifoam effect. Air–water–solid films are of course analogous to air–water–oil pseudoemulsion films, the stability of which, we remember, determines the ease of entry of oil drops into air–water surfaces. The stability of the air–water–solid film presumably plays an analogous role in determining the emergence of particles into that surface.

In summary then, theory suggests that hydrophobic particles decorated with many rugosities can show antifoam effects. However, they require contact angles high enough to ensure that air–water surfaces hinge on the rugosity that requires most of the particle to be dewetted from the aqueous phase. This then permits antifoam action as the second air–water surface in a foam film attempts to satisfy the local contact angle (see Figure 4.59). Effective antifoam action is therefore favored if contact angles are at least close to 90° regardless of the contact angle requirements at the rugosities. Any difference in antifoam effectiveness between a smooth spherical particle and one decorated by rugosities must then be enhanced under foam generation conditions involving intense agitation, which favor high dynamic contact angles and the formation of foam collapse configurations by rough particles at the air–water surface. The empirical evidence also suggests that regardless of these factors, the presence of rugosities should also facilitate the rupture of the air–water–solid film, which in turn determines the ease of emergence of the particle into the air–water surface—a key initial step in the realization of antifoam action.

4.7.5 Rugosities and Stability of Air–Water–Solid Films

The rupture of the aqueous film separating a particle from the air–water surface is usually believed to occur by catastrophic growth of random fluctuations in the air–water surface of the film at a certain critical film thickness in a manner analogous to that described by Vrij and Overbeek [180, 181] for free aqueous foam films (see Chapter 1). Here we remember that two factors influence the growth of those fluctuations. The first is the work done against the air–water surface tension in perturbing the surface, and the second is the work of interaction due to the disjoining force acting across the film. A recent review of this subject in the present context, where one side of the film is a solid surface, can be seen, for example, in reference [182].

As with free foam and pseudoemulsion films, both attractive and repulsive disjoining forces will exist in the films formed between hydrophobic particles and air–water surfaces in aqueous surfactant solutions. Repulsive forces will often be electrostatic in origin and are of relatively long range. They may be due to overlapping electrostatic double layers caused by adsorbed charged surfactant. They may even be a consequence of charge intrinsic to the solid surface in an aqueous environment. Thus, strong electrostatic repulsive forces stabilize films of pure water on silica even when the latter is hydrophobed with methyl silanes [183].

The Hamaker constants for the aqueous films separating hydrophobic materials such as n-alkanes (at least up to C_{16}) and various polymers (such as polystyrene and polmethylmethacrylate) from air are all negative [184]. This implies repulsive van der Waals interactions favoring stability of the films. That these materials are hydrophobic has therefore been attributed to ultra short range, so-called hydrophobic forces [28]. By contrast, the relevant Hamaker constant for PTFE is positive, which implies repulsive van der Waals forces. The relevant Hamaker constants for minerals such as fused silica and calcite all tend to be negative [184]. However, these materials can be rendered hydrophobic by chemical adsorption of hydrocarbon groups on their surfaces. It is difficult to see why such a limited molecular change should reverse the sign of the Hamaker constants. Again, this tends to suggest a role for short-range hydrophobic forces.

Despite these complications, it seems reasonable to suppose that the disjoining forces across aqueous films on hydrophobic solids (or those rendered hydrophobic) are expected to be composed of both attractive and repulsive components, which can conspire to produce maxima in the relevant disjoining pressure isotherms. If a particle is to emerge into an air–water surface of a foam film, a sufficiently high capillary pressure must presumably be applied to overcome that disjoining pressure. Film rupture would then follow by growth of fluctuations provided, as we have seen in other contexts, the derivative of the disjoining pressure with respect to film thickness is positive (see Chapters 1 and 3).

Froth flotation is a process in which particles are collected by gas bubbles according to the ease with which they emerge into air–water surfaces. Therefore, this process clearly has something in common with the antifoam process we are considering here. Anfruns and Kitchener [185] have reported a flotation study that has in fact direct relevance to that process. This study concerned the efficiency of the capture of particles dispersed in an aqueous medium by rising bubbles under carefully controlled conditions. Both glass microspheres and rough quartz particles were used, which were rendered hydrophobic with methyl silanes. They were of similar size and contact angle. Bubbles were in the size range 0.05–0.1 cm, and particles were of size about 30 μm. These experiments enabled the collection efficiency C_e to be obtained as

$$C_e = \frac{\text{Number of particles collected}}{\pi r_b^2 L N} \tag{4.64}$$

where r_b is the radius of the bubble, L is the path length of the bubble, and N is the number of particles per unit volume suspended in the medium through which the bubble passes. Here $C_e \ll 1$ because most particles ahead of a rising bubble are swept along the fluid streamlines remote from the bubble surface. The process is depicted schematically in Figure 4.61.

The conditions of these experiments were such that the relevant streamlines could be calculated so that C_e could in turn be calculated. In Figure 4.62a and b, the

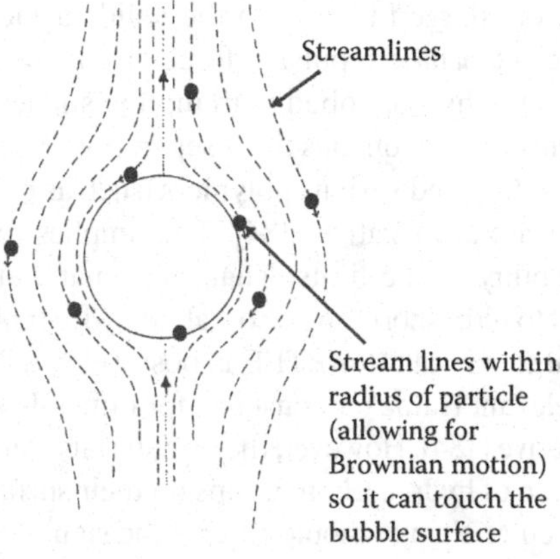

FIGURE 4.61 Bubble rising through suspension of hydrophobic particles under laminar flow illustrating inefficient interception of particles by bubbles (schematic).

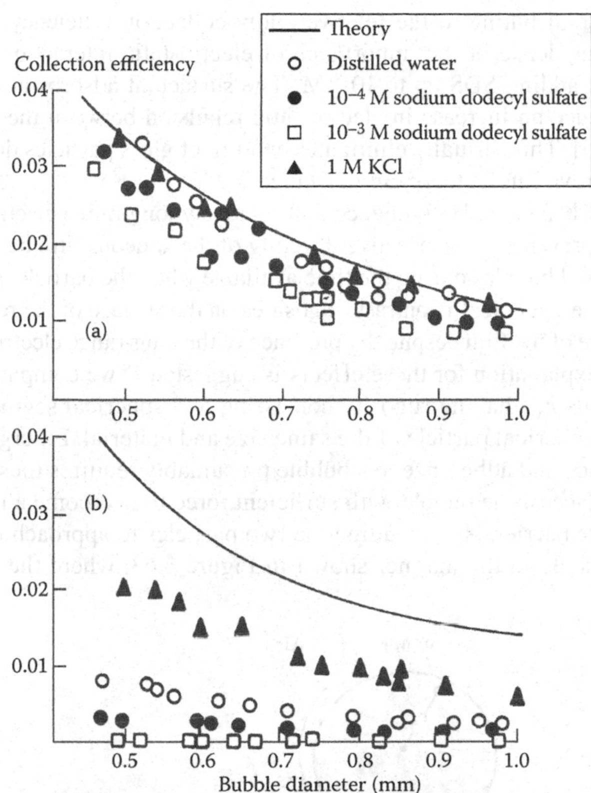

FIGURE 4.62 Collection efficiencies for rough quartz and spherical glass particles. (a) Quartz particles of 31 microns Stokes equivalent diameter. (b) Spherical glass particles of 32-micron diameter. (From Anfruns, J.F., Kitchener, J.A., *Trans. Inst. Min. Metall., London C*, 86, 9, 1977, reproduced by permission of the Institute of Mining and Metallurgy, London.)

calculated and experimental values of C_e for both smooth glass microspheres and rough quartz particles are compared for various solution conditions. The efficiency of collection of the glass microspheres is seen to be considerably less than that of the quartz particles under all these conditions. Indeed, the efficiency of collection of the quartz particles generally approaches theoretical values. This means that every quartz particle following a streamline that takes the particle within a particle radius of the surface of a bubble is captured by the bubble. These differences between the quartz and glass particles occur despite the similarity of both the surface treatment and contact angles measured on representative smooth surfaces (~85–90°) in distilled water. Similar observations concerning the effectiveness of surface roughness in promoting the efficiency of capture of particles by bubbles are reported by Ducker et al. [186] and Strnad et al. [187].

Interactions due to electrical double layers are essentially suppressed by the addition of 0.1 M KCl. This is seen from Figure 4.62b to more than triple the collection efficiency of glass microspheres. Electrical double layers are therefore clearly implicated in contributing to the relatively low collection efficiency of these particles. Further evidence of the importance of electrostatic interactions is provided by the effect of adding SDS up to 10^{-3} M. The surfactant adsorbs at the air–water surface to produce an increase in electrostatic repulsion between the particles and the bubbles [185]. This virtually eliminates capture of glass particles despite contact angles of 30° prevailing in these circumstances.

Anfruns and Kitchener [185] deduce that relatively long-range electrostatic repulsion forces are preventing the effective thinning of the aqueous film between smooth glass particles and bubbles during the time available when the particle is in near contact with the rising bubble. In contrast, rugosities on the surface of the rough particles facilitate rupture of the film despite the presence of the long-range electrostatic forces.

A possible explanation for these effects is suggested if we compare a spherical particle of radius r_{AF}, having rugosity consisting of a spherical segment radius r_a, with a smooth spherical particle of the same size and material. Emergence into the air–water surface and adherence to a bubble presumably requires these particles to impact the surface of the bubble with sufficient force to overcome a repulsive disjoining pressure barrier. Now imagine the two particles to approach the air–water surface of a bubble in the manner shown in Figure 4.63, where the impact force

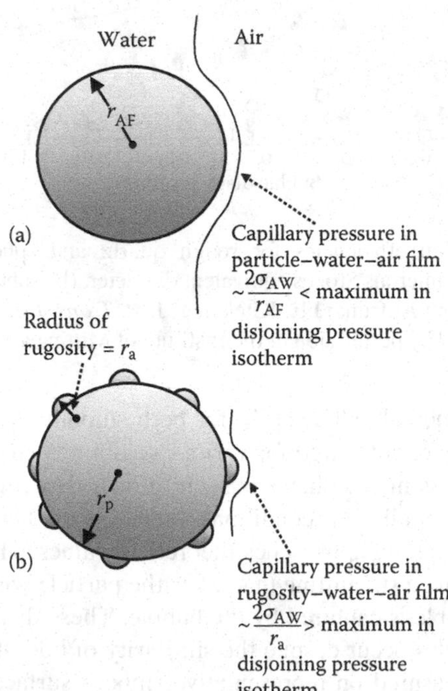

FIGURE 4.63 Effect of rugosities on capillary pressure in film formed between particle and air–water surface as it intercepts a bubble in flotation experiment. (a) Smooth spherical particle of radius r_{AF}. (b) Spherical particle also of radius r_p but with rugosities of radius r_a where $r_a \ll r_p$.

produces deformation of the air–water surface. In the case of the smooth particle, a capillary pressure of ~$2\sigma_{AW}/r_p$ is induced by that force in the solid–water–air film. However, in the case of the rough particle, a capillary pressure of ~$2\sigma_{AW}/r_a$ is induced in the vicinity of a rugosity. The force required to produce these deformations of the air–water surface is, however, approximately proportional to the radius of either the particle or the rugosity. The maximum in the disjoining pressure isotherm must be exceeded by the capillary pressures if the solid–water–air film is to rupture and the particles are to adhere to the air–water surface. The ratio of the capillary pressure in the vicinity of the rugosity to that in the smooth particle will be r_p/r_a, where $r_p \gg r_a$, and it is possible that the capillary pressure in the vicinity of the rugosity is high enough to exceed the maximum in the disjoining pressure isotherm but not in the absence of the rugosity. Rupture of the solid–water–air film and therefore adherence of the particle to the air–water surface would therefore require the presence of rugosities. Adherence of the rugosity to the air–water surface would, however, still leave an aqueous film separating most of the particle from the air bubble. Presumably the rugosity could then have a similar effect on the overall stability of that film to that shown by particles bridging free foam films. Provided the spherical-segment rugosity protrudes from the particle by a distance less than its radius, then it can dewet to rupture the aqueous film separating the whole particle from the air bubble even though the air–water contact angle is <90°.

Similar considerations could concern particles in foam films subject to a capillary pressure. To this end, we may adapt the argument of Zhang et al. [188]. In Figure 4.64a, we depict a smooth spherical particle in a thin foam film. If the Plateau border capillary pressure in the foam film is p_c^{PB}, then the total capillary pressure in the solid–water–air film in the vicinity of the particle is equal to $p_c^{PB} + 2\sigma_{AW}/r_{AF}$. Similarly, in Figure 4.64b, we depict a particle with rugosities where the total capillary pressure in the vicinity of the rugosity is then equal to $p_c^{PB} + 2\sigma_{AW}/r_a$. Again, then, since $r_{AF} \gg r_a$, we find that the capillary pressure in the vicinity of the rugosity is higher than that in the vicinity of the smooth particle leading to the possibility of rupture of the solid–water–air film, as the maximum in the disjoining pressure isotherm is exceeded only on the rugosity. However, the critical thickness for rupture of the solid–water–air film is likely to be smaller for the rugosity because of the requirement that the critical wavelength is less than the dimensions of that rugosity where that wavelength is a function of the film thickness (see, e.g., Chapter 1 and references [180, 181]). This factor would increase the time taken for that film to rupture.

Frye and Berg [152] in their study of the effect of particle morphology on antifoam effectiveness were apparently unaware of this work of Anfruns and Kitchener [185]. They selected similar systems, however, to those selected by the latter. The superior antifoam action of hydrophobed ground-glass particles relative to the effect of spherical glass particles, shown in Figure 4.53, may therefore owe something to the more ready attachment of rough particles to bubble surfaces. The importance of roughness in determining dewetting behavior does hint that the contribution to the weak antifoam behavior of the spherical particles by a contaminating proportion of

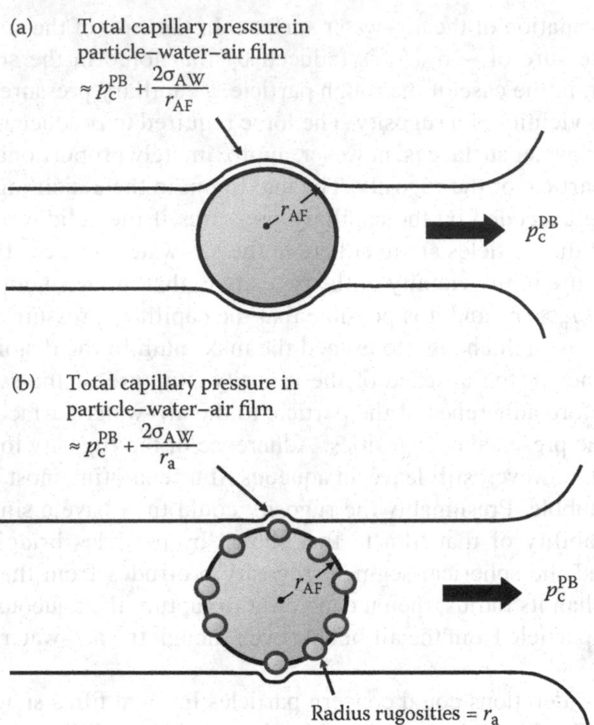

FIGURE 4.64 Effect of rugosities on total capillary pressure in film formed between particle and air–water surface of a foam film subject to Plateau border suction, p_c^{PB}. (a) Smooth spherical particle of radius r_{AF}. (b) Spherical particle also of radius r_{AF} but with rugosities of radius r_a where $r_a \ll r_{AF}$.

particles with sharp edges is likely to be even more important than Frye and Berg [152] indicate.

We conclude then that the presence of rugosities facilitates rupture of the metastable aqueous film separating particles from the air–water surface. As we show in Figure 4.62, enhancing the electrostatic repulsive component of the disjoining pressure by addition of a charged surfactant has little effect on the collection efficiency of rough quartz particles but significantly diminishes that of smooth glass particles. Similarly, diminution of the electrostatic component of the disjoining pressure by addition of a simple electrolyte, in the absence of surfactant, enhances the relative collection efficiency of the smooth glass particles but has little effect on the already high collection efficiency of rough quartz particles. Particle collection efficiencies, and therefore their ease of emergence into air–water surfaces, clearly concern both the presence of rugosities and the electrostatic component of the disjoining pressure isotherm. This may well therefore offer an explanation for the apparent enhanced antifoam effect, under dynamic conditions, of rough particles with respect to smooth spherical particles with similar contact angles.

4.7.6 Particle Size and Kinetics of Foam Film Rupture

An essential feature of the bridging mechanism for antifoam action by particles is that the particle size be of the same order as the thickness of the foam film. Foam films drain and stretch during foam generation and after foam generation has ceased. The rate of such processes will clearly play an important role in determining the frequency of foam film collapse if the film must first drain to the dimensions of any particles present.

Dippenaar [77] has developed a simple model to describe foam film collapse as a function of particle size for the case of foam generation by shaking cylinders. With this method of foam generation, repeated shaking will often eventually produce a constant foam volume. Here, presumably, the rate of production of foam equals the rate of destruction due to the presence of hydrophobic particles. The rate of destruction according to Dippenaar [77] will, however, be proportional to the amount of foam present. Therefore, the final foam volume is proportional to the volume rate of foam destruction. That volume may be maintained constant for different particle sizes by adjusting the total mass M of the particles present. Then M becomes the mass of particles required to reduce the final foam volume to the chosen reference value. All of this implies a relation between particle size and the total mass of particles required to achieve a constant volume rate of foam destruction (which is proportional to the rate of film rupture).

At the supposedly constant final foam volume, the frequency ν_r with which a single particle ruptures a film is simply the rate of film destruction/total number of particles. Therefore, ν_r is given by [77]

$$\nu_r = \frac{k_1 r_{AF}^3}{M} \tag{4.65}$$

where r_{AF} is the antifoam particle radius and k_1 is a constant.

As we have seen, for particles to rupture thin films, they must first thin down to a thickness proportional to the particle dimension (dependent on particle size, shape, and contact angle). The time taken for this to occur is difficult to assess with certainty. Foam films prepared from aqueous solutions of typical surfactants are "mobile" and exhibit rapid drainage by processes involving hydrodynamic instabilities (see Chapter 1). These instabilities give rise to marginal regeneration [189] in the case of vertical foam films and asymmetric drainage in the case of horizontal films [190, 191]. No analytical expression is available for describing film drainage in either case. Mobile film behavior is usually associated with air–water surfaces of low surface shear viscosity and/or low surface dilational modulus. In the case of surfactant solutions with air–water surfaces of high surface shear viscosity and/or high surface dilational modulus, the hydrodynamic instability is suppressed [64, 190, 192]. The resulting foam films are "rigid" and exhibit simple drainage behavior for which analytical solutions are available. In the exceptional case of, for example, rigid vertical films, drainage is by Poiseulle flow under gravity [189]. It seems unlikely, however, that the aqueous solutions of 1.6×10^{-4} M 1,1,3-triethoxy butane used by Dippenaar [77] would have the required surface rheological properties to ensure such drainage behavior. Nevertheless, Dippenaar [77] assumes Poiseulle flow. According to Mysels et al. [189], the drainage time t for a vertical film to thin to thickness h at a distance y from the top of the film is then given by

$$t = \frac{4\eta_W y}{\rho_W g h^2} \tag{4.66}$$

where η_W is the viscosity. In the case of small (<100 microns radius) rigid, circular, horizontal films, drainage is driven by capillary suction from adjacent Plateau borders so that the time for a film to thin is given by the Reynolds equation [193]

$$t = \frac{3\eta_W r_f^2}{4 p_c^{PB} h^2} \tag{4.67}$$

where r_f is the radius of the film and p_c^{PB} is the capillary pressure in the Plateau border. Comparison of Equations 4.66 and 4.67 reveals that the time taken for a film to thin to a given thickness is proportional to the inverse square of that thickness irrespective of whether drainage is driven by gravity or Plateau border capillary pressure. However, if, as seems likely, films of this surfactant solution exhibit mobile behavior, both Equations 4.66 and 4.67 will overestimate drainage times.

Dippenaar [77] supposes that the time taken for a film to thin to the dimension required for rupture by a particle is given by Equation 4.66 and is therefore simply inversely proportional to the square of that dimension. This would be true if drainage is described by either Equation 4.66 or 4.67, where under the conditions of constant final foam volume, either y or r_f^2/p_c^{PB} is constant. Therefore, combining Equations 4.65 and either 4.66 or 4.67 yields the relation

$$\nu_r = k_2\left(\frac{\eta_W}{t_p}\right)\left(\frac{r_{AF}}{M}\right) \tag{4.68}$$

where k_2 is another constant and t_p is the time taken for the film to thin to the thickness where rupture by a particle can occur. Using a range of hydrophobic quartz particles differing by more than two orders of magnitude in size, Dippenaar [77] shows that r_{AF}/M is essentially constant for foams generated from aqueous solutions of triethoxybutane. This implies, if we take account of Equation 4.68, that the frequency of rupture of foam films is inversely proportional to the time taken for the films to thin to a thickness where rupture by a particle can occur. Dippenaar [77] then concludes that the "rate-determining step" for film collapse by particles is "natural" thinning of films. This result is, however, obtained by using expressions for that natural thinning process, which are only true for rigid film behavior and which seems unlikely to be realistic in the case of the surfactant solution used for this work.

Equation 4.68 also means that if ν_r is proportional to t_p^{-1}, then for a constant particle size the ratio η_W/M must be constant. Dippenaar [77] has verified this using glycerol to modify solution viscosity. It implies that the mass of antifoam required to achieve a given final volume of foam is proportional to the viscosity of the solution.

If the rate-determining step for film rupture is generally the time taken for films to thin, then the actual process of film rupture, once particle bridging has occurred, must be relatively rapid. Frye and Berg [152] have developed a model of that process that yields estimates of the rupture time. The model assumes an axially symmetrical configuration similar to that shown in Figure 4.65. Here the air–water surface in the vicinity of the particle is assumed to have a circular cross section between the contact line on the particle and flat surface of the plane parallel film. The capillary pressure Δp_c^{part} in the vicinity of the particle is crudely simplified by considering only the curvature in the plane orthogonal to the plane of the foam film (i.e., Δp_c^{part} is given by Equation 4.46). The rate of flow of fluid $\tilde{u}_p$ from the vicinity of the particle under the influence of Δp_c^{part} is then calculated under the assumption that all the displaced liquid is squeezed out of the film so that it remains plane parallel except in the vicinity of the particle. A rigid circular film is assumed, so that $\tilde{u}_p$ is given by

$$\tilde{u}_p = \frac{\pi \Delta p_c^{part} h^3}{3\eta_W} \tag{4.69}$$

This expression is readily obtained by differentiating Equation 4.67, replacing p_c^{PB} by Δp_c^{part} and noting that $\tilde{u}_p = \pi r_f^2 .\mathrm{d}h/\mathrm{d}t$.

The position of the three-phase contact line is adjusted by small increments Δh^* from $h^* = 0$ to $h^* = h/2$, where h is the thickness of the film (see Figure 4.65). Here, h^* is the distance from the plane of the film surface to the plane of the three-phase

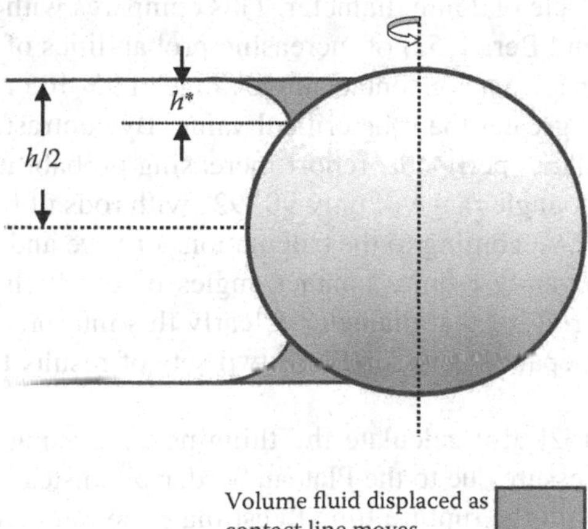

FIGURE 4.65 Illustration of procedure for estimation of time for antifoam action by hydrophobic spherical particle. (Reprinted from *J. Colloid Interface Sci.*, 127, Frye, G.C.C., Berg, J.C., 222. Copyright 1989, with permission from Elsevier.)

contact line. The time increment corresponding to each increment Δh^* is calculated from the relation

$$\Delta t = \frac{V}{\tilde{u}_p} \tag{4.70}$$

where ΔV is the corresponding incremental change in volume of fluid removed from the shaded area in Figure 4.65. The time to rupture is then $N\Delta t$, where the number of increments N is given by $N\Delta h^* = h/2$.

The assumption that all the displaced fluid is completely removed from the film so that it remains plane parallel facilitates calculation but may not be entirely realistic. In practice, liquid may migrate only a small distance from the particle to form a bulge. Calculated rupture times may therefore be overestimates.

Rupture times calculated with these procedures are dependent on contact angle and particle size. Thus, as the contact angle declines to the appropriate critical value, the rupture time asymptotically approaches infinity. The larger the particle, the greater the volume of liquid that must be squeezed out of the film before rupture can occur. Rupture times therefore increase with particle size. However, Frye and Berg [152] show that a three-orders-of-magnitude increase in particle size for rod-shaped particles results in a difference of only about 10° in the required contact angle if film rupture is always to occur in less than ~10^{-3} s.

In the case of experiments with rod-shaped particles, the total time for antifoam action is given by the rupture time—foam film drainage is then irrelevant because rods are directly inserted into an existing film. Using this technique, Frye and Berg [152] report a perception of instantaneous rupture, which presumably implies rupture times of ≤0.1 s. According to their theoretical estimates, such rupture times would apparently require contact angles ≥ 5° higher than the critical value of 90° for a rod-shaped particle of 2 mm diameter. This compares with experimental measurements by Frye and Berg [152] of increasing probabilities of rupture for rods of that diameter over an advancing contact angle range of 95–105°, as shown in Figure 4.46, which is 5–15° greater than the critical value. By contrast, however, Aveyard et al. [24] with similar experiments report increasing probabilities of rupture over an advancing contact angle range of only 90–92° with rods of 0.2 mm diameter, as shown in Figure 4.47. According to the calculations of Frye and Berg [152], rupture times of ≤0.1 s apparently require contact angles of ≥3° higher than the critical value in the case of rods of that diameter. Clearly this interpretation does assist in reconciling the discrepancy between these two sets of results that we highlighted in Section 4.7.2.

Frye and Berg [152] also calculate the thinning time using Equation 4.69, but with the capillary pressure due to the Plateau border p_c^{PB} instead of Δp_c^{part}. They then compare that time with the rupture time to estimate the rate-determining step. For small particles (≪100 μm), rupture times are minimal and thinning is rate determining in agreement with the conclusion of Dippenaar [77]. However, if the particles are sufficiently large (≫100 μm), then thinning times decline and rupture becomes rate determining. Frye and Berg [152] calculate an optimum particle size for which the total time of antifoam action is minimal. Rates are dominated by rupture times for sizes larger than the optimum and by thinning times for sizes smaller than the optimum. We should note here, however, that particles of optimum size for minimal antifoam action time are not necessarily of optimum size for overall effectiveness. A dominating factor is that the smaller the particle size, the greater the number of particles for a given antifoam mass and the higher the probability of film rupture. This is shown empirically by Dippenaar [77], where as we have seen the mass M of antifoam required to reduce the foam volume to a chosen reference value is proportional to the particle size.

Frye and Berg [152] provide no detailed comparison with experiment for these considerations. However, they note that the deviation from constant r_{AF}/M recorded by Dippenaar [77] for extremely large particles is consistent with rupture times becoming rate determining.

These calculations concerning total time for antifoam action rely essentially on the relevance of the Reynolds equation [193] and require knowledge of p_c^{PB}, which is given approximately by the hydrostatic head for a foam only after drainage has ceased (see Section 1.4.1). We havc already noted the limitations of the applicability of the Reynolds equation in this context. At most we can state that arguments about the overall time for antifoam action based on the use of that equation are likely to be overestimates except in the exceptional cases where the air–water rheology is consistent with rigid immobile film behavior and film sizes are so small that dimple formation is not possible (i.e., <100-micron diameter) [194].

4.7.7 Antifoam Effects of Calcium Soaps

Soaps have often been incorporated in detergents to control levels of foam, especially in the context of products designed to be used in foam-intolerant washing machines. Utility as antifoam additives in this context alone invites attention with respect to their mode of action. We therefore explore the possibility that they function as hydrophobic particles according to the principles outlined here. Of particular interest is the possibility that formation of highly insoluble calcium soaps represent the key to their mode of action.

More than half a century ago, Peper [195] considered the effect of sodium soaps and fatty acids on the foamability and foam stability of dilute solutions of anionic surfactant solutions. In this work, these soaps and fatty acids were present at a concentration of only 0.02 g dm^{-3} (in the range of 0.6×10^{-4}–1.0×10^{-4} M depending on molecular weight). Peper does not make clear the state of these materials dispersed in the surfactant solution. However, the concentrations of surfactant appeared to exceed the CMC in most cases and it seems likely that both sodium soaps and fatty acids were solubilized. Foam measurements were made using hand shaking of measuring cylinders at 25°C.

Peper showed [195] that in the presence of 1.8×10^{-5} M calcium chloride, the adverse effect of these sodium soaps on both foamability and foam stability of a ~3×10^{-3} M (1.0 g dm^{-3}) sodium dodecylbenzene sulfonate solution was in the order sodium stearate ≈ sodium palmitate > sodium oleate > sodium laurate, which follows the order of solubility products K_{sp} of the corresponding calcium soaps (calcium stearate, K_{sp} ~2×10^{-20}; calcium palmitate, K_{sp} ~6×10^{-18}; calcium oleate, K_{sp} ~10^{-15}; calcium laurate, K_{sp} ~6×10^{-13} at 25°C [151]). Measurements of foam volume as a function of time are reproduced in Figure 4.66. With the exception of sodium laurate, the solubility products for calcium soap precipitation were exceeded by from two to six orders of magnitude (depending on the soap), which implies that precipitation is likely irrespective of the effect of solubilization. With sodium laurate, no precipitation of calcium laurate would be expected based on these solubility products even if we again ignore micellar solubilization. This soap had a negligible effect on foam behavior under these conditions. The effect on foam stability for the other soaps was more marked than the effect on foamability—in the case of sodium palmitate, for example, the foamability was reduced by 40% and by a further 57%, 60 s after foam generation had ceased (see Figure 4.66). These effects were largely absent in deionized water. Antifoam effects were also observed with palmitic acid in the presence of calcium chloride at pH 6.5. However, at pH 3, palmitic acid remains un-ionized, soaps are not formed, and no antifoam effects were observed.

It would therefore seem likely that these antifoam effects are largely associated with the formation of calcium soap precipitates. Curiously, Peper [195] ignores this possibility and interprets his results in terms of formation of rigid islands of calcium soap monolayer interspersed with gaseous film of adsorbed surfactant. Peper [195] asserts that these islands will make the film unstable because of their "inflexible brittle nature." No theoretical arguments are given for why these should be unstable.

Peper [195] found that fatty acids reduce the surface tension of solutions of sodium dodecylbenzene sulfonate containing calcium chloride. The effect was found to be least pronounced for the fatty acid that forms the least soluble calcium soaps and the best antifoam. It is difficult to reconcile this finding with a Marangoni spreading mechanism for antifoam action (see Section 4.4.2).

If, by contrast, these antifoam effects with calcium soaps are a consequence of the formation of unstable bridging configurations in foam films, then we would expect

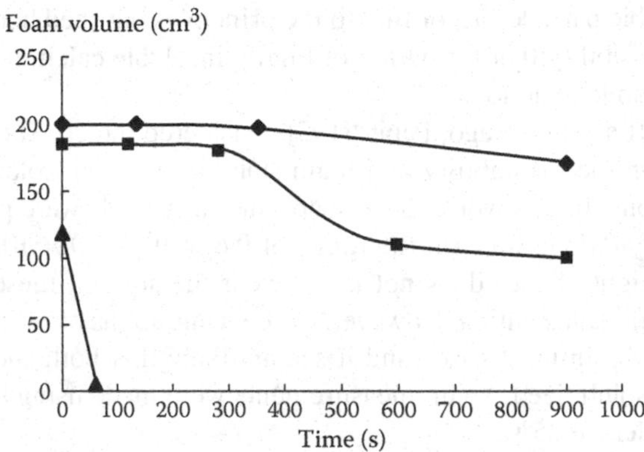

FIGURE 4.66 Foam volume as function of time of ~3 mM aqueous sodium dodecylbenzene sulfonate solution containing 0.018 mM $CaCl_2$ and 0.02 g dm^{-3} sodium soap. Foam generation by hand shaking cylinders at 25°C. ◆, Sodium laurate; ■, sodium oleate; ▲, sodium palmitate and sodium stearate (identical results). (After Peper, H., *J. Colloid Sci.*, 13, 199, 1958.)

the particles to exhibit finite contact angles at the air–water surface of the relevant foaming surfactant solutions. There are, however, few measurements of such contact angles except those reported by Scamehorn and coworkers [36, 196, 197]. Although calcium soaps invariably exhibit finite contact angles at air–water surfaces, values, both advancing and receding, are found by these workers to be usually <90°. Scamehorn and coworkers then argue [196] that this implies that the particles cannot function by bridging and dewetting in foam films. This conclusion is, however, demonstrably incorrect, as revealed by both the experimental evidence exemplified in Figures 4.42, 4.48, 4.51, and 4.53 and theoretical considerations in Section 4.7.3.2, where it is clear that both dynamic factors and geometry of particles can give rise to antifoam effects by bridging even if static contact angles are <90°.

Selection of calcium laurate as the model calcium soap in the presence of various surfactant solutions means that direct comparison of the contact angles measured by Scamehorn and coworkers [36, 196, 197] with the results of Peper [195] is not possible. However, Peper [195] observes an antifoam effect with calcium palmitate in submicellar 3.5×10^{-3} M SDS solution, where both foamability and foam stability are significantly diminished. The advancing contact angle measured by Scamehorn and coworkers [196] for this solution against calcium laurate is ~60°, with an expectation that the receding angle is even smaller. If, as seems likely, the corresponding angle for calcium palmitate is similar to that of calcium laurate, then we would expect antifoam effects by bridging only under dynamic conditions where the actual contact angle may be significantly higher and/or if the geometry of the calcium palmitate particles has sharp edges. That the calcium palmitate is effective even after foam generation has ceased, however, suggests that it can still function as air–water surface tensions tend toward equilibrium values. This argues against a role for dynamic contact angles under those conditions.

Another study of the antifoam behavior of calcium soaps has been presented by Zhang et al. [198]. They describe the effect of calcium oleate precipitation on the stability of the foam of both sodium alkyl ethoxy sulfate (C_{12}–C_{15}.EO_3.SO_4Na) and alcohol ethoxylate (C_{12}–C_{15}.EO_7) solutions. Here precipitation of the calcium soap was achieved by adding calcium ions to aqueous solutions of the surfactants containing sodium oleate at pH 9. Measurements concerned only the stability of foam generated to a fixed volume by an air entrainment method (see Section 2.2.6) [188]. Results are exemplified by those for the non-ionic alcohol ethoxylate presented in Figure 4.67a. Decline in foam stability correlated with an increase in turbidity associated with precipitation of calcium oleate. The effect was apparent with solutions of both surfactants. The decline in foam stability is seen to occur over several minutes and is clearly occurring over a time scale similar to that of Peper [195] for oleate soap shown in Figure 4.66.

A curious feature of these experiments concerns a decrease in the antifoam effect as the solutions aged over a few days. This occurred despite a concomitant increase in turbidity associated with increase in particle size of the precipitate. The effect was apparent with both surfactants and is exemplified in Figure 4.67b for the alcohol ethoxylate. More or less total elimination of the antifoam effect is seen after about 2 days of aging. We are therefore forced to conclude that the observed antifoam effect concerns some process of interaction between calcium and oleate ions

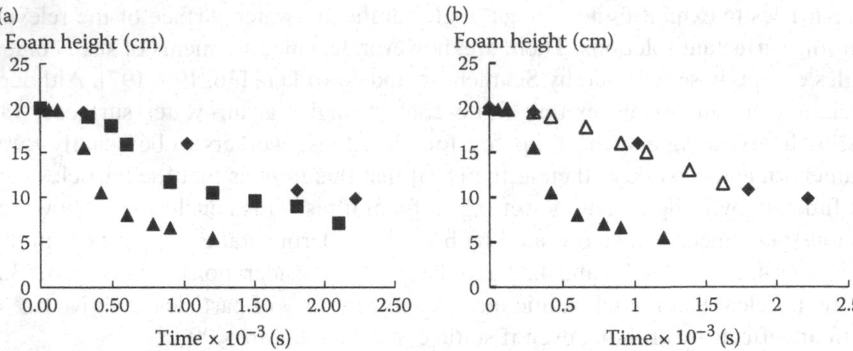

FIGURE 4.67 Effect of *in situ* formation of calcium oleate from dissolved sodium oleate on foam stability of 0.1 g dm^{-3} aqueous solution of commercial alcohol ethoxylate ($C_{12-15}.EO_7$) in 3 mM Ca^{2+} at pH 9. (a) Immediate effect of calcium oleate formation. ◆, absence of oleate; ■, 10^{-2} mM oleate; ▲, 3 × 10^{-2} mM oleate. (b) Comparison of effect of *in situ* formation of calcium oleate immediately and after 60 h aging. ◆, absence of oleate; ▲, immediate effect of 3 × 10^{-2} mM oleate; △, effect of 3 × 10^{-2} mM oleate after 60 h. (After Zhang, H., Miller, C.A., Garrett, P.R., Raney, K., *J. Colloid Interface Sci.*, 279, 539, 2004.)

because it is largely eliminated as the system equilibrates. Similar observations had been made earlier by Raghavachari et al. [199]. Whether this is a general aspect of the action of calcium soap formation or specific to calcium oleate formation has not of course been established by this work. A possible explanation concerns the diminution of surface charge in the adsorbed surfactant under initial conditions of supersaturation with enhanced calcium and oleate adsorption leading to diminished disjoining pressures and film stability. However, such an effect would surely mean that equilibration would lead to an increase in surface tension as the oleate and calcium ion activities decline with precipitation of calcium oleate. In fact, surface tensions actually decrease slightly [198] on aging of these solutions (at least in the case of the sodium alkyl ethoxy sulfate).

Observation of small horizontal foam films using a Scheludko cell (see Chapter 2) revealed many more particles to be present in fresh solutions than were present in aged solutions [198]. It is tempting therefore to suppose that the relevant process is actual precipitation of calcium oleate inside foam films as they drain. This may include nucleation on foam film surfaces. In turn, this could eliminate any issues concerning emergence of the particles into the air–water surface and also mean that particles are less readily flushed out of the draining film than is the case with particles that have nucleated and grown elsewhere. Destabilization of the foam films by bridging would then be favored, possibly also even by the geometry of the resulting particles.

In summary then, it seems probable that precipitated calcium soap particles are responsible for antifoam effects. The only available experimental evidence suggests that such particles appear to be characterized by static contact angles < 90° in surfactant solution. This suggests that any bridging foam film instabilities will require that the particles posses either certain geometrical features or be confined to the non-equilibrium conditions prevailing during foam generation by methods involving rapid air–water surface generation such as hand shaking, where dynamic contact angles could be significantly higher than static angles. However, no foam studies have been combined with study of particle size and geometry so it is difficult to make firm conclusions about the mode of action of these calcium soaps. Discovery of transient antifoam effects involving the process of precipitation of calcium oleate may even be representative of a general phenomenon involving precipitation of hydrophobic calcium soaps (or even other types of particles) in foam films. All of these considerations indicate that there is clearly a need to develop a firmer understanding of the calcium soap antifoam effects described briefly here.

4.7.8 Melting of Hydrophobic Particles and Antifoam Behavior

Certain hydrophobic materials that exhibit antifoam effects have melting points at temperatures < 100°C. Examples include hydrocarbon waxes, triglycerides, and long-chain fatty acids.

Dispersions of hydrophobic particles can be prepared by first emulsifying the molten materials and then cooling the emulsions below the melting point. Observations by Davis and Garrett [200] of the antifoam behavior of the resulting entities reveal some interesting behavior. By way of example, in Figure 4.68, we plot the ratio F, where

$$F = \frac{\text{volume of air in foam in presence of antifoam}}{\text{volume of air in foam in absence of antifoam}} \tag{4.71}$$

against temperature for 1.2 g dm^{-3} dispersion of n-docosane, n-eicosane, and paraffin wax in 0.5 g dm^{-3} (~7 × 10^{-3} M) sodium (C_{10}–C_{14}) alkylbenzene sulfonate solution. Foam measurements were made in a static Ross–Miles apparatus [201]. Particle sizes were in the range 0.5–8 microns. The melting ranges of these hydrocarbons were determined by nuclear magnetic resonance (NMR) spectroscopy. They showed no evidence of spreading on the surfaces of distilled water or the surfactant solution at 25°C as indicated by equilibrium surface tension measurements.

The most striking feature of the behavior of n-docosane is the sharp deterioration of antifoam effect in the region of the melting point at 44°C. An extremely pure sample was used, and therefore a sharp melting transition was observed. Similar observations have been made by Aronson [37] for hydrocarbons and fatty acids in solutions of sodium tridecylbenzene sulfonate.

Neither the n-eicosane nor the paraffin wax used in this work were pure materials. Therefore, both exhibit a relatively wide temperature range over which both solid and liquid phases coexist. The antifoam behavior of these materials, shown in Figure 4.68, is therefore complex. At low temperatures in region AB, an antifoam effect associated with solid hydrocarbon particles is present. Curiously, that effect appears to deteriorate with increase in temperature. However, in region BC, where both solid and liquid hydrocarbon coexist, a pronounced enhancement of the antifoam effect is seen to occur. This type of behavior has apparently also been observed by Kulkarni et al. [53] for impure triglycerides and by Joshi et al. [202] for an unspecified impure long-chain alcohol in a solution of a non-ionic surfactant. Here Joshi et al. [202] have

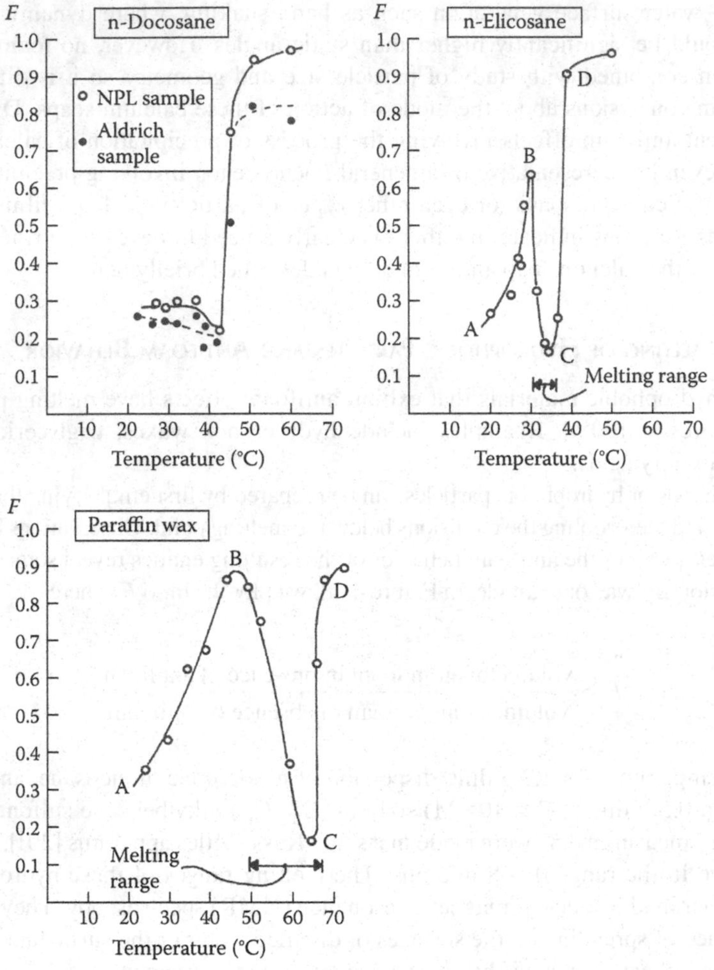

FIGURE 4.68 Effect of temperature on antifoam behavior of finely divided hydrocarbons [1.2 g dm^{-3} hydrocarbon in aqueous 0.5 g dm^{-3} commercial sodium alkyl (C_{10-14}) benzene sulfonate solution]. (After Garrett, P.R., Mode of action of antifoams, in *Defoaming, Theory and Industrial Applications*, Garrett, P.R., ed., Marcel Dekker, New York, Chpt 1, p 1, 1993; Davis, J., Garrett, P.R. unpublished work.)

also used NMR to determine the solid content as a function of temperature. As with paraffin wax, they find a pronounced minimum in foam generated by sparging at a temperature a few degrees below complete melting where the solid content is only ~10 wt.%. At sufficiently high temperatures where all solid is converted to liquid, the antifoam effect is seen to deteriorate more or less totally in much the same manner as for paraffin wax.

That antifoam effects may be associated with solid wax or fatty acid particles is not altogether surprising. Under microscopic observation, the particles are observed to exhibit more irregular shapes than the emulsion spheres from which they are derived. As we have seen (Section 4.7.4), Aronson [37] observes particularly marked roughness with many sharp edges for particles of stearic and palmitic acid. These particles are also found to be more effective antifoams than the more symmetrical hydrocarbon particles, for which similar contact angles would be expected to prevail. The equilibrium receding contact angles for the hydrocarbons used to obtain the results shown in Figure 4.68 are about 105° against distilled water, which decrease to about 30° against the relevant alkylbenzene sulfonate solutions [200]. It seems probable then that a combination of edges/asperities, finite contact angles, and dynamic effects (where the dynamic contact angle is higher than the equilibrium contact angle due to slow surfactant transport to the relevant surfaces) will mean that antifoam behavior by a bridging mechanism will occur. Aronson [37] presents evidence that suggests that dynamic effects are important for these systems.

Gradual deterioration of antifoam effectiveness of the solid particles with increase in temperature in region AB for the impure hydrocarbons in Figure 4.68 is difficult to explain. No significant change occurs in contact angles over the relevant temperature range [200]. It is, however, possible that the hydrocarbon softens with increasing temperature, so that asperities and edges are gradually removed. It may, in fact, concern the formation of the so-called plastic-crystalline rotator phases, which exist between the low temperature crystalline phases and the melting temperatures of n-alkanes [203]. The range of stability of those phases is increased by the chain mixing [203], which could be associated with paraffin wax blends and relatively impure n-alkanes such as the n-eicosane sample used here. All of this at least correlates with the relative range of the regions AB for n-eicosane and paraffin wax shown in Figure 4.68.

That the liquid hydrophobic oils formed above their respective melting temperatures are ineffective is a consequence of the metastability of the relevant pseudoemulsion films, as we have discussed elsewhere (see, e.g., Section 3.3). In the case of the impure materials where a region of mixed solid and liquid phase exists, it seems likely that the solid component is able to destabilize those films. Similar effects have been described in oil-in-water emulsion systems where partial melting leads to instability of the oil–water–oil films due the presence of crystals at the oil–water surface [204–206]. We address the effect of particle destabilization of pseudoemulsion films on antifoam action in the following section.

4.8 MIXTURES OF HYDROPHOBIC PARTICLES AND OILS AS ANTIFOAMS FOR AQUEOUS SYSTEMS

4.8.1 ANTIFOAM SYNERGY

One of the most striking aspects of antifoam behavior is the synergy shown by mixtures of insoluble hydrophobic particles and hydrophobic oils when dispersed in aqueous media. We illustrate this behavior in Figure 4.69 with a plot of the ratio F (defined by Equation 4.71) against antifoam composition for a mixture of methylsilane hydrophobed silica and liquid paraffin. Foam from a solution of 0.5 g dm^{-3} (~1.4 × 10^{-3} M) commercial sodium (C_{10}–C_{14}) alkylbenzene sulfonate was generated by cylinder shaking [43]. The liquid paraffin is seen to be virtually without effect on

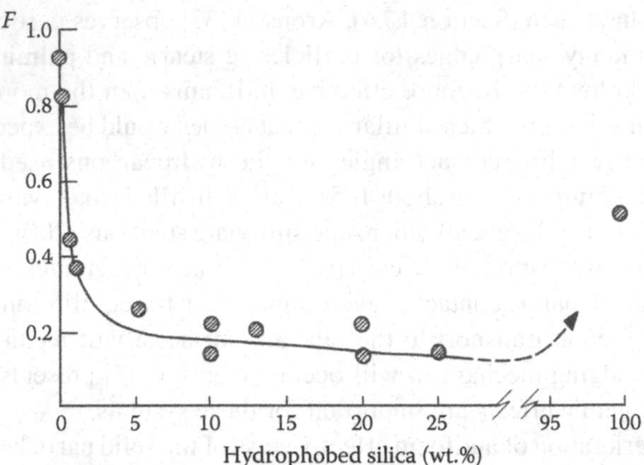

FIGURE 4.69 Antifoam effectiveness F (= volume of air in foam with antifoam/volume of air in foam absent antifoam) in aqueous surfactant solution of dispersion of hydrophobed silica–liquid paraffin antifoam as function of antifoam composition, illustrating antifoam synergy. Surfactant solution, 0.5 g dm^{-3} commercial sodium alkyl (C_{10-14}) benzene sulfonate. Hydrophobed silica, D17 ex-Degussa. Concentration of antifoam, 1.2 g dm^{-3}. Foam was generated by cylinder shaking at ambient temperature (22 ± 2°C). (After Garrett, P.R., Mode of action of antifoams, in *Defoaming, Theory and Industrial Applications*, Garrett, P.R., ed., Marcel Dekker, New York, Chpt 1, p 1, 1993; Garrett, P.R. et al., *Colloids Surf. A*, 85, 159, 1994.)

foam behavior, and the silica particles exhibit only a weak effect. Adding only a few percent of hydrophobed silica to the liquid paraffin is, however, enough to produce a significant enhancement of antifoam performance. Similar observations have been reported for mixtures of calcium stearyl phosphate and liquid paraffin [43], and by others for mixtures of hydrophobed glass with hexadecane [75]; dodecane with paraffin wax, ethylene bistearamide, or PTFE [44]; and often for hydrophobed silica with PDMSs [1, 53, 67, 70, 98, 207–209]. With the latter, silica may be first hydrophobed by use of a suitable agent, such as a silane or alcohol, before mixing with the PDMS. Alternatively, the silica may be hydrophobed *in situ* by heating with PDMS to temperatures in excess of 150°C (possibly in the presence of a catalyst) [210]. Antifoams produced by this process are often described as silicone "compounds." *In situ* hydrophobic particle–oil antifoam formation can also occur as a result of reactions between components of the oil phase and the aqueous phase to form particulate precipitates at the oil–water interface. Examples include the reaction between fatty acids dissolved in various hydrocarbon or triglyceride oils and calcium in the aqueous phase at high pH to form calcium soap particles at the interface [188].

This antifoam synergy appears to be quite general for all manner of hydrophobic particles and oils. For example, intrinsically hydrophobic organic particulates, such as precipitates of polyvalent metals with long-chain alkyl phosphates or carboxylates, may be combined with either hydrocarbons [43, 71, 211, 212] or PDMSs [213] to produce synergistic antifoam behavior. Moreover, antifoam synergy has also been shown to be effective with many different surfactant types, including anionics [43, 53, 67, 70, 98, 207–209, 214], cationics [70], and non-ionics [60, 188].

As we have shown in Section 4.6, emulsified oils are usually relatively ineffective antifoams for aqueous solutions despite possession of the properties necessary for participation in, for example, the bridging–stretching mechanism for antifoam action as described in Section 4.5.1. It is now widely accepted that this limitation derives from the stability of oil–water–air pseudoemulsion films. It is also now firmly established that the role of the particles concerns rupture of the otherwise metastable oil–water–air pseudoemulsion films [1, 22, 43, 44, 70, 71, 98, 215]. Here we give an account of the evidence that has led to these conclusions after briefly summarizing the many examples of hydrophobic oil–hydrophobic particle mixtures described in the early patent literature, which in most cases predated any reference to the phenomenon in the scientific literature.

4.8.2 Early Patent Literature

Mechanistic studies of oil–particle antifoam synergy did not appear in the scientific literature until the mid-seventies [53, 67, 71, 207, 209] of the last century. Indeed only a few references to it appeared before 1970 [39, 106]. However, descriptions of many examples of mixtures of hydrophobic particles and oils have been appearing in the patent literature since the early fifties of that century. Examples of those early patents, filed before 1980, are listed in Appendix 4.1 as Tables 4.A1 through 4.A3. Here they are categorized as mixtures of hydrophobed mineral particles and silicone oils (Table 4.A1), mixtures of hydrophobed mineral particles and organic liquids (Table 4.A2), and mixtures of intrinsically hydrophobic organic particles and organic liquids (Table 4.A3).

The finely divided nature of the particles is often stressed in these patents, and methods of milling are sometimes described [216]. Whenever particle size is quoted, it usually falls in the range of 0.001–1.0 microns. The preferred particle concentration in the oil ranges from 1% to about 30% by weight.

As we have seen, the antifoam performance of both hydrocarbons and PDMS oils may be considerably enhanced by the addition of finely divided silica. This suggests an obvious phenomenological similarity. However, when hydrophobed silica is added to PDMSs, it is often referred to as a "filler" [209, 217–219] or "activator" [220], implying that the oil plays the active role in the antifoam mechanism. On the other hand, when hydrophobed silica is added to hydrocarbon oils, the oil is sometimes referred to as the "carrier," implying that the silica plays the active role in the antifoam mechanism. Similarly, when long-chain organic materials are added to hydrocarbons, the particles are often referred to as the "active" ingredients and the oils as the "carriers" (see, e.g., reference [216]), which once again implies that the particles play the active role in the antifoam mechanism. This terminological distinction between PDMS and hydrocarbon-based systems suggests a complete mechanistic dissimilarity that seems unlikely. It is more probable that the terminology simply reflects the relative costs of the oil and particulate ingredients!

It is clear from the patent literature that an antifoam mechanism invoking Marangoni spreading of the oil (see Figure 4.10) underlies some of the thinking behind the development of these oil-based antifoams. It is, however, equally clear that it represents an inadequate explanation for all reported phenomena. Thus, Boylan [221], in a patent concerning the addition of hydrophobed silica to organic liquids, claimed that the addition of a spreading agent (e.g., a soap) to the oil is necessary to allow the water-insoluble organic liquid to spread at the air–water surface. In later patents concerning the same system, however, it is claimed that the addition of a spreading agent is unnecessary. Buckman [222], for example, states that Boylan's product is not entirely satisfactory because the presence of a surface-active agent in the system tends to stabilize the foam. In another patent also concerning this system, Miller [223] states that the addition of spreading agents may give an improvement only if the surface treatment of the silica is inadequate. Boylan apparently realized the error of his ways and eventually issued a patent [224] in which the spreading agent was claimed as an optional additive.

4.8.3 Role of Oil in Synergistic Oil–Particle Antifoams

As we have discussed in Chapter 3, oils typically used as antifoams can exhibit different types of behavior when emerging into the surfaces of surfactant solutions. Essentially this means that they can exhibit complete wetting, pseudo-partial wetting, or partial wetting, depending on the nature of the oil and the surfactant solution. We have shown in Section 4.5.1 that, provided the bridging coefficient B satisfies condition 4.31, all of this wetting behavior could in principle give rise to the formation of unstable bridging configurations in foam films and therefore foam film rupture. However, this simple proposition is complicated by the metastability of pseudoemulsion films and the possibility of significant deviations from equilibrium surface tensions during foam generation. In the absence of hydrophobic particles, the former can mean that antifoam action is only possible when capillary pressures in the Plateau borders are sufficiently high to rupture those pseudoemulsion films. In turn, this requires drainage from the foam to take place, which will mean that antifoam drops are removed from the foam films before they can cause foam film rupture. Antifoam action is then confined to Plateau borders, as we have discussed in Section 4.5.3. As we will show in Section 4.8.4, the presence of effective hydrophobic particles can, on the other hand, mean rapid rupture of the pseudoemulsion films leading to effective antifoam action in foam films. If such particles are present, then antifoam action should then occur in foam films rather than in Plateau borders and be determined largely by the sign of the bridging coefficient. Here we explore the relevance of that proposition.

Partial and pseudo-partial wetting behaviors represent the most common wetting behavior of hydrocarbons at the air–water surfaces of aqueous surfactant solutions, as we have shown in Chapter 3. In the case of the system represented in Figure 4.69, the liquid paraffin has been shown by Garrett et al. [43, 71] to exhibit partial wetting behavior on the sodium alkylbenzene sulfonate solution. Measurement of the relevant surface tensions and particle contact angles revealed the nature of the antifoam entities in this system as shown in Figure 4.70 [3, 71]. The particles are seen to adhere to the oil-water interface with a contact angle $\theta_{OW} > 90°$ (measured through the aqueous phase). Lens formation occurred at the air–water surface with the dimensions summarized in the

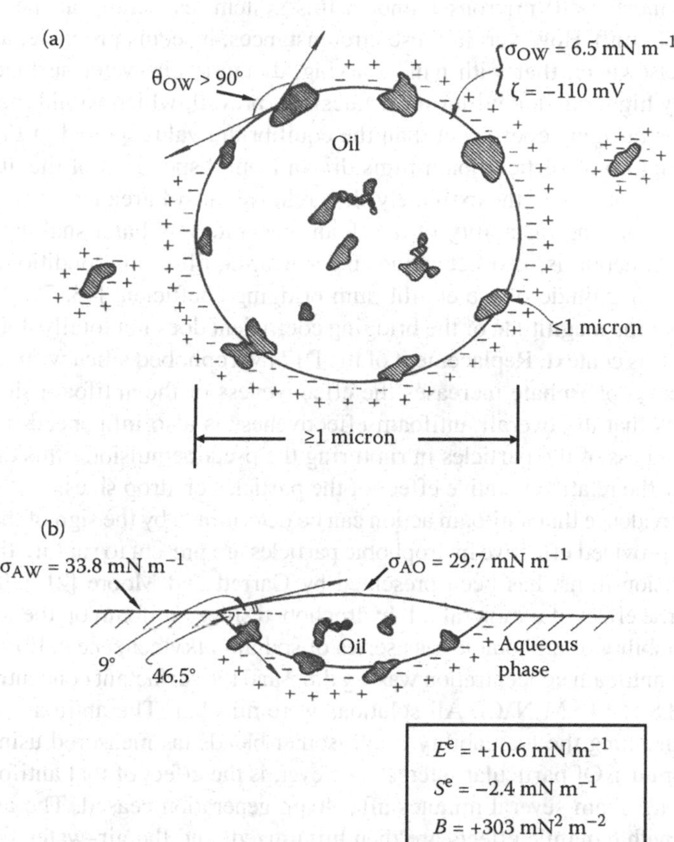

FIGURE 4.70 Nature of entities formed upon dispersal of mixtures of hydrophobed silica–liquid paraffin in aqueous 0.5 g dm^{-3} sodium alkyl (C_{10-14}) benzene sulfonate solution to yield the antifoam effects depicted in Figure 4.69. (a) Dispersed drops. (b) Lenses formed at air–water surface. (Reprinted from *Colloids Surf. A*, 85, Garrett, P.R., Davis, J., Rendall, H.M., 159. Copyright 1994, with permission from Elsevier.)

figure. The bridging coefficient, calculated from equilibrium values of the relevant surface tensions, is positive and the angle θ* is clearly >90°. An exhaustive check of the spreading behavior of this oil on water and on solutions of the surfactant at various concentrations under both equilibrium and dynamic conditions failed to detect any contamination of the air–water surface by the oil. For the relevant spreading coefficients, we therefore have $S^i < 0$, $S^{si} < 0$, $S^e < 0$, $S^i = S^{si} = S^e$, and for the spreading pressure $\Delta\sigma^{sp}_{AW} = 0$. The hydrophobed silica was equally inert in this respect. Clearly we have a system that cannot exhibit Marangoni spreading.

Increasing the concentration of this hydrophobed silica–liquid paraffin antifoam essentially eliminates foamability when attempting to generate foam by hand shaking [43, 71]. This strongly implies that the antifoam effect is occurring in foam films rather than in the Plateau borders because in these circumstances, the Plateau border dimensions are likely to be too large to permit accumulation of antifoam drops of sizes of mostly <10 microns found in this system (especially at the bottom of the foam column!). However, in these circumstances, it seems probable, as we have suggested elsewhere, that with hand shaking, dynamic air–water surface tensions significantly higher than equilibrium values will prevail, which would mean higher values of the bridging coefficient than the equilibrium value quoted in Figure 4.70. However, rupture of vertical foam films drawn from dispersions of the antifoam in the surfactant solution at the extremely slow relative rate of area generation of $<10^{-3}$ s^{-1}, together with the instability of any foam generated by hand shaking, suggests that antifoam action is also occurring under near-equilibrium conditions, as indicated by the magnitude of the equilibrium bridging coefficient [43, 71]. We should note also that the magnitude of the bridging coefficient does not totally dominate the outcome in this context. Replacement of the D17 hydrophobed silica with particles of calcium stearyl phosphate increases the effectiveness of the antifoam significantly. This implies that the overall antifoam effectiveness is also influenced by the relative effectiveness of the particles in rupturing the pseudoemulsion films or even, for example, by the relative putative effect of the particles on drop sizes.

Further evidence that antifoam action can be determined by the sign of the bridging coefficient, provided effective hydrophobic particles are present to rupture the relevant pseudoemulsion films, has been presented by Garrett and Moore [214]. This study concerned the effect of a mineral oil–hydrophobed silica antifoam on the foamability and foam stability of an homologous series of sodium alkylbenzene sulfonate isomer blends. The antifoam concentration was 1 g dm^{-3} and the surfactant concentration $\sim 5 \times 10^{-3}$ M in 3.8×10^{-2} M NaCl. All solutions were micellar. The antifoam was effective in diminishing the foamability of all isomer blends (as measured using a Ross–Miles apparatus). Of particular interest, however, is the effect of that antifoam on the stability of the foam several minutes after foam generation ceased. The ambiguities associated with dynamic effects are then minimized, and the air–water surface tension of foam films is reasonably supposed to approximate the equilibrium surface tension of a free air–water surface. In Table 4.8, we present the entry coefficient E^e, the spreading coefficient S^e, and the bridging coefficient B, for each of the chain lengths of sodium alkylbenzene sulfonates calculated by using the equilibrium values of σ_{AW} and σ_{OW}. Here it is clear that everywhere $E^e > 0$ and $S^e < 0$ so that again we have a system that cannot exhibit Marangoni spreading. However, for chain lengths $< C_{12}$ at 25°C and $<C_{13}$ at 50°C, the bridging coefficient is > 0 and therefore $\theta^* > 90°$ so that foam film rupture can occur as a result of formation of unstable oil bridging configurations.

In the absence of antifoam, the foam volume of each of these solutions of alkylbenzene sulfonate was stable for up to 1800 s after foam generation ceased (although bubble disproportionation and drainage significantly affected the appearance of the foam). A measure of the effectiveness of the antifoam after foam generation has ceased is the ratio $F(t = 960\text{ s})/F(t = 0)$, where t is the foam age and where we remember F = volume of air in foam in the presence of antifoam/volume of air in foam in the absence of antifoam. In effect, the ratio $F(t = 960\text{ s})/F(t = 0)$ is the ratio of the volume of air in the foam in the presence of antifoam after 960 s to the volume of air in the foam in the presence of antifoam immediately after foam generation has ceased. This follows because the foam volume is stable in the absence of antifoam. The ratio $F(t = 960\text{ s})/F(t = 0)$ has the value of unity if F does not change with the age

TABLE 4.8
Comparison of Entry E^e, Spreading S^e, and Bridging B Coefficients with Antifoam Effectiveness after Cessation of Foam Generation for Mineral Oil/Hydrophobed Silica Antifoam in Solutions of Homologues of Sodium Alkylbenzene Sulfonates

Chain Length	σ_{AW} (mN m^{-1})	σ_{OW} (mN m^{-1})	E^e (mN m^{-1})	S^e (mN m^{-1})	B ((mN m^{-1})^2)	Antifoam Effectiveness, $F(t = 960\ s)/F(t = 0\ s)$
			25°C			
C_9	33.0	6.5	+8.5	−4.5	+170	0.50
C_{10}	34.0	6.8	+9.8	−3.8	+241	0.33
C_{11}	31.7	6.75	+7.5	−6.1	+89	0.62
C_{12}	30.2	4.9	+4.1	−5.7	−25	0.94
C_{13}	28.8	4.0	+1.8	−6.2	−115	1.00
C_{14}	27.7	3.3	0.0	−6.6	−183	0.90
			50°C			
C_{10}	33.8	8.6	+12.9	−4.3	+346	0.00
C_{11}	31.2	8.1	+9.8	−6.4	+169	0.28
C_{12}	29.5	5.8	+5.8	−5.8	+34	0.79
C_{13}	28.0	5.9	+4.4	−7.4	−61	0.92
C_{14}	27.1	3.9	+1.5	−6.3	−119	0.95

Source: Reprinted from *J. Colloid Interface Sci.*, 159, Garrett, P.R., Moore, P., 214. Copyright 1993, with permission from Elsevier.

Note: Foam generation was by Ross–Miles. Mineral oil: air–oil surface tension = 31.0 mN m^{-1} at 25°C and 29.5 mN m^{-1} at 50°C. Homologues: each is a mixture of all possible isomers except 1-phenyl isomer.

of the foam because the antifoam does not function under the near-equilibrium conditions then prevailing. It is compared with S^e and B in Table 4.8. Here we see that $F(t = 960\ s)/F(t = 0) \ll 1$ for wherever $B > 0$. Therefore, significant foam film collapse due to the antifoam after foam generation has ceased only occurs for solutions where $B > 0$, and therefore $\theta^* > 90°$. Conversely negative values of the bridging coefficient, for which $\theta^* < 90°$, mean absence of antifoam effects.

It is, however, not clear whether antifoam action after cessation of foam generation with the lower homologues occurred in the foam films or in the Plateau borders. It is possible that most of the foam collapse occurred in the foam films soon after foam generation ceased while antifoam drops were still present. As we argue in Chapter 5, antifoam drops are likely to be flushed out of foam films rapidly because of an excluded volume effect. If the antifoam drops removed are too small to bridge Plateau borders at that stage, then foam collapse will cease. These effects are dependent on the size of antifoam drops and the rapidity with which the hydrophobed particles rupture the pseudoemulsion films.

Triglycerides that are liquid at ambient temperatures, such as triolein, are characterized by relatively high air–oil surface tensions (e.g., σ^{e}_{AO} = 31.6 nM m^{-1} [27] for triolein and 29.7 mN m^{-1} for liquid paraffin [43, 71] at 25°C). In consequence, entry, spreading, and bridging coefficients tend to be relatively low and often negative with typical micellar hydrocarbon chain–based surfactant solutions. For example, mixtures of triolein and oleic acid dispersed in aqueous solutions at high pH and in the presence of calcium ions form oil drops decorated with calcium oleate particles where the latter can be effective in rupturing the relevant pseudoemulsion films. However, Zhang et al. [188] show that the bridging, entry, and spreading coefficients of this antifoam are all negative in the case of a 0.01 wt.% solution of a commercial alkyl ethoxylated sulfate at pH 9 in the presence of calcium ions. The relevant coefficients are claimed to be equilibrium values but unfortunately violate conditions 3.11, 3.12, and 4.31, which suggests some failure to reach true equilibrium. Despite this, the antifoam is effective in reducing the volume of foam generated from that solution. Any foam generated is, however, stable. Zhang et al. [188] present some evidence that under dynamic conditions, a dynamic air–water surface tension higher than the equilibrium value will prevail so that all three coefficients can become positive (with, however, a near-zero value for the spreading coefficient). It is then argued that positive values will prevail during foam generation so that antifoam effects are observed. However, as equilibrium is approached after foam generation has ceased, the relevant coefficients become negative, the antifoam effect switches off, and the foam is stable as observed. This finding is similar to that reported by Garrett and Moore [214] in the case of the mineral oil–hydrophobed silica antifoam in saline solutions of the higher homologues of the sodium alkylbenzene sulfonate blends summarized here. Similar behavior has also been reported for mixtures of tristearin particles with triolein in saline solutions of sodium dodecyl 4-phenyl sulfonate where despite positive equilibrium values of the entry coefficient, both equilibrium spreading and bridging coefficients are negative. Again the antifoam is effective in reducing the amount of foam generated but is ineffective after foam generation has ceased [27]. Such systems do not, however, represent suitable models for establishing the role of oils in antifoam mechanism. The absence of exact knowledge of the values of dynamic surface tensions prevailing during foam generation precludes calculation of the relevant bridging coefficients with any certainty.

Both Koczo et al. [98] and Wang et al. [215] argue that even in the case of hydrophobic particle–oil mixtures, antifoam action occurs in Plateau borders rather than in foam films. Kozco et al. [98] give no direct evidence for this, save statements about the relative size of effective antifoam drops and the effective radius of a sphere inscribed by Plateau borders. They also seem to miss the possibility that the capillary pressure in the Plateau border can cause an oil drop to emerge into the air–water surface in a foam film—it is, after all, the basis of the interpretation of results with the FTT described by Denkov and coworkers [1]. Koczo et al. [98] do, however, state that when "the foam is under dynamic conditions, such as during vigorous shaking, there is probably not enough time for the formation of thin liquid films (only thick lamellae exist) and also for the antifoam drops to flow to the Plateau borders. In this case the antifoam drops/particles probably directly bridge these thick lamellae." Wang et al. [215], on the other hand, have made an experiment using video-microscopy where large (>50 micron radius) mineral oil–hydrophobed silica antifoam lenses on adjacent bubbles coalesce to cause foam film rupture where the event is clearly occurring outside any foam films present between the bubbles. The relevant video frames are reproduced in Figure 4.71. However, the point here is not that bridging oil drops in such circumstances can lead to foam collapse but rather whether it does happen with hydrophobic particle–oil antifoams in a real foam.

In contrast to hydrocarbon oils, PDMS oils usually exhibit either complete wetting or pseudo-partial wetting on the surfaces of surfactant solutions (see, e.g., Table 3.3). Although the initial spreading coefficients are generally positive, equilibrium values are zero or close to zero. In the case of pseudo-partial wetting this clearly implies the presence of oil lenses with extremely small dihedral angles. Indeed it is occasionally difficult to distinguish pseudo-partial wetting from complete wetting behavior in the case of PDMS oil spread on the surface of surfactant solutions.

Kulkarni et al. [53, 207] claim that hydrophobic particle–oil antifoam action usually requires a positive value of the initial or semi-initial spreading coefficient of the oil. As we have shown in Section 4.5.1, the bridging coefficient is meaningless in those circumstances. However, Kulkarni et al. [207] show that antifoam efficiency does not correlate with the magnitude of the semi-initial spreading coefficient S^{si}. A comparison of antifoam efficiency with S^{si} for mixtures of hydrophobed silica–PDMS is presented in Figure 4.72. Here the increase in initial spreading coefficient

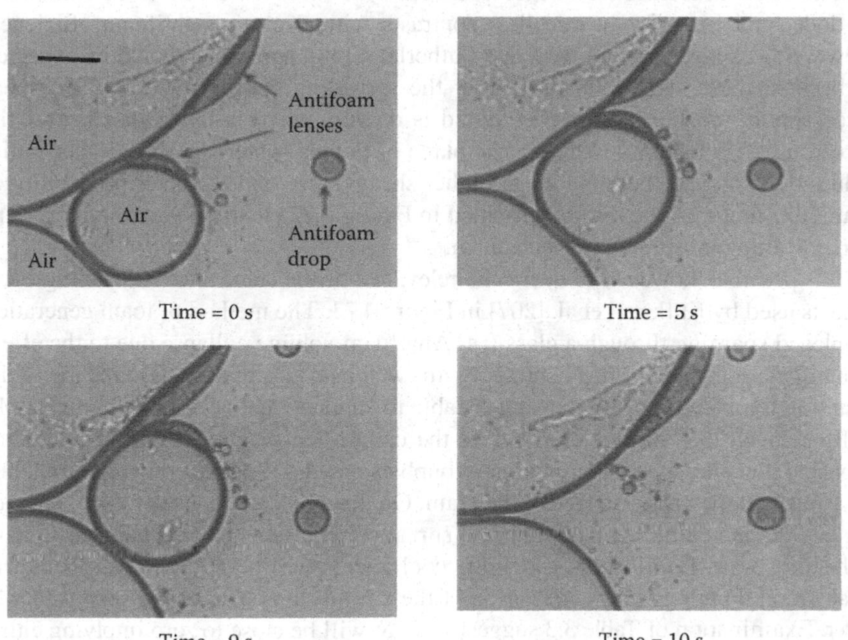

FIGURE 4.71 Sequence of video frames where hydrophobed silica–mineral oil antifoam lenses on adjacent bubbles coalesce to cause film rupture where the event is apparently occurring outside actual foam films. Size bar, 100 microns. (Reprinted with permission from Wang, G. et al. *Langmuir*, 15, 2202. Copyright 1999 American Chemical Society.)

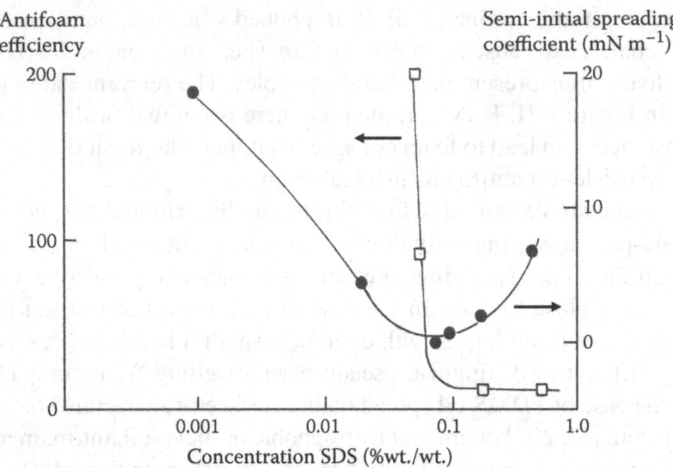

FIGURE 4.72 Comparison of semi-initial spreading coefficient, S^{si}, and antifoam efficiency for PDMS–hydrophobed silica antifoam in SDS solution. (After Kulkarni, R.D., Goddard, E.D., *Croatica Chem. Acta*, 50, 163, 1977.)

at high SDS concentration (where presumably S^{si} increases due to solubilization of dodecanol impurity in micelles) contrasts with declining antifoam efficiency. However, as suggested by Ewers and Sutherland [38], spreading would be expected to become more significant the higher the spreading coefficient. Thus, the shear force applied to the intralamellar liquid is dS^{si}/dy, where y is the distance of the spreading layer from its source in the plane of the air–water surface. Higher initial values of S^{si} should therefore mean higher shear forces and a greater probability of foam film rupture. The results presented in Figure 4.72 clearly do not therefore support a Marangoni spreading mechanism.

It is possible, however, to query the relevance of the semi-initial spreading coefficients used by Kulkarni et al. [207] in Figure 4.72. The method of foam generation employed sparging through a glass frit. Any foam volume collapse due to the effect of antifoam must involve the continuous air–water surface that exists at the top of the resulting foam column. It seems reasonable to suppose that accumulation of PDMS will occur on that surface even before the experiment begins. Moreover, it seems probable that silicone contamination of bubbles may have occurred by the time they have reached the upper surface of the foam. On the whole, then, use of initial spreading coefficients rather than equilibrium spreading coefficients would appear to have a dubious basis. Equilibrium spreading coefficients will be ≤ 0. However, Kulkarni et al. [207] do not give any indication of the equilibrium state of the spread PDMS layer. Examination of Table 3.3 suggests that S^e will be close to zero implying either a duplex film or a film with oil lenses of small dihedral angle. In either case, the bridging coefficient is likely to be positive with $\theta^* \rightarrow 180°$.

Bergeron et al. [70] have also made a study of the mode of action of a PDMS oil–particle antifoam using micellar solutions of a variety of surfactants. The latter included AOT and some homologues of alkyl trimethylammonium bromides, the spreading properties of which we list in Table 3.3 and discuss in Section 3.6.2.1. The equilibrium entry coefficients were all positive and the equilibrium spreading coefficients for the PDMS oil on each of these surfactants were all close to zero. As we discuss in Section 3.6.2.1, it seems likely that duplex film formation prevails on the alkyl trimethylammonium bromide solutions and pseudo-partial wetting on the AOT solution with lenses having extremely low dihedral angles. In the case of the latter, we remember that once oil lenses have formed on the surface any additional oil simply increases the amount of material in the lenses without any change in the oil film thickness. By contrast, addition of more oil to a duplex film continuously increases the thickness of the film. Despite these differences, Bergeron et al. [70] find effective synergistic antifoam action of mixtures of the chosen PDMS oil with hydrophobed silica on all of these surfactant solutions. Using an automated shake test to generate foam in the presence of antifoam, total collapse of the foam occurred in 20 s for all these solutions. Again then we find effective antifoam action with a positive bridging coefficient in all cases with $\theta^* \rightarrow 180°$. However, Bergeron et al. [70] also showed that this oil–particle antifoam was totally ineffective in a solution of a commercial zwitterionic fluorinated betaine surfactant for which the equilibrium entry coefficient of the oil was positive but both the initial and equilibrium spreading coefficients were negative, implying partial wetting. The bridging coefficient was also negative. The values of these coefficients imply the presence of oil lenses where $\theta^* < 90°$. As with the mineral oil–hydrophobed silica antifoam with various sodium alkylbenzene sulfonate isomer blends (for which results are shown in Table 4.8), negative values of the bridging coefficient for the oil correlate with absence of an antifoam effect despite positive values of the entry coefficient.

Jha et al. [225] have also made a study of the antifoam effect of a PDMS-based antifoam on the foam behavior of several different surfactant solutions, including of some of the same surfactants considered by Bergeron et al. [70]. However, Jha et al. [225] used a commercial PDMS antifoam that was not characterized at all. The presence or absence of hydrophobed silica was not recorded; moreover, the antifoam was described as a "dry powder," which implies the presence of some carrier material because PDMS oil–hydrophobed silica antifoam is a fluid mixture not a powder! Foam heights were measured within 10 s of hand shaking volumetric cylinders 10 times. Jha et al. [225] compared the spreading pressure $\Delta\sigma^{sp}_{AW}$ of the antifoam with the antifoam effect as measured by the ratio $\Delta H_{rel} = (H_o - H_{AF})/H_o$, where H_o is the foam height in the absence of antifoam and H_{AF} is the foam height in the presence of antifoam measured within 10 s of the cessation of foam generation. Results are presented in Figure 4.73. A linear correlation between ΔH_{rel} and spreading pressure is apparent for surfactant solutions with $\Delta\sigma^{sp}_{AW} \leq \sim 8$ nM m^{-1}. Jha et al. [225] interpret this as evidence for a Marangoni spreading mechanism. The apparent correlation appears to contradict the findings of Kulkarni et al. [207], depicted in Figure 4.72, where no correlation is observed between the semi-initial spreading coefficient and antifoam efficiency. Here we note that $\Delta\sigma^{sp}_{AW}$ is equal to the difference between the semi-initial spreading coefficient, S^{si}, and the equilibrium spreading coefficient (see Section 3.5, Equation 3.29) where the latter is expected, as shown in Table 3.3, to be near zero in the case of PDMS-based antifoams. Therefore, use of $\Delta\sigma^{sp}_{AW}$ by Jha et al. [225] and Ssi by Kulkarni et al. [207] should be essentially equivalent.

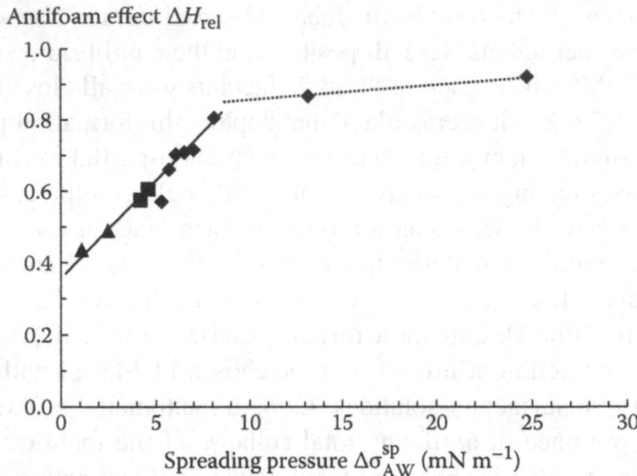

FIGURE 4.73 Plot of antifoam effect ΔH_{rel} (see text for definition) of PDMS-based antifoam on foamability of various aqueous surfactant solutions against spreading pressure $\Delta\sigma^{sp}_{AW}$ for a PDMS oil. Surfactant concentration, 1.55 g dm^{-3} in solution of ionic strength 12 mM at pH 8.5. Foamability determined by hand shaking cylinders. ◆, pure surfactants; ■, mixed commercial surfactants; ▲, mixed pure surfactants. (Reprinted with permission from Jha, B. et al. *Langmuir*, 16, 9947. Copyright 2000 American Chemical Society.)

There are other difficulties with the interpretation by Jha et al. [225] of the correlation shown in Figure 4.73. The first is clear from the figure—as pointed out by the authors, extrapolation to $\Delta\sigma^{sp}_{AW} \rightarrow 0$ suggests an antifoam effect even at zero spreading pressure. Another difficulty concerns results of measurements of foam heights after 120 s that are not cited. Apparently these "show no obvious correlation with any measurable parameter" for the antifoam–surfactant solution combinations and (presumably) techniques used in this study. Jha et al. [225] suggest that the reason foam measurements at 120 s do not correlate with measurements of $\Delta\sigma^{sp}_{AW}$ while those at ≤10 s do correlate concerns bias of the latter toward measurements of "antifoam performance defined by events that occur on a relatively short time scale (≈seconds)." However, the correlation shown in Figure 4.73 between ΔH_{rel} and $\Delta\sigma^{sp}_{AW}$ must therefore concern two quantities that relate to markedly different timescales. Thus, the authors measure $\Delta\sigma^{sp}_{AW}$ as the difference between the equilibrium air–water surface tension of a surfactant solution (by the Wilhelmy plate method) and the air–water surface tension of the same solution after equilibration for up to 10 min with the antifoam "powder." Therefore, the proposition, based on the correlation between ΔH_{rel} and $\Delta\sigma^{sp}_{AW}$ shown in Figure 4.73, that this antifoam functions by Marangoni spreading remains unproven.

Denkov et al. [54] have made a detailed study of the role of PDMS oil in the mode of action of mixtures of such oils with hydrophobed silica. The work concerned the effect of such an antifoam on micellar (10 mM) aqueous solutions of AOT. The chosen system was characterized by a positive initial spreading coefficient and a slightly negative equilibrium spreading coefficient for the oil on the surfactant solution (see Table 3.3). Both the equilibrium entry coefficient and the bridging coefficient were positive with the latter close to the limit permitted by condition 4.31. The system

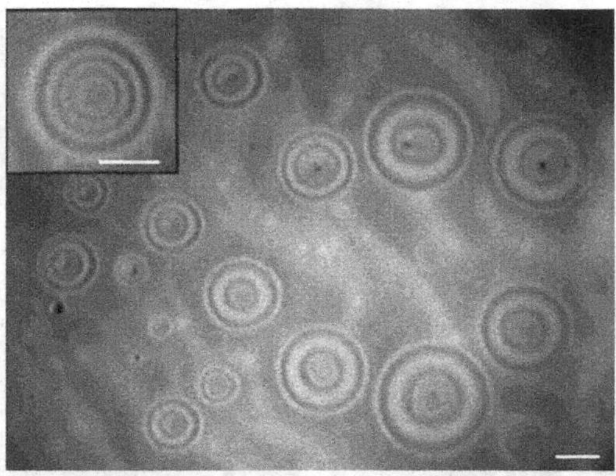

FIGURE 4.74 Lenses of PDMS–hydrophobed silica antifoam on surface of 10 mM aqueous solution of AOT. Lenses are viewed in reflected monochromatic light and exhibit interference patterns in form of Newton's rings. Relatively large silica particles are clearly visible in most lenses. Presence of large silica particles appears to increase the penetration depth of lenses. Inset shows lens without visible silica particles. However, size range of silica particles extends into suboptical range (<~0.5 microns), which means particles may be present even though not revealed in image. Size bars, 100 microns. (Reprinted with permission from Denkov, N.D. et al. *Langmuir*, 15, 8514. Copyright 1999 American Chemical Society.)

exhibited pseudo-partial wetting with oil lenses in equilibrium with a thin oil film. The oil lenses, however, had extremely low dihedral (water–oil–air) angles of only 0.4° so that $\theta^* \rightarrow 180°$. An image of lenses of the antifoam on the AOT solution is shown in Figure 4.74. The presence of large hydrophobed silica agglomerates is seen to be present in some lenses (visible as black dots). However, we should note that suboptical silica particles of submicron size may also be present but would not be revealed in such images.

This study involved the use of high-speed video-microscopy to examine the effect of antifoam drops on the stability of foam films generated in either Scheludko cells or in large vertical frames (see Chapter 2 for detailed description of these techniques) [54]. Unless special precautions were taken, filling the Scheludko cell, for example, with a dispersion of an emulsion of the antifoam in AOT solution meant the presence of a spread layer of oil on the solution surface. Under these circumstances, the films were extremely unstable—rupturing in 1–10 s. Rupture events were, however, always preceded by the appearance of the so-called fish eyes in the interference pattern made by the film in reflected incident illumination. Images of such fish eyes, present at various stages during the drainage of the foam film (see Chapter 1), are reproduced in Figure 4.75. Denkov et al. [54] show that these fish eyes are in fact caused by bridging oil drops surrounded by a deformed air–water surface. Actual film rupture followed the expansion of the fish eyes, which seemed to occur within a few milliseconds. Such expansion is of course consistent with the bridging–stretching mechanism described in Section 4.5.1 and shown in Figure 4.20. However, in some cases, this process was preceded by a relatively quiescent period lasting of the order of a second. Denkov et al. [54] suggest that this induction period is required for the ratio of the bridging drop diameter to the film thickness to increase by drainage

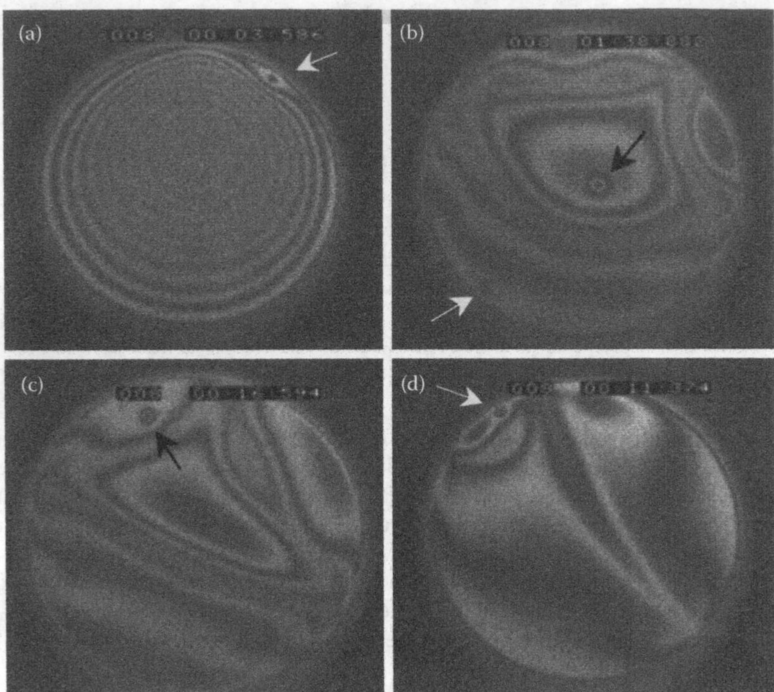

FIGURE 4.75 Interference patterns formed by films of emulsion of PDMS–hydrophobed silica antifoam in aqueous 10 mM solution of AOT solution in a Scheludko cell [79, 80]. In each case, a prespread layer of PDMS oil was present. Image a concerns the initial stages of the foam film with formation of a dimple. Other images concern older films exhibiting asymmetric drainage (see Section 1.2.2). Arrows indicate "fish-eye" interference patterns formed by bridging oil lenses. Appearance of these patterns was always followed by film rupture. (Reprinted with permission from Denkov, N.D. et al. *Langmuir*, 15, 8514. Copyright 1999 American Chemical Society.)

of the foam film or transfer of oil from the spread layer. In turn, this should move the bridging drop from a metastable state to an unstable state as shown in Figure 4.17a (again see Section 4.5.1). Absence of such an induction period is therefore presumably due to the formation of a bridge from an oil lens with a large volume augmented by the effect of the presence of silica. Nevertheless, direct observation and the relatively small size of the drops present in these dispersions precluded the possibility that foam film rupture was initiated in the Plateau borders.

Careful loading of the Scheludko cell to avoid contamination of the air–water surface with a prespread layer of oil surprisingly eliminated the formation of unstable bridges and the accompanying foam film instability. Denkov et al. [54] attributed this observation to the effect of the spread layer on the emergence of the PDMS oil into the air–water surface, which they have later shown to augment the effect of the hydrophobed silica. We return to this issue in Section 4.8.4.

These observations concerning the role of the oil in foam film rupture are clearly not consistent with a Marangoni spreading mechanism. As we have seen, presence of a spread film of oil means that the spreading coefficient for an additional drop that emerges into the air–water surface is slightly negative. Marangoni spreading cannot therefore occur. Indeed, Denkov et al. [54] report the occasional observation of a PDMS drop emerging into the air–water surface of a film without a prespread layer (a condition that is difficult to arrange in practice). The drop was then observed to spread as revealed by changes in the interference fringes caused by changes in film thickness due presumably to the Marangoni effect. However, foam film rupture did not usually then occur—the drop simply spreads to leave behind a residue of silica and presumably some oil.

It is clear that for entering oils with wetting properties ranging from partial wetting [43, 71] to pseudo-partial wetting (albeit with lenses of very small dihedral angle) [54] through to complete wetting [70], Marangoni spreading has been shown to be an unnecessary aspect of antifoam action. We can also in fact conclude that the available experimental evidence indicates that the role of the oil in oil–hydrophobic particle mixtures usually concerns events in foam films rather than Plateau borders, which indeed appear to be largely determined by the sign of the bridging coefficient for sufficiently large oil bridges. This evidence is clearly at variance with respect to the opinion of Aveyard and Clint [85] that "the bridging coefficient, which relates to an already formed and originally static bridge, may not be a good guide as to whether the bridge will rupture a film or not." However, the need for a spread oil film in order to augment the role of the hydrophobic particle in rupturing the relevant pseudoemulsion films can in some important circumstances represent an additional complication. Ensuring the presence of such a film can paradoxically, in principle, favor oils with high values of the initial (or semi-initial) spreading coefficient [54].

The bridging coefficient should be determined under appropriate circumstances, which can clearly mean, for example, that the sign is dependent on the extent of deviation from equilibrium during foam generation due to the effect of relatively slow surfactant transport to air–water surfaces. The nature of the actual foam film rupture process caused by the presence of unstable bridging configurations with positive bridging coefficients has not always been established. In the case of the hydrocarbon-based antifoams studied by Garrett et al. [4, 71, 214], no direct evidence has been given concerning that process. However, it seems that in the case of the PDMS-based antifoam studied by Denkov et al. [54], a bridging–stretching mechanism gives rise to the actual foam film rupture.

4.8.4 Early Hypotheses Concerning Role of Particles in Synergistic Oil–Particle Antifoams

Before the first publication by Garrett [22] in the early nineties of the last century* of the hypothesis that the role of the particles in hydrophobic oil–particle antifoams concerned rupture of metastable pseudoemulsion films, a number of different hypotheses had been advanced in explanation for that role. Here we summarize that work. In the main, it concerned mixtures of oil with silica where the requirement that the silica be surface treated to render it hydrophobic had been amply demonstrated.

* Actually earlier—during the late 1970s by Garrett et al. [71].

Some of these early hypotheses are essentially naive and are easily disposed of. Thus, Sinka and Lichtman [226] attributed the role of the particle to inhibition of solubilization of the oil. However, it is easily shown that the presence of concentrations of antifoam oil alone, well in excess of the (usually extremely low) amounts that are solubilized, may still produce a negligible antifoam effect with aqueous solutions of, say, anionic surfactants [43, 71]. Povich [209] showed that the increase in bulk shear viscosity of PDMS due to the presence of the silica is not responsible for the 7-fold increase in antifoam effectiveness accompanying the presence of that material in the oil. Moreover, Ross and Nishioka [210] showed that the presence of hydrophobed silica (at concentrations of the ≤5 wt.% normal in antifoams) has no effect on the surface shear viscosity of PDMS. Garrett et al. [43, 71] have shown that although the presence of hydrophobed silica particles can facilitate dispersal of the antifoam oil, that is not the principal role of the particles. Thus, mineral oil dispersions of essentially the same size distribution as hydrophobed silica–liquid paraffin dispersions reveal a markedly different antifoam effectiveness. This is exemplified in Figure 4.76, where the relevant size distributions are compared with the respective antifoam effectiveness as measured by F for a solution of a commercial sodium alkylbenzene sulfonate.

The possibility that the spreading coefficient of PDMS oils is modified by the addition of hydrophobic particles had been explored by Povich [209]. Here the initial spreading coefficient for PDMS was shown to slightly decrease upon the addition of hydrophobed silica (despite the effectiveness of this silica in promoting the antifoam behavior of the oil). Povich [209] attributed this to adsorption of oil-soluble surface-active impurities on the silica. Hydrophobed silica has also been shown to be without significant effect on the spreading behavior of a liquid paraffin [43, 71].

An often-quoted mechanism for mixtures of hydrophobic oils and particles is that due to Kulkarni et al. [53, 207]. These authors claimed that the oil spreads over the air–water surface exposing the particles to the aqueous solution. Adsorption of surfactant onto the surface of the particles is then supposed to occur, rendering the particle hydrophilic so that particles are progressively extracted from the oil into the aqueous phase. Rapid local depletion of surfactant in the aqueous film is then, in turn, supposed to produce a "surface stress" that renders the foam film unstable so that rupture occurs.

There are a number of problems with this mechanism. The first problem clearly concerns the requirement that the oil spread at the air–water surface. We have already established that this is not a necessary property of antifoam oils. A second problem is conceptual—depletion of surfactant in the foam film by adsorption on particles will give rise to an increase in surface tension, which will in turn produce a Marangoni flow in the direction of the foam film. This will tend to enhance film stability. Finally it is clear that an aspect of the mechanism is an intrinsic tendency for the particle to be removed from the oil phase to the aqueous phase. However, it has been shown [43, 71] that addition of liquid paraffin to a dispersion of hydrophobed silica in sodium alkylbenzene sulfonate solution produces an enhancement of antifoam effectiveness. The duplicated results are given in Table 4.9 where the antifoam effectiveness is indicated by the ratio F (given by Equation 4.71). This is clearly not consistent with a process of removal of particles from oil having a central role in the

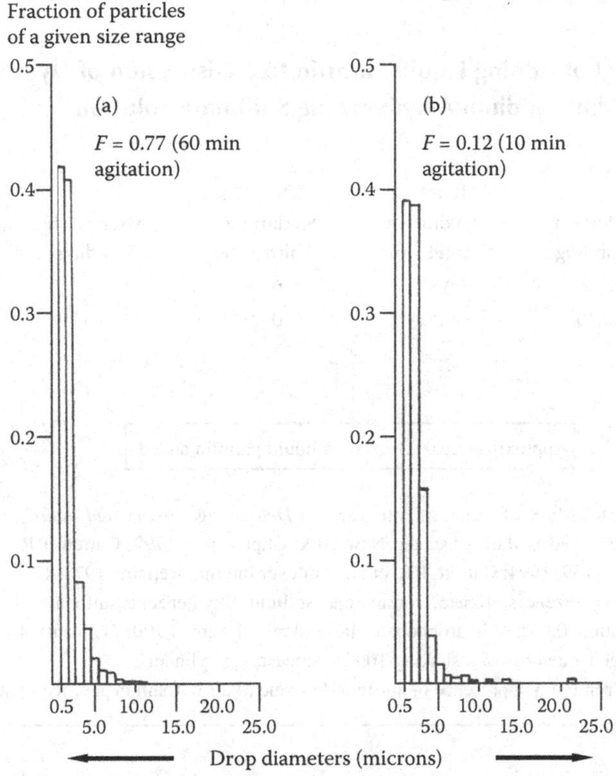

FIGURE 4.76 Comparison of antifoam effectiveness F (= volume air in foam with antifoam/volume of air in foam without antifoam) with size distribution of (a) liquid paraffin; (b) liquid paraffin–hydrophobed silica (90/10 by weight). Surfactant solution: 0.5 g dm^{-3} commercial sodium alkylbenzene sulfonate; antifoam concentration: 1.2 g dm^{-3} (emulsified using a high speed mixer). Here size distributions were adjusted by altering agitation time so that they are approximately equal. (After Garrett, P.R., Mode of action of antifoams, in *Defoaming, Theory and Industrial Applications*, Garrett, P.R., ed., Marcel Dekker, New York, Chpt 1, p 1, 1993; Garrett, P.R. et al., *Colloids Surf. A*, 85, 159, 1994.)

synergistic mode of action of the antifoam. On the whole, then, we are forced to conclude that the mechanism proposed by Kulkarni et al. [53, 207] is probably wrong.

Dippenaar [77] has suggested that the particle may have a central role in the mode of action of the antifoam. He supposes that the particle functions by a dewetting mechanism similar to that outlined in Section 4.7. The oil is supposed to adhere to the particle to yield a higher contact angle than would otherwise prevail. Dippenaar [77] illustrated this mechanism by comparing the contact angle and effectiveness at film rupture of sulfur particles before and after contamination with liquid paraffin. This mechanism would, however, be most effective in the case of rough particles where the oil may adhere in the rugosities to increase the contact angle at the air–water surface. It is, however, difficult to see how this mechanism could function where the oil forms the overwhelming proportion in the antifoam and where the particle is preferentially

TABLE 4.9
Antifoam Effect of Adding Liquid Paraffin to a Dispersion of Hydrophobed Silica in 0.5 g dm^{-3} Sodium Alkylbenzene Sulfonate Solution

F[a]					
After 15 s Shaking	After 2 h Standing	After 15 s Shaking for Second Time	After 15 s Shaking for Third Time	After 1.5 h Standing	After 15 s Shaking for Fourth Time
0.41	0.36	0.55	0.18	0.0	0.15
0.43	0.40	0.52	0.10	0.0	0.11

Approximately 0.03 ± 0.01 g liquid paraffin added

Sources: Garrett, P.R., Mode of action of antifoams, in *Defoaming, Theory and Industrial Applications* (Garrett, P.R., ed.), Marcel Dekker, New York, Chpt 1, p 1, 1993; Garrett, P.R. et al., *Colloids Surf. A*, 85, 159, 1994; Garrett, P.R. et al., Unilever Internal Reports, 1978–1979.

Note: Sodium alkylbenzene sulfonate, commercial sodium alkylbenzene sulfonate of C_{10}–C_{14} chain length. Solution, 0.03 g of hydrophobed silica (silanized Aerosil 200) in 25 cm^3 of surfactant solution; foam generated by hand shaking 100 cm^3 measuring cylinder.

[a] F = volume of air in foam in presence of antifoam/volume of air in foam in absence of antifoam.

wetted by the oil. In this case, entities where a rough particle is embedded in an oil drop with only minimal exposure of the particles to the aqueous phase will be favored. This issue is addressed in more detail below (see Section 4.8.4).

Frye and Berg [75] proposed that if the oil completely wets the particles, then the size of those oil drops containing particles will be determined by the size of the particles. Sufficiently large particles should therefore reduce the time required for foam films to thin to the point where bridging and foam film collapse can occur according to a mechanism suggested by a positive bridging coefficient and shown in Figure 4.18. However, the presence of particles that produce synergistic foam behavior does not necessarily mean larger oil drops, as we have shown in Figure 4.76. Moreover, reduced thinning time does not necessarily imply increased frequency of foam film collapse because larger antifoam entities means fewer entities. Thus, we remember the finding of Dippenaar [77] that the mass of antifoam required to remove a given proportion of foam is proportional to the size of the antifoam—smaller entities therefore mean a higher overall antifoam efficiency. The hypothesis of Frye and Berg [75] also ignores the usual observations [43, 44, 71] that the hydrophobic particles are not completely wetted by the oil but adopt a finite contact angle θ_{OW} at the oil–water surface so that

$$90° < \theta_{OW} < 180° \tag{4.72}$$

where the angle is measured through the aqueous phase.

4.8.5 Role of Particles in Synergistic Oil–Particle Antifoams

4.8.5.1 Experimental Observation

Mixtures of the oils and hydrophobic particles used in antifoams disperse in aqueous surfactants as composite entities. Examples of optical and electromicrographs of various antifoam drops are depicted in Figure 4.77. Particles include hydrophobed silica, intrinsically hydrophobic calcium stearyl phosphate and *in situ* formed calcium oleate. Oils include triolein, liquid paraffin, hexadecane, and PDMS, listed in order of declining air–oil surface tensions. Other images of a similar nature are to be found elsewhere (see, e.g., references [56, 215]). Regardless of the chemical nature of the entities, the particles are seen to adhere to the oil–water interface. Measurement of contact angles of compressed discs of the relevant particles or of representative smooth plates of the same material [43, 44, 70, 71, 226] suggests that condition 4.72 is normally satisfied by the relevant advancing angles of the adhering particles. This is in fact the condition required for water-in-oil Pickering emulsion formation where particles adhering to the oil–water surface ensure high stability. Indeed, a characteristic of oil–particle antifoams is that if equivolume amounts of the antifoam and the solution to be defoamed are shaken together, a water-in-oil emulsion invariably forms [27, 43, 71, 226]. If, however, the oil alone is shaken with the same solution, an oil-in-water emulsion usually results. Clearly, then, there is strong evidence to suggest that the particles in the oil–particle antifoam adhere to the oil–water surface with a contact angle that satisfies condition 4.72.

It seems probable that the location of the particles at the oil–water interface is the key to their role. Kulkarni et al. [227] have emphasized the importance of long-range electrostatic repulsion forces between antifoam entities and bubbles in contributing to antifoam behavior with hydrophobed silica–PDMS antifoams. These authors present some evidence of diminished antifoam effectiveness accompanying increasing zeta potentials in sodium lauryl sulfate solution with increasing concentration. This work then suggests that the role of particles at the oil–water surface will concern the (often electrostatic) forces between bubbles and antifoam entities.

Further evidence concerning this proposition is given if we remember (see Section 4.6.2) that hydrophobic particles are often not added to PDMS oils when these materials are used as antifoams for non-aqueous systems. For example, Shearer and Akers [5], in a study of PDMS antifoam behavior in lube oils, have shown that electrostatic interactions are essentially absent in that system.

As we have seen (Section 4.7.5), rugosities facilitate the emergence of hydrophobic particles into air–water surfaces. This effect appears to concern the reduced force necessary for a rugosity to penetrate the electrostatic double layer in order to initiate rupture of the relevant metastable air–water–solid film. It seems reasonable then to attribute a similar role to particles at the oil–water surface. Thus, in effect, the particles should facilitate rupture of the aqueous pseudoemulsion films separating oil drops from the air–water surfaces.

That presence of particles adhering to the oil–water surface could destabilize pseudoemulsion films was first suggested and tested by Garrett et al. [22, 43, 71] in examining the effect of hydrophobed silica on the time of emergence of liquid paraffin drops into the air–water surface of the solution of a sodium (C_{10}–C_{14}) alkylbenzene sulfonate [43, 71] used to obtain the foam results shown in Figure 4.69. The

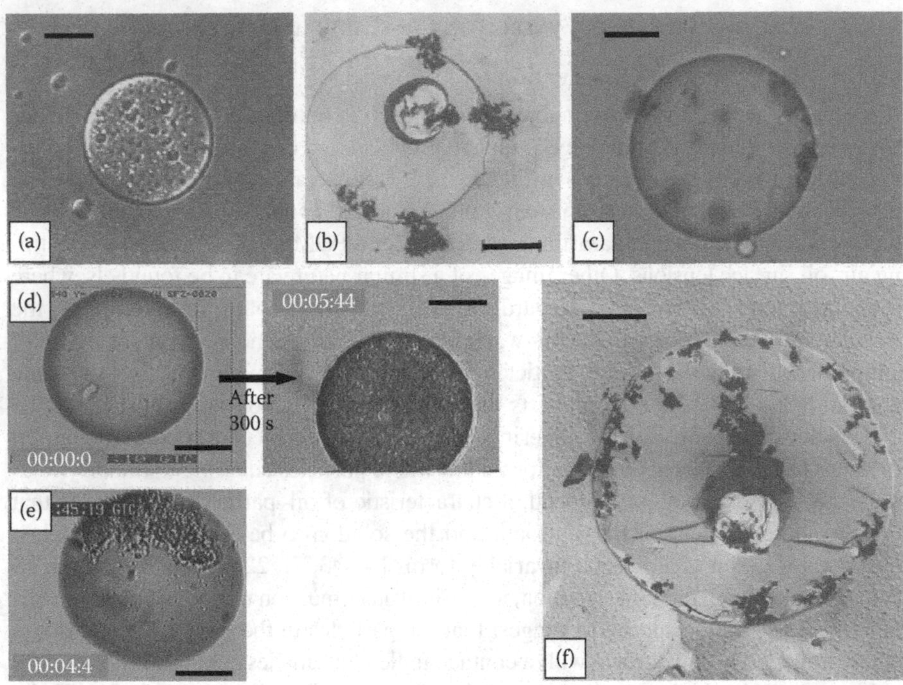

FIGURE 4.77 Examples of images, using various techniques, of hydrophobic particle–oil antifoam drops dispersed in aqueous surfactant solution. (a) Optical photomicrograph of hydrophobed silica (D17 ex-DeGussa)–liquid paraffin in 5 mM sodium dodecylbenzene sulfonate in 38 mM NaCl (size bar, 30 microns). (After Curtis, R. et al. [235], Preliminary observations concerning the use of perfuoroalkyl alkanes in hydrocarbon-based antifoams in *Emulsions Foams and Thin Films* (Mittal, K., Kumar, P., eds.), Marcel Dekker, New York, p 177, 2000.) (b) Electron micrograph (using freeze fracturing/etching) of hydrophobed silica (D17 ex-DeGussa)–liquid paraffin in 0.5 g dm^{-3} (~0.7 mM) sodium alkylbenzene sulfonate (size bar, 1 micron). (Reprinted from *Colloids Surf. A*, 85, Garrett, P.R., Davis, J., Rendall, H.M. [43], 159, Copyright 1994, with permission from Elsevier.) (c) Optical photomicrograph of hydrophobed silica–PDMS oil in 10 mM AOT solution (size bar, ~50 microns). (After Denkov, N.D. [1], *Langmuir*, 20, 9463, 2004.) (d) Optical photomicrograph showing formation of calcium oleate particulate "skin" at hexadecane–oleic acid–water interface. Surfactant solution, 0.1 g dm^{-3} of commercial alcohol ethoxylate ($C_{12-15}EO_7$) in 3 mM Ca^{2+} at pH 9. Size bar, ~50 microns. (Reprinted from *J. Colloid Interface Sci.*, 263, Zhang, H., Miller, C.A., Garrett, P.R., Raney, K. [188], 633, Copyright 2003, with permission from Elsevier.) (e) As for (d) but replacing hexadecane with triolein. (f) Electron micrograph (using freeze fracturing/etching) of calcium stearyl phosphate–liquid paraffin in 0.5 g dm^{-3} (~0.7 mM) sodium alkylbenzene sulfonate (size bar, ~2 microns). (Reprinted from *Colloids Surf. A*, 85, Garrett, P.R., Davis, J., Rendall, H.M. [43], 159, Copyright 1994, with permission from Elsevier.)

drops were of volume ~0.01 cm^3. Results are presented in Figure 4.78. Here it is clear that the particles significantly reduce the emergence time for the oil drops. A similar result has been obtained for oil drops coalescing with a layer of oil on the surface of a solution [43, 71]. This then implies that the emulsion behavior found with these particle–oil mixtures is determined by a tendency of the particles to rupture the aqueous film

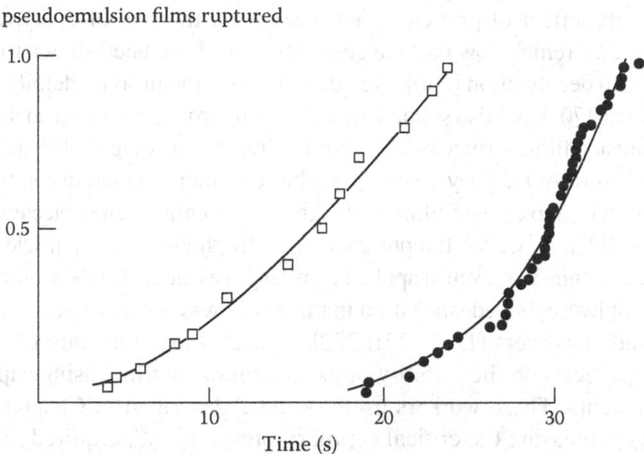

FIGURE 4.78 Effect of hydrophobed silica on time of emergence of liquid paraffin oil drops into air–water surface (thereby rupturing the pseudoemulsion film) of 0.5 g dm^{-3} (~0.7 mM) sodium (C_{10-14}) alkylbenzene sulfonate solution. Drop volumes, 0.01 cm^3. □, Liquid paraffin–hydrophobed silica (90/10 by weight); ●, liquid paraffin. (After Garrett, P.R., Mode of action of antifoams, in *Defoaming, Theory and Industrial Applications*, Garrett, P.R., ed., Marcel Dekker, New York, Chpt 1, p 1, 1993; Garrett, P.R. et al., *Colloids Surf. A*, 85, 159, 1994.)

between oil drops. Essentially the same explanation for the effect of particles on emulsion behavior has, for example, been offered by van Boekel and Walstra [228] and by Mizrahi and Barnea [229].

These considerations have pushed much of the experimental effort concerning the mode of action of hydrophobic particle–oil mixtures, during the past couple of decades, to focus on the effect of the particles on the stability of the relevant pseudoemulsion films. For example, Koczo et al. [98] have studied the effect of the concentration of hydrophobed silica particles on the stability of pseudoemulsion films formed by PDMS oils and 0.06 M SDS solution at the tips of capillaries. Concentrations as low as only 0.001–0.1 wt.% were required to rupture the pseudoemulsion films that were otherwise stable. Not surprisingly, Koczo et al. [98] find that the greater the size of the capillary, the larger the pseudoemulsion film, and the greater the probability of a particle causing film rupture as indicated by the decreasing minimum concentrations of particles required. Such concentrations are, however, more than an order of magnitude lower than is usually found in commercial oil–particle antifoam mixtures (see, e.g., the tables in Appendix 4.1). The discrepancy presumably arises, at least in part, because there is an obvious requirement during practical application that at least one particle be present in each drop of emulsified antifoam. Thus, if for example, at least one spherical silica particle of 2-micron diameter is always to be present in PDMS oil drops of 10-micron diameter, a concentration of silica of >~1.6 wt.% is required assuming that the density of the silica is about double that of the PDMS. This compares with critical concentrations reported by Koczo et al. [98] for pseudoemulsion film rupture of only 0.01–0.03 wt.% for such particles. Another factor contributing to the discrepancy between these critical concentrations and practical application is the effect of particle concentration on the rate of deactivation of the antifoam where extremely low particle concentrations have been shown to mean rapid rates of antifoam deactivation [230]. We address this topic in some detail in Chapter 6.

Bergeron et al. [70] have also studied the effect of hydrophobed silica on the stability of the pseudoemulsion films formed by PDMS oil (of molecular weight ~10^4), but in aqueous micellar solutions of AOT. They showed that the relevant pseudoemulsion films drained to form indefinitely stable white films of thickness <100 nm when subjected to capillary pressures of ~60 Pa. However, the presence of hydrophobed silica particles caused the films to rupture within 30 s, even at applied capillary pressures <60 Pa. Unfortunately, the concentration of hydrophobed silica used in this study was not reported.

Denkov and coworkers [1, 62, 231, 232] have also made a study of the effect of hydrophobic particles on the stability of pseudoemulsion films using capillary pressure measurements. These workers used the FTT described in Chapter 2 (also see Section 3.4.1) to measure the critical capillary pressure, p_c^{crit}, required to rupture the pseudoemulsion films, both in the presence and absence of particles. They also studied the effect of spread films formed by oils exhibiting pseudo-partial wetting behavior, both in the presence and absence of hydrophobed particles. Illustrative results comparing PDMS oil and neat hydrophobed silica–PDMS oil mixtures for micellar solutions of (10 mM) AOT and a non-ionic surfactant, octylphenol ethoxylate (OP. EO_{10}) are shown in Table 4.10 [62]. The hydrophobed silica was present at 4.2 wt.%

TABLE 4.10
Effect of Hydrophobed Silica and Spread Layers on Critical Capillary Pressures, p_c^{crit}, for Hydrophobed Silica–PDMS Oil Antifoams in Micellar AOT and Triton X-100 Solutions

		p_c^{crit} (Pa)	
Antifoam	**Spread Layer?**	**AOT[a]**	**Triton X-100[b]**
Neat PDMS oil[c]	No	28 ± 1	>200
	Yes	19 ± 2	>200
PDMS/h.silica[d]	No	8 ± 1	30 ± 1
	Yes	3 ± 2	5 ± 2
PDMS/h.silica emulsion[e]	No	20 ± 5	22 ± 1
	Yes	4 ± 1	7 ± 1

Source: Reprinted with permission from Denkov, N.D. et al. *Langmuir*, 18, 5810. Copyright 2002 American Chemical Society.

[a] At 10 mM.
[b] At 1 mM.
[c] Of dynamic viscosity 1000 mPa s.
[d] 4.2 wt.% hydrophobed pyrogenic silica.
[e] Emulsion of PDMS–h.silica antifoam stabilized by sorbitan monostearate (Span 60) and a stearic acid ethoxylate (EO_{40}). Span 60 is a particulate solid at ambient temperatures.

in the oil and had a typical fractal structure with a size range of 0.1–5 microns. Also included in the table are results with emulsions of the same hydrophobed silica–PDMS oil mixtures stabilized with suitable emulsifiers.

The results presented in Table 4.10 reveal that the presence of hydrophobed silica clearly diminishes the critical capillary pressure p_c^{crit} required to rupture the relevant pseudoemulsion films. The effect appears to be general since two radically different surfactants are considered. However, p_c^{crit} is increased if precautions are taken to remove the spread oil layer that is usually present at the air–water surfaces of these solutions in the presence of dispersed antifoam drops. The effect of drop size on p_c^{crit} is shown in Figure 4.79. It is clear from the figure that the spread layer diminishes p_c^{crit} irrespective of the size of the drops. There would appear then to be a synergistic effect between the presence of the hydrophobic particles and the spread layer leading to diminished stability of the relevant pseudoemulsion films, as indicated by diminished magnitudes of the critical capillary pressures. Denkov et al. [62] show that absence of the spread layer can in fact diminish significantly the antifoam effect in foam generation by an automatic shake test of the emulsion listed in Table 4.10. We consider the causes of this synergy in Section 4.8.4.2.

Here we should note that the p_c^{crit} for PDMS oil drops in the absence of particles with micellar solutions of AOT is unusually low. Not only is p_c^{crit} much higher for micellar TX-100 solutions but Hadjiiski et al. [102] report values of >3000 Pa for PDMS oil drops in the absence of particles in saline micellar solutions of sodium dodecylbenzene sulfonate.

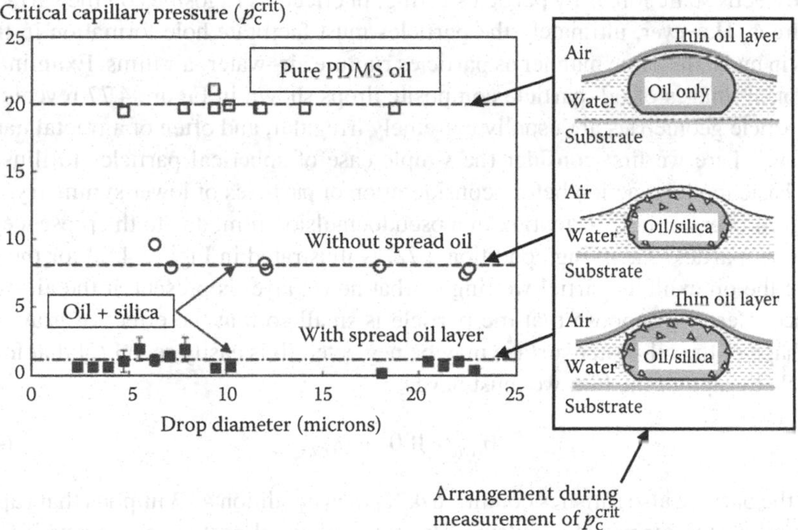

FIGURE 4.79 Effect of spread PDMS oil layers on ease of rupture of air-water-oil pseudoemulsion film, even in presence of hydrophobed silica particles, as revealed by measurements of critical capillary pressure p_c^{crit}. Effect is seen to be insensitive to changes in drop diameter. (Reprinted with permission from Denkov, N.D. *Langmuir*, 20, 9463. Copyright 2004 American Chemical Society.)

Other direct observations of the adverse effect of hydrophobic particles on the stability of pseudoemulsion films have concerned *in situ* precipitation of calcium oleate particles at the hexadecane–water surface [188]. Such precipitation is the result of a reaction between oleic acid dissolved in the oil phase and calcium ions in the aqueous phase. The reaction requires a sufficiently high pH (at least pH 9)—at lower pHs, formation of calcium oleate precipitate is suppressed and antifoam action is similar to that of hexadecane alone (which is weak—see Section 4.6.1.3). Synergistic antifoam action was found using this system in micellar solutions of both a commercial sodium alkyl ethoxy sulfate and an alcohol ethoxylate. The initial spreading coefficient of hexadecane on solutions of the latter was positive. Examination of Figure 3.18 suggests that hexadecane generally exhibits pseudo-partial wetting behavior on micellar solutions of solutions of alcohol ethoxylates. It is therefore reasonable to suppose that a spread film of hexadecane was present on the air–water surface of the relevant pseudoemulsion film. It would be interesting to establish whether that spread film had any role in determining antifoam effectiveness in this case especially since a spread layer of hexadecane in the absence of particles is known to actually increase p_c^{crit} in the case of saline micellar solutions of a sodium dodecylbenzene sulfonate (see Table 3.4 and references [61, 102]).

4.8.5.2 Spherical Particles, Spread Oil Layers, and Rupture of Pseudoemulsion Films

The process of emergence of oil drops into the air–water surface will clearly be initiated by penetration of the energy barrier in the pseudoemulsion films due to long-range electrostatic forces by particles acting, in effect, as rugosities on the surface of the drops. However, ultimately the particles must facilitate hole formation in those films in much the same manner as particles rupture air–water–air films. Examination of typical images of oil–particle composite drops shown in Figure 4.77 reveals that the particle geometries are usually extremely irregular, and often of a fractal nature. However, here we first consider the simple case of spherical particles to illustrate their basic mode of action before consideration of particles of lower symmetry.

The process of hole formation in a pseudoemulsion film, due to the presence of a spherical particle satisfying condition 4.72, is illustrated in Figure 4.80 for the case where the oil exhibits partial wetting so that no oil layer is present at the air–water surface. Here we suppose that the particle is small so that the effect of gravity on the shape of the oil–water surface may be neglected. It is easily seen [22] that for the particle to rupture the film we must have

$$\theta_{AW} > 180° - \theta_{OW} \tag{4.73}$$

If the particle also satisfies condition 4.72, then condition 4.73 implies that rupture of an oil–water–air pseudoemulsion film by a spherical particle can occur if $\theta_{AW} <$ 90°. Such a particle will not, as we have seen, rupture symmetrical air–water–air films (for which we generally require $\theta_{AW} > 90°$—but see Section 4.7). We therefore find that spherical particles that satisfy conditions 4.72 and 4.73 will promote the emergence of an oil drop into the air–water surface without exhibiting any antifoam behavior when used alone. If the oil satisfies condition 4.31 for formation of unstable

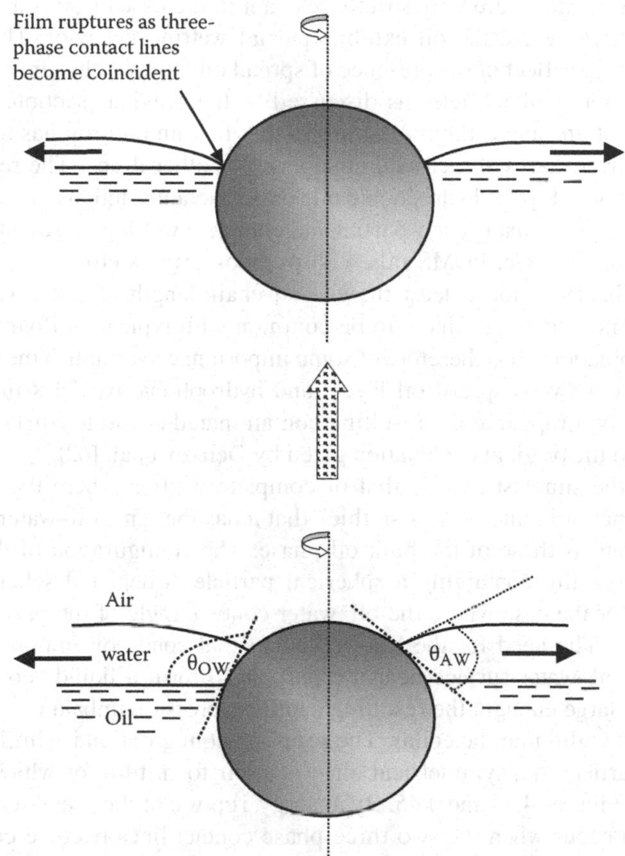

FIGURE 4.80 Rupture of air–water–oil pseudoemulsion film by a spherical particle with $90° < \theta_{OW} < 180°$ and $\theta_{AW} > 180° - \theta_{OW}$ so that $\theta_{AW} < 90°$.

bridges, then we have a clear explanation for the antifoam synergy that Frye and Berg [75] describe for mixtures of hydrophobed glass spheres and hexadecane where the reported receding contact angles clearly satisfy condition 4.73. Thus, the particle alone will not function as an antifoam if the contact angle at the air–water surface is <90° and the oil will not function because of low probability of emergence into the air–water surface. Particles in the oil facilitate the latter process, enabling the resultant oil lens (oil drop if sufficiently large) to bridge an aqueous film and participate in, for example, the bridging–stretching foam film rupture mechanism [54] shown in Figure 4.20.

Inversion of the tendency of the oil to form oil-in-water emulsions by the presence of spherical particles is also easily understood. Thus, if the particles are preferentially wetted by the oil so that they satisfy condition 4.72, then they may rupture aqueous films between oil drops in much the same manner as spherical particles, with $\theta_{AW} > 90°$, rupture aqueous films between air bubbles (Figures 4.43 and 4.45). Water-in-oil emulsions are therefore formed.

These considerations are only strictly relevant in the case of oil–surfactant solution combinations where the oil exhibits partial wetting behavior. They offer no explanation for the effect of the presence of spread oil layers at the air–water surface on the effectiveness of particles as discussed in the previous section. This effect appears to be of an "autocatalytic" nature in that first an oil drop has to dewet and form such a layer that will facilitate emergence of other drops. The review of the wetting behavior of typical hydrophobic oils on surfactant solutions given in Chapter 3 does in fact suggest that pseudo-partial and complete wetting represent the normal behavior of, for example, PDMS oils with pseudo-partial wetting representing the predominant behavior for at least the shorter-chain-length alkanes. The presence of spread layers is therefore likely to be common with typical antifoam–surfactant solution combinations. It is therefore of some importance to establish the likely cause of the synergy between spread oil layers and hydrophobic particles in facilitating emergence of oil drops into the resulting contaminated air–water surface. Here we slightly extend the original explanation given by Denkov et al. [62].

Arguably the simplest case is that of complete wetting where the oil forms a duplex layer that, we remember, is so thick that it has the same oil–water and oil–air surface tensions as those of the bulk oil phase. The configuration of the resulting pseudoemulsion film containing a spherical particle is depicted schematically in Figure 4.81a for the case where the oil–water contact angle of the particle satisfies condition 4.72. The need for the film to satisfy that condition imposes a concave shape on the oil–water surface near the particle to form a liquid "collar" [62]. If the particle is large enough, the resulting capillary pressure imbalance will suck oil from the duplex film into the collar. The resulting configuration is similar to that of a spherical particle in a symmetrical air–water–air foam film for which $\theta_{AW} > 90°$ and shown in Figures 4.43 and 4.45. By analogy, rupture of the pseudoemulsion film will therefore occur when the two three-phase contact lines become coincident as the film drains provided the particle is large enough. The particle would then either remain at the air–oil surface or be totally engulfed within the oil phase, depending on the magnitude of the relevant contact angle. We can in fact extend this argument to the case where a large enough particle bridges an air–water–air foam film contaminated by duplex oil films on both surfaces. If the contact angle satisfies condition 4.72 in that $\theta_{OW} > 90°$, then an unstable oil bridge across the whole foam film will be formed. Film rupture by the bridging–stretching mechanism is then inevitable as an unbalanced capillary pressure at the oil–water interface sucks oil into the bridge and drives the formation of an oil-filled hole in the foam film. This eventuality is shown schematically in Figure 4.81b. Indeed, it is possible that it is only by this means that large enough oil bridges can be formed in foam films contaminated by duplex layers because lenses protruding into the foam film will be absent and drops large enough to bridge a film may be too small, as indicated by the calculations presented in Figure 4.17a.

Pseudo-partial wetting is also often found with both PDMS and short-chain alkane oils. However, as we have described in Section 3.6.1, in the case of such spreading on aqueous micellar surfactant solutions, equilibrium spreading coefficients are usually close to zero. This indicates that the films in equilibrium with oil lenses are relatively thick and are close to duplex in character. In turn, this means

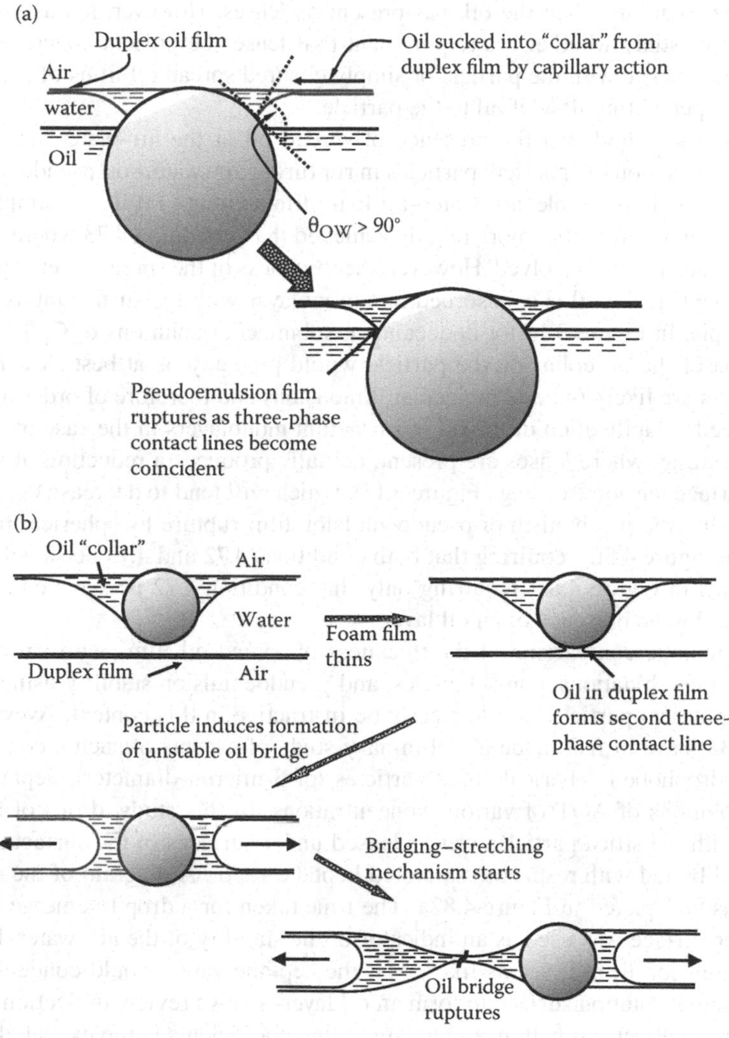

FIGURE 4.81 Schematic showing potential roles of spherical particle in case of a duplex film-forming oil for which $90° < \theta_{OW} < 180°$. (a) Particle may facilitate rupture of air–water–oil pseudoemulsion film by forming an unstable oil bridge (regardless of magnitude of air–water contact angle, θ_{AW}). (b) In case of large particle, it may penetrate an air–water–air foam film, facilitating formation of oil bridge and subsequent foam film rupture by the bridging–stretching mechanism.

extremely low lens dihedral angles, Denkov et al. [54], for example, quoting a value of only 0.4° for PDMS oil on a micellar AOT solution where the film has a thickness of 250 nm according to Marinova et al. [230]. It seems reasonable, therefore, to suppose that such systems will closely resemble those exhibiting complete wetting. Indeed Denkov et al. [62] demonstrate the formation of a "collar" of PDMS oil around a spherical glass particle present in the air–water surfaces of micellar surfactant solutions when the oil was present as lenses. However, it was not made clear in this study whether the requirement that lenses be present concerned their direct coalescence with the particle or simply ensured spread oil films of maximum thickness, permitting flow of oil to the particle.

We therefore find that the presence of spread oil at the air–water surface can facilitate the action of spherical particles in rupturing air–water–oil pseudoemulsion films (or even, in principle, air–water–air foam films directly) if θ_{OW} is simply >90°. This condition is arguably more readily achieved than condition 4.73 where the air–water contact angle is involved. However, the thickness of the spread layer is presumably relevant. If the oil is coadsorbed in a monolayer with the surfactant as shown, for example, in Figure 3.16 for dodecane on submicellar solutions of C_{14}TAB, then formation of the oil collar on the particle would probably be at best slow because such layers are likely to be of molecular dimensions and therefore of order only 2–3 nm. Indeed solubilization of the oil in surfactant monolayers in the case of pseudo-partial wetting, where lenses are present, actually produces a reduction of the air–water surface tension (see, e.g., Figure 3.13), which will tend to decrease θ_{AW}. In this case, neither the mechanism of pseudoemulsion film rupture by spherical particles shown in Figure 4.80 (requiring that both conditions 4.72 and 4.73 be satisfied), nor that shown in Figure 4.81 (requiring only that condition 4.72 be satisfied), will be facilitated by the presence of an oil layer.

A systematic comparison of the thickness of spread oil films with various oil–surfactant combinations, contact angles, and pseudoemulsion stability using hydrophobic spherical particles would clearly be instructive in this context. Aveyard and Clint [233] have in fact made a preliminary study of aspects of such a comparison using hydrophobed spherical silica particles (of 3-micron diameter), heptane, and saline solutions of AOT of various concentrations. In this study, drops of heptane coated with the silica particles were released under surfaces of the surfactant solution equilibrated with respect to saturated heptane vapor. A diagram of the relevant apparatus is depicted in Figure 4.82a. The time taken for a drop to emerge into the air–water surface was used as an indicator of the stability of the air–water–heptane pseudoemulsion film. It seems likely that the heptane vapor would condense upon the surfactant solution surface to form an oil layer—as we review in Section 3.6.1.2, pseudo-partial wetting with near-zero spreading coefficients is the expected behavior for such a system (although Aveyard and Clint [233] report no relevant measurements with the actual system considered). A plot of mean drop lifetime against the sum of the contact angles $\theta_{AW} + \theta_{OW}$ is shown in Figure 4.82b. It is not, however, made clear by these authors whether these are advancing or receding angles or how the angles were measured. Also apparently no comparison was made with the lifetimes of oil drops in the absence of silica particles. If, despite this omission, we suppose that the low lifetimes for the sum of contact angles in the range $50° < (\theta_{AW} + \theta_{OW}) < 180°$ reveal pseudoemulsion film instability induced by particles, then it is clear that neither condition 4.72 nor 4.73 is satisfied. The discrepancy is attributed both by Aveyard and Clint [233] and later by Denkov et al. [62], rather hopefully, to "dynamic wetting effects." There would appear to be much more to do in this area!

Another aspect of the role of particles in rupturing air–water–oil pseudoemulsion films concerns the extent to which the particles protrude from the oil–water

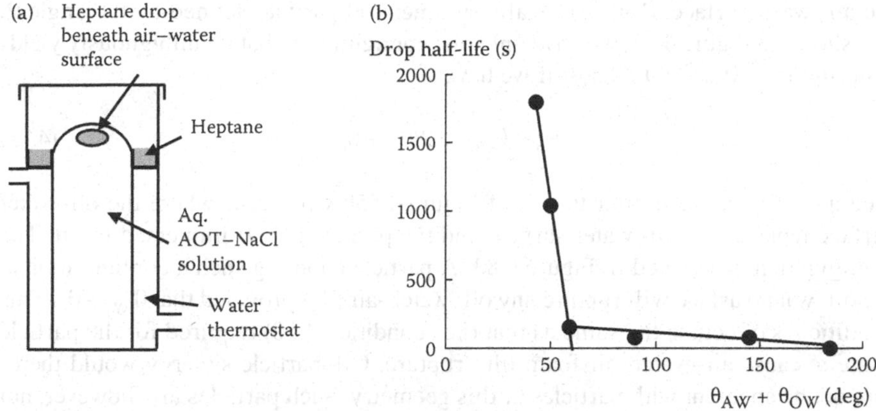

FIGURE 4.82 Effect of spherical hydrophobed silica particles on stability of pseudoemulsion film formed by heptanes in saline aqueous solutions of AOT. (a) Apparatus used. Contact angles were varied by changing AOT concentration. (b) Plot of heptane drop half-life against $\theta_{AW} + \theta_{OW}$ (see text). (Aveyard, R., Clint, J.H., *JCS Faraday Trans.*, 91, 2681, 1995. Reproduced by permission of The Royal Society of Chemistry.)

surface. The overall thickness of a metastable pseudoemulsion film separating an oil drop from the air–water surface will be determined by the requirement that the disjoining pressure in the film be equal to the sum of the capillary pressure due to the curved surface of the film and that in the adjacent air–water–air foam film. If any particles present at the oil–water surface do not protrude into the pseudoemulsion film at least a distance equal to the thickness of that film, then their presence will be essentially irrelevant for the stability of the film at that thickness regardless of their other properties. Simple geometrical arguments show [232] that the relation between the contact angle θ_{OW}, the film thickness h_{pef}, and the radius of the particle R_p where the particle protrudes into the pseudoemulsion film, a distance equal to the film thickness is given by

$$\cos \theta_{OW} = h_{pef}/R_p - 1 \tag{4.74}$$

where θ_{OW} represents the optimum contact angle. Higher values of θ_{OW} will mean that the particles do not protrude far enough into the pseudoemulsion film to cause rupture until it thins as a result of increasing capillary pressures in the adjacent foam film. Conversely, lower values may mean that conditions 4.72 and 4.73 may not be satisfied. Here of course we remember condition 4.73 is only relevant in the absence of thick spread oil films at the air–water surface.

4.8.5.3 Smooth Particles with Edges in Absence of Spread Oil Layers

In the case of smooth particles with edges then, as with air–water–air films, a more complex set of conditions for rupture of oil–water–air films arises. Here we consider the case where the oil exhibits partial wetting so that no spread oil layer is present at the air–water surface. With an axially symmetrical particle defined by the angle θ_p and shown in Figure 4.83, we have only one configuration that unambiguously yields film rupture if $\theta_{AW} < 90°$. Thus, if we have

$$\theta_p < \theta_{OW} < 180° - \theta_p \tag{4.75}$$

then a configuration similar to that of Figure 4.55b will occur where the oil–water surface replaces the air–water surface and the particle is half immersed in oil. The configuration is depicted in Figure 4.83. A particle adopting such a configuration at the oil–water surface will rupture any oil–water–air film provided that $\theta_{AW} > \theta_p$. This condition is of course the same as that (i.e., condition 4.50) required for the particle alone to cause air–water–air foam film rupture. Oil–particle synergy would therefore not be apparent with particles of this geometry. Such particles are, however, not readily prepared!

We now consider the case of hypothetical crystal-like particles by again selecting a geometry identical to that depicted in Figure 4.56a. The aspect ratio is, we recall, assumed to be so large that the curved surfaces of catenoid section present at surfaces that do not permit the relevant fluid–fluid surfaces to hinge on edges can be neglected. It will therefore be instructive to compare the conditions required for such

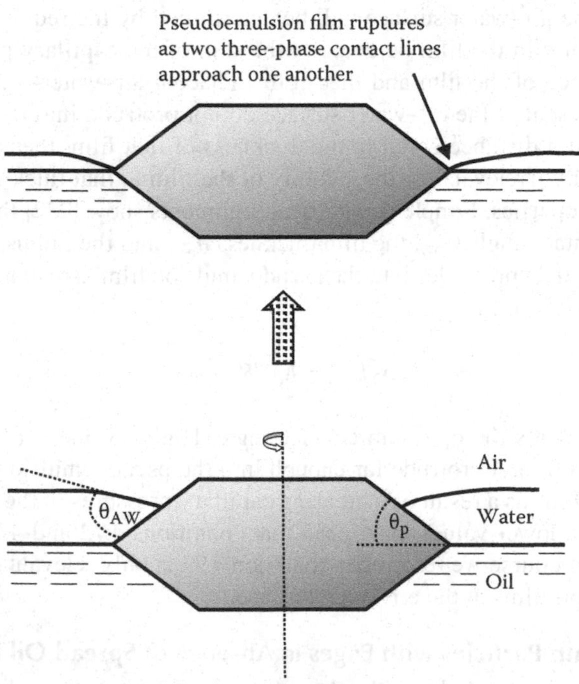

FIGURE 4.83 Effect of axially symmetrical particle with edges on stability of oil–water–air pseudoemulsion film with $\theta_p < \theta_{OW} < 180° - \theta_p$ and $\theta_{AW} > \theta_p$ with $\theta_p < 90°$.

a hypothetical particle to cause rupture of a pseudoemulsion film with those required for rupture of an air–water–air foam film.

In Table 4.11 we present the configurations such a particle could adopt at the oil–water interface if condition 4.72 is satisfied and if $\theta_p = 30°$. The conditions required for θ_{AW} if the particle is to cause rupture of oil–water–air films and facilitate emergence of the oil drop into the air–water surface are also deduced. Here we see that three configurations give rise to unambiguous rupture of oil–water–air films with $\theta_{AW} < 90°$. The remaining configurations will only affect rupture of such films if the critical thickness of rupture of the aqueous film on the solid is greater than that of the aqueous film on the oil.

Synergy between an oil that can form unstable bridging configurations in foam films where condition 4.31 is satisfied and a particle that satisfies condition 4.75 requires that the air–water contact angle conditions for the particle to rupture air–water–air foam films be more severe than those for rupture of air–water–oil pseudoemulsion films. This situation can, however, only occur for the particle considered in Table 4.11 if the configuration $j = 4$ occurs because the conditions for θ_{AW} for all other configurations that yield rupture of air–water–oil pseudoemulsion films would also mean rupture of air–water–air films (see Table 4.6). It is clear from Table 4.11 that the configuration $j = 4$ can coexist with both $j = 5$ and 6, where θ_{OW} lies between 135° and 150°. In these circumstances, the work of emergence of the particle into the oil–water surface from the oil phase will largely determine the relative probabilities of the relevant configurations (ignoring the effect of "degeneracy"). The work of emergence W_j^0 of each of the configurations at the oil–water surface is given (by analogy with the corresponding Equation 4.54 for the air–water surface)

$$W_j^0 = \sigma_{OW}\left(\Delta A_{SW}\cos(180° - \theta_{OW}) - \Delta A_{OW}\right) \tag{4.76}$$

where ΔA_{SW} and ΔA_{OW} are the changes in solid–water and oil–water surface areas accompanying emergence of the particles into the oil–water surface from the oil phase, respectively. If we consider the emergence of a particle into the oil–water surface of an oil drop where the radius of the oil drop is large compared with the size of the particle, then we may neglect the reduction in surface area of the oil drop due to changes in radius as the particle emerges into the oil–water surface. The geometry of the problem of estimating W_j^0 then becomes exactly the same as that involved in estimating the work W_i of emergence of the same particle from the aqueous phase into the air–water surface (where $180° - \theta_{OW}$ is substituted for θ_{AW} and σ_{OW} for σ_{AW} in the relevant expressions given in Table 4.6).

Whether configuration $j = 4$ has the largest negative work of emergence depends on θ_{OW}. Thus, it is easy to deduce from the data presented in Table 4.11 that synergy may occur for these hypothetical particles if $\theta_{OW} = 140°$ and $15° < \theta_{AW} < 30°$. This follows because under those conditions, configuration $j = 4$ is the most probable configuration (because it has the largest negative work of emergence when coexistence with other configurations is possible) and the particle alone cannot function as an antifoam (because this requires $\theta_{AW} > 30°$; see Table 4.6). In the case of $\theta_{OW} = 145°$, however, configuration $j = 5$ at the oil–water surface is most probable and collapse

TABLE 4.11
Air–Water–Oil Pseudoemulsion Film Rupture and Configurations at Oil–Water Surface of Particle with Several Edges

Possible Configurations at Air–Water Surface if $\theta_{OW} < 90°$	Conditions for Realization of Configurations	"Degeneracy"	Air–Water–Oil Pseudoemulsion Film Rupture if $\theta_{AW} < 90°$	Work of Emergence[a] of Particle into Oil–Water Surface from Oil $W_j/\sigma_{AW}X^2$ for $Z = 20X$: $\theta_{OW} = 145°$	$\theta_{OW} = 140°$
water, oil, $j = 1$	$150° < \theta_{OW} < 180°$	2	–	–	–
$j = 2$	$150° < \theta_{OW} < 180°$	4	–	–	–
$j = 3$	$90° < \theta_{OW} < 120°$	2	Yes provided $\theta_{AW} > 60°$	–	–
$j = 4$	$135° < \theta_{OW} < 165°$	4	Yes provided $\theta_{AW} > 15°$	–5.9	–8.0
$j = 5$	$30° < \theta_{OW} < 150°$	2	Yes provided $\theta_{AW} > 30°$	–5.5	–8.68
$j = 6$	$90° < \theta_{OW} < 180°$	2	–	–1.3	–0.44

Source: Adapted from Garrett, P.R., Mode of action of antifoams, in *Defoaming, Theory and Industrial Applications* (Garrett, P.R., ed.), Marcel Dekker, New York, Chpt 1, p 1 1993.

Note: Particle geometry as shown in Figure 4.56a with $\theta_p = 30°$ and $Y = X$. Z is the length of the particle as defined in the figure.

[a] Calculated using expressions shown in Table 4.6 for emergence into air–water surface from water, which is exactly analogous except that σ_{OW} replaces σ_{AW} and $180° - \theta_{OW}$ replaces θ_{AW}.

of the air–water–oil film will require $\theta_{AW} > 30°$. This latter is, however, the same condition as that required for the particle to collapse air–water–air films, so synergy will be absent in this case. Clearly, then, for this hypothetical particle, we have potentially complex behavior where synergy may or may not occur, depending on the magnitudes of θ_{OW} and θ_{AW}, which will in turn be influenced by the type and concentration of the surfactant present.

No experimental study of the antifoam effect of mixing hydrophobed crystals of well-characterized habit with oils has been reported. Experimental confirmation of the expectation of synergy only under restricted circumstances is therefore not available. However, Frye and Berg [75] have shown evidence of only weak antifoam synergy when hydrophobed ground-glass particles, which possess sharp edges (and are therefore relatively effective antifoams when used alone), are mixed with hexadecane.

4.8.5.4 Smooth Particles with Edges in Presence of Spread Oil Layers

We can now extend consideration of the effect of the possible orientations of the particle of crystal-like geometry shown in Figure 4.56a and Table 4.11 to the case where a relatively thick oil film is present at the air–water surface (corresponding to complete wetting or pseudo-partial wetting with a near-zero spreading coefficient). Assuming that the oil–water contact angle $\theta_{OW} > 90°$, then all configurations $j = 3–5$, which can lead to pseudoemulsion film rupture, are in principle realizable, albeit not simultaneously (e.g., configurations $j = 3$ and $j = 5$ can in principle coexist if $\theta_{OW} > 90°$ but $j = 4$ requires $\theta_{OW} > 135°$). The condition $\theta_{OW} > 90°$ of necessity implies the formation of a concave meniscus against the particle as it emerges into the oil-contaminated air–water surface with consequences analogous to those proposed for the equivalent "collar" around a spherical particle depicted in Figure 4.81. This in turn must result in pseudoemulsion film rupture irrespective of the magnitude of the air–water contact angle in the absence of an oil layer.

Examination of Table 4.11 reveals that, for example, if $\theta_{AW} < 15°$ in the absence of an oil layer, then none of the configurations $j = 3–5$ will result in pseudoemulsion film rupture. The presence of the oil layer would, however, result in pseudoemulsion film rupture in all cases provided $\theta_{OW} > 90°$ (where of course we must have $\theta_{OW} > 135°$ if configuration $j = 4$ is to occur at all). Moreover, if $\theta_{AW} < 15°$, then as we have seen, there exists no configuration at the air–water surface that would permit rupture of air–water–air foam films (see Table 4.6). The presence of a spread oil layer is therefore seen to expand the circumstances in which oil–particle antifoam synergy is to be expected relative to those expected in the absence of a spread layer.

4.8.5.5 Rough Particles with Many Edges

As we have illustrated in Figure 4.77, the particles present in synergistic oil–particle antifoams are usually highly irregular and even fractal in nature. We should therefore extend our analysis to include at least some aspects of that irregularity.

The case of rough particles with many edges can be modeled by selecting the axially symmetrical particle with regular rugosities shown in Figure 4.59. In the interests of clarity, the particle is again depicted in Figure 4.84 for the case of interaction

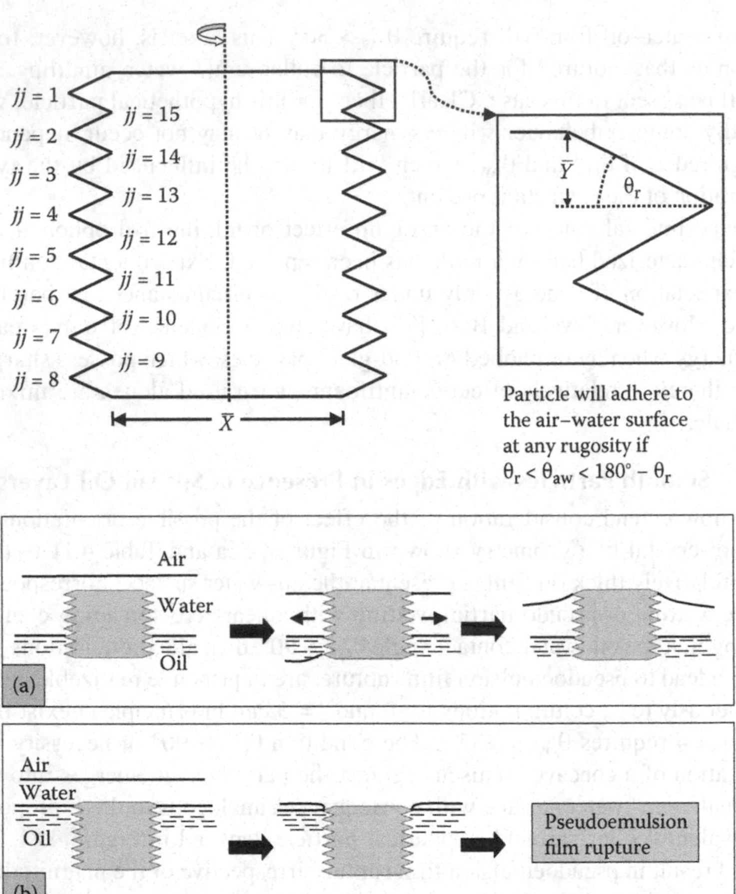

FIGURE 4.84 Axially symmetrical particle with many edges at oil–water interface. (a) Example of configuration that stabilizes air-water–oil pseudoemulsion film where oil–water interface hinges on any rugosity $jj \neq 1$. (b) Configuration where oil–water interface hinges on rugosity $jj = 1$ will cause air-water–oil pseudoemulsion film rupture provided $\theta_{AW} > \theta_r$. (After Garrett, P.R., Chap 1, in *Defoaming, Theory and Industrial Applications*, Garrett, P.R., ed., Marcel Dekker, New York, 1993, p 1.)

with an air–water–oil film. At the oil–water surface, such a particle may adopt configurations $jj = 1$ to 15 where that surface hinges at the rugosities provided

$$\theta_r < \theta_{OW} < 180° - \theta_r \tag{4.77}$$

where $\theta_r < 90°$ and is defined in Figure 4.84. This particle can only cause rupture of oil–water–air films when $\theta_{AW} < 90°$ if configuration $jj = 1$ is adopted (where most of the particle is immersed in the oil) and $\theta_{AW} > \theta_r$. A particle with such a configuration would also cause rupture of oil–water–oil films and would therefore cause inversion of emulsion behavior so that water-in-oil emulsions would be found.

The work of emergence from the oil phase, W_{jj}^{+0}, of each of the configurations of this rough particle at the oil–water surface may be calculated in a similar manner to W_j^0 by using Equation 4.76 if due allowance is made for the relevant geometry. If again we assume that the particles are small relative to the size of the oil drops, then we find

$$\frac{W_{jj}^{+0}}{\sigma_{OW}\bar{X}^2} = \pi\left[p_{jj}\left(\frac{\bar{Y}/\bar{X}}{\sin\theta_r}\right)\left(1+\frac{\bar{Y}/\bar{X}}{\tan\theta_r}\right)+0.25\right]\cos(180° - \theta_{OW}) - \pi\left[0.5+\frac{\bar{Y}/\bar{X}}{\tan\theta_r}\right]^2 \tag{4.78}$$

where $\bar{X}$ is defined in Figure 4.84 and p_{jj} is the number of segments of height $\bar{Y}$ (see Figure 4.84) transferred from the oil phase to the aqueous phase.

Calculations of $W_{jj}^{+0}/\sigma_{OW}\bar{X}^2$ for a particle that satisfies both conditions 4.72 and 4.77, so that

$$90° < \theta_{OW} < 180° - \theta_r \tag{4.79}$$

are given in Table 4.12. From this table, it is clear that the configuration $jj = 1$, with the particle mainly in the oil phase, yields the largest negative value of W_{jj}^{+0}. That is therefore the most probable configuration.

TABLE 4.12
Work of Emergence into Oil–Water Surface W_{jj}^{+0} of an Axially Symmetrical Particle with Regular Rugosities

Configurations with Oil–Water Surface Hinging on Rugosity jj	$W_{jj}^{+0}/\sigma_{OW}\bar{X}^2$	
	$\theta_{OW} = 135°$	$\theta_{OW} = 95°$
$jj = 1$[a]	−0.35	−1.29
2	+0.70	−1.16
3	+1.74	−1.03
4	+2.78	−0.91
5	+3.82	−0.78
6	+4.87	−0.65
7	+5.91	−0.52
8	+6.95	−0.39

Source: Adapted from Garrett, P.R., Mode of action of antifoams, in *Defoaming, Theory and Industrial Applications* (Garrett, P.R., ed.), Marcel Dekker, New York, Chp 1, p 1 1993.

Note: Calculated from Equation 4.78 where $\theta_r = 30°$ and $\bar{Y}/\bar{X} = 0.1$: see Figure 4.84 for definition of θ_r, $\bar{Y}$ and $\bar{X}$.

[a] $jj = 1$ (see Figure 4.84) is the only configuration that could lead to pseudo-emulsion film rupture.

An alternative configuration may occur if the particle is completely removed from the bulk of the oil so that the oil just fills the rugosities. This is illustrated in Figure 4.85. It is the configuration considered by Dippenaar [77], where the effective contact angle of the oil-filled particle at the air–water surface is supposed higher than that of the uncontaminated particle. This configuration can occur if the oil–water contact angle satisfies the condition

$$90° + \theta_r < \theta_{OW} < 180° \tag{4.80}$$

Conditions 4.79 and 4.80 overlap so that the configuration depicted in Figure 4.85 can coexist with the configurations $jj = 1$ to 15 if

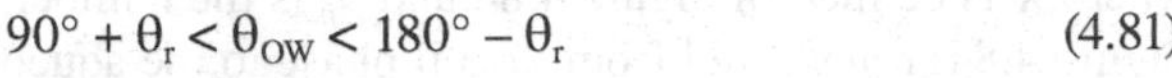

$$90° + \theta_r < \theta_{OW} < 180° - \theta_r \tag{4.81}$$

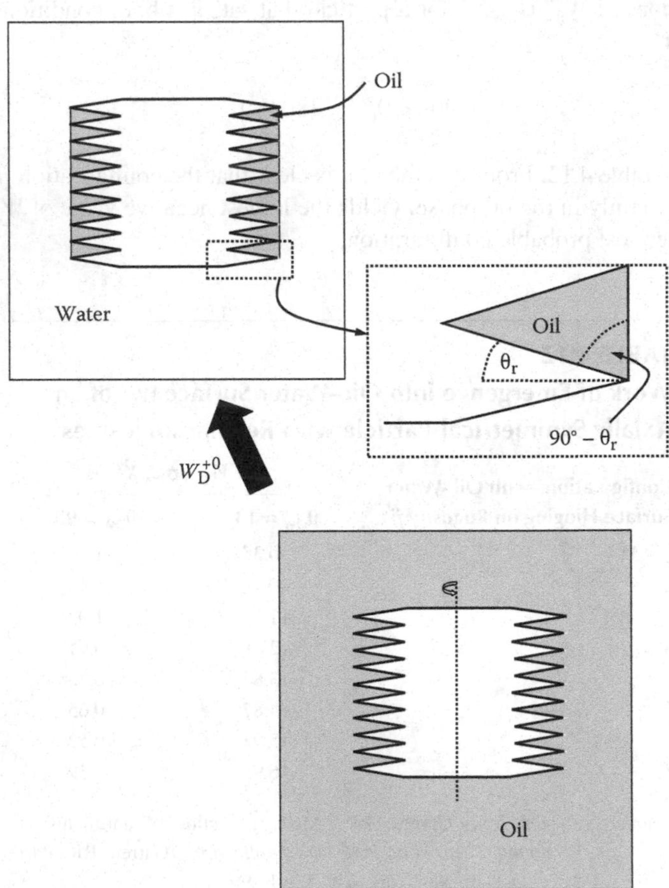

FIGURE 4.85 Formation of hydrophobic particle with oil-filled rugosities after dispersal of oil–particle mixture in aqueous phase. Essential features of particle geometry are as defined in Figure 4.84. (After Garrett, P.R., Chap 1, in *Defoaming, Theory and Industrial Applications*, Garrett, P.R., ed., Marcel Dekker, New York, 1993, p 1.)

However, the configuration where the rugosities are just filled with oil (shown in Figure 4.85) will have a higher work of formation W_D^{+0} than that of the most probable configuration at the oil–water surface where jj = 1 and where most of the particle is immersed in the oil This will arise because of both the greater oil–water surface area and the greater particle–water surface area. Thus, if we assume the particles are small compared with any oil drop in which they are initially dispersed, then the difference in work of formation of the two configurations is

$$\frac{W_D^{+0} - W_{jj=1}^{+0}}{\pi\sigma_{OW}\bar{X}^2} = \left[\left(\frac{\bar{Y}/\bar{X}}{\sin\theta_r}\right)\left(1+\frac{\bar{Y}/\bar{X}}{\tan\theta_r}\right)+0.25\right]\cos(180°-\theta_{OW}) + \left[0.5+\frac{\bar{Y}/\bar{X}}{\tan\theta_r}\right]^2 + \left[14\left(\frac{\bar{Y}}{\bar{X}}\right)\left(1+\frac{\bar{Y}/\bar{X}}{\tan\theta_r}\right)\right] \tag{4.82}$$

so that we always have $W_D^{+0} > W_{jj=1}^{+0}$ provided $\theta_{OW} > 90°$ since $\theta_r < 90°$.

We therefore find for a rough particle, which satisfies conditions 4.72 and 4.77, that the most probable configuration is one where the particle is mostly immersed in the oil phase. The oil–water surface hinges on an edge so that exposure of particle surface to the aqueous phase is minimized. As we have seen, this is also the only configuration that can cause rupture of the oil–water–air film (provided condition 4.62 is also satisfied so that $\theta_{AW} > \theta_r$). By contrast, a particle of the same geometry that satisfies condition 4.62, and for which $\theta_{AW} < 90°$, will have a low probability of rupturing air–water–air foam films. This follows because the only configuration that such a particle may adopt at the air–water surface that can cause rupture has the lowest probability of occurrence (see Section 4.7.4). Therefore, a rough particle of geometry shown in Figure 4.84 (or 4.59) has a high probability of rupturing an oil–water–air film if $\theta_{OW} > 90°$ and $\theta_r < \theta_{AW} < 90°$ but a low probability of rupturing an air–water–air foam film. Again, then, we would expect synergy if such particles are mixed with oils of low σ_{AO} so that the bridging condition 4.31 is satisfied. Since most particles used as antifoam promoters are best considered to be rough with many edges, then the ubiquitous occurrence of such synergy is clearly consistent with the argument outlined here.

These considerations obviously do not concern oils that form duplex films over the air–water surface. With such oils, and by analogy with the case of spherical particles, a "collar" of oil will form at the upper surface of the particle if $\theta_{OW} > 90°$ for which the most probable configuration is jj = 1 shown in Figure 4.84. We therefore expect the situation depicted in Figure 4.86 to develop so that the pseudoemulsion film will rupture irrespective of the magnitude of the air–water contact angle in the absence of the oil layer. In this case, condition 4.79 completely determines the role of the particle. Again it is to be expected that this will also apply to oils that exhibit pseudo-partial wetting for which $S^e \to 0$.

We therefore have two sets of conditions for the effectiveness of a rough particle in rupturing air–water–oil pseudoemulsion films while not rupturing air–water–air foam films. For partial wetting oils with no spread oil layer at the air–water surface,

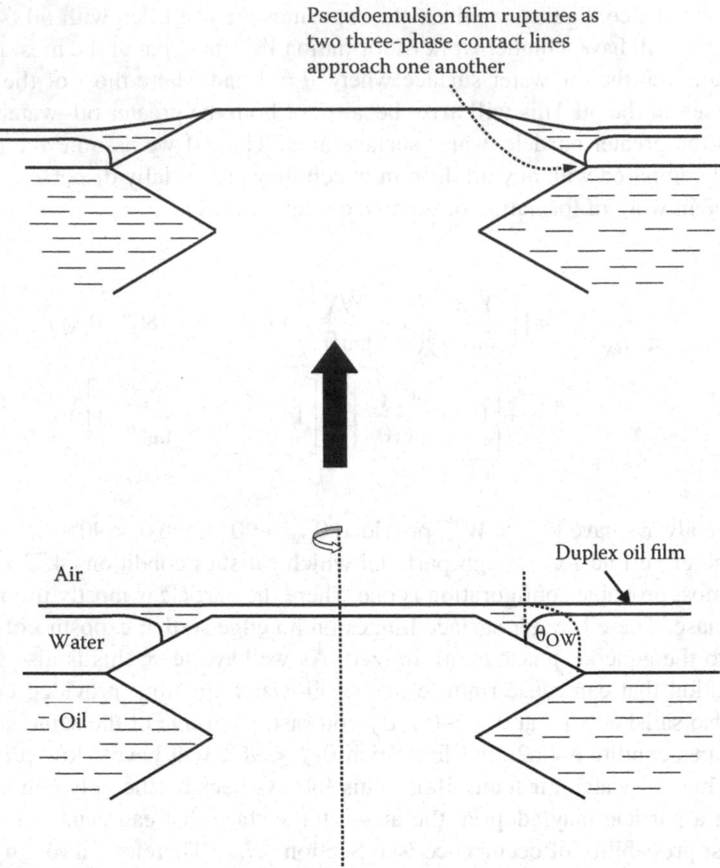

FIGURE 4.86 Effect of duplex oil film on rupture of pseudoemulsion film by an axially symmetrical particle with many edges hinging in oil–water interface at rugosity $jj = 1$ (see Figure 4.84 for definition of particle geometry). Film rupture only requires that $\theta_{OW} > 90°$.

the oil must preferentially wet the particle with $\theta_{OW} > 90°$ (satisfying condition 4.79 for the present geometry), hinging on rugosities that minimize exposure to the aqueous phase. In addition, the particle should be sufficiently hydrophobic to emerge into the air–water surface with a contact angle, θ_{AW}, high enough to rupture the pseudoemulsion film (in the case of the geometry selected here, this means $\theta_{AW} > \theta_r$). The geometry of the relevant rugosities mean that we should have $\theta_{AW} < 90°$. The latter will in turn mean that the particle would have an extremely low probability of adopting a configuration at the air–water surface, which could give rise to rupture of air–water–air foam films (see Table 4.7 and Figure 4.59).

In the case of complete wetting of the air–water surface by the oil (or even pseudo-partial wetting with thick spread layers at the air–water surface), we must simply have $\theta_{OW} > 90°$ (satisfying condition 4.79 for the present geometry). Again, however, if the air–water contact angle of the particle in the absence of oil is <90°, configurations at the air–water surface, which could give rise to air–water–air film rupture, would have extremely low probabilities of occurrence.

Quantitative comparison of these considerations with experiment is unfortunately fraught with difficulty. Rigorous measurement of the contact angles at fluid–fluid surfaces of irregular particles, replete with rugosities, in the relevant size range of a few microns is not at present feasible. Such contact angles are usually inferred from measurements on either compressed discs prepared from the relevant particles [234] or on smooth flat surfaces deemed representative of those of the relevant particles. Apart from the obvious objection that such measurements may not be representative, another difficulty concerns contact angle hysteresis.

Despite these difficulties, there have been a number of attempts to measure both oil–water and air–water contact angles supposedly representative of those relevant for the effectiveness of oil–particle antifoams in aqueous surfactant solutions. Typical results of such contact angle measurements are reproduced in Table 4.13 for a variety of effective antifoams (the antifoam effectiveness of each of these examples is to be found in the papers cited in the table). These antifoams represent examples of practical systems where we would expect irregular particles decorated with rugosities—images of the oil–particle composite entities formed upon dispersal of some of the antifoams listed in the table are in fact shown in Figure 4.77. These results all concern advancing contact angles measured through the relevant aqueous phase. Unfortunately, the role of the particle in rupturing the relevant pseudoemulsion film has only been directly established in the case of the hydrophobed silica–PDMS [70] and the D17 hydrophobed silica–liquid paraffin [43] antifoams listed in the table.

It is clear from the table that, regardless of the chemical composition of the oil–particle antifoams, we have $\theta_{OW} > 90°$ and $0 < \theta_{AW} < 90°$. Here we note that it seems likely that the roughness of some of the surfaces used to make these measurements is in general less than that of the particles. Application of the Wenzel equation [172, 173] (Equation 4.61 for air–water contact angles) would then suggest that roughness would mean that the actual values of θ_{OW} for rough particles would be higher than quoted and the actual values of θ_{AW} lower than quoted.

The measurements of the contact angles quoted in Table 4.13 are apparently at least consistent with the theoretical considerations given here. In all cases, for example, $\theta_{OW} > 90°$. In the case of partial wetting where a spread layer at the air–water surface is absent, the air–water contact angle is relevant and we would expect $0 < \theta_{AW} < 90°$. However, the relevant contact angle should be the receding angle, not the measured advancing angle quoted in the table. By contrast, we would expect the advancing angle $\theta_{OW} > 90°$ to be relevant in the absence of spread oil layers at the oil–water surface. If spread oil layers are present, then two oil–water–solid three-phase contact lines are supposed involved in pseudoemulsion film rupture. At one line, the contact angle is advancing and at the other it is receding. The relevance of this supposed contact angle hysteresis, if at all, in the process of pseudoemulsion film rupture is of course not established. However, the presence of such hysteresis is of course relevant in establishing the intrinsic contact angle from measurements on plates or compressed discs.

Of some interest is the result of Bergeron et al. [70] concerning the effect of a hydrophobed silica–PDMS antifoam on solutions of a commercial zwitterionic

TABLE 4.13
Advancing Contact Angles Representative of Antifoam Particles

Oil–Particle Antifoam Mixture	Surfactant Solution[a]	Spread Layer?	θ_{OW} (°)	θ_{AW} (°)	Nature of Solid Surface	Ref.
Calcium stearyl phosphate/liquid paraffin	1.4 mM Sodium alkylbenzene sulfonate	No	122	68	Compressed disc	[43]
D17 hydrophobed silica/liquid paraffin	1.4 mM Sodium alkylbenzene sulfonate	No	143	39	Compressed disc	[43]
Aerosol hydrophobed silica/liquid paraffin	1.4 mM Sodium alkylbenzene sulfonate	No	101	31	Compressed disc	[43]
Ethylene bis-stearamide/dodecane	0.01 mM SDS	Yes	140	~75	Cooled melt on glass	[44]
Ethylene bis-stearamide/dodecane	0.01 mM $C_{16}TAB$	Yes	140	~70	Cooled melt on glass	[44]
Ethylene bis-stearamide/dodecane	0.01 mM AOT	Yes	140	~45	Cooled melt on glass	[44]
Paraffin wax/dodecane	0.01 mM SDS	Yes	140	~75	Compressed disc	[44]
Paraffin wax/dodecane	0.01 mM $C_{16}TAB$	Yes	140	~70	Compressed disc	[44]
Paraffin wax/dodecane	0.01 mM AOT	Yes	140	~45	Compressed disc	[44]
PTFE/dodecane	0.01 mM SDS	Yes	~160	~75	Smooth sheet	[44]
PTFE/dodecane	0.01 mM $C_{16}TAB$	Yes	~160	~70	Smooth sheet	[44]
PTFE/dodecane	0.01 mM AOT	Yes	~160	~45	Smooth sheet	[44]
Hydrophobed silica/PDMS oil	3 × CMC $C_{12}EO_5$	Yes	150	60	Hydrophobed silica simulated using PDMS elastomer	[70]
Hydrophobed silica/PDMS oil	3 × CMC $C_{12}TAB$	Yes	151	71	Hydrophobed silica simulated using PDMS elastomer	[70]
Hydrophobed silica/PDMS oil	3 × CMC $C_{14}TAB$	Yes	152	68	Hydrophobed silica simulated using PDMS elastomer	[70]
Hydrophobed silica/PDMS oil	3 × CMC $C_{16}TAB$	Yes	151	80	Hydrophobed silica simulated using PDMS elastomer	[70]
Hydrophobed silica/PDMS oil	3 × CMC AOT	Yes	155	48	Hydrophobed silica simulated using PDMS elastomer	[70]
Hydrophobed silica/PDMS oil	3 × CMC Zonyl FSK	No	125	75	Hydrophobed silica simulated using PDMS elastomer	[70]

Note: All contact angles were measured through the aqueous phase; air–water contact angles, θ_{AW}, were measured in the absence of oil layers.

[a] Abbreviations for surfactants: SDS, sodium dodecyl sulfate; AOT, sodium bis-diethylhexyl sulfosuccinate; C_nTAB, $C_nH_{2n+1}.(CH_3)_3N+Br-$; $C_{12}EO_5$, $C_{10}H_{21}(OCH_2CH_2)_5OH$; Zonyl FSK, a commercial zwitterionic fluorinated betaine; D17, hydrophobed silica, ex-Degussa.

fluorinated betaine surfactant solution ("Zonyl FSK" in Table 4.13). We have already noted (see Section 4.8.3) that this antifoam is totally ineffective in that solution and exhibits partial wetting, forming oil lenses with negative bridging coefficients and $\theta^* < 90°$. The contact angles given for this system in Table 4.13 indicate that the hydrophobed silica should rupture the relevant pseudoemulsion film. In which case, this system tends to confirm the theoretical expectation that oil lenses with negative bridging coefficients should be ineffective antifoam components. Any ambiguity concerning the role of the particles would in this case appear to be absent.

Another aspect of the role of the particle in rupturing pseudoemulsion films concerns the possibility of an optimum hydrophobicity for a given particle size. We have already discussed this issue in the context of spherical particles (see Section 4.8.42). If, for example, the particle is too hydrophobic (so that $\theta_{OW} \rightarrow 180°$), it could protrude into the pseudoemulsion film a distance less than the equilibrium film thickness at a given capillary pressure. The particle will not therefore bridge the film and cause pseudoemulsion film rupture. Conversely if the particle is too hydrophilic so that, for example, $\theta_{OW} < 90°$, then again it will not rupture the pseudoemulsion film regardless of the presence or absence of a spread oil layer. We illustrate these extremes for the present geometry in Figure 4.87.

Some evidence that an optimum hydrophobicity exists has been obtained by Marinova et al. [231, 232] using hydrophobed silica–PDMS oil antifoams. These were prepared by *in situ* reaction of the oil with silica at room temperature for prolonged periods of time under continuous stirring to minimize the formation of three-dimensional bridged gels. This procedure results in a progressively increasing adsorption of polydimethysiloxane molecules onto the silica (presumably as a result of chemical reaction between silica surface hydroxyl groups and the PDMS) [231]. The effectiveness of antifoams prepared for various periods of time in this manner was compared as an indication of the effect of increasing hydrophobicity of the silica, and therefore contact angle, θ_{OW}. Unfortunately, differences in effectiveness were only apparent after repeated foam generation until some measure of antifoam deactivation had occurred. The resulting "durability" was then used as an indicator of effectiveness. Implicit in this equivalence is the assumption that deactivation is more rapid if the antifoam is intrinsically less effective. This may well concern the effectiveness of the antifoam in maintaining a spread oil film at the air–water surface. Absence of that film in the case of hydrophobed silica–PDMS oil antifoam is a symptom of antifoam deactivation as we discuss in detail in Chapter 6.

A plot of antifoam "durability" against time of hydrophobization (and therefore hydrophobicity) of the silica obtained by Marinova et al. [231] is given in Figure 4.88 for solutions of 0.6 mM of a commercial $C_{12/14}$ glucopyranoside. Included in the figure is a plot of the critical applied capillary pressure, p_c^{crit}, for emergence of a drop of the antifoam into the air–water surface as measured using the FTT. A pronounced maximum in the antifoam durability plot correlates exactly with a minimum in the p_c^{crit} plot. Similar plots are reported for micellar solutions of AOT and Triton X-100. Marinova et al. [231] attribute the maximum in durability to an optimum hydrophobicity using an argument similar to that given here. The correlation of

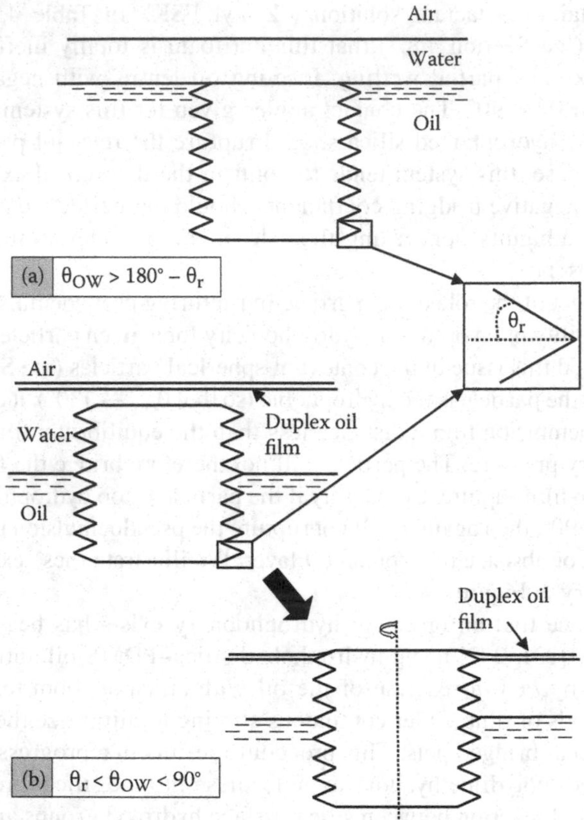

FIGURE 4.87 Extremes of hydrophobicity and non-rupture of pseudoemulsion film by an axially symmetrical particle with many edges. (a) Case where $\theta_{OW} > 180° - \theta_r$ and particle is too hydrophobic so that it does not penetrate the pseudoemulsion film irrespective of presence or absence of spread oil films (layers). (b) Case where $\theta_r < \theta_{OW} < 90°$; while the particle can hinge on edges, it is not sufficiently hydrophobic to hinge on edge $jj = 1$ (see Figure 4.84 and text). Therefore, coincidence of the three-phase contact lines of bulk phase oil and duplex film oil on one edge is improbable and pseudoemulsion film rupture will not occur. Similar considerations apply in case of partial wetting oils if also $\theta_r < \theta_{AW} < 90°$.

declining durability with increasing values of p_c^{crit} does of course strongly suggest the expected link between hydrophobicity and effectiveness of rupture of pseudoemulsion films.

4.8.6 Antifoam Dimensions and Kinetics

There have been no systematic studies of the effect of particle size on the efficiency of particle–oil antifoams. However, if part of the role of the particle is to penetrate any electrical double layer between oil drops and air bubbles, then clearly the

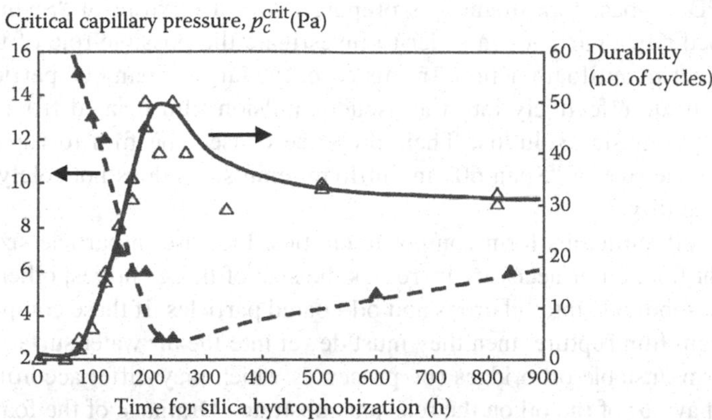

FIGURE 4.88 Applied capillary pressure, p_c^{crit} (measured using the FTT), required for emergence of PDMS–hydrophobed silica antifoam into air–water surface of a surfactant solution is plotted against time for hydrophobization of silica as an indication of hydrophobicity. Durability of the resulting antifoam is also plotted against the latter. Surfactant solution is 0.6 mM of non-ionic commercial alkyl (C_{12-14}) glucopyranoside. (Reprinted with permission from Denkov, N.D. *Langmuir*, 20, 9463. Copyright 2004 American Chemical Society.)

particle size should be at least of the same order as the double layer thickness, i.e., ~10–100 nm. If the particle size is appreciably smaller than this, then the oil drop will in effect be smooth so that the presence of particles will be irrelevant (except in that they modify the overall interaction forces between bubbles and antifoam entities). This is therefore essentially the corollary of the argument concerning excessively hydrophobic particles that protrude into pseudoemulsion films to a minimal extent.

Marinova and Denkov [60] have published an interesting illustration of this aspect of antifoam mechanism. The study involved a comparison of the effectiveness of hydrophobed silica in rupturing the pseudoemulsion films formed by PDMS oil in an aqueous solution of an anionic surfactant with that in an aqueous solution of a non-ionic surfactant. The anionic surfactant solution was 10 mM AOT and the non-ionic solution was 0.45 mM of a commercial $C_{12/14}$ glucopyranoside. Equilibrium thicknesses of pseudoemulsion films of the non-ionic surfactant were significantly greater than those of the anionic surfactant largely as a result of the lower ionic strength present in the former leading to large Debye screening lengths. As a consequence, the hydrophobed silica particles were seen to be too small to protrude significantly into the pseudoemulsion films of the non-ionic surfactant solution, which were therefore stable. By contrast, the silica particles were easily visible, protruding into the pseudoemulsion films of the anionic surfactant, leading to film rupture. Hydrophobed silica–PDMS antifoams are often prepared as emulsions stabilized by particles of sorbitan monostearate ("Span 60") where the particle size of the latter is significantly larger than that of the hydrophobed silica. A PDMS-based antifoam was prepared with a mixture of "Span 60" and hydrophobed silica particles in order to investigate the possible role of the emulsion stabilizer in antifoam action. In this case, the large "Span 60" particles were seen to protrude effectively into the pseudoemulsion film formed from the nonionic glucopyranoside solution. Their presence caused the film to rupture, suggesting that the role of "Span 60" in antifoam emulsions does not solely concern emulsion stability.

Particle–oil antifoams form composite entities. Increase in particle size at constant weight fraction of necessity increases the size of those entities; otherwise, the system will subdivide into oil drops and oil-coated particles. If these composites are to cause foam film rupture, then they must dewet into the air-water surface of foam films to form unstable oil bridges (by processes which may differ according to the wetting behaviour of the oil on the relevant substrate). Drainage of the foam film to dimensions of the same order as that of the antifoam could therefore be expected to play a similar role with oil–particle antifoams as with particle antifoams. We would therefore expect that the considerations of Dippenaar [77] and Frye and Berg [152] to be relevant for particle–oil antifoams so that foam film draining could be the rate-determining step for antifoam action. Sizes of the antifoam particle–oil composites should therefore be small enough to ensure a high probability of presence in a given foam film but not so small that foam film drainage to the dimensions of the antifoam entity is too slow. Observations with both mineral oil–hydrophobed silica [43] and silicone [201] antifoams suggest that improvements in antifoam effectiveness of a given weight concentration of antifoam may be achieved by decreasing antifoam entity sizes down to at least 1–2 μm. However, we return to the issues associated with antifoam drop size, foam film drainage, and the kinetics of antifoam action and deactivation in Chapters 5 and 6.

4.9 SUMMARIZING REMARKS

In summarizing, it appears appropriate to be mainly concerned with outstanding issues concerning the mode of action of antifoams that remain to be addressed.

Antifoam effects are usually attributed to undissolved materials. In general, much of the evidence concerning the supposed adverse effect of solubilized oils on the foamability of aqueous micellar surfactant solutions may simply concern the presence of undissolved oil drops. Evidence that solubilized alkanes adversely affect the stability of foams of certain aqueous micellar solutions [15, 21] is, however, difficult to simply dismiss in this manner. These observations have been attributed to the effect of the solubilizate on the repulsion forces between micelles. In turn, this is supposed to reduce stratification and therefore foam film stability. Direct evidence of this supposed reduced stability, using, for example, a suitable film balance method, is lacking.

Precipitation of anionic surfactant from aqueous solution as either crystalline or liquid crystalline particles by addition of inorganic electrolytes generally causes diminished foamability. This does not, however, always imply antifoam action by the precipitate. It is usually largely due to the slow rates of surfactant transport by the precipitated particles to the rapidly formed air–water surfaces during foam generation. There is, however, some evidence that if the precipitate is crystalline, then antifoam effects are superimposed upon this transport effect as a consequence of favorable geometry and finite contact angles at the air–water surface. Conversely, liquid crystalline precipitates are expected to be hydrophilic and unlikely to give rise to antifoam effects. Placing these generalizations on a firmer basis with more examples would, however, appear to be desirable.

Ewers and Sutherland [38] first proposed the Marangoni spreading mechanism for antifoam action more than six decades ago. Claims to wide generality have been undermined by experimental observations of antifoam phenomena that cannot function by this mechanism [9, 12, 39, 40–44, 54]. Indeed, the Marangoni spreading mechanism is limited in applicability because it is necessarily a transient effect. Thus, as foam generation ceases, then the air–liquid surfaces of the foam approach equilibrium and the driving force for spreading, the spreading coefficient, becomes either zero or negative. Yet antifoams still function after foam generation ceases.

Arguably the most recent theoretical treatment concerning antifoam action by Marangoni spreading is that due to Prins [45] which was published more than two decades ago. This theory predicts that antifoam effectiveness should increase as the initial spreading coefficient decreases. Perhaps therefore a serious reappraisal of the theoretical implications of Marangoni spreading for antifoam action is now necessary.

Effective antifoam action by oils requires that the relevant pseudoemulsion film be first ruptured by either the action of particles adsorbed at the oil–water interface or by the application of a sufficiently high capillary pressure. Oils may be active antifoam components regardless of whether they exhibit complete wetting (as duplex layers with no lenses), pseudo-partial wetting (lenses with oil layers), or partial wetting (oil lenses and no oil layer) at the surface of the foaming liquid. Regardless of these different wetting behaviors, oil bridges in foam films have been shown to lead to foam film rupture provided the bridging coefficient is positive and the volume of the oil bridge is large enough. The actual process of foam film rupture is usually attributed to the bridging–stretching mechanism proposed by Denkov [54]. Both the precise fate of a bridging drop after film rupture and the role of the magnitude of the bridging coefficient in the process remain to be resolved.

In the absence of particles adsorbed at the oil–water surface, oil drops tend to be removed from draining foam films into the Plateau borders before they can emerge into the air–water surface. Unstable bridging configurations do not then form in foam films and rupture of those films does not occur. Antifoam action can then only occur when oil drops trapped in Plateau borders are subjected to a capillary pressure, sufficiently high to rupture the pseudoemulsion films, as liquid drains out of the foam. Any resulting antifoam effect tends therefore to occur only after an induction period, often of several minutes. The nature of the foam rupture process induced by oil drops bridging Plateau borders has not, however, been established and it is therefore uncertain whether the bridging coefficient determines the outcome.

We have reviewed the antifoam behavior in aqueous surfactant solution of neat oils represented by short-chain alcohols, alkanes, and polydimethylsiloxanes, all without the presence of particles. In all cases, the antifoam behavior appears to be dominated by the critical applied capillary pressure necessary to rupture the relevant pseudoemulsion films. Antifoam effects therefore tended to be weak and mainly confined to Plateau borders. By contrast, neat PDMS oils are extremely effective antifoams for foaming non-aqueous liquids (see Chapter 10). Such liquids are characterized by low dielectric constants, minimal ionic dissociation, and the absence of electrostatic effects. This presumably means that the relevant pseudoemulsion films are easily ruptured at low capillary pressures despite the absence of particles adsorbed at the PDMS–foaming liquid interface. Effective antifoam action is then possible in foam films rather than confined to Plateau borders. The literature concerning the antifoam effect of neat oils on the foam behavior of non-polar liquids is, however, characterized by a lack of details concerning basic properties such as entry, spreading, and bridging coefficients together with no studies at all of the behavior of the relevant pseudoemulsion films or attempts to visualize antifoam action in foam films.

The onset of partial miscibility as a system passes through the coexistence curve produces a marked reduction in foamability. In effect, one conjugate liquid then acts as an antifoam for the other. This is exemplified by the behavior of aqueous solutions of ethoxylated non-ionic surfactants. At temperatures above the cloud point, phase separation occurs to produce a concentrated micellar conjugate phase and a dilute phase. The former acts as an antifoam for the latter. That the effect is a true antifoam effect rather than an effect due to slow transport of surfactant by drops of the concentrated conjugate is revealed by removal of the concentrated micellar conjugate—foamability is restored. The concentrated conjugate is an effective antifoam for the dilute conjugate—effects would appear to be occurring in foam films rather than Plateau borders. This tends to suggest that the pseudoemulsion films are intrinsically unstable, requiring only low values of an applied capillary pressure to rupture. Direct evidence for this supposition is, however, lacking. This antifoam effect can be employed in systems where a second surfactant component is present—the concentrated conjugate is then acting as a "cloud point antifoam" for the mixed system. Few studies have addressed the behavior of such systems despite their importance for some practical applications. Many issues deserve attention, apart from the stability of the pseudoemulsion films, such as the potential role of spinodal decomposition in phase separation, and therefore antifoam behaviour, with these systems.

It is now well known that hydrophobic particles can form unstable bridging configurations in foam films and therefore act as antifoams for aqueous foaming surfactant solutions. The role of contact angle and particle geometry has been explored by several authors. The main issues concern accurate determination of the relevant contact angles and the determination of the orientation of particles of complex geometry in the air–water surface. One possible approach to the latter involves the use of the Surface Evolver software [81] to determine the relevant minimum energy configurations. A limitation of this approach, however, concerns incorporation of disjoining potentials in the calculation—this means that realistic calculation of the minimum energy route to foam film rupture by a particle adopting a particular orientation at the surface of a foam film is not at present possible.

The role of calcium soap particles as antifoam additives is surprisingly poorly understood. Available evidence (which is scant) suggests that such particles may function simply as hydrophobic particles. However, measurements of the relevant contact angles are few and never combined with imaging and sizing of the particles. Moreover, there is evidence of transient antifoam effects accompanying the precipitation of such particles by reaction of calcium ions with soaps [198]. There is as yet no explanation for such transient effects.

Effective antifoams for aqueous surfactant solutions usually consist of mixtures of hydrophobic particles and oils. The role of the particles in rupturing the relevant pseudoemulsion films is now well established. As with hydrophobic particles alone, issues in the main concern accurate measurement of the relevant contact angles and the determination of the orientation of particles of complex geometry at a surface (in this case, the oil–water surface). An additional complication concerns the effect of spread layers of oil on the effectiveness of the particles. The role of the thickness of that layer in this respect does not appear to have been systematically studied. Unstable bridging configurations in the foam film by the oil give rise to foam film rupture. There is clear evidence of examples where Marangoni spreading cannot occur in both the case of an oil that exhibits partial wetting at an air–water surface and an oil that exhibits pseudo-partial wetting (albeit with equilibrium spreading coefficients close to zero). Effective antifoam action is observed in all cases.

One general issue concerns the effect of the extent of the likely deviations from equilibrium surface properties during foam generation on antifoam action. In principle, such effects could give rise to significant transient deviations from equilibrium values of entry, spreading, and bridging coefficients together with deviations in disjoining pressure isotherms and the critical applied capillary pressure for pseudoemulsion film rupture. Allowing for such effects can at best be qualitative until some means can be established for characterizing the extent of deviation from equilibrium of the continuous air–liquid surface at the top of a foam column immediately after foam generation has ceased.

ACKNOWLEDGMENTS

The author is grateful to Drs. Stephen Neethling and Gareth Morris (of Imperial College) for discussions concerning the application of Surface Evolver software for determination of particle orientations at surfaces and for supplying copies of the illustrations in Figures 4.26 and 4.58. The author is also grateful to Prof. Nikki Denkov (of Sofia University) for granting general permission to use figures from his publications and for supplying copies of the photomicrographs in Figures 4.74 and 4.75.

APPENDIX 4.1

Tables Listing Examples of Early Patents Concerning Mixtures of Hydrophobic Particles and Oils as Antifoams for Aqueous Systems

TABLE 4.A1
Some Examples of Early Antifoam Patents Concerning Mixtures of Hydrophobed Mineral Particles and Polydimethylsiloxane (Silicone) Oils

Oils Claimed To Be Effective	Particles Claimed To Be Effective	Actual Examples Given	Preferred Conc. of Particles (wt.%)	Particle Size	Any Other Additives	Application	Year of Publn.	Ref.
Partially oxidized methylsiloxane polymer. May be diluted with dimethylsiloxane polymer and dispersed in benzene	Silica aerogel presumably rendered hydrophobic by reaction *in situ* with the siloxane	As claimed	7.5	–	–	Aqueous alkaline solutions for paper industry, rubber industry, metal working industry	1953	[236]
Methyl polysiloxane	Finely divided silica presumably rendered hydrophobic by reaction *in situ* with the siloxane	Silica + methyl polysiloxane + emulsifiers	2–10	"Finely divided"	Patent really concerns emulsifiers, which are monostearic acid ester of polyethylene glycol + monostearic acid ester of sorbitol ("Span 60")	Particularly concerns emulsifiers for silicone antifoams for cosmetics and drug application	1958	[237][a]
Silicone oils (Product may be diluted by hydrocarbons, ethers, ketones, or chlorohydrocarbons. This does not, however, appear as a specific claim)	Aluminum oxides, titanium dioxide, particularly all manner of silicas. These react with silicone oil *in situ* catalyzed by an acid condensation reagent	1. Dimethyl polysiloxane + silica aerogel + $AlCl_3$ diluted by toluene 2. Dimethyl polysiloxane + alumina + $SnCl_4$ diluted by toluene 3. Dimethyl polysiloxane + precipitated silica + phosphorus nitrile chloride + polyethylene glycol stearate	1–30	0.01–25 microns	1. Acid condensation catalyst necessary (e.g., $AlCl_3$, $SnCl_4$, BF_3 etc.). 2. Emulsifiers such as methyl cellulose, polyethylene glycol monstearate, polyethylene glycol trimethyl nonyl ether added to prepare stable O/W emulsions of the antifoam if desired	Alkaline aqueous solutions. Particularly alkaline cleaning products for automatic washing machines	1996	[217]
Silicone oils	Pyrogenic or precipitated silicas hydrophobed with "chemically bound methyl groups" (probably silanized) Special reference is made to the relative ineffectiveness of the oil alone	As claimed	2–8 (preferably 5)	0.015–0.05 micron	–	Aqueous surfactant solutions	1968	[220]

Organosiloxane liquid polymers, which may be partly or entirely replaced by other non-aqueous fluids such as hydrocarbon oils and polyalkylene glycols	Hydroxyl-containing inorganic "fillers" TiO_2, Al_2O_3, and preferably SiO_2 reacted with a dialkyl-amino-organosilicone. Special provision of patent is that necessity for catalysts is obviated. Heating and cooling cycles are also obviated	1. Dialkyl amino-silicone treated silica in dimethylsiloxane polymer diluted with polypropylene glycol 2. Dialkyl amino- organosilicone – treated silica in polypropylene glycol 3. Dialkyl amino-organosilicone–treated silica in mineral oil diluted with polypropylene glycol	1–30	0.007–0.025 micron	–	Alkaline aqueous solutions, particularly aqueous paints, latex systems, laundry and detergent products	1971	[218]
Dimethyl polysiloxane or about equal proportions of mixture of dimethyl polysiloxane + diorganosiloxane (containing silicon-bonded Me, Et, and 2-phenylpropyl groups)	Silica of surface area >50 $m^2 gm^{-1}$ (List of previous silicone antifoam patents given as examples of the type of materials considered.)	Silica + polydimethyl siloxane	1–10	–	1. Whole mixed with sodium tripolyphosphate to produce free-flowing powder. This is the particular provision of the patent 2. Emulsifying agents, mold inhibitors, etc.	Laundry washing formulations	1974	[219]
Alkylated polysiloxane materials of various types	Silicas rendered hydrophobic by a variety of methods	Dimethylsiloxane + trimethyl silanated silica	10–50	Not >0.1 micron, preferably 0.01–0.02 micron	Antifoam incorporated in detergent-impermeable water-soluble or dispersible carrier material such as gelatin and polyethylene glycol. This material is in turn coated with a water-soluble granular material	Detergent formulations for automatic clothes and dishwashing machines	1975	[238]

Source: Garrett, P.R., Mode of action of antifoams, in *Defoaming, Theory and Industrial Applications* (Garrett, P.R., ed.), Marcel Dekker, New York, Chpt 1, p 1, 1993.

[a] Also used in a recent publication by Marinova and Denkov [60].

TABLE 4.A2
Some Examples of Early Antifoam Patents Concerning Mixtures of Hydrophobed Mineral Particles and Organic Liquids

Oils Claimed To Be Effective	Particles Claimed To Be Effective	Actual Examples Given	Preferred Conc. of Particles (wt.%)	Particle Size	Any Other Additives	Application	Year of Publn.	Ref.
Kerosene, naphthenic mineral oil, paraffinic mineral oil, chlorinated naphthenic mineral oil, chlorinated paraffinic mineral oil, liquid trifluorovinyl chloride polymer, fluorinated hydrocarbons	Aerogel, fume, or precipitated silicas hydrophobed by any suitable method	Silicas hydrophobed with silicone oil or with alkyl chlorosilanes + most of the oils claimed	2–20	Most preferred from 0.02 to 1 micron	Spreading agent claimed to be essential (anionic, cationic, or non-ionic surfactant)	Probably mainly intended for paper pulp mills although patent not restrictive in this respect	1963	[221]
Aliphatic or aromatic hydrocarbons with at least six carbon atoms	Precipitated silica having pH from 8 to 10. Rendered hydrophobic by any suitable method but only polysiloxane or alkyl (aryl or alicyclic) silanes claimed. Type of silica employed is a special provision of the patent	Precipitated silica hydrophobed with polymethylsiloxane and mixed with various mineral oils	3–30	0.005–0.05 micron	Need for a spreading agent is eliminated—a special feature of this patent. Use of a spreading agent is claimed to reduce rather than increase the effectiveness of the antifoam	Especially adopted for paper pulp mill application although patent apparently not restrictive in this respect	1965	[239]
Water-insoluble polyalkylene glycol	Silica subjected to heat and shear treatment in oil. Silica may also be hydrophobed before mixing with oil by treatment with silane or alcohol	1. Polypropylene glycol with silica (hydrophobed with trimethylsilane or isobutanol) 2. Polybutylene glycol with silica	1–10	Finely divided high surface area (>50 m^2 g^{-1})	–	No specific application quoted	1967	[240]
Aliphatic hydrocarbon or paraffin oil, including mineral seal oil, kerosene, various light aliphatic fuel oils, gas oils, paraffin waxes, etc.	Any type of silica possessing reactive surface hydroxyl groups rendered hydrophobic with dialkyl dihalosilane. Reaction made in the oil. Special claim of patent concerns reaction conditions giving rise to dialkyl substituted cyclic siloxanes on the surface of silica	Mineral seal oil + fume silica + dimethyl-dichlorosilane	Preferred 1–40 (but reference made to use of the particles by themselves, i.e., 100%)	0.005–0.15 micron (prefer non-aggregated silicas)	Emulsifiers and dispersants may be added	Wide variety of industrial applications quoted. A specific claim concerns paper pulp stock	1968	[241]

Paraffinic and napthenic mineral oils, cutting oils, kerosene, similar petroleum fractions, including food-grade mineral oils and halogenated hydrocarbons. Synthetic oils also claimed; aliphatic diesters, silicate esters, and polyalkylene glycol or their derivatives (detailed description of viscosity, etc., of oils specified)	Any type of silica possessing reactive surface hydroxyls rendered hydrophobic by reacting with alkoxysilicon chloride $SiCl_m(OR)_n$. Reaction made in the oil	Mineral seal oil with silica treated with various alkoxysilicon chlorides	7–45	<0.1 micron but prefer <0.05 micron	1. Need for spreading agent is eliminated—a special feature of this patent. Surface-active agent claimed to stabilize the foam. 2. Alkylene oxide may be added to react with HCl formed when alkoxy silicon chloride reacts with silica (in the oil)	Paper pulp mill application	1969	[222]
Halocarbon or hydrocarbon fluid	Synthetic alkali metal or alkaline earth metal silicoaluminate rendered hydrophobic by reaction with halosilane. Reaction made in the oil	Sodium or calcium silicoaluminates hydrophobed with methyl chlorosilanes in paraffinic mineral oil, chloronaphthalene or kerosene	5–30	<0.2 micron; is a particular provision of the patent	Lewis base may be added to neutralize haloacid formed in reaction of halosilane with silicoaluminate	Paper pulp mills are preferred application	1970	[242]
Water-insoluble organic liquid selected from vegetable oils, aliphatic, alicyclic, or halogenated aromatic hydrocarbons, long-chain alcohols, long-chain esters, and amines (must be organic liquid at ambient with boiling pt >65°C). Dependent claim states that liquid is hydrophobic	Aluminum oxide reacted in situ with alkali or alkaline earth hydroxide and fatty acid (with 6–24 carbon atoms)	Colloidal aluminum oxide prepared by hydrolysis of aluminum chloride in a flame + paraffinic hydrocarbon oil + $Ca(OH)_2$ + variety of fatty acids	Most preferred from 6 to 16	Alumina <15 microns; preferred 0.01–1.3 microns	1. Water may be added (forming W/O emulsion) 2. Surfactant may be added	Probably intended for paper pulp mills but no explicit reference to such an application	1972	[243]
Water-insoluble organic liquid selected from kerosene, naphthenic paraffinic, chlorinated naphthenic, or chlorinated paraffinic mineral oils and liquid difluorovinylchloride polymer (detailed descriptions of viscosity, volatility etc.) also specified	Exclusively claims the preparation of an alkalized microfine precipitated silica (which may be hydrophobed by any method)	Silica hydrophobed with dimethylpolysiloxane and dispersed in naphthenic, mineral, or mineral seal oil	3–30	<0.05 micron (ultimate particle size)	Catalyst to promote reaction of silicone with silica (e.g., tin octanoate) Addition of spreading agents or other surfactants usually not helpful except occasionally when silicone oil loading of silica low	Particularly paper pulp mill application	1973	[223]

Source: Garrett, P.R., Mode of action of antifoams, in *Defoaming, Theory and Industrial Applications* (Garrett, P.R., ed.), Marcel Dekker, New York, Chpt 1, p 1, 1993.

TABLE 4.A3
Some Early Examples of Antifoam Patents Concerning Mixtures of Intrinsically Hydrophobic Particles and Organic Liquids

Oils Claimed To Be Effective	Particles Claimed To Be Effective	Actual Examples Given	Preferred Conc. of Particles (wt.%)	Particle Size	Any Other Additives	Application	Year of Publn.	Ref.
Water-immiscible organic liquid boiling above 100°C, e.g., a hydrocarbon	Polyamide or polymethylene polyamine having 2–12 methylene groups and a carboxylic acid of the group consisting of aliphatic and cycloaliphatic monocarboxylic acids, each acyl group containing 11–18 carbon atoms	*N,N'*-distearyl ethylene diamide + white spirit	0.05–10	"Fine particles" prepared by heating till dissolved + rapid cooling or milling	–	Anionic detergent solutions (has no detrimental effect on wetting or detergent power of the solution or on its content of active material)	1955	[244]
Organic liquid is nonsolvent for particle and is immiscible with water. Nonpolar are saturated and unsaturated aliphatic hydrocarbons. Polar include alcohols, esters, ketones, chlorinated aromatic hydrocarbons, fluorinated hydrocarbons	Finely divided poly-α-olefin polymers such as polypropylene, polyisobutylene etc. Also claims thermoplastic polyesters such as nylon	1. Polypropylene + xylene 2. Polypropylene + commercial aromatic hydrocarbon mixture 3. Polypropylene + mineral oil + surfactant 4. Polyethylene tetraphthalate + mineral oil, etc.	2–25	0.02–50 microns but preferred size 0.2–5.0 microns	Surface-active agent may be added in order to provide "an improvement in the property of these dispersions to spread at the air–water surface"	Reference made to paper pulp mill application only	1972	[245]
Hydrophobic organic solvents, e.g., aliphatic, cycloaliphatic, hydroaromatic, aromatic hydrocarbons, Natural fatty oils (e.g., olive oil), silicone oils, phosphoric acid esters	Hydrates of fatty acid mixed salts of polyvalent metals and/or of lower dibasic amines	Mineral oil + 1. Aluminum magnesium stearate 2. Zinc-ethylene diamine stearate 3. Magnesium zinc stearate	2–20	–	Non-ionic emulsifier	General industrial application	1972	[246]

Mineral oil of specified viscosity	Aliphatic diamide derivative of polymethylene diamine	*N,N'*–distearylethylene diamine + paraffin oil	4–12	15–20 microns (by Hegman grind gauge)	1. Spreading agent 2. Silicone oil (dependent claims)	Paper pulp mill application	1973	[216]
Hydrocarbon oil	Amide, which is the reaction product of a polyamine containing at least one CH_2-CH_8 alkylene group and a CH_6-CH_{18} fatty acid	Stearic diamide of ethylene diamine + paraffin oil	1–20	Heat till particles dissolve followed by rapid cooling + homogenizing = "smaller particles"	1. Oil-soluble organic polymer 2. A fat 3. Silicone oil	General industrial application	1973	[247]
Mineral oil or esters of unsaturated fatty acids with mono- or polyhydric alcohols, liquid fatty acids or alcohols, terpene hydrocarbons	Mono- or diester of hydroxystearyl alcohol with saturated fatty acid or hydroxyl-fatty acid	Mineral oil + hydroxylstearyl monobehenate + various other additives	5–15 Particles may, however, show antifoam behavior alone, i.e., 100%	"Finely divided"	Silicone oils, non-ionic ethoxylated surfactants, metal stearates, etc.	Paper pulp mills, manufacture of dispersions of plastics, etc.	1975	[248]
Mineral oils, fatty oils, fatty acids, tetraisobutylene	Polyethylene having molecular weight from 500 to 25,000	As claimed	0.5–15	Of order 0.1 micron. Dissolve in oil at high temp. + cool rapidly to optimize	Emulsifier	General application for aqueous systems	1975	[249]
Liquid hydrocarbon or solid hydrocarbon which melts between 20°C and 120°C	Oil- and water-insoluble polyvalent metal salt of an alkyl phosphoric acid	Mineral oil + various polyvalent salts of alkyl phosphoric acids (of various alkyl chain lengths)	>5% Particles may show antifoam behavior alone, i.e., 100%	–	Incorporation in detergent formulation described	Laundry detergent compositions	1977	[250]

Source: Garrett, P.R., Mode of action of antifoams, in *Defoaming, Theory and Industrial Applications* (Garrett, P.R., ed.), Marcel Dekker, New York, Chpt 1, p 1, 1993.

REFERENCES

1. Denkov, N.D. *Langmuir*, 20(22), 9463, 2004.
2. Fiske, C.H. *J. Biol. Chem.*, 35, 411, 1918.
3. Sasaki, T. *Bull. Chem. Soc. Jpn.*, 11, 797, 1936.
4. Sasaki. T. *Bull. Chem. Soc. Jpn.*, 13, 517, 1938.
5. Shearer L.T., Akers, W.W. *J. Phys. Chem.*, 62, 1264 and 1269, 1958.
6. McBain, J.W., Ross, S., Brady, A.P., Robinson, J.V., Abrams, I.M., Thorburn, R.C., Lindquist, C.G. National Advisory Committee for Aeronautics A.R.R. No. 4105, 1944.
7. Arnaudov, L., Denkov, N.D., Surcheva, I., Durbut, P., Broze, G., Mehreteab, A. *Langmuir*, 17(22), 6999, 2001.
8. Patist, A., Axelberd, T., Shah, D. *J. Colloid Interface Sci.*, 208, 259, 1998.
9. Robinson, J.V., Woods, W.W. *J. Soc. Chem. Ind.*, 67, 361, 1948.
10. Ross, S. *J. Phys. Colloid Chem.*, 54, 429, 1950.
11. Ross, S., Haak, R.M. *J. Phys. Chem.*, 62, 1260, 1958.
12. Okazaki, S., Hayashi, K., Sasaki, T. *Proceedings of the IV International Congress on SAS, V3*, Brussels, 67, 1964.
13. Koretskaya, T.A., Kruglyakov, P.M. *Izv. Sib. Otd. An SSSR, Ser. Khim. Nauk.*, 7(3), 129, 1976.
14. Ross, S., Bramfitt, T.H. *J. Phys. Chem.*, 61, 126, 1957.
15. Koczo, K., Lobo, L., Wasan, D. *J. Colloid Interface Sci.*, 150(2), 492, 1992.
16. Binks, B.P., Fletcher, P.D.I., Haynes, M. *Colloids Surf. A*, 216, 1, 2003.
17. Aveyard, R., Binks, B.P., Clark, S., Fletcher, P.D.I. *JCS Faraday Trans.*, 86(18), 3111, 1990.
18. Aveyard, R., Binks, B.P., Fletcher, P.D.I. *Langmuir*, 5, 1210, 1989.
19. Rosen, M.J., Cohen, A.W., Dahanayake, M., Hua, X. *J. Phys. Chem.*, 86, 541, 1982.
20. Aveyard, R., Binks, B.P., Fletcher, P.D.I., MacNab, J.R. *Langmuir*, 11(7), 2515, 1995.
21. Lobo, L., Nikolov, N., Wasan, D. *J. Disper. Sci. Technol.*, 10(2), 143, 1989.
22. Garrett, P.R. Mode of action of antifoams, in *Defoaming, Theory and Industrial Applications* (Garrett, P.R., ed.), Marcel Dekker, New York, 1993, Chpt 1, p 1.
23. Bergeron, V., Radke, C. *Colloid Polym. Sci.*, 273, 165, 1995.
24. Aveyard, R., Binks, B.P., Fletcher, P.D.I., Peck, T.G., Rutherford, C.E. *Adv. Colloid Interface Sci.*, 48, 93, 1994.
25. Peck, T.G. The mechanisms of foam breakdown by oils and particles, PhD Thesis, University of Hull, 1994.
26. Farquhar, K.D. A study of the relationship between the kinetics of micelle/vesicle breakdown and foaming ability, PhD Thesis, University of East Anglia, 1994.
27. Ran, L. Foaming of anionic surfactant solutions in the presence of calcium ions and triglyceride-based antifoams, PhD Thesis, University of Manchester, 2011.
28. Isrealachvilli, J.N. *Intermolecular and Surface Forces with Applications to Colloidal and Biological Systems*, Academic Press, London, 1985.
29. Farquhar, K.D., Misran, M., Robinson, B.H., Steytler, D.C., Morini, P., Garrett, P.R., Holzwarth, J.F. *J. Phys. Condens. Matter*, 8, 9397, 1996.
30. Ran, L., Jones, S., Embley, B., Tong, M., Garrett, P.R., Cox, S., Grassia, P., Neethling, S. *Colloid Surf. A*, 382, 50, 2011.
31. Garrett, P.R., Gratton, P. *Colloids Surf. A*, 103, 127, 1995.
32. Nakayama, H., Shinoda, K. *Bull. Chem. Soc. Jpn.*, 40, 1797, 1967.
33. Moroi, Y., Matuura, R. *Bull. Chem. Soc. Jpn.*, 61, 333, 1988.
34. Rodriguez, C., Scamehorn, J. *J. Surfactants Deterg.*, 2(1), 17, 1999.
35. Vautier-Giongo, C., Blaes, B. *J. Phys. Chem. B*, 107, 5398, 2003.
36. Luangpirom, N., Dechabumphen, N., Saiwan, C., Scamehorn, J. *J. Surfactants Deterg.*, 4(4), 367, 2001.

37. Aronson, M. *Langmuir*, 2(5), 653, 1986.
38. Ewers, W.E., Sutherland, K.L. *Aust. J. Sci. Res.*, 5, 697, 1952.
39. Ross, S., Young, G.J. *Ind. Eng. Chem.*, 43(11), 2520, 1951.
40. Kruglyakov, M., Taube, P.R. *Zh. Prikl. Khim.*, 44(1), 129, 1971.
41. Kruglyakov, P.M. Equilibrium properties of free films and stability of foams and emulsions, in *Thin Liquid Films, Fundamentals and Applications* (Ivanov, I.B., ed.), Marcel Dekker, New York, 1988, p 767.
42. Garrett, P.R. *J. Colloid Interface Sci.*, 69(1), 107, 1979.
43. Garrett, P.R., Davis, J., Rendall, H.M. *Colloids Surf. A*, 85, 159, 1994.
44. Aveyard, R., Cooper, P., Fletcher, P.D.I., Rutherford, C.E. *Langmuir*, 9, 604, 1993.
45. Prins, A. Theory and practice of formation and stability of food foams, in *Food Emulsions and Foams* (Dickinson, E., ed.), Royal Society of Chemistry Special Publication 58, 1986, p 30.
46. Prins, A. Foam stability as affected by the presence of small spreading particles, in *Surfactants in Solution*, Vol 10, (Mittall, K.L., ed.), 1989, p 361.
47. Fay, J.A. In *Oil on the Sea* (Hoult, D.P., ed.), Plenum Press, New York, 1969, p 53.
48. Hoult, D.P. *Ann. Rev. Fluid Mech.*, 4, 341, 1972.
49. Joos, P., Pintens, J. *J. Colloid Interface Sci.*, 60, 507, 1977.
50. Huh, C., Inoue, M., Mason, S.G. *Can. J. Chem. Eng.*, 53, 367, 1975.
51. Bergeron, V., Langevin, D. *Phys. Rev. Lett.*, 76(17), 3152, 1996.
52. Lucassen, J. *Trans. Faraday Soc.*, 64(8), 2221, 1968.
53. Kulkarni, R.D., Goddard, E.D., Kanner, B. *Ind. Eng. Chem. Fundam.*, 16(4), 472, 1977.
54. Denkov, N.D., Cooper, P., Martin, J. *Langmuir*, 15(24), 8514, 1999.
55. Denkov, N.D. *Langmuir*, 15(24), 8530, 1999.
56. Denkov, N.D., Marinova, K.G., Christova, C., Hadjiiski, A., Cooper, P. *Langmuir*, 16(6), 2515, 2000.
57. Denkov, N.D., Marinova, K.G., Tcholakova, S., Deruelle, M. Mechanism of foam destruction by emulsions of PDMS-silica mixtures, in *Proceedings of 3rd World Congress on Emulsions*, 24–27 September 2002, Lyon, France, paper 1-D-199.
58. Basheva, E.S., Ganchev, D., Denkov, N.D., Kasuga, K., Satoh, N., Tsujii, K. *Langmuir*, 16(3), 1000, 2000.
59. Basheva, E.S., Stoyanov, S., Denkov, N.D., Kasuga, K., Satoh, N., Tsujii, K. *Langmuir*, 17(4), 969, 2001.
60. Marinova, K.G., Denkov, N.D. *Langmuir*, 17(8), 2426, 2001.
61. Hadjiiski, A., Tcholakova, S., Denkov, N.D., Durbut, P., Broze, G., Mehreteab, A. *Langmuir*, 17(22), 7011, 2001.
62. Denkov, N.D., Tcholakova, S., Marinova, K.G., Hadjiiski, A. *Langmuir*, 18(15), 5810, 2002.
63. Kitchener, J.A., Cooper, C.F. *Quart. Rev.*, 13, 71, 1959.
64. Lucassen, J. Dynamic properties of free liquid films and foams, in *Anionic Surfactants–Physical Chemistry of Surfactant Action* (Lucassen-Reijnders, E.H., ed.), Marcel Dekker, New York, 1981, p 217.
65. Burcik, E.J. *J. Colloid Sci.*, 5, 421, 1950.
66. Aveyard, R., Binks, B.P., Fletcher, P.D.I., Peck, T., Garrett, P.R. *JCS Faraday Trans.*, 89(24), 4313, 1993.
67. Ross, S., Nishioka, G. Experimental researches on silicone antifoams, in *Emulsions, Lattices and Dispersions* (Becker, P., Yudenfreud, M., eds.), Marcel Dekker, New York, 1978, p 237.
68. Roberts, K., Axberg, C., Osterlund, R. Emulsion foam killers in foams containing fatty and rosin acids, in *Foams* (Akers, R., ed.), Academic Press, London, 1976, from original conf. proceedings p39.
69. Mansfield, W. *Aust. J. Chem.*, 16(1), 76, 1963.

70. Bergeron, V., Cooper, P., Fisher, C., Giermask-Kahn, J., Langevin, D., Pouchelon, A. *Colloids Surf. A*, 122, 103, 1997.
71. Garrett, P.R., Davis, J., Rendall, H. Unilever Internal Reports, 1978–1979.
72. Princen, H.M. The equilibrium shape of interfaces, drops, and bubbles. Rigid and deformable particles at interfaces, in *Surface and Colloid Science* Vol. 2 (Matijevic, E., ed.), Wiley, New York, 1969, p 1.
73. Garrett, P.R. *J. Colloid Interface Sci.*, 76(2), 587, 1980.
74. Lobo, L., Wasan, D. *Langmuir*, 9, 1668, 1993.
75. Frye, G.C., Berg, J.C. *J. Colloid Interface Sci.*, 130(1), 54, 1989.
76. de Vries, A.J. *Rec. Trav. Chim.*, 77, 383, 1958.
77. Dippenaar, A. *Int. J. Miner. Process.*, 9, 1–22, 1982.
78. Denkov, N.D., Marinova, K.G. Antifoam effects of solid particles, oil drops and oil-solid compounds in aqueous foams, in *Colloidal Particles at Liquid Interfaces* (Binks, B.P., Horozov, T., eds.), Cambridge University Press, Cambridge, UK 2006, p 383.
79. Scheludko, A., Exerowa, D. *Commun. Dept. Chem (Bulg. Acad. Sci.)*, 7, 123, 1959.
80. Scheludko, A. *Adv. Colloid Interface Sci.*, 1, 391, 1967.
81. Brakke, K. *Exp. Math.* 1, 141, 1992, and www.susqu.edu/facstaff/b/brakke/evolver/html/default.htm, 1999.
82. Gibbs, J.W. *The Collected Works of J Willard Gibbs*, Longmans Green, New York, 1928, Vol 1, p 258.
83. Rowlinson, J.S., Widom, B. *Molecular Theory of Capillarity*, Oxford University Press, Oxford, 1982.
84. Pujado, P.R., Scriven, L. *J. Colloid Interface Sci.*, 40, 82, 1972.
85. Aveyard R., Clint J. *JCS Faraday Trans.*, 93(7), 1397, 1997.
86. Garrett, P.R., Wicks, S.P., Fowles, E. *Colloids Surf. A*, 282–283, 307, 2006.
87. Scheludko, A., Chakarov, V., Toshev, B. *J. Colloid Interface Sci.*, 82(1), 83, 1981.
88. Dussaud, A., Vignes-Adler, M. *Langmuir*, 13, 581, 1997.
89. Chen, P., Susnar, S., Amirfazli, A., Mak, C., Neumann, A. *Langmuir*, 13, 3035, 1997.
90. Aveyard, R., Clint, J.H., Nees, D., Paunov, V. *Colloids Surf. A*, 146(1–3), 95, 1999.
91. Takata, Y., Matsubara, H., Kikuchi, Y., Ikeda, N., Matsuda, T., Takiue, T., Aratono, M. *Langmuir*, 21, 8594, 2005.
92. Takata, Y., Matsubara, H., Matsuda, T., Kikuchi, Y., Takiue, T., Law, B., Aratono, M. *Colloid. Polym. Sci.*, 286, 647, 2008.
93. Harkins, W.D. *J. Chem. Phys.*, 5, 135, 1937.
94. Drelich, J. *Colloids Surf. A*, 116, 43, 1996.
95. Gaydos, J., Neumann, A. *J. Colloid Interface Sci.*, 120(1), 76, 1987.
96. Li, D., Neumann, A. *Colloids Surf. A*, 43, 195, 1990.
97. Duncan, D., Li, D., Gaydos, J., Neumann, A. *J. Colloid Interface Sci.*, 169, 256, 1995.
98. Koczo, K., Koczone, J., Wasan, D. *J. Colloid Interface Sci.*, 166, 225, 1994.
99. Miller, C.A., Neogi, P. *Interfacial Phenomena, Equilibrium and Dynamic Effects*, Marcel Dekker, Surfactant Science Series Vol 17, New York, 1985, p 165.
100. Nikolov, A.D., Wasan, D.T., Huang, D.D.W., Edwards, D.A. Paper SPE 15443, presented at 61st Annual Technical Conference and Exhibition of Soc Pet Engrs., New Orleans, 1986.
101. Wasan, D.T., Nikolov, A.D., Huang, D.D., Edwards, D.A. Foam stability: Effects of oil and film stratification, in *Surfactant Based Mobility Control—Progress in Miscible Flood Enhanced Oil Recovery* (Smith, D., ed.), ACS Symp series 373, 1988, Chpt 7, p 136.
102. Hadjiiski, A.D., Denkov, N.D., Tcholakova, S.S., Ivanov, I.B. Role of entry barriers in foam destruction by oil drops, in *Adsorption and Aggregation of Surfactants in Solution* (Mittal, K., Shah, D., eds.), Marcel Dekker, Surfactant Science Series Vol 109, New York, 2003, Chpt 23, p 465.
103. Princen, H.M. *Langmuir*, 4(1), 164, 1988.

104. Neethling, S., Morris, G., Garrett, P.R. *Langmuir*, 27(16), 9738, 2011.
105. Pattle, R. *J. Soc. Chem. Ind.*, 69, 363, 1950.
106. Ross, S. *Chem. Eng. Progr.*, 63(9), 41, 1967.
107. Kruglyakv, P.M., Koretskaya, T.A. *Kolloid Zh.*, 36(4), 682, 1974.
108. Abe, Y., Matsumura, S. *Tenside Deterg.*, 20(5), 218, 1983.
109. Kellay, H., Meunier, J., Binks, B.P. *Phys. Rev. Lett.*, 69(8), 1220, 1992.
110. Kellay, H., Binks, B.P., Hendrikx, Y., Lee, L., Meunier, J. *Adv. Colloid Interface Sci.*, 49, 85, 1994.
111. Varaprath, S., Frye, C.L., Hamelink, J. *Environ. Toxicol. Chem.*, 15(8), 1263, 1996.
112. Ross, S., Nishioka, G. *Colloid Polym. Sci.*, 255, 560, 1977.
113. Callaghan, I.C., Hickman, S., Lawrence, F., Melton, P. Antifoams in gas-oil separation, in *Industrial Applications of Surfactants* (Karsa, D., ed.), Royal Society of Chemistry Special Publication, Cambridge, UK 59, 1987, p 48.
114. Callaghan, I.C., Gould, C.M., Hamilton, R.J., Neustadter, E.L. *Colloids Surf.*, 8, 17, 1983.
115. McKendrick, C.B., Smith, S.J., Stevenson, P.A. *Colloids Surf.*, 52, 47, 1991.
116. Wu, F., Cai, C., Yi, W., Cao, Z., Wang, Y. *J. Appl. Polym. Sci.*, 109(3), 1950, 2008.
117. Lang, J., Morgan, R. *J. Chem. Phys.*, 73(11), 5849, 1980.
118. Guggenheim, E.A. *Thermodynamics*, Fifth Edition, North Holland, Amsterdam, the Netherlands 1967, p 195.
119. Dill, K.A., Bromberg, S. *Molecular Driving Forces*, Second Edition, 2011, Chpt 25, Taylor and Francis, New York. p 489.
120. Strey, R. *Ber. Bunsenges Phys. Chem.*, 100(3), 182, 1996.
121. Fineman, M., Brown, G., Myers, R. *J. Phys. Chem.*, 56, 963, 1952.
122. Koretsky, A.P., Smirnova, A.V., Koretskaya, T.A., Kruglyakov, P.M. *Zh. Prikl. Khim.*, 50, 84, 1977.
123. Koretskaya, A.P. *Kolloid. Zh.*, 39(3), 571, 1977.
124. Bonfillon-Colin, A., Langevin, D. *Langmuir*, 13(4), 599, 1997.
125. Chaisalee, R., Soontravanich, S., Yanumet, N., Scamehorn, J. *J. Surfactants Deterg.*, 6(4), 345, 2003.
126. Colin, A., Giermanska-Kahn, J., Langevin, D. *Langmuir*, 13(11), 2953, 1997.
127. Tan, S., Fornasiero, D., Sedev, R., Ralston, J. *Colloids Surf. A*, 250, 307, 2004.
128. Blease, T.G., Evans, J.G., Hughes, L. Surfactant antifoams, in *Defoaming, Theory and Industrial Applications* (Garrett, P.R., ed.), Marcel Dekker, New York, 1993, Chpt 8, p 299.
129. Mukherjee, P., Padhan, S., Dash, S., Patel, S., Mishra, B. *Adv. Colloid Interface Sci.*, 162, 59, 2011.
130. Jonsson, B., Lindman, B., Holmberg, K., Kronberg, B. *Surfactants and Polymers in Aqueous Solution*, Wiley, New York, 1998.
131. Nemeth, Z., Racz, G., Koczo, K. *J. Colloid Interface Sci.*, 207, 386, 1998.
132. Nemeth, Z., Racz, G., Koczo, K. *Colloids Surf. A*, 127, 151, 1997.
133. Joshi, K., Jeelani, A., Blickenstorfer, C., Naegeli, I., Oliviero, C., Windhab, E. *Langmuir*, 22, 6893, 2006.
134. Ross, S., Nishioka, G. Foaming behaviour of partially miscible liquids as related to their phase diagrams, in *Foams; Proceedings of a Symposium Organized by the Soc. Chem. Ind., Colloid and Surface Chem. Group, Brunel Univ., 1975* (Akers, R.J., ed.), Academic Press, London, 1976, p 17.
135. Ross, S., Nishioka, G. *J. Phys. Chem.*, 79(15), 1561, 1975.
136. Ross, S., Patterson, R. *J. Phys. Chem.*, 83(17), 2226, 1979.
137. Ross, S., Nishioka, G. *Chem. Ind.*, *Jan*, 47, 1981.
138. Andrew, S.P.S. International Symposium on Distillation 1960, Inst. Chem. Engrs., p 73.
139. Hildebrand, J., Prausnitz, J., Scott, R. *Regular and Related Solutions*, Van Nostrand Reinhold, New York, 1970.

140. Ramsden, W. *Proc. Royal Soc.*, 71, 156, 1903.
141. Pickering, S.U. *J. Chem. Soc.*, 91, 2001, 1907.
142. Bartsch, O. *Kolloidchem. Beihefte*, 20, 1, 1924.
143. Lovell, V.M. In *Flotation; A. M. Gaudin Memorial Volume* (Feurstenau, M.C., ed.), American Institute of Mining, Metallurgical and Petroleum Engineers Inc., 1976, Vol 1, p 597.
144. Tate, J.R., McRitchie, A.C. (assigned to Procter and Gamble), GB 1,492,938, November 23, 1977, filed January 1, 1974.
145. Mokrushin, S.G. *Kolloidn. Zh.*, 12, 448, 1950.
146. Dombrowski, N., Fraser, R.P. *Phil. Trans. Royal Soc. London, Ser. A*, 247, 13, 1954.
147. Livshitz, A.K., Dudenkov, S.V. *Tsvet. Metally*, 30(1), 14, 1954.
148. Schulman, J.H., Leja, J. *Trans. Faraday Soc.*, 50, 598, 1954.
149. Livshitz, A.K., Dudenkov, S.V. *Proceedings of 7th IMPC New York*, 1965, p 367.
150. Dudenkov, S.V. *Tsvet. Metally*, 40, 18, 1967.
151. Irani, R.R., Callis, C.F. *J. Phys. Chem.*, 64, 1741, 1960.
152. Frye, G.C.C., Berg, J.C. *J. Colloid Interface Sci.*, 127(1), 222, 1989.
153. Livschitz, A.K., Dudenkov, S.V. *Tsvet. Metally*, 33, 24, 1960.
154. Tang, F., Xiao, A., Tang, J., Jiang, L. *J. Colloid Interface Sci.*, 131(2), 498, 1989.
155. Murray, B., Ettelaie, R. *Curr. Opin. Colloid Interface Sci.*, 9, 314, 2004.
156. Hunter, T.N., Pugh, R., Franks, G., Jameson, G. *Adv. Colloid Interface Sci.*, 137, 57, 2008.
157. Binks, B.P. *Curr. Opin. Colloid Interface Sci.*, 7, 21, 2002.
158. Binks, B.P., Horozov, T. *Angew. Chem. Int. Ed.*, 44, 3722, 2005.
159. Binks, B.P., Kirkland, M., Ridrigues, J. *Soft Matter*, 4, 2373, 2008.
160. Dickinson, E., *Curr. Opin. Colloid Interface Sci.*, 15, 40, 2010.
161. Abkarian, M., Subramaniam, A., Kim, S., Larsen, R., Yang, S., Stone, H. *Phys. Rev. Lett.*, 99, 188301, 2007.
162. Aveyard, R., Binks, B.P., Fletcher, P.D.I., Rutherford, C.E. *J. Disper. Sci. Technol.*, 15(3), 251, 1994.
163. Johnson, R.E., Dettre, R.H. Wettability and contact angles, in *Surface and Colloid Science Vol. 2* (Matijevic, E., ed.), Wiley, New York, 1969, p 85.
164. Johansson, G., Pugh, R. *Int. J. Miner. Process.*, 34, 1, 1992.
165. Oliver, J. F., Huh, C., Mason, S.G. *J. Colloid Interface Sci.*, 59(3), 568, 1977.
166. Xu, H., Goedel, W. *Langmuir*, 19(12), 4950, 2003.
167. Hopkins, A.J., Woodcock, L.V. *JCS Faraday Trans.*, 86(12), 2121, 1990.
168. Morris, G., Neethling, S., Cilliers, J. *Miner. Eng.*, 23, 979, 2010.
169. Morris, G., Neethling, S., Cilliers, J. *J. Colloid Interface Sci.*, 354, 380, 2011.
170. Zhou, L., Kambhamettu, C. *Graph Models*, 63(1), 1, 2001.
171. Vacher, R., Woignier, T., Pelous, J., Courtens, E. *Phys. Rev. B.*, 37(11), 6500, 1988.
172. Wenzel, R.N. *Ind. Eng. Chem.*, 28, 988, 1936.
173. Wenzel, R.N. *J. Phys. Chem.*, 53, 1466, 1949.
174. Shuttleworth, R., Bailey, G.L.J. *Disc. Faraday Soc.*, 3, 16, 1948.
175. Good, R.J. *J. Am. Chem. Soc.*, 74, 5041, 1952.
176. Johnson, R.E., Dettre, R.H. *Adv. Chem.*, 43, 112, 1964.
177. Tamura, T., Kageyama, M., Kaneko, Y., Kishino, T., Nikaido, M. *J. Colloid Interface Sci.*, 213, 179, 1999.
178. Tamura, T., Kageyama, M., Nikaido, M. Preparation of novel silicone-based antifoams having a high defoaming performance, in *Emulsions Foam and Thin Films* (Mittal, K. Kumar, P., eds.), Marcel Dekker, New York, 2000, p 161.
179. Tamura, T., Kaneko, Y. Foam film stability in aqueous systems, in *Surface and Interfacial Tension, Measurement, Theory and Applications* (Hartland, S., ed.), Marcel Dekker, New York, Surfactant Science Series Vol 119, 2004, p 91.
180. Vrij, A., Overbeek, J.Th.G. *J. Am. Chem. Soc.*, 90, 3074, 1968.
181. Vrij, A. *Disc. Faraday Soc.*, 42, 23, 1966.

182. Nguyen, A.V., Schultz, H.J. *Colloidal Science of Flotation*, Marcel Dekker, Surfactant Science Series Vol 118, New York 2004, pp 415 and 445.
183. Blake, T.D., Kitchener, J.A. *JCS Faraday Trans. I*, 68, 1435, 1972.
184. Hough, D., White, L. *Adv. Colloid Interface Sci.*, 14, 3, 1980.
185. Anfruns, J.F., Kitchener, J.A. *Trans. Inst. Min. Metall., London C*, 86, 9, 1977.
186. Ducker, W.A., Pashley, R.M., Ninham, B. *J. Colloid Interface Sci.*, 128(1), 66, 1989.
187. Strnad, J., Kohier, H., Heckmann, K., Pitsch, M. *J. Colloid Interface Sci.*, 132(1), 283, 1989.
188. Zhang H., Miller, C.A., Garrett, P.R., Raney, K. *J. Colloid Interface Sci.*, 263, 633, 2003.
189. Mysels, K.J., Shinoda, K., Frankel, S. *Soap Films-Studies of Their Thinning*, Pergamon Press, London, 1959.
190. Joye, J., Hirasaki, G., Miller, C.A. *Langmuir*, 10, 3174, 1994.
191. Joye, J., Hirasaki, G., Miller, C.A. *J. Colloid Interface Sci.*, 177, 542, 1996.
192. Prins, A., van Voorst Vader, F. *Proceedings of VI Int. Congress on Surfactants, Zurich,* 11(15), September 1972, p 441.
193. Reynolds, O. *Phil. Trans. Royal Soc. London, Ser. A*, 177, 157, 1886.
194. Exerowa, D., Kruglyakov, P. *Foam and Foam Films*, Elsevier, Amsterdam, 1998, p 108.
195. Peper, H. *J. Colloid Sci.*, 13, 199, 1958.
196. Balasuwatthi, P., Dechabumphen, N., Saiwan, C., Scamehorn, J. *J. Surfactants Deterg.*, 7(1), 31, 2004.
197. Luepakdeesakoon, B., Saiwan, C., Scamehorn, J. *J Surfactants Deterg.*, 9(2), 125, 2006.
198. Zhang, H., Miller, C.A., Garrett, P.R., Raney, K. *J. Colloid Interface Sci.*, 279, 539, 2004.
199. Raghavachari, R., Narayan, K.S., Nayyar, N., Srisankar, S. in Abstracts of Eufoam conference (Garrett, P.R., ed.), 2002.
200. Davis, J., Garrett, P.R. unpublished work.
201. Veber, V., Pauchek, M. *Acta Fac. Pharm. Univ. Comenianae*, 26, 221, 1974.
202. Joshi, K., Jeelani, S., Blickenstorfer, C., Naegeli, L., Windhab, E. *Colloids Surf. A*, 263, 239, 2005.
203. Sirota, E.B., King, H., Shao, H., Singer, D. *J. Phys. Chem.*, 99, 798, 1995.
204. Boode, K., Walstra, P. *Colloids Surf. A*, 81, 139, 1993.
205. Hindle S., Povey, M., Smith K. *J. Colloid Interface Sci.*, 232, 370, 2000.
206. Golemanov, K., Tcholakova, S., Denkov, N.D., Gurkov, T. *Langmuir*, 22, 3560, 2006.
207. Kulkarni, R. D., Goddard, E.D. *Croatica Chem. Acta*, 50(1–4), 163, 1977.
208. Birtley, R., Burton, J., Kellett, D., Oswald, B., Pennington, J. *J. Pharm. Pharmacol.*, *25*, 859, 1973.
209. Povich, M.J. *Am. Inst. Chem. Eng. J.*, 21(5), 1016, 1975.
210. Ross, S., Nishioka, G. *J. Colloid Interface Sci.*, 65(2), 216, 1978.
211. Carter, M.N.A., Garrett, P.R. (assigned to Unilever Ltd.), GB 1,571,502; July 16, 1980, filed January 23, 1976.
212. Schweigl, O.F., Best, G.P. (assigned to Unilever Ltd.), GB 1,099,502; January 17, 1968, filed July 8, 1965.
213. Garrett, P.R. (assigned to Unilever Ltd.), EP 75,433; March 30, 1983, filed September 16, 1981.
214. Garrett, P.R., Moore, P. *J. Colloid Interface Sci.*, 159, 214, 1993.
215. Wang, G., Pelton, R., Hrymak, A., Shawafaty, N., Hen, Y. *Langmuir*, 15, 2202, 1999.
216. Shane, H.J., Schill J.E., Lilley, J.W. (assigned to Hart Chemical Ltd.), Canada 922,456; March 13, 1973, filed March 31, 1971.
217. Nitzsche, S., Firson, E. (assigned to Wacker-Chemie GmbH), US 3,235,509; February 15, 1966, filed October 3, 1962.
218. O'Hara, M.J., Rink, D.R. (assigned to Union Carbide Corp.), GB 1,247,690; September 29, 1971, filed August 11, 1987.
219. Farminer, K.W., Brooke, C.M. (assigned to Dow Corning Ltd.), US 3,843,558; October 22, 1974, filed June 16, 1972.

220. (Assigned to Degussa), Fr 1,533,825; July 19, 1968, filed August 8, 1967.
221. Boylan, F.J. (assigned to Hercules Powder Co.), US 3,076,768; February 5, 1963, filed April 5, 1960.
222. Buckman, H. (assigned to Buckman Laboratories Ltd.), GB 1,166,877; October 15, 1969, filed December 8, 1967.
223. Miller, J.R., Pierce, R.H., Linton, R.W., Wills, J.H. (assigned to Philadelphia Quartz Co.), US 3,714,068; January 30, 1973, filed December 28, 1970.
224. Boylan, F.J. (assigned to Hercules Inc.), US 3,408,306; October 29, 1968, filed July 7, 1961.
225. Jha, B., Christiano, P., Shah, D. *Langmuir*, 16(26), 9947, 2000.
226. Sinka, J., Lichtman, I. *Int. Dyer Textile Printer, May*, 489, 1976.
227. Kulkarni, R.D., Goddard, E.D., Kanner, B. *J. Colloid Interface Sci.*, 59(3), 468, 1977.
228. van Boekel, M.A.J.S., Walstra, P. *Colloids Surf.*, 3, 109, 1981.
229. Mizrahi, J., Barnea, E., *Br. Chem. Eng.*, 15(4), 497, 1970.
230. Marinova, K., Tcholakova, S., Denkov, N.D., Roussev, S., Deruelle, M. *Langmuir*, 19(7), 3084, 2003.
231. Marinova, K., Denkov, N.D., Tcholakova, S., Deruelle, M. *Langmuir*, 18(23), 8761, 2002.
232. Marinova, K., Denkov, N.D., Branlard, P., Giraud, Y., Deruelle, M. *Langmuir*, 18(9), 3399, 2002.
233. Aveyard, R., Clint, J.H. *JCS Faraday Trans.*, 91(17), 2681, 1995.
234. Kossen, N., Heertjes, P. *Chem. Eng. Sci.*, 20, 593, 1965.
235. Curtis, R., Garrett, P.R., Nicholls, M., Yorke, J. Preliminary observations concerning the use of perfuoroalkyl alkanes in hydrocarbon-based antifoams, in *Emulsions Foams and Thin Films* (Mittal, K., Kumar, P., eds.), Marcel Dekker, New York, 2000, p 177.
236. Currie, C.C. (assigned to Dow Chemical Co.), US 2,632,736, March 24, 1953, filed August 22, 1946.
237. Solomon, M.M. (assigned to General Electric Co.), US 2,829,112; April 1, 1958, filed September 22, 1955.
238. Bartolotta, G., de Oude, N.T., Gunkel, A.A. (assigned to Procter and Gamble Co.), GB 1,407,997; October 1, 1975, filed August 1, 1972.
239. Leibling, R., Canaris, N.M. (assigned to Nopco Chemical Co.), US 3,207,698; September 21, 1965, filed February 13, 1963.
240. Sullivan, R.E. (assigned to Dow Corning Corp.), US 3,304,266; February 14, 1967, filed May 6, 1963.
241. Domba, E. (assigned to Nalco Chemical Co.), US 3,388,073; June 11, 1968, filed December 16, 1963.
242. Harrison, G.C., Stumpo, A.J. (assigned to Pennsalt Chemicals Corp.), GB 1,195,589; June 17, 1970, filed August 10, 1967.
243. Lieberman, H., Duharte-Francia, C.A., Henderson, J. W. (assigned to Betz Laboratories Inc.), GB 1,267,479; March 23, 1972, filed May 26, 1969.
244. Caviet, M. (assigned to Shell Development Co.), Canada 508,856; April 4, 1955, filed November 28, 1949.
245. Boylan, F.J. (assigned to Hercules Inc.), US 3,705,859; December 12, 1972, filed December 30, 1970.
246. Boehinke, G., Quaedrlieg, M., Kolla, G. (assigned to Farbenfabriken Bayer Aktiengesellschaft), GB 1,267,482; March 22, 1972, filed July 14, 1970.
247. Lichtman, I.A., Rosengart, A.M. (assigned to Diamond Shamrock Corp.), Canada 927,707; June 5, 1973, filed September 3, 1971.
248. (assigned to Henkel and Cie GmbH), GB 1,386,042; March 5, 1975, filed January 25, 1973.
249. Ernst, F.M. (assigned to Mobil Oil Corp.), US 3,909,445; September 30, 1975, filed September 21, 1972.
250. Carter, M.N.A., Garrett, P.R. (assigned to Unilever Ltd.), Belgium 850,458, 1977.

5 Effect of Antifoam Concentration on Volumes of Foam Generated by Air Entrainment in Aqueous Solutions

5.1 PHENOMENOLOGY

5.1.1 INTRODUCTION

For a given antifoam–surfactant solution combination, the main practical concern is simply to ensure that sufficient antifoam is present to produce the desired diminution in foam volume given the nature and intensity of the relevant aeration processes involved. If this is to be achieved with a minimal level of empiricism, some knowledge of the relation between antifoam concentration and foam volume under given conditions of foam generation is required. Despite the evident importance of this fundamental aspect of antifoam behavior, there have been relatively few attempts to understand it [1]. This of course derives from the complexity that we have shown characterizes the overall process of antifoam action on foam volume. We may, however, derive useful generalizations if we make some simple, but realistic, assumptions.

Let us confine our generalizations to the effect of antifoam on the generation of foam (i.e., to measurements of foamability) without making reference to the stability of the resulting foam. Control of foamability is after all often the primary function of antifoams. Inconvenient consequences accompanying the formation of large quantities of unwanted foam are not always effectively ameliorated by a gradual collapse of that foam over a period of several minutes (or even seconds)! We are therefore particularly concerned here with foam generation by air entrainment where foam is produced during shaking, rotation, stirring, plunging, or by impinging jets. It is arguable that many, if not most, foam control problems are caused by such air entrainment.

Antifoams are often conveniently dispersed into foaming systems in the form of emulsion drops or individual particles. Here we assume that the concentration, state of dispersion and the effectiveness of each antifoam entity in such antifoam dispersions remain constant during foam generation. These assumptions therefore ignore the process of antifoam deactivation, which often occurs after prolonged aeration [2–8]. We therefore confine experimental measurements to short time intervals where such deactivation is reasonably assumed to be negligible. The phenomenon of antifoam deactivation is addressed in detail in Chapter 6 where some of the arguments given here are extended to include it.

Let us make the additional assumption that the effectiveness of any individual antifoam entity in this context is unaffected by the concentration of such entities in a dispersion. Each foam film collapse event is therefore supposed caused by a single antifoam entity. We therefore assume the absence of, for example, any "avalanche" effect that is seen during some foam stability measurements where the shock of foam film rupture by a single antifoam entity is supposedly propagated through the foam to induce many other foam film collapses by other antifoam entities [9, 10]. Such effects have been associated with the buildup of high capillary pressures in Plateau borders containing antifoam drops as the foam drains [11]. They are, therefore, unlikely to occur during the generation of foam.

It is known that different antifoam dispersions, based, for example, upon different chemistries, can differ significantly in effectiveness whilst both still remain effective. A given degree of foam control, as measured by the amount of foam generated in the presence of antifoam, can usually be achieved in the case of two different antifoam dispersions by simply increasing the relative concentration of the least effective [1]. This suggests a stochastic process for the observation because otherwise we would observe one dispersion as effective and the other as totally ineffective. We will show below, in Section 5.2, that it is in fact possible to derive a statistical theory of antifoam action, which is consistent with the simple phenomenological approach described here.

We first deduce the effect of antifoam concentration on generated foam volumes, which should result if we make the simple assumption that the relative effectiveness of different antifoams is independent of both antifoam concentration and foam volume for a given method of foam generation and surfactant solution. Comparison with the experiment is made.

5.1.2 Dispersion of Single Antifoam

If, in the case of a given method of foam generation and surfactant solution, the relative effectiveness of an antifoam dispersion can be adjusted by simply increasing the relative concentration, then we can write

$$\tilde{c}_1^0 = \tilde{n}\tilde{c}_2^0 \tag{5.1}$$

where $\tilde{c}_1^0$ and $\tilde{c}_2^0$ are the concentrations of two different antifoam dispersions that are adjusted to give the same foam volume, V_{AF}^0. The dispersions differ in effectiveness so that this can be achieved if the concentration of the least effective antifoam 1 is higher by a factor $\tilde{n}$, where $\tilde{n} > 1$. Let us assume that if the mode of action of these antifoams does not involve cooperative events between different entities in a given dispersion, then $\tilde{n}$ should be constant, independent of the foam volume and antifoam concentrations. Therefore, it should also be independent of the ratio F (= volume foam in presence of antifoam/volume of foam in the absence of antifoam).

Equation 5.1 means that in any plot of foam volume or F against antifoam concentration, the gradient at a given value of F (= F^0), or foam volume V^0_{AF}, should be greater by a factor $\tilde{n}$ for the most effective antifoam. This is illustrated schematically in Figure 5.1 from which it is easy to see that

$$\left(\frac{dF}{d\tilde{c}_2}\right)^0 \Bigg/ \left(\frac{dF}{d\tilde{c}_1}\right)^0 = \left(\frac{-\Delta F}{\Delta\tilde{c}_2^0}\right) \Bigg/ \left(\frac{-\Delta F}{\Delta\tilde{c}_1^0}\right) = \tilde{n} \text{ (at } F = F^0) \tag{5.2}$$

because $\Delta\tilde{c}_1^0 = \tilde{n}\Delta\tilde{c}_2^0$. Such plots should not, however, cross despite differences in gradient. Rather they should converge on $F = 1$ (or the foam volume in the absence of antifoam) where both $\tilde{c}_1^0 = 0$ and $\tilde{c}_2^0 = 0$. However, if we now consider a plot of F (or foam volume) against the logarithm of the antifoam dispersion concentration, then we must have

$$\left(\frac{dF}{d\ln\tilde{c}_2}\right)^0 \Bigg/ \left(\frac{dF}{d\ln\tilde{c}_1}\right)^0 = \left(\frac{-\tilde{c}_2^0\Delta F}{\Delta\tilde{c}_2^0}\right) \Bigg/ \left(\frac{-\tilde{c}_1^0\Delta F}{\Delta\tilde{c}_1^0}\right) = 1 \text{ (at } F = F^0) \tag{5.3}$$

This means that plots of F or foam volume against log(antifoam concentration) for antifoams of different effectiveness should all have the same gradient at the same value of F (= F^0) or foam volume provided $\tilde{n}$ is constant. The plots should therefore be isomorphous.

Surprisingly, there are very few reported experimental results in the literature concerning measurements of the effect of the concentration of different antifoams on the foamability of a given surfactant solution using a given, precisely defined, foam generation method. Perhaps the only examples are those presented by Garrett [1], Ran [12], and Cubero et al. [13]. In making such measurements, it should be

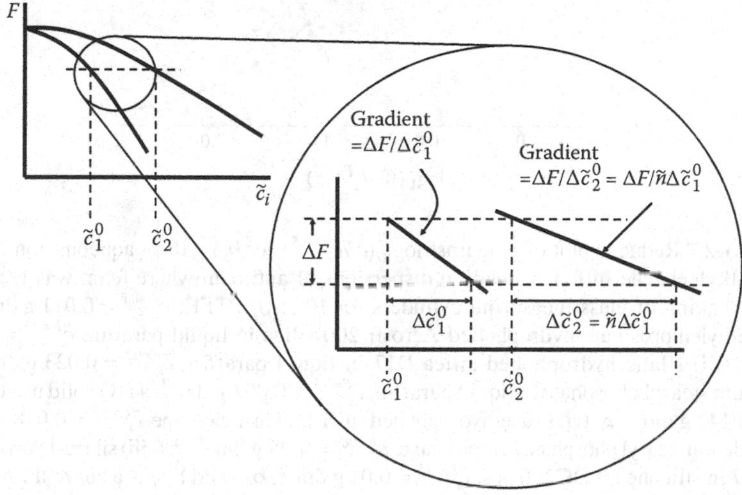

FIGURE 5.1 Derivation of Equation 5.2.

emphasized that the assumption that the state of dispersion of the antifoam be constant has implication for the preparation of antifoam dispersions of varying concentration. This means, for example, that the antifoam should preferably be dispersed using a high shear mixer so that the subsequent effect of aeration on the state of dispersion is minimal. It is also desirable that the antifoam dispersion be prepared at the highest concentration to be considered so that dispersions of lower concentration can be prepared by simply diluting with the relevant surfactant solution with minimal agitation to ensure that the state of dispersion is independent of concentration. Finally, observations of foam generation should concern short periods to minimize any deactivation of antifoam effectiveness.

Garrett [1] has presented measurements of F against log(antifoam concentration) for a variety of solid hydrophobic particulate antifoams and mixed hydrophobic particle–hydrocarbon oil antifoams. In this study, methods included both cylinder shaking and a static Ross–Miles apparatus [14]. The surfactant was a commercial sodium alkylbenzene sulfonate. Despite the chemical dissimilarity of the antifoams and large differences in their effectiveness, plots of F against the logarithm of the antifoam concentration were in fact found to be essentially isomorphous. A plot of F against the logarithm of the reduced antifoam concentration, $\tilde{c}_i/\tilde{c}_i^{F=0.5}$, is given in Figure 5.2 by way of example. Here $\tilde{c}_i$ is the concentration of antifoam i and $\tilde{c}_i^{F=0.5}$ is

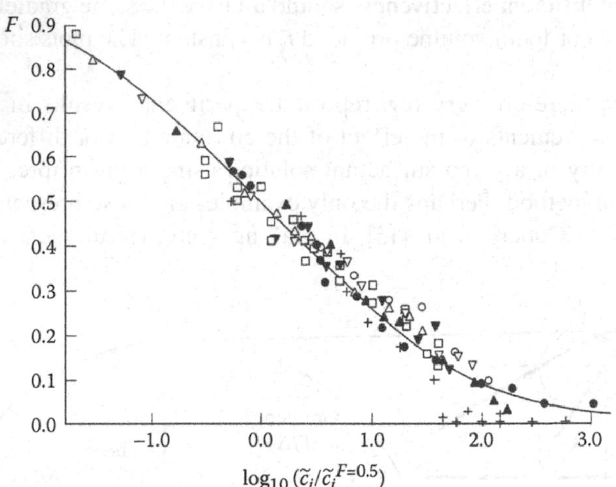

FIGURE 5.2 Reduced plot of F against $\log_{10}(\tilde{c}_i/\tilde{c}_i^{F=0.5})$ for 0.5 g dm^{-3} aqueous commercial sodium alkylbenzene sulfonate solution dispersions of antifoam where foam was generated by hand shaking of glass measuring cylinders for 10 s: (i) PTFE, $\tilde{c}_i^{F=0.5} = 0.011$ g dm^{-3}, ●; (ii) trimethylchlorosilane-hydrophobed Aerosil 200 silica in liquid paraffin, $\tilde{c}_i^{F=0.5} = 0.10$ g dm^{-3}, □; (iii) silane hydrophobed silica D17 in liquid paraffin, $\tilde{c}_i^{F=0.5} = 0.023$ g dm^{-3}, ▲; (iv) calcium stearyl phosphate in liquid paraffin, $\tilde{c}_i^{F=0.5} = 0.007$ g dm^{-3}, +; (v) solid n-eicosane, $\tilde{c}_i^{F=0.5} = 0.141$ g dm^{-3}, △; (vi) silane-hydrophobed silica D17 in n-eicosane, $\tilde{c}_i^{F=0.5} = 0.078$ g dm^{-3}, ▼; (vii) calcium stearyl phosphate in n-eicosane, $\tilde{c}_i^{F=0.5} = 0.05$ g dm^{-3}, ▽; (viii) silane-hydrophobed silica D17 in silicone oil DC200 5cs, $\tilde{c}_i^{F=0.5} = 0.01$ g dm^{-3}, o. Solid line is a curve fit. All mixtures were 90% by weight oil. (Reprinted with permission from Garrett, P.R., *Langmuir*, 11, 3576. Copyright 1995 American Chemical Society.)

the concentration of that antifoam required to give $F = 0.5$. Foamabilities were measured using cylinder shaking. The antifoams exhibited a wide range of effectiveness with $\tilde{c}_i^{F=0.5}$ varying by more than an order of magnitude [1]. That all the points lie on the same curve is clear in the figure where it should be noted that the precision of measurement of F by cylinder shaking is no better than ±0.04. The specific form of the curve is of course irrelevant for the present argument. It is in fact just a fit with a single adjustable parameter [1]. However, results for the system calcium stearyl phosphate–liquid paraffin at high concentration appear to lie below that curve for values of $F < 0.1$. This apparent anomaly could concern the marked effect of this antifoam on foam stability at high antifoam concentrations. Rapid foam collapse after foam generation has ceased could therefore preclude accurate measurement of foamability.

Isomorphous plots of F against log(antifoam concentration) have also been found for foam generated by mechanically rotating cylinders (see Section 2.2.4) containing an antifoam dispersion in a solution of a commercial sodium alkylbenzene sulfonate at different pH and calcium concentrations [12]. Here a comparison is made between two antifoams one of which is artificial sebum, consisting of a complex mixture of fatty acids, triglycerides, hydrocarbons, and cholesterol. The other is a simple mixture of stearic acid and triolein. The relevant plots are shown in Figure 5.3 where we distinguish between surfactant solutions at pH 3 without calcium and surfactant solutions at pH 10.5 with calcium. These different conditions would be expected to not only influence antifoam action as a result of formation of calcium soaps but also the intrinsic surface chemistry and therefore foam behavior of the surfactant solutions. Therefore, direct comparison between antifoam effects under these different conditions would violate the assumption that $\tilde{n}$ in Equation 5.1 is a constant for a given method of foam generation and surfactant solution. We find in fact that plots of F against log(antifoam concentration) are linear with essentially the same gradients under given conditions of pH and calcium content despite the marked difference in effectiveness of artificial sebum and mixtures of stearic acid and triolein. However, the gradients appear to be systematically different if comparison is made for a given antifoam at different conditions of pH and calcium content.

Cubero et al. [13] present similar reduced plots to that shown in Figure 5.2 for various unspecified antifoams of different effectiveness dispersed in sugar beet juices. Foamability was measured using a modified Bikerman method [15], which resembles sparging except that bubbles are formed at a porous plate using a vacuum. Results are summarized in Figure 5.4. Again, despite differences in antifoam effectiveness, all plots of F against logarithm of antifoam concentration appear to lie on the same curve.

5.1.3 Mixed Dispersions of Two Antifoams

We may extend this argument to mixtures of antifoam dispersions. Here we suppose, for example, that two different antifoams are separately dispersed in the same surfactant solution and the resultant dispersions are mixed in known ratios. Again we assume that the antifoam effects are caused by single entities and that therefore the effects are not cooperative. The argument is illustrated in terms of the ratio F but could of course be equally well illustrated using the relevant foam volumes since

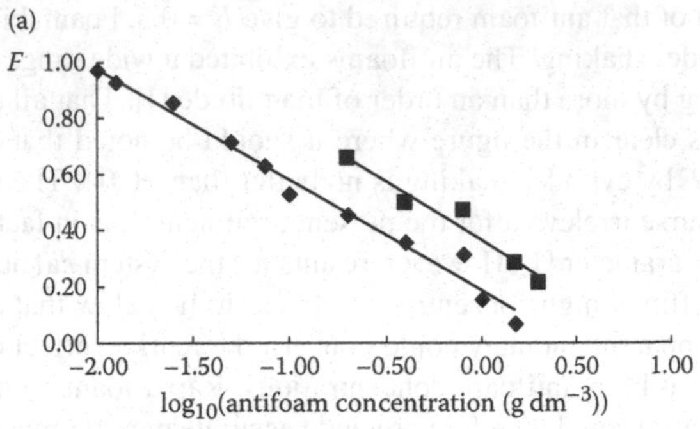

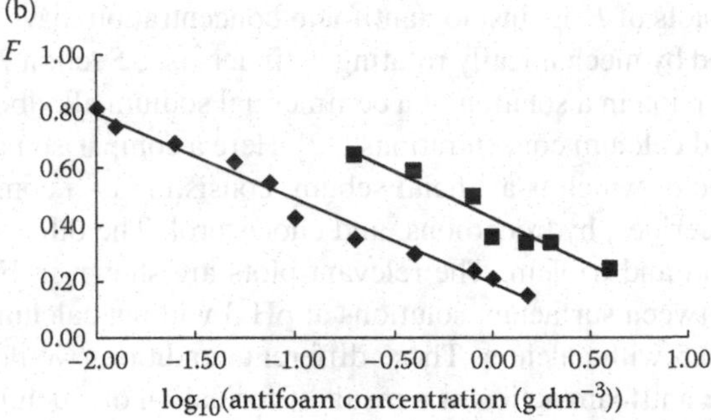

FIGURE 5.3 Plots of F against log(antifoam concentration) for 2×10^{-3} M aqueous commercial sodium alkylbenzene sulfonate solution dispersions where foam was generated by mechanical rotation of cylinders at 0.5 Hz for 15 s and antifoam was dispersed using high-speed emulsifier. Lines represent linear least squares fits. (a) pH 3, no calcium. Antifoams: (i) artificial sebum (a mixture of triglycerides, fatty acids, liquid hydrocarbons, and cholesterol, ◆; gradient of plot 0.40 ± 0.02); (ii) mixture of stearic acid and triolein (ratio 90/5 by weight), ■; gradient of plot 0.42 ± 0.05. (b) pH 10.5, 5.3×10^{-4} M calcium. Antifoams: (i) artificial sebum, ◆; gradient of plot 0.30 ± 0.01; (ii) mixture of stearic acid and triolein (ratio 90/5 by weight), ■; gradient of plot 0.33 ± 0.04. (After Ran, L. Foaming of anionic surfactant solutions in the presence of calcium ions and triglyceride-based antifoams, PhD Thesis, University of Manchester, 2011.)

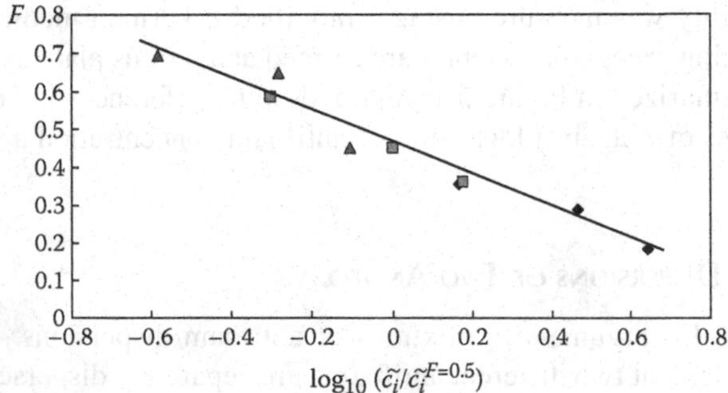

FIGURE 5.4 Reduced plot of F against $\log_{10}(\tilde{c}_i/\tilde{c}_i^{F=0.5})$ for "thin" sugar beet juice containing three unspecified different antifoams of significantly different effectiveness [13]. Foam was generated by the Bikerman method [15].

the surfactant solution and foam generation methods are assumed to be the same throughout (so that the volume of foam generated in the absence of antifoam is therefore the same throughout).

If $\tilde{c}_{\mathrm{T}}$ is the concentration of the mixed antifoam dispersion, then we must have

$$\tilde{c}_{\mathrm{T}} = \tilde{c}_{1\mathrm{m}} + \tilde{c}_{2\mathrm{m}} \tag{5.4}$$

$$\tilde{c}_{1\mathrm{m}} = \tilde{x}_{1\mathrm{m}}\tilde{c}_{\mathrm{T}} \tag{5.5}$$

and

$$\tilde{c}_{2\mathrm{m}} = \tilde{x}_{2\mathrm{m}}\tilde{c}_{2} \tag{5.6}$$

where $\tilde{c}_{1\mathrm{m}}$, $\tilde{c}_{2\mathrm{m}}$, $x_{1\mathrm{m}}$, and $x_{2\mathrm{m}}$ are the concentrations and weight fractions of antifoam species 1 and 2 in the mixed dispersion, respectively.

The ratio F for the mixed antifoam dispersions is obviously some function of the concentration of each antifoam species so that

$$F = f(\tilde{c}_{1\mathrm{m}}, \tilde{c}_{2\mathrm{m}}) \tag{5.7}$$

which means that we can write, using Equations 5.5 and 5.6

$$\begin{aligned}\left[\frac{\mathrm{d}F}{\mathrm{d}\tilde{c}_{\mathrm{T}}}\right]^{0} &= \left(\frac{\partial F}{\partial \tilde{c}_{1\mathrm{m}}}\right)^{0}_{\tilde{c}_{2\mathrm{m}}}\frac{\mathrm{d}\tilde{c}_{1\mathrm{m}}}{\mathrm{d}\tilde{c}_{\mathrm{T}}} + \left(\frac{\partial F}{\partial \tilde{c}_{2\mathrm{m}}}\right)^{0}_{\tilde{c}_{1\mathrm{m}}}\frac{\mathrm{d}\tilde{c}_{2\mathrm{m}}}{\mathrm{d}\tilde{c}_{\mathrm{T}}} \\ &= \left(\frac{\partial F}{\partial \tilde{c}_{1\mathrm{m}}}\right)^{0}_{\tilde{c}_{2\mathrm{m}}}\tilde{x}_{1\mathrm{m}} + \left(\frac{\partial F}{\partial \tilde{c}_{2\mathrm{m}}}\right)^{0}_{\tilde{c}_{1\mathrm{m}}}\tilde{x}_{2\mathrm{m}} \ (\text{at } F = F^{0})\end{aligned} \tag{5.8}$$

where all quantities except the mixture compositions $\tilde{x}_{1\mathrm{m}}$ and $\tilde{x}_{2\mathrm{m}}$ refer to the same given value of F (= F^0). Since the antifoam effects are assumed not to be cooperative, then it seems reasonable to assume that

$$\left(\frac{\partial F}{\partial \tilde{c}_{1\mathrm{m}}}\right)^{0}_{\tilde{c}_{2\mathrm{m}}} = \left[\frac{\mathrm{d}F}{\mathrm{d}\tilde{c}_{1}}\right]^{0} \ (\text{at } F = F^{0}) \tag{5.9}$$

where $[\mathrm{d}F/\mathrm{d}\tilde{c}_1]^0$ is the corresponding derivative for the separate constituent antifoam dispersion of species 1 at the same value of F (= F^0) as that of the mixture. This equality means that the incremental decrease in F, at a given value of F^0, due to the incremental increase in concentration of species 1 is necessarily the same irrespective of whether species 2 is present, provided the concentration of the latter, if present, remains constant and provided the antifoam entities are non-cooperative in their function. The same assumption can be extended to antifoam species 2 so that

$$\left(\frac{\partial F}{\partial \tilde{c}_{2m}}\right)^0_{\tilde{c}_{1m}} = \left[\frac{dF}{d\tilde{c}_2}\right]^0 \text{ (at } F = F^0\text{)} \tag{5.10}$$

If we now allow the relative effectiveness of the mixed antifoam dispersion to be adjusted to give the same value of $F = F^0$ as that of the dispersions of the constituent antifoams alone by simply adjusting the overall concentration (at constant composition), then we can write by analogy with Equation 5.1

$$\tilde{c}_T^0 = \tilde{n}_1(\tilde{x}_{1m}, \tilde{x}_{2m})\tilde{c}_1^0 \tag{5.11}$$

and

$$\tilde{c}_T^0 = \tilde{n}_2(\tilde{x}_{1m}, \tilde{x}_{2m})\tilde{c}_2^0 \tag{5.12}$$

where $\tilde{c}_T^0$, $\tilde{c}_1^0$, and $\tilde{c}_2^0$ are the concentrations of mixed antifoam dispersion, dispersion of antifoam 1 alone, and dispersion of antifoam 2 alone required to give the same value of F (= F^0), respectively. $\tilde{n}_1(\tilde{x}_{1m}, \tilde{x}_{2m})$ and $\tilde{n}_2(\tilde{x}_{1m}, \tilde{x}_{2m})$ are dependent on composition of the mixed antifoam dispersion but are independent of the value of F (or foam volume) and the overall antifoam concentrations. By analogy with Equation 5.3, we can therefore deduce from Equations 5.11 and 5.12 that

$$\left(\frac{dF}{d\ln \tilde{c}_T}\right)^0 = \left(\frac{dF}{d\ln \tilde{c}_1}\right)^0 = \left(\frac{dF}{d\ln \tilde{c}_2}\right)^0 \text{ (at } F = F^0\text{)} \tag{5.13}$$

which means again that plots of F against log(antifoam concentration) for the mixed antifoam dispersion and for the dispersions of the constituent antifoams alone should all be isomorphous.

From Equation 5.13 it is easy to see that

$$\left(\frac{dF}{d\tilde{c}_1}\right)^0 = \frac{\tilde{c}_T^0}{\tilde{c}_1^0}\left(\frac{dF}{d\tilde{c}_T}\right)^0 \text{ and } \left(\frac{dF}{d\tilde{c}_2}\right)^0 = \frac{\tilde{c}_T^0}{\tilde{c}_2^0}\left(\frac{dF}{d\tilde{c}_T}\right)^0 \text{ (at } F = F^0\text{)} \tag{5.14}$$

If now we combine Equations 5.8 through 5.10 and 5.14, we can deduce that

$$\frac{1}{\tilde{c}_T^0} = \frac{\tilde{x}_{1m}}{\tilde{c}_1^0} + \frac{\tilde{x}_{2m}}{\tilde{c}_2^0} \text{ (at } F = F^0\text{)} \tag{5.15}$$

where we again emphasize that all quantities except composition refer to the same value of F (= F^0). That this equation is reasonable is revealed if we substitute $\tilde{c}_2^0 = \infty$.

This corresponds to the dilution of an antifoam dispersion with a material that exhibits no antifoam behavior; $1/\tilde{x}_{1m}$ is then simply the dilution factor.

Garrett [1] has reported what is apparently the only observation of the effect of the concentration of a mixed antifoam dispersion on F (or foam volume). Results are shown in Figure 5.5 where F is plotted against log(antifoam concentration). Here dispersion C was prepared by mixing dispersions A and B in a fixed proportion. Dispersions A and B were each prepared using only one antifoam species. The two different antifoams are both mixtures of hydrophobic particles and hydrocarbon oils—a commercial calcium alkyl phosphate–liquid paraffin (dispersion A) and a hydrophobed silica–liquid paraffin (dispersion B). All the dispersions were made using a solution of a commercial sodium alkylbenzene sulfonate. Foam measurements were done using a static Ross–Miles apparatus [1].

The hydrophobed silica–liquid paraffin dispersion B is seen to be almost an order of magnitude less effective than the calcium alkyl phosphate dispersion A. Plots of F against $\log_{10}$(antifoam concentration) for the three dispersions are seen to be approximately linear over the concentration range used. They are also of approximately the same gradient and are therefore isomorphous. This is illustrated in the figure where a fitted straight line is drawn through the plot for dispersion B and a line of the same gradient is drawn through the plot for dispersion A using one of the experimental points. A line of the same gradient is also drawn on the experimental plot for the mixed dispersion C but using Equation 5.15 to calculate its position. The agreement of experiment with Equation 5.15 is seen to be good. It should be noted

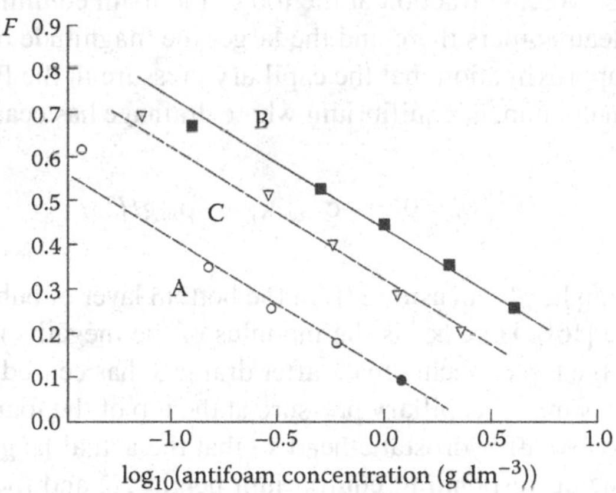

FIGURE 5.5 Plot of F against $\log_{10} \tilde{c}_i$ for dispersions of antifoam in a solution of 0.5 g dm^{-3} aqueous commercial sodium alkylbenzene sulfonate. Foam was generated at 25°C using a static Ross–Miles apparatus [14]. (i) Calcium alkyl acid phosphate in liquid paraffin, A, ○; (ii) silane-hydrophobed silica D17 in liquid paraffin, B, ■; (iii) mixed dispersion (0.13A + 0.87B weight proportions), ▽. Dashed lines are calculated from the experimental result marked ● and from theory using Equation 5.15 for the mixed antifoam dispersion and assuming isomorphous F against log(antifoam concentration) for all antifoams. (Reprinted with permission from Garrett, P.R., *Langmuir*, 11, 3576. Copyright 1995 American Chemical Society.)

that this account of the findings corrects an error in the text of the original paper [1] concerning the labeling of the dispersions.

5.1.4 Relative Effectiveness of Antifoam Entities and Foam Structure

We find that Equation 5.15 permits calculation of the foam volume of solutions containing mixed antifoam dispersions from knowledge of the concentration dependence of dispersions of the constituent species. Again this equation is derived from an assumption that the relative effectiveness of any antifoam entity is unaffected, either by the concentration of such entities or the concentration of other types of antifoam entity that exhibit different levels of effectiveness. The success of this assumption in predicting the observed behavior, as revealed by Figures 5.2 through 5.5, invites comparison with prevailing views concerning the interaction of antifoam entities with foam structure.

Antifoam emergence into air–water surfaces during foam drainage has been rather convincingly demonstrated to be controlled by the balance between the critical capillary pressure required for this to occur (see Chapter 4 and, e.g., reference [11]) and the capillary pressure exerted by adjacent Plateau borders. As we have described in Section 1.4.1 for the case of an equilibrium foam in which drainage has ceased, Princen [16] has stated that the capillary pressure $p_c^{PB}(y)$ in the Plateau borders at a distance y from the top of the foam column is determined by a combination of the capillary pressure at the top of the foam column and the hydrostatic head [16]. The capillary pressure at $y = 0$ is in turn determined by the curvature κ_t of the free air–water surface of the Plateau borders at the top of the foam column. Therefore, the larger the gas phase volume fraction at the top of the foam column, the smaller the volume of the Plateau borders there and the larger the magnitude of κ_t. These arguments lead to the approximation that the capillary pressure in the Plateau borders at the top of the foam column, at equilibrium where drainage has ceased, is given by

$$p_c^{PB}(y=0) = -\sigma_{AW}\left|\kappa_t\right| \approx -\rho_W g H_0^e \tag{5.16}$$

where H_0^e is the foam height measured from the bottom layer of bubbles (see Section 1.4.1 and reference [16]). Here $\left|\kappa_t\right|$ is the modulus of the (negative) curvature of the Plateau borders at the top of such a foam after drainage has ceased. However, in the case of a draining foam, the capillary pressure at the top of the foam, $p_c^{PB}(y=0)$, no longer matches the overall hydrostatic head so that the actual height H of the foam column exceeds the corresponding equilibrium height H_0^e and the modulus of the Plateau border capillary pressure $\left|p_c^{PB}(y=0)\right| < \rho g(H > H_0^e)$. Such a foam will continue to drain until $H = H_0^e$ as a result of increases in κ_t and reduction of H following increases in the gas phase volume fraction Φ_G^{foam}.

Equation 5.16 implies that the chance of the Plateau border capillary pressure at the top of the foam column exceeding the critical capillary pressure for emergence of antifoam entities into air–water surfaces (and in consequence for foam film rupture) therefore increases with increasing foam height in the case of an equilibrium foam. In principle, this could mean that a given antifoam dispersion will be totally ineffective below a certain limiting foam height (or value of F). This type of behavior has in fact been presented by Denkov and coworkers [10, 11]. However, as described in Section 4.5.3, this work involved measurements of the effect of neat oils on foam stability under conditions where equilibrium with respect to drainage and foam height is at least approximated. In this respect, the conditions differed markedly from the non-equilibrium conditions prevailing during foam generation with which we are concerned here.

Consider then a comparison between a hypothetical antifoam dispersion that is only effective above a certain capillary pressure (and therefore foam height) and another that is effective at all foam heights. The ratio $\tilde{n}$ will then be finite for the two antifoam dispersions at foam heights above the limiting foam height for the least effective antifoam dispersion but infinite at foam heights below that limiting foam height. This obviously represents an extreme comparison but it illustrates that, in general, if antifoam effectiveness during foam generation is in part determined by a Plateau border capillary pressure, which changes with foam height, then $\tilde{n}$ will also become a function of foam height (or F) and therefore antifoam concentration. Such behavior would invalidate the treatment represented by Equations 5.2 through 5.15 and mean that in general plots of F against log(antifoam concentration) in a given solution using a given foam generation method should not be isomorphous. However, as we have seen, the available experimental evidence suggests rather that such plots are isomorphous, at least in the cases of foam generation by the Ross–Miles method, cylinder shaking, and tumbling. It seems likely that this finding concerns the nature of the capillary pressure gradients existing during the process of foam generation using such methods.

In the case of foam generation by the Ross–Miles method [14, 15], a jet of surfactant solution containing dispersed antifoam, ejected from an upper reservoir, impinges upon already generated foam in a lower reservoir containing the same dispersion. That jet is usually directed at the center of the lower reservoir so that it continuously irrigates the center and especially the upper surface of any foam column. Aeration presumably occurs at the surface of that foam. The gas phase volume fraction then varies, mainly radially, in a direction perpendicular to the direction of the impinging jet. Any capillary pressure gradient at the top of the foam column will therefore be largely independent of foam height and therefore foam volume. Additionally, the effect of the jet on the foam is to stir the foam, which would in any case tend to reduce such gradients. Antifoam action, which reduces the overall foam volume, must of course concern rupture of foam films at the top of the foam column rather than rupture of foam films between bubbles (which of course only have consequence for the bubble size distribution). It would therefore seem unsurprising that in the case of the process of foam generation by the Ross–Miles method, $\tilde{n}$ is independent of foam height, volume, or indeed F so that plots of F against log(antifoam concentration) are isomorphous for antifoams of different effectiveness as shown in Figure 5.5. Clearly, however, after foam generation has ceased and drainage commences, a vertical capillary pressure gradient will be established, which will eventually balance the hydrostatic head to establish mechanical equilibrium. Plots of F against log(antifoam concentration) for antifoams of differing effectiveness, measured in such circumstances, would not therefore then be expected to be necessarily isomorphous.

Cylinder shaking by hand and mechanically driven cylinder rotation involves intensive mixing of air and liquid, which will tend to produce large fluctuations in the gas phase volume fraction regardless of foam height during the process of foam generation. The hydrostatic head is then irrelevant. That $\tilde{n}$ is independent of foam height, volume, or F, as implied by the results shown in Figures 5.2 through 5.3, suggests therefore that the Plateau border capillary pressure involved in antifoam action during the process of foam generation is also independent of foam height using this method. Somewhat surprisingly, the results of Cubero et al. [13] shown in Figure 5.4, using a method that resembles sparging, appear to suggest a similar conclusion.

We therefore find that if the relative effectiveness of an antifoam dispersion in a given surfactant solution can be adjusted by simply changing the relative concentration so that $\tilde{n}$ is constant, then this implies that the antifoam effectiveness should not be a function of foam height. This situation would seem likely to prevail only during the process of foam generation. By contrast, during foam drainage, Plateau border capillary pressures, and therefore antifoam action, depend on the hydrostatic head so that $\tilde{n}$ could vary with foam height.

We now explore whether we can provide a theoretical basis, using a physical model, for the phenomenology described here.

5.2 STATISTICAL THEORY OF ANTIFOAM ACTION

5.2.1 Assumptions

Practical antifoam dispersions are characterized by the heterogeneity of the constituent particle-containing drops. Individual drops vary in size and may contain different numbers of hydrophobic particles, which in turn usually vary in size, shape, and probably even contact angle (see reference [17]). In turn, this variability will cause a variation in effectiveness of different drops. Antifoam dispersions consisting of hydrophobic particles alone will also exhibit heterogeneity. We have seen that even identical hydrophobic particles with edges (see Chapter 4) may adopt different configurations at both air–water and oil–water surfaces, many of which do not permit film rupture. It is possible to write a Boltzmann-type expression for the relative probabilities of such configurations, and therefore for the relative probability of a given particle being ineffective in causing film rupture (see Chapter 4). Even in the case of perfect monodisperse spherical particles alone, each with identical contact angles, there will be variations in the state of the air–water surface seen by each of these particles due to the presence of surface tension gradients at the relevant air–water surfaces during foam generation.

These considerations suggest that antifoam action should have some of the aspects of a stochastic process. Such an approach has therefore been adopted by Garrett [1]. Here we outline that theory but include a number of significant modifications in order to take account of recent developments in the relevant science. The basis of the theory is an expression for the probability of a bubble surviving despite having several antifoam entities associated with it. Thus, if the probability of each antifoam entity in a dispersion of antifoam species i being ineffective is q_i, and there are $N_i(r_b)$ such entities associated with a given bubble of equivalent spherical radius r_b, then the probability of the bubble surviving without coalescence is $q_i^{N_i(r_b)}$. Therefore, the greater the number of antifoam entities, $N_i(r_b)$, for which $q_i < 1$, the smaller the chance of a given bubble surviving. However, if the total number of antifoam entities associated with bubbles of a given equivalent radius r_b is less than the number of bubbles, then the quantity $q_i^{N_i(r_b)}$ no longer describes the probability of bubbles surviving because some bubbles must of necessity be without antifoam entities. A more sophisticated statistical treatment is then required. Here we therefore confine the treatment to situations where $N_i(r_b) \gg 1$ for all bubble sizes, where $N_i(r_b)$ is a mean so that non-integer values are allowed. However, we also suppose that any fluctuations in the number of antifoam entities, associated with bubbles of a given size, is always extremely small so that deviations from the mean for any given bubble can be neglected. As we will show (see Section 5.2.4), allowing for significant fluctuations requires a more sophisticated statistical treatment leading to significantly greater complexity.

If a theory is to be developed from all this, it is necessary to both relate $N_i(r_b)$ to the antifoam concentration and relate the probability of a bubble surviving to the overall foam volume. This volume will of course be unaffected by any interbubble coalescence events but will be determined only by bubble coalescence with ambient air at the top of the foam column. We will, however, assume that such interbubble coalescence is in any case negligible during foam generation in the presence of antifoam. This will mean that the antifoam simply induces coalescence with ambient air of a proportion of the bubbles that would otherwise be present in the absence of antifoam. If, for example, $N_i(r_b)$ is independent of bubble size, the effect of the antifoam will then be to reduce the total number of bubbles, and therefore foam volume, without having any effect on the bubble size distribution. If, on the other hand, $N_i(r_b)$ is larger for larger bubbles, then the resulting bubble size distribution would be skewed toward smaller sizes relative to that prevailing in the absence of antifoam. This approach therefore clearly differs from a model of Pelton and Goddard [18, 19] where interbubble coalescence is assumed to be induced by antifoam.

Bubble coalescence only with ambient air without interbubble coalescence is in fact often observed in experiments. This behavior occurs in foam generation where foam films tend to rupture at a certain critical thickness, or when antifoams are present at a thickness equal to the characteristic dimension of antifoam entities such as the diameter of spherical drops. That foam film rupture tends to occur mostly in ambient air–bubble films is to be expected because they are the thinnest films in the foam. It is in fact found with intrinsically unstable foams formed by sparging aqueous alcohol [20, 21], sodium dodecyl sulfate, or "Teepol" solutions in the absence of antifoam where film rupture is essentially confined to the top of the foam column [22]. Absence of interbubble coalescence in the case of intrinsically unstable foams, generated by sparging, is in fact a basic assumption of the model of foam growth developed by Neethling [22]. Similarly, Denkov et al. [23] find that foam bubble radii are relatively unaffected by the presence of predispersed hydrophobed silica–polydimethylsiloxane antifoam when monodisperse bubble foam is formed by bubbling through an array of uniform capillaries. That interbubble coalescence in the bulk of the freshly generated foam formed by shaking a closed vessel is often not apparent is illustrated in Figure 5.6. Here polytetrafluoroethylene (PTFE) particles

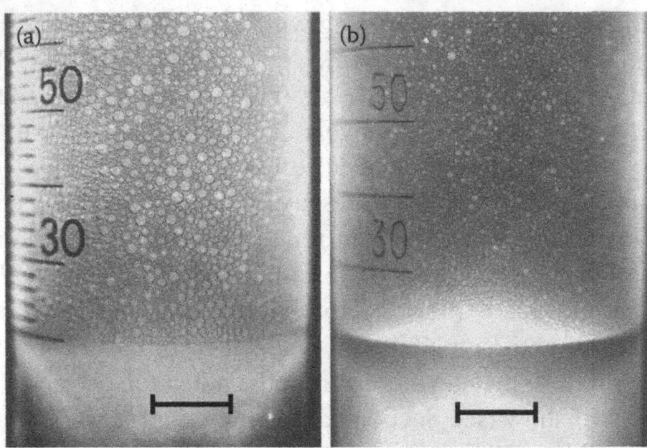

FIGURE 5.6 Foam generated by shaking 0.05 dm^3 solution in a 0.25 dm^3 graduated glass cylinder for 15 s. (a) 0.01 M aqueous sodium dodecyl sulfate solution—volume of air in foam ~0.27 dm^3; (b) 0.01 M aqueous sodium dodecyl sulfate with 2 g dm^{-3} of PTFE powder [25] of particle size range 5–10 microns—volume of air in foam ~0.17 dm^{-3} representing a reduction of ~38% due to the antifoam action of the PTFE [24]. Size bars, 0.1 dm.

were shaken with 0.01 M sodium dodecyl sulfate solution and the resulting foam compared with that formed from the solution without the particles. Images of the foams are reproduced in the figure where it is obvious that bubbles are approximately spherical, implying relatively low gas phase volume fractions [24]. Bubble sizes of the foam containing particles are seen to be apparently smaller than those formed in the absence of particles despite statistical sampling bias that exaggerates the relative proportion of larger bubbles [26, 27]. Such differences imply that the PTFE particles preferentially remove the larger bubbles from the foam during shaking. Similar observations have been reported elsewhere with foam generated by shaking sodium alkylbenzene sulfonate solution in both the presence and absence of hydrophobed silica–liquid paraffin antifoam [1]. The available experimental evidence therefore suggests that it is reasonable to assume that the dominant coalescence phenomena during foam generation, caused by the presence of antifoam, are confined to the region near the continuous air–water surface at the top of the foam. There is, however, a need for systematic studies of the effect of antifoam action on bubble size distributions in foam generated by various methods in order to assess the general applicability of this assumption.

The magnitude of q_i will be determined by the nature of the antifoam. In this, as we have seen (see Chapter 4), there are two dominant factors—whether the capillary pressure in Plateau borders is sufficiently high to permit the antifoam entity to emerge into the air–water surface and whether, once emerged, the antifoam entity has the correct characteristics to cause foam film rupture. The factors intrinsic to the antifoam in this context depend on whether the antifoam consists of particles alone, oils alone, or mixtures of particles and oils. In the case of particles alone, the relevant factors are the particle size, geometry, and wettability together with the various colloidal interaction forces that conspire to produce the magnitude of the disjoining forces between the particle and the air–water surface. In the case of particle–oil mixtures, there are additional factors such as the particle concentration in the oil–water interface and wettabilities at both oil–water and air–water interfaces. Clearly then, the possibilities for variation from antifoam entity to antifoam entity (which means that generally $0 < q_i < 1$) are greater when solid particles are involved.

Obviously q_i will also depend on the capillary pressure in the Plateau borders. This pressure will be determined in part by the surface tension at the ambient air–water surface at the top of the foam column. This surface tension could differ significantly from equilibrium during foam generation if air–water surfaces are formed at a rate such the rate of transport of surfactant is too slow to maintain that equilibrium. The capillary pressure will also be determined by the overall structure of the foam, including both gas phase volume fraction and bubble sizes. However, as we have seen, Princen [16] has pointed out that in the case of an equilibrium foam, the curvature of the top air–water surface κ_t of Plateau borders will be everywhere the same despite the presence of bubbles of different sizes (see reference [16]). During generation of a non-equilibrium draining foam where bubbles are well mixed, it seems reasonable to expect that any capillary pressure differences between adjacent Plateau borders at the top of the foam will be rapidly eliminated by flow-induced changes in local liquid saturation. If such flows are rapid compared with the drainage rates of foam films, then the capillary pressure exerted by those borders on a given film at the top of a foam column will be the same regardless of size of the adjacent bubbles. This means that q_i will not be a function of the radius, r_b, of any particular bubble but rather of the overall capillary pressure, which, although determined by bubble sizes, will be essentially constant for all Plateau borders at the top of the foam column. Following the discussion in Section 5.1.4, we will also suppose that the method of foam generation is such that neither the capillary pressure at the top of the foam column nor therefore q_i are functions of foam height (or therefore F). Clearly then, q_i is not just an intrinsic property of antifoam entities—it also depends on the properties of the foaming liquid (especially the adsorption behavior of any surfactant present) and the method of foam generation (which will influence intrinsic gas phase volume fractions and bubble sizes distributions and therefore Plateau border capillary pressures).

As we have seen, in developing the theory, it is necessary to relate $N_i(r_b)$ to the antifoam concentration. In the original version of the theory [1], $N_i(r_b)$ was supposed determined by the attachment of antifoam entities to bubbles during their passage through the foaming liquid before they form films at the top of the foam. The envisaged process therefore resembles that occurring in mineral separation by "froth flotation." The presence of unattached antifoam entities in the films was ignored. This is tantamount to assuming that any such entities would be removed from foam films before they could attach to the relevant air–water surfaces. The main problem with this approach concerns the nature of the attachment process. It clearly requires that the pseudoemulsion film between antifoam entities and the air–water surface rupture when bubbles intercept those entities during movement through the foaming liquid. However, the stability of the pseudoemulsion film is one of the main factors determining the probability that a given antifoam entity causes rupture of a foam film as it forms a bridging configuration. The approach then has both q_i and $N_i(r_b)$ largely dependent on the same factor. This tends to undermine the legitimacy of supposing that the probability of a bubble surviving without coalescence is $q_i^{N_i(r_b)}$. It would tend to imply that if $N_i(r_b) > 1$, then always $q_i \to 0$.

It is more than 30 years since the froth flotation approach for estimating $N_i(r_b)$ was published in private corporate literature [24] and more than 15 years since it appeared in the published literature [1]. During that time, our understanding of antifoam behavior has advanced significantly in many respects (without, however, any marked change in the understanding of the basic mode of action of such materials). One such advance by Hadjiiski et al. [10] has been measurement of the critical capillary pressure required to force an antifoam entity to emerge into the air–water surface of a film. That capillary pressure is in turn identified with the pressure exerted by the Plateau borders in a foam film. All this suggests that antifoam entities, already present in a foam film, are forced by that pressure to emerge into both air–water surfaces of a foam film to form unstable bridging configurations that give rise to foam film rupture. In this view of antifoam action, the antifoam concentration in foam films is independent of processes associated with antifoam action. It is the intrinsic concentration possibly, as we shall see, modified by the draining processes occurring in the foam film. In this case, $N_i(r_b)$ is simply the antifoam concentration in the film multiplied by the volume of the foam film. It is therefore clearly independent of the effectiveness of the antifoam entities. This then removes one of the fundamental objections to the original statistical theory of antifoam action. We therefore proceed on that basis here.

5.2.2 Factors Determining Number of Antifoam Entities in Foam Film

Let us suppose then that antifoam entities do not adhere to bubble surfaces during their passage through the foaming liquid. Then $N_i(r_b)$ becomes equal to the number of antifoam entities present in the films separating bubbles from the ambient air under conditions where they can, in principle, cause film rupture. As we have seen, such entities should bridge those films so that their diameters then approximate the thicknesses of the films at the point of rupture.

A preliminary analysis of the factors contributing to the magnitude of $N_i(r_b)$ may be readily obtained if we make some further simplifying assumptions. Let us assume therefore that the antifoam consists of monodisperse spheres, distributed throughout the foaming liquid. The total number of monodisperse antifoam entities, N_T, in a film of thickness h separating a bubble from ambient air is then given by the product of the number of antifoam entities per unit volume and the volume of the film. Thus, we can write

$$N_T = \frac{3S(r_b)\pi r_b^2 h\tilde{c}_i}{4\rho_{AF} r_{AF}^3} \tag{5.17}$$

where $\tilde{c}_i$ is the antifoam concentration of species i by weight, ρ_{AF} is the average antifoam density, and r_{AF} is the radius of the antifoam drop. r_b is the radius of a spherical bubble having the same volume as that of a given bubble at the top of the foam column. $S(r_b)$ is a factor relating the area of the film to the cross-sectional area, πr_b^2, where it has been shown by simulation that $S(r_b)$ is of order unity [28].

It is reasonable to assume that $N_i(r_b)$ is given by the number of antifoam entities remaining in foam films after they have drained to a thickness equal to the diameter of those entities. If foam film drainage occurs at constant antifoam concentration, $\tilde{c}_i$, then we must have $N_i(r_b) = N_T$ if $h = 2r_{AF}$ in Equation 5.17 so that

$$N_i(r_b) = \frac{3S(r_b)\pi r_b^2 \tilde{c}_i}{2\rho_{AF} r_{AF}^2} \tag{5.18}$$

In using this equation to make numerical estimates of the magnitude of $N_i(r_b)$, we note that mean antifoam drop radii in practical applications are typically ~1 micron (see, e.g., references [7, 29]) and the average density of antifoams is reasonably approximated to within ±10% as 10^3 g dm^{-3}.

In Table 5.1, a range of values of $N_i(r_b)$, corresponding to a typical antifoam concentration range of 0.1 to 1.0 g dm^{-3} [11], are calculated from Equation 5.18 (setting

TABLE 5.1
Estimated Number of Antifoam Entities in Foam Films as Function of Bubble Radius

Method	Aqueous Solutions	Bubble Radii (microns)[a]	Range of Mean Values for $N_i(r_b)$; Corresponding to Values of $\tilde{c}_i$ from 0.1 to 1.0 g dm^{-3}	Range of Probabilities of Bubble Collapse = $(1-q_i^{N_i(r_b)})$ with $q_i = 0.9$
Shaking cylinders	0.01 M Sodium dodecyl sulfate	≥500[b] [24]	38–375	0.98–1.0
Ross–Miles	0.005 M Sodium alkylbenzene sulfonate	Mean ~150 [30]	3–34	0.30–0.97
		Maximum ~500 [30]	38–375	0.98–1.0
Sparging with glass frits	0.017 M Sodium dodecyl sulfate	Course frits ~500 [19]	38–375	0.98–1.0
		"Regular" frit ~150 [19]	3–34	0.30–0.97
Whipping	Teepol + sodium alginate	Minimum ~12.5 [27]	0.024–0.24	Meaningless because $N_i(r_b) < 1$
		Maximum ~200 [27]	6–60	0.47–1.0

Note: Calculated using Equation 5.18 with $r_{AF} = 1$ micron and $\tilde{c}_i = 0.1-1.0$ g dm^{-3}.

[a] Initial values measured immediately after foam generation ceased.

[b] See Figure 5.6.

$S(r_b)$ at unity [28]) for bubble sizes observed for various methods of foam generation. The polydispersity of real antifoam dispersions is of course neglected in this calculation. In the case of extremely small bubbles, with radii in the region of 12.5 microns, we find that $N_i(r_b) < 1$, which implies that the number of antifoam entities is less than the number of films. Total rupture of all bubbles in such a foam is therefore in principle not then possible. As we have seen, this situation must also, in turn, render $q_i^{N_i(r_b)}$ an invalid expression for the probability of bubble survival. However, for most situations, it is clear from the table that $N_i(r_b) > 1$. Indeed, for bubbles of ~500-micron radius, which can be seen in the table to be typical of many foam generation methods, $N_i(r_b) \gg 1$, which means that even values of q_i close to unity are predicted by Equation 4.19 to produce low probabilities $q_i^{N_i(r_b)}$ of bubble survival. Thus, for example, we calculate that for a bubble of 500-micron radius, we need only that 10% (i.e., $q_i = 0.9$) of antifoam entities, of 1-micron radius at a concentration of 0.1 g dm^{-3}, be effective to produce a 98% chance of bubble collapse.

We may compare the estimates shown in Table 5.1 with actual experimental observation if we consider the results shown in, for example, Figure 5.5. Here results are presented for hydrocarbon-based antifoams where foam was generated using a Ross–Miles apparatus. A mixture of calcium alkyl phosphate and liquid paraffin is seen to be the most effective of those for which results are presented in the figure. For this antifoam, it can be inferred from the values of F given therein that at an antifoam concentration of 0.1 g dm^{-3} about 55% of the foam is destroyed and at a concentration of 1 g dm^{-3} about 90% of the foam is destroyed. The mean bubble size for Ross–Miles foam generation with the same apparatus has been found to be only about 150 microns [30], which is shown in Table 5.1 to produce a calculated probability of bubble collapse of 30% at an antifoam concentration of 0.1 g dm^{-3} and 97% at a concentration of 1 g dm^{-3} if it is assumed that only 10% of antifoam entities are effective. Clearly such low probabilities of antifoam effectiveness have produced estimates of bubble collapse that are of the same order as experimental observations of foam collapse at the same antifoam concentrations. Even lower probabilities of antifoam effectiveness would obviously be required to produce estimates of foam film collapse in the case of the less effective antifoam shown in Figure 5.5. These estimates ignore, of course, the polydispersity of the foam and are dependent on the assumption that antifoam drop radii equal 1 micron (which seems reasonable in view of the means of dispersal by stirring with a high-speed emulsifier for several minutes). However, even if the drop radii are increased to 3 microns at a concentration of 1 g dm^{-3}, a 90% probability of foam film rupture would only require that 46% of the antifoam entities are effective. Further increases in drop radii to, say, 5 microns would mean that at an antifoam concentration of 1 g dm^{-3}, a 90% probability of foam film collapse would require that 93% of antifoam entities to be effective. However, with drops of 5 microns radii at 0.1 g dm^{-3}, $N_i(r_b) = 0.14$ and most foam films would be without antifoam regardless of effectiveness so that no more than 14% (regardless of the statistical approach employed) of foam films would be ruptured. That is significantly lower than the 55% implied by the experimental observations. Clearly better agreement with the experiments summarized in Figure 5.5 is obtained if it is assumed that antifoam drops are smaller and therefore more numerous but less effective.

Although somewhat surprising, then it seems possible that a high proportion of antifoam entities are intrinsically ineffective even in the case of mixed oil–particle drops where the bridging coefficient is positive and sufficient particles are present to saturate the oil–water surface. This possibility, however, arises only if Equation 5.18 does not overestimate the effective value of $N_i(r_b)$. Any such overestimation could derive from neglect of the effect of film thinning on the number of antifoam entities present in a foam film. Thus, if thinning results in a significant reduction of antifoam concentration $\tilde{c}_i$ by the time the film assumes a thickness of the order of the diameter of the antifoam entities, then $N_i(r_b)$ would also be significantly reduced and realistic probabilities of bubble collapse could then require relatively high probabilities of individual antifoam entity effectiveness.

In assessing the possible effect of film thinning on antifoam concentration, we will suppose that drainage is not influenced by the presence of the antifoam. The latter must be unambiguously true in those circumstances where the hydrodynamics of foam generation and film thinning, together with the relevant surface and bulk rheological properties, are not themselves influenced by the presence of antifoam. In typical applications, the volume fraction of antifoam is <0.001 and any influence on bulk rheology is therefore necessarily negligible. We are of course considering a situation where antifoam entities are not initially present in the air–water surfaces of foam films, only emerging as a result of the action of capillary forces. The absence of any influence on surface rheology by antifoam entities during film thinning is therefore implicit in the system under consideration. In reality, only situations involving contamination of air–water surfaces by antifoam material that has originated from emergence of antifoam drops into air–water surfaces remote from a given film would appear to potentially involve changes to its surface rheology. Such behavior could influence the thinning behavior of a foam film containing antifoam entities that are not emerged into the relevant air–water surfaces. Polydimethylsiloxane-based antifoams would appear to exhibit this type of behavior [7], whereas certain hydrocarbon-based antifoams do not [31].

Relatively simple drainage behavior has sometimes been observed in the case of small (<100 μm radius) plane-parallel foam films with immobile surfaces [32]. Such films can drain by laminar flow so that the rate of thinning is given by the Reynolds equation [33]. In the case of larger films where the surface shear viscosity is extremely high ($\gg 10^{-3}$ g s^{-1}), then drainage can be slower than predicted by the Reynolds equation. In the case of such films, a thin region is usually formed near the Plateau border, which restricts drainage from the center of the film leading to the formation of an axisymmetrical "dimple" (see Section 1.3.2 and reference [34]). In the case of films of >100 μm radius where the surface shear viscosity is low ($<10^{-3}$ g s^{-1}), then the dimple is unstable and a hydrodynamic instability results in complex asymmetrical drainage where the dimple disgorges directly into the Plateau border [35]. It would seem that low surface shear viscosities mean that most aqueous foam films formed from simple surfactant solutions exhibit this complex and relatively rapid drainage behavior. No analytical expression exists for such asymmetrical drainage. Indeed it is only in the case of Reynolds drainage that there is a simple analytical expression for the time dependence of film thickness. We will therefore conveniently consider films that exhibit such behavior in assessing the likely effect of drainage on antifoam concentration. Obviously, any conclusions resulting from such an exercise can only have qualitative significance because of the limited relevance of the Reynolds equation [34].

The Reynolds equation [34] for the rate of film thinning in the case of rigid plane parallel foam films is given

$$-\frac{dh}{dt} = \frac{2h^3 p_c^{PB}}{3\eta_W \pi r_f^2} \tag{5.19}$$

where p_c^{PB} is the pressure drop across the film due to the capillary pressure at the Plateau border, η_W is the viscosity, t is time, and r_f is the radius of the film. The latter is a function of the bubble size because $r_f = r_b\sqrt{S(r_b)}$ where as we have seen $S(r_b)$ is of order unity [28]. For a given film thickness and p_c^{PB}, the larger the film the slower the drainage rate because the pressure drop is applied over a larger distance. In deriving this equation, it is assumed that the air–water surfaces are immobile so that the fluid velocity at those surfaces is zero. This results in a parabolic fluid velocity profile across the film. If a supposedly spherical antifoam entity is present in such a film, the center of the entity will be unable to approach the air–water surface at a closer distance than its radius. As a result, the entity will not fully sample the low fluid velocities in the neighborhood of the air–water surface. The mean velocity of the antifoam entity will therefore be greater than the mean fluid velocity. This excluded volume effect is illustrated in Figure 5.7. It finds practical application in the process of hydrodynamic chromatography (see, e.g., reference [36]) where particles are eluted at a greater rate from a porous medium than the suspending fluid and

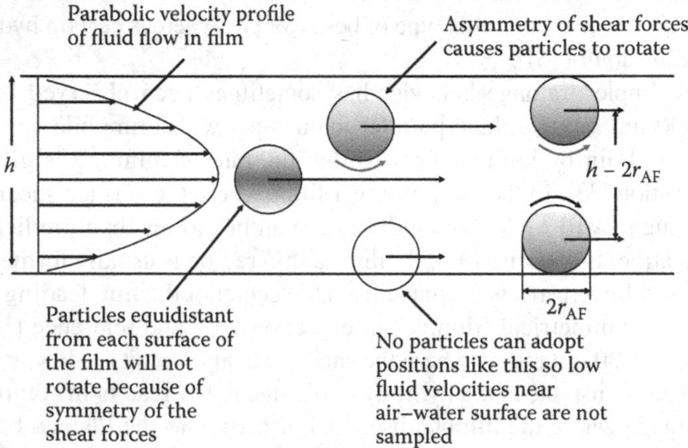

FIGURE 5.7 Schematic illustration of excluded volume effect on movement of spherical antifoam entity in a draining plane-parallel foam film (of 100 microns radius) with immobile surfaces. Film would exhibit drainage according to Reynolds equation (Equation 5.20) with characteristic parabolic velocity profile. Asymmetry of shear forces means that particles will also rotate except when equidistant from each surface of film.

where larger particles are eluted faster than smaller particles. We show in Appendix 5.1 that the mean velocity $\overline{u}_{AF}$ of antifoam entities in a film draining according to the Reynolds equation [33], after making a number of simplifying assumptions, including neglect of electrostatic effects, is given by

$$\overline{u}_{AF} = u_{mean}(1 + \Omega - \Omega^2/2) \tag{5.20}$$

where u_{mean} is the mean fluid velocity in the film and Ω is the ratio of the diameter of the antifoam entity to the thickness of the film. Here both $\overline{u}_{AF}$ and u_{mean} vary with both radial position in the film and film thickness (and therefore time). It is easy to see from Equation 5.20 that if the antifoam entity is extremely small so that $\Omega \to 0$, then the excluded volume is negligible and $\overline{u}_{AF} = u_{mean}$. However, if $\Omega \to 1$ so that the entity is of comparable size to the thickness of the film, then we have a maximum excluded volume effect so that $\overline{u}_{AF} = 1.5u_{mean}$ and the particle moves more rapidly than the mean fluid velocity. In fact, in these circumstances, the particle must move at the maximum fluid velocity, which occurs in the plane equidistant from both air–water surfaces of the film. We must conclude then that antifoam entities may in general flow out of this type of foam film at a faster rate than the surrounding fluid where the difference in rates becomes more pronounced as the film thins toward the dimensions of the antifoam. Of necessity, this will mean that the concentration, $\tilde{c}_i$, of antifoam in the film will decline as the film drains. Calculations given in Appendix 5.1 reveal that this excluded volume effect produces at most a 4-fold reduction in antifoam concentration as a film exhibiting Reynolds drainage thins to a thickness where $\Omega = 1$ and $h = 2r_{AF}$. This should obviously lead to diminished probabilities of foam film rupture, the magnitude of which can be illustrated by taking an example from Table 5.1. Here again then we suppose $q_i = 0.9$. Thus, a 4-fold reduction in antifoam concentration as films drain leads to a reduction in the probability of foam film collapse with foam films of 500 μm radius from 0.98 to 0.63 in the case of initial antifoam concentrations of 0.1 g dm^{-3} and negligible change from 1.0 in the case of initial antifoam concentrations of 1.0 g dm^{-3}. This dilution will therefore make only a partial contribution to an explanation for the apparent overestimation of $N_i(r_b)$ given by Equation 5.18 and shown in Table 5.1.

Clearly allowing for dilution due to the excluded volume effect with Reynolds drainage does not significantly alter the condition $N_i(r_b) \gg 1$ found in most situations summarized in Table 5.1. This may mean that the probability of causing foam film rupture by individual entities in typical antifoam dispersions really is extremely low! Alternatively, it may simply concern the likelihood that Reynolds drainage is not relevant for typical foam films. As we have seen in Section 1.3.2, rapid asymmetrical drainage, where a dimple disgorges directly into the Plateau borders, is the most likely process of thinning with films formed from typical surfactant solutions. As we describe in Section 1.3.2 most explanations of the origin of this effect suggest it is a hydrodynamic instability in which surface tension gradient–driven flows cause thin film to be formed at the expense of thick film (the dimples) [35]. If we suppose that both thick and thin film elements are draining, then the excluded volume effect will obviously be more marked for the latter than the former. Thin film is also expanding at the expense of thick film and we would therefore expect this effect to largely dominate in causing an overall concentration depletion effect in such a film. Since the hydrodynamic instability probably involves a critical dimension, it seems likely that, unlike Reynolds drainage, the dilution of antifoam particle concentration in asymmetrical drainage is some function of film and therefore bubble size. However, it is not possible to readily estimate the magnitude of the depletion effect because no simple analytical solution exists to describe asymmetrical drainage [35]. Estimates of the magnitude of the extent of that depletion probably await relevant numerical simulations.

We must therefore write as the best estimate for $N_i(r_b)$

$$N_i(r_b) = \frac{3X(r_b).S(r_b)\pi r_b^2 \tilde{c}_i}{2\rho_{AF} r_{AF}^2} \tag{5.21}$$

where $X(r_b) < 1$ is the excluded volume dilution effect, which is equal to a pure number, 0.25, in the case of Reynolds drainage (see Appendix 5.1) but some unknown function of bubble size r_b in the case of asymmetrical drainage. Therefore, we can simplify

$$N_i(r_b) = A_i W(r_b)\tilde{c}_i \tag{5.22}$$

where $A_i = 3/2\rho_{AF} r_{AF}^2$ is a constant only dependent on the antifoam properties, $W(r_b) = X(r_b).S(r_b)\pi r_b^2$ is a function of bubble size and is independent of the antifoam properties. In the case of a polydisperse antifoam consisting of spherical drops, then we can write for a mean value of A_i

$$A_i = 3/2\rho_{AF} \int_0^\infty G(r_{AF}) r_{AF}^{-2} \, dr_{AF} \tag{5.23}$$

where $G(r_{AF})$ is the normalized distribution of antifoam drop sizes.

5.2.3 Calculation of Volume of Air in Foam in Presence of Antifoam

Expressions can now be derived for the volume of air present in a freshly generated foam where bubbles are approximately spherical, both in the absence and presence of antifoam where we assume $N(r_b) > 1$ for all bubbles.

Air entrainment processes inevitably produce a polydisperse foam. Therefore, in the absence of antifoam, we can write for the volume V_T of air in the foam

$$V_T = 1.33\pi N_b \int_0^\infty P(r_b) r_b^3 \, dr_b \tag{5.24}$$

where $P(r_b)$ is the normalized bubble size distribution (where we remember r_b is the equivalent spherical radius) and N_b is the total number of bubbles present. Using the definition of q_i and Equation 5.22, we can write for the volume of air in the foam V_{AF} in the presence of antifoam

$$V_{AF} = 1.33\pi N_b \int_0^\infty P(r_b) r_b^3 q_i^{N_i(r_b)}\, dr_b = 1.33\pi N_b \int_0^\infty P(r_b) r_b^3 q_i^{A_i W(r_b)\tilde{c}_i}\, dr_b \qquad (5.25)$$

Therefore, the ratio F of volume of air in the foam in the presence of antifoam to that in the absence of antifoam is given

$$F = \frac{V_{AF}}{V_T} = \frac{\int_0^\infty P(r_b) r_b^3 q_i^{N_i(r_b)}\, dr_b}{\int_0^\infty P(r_b) r_b^3\, dr_b} = \frac{\int_0^\infty P(r_b) r_b^3 q_i^{A_i W(r_b)\tilde{c}_i}\, dr_b}{\int_0^\infty P(r_b) r_b^3\, dr_b} \qquad (5.26)$$

which is an equation mathematically exactly analogous to that derived earlier [1] in the original version of the theory for the case where $N_i(r_b)$ is determined by antifoam entities that attach to bubbles during their passage through the foaming liquid. The physical meanings of both A_i and $W(r_b)$ are, however, different.

Equation 5.26 predicts that as the antifoam concentration $\tilde{c}_i$ of species i becomes very large, V_{AF}, and therefore F, both tend to zero because we must have by definition $q_i < 1$. This is consistent with the usual behavior of antifoams—at sufficiently high concentrations, they eliminate any tendency for foam to form during aeration (see, e.g., Figure 4.2). On the other hand, if $\tilde{c}_i \to 0$, then $F \to 1$ even if $q_i \to 0$. However, it should be noted that at extremely low concentrations of antifoam $N_i(r_b)(= A_i W(r_b)\tilde{c}_i)$ will become <1. The latter implies that we can in principle have $q_i^{N_i(r_b)} \to 0$, and therefore $F \to 0$, if $q_i \to 0$ and $0 < N_i(r_b) < 1$ even though the latter means that some films will be without antifoam! Clearly then the physical basis of supposing that the probability of a bubble surviving without rupture is $q_i^{N_i(r_b)}$ (and therefore of Equations 5.25 and 5.26) will be invalidated if $N_i(r_b) < 1$.

The significance of Equation 5.26 can now be compared with the conclusions derived from the simple phenomenological approach described in Section 5.1. In the latter approach, the main assumption, we recall, was that the relative effectiveness of different antifoam dispersions is independent of antifoam concentration. This is expressed by Equation 5.1. It leads to the conclusion that plots of F against log(antifoam concentration) for different antifoams should be isomorphous. Experimental evidence that this is often observed is presented in Figures 5.2 through 5.5 (also see references [1, 12, 13]).

Since the mathematical form of Equation 5.26 is the same as that presented in the earlier statistical theory [1], then the same analysis can be applied. We present this in outline here but details are to found in the original paper. Consider then dispersions of two different antifoams, 1 and 2, in the same surfactant solution and undergoing the same method of foam generation where $q_1 \neq q_2$ and $A_1 \neq A_2$ where the concentrations of each are adjusted to $\tilde{c}_1^0$ and $\tilde{c}_2^0$ to give the same value of $F = F^0$, respectively.

Since the surfactant solution and foam method are identical, then the bubble size distribution $P(r_b)$ in the absence of antifoam and $W(r_b)$ will be the same in both cases. Then Equation 5.26 implies we can write

$$\int_0^\infty P(r_b) r_b^3 q_1^{A_1 W(r_b)\tilde{c}_1^0} \mathrm{d}r_b = \int_0^\infty P(r_b) r_b^3 q_2^{A_2 W(r_b)\tilde{c}_2^0} \mathrm{d}r_b \tag{5.27}$$

which means that

$$A_1 \tilde{c}_1^0 \ln q_1 = A_2 \tilde{c}_2^0 \ln q_2 \tag{5.28}$$

(where proof is to be found in reference [1]). If Equations 5.27 and 5.28 are combined, then it clear that the probability of bubble rupture is the same function of bubble size for both antifoam dispersions at $F = F^0$ because we must have

$$q_1^{A_1 W(r_b)\tilde{c}_1^0} = \exp\{A_1 W(r_b)\tilde{c}_1^0 \ln q_1\} = \exp\{A_2 W(r_b)\tilde{c}_2^0 \ln q_2\} \tag{5.29}$$

Equation 5.26 can now be differentiated to obtain for antifoam 1 at $F = F^0$

$$\left(\frac{\mathrm{d}F}{\mathrm{d}\ln \tilde{c}_1}\right)^0 = \frac{\int_0^\infty P(r_b) r_b^3 (A_1 W(r_b)\tilde{c}_1^0 \ln q_1) q_1^{A_1 W(r_b)\tilde{c}_1^0} \mathrm{d}r_b}{\int_0^\infty P(r_b) r_b^3 \mathrm{d}r_b} \tag{5.30}$$

This equation can be combined with the equivalent expression for antifoam 2 together with Equations 5.28 and 5.29 to establish that at $F = F^0$

$$\left(\frac{\mathrm{d}F}{\mathrm{d}\ln \tilde{c}_1}\right)^0 = \left(\frac{\mathrm{d}F}{\mathrm{d}\ln \tilde{c}_2}\right)^0 \tag{5.31}$$

which is of course equivalent to Equation 5.3. We have therefore shown that the statistical theory developed here provides a theoretical basis for the conclusion, derived from a simple phenomenological approach, that plots of F against log(antifoam concentration) for different antifoams should be isomorphous.

We should note that Equation 5.31 follows because of the identity represented by Equation 5.28. Differences between the bubble size distribution function $P(r_b)$ are, however, to be expected if different surfactant solutions are to be considered. This in turn means that the identity represented by Equation 5.28 will not hold (see reference [1] for proof) so that the isomorphous behavior represented by Equation 5.31 (and Equation 5.3) will not in general be found when comparison is made between F

vs. log(antifoam concentration) plots in different surfactant solutions. The effect of pH and calcium content on plots of F against log(antifoam concentration) for a given antifoam shown in Figure 5.3 exemplifies this effect.

Comparison of this theory with experiment is at present largely limited to the prediction of Equation 5.31, which is of course no advance on the simple phenomenological approach. However, in principle, some of the quantities in Equation 5.26 are either directly accessible by experiment or could be calculated. Clearly the effect of antifoam on bubble size distribution $P(r_b)$ could be measured but rarely is—no reliable data are apparently to be found in the published literature. Alternatively the effect of antifoam concentration on monodisperse foam volumes would be instructive. In this case, Equation 5.27 reduces to

$$F = \exp(A_i W(r_b)\tilde{c}_i \ln q_i) \tag{5.32}$$

which predicts that F is then simply an exponential function of antifoam concentration. Observations by Kulkarni et al. [37], concerning the effect of commercial polydimethylsiloxane antifoam on the foam generated from sodium oleate solution by sparging, have been shown by Garrett [1] to exhibit such behavior. However, Kulkarni et al. [37] do not give any indication of bubble size distribution, although a porous frit was used, which can mean relatively narrow bubble size distributions. We should note as well that a significant gradient of capillary pressure could exist during foam generation by sparging, which means that q_i will become a function of foam height. That would of course violate one of the assumptions upon which this theory is based. In contrast to foam generation by air entrainment there is also evidence that antifoam induced interbubble coalescence occurs during foam generation by sparging [18, 19]. This would violate another assumption of the theory which supposes that antifoam action only concerns coalescence of the uppermost layer of bubbles with ambient air. If prevalent such a process could mean antifoam action producing a bubble size distribution skewed towards sizes larger than those present in the intrinsic foam.

The function $W(r_b)$ is mainly determined by bubble sizes and the nature of the drainage processes occurring in foam films. The efficiency with which film drainage, for example, reduces the antifoam population in foam films should be, in principle, accessible to theoretical estimation if that is combined with simulations of film drainage in the absence of antifoam (as we have seen, theory is not yet available for the prediction of the drainage rates of the most common mobile asymmetrically draining films [35]). Estimation of $W(r_b)$ is of course of some importance because along with antifoam concentration, it determines whether $N_i(r_b)$ is greater than or less than unity. In turn, that determines the relevance of the statistical theory presented here.

A_i can be obtained from the drop size distribution. However, direct measurement or calculation of q_i would seem to be problematic. It is, of course, as we have seen, determined by many factors most of which cannot be independently estimated. These factors include those that are intrinsic to a given antifoam dispersion, as well as those that are determined by the properties of both the surfactant solution, the structure of the foam, and the method of foam generation. That both the phenomenological approach and this statistical approach are mutually consistent and also consistent with the experimental evidence does, however, tend to validate the assumption that q_i is independent of bubble radii and foam height (and therefore F), at least in the case of the methods considered here.

This statistical approach can be readily extended to mixed dispersions of different antifoams. Thus, the probability of a bubble surviving in the presence of two different antifoam dispersions of species, 1 and 2, respectively, is obviously $q_1^{\tilde{N}_1(r_b)} q_2^{\tilde{N}_2(r_b)}$ from which we may deduce by analogy with Equation 5.26 that we can write

$$F = \frac{\int_0^{\infty} P(r_b) r_b^3 q_1^{A_1 W(r_b)\tilde{c}_{1m}} q_2^{A_2 W(r_b)\tilde{c}_{2m}} \, dr_b}{\int_0^{\infty} P(r_b) r_b^3 \, dr_b} \quad (5.33)$$

where A_1 and A_2 have the same meaning as in Equations 5.27 through 5.30, and $\tilde{c}_{1m}$ and $\tilde{c}_{2m}$ are the weight concentrations of antifoam 1 and antifoam 2 in the mixed dispersion, respectively. Equation 5.33 has the same mathematical form as that derived for the earlier flotation approach for a statistical theory of antifoam action [1]. It may therefore be deduced by analogy with the earlier treatment that Equation 5.33 implies that plots of F against log(antifoam concentration) for both the mixed antifoam dispersion and the constituent antifoam are isomorphous (i.e., consistent with Equation 5.13). It is moreover also possible to derive Equation 5.15, which Figure 5.5 reveals can be used to predict the foam behavior of mixed antifoam dispersions, from those of the constituent antifoam dispersions [1].

Finally we note that Equation 5.26 can in fact be fitted to reduced experimental plots of F against $\log(\tilde{c}_i/\tilde{c}_i^{F=0.5})$ using a single fitting parameter if we substitute the bubble size distribution of de Vries [26] for $P(r_b)$ and simplify by setting $W(r_b) = r_b^{\varphi}$ where φ is some constant (see reference [1] for details). The curve that fits the results in Figure 5.2 was obtained using $\varphi = 6$ as the only fitting parameter. Similarly the results obtained by Ross–Miles foam measurement shown in Figure 5.5 may be replotted in a reduced form and fitted successfully using $\varphi = 4$ [1].

5.2.4 Limitations of Theory

The simple statistical theory presented here is clearly consistent with the phenomenological approach given in Section 5.1 in that Equations 5.3 and 5.15 can be derived from it. That theory therefore affords some explanation for the success of the assumption, expressed in Equation 5.1, that the relative effectiveness of different antifoam dispersions can be adjusted to give the same foam volume, and therefore F, if their concentrations are adjusted in a fixed ratio, which is independent of foam volume and antifoam concentration. However, this theory uses a relatively simplistic statistical treatment that cannot effectively account for any fluctuations in the number of antifoam entities associated with films of a given size. In consequence, it is not relevant for low concentrations of antifoam entities, where the total number of such entities is less than the number of relevant bubbles (so that the mean value of $N_i(r_b)$ is <1) because the probability of a bubble surviving without rupture is then no longer $q_i^{N_i(r_b)}$. It is easy to see, therefore, that in the case of two different antifoam dispersions, each with different values of q_i, the concentrations at which mean values of $N_i(r_b)$ become <1 will not be identical. Plots of F against log(antifoam concentration) are therefore predicted to be no longer isomorphous at the point where $N_i(r_b) < 1$ for even part of the relevant bubble size distribution of one of the dispersions. This follows because the mathematical expression for the probability of survival of a bubble would have to change when $N_i(r_b) < 1$.

These limitations of the theory suggest that a more general statistical theory is necessary if the artificial constraint that applicability is limited to situations where $N_i(r_b) < 1$ is to be removed. In making such a theory, we retain the assumption that $N_i(r_b)$ is an average value given by Equation 5.22. However, we now explicitly introduce a binomial distribution to account for the variation from the mean number of antifoam entities associated with any given bubble size.

We can write for the total number, $N_i^T(r_b)$, of antifoam entities of species i associated with bubbles of radius r_b

$$N_i^T(r_b) = P(r_b)\, N_b\, N_i(r_b) = N_b(r_b)\, N_i(r_b) \tag{5.34}$$

where N_b and $N_b(r_b)$ are the total number of bubbles in the foam and the total number of bubbles of radius r_b in the foam, respectively. Here $N_i(r_b)$ is the average number of antifoam entities of species i associated with each bubble of radius r_b and is given by Equation 5.22. However, we place no constraint on $N_i(r_b)$ so that $N_i(r_b) < 1$ is allowed.

We suppose that the distribution of antifoam entities associated with bubbles of a given radius is represented by a binomial distribution (see, e.g., reference [38]). We assume that the probability of an antifoam being associated with a bubble of radius r_b is $1/N_b(r_b)$ and that the probability of it being elsewhere among such bubbles is $(1 - 1/N_b(r_b))$. We can therefore write for the probability, $Q(r_b,m)$, of having m antifoam entities associated with bubbles of radius r_b as

$$Q(r_b,m) = \frac{N_i^T(r_b)!}{[N_i^T(r_b) - m]!\; m!}\left(\frac{1}{N_b(r_b)}\right)^m \left(1 - \frac{1}{N_b(r_b)}\right)^{N_i^T(r_b)-m} \tag{5.35}$$

where $N_i^T(r_b)!/[N_i^T(r_b) - m]!\; m!$ is the number of ways of achieving the arrangement and where

$$\sum_{m=0}^{N_i^T(r_b)} Q(r_b,m) = 1 \tag{5.36}$$

A similar approach has been adopted in Section 6.4.2 for calculating the probability of finding antifoam drops with a given number of particles.

The probability of a bubble of radius r_b surviving despite the presence of m antifoam particles in the relevant films is, of course, the product $Q(r_p,m).q_i^m$. We can therefore write that

$$F = \frac{\int_0^\infty P(r_b).r_b^3 \left(\sum_{m=0}^{N_i^T(r_b)} Q(r_b,m).q_i^m \right) dr_b}{\int_0^\infty P(r_b).r_b^3 dr_b} \tag{5.37}$$

which is clearly a significantly less convenient expression than Equation 5.26. That Equation 5.36 is, however, physically reasonable is revealed if we substitute $N_i^T(r_b) = N_i(r_b) = m = 0$ to represent the absence of any antifoam entities whereupon we find $F = 1$ even if we set $q_i = 0$.

As we have seen, a problem with Equation 5.26 concerns the situation where $q_i = 0$ (representing an extremely effective antifoam) and $0 < N_i(r_b) < 1$—that theory predicts $F = 0$ despite the presence of bubbles without antifoam entities in the relevant films! That Equation 5.37 avoids this limitation is easily revealed if we consider a simple numerical example. From Equation 5.35, it is clear that the probability of a bubble of radius r_b being without antifoam entities (i.e., where m = 0) is

$$Q(r_b, m=0) = \left(1 - \frac{1}{N_b(r_b)}\right)^{N_i^T(r_b)} \approx \exp\left(-N_i^T(r_b)/N_b(r_b)\right) \tag{5.38}$$

where the approximation follows if $N_b(r_b) \gg 1$. Therefore, we can calculate, for example, that if the ratio of the total number of antifoam entities associated with bubbles of radius r_b to the total number of such bubbles (= $N_i(r_b)$) is 0.5, then we will have ~60% of those bubbles surviving even if $q_i = 0$ because $q_i^{m=0} = 1$. The corollary of this argument is, of course, that Equation 5.37 means that we must have $N_i(r_b) > 1$ for all r_b to yield $F = 0$ even if $q_i = 0$.

Unfortunately, the complexity of Equation 5.37 means that consistency with the experimental finding that plots of F against log(antifoam concentration) for antifoams of different effectiveness should be isomorphous even if the restriction $N_i(r_b) > 1$ is relaxed can only be established by numerical analysis, which has not yet been made. It is clear, however, that the equation does retain an expectation that the probability of foam bubble collapse by antifoam action should increase with increase in bubble size because we retain implicit use of Equation 5.22 to relate $N_i(r_b)$ to bubble size.

5.3 SUMMARIZING REMARKS

It should be emphasized that this chapter is concerned exclusively with the effect of antifoam concentration on foam generation. We are mainly concerned here with foam generation by air entrainment using, for example, shaken, rotated, or stirred vessels and impinging jets, etc. Perhaps the majority of foam control problems for which antifoams are employed concern foam generation by such methods. They are characterized by the polydispersity of the foam produced—it is difficult in most cases to even envisage design of an experimental technique that could produce monodisperse foam by such methods.

We have developed a simple phenomenology for the relative effectiveness of different antifoams in determining the amount of foam generated by such air entrainment method. Thus, if we make the simple assumption that the relative concentrations of different antifoams required to ensure a given foam volume is a constant $\tilde{n}$, independent of both antifoam concentration and foam volume, we can derive simple relations that can be compared with experiment. This leads to the prediction that plots of foam height, volume, or F against log(antifoam concentration) should be isomorphous for different antifoams with the same solution and foam generation method. It is even in principle possible to quantitatively predict the overall foam behavior to be expected from mixed dispersions of different antifoams from the behavior of the individual antifoam dispersions.

The available experimental evidence is consistent with the conclusions derived from this phenomenology. That $\tilde{n}$ is independent of foam height is perhaps surprising since antifoam action is considered to be, in part, determined by Plateau border capillary pressures at the top of foam columns where antifoam action directly reduces foam volumes. If changes in foam height resulted in differences in Plateau border capillary pressure, then it would be expected that the responses to such changes by different antifoams would be different, which would mean that $\tilde{n}$ would vary. That it does not vary suggests that capillary pressures in the Plateau borders at the top of a column of foam being generated by air entrainment are not a function of foam height. Perhaps this is to be expected in the case of shaken or rotated vessels where hydrostatic head has little relevance. It seems likely that this is also true in the case of foam generation by impinging jet(s) (such as by Ross–Miles method [14, 39, 40]) where the upper surface of the foam is simultaneously irrigated and aerated regardless of foam height.

We have derived some theoretical basis for this phenomenology using a statistical approach. That approach assumes that we can define a probability q_i that a given antifoam entity will be ineffective in a given dispersion and foam generation method. If there are $N_i(r_b)$ antifoam entities in a foam film formed by a bubble at the top of a foam column, then the probability of the bubble surviving is $q_i^{N_i(r_b)}$. Since larger bubbles will mean larger films with higher $N_i(r_b)$, the probability of a bubble surviving is predicted to be lower. This approach therefore predicts that bubble size distributions will be skewed toward smaller sizes as a result of antifoam action. Provided $N_i(r_b) \gg 1$, this approach yields the same basic relations as the simple phenomenology. However, if $N_i(r_b) < 1$, then the number of antifoam entities is less than the number of relevant bubbles and this statistical treatment becomes invalid. An alternative more rigorous statistical theory has also been presented, which does not have this limitation. However, the complexity of the relevant equations implies that consistency with the experimental finding that plots of F against log(antifoam concentration) for different antifoams are isomorphous can only be established numerically. Such a numerical analysis has not yet been made.

The prediction of these statistical theories that bubble size distributions should become more skewed to small sizes as the antifoam concentration increases has not been generally validated. Some experimental evidence is of course shown in Figure 5.7 that this can in fact occur. There is, however, experimental evidence that, in the apparently exceptional case of foam generation by sparging, bubble size distribution is skewed to large sizes by antifoam action as a result of interbubble coalescence [18, 19, 41]. All of this underlies the need for systematic knowledge of the effect of antifoam action on bubble size distribution with different methods of foam generation. Lack of such knowledge represents a surprising deficiency which should be remedied.

APPENDIX 5.1 EFFECT OF EXCLUDED VOLUME ON ANTIFOAM CONCENTRATION IN A FILM EXHIBITING REYNOLDS DRAINAGE

A5.1.1 Mean Flow Velocity of Antifoam Entities

We will suppose that a foam film drains according to lubrication theory, which means that the drainage is described by the Reynolds equation [33]. For such a cylindrical film of radius r_f and thickness h with two immobile surfaces, the fluid flow velocity profile $u(z)$ is given [33, 42]

$$u(z) = \frac{3\varphi_h r_f}{h^3}(zh - z^2) \qquad \text{(A5.1.1)}$$

where z is the distance from the lower surface of the film at $z = 0$ and where the upper surface is at $z = h$. The overall rate of thinning of the film is $\varphi_h = dh/dt$. These quantities are defined in Figure 5.8. The mean fluid velocity, u_{mean}, out of the film is therefore given

$$u_{mean} = \frac{1}{2\pi r_f h}\int_0^h 2\pi r_f . u(z)\,dz = \frac{3\varphi_h r_f}{h^4}\int_0^h (zh - z^2)\,dz = \frac{\varphi_h r_f}{2h} \qquad \text{(A5.1.2)}$$

As illustrated in Figure 5.7, the excluded volume effect means that spherical antifoam entities of radius r_{AF} can only experience flow between $z = r_{AF}$ and $z = h - r_{AF}$.

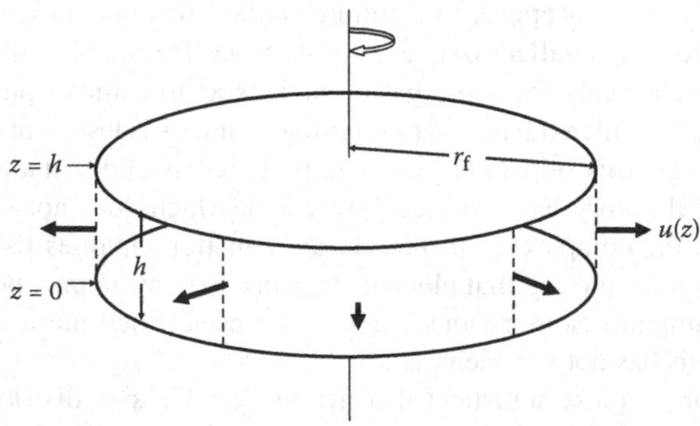

FIGURE 5.8 Schematic illustration of drainage in cylindrical, plane-parallel film with immobile air–water surfaces.

Here we suppose that the concentration of antifoam entities is sufficiently high so that entities sample the complete velocity profile. If then we neglect interactions between antifoam entities and their interactions with the air–water surface (including electrostatic interactions), we can write for the mean velocity $\bar{u}_{AF}$ of the antifoam entities

$$\bar{u}_{AF} = \frac{3\varphi_h r_f}{h^3(h-2r_{AF})}\int_{r_{AF}}^{h-r_{AF}}(zh-z^2)\,dz \tag{A5.1.3}$$

where we assume their centers of mass move at the same velocity as the fluid so that inertial effects are neglected as are those due to the distribution of fluid velocities over the antifoam entities [43]. If we now substitute both Equation A5.1.2 and $\Omega = 2r_{AF}/h$ in Equation A5.1.3 and integrate we find

$$\bar{u}_{AF} = \frac{u_{mean}}{(1-\Omega)}\left(1-3\frac{\Omega^2}{2}+\frac{\Omega^3}{2}\right) = u_{mean}\left(1+\Omega-\frac{\Omega^2}{2}\right) \tag{A5.1.4}$$

A5.1.2 Effect of Excluded Volume on Antifoam Concentration in a Draining Film

The rate, dN_0/dt, of the total number of entities leaving a film without any excluded volume effect (where h is large so that $\Omega \ll 1$ and $\bar{u}_{AF} \approx u_{mean}$) is given

$$\frac{dN_0}{dt} = \tilde{c}_0\frac{dV_f}{dt} = \tilde{c}_0 u_{mean}.2\pi r_f h \tag{A5.1.5}$$

where $\tilde{c}_0$ is the initial antifoam concentration and V_f is the volume of fluid in the film and t is the time. Similarly, if there is an excluded volume effect so that the mean velocity $\bar{u}_{AF} > u_{mean}$, then the rate, dN_{ev}/dt, of the total number of entities leaving a film is given

$$\frac{dN_{ev}}{dt} = \tilde{c}_0\bar{u}_{AF}.2\pi r_f h \tag{A5.1.6}$$

Therefore, the rate of depletion of the total number of antifoam entities $d\Delta N/dt$ in the film due to the excluded volume effect is given by combining Equations A5.1.5 and A5.1.6

$$\frac{d\Delta N}{dt} = \frac{dN_{ev}}{dt} - \frac{dN_0}{dt} = \tilde{c}_0(\bar{u}_{AF} - u_{mean}).2\pi r_f h \tag{A5.1.7}$$

The rate of reduction in antifoam concentration $\mathrm{d}\tilde{c}/\mathrm{d}t$ is then simply the rate of depletion of the total number of antifoam entities per unit volume of the film so that

$$\frac{\mathrm{d}\tilde{c}}{\mathrm{d}t} = \frac{1}{\pi r_\mathrm{f}^2 h} \cdot \frac{\mathrm{d}\Delta N}{\mathrm{d}t} = 2\tilde{c}_0(\overline{u}_\mathrm{AF} - u_\mathrm{mean})/r_\mathrm{f} \tag{A5.1.8}$$

Conservation of volume means that we can write for the rate of film drainage $\mathrm{d}h/\mathrm{d}t$

$$\pi r_\mathrm{f}^2 \frac{\mathrm{d}h}{\mathrm{d}t} = 2\pi r_\mathrm{f} h u_\mathrm{mean} \tag{A5.1.9}$$

Combining Equations A5.1.4, A5.1.8, and A5.1.9 yields a relation between the rate of reduction in antifoam concentration, due to the excluded volume effect, and the rate of film thinning so that

$$\frac{\mathrm{d}\tilde{c}}{\mathrm{d}t} = \frac{\tilde{c}_0}{h}\left(\frac{2r_\mathrm{AF}}{h} - \frac{2r_\mathrm{AF}^2}{h^2}\right) \cdot \frac{\mathrm{d}h}{\mathrm{d}t} \tag{A5.1.10}$$

Equation A5.10 can be rearranged and integrated between the limits $h = \infty$ and $h = 2r_\mathrm{AF}$ to give the total change in antifoam concentration, due to the excluded volume effect, as the film drains from one of infinite thickness to one of a thickness equal to that of the diameter of the antifoam entities. We therefore have

$$\tilde{c}_{h=\infty} - \tilde{c}_{h=2r_\mathrm{AF}} = \tilde{c}_0 \int_{h=2r_\mathrm{AF}}^{h=\infty} \left(\frac{2r_\mathrm{AF}}{h^2} - \frac{2r_\mathrm{AF}^2}{h^3}\right) \mathrm{d}h \tag{A5.1.11}$$

Since we must have $\tilde{c}_0 = \tilde{c}_{h=\infty}$ (because the excluded volume effect will not be present as $\Omega \ll 1$ at $h = \infty$), the integral reduces to $\tilde{c}_{h=2r_\mathrm{AF}} = 0.25\tilde{c}_0$. Thus, the excluded volume effect produces a 4-fold reduction in antifoam concentration during drainage of an immobile foam film (of the type considered here) until the film thickness approximates the diameter of an antifoam entity. As we have stated, this analysis is, however, subject to a number of simplifying assumptions, which include neglect of electrostatic effects and the effect of the liquid layer between the rigid air–water surface and antifoam entities. The latter is likely to slow the motion of those entities. If the air–water surface is mobile, then this too will tend to reduce the significance of the excluded volume effect. The estimate of a 4-fold reduction in concentration due to that effect should therefore be regarded as an approximate upper limit for a foam film exhibiting Reynolds drainage.

REFERENCES

1. Garrett, P.R. *Langmuir*, 11(9), 3576, 1995.
2. Pouchelon, A., Araud, C. *J. Disper. Sci. Technol.*, 14(4), 447, 1993.
3. Koczo, K., Koczone, J., Wasan, D. *J. Colloid Interface Sci.*, 166, 225, 1994.
4. Racz, G., Koczo, K., Wasan, D. *J. Colloid Interface Sci.*, 181, 124, 1996.
5. Bergeron, V., Cooper, P., Fischer C., Giermanska-Kahn, J., Langevin, D., Pouchelon, A. *Colloids Surf. A*, 122, 103, 1997.
6. Denkov, N., Cooper, P., Martin, J. *Langmuir*, 15, 8514, 1999.
7. Denkov, N., Marinova, K.G., Christova, C., Hadjiiska, A., Cooper, P. *Langmuir*, 16, 2515, 2000.
8. Marinova, K.G., Tcholakova, S., Denkov, N., Roussev, S., Deruelle, M. *Langmuir* 19, 3084, 2003.
9. Arnaudov, L., Denkov, N, Surcheva, I., Durbut, P., Broze, G., Mehreteab, A. *Langmuir*, 17, 6999, 2001.
10. Hadjiiski, A., Tcholakova, S., Denkov, N., Durbut, P., Broze, G., Mehreteab, A. *Langmuir*, 17, 7011, 2001.
11. Denkov, N. *Langmuir*, 20(22), 9463, 2004.
12. Ran, L. Foaming of anionic surfactant solutions in the presence of calcium ions and triglyceride-based antifoams, PhD Thesis, University of Manchester, 2011.
13. Cubero, M.T.G., Benito, G.G., Fernandez, O., Blanco J.F.H. *Zuckerindustrie*, 125(7), 524, 2000.
14. Veber, V., Pauchek, M. *Acta Fac. Pharm. Univ. Comenianae*, 26, 221, 1974.
15. Bikermann, J.J. *Foams*, Springer, Berlin, 1973, Chapter 3, p 80.
16. Princen, H. *Langmuir*, 4(1), 164, 1988.
17. Mingins, J., Scheludko, A. *J.C.S Faraday I*, 75, 1, 1976.
18. Pelton, R.H., Goddard, E.D. *Colloids Surf.*, 21, 167, 1986.
19. Pelton, R.H. *Chem. Eng. Sci.*, 51(19), 4437, 1996.
20. Malysa, K., Cohen, R., Exerowa, D., Pomianowski, A. *J. Colloid Interface Sci.*, 80(1), 1, 1981.
21. Malysa, K., Lunkenheimer, K., Miller, R., Hempt, C. *Colloids Surf.*, 16, 9, 1985.
22. Neethling, S. *Colloids Surf. A*, 263, 184, 2005.
23. Denkov, N., Tcholakova, S., Marinova, K., Vankova, N., Christov, N., Russev, S. unpublished work.
24. Garrett, P.R. unpublished work.
25. Garrett, P.R. *J. Colloid Interface Sci.*, 69(1), 107, 1979.
26. de Vries, A. *Rec. Trav. Chim.*, 77, 209 and 283, 1958.
27. Lemlich, R. *Ind. Eng. Chem. Fundam.*, 17(2), 89, 1978.
28. Neethling, S. personal communication.
29. Garrett, P.R. Mode of action of antifoams, in *Defoaming: Theory and Industrial Applications*, First Edition (Garrett, P.R., ed.), Marcel Dekker, New York, 1993, p 95.
30. Hines, J. unpublished work.
31. Garrett, P.R., Davis, J., Rendall, H. *Colloids Surf A*, 85, 159, 1994.
32. Exerowa, D., Kruglyakov, P. *Foam and Foam Films*, Elsevier, Amsterdam, 1998, p 108.
33. Reynolds, O. *Phil. Trans. Roy. Soc., London, A* 177, 157, 1886.
34. Joye, J., Miller, C.A., Hirasaki, G.J. *Langmuir*, 8, 3083, 1992.
35. Joye, J., Hirasaki, G.J., Miller, C.A. *Langmuir*, 10, 3174, 1994.
36. Probstein, R. *Physicochemical Hydrodynamics, An Introduction*, Second Edition, Wiley Interscience, Hoboken, NJ, 2003, p 158.
37. Kulkarni, R., Goddard, E.D., Kanner, B. *J. Colloid Interface Sci.*, 59(3), 468, 1977.
38. Pugh, E.M., Winslow, G.H. *The Analysis of Physical Measurements*, Addison-Wesley, Reading, MA, 1966, p 41.

39. Testing of Surfactants. Determination of foaming power: Modified Ross–Miles method, DIN 53902, Sheet 2, January 1971.
40. Pasztor-Rozzo, F. *Fette, Seifen, Anstrichmittel.*, 67(9), 688, 1965.
41. van der Zon, M., Hammersma, P.J., Poels, E.K., Bliek, A. *Chem. Eng. Sci.*, 57, 4845, 2002.
42. Barber, A.D. A model for a cellular foam, PhD Thesis, University of Nottingham, 1973, p 12.
43. Faxen, H. *Ann. Phys.*, 68(4), 89, 1922.

6 Deactivation of Mixed Oil–Particle Antifoams During Dispersal and Foam Generation in Aqueous Media

6.1 INTRODUCTION

It is well known that the effectiveness of hydrophobed silica–polydimethylsiloxane antifoams in aqueous solutions deteriorates markedly during both use [1–7] and dispersal as an emulsion [6]. Results of Marinova et al. [7] illustrating this effect are reproduced in Figure 6.1. Here the time taken for foam collapse after repeated shake–quiescent cycles of a solution of 0.01 M aerosol OT (sodium bis-octyl sulfosuccinate) containing ~0.1 g dm^{-3} of hydrophobed silica–polydimethylsiloxane antifoam is plotted against number of cycles. Antifoam was added neat to the solution without predispersal. Marked deactivation of the antifoam is observed after 50–60 cycles.

There is some evidence that this deactivation is apparent with other hydrophobic particle–hydrophobic oil mixed antifoams [8–10]. It also seems probable that the prolongation of antifoam effectiveness accompanying increase in viscosity of the oil [2, 3] concerns the effect of that increase on the process of antifoam deactivation rather than on the intrinsic effectiveness of the antifoam. Moreover, partial deactivation of antifoams during dispersal to form emulsions appears to result from the same cause as deactivation during foam generation [6]. This deactivation is commercially important because antifoam products for certain applications are often necessarily prepared as emulsions.

Deactivation is clearly a large effect and can involve complete loss of antifoam effectiveness after prolonged exposure to foam generation. The evident commercial significance and apparent generality of this deactivation phenomenon therefore clearly suggest that it merits detailed study. Indeed, its importance was recognized by Ross [11] who wrote more than half a century ago that "not enough studies have been made of the effect of the passage of relatively long periods of time on systems containing antifoam agents." It is, however, only in the past 20 years that significant attention has been given to this topic.

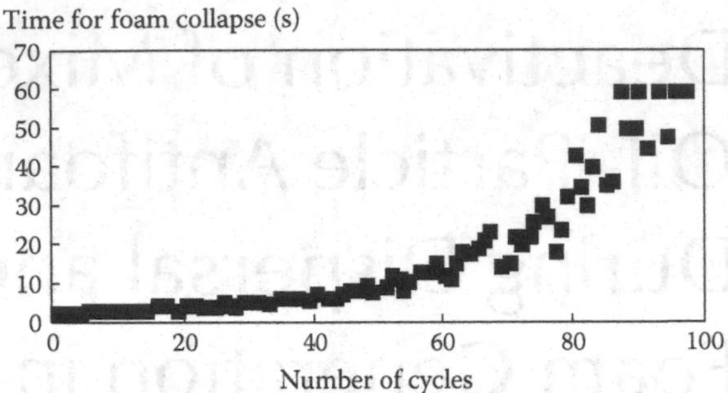

FIGURE 6.1 Deactivation of hydrophobed silica–polydimethylsiloxane antifoam. Neat antifoam (~0.1 g dm^{-3} added to 0.01 M AOT solution and subjected to repeated shake–quiescent cycles). (Adapted with permission from Marinova, K.G., Tcholakova, S., Denkov, N., Roussev, S., Deruelle, M. *Langmuir* 19, 3084. Copyright 2003 American Chemical Society.)

In this chapter, we present a detailed review of deactivation of particle–oil antifoams, with particular emphasis on hydrophobed silica–polydimethylsiloxane antifoams about which most studies have been made. The available evidence suggests that the phenomenon concerns disproportionation of antifoam drops during processes of splitting and coalescence, which occur as the antifoam is either dispersed or interacts with foam films. We speculate here about the likely causes of this phenomenon and describe theories that attempt to predict foam volume growth in the presence of deactivating antifoam.

6.2 DEACTIVATION OF THE ANTIFOAM EFFECT OF POLYDIMETHYLSILOXANE OILS WITHOUT PARTICLES

Hydrophobic oils such as polydimethylsiloxanes can exhibit significant antifoam effects even in the absence of hydrophobic particles. The antifoam efficiency of such oils is of course much enhanced by the presence of hydrophobic particles admixed with the oil. However, the oils alone are known to lose antifoam effectiveness during foam generation, albeit more rapidly than is the case with oil–particle mixtures [3, 4]. It is argued by Basheva et al. [12] that this deactivation may be indirectly attributed to the relatively high stability of the relevant polydimethylsiloxane–water–air pseudoemulsion films in the absence of hydrophobic particles. As we have seen (see Chapter 4), stable pseudoemulsion films mean that drops of oil do not have time to emerge into the air–water surfaces of foam films before they are swept out by the draining flux present in those films. If, however, such drops are sufficiently large (≫10 microns), they may subsequently accumulate in Plateau borders where the capillary pressure may become high enough to overcome the disjoining forces stabilizing the pseudoemulsion films. Drops may then emerge into the air–water surfaces of the Plateau borders, forming bridging configurations that are unstable [2] and which lead to foam collapse as we discuss in Section 4.5.3. Continuous agitation accompanying foam generation will, however, mean drop breakup and progressive disappearance of drops large enough to produce these Plateau border collapse phenomena. This will lead to deactivation of the antifoam effect. Unpublished results of Yorke [13], illustrating this deactivation, are shown in Figure 6.2 for a 50 mPa s polydimethylsiloxane oil in a solution of 5×10^{-3} M sodium nonyl benzene sulfonate in 3.8×10^{-2} M NaCl. Foam was generated by shaking a graduated measuring cylinder. The foam height is seen to initially fall as the oil is dispersed. However, it subsequently increases as the antifoam effect is deactivated with an increasing number of shakes. Deactivation correlates, as expected, with the decrease in the proportion of large (≫10 microns) drops also shown in the figure. We would expect therefore that increase in viscosity of the oil would diminish the rate of drop breakup and therefore decrease the rate of deactivation. Racz et al. [3] present some evidence to show that this may happen, at least in the case of repeated foam film formation using a film frame.

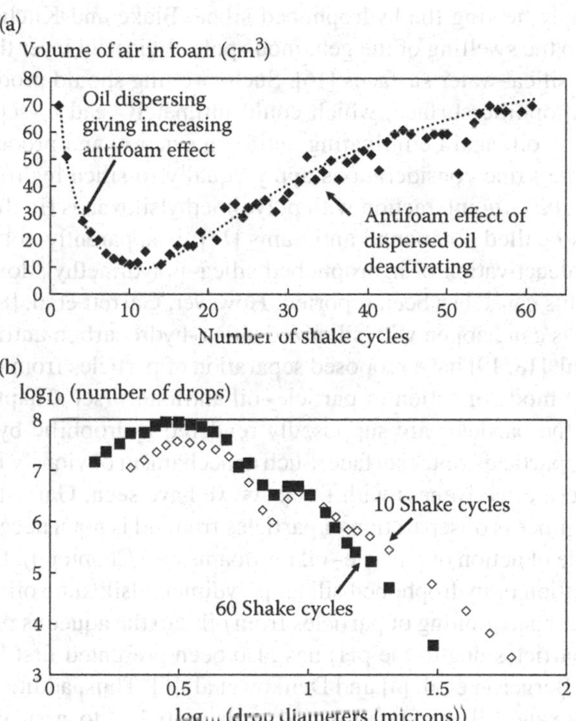

FIGURE 6.2 Deactivation of antifoam effect of polydimethylsiloxane oil alone. Polydimethylsiloxane, of viscosity 50 mPa s, was added neat to solution of 5×10^{-3} M sodium nonyl benzene sulfonate in 3×10^{2} M NaCl. (a) Volume of air in foam generated by shaking 0.025 dm³ of a surfactant solution containing 0.63 g dm⁻³ of the oil in a 0.1 dm³ measuring cylinder—each cycle consisting of 30 shakes. (b) Effect of several shake cycles on drop size distribution of the oil (by Malvern Sizer). (After Yorke, J.W.H., unpublished work.)

6.3 EARLY WORK WITH HYDROPHOBED SILICA–POLYDIMETHYLSILOXANE ANTIFOAMS

6.3.1 Separation of Silica from Oil

Arguably, the simplest explanation for the phenomenon of deactivation of hydrophobed silica–polydimethylsiloxane antifoams concerns the chemical and physical stability of the hydrophobed silica. At sufficiently high pH, the silica will in fact dissolve, effectively eliminating the synergy characteristic of these antifoams (see, e.g., reference [14]). Indeed, at a sufficiently high pH, the chemical stability of the polydimethylsiloxane will be compromised. No evidence of deactivation attributable to such causes has, however, been reported. Another possible cause of instability of the hydrophobed silica concerns the nature of the hydrophobed layer. Blake and Kitchener [15] have shown that the contact angle at the air–water surface of silica hydrophobed with trimethylchlorosilane is unstable—it gradually declines with time of exposure to the aqueous phase. The effect is reversible—the contact angle can be restored by simply heating the hydrophobed silica. Blake and Kitchener [15] attribute the effect to the swelling of the gelatinous polysilicic acid layer that is supposed to be present at silica–water surfaces [16]. Such swelling should produce a progressively more hydrophilic surface, which could ultimately lead to separation of the particles from the oil, again eliminating antifoam synergy and producing deactivation. Whether the same considerations apply equally to silica hydrophobed by, for example, thermal *in situ* interaction with polydimethylsiloxanes (in the manner used to prepare the so-called compound antifoams [17]) is apparently not known. Again no evidence of deactivation of hydrophobed silica–polydimethylsiloxane antifoams attributable to this cause has been reported. However, Garrett et al. [8] present some evidence that this can happen with silanized silica–hydrocarbon antifoams.

Kulkarni et al. [18, 19] have proposed separation of particles from oil as an intrinsic aspect of the mode of action of particle–oil antifoams (see Chapter 4). Here we remember that the particles are supposedly rendered hydrophilic by adsorption of surfactant at the particle–water surface. Such a mechanism obviously implies that the antifoam will lose effectiveness with time. As we have seen, Garrett et al. [8] have shown that this process of separation of particles from oil is not a necessary aspect of the general mode of action of particle–oil antifoams (see Chapter 4). Direct evidence that the deactivation of hydrophobed silica–polydimethylsiloxane oil antifoam is not necessarily due to partitioning of particles from oil into the aqueous phase (or indeed dissolution of particles due to the pH) has also been presented first by Koczo et al. [2] and later by Bergeron et al. [4] and Denkov et al. [6]. Thus, addition of fresh bulk polydimethylsiloxane oil, uncontaminated with particles, to a surfactant solution containing antifoam deactivated by many foam-generating shake cycles produces a marked reactivation of antifoam activity. Here Koczo et al. [2] used 0.06 M sodium dodecyl sulfate (SDS) solution, Bergeron et al. [4] used 0.008 M aerosol OT (sodium bis-ethylhexyl sulfosuccinate) solution and Denkov et al. [6] used 0.0113 M aerosol OT solution. The results of Denkov et al. [6] are reproduced in Figure 6.3. Clearly the fresh oil is interacting in some way with the deactivated antifoam to produce effective material. This implies that the fresh oil is acquiring silica particles because

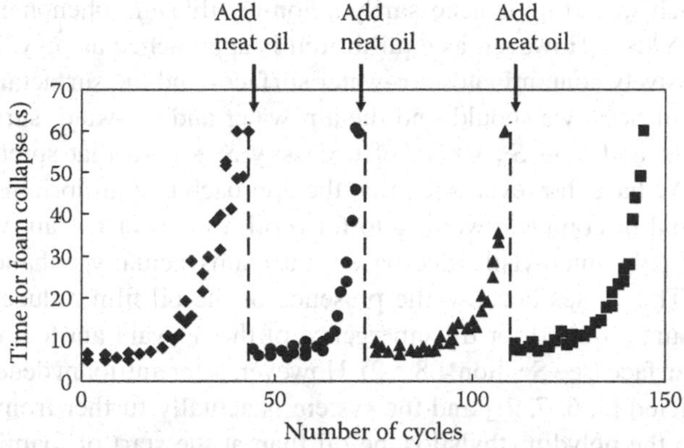

FIGURE 6.3 Successive deactivation and reactivation events accompanying addition of aliquots of neat polydimethylsiloxane oil (without particles) to hydrophobed silica–polydimethylsiloxane antifoam dispersion. Addition of neat oil is seen to restore antifoam activity as measured by foam collapse times. However, that activity is progressively lost with succeeding foam generation shake–quiescent cycles only to be restored by further additions of neat oil. Original antifoam added neat at concentration of 0.005 g dm^{-3} in 0.0113 M AOT solution. Added aliquots of polydimethylsiloxane oil each equivalent to ~5×10^{-5} g dm^{-3}. (Reprinted with permission from Denkov, N., Marinova, K.G., Christova, C., Hadjiiska, A., Cooper, P. *Langmuir*, 16, 2515. Copyright 2000 American Chemical Society.)

according to Denkov et al. [6], the oil alone is without effect on this solution under the selected conditions. However, we should note that Bergeron et al. [4], using a similar method to that of Denkov et al. [6], and 0.008 M solutions of AOT, find that polydimethylsiloxane oil alone is "rather effective" but that it deactivates in three cycles, which compares with >50–60 cycles for hydrophobed silica–polydimethylsiloxane mixtures (see Figure 6.1). Irrespective of this apparent discrepancy, the reactivation reproduced in Figure 6.3 yields an antifoam effect, which deactivates over a time scale comparable to that of hydrophobed silica–polydimethylsiloxane mixtures rather than the oil alone. Such reactivation could not happen if the silica particles had become hydrophilic during deactivation. Separation of particles from oil as a result of the particles becoming hydrophilic, by either surface polysilicic acid gel formation, by adsorption of surfactant or by a combination of both, is therefore clearly not a necessary aspect of the process of deactivation of hydrophobed silica–polydimethylsiloxane antifoams.

6.3.2 Equilibration and Deactivation

It could be argued that an explanation for the deactivation of hydrophobed silica–polydimethylsiloxane antifoams concerns the mutual equilibration of the oil and the relevant surfactant solution rather than the role of the particles [4, 20]. This argument is of course derived from the assumption that the mode of action of these antifoams concerns Marangoni spreading (see Section 4.4.2). As we have described in

Section 3.2, such spreading is necessarily a non-equilibrium phenomenon deriving from $S^i > 0$ or $S^{si} > 0$. However, as equilibrium is approached as polydimethylsiloxane oil progressively contaminates air–water surfaces and the surfactant adsorbs on the oil–water surfaces, we should find the air–water and oil–water surface tensions decline, $E \rightarrow E^e$ and $S \rightarrow S^e$, where of necessity $S^e \leq 0$ so that spreading can no longer occur. We have, however, seen that the approach to equilibrium by a process of pseudo-partial or complete wetting to form oil films over the air–water surface in the case of polydimethylsiloxane-based antifoams actually enhances antifoam effectiveness. This arises because the presence of the oil film reduces the critical capillary pressure required for the emergence of the relevant antifoam entities into the air–water surface (see Section 4.8.5.2). However, after antifoam deactivation, this oil film is depleted [3, 6, 7, 21] and the system is actually further from equilibrium with respect to the polydimethylsiloxane oil than at the start of foaming when the antifoam is most effective. Clearly then it would seem that mutual equilibration of the antifoam oil with the surfactant solution and the relevant surfaces is not the cause of polydimethylsiloxane–hydrophobed silica antifoam deactivation.

6.3.3 Deactivation, Emulsification, and Drop Sizes

While not invoking a Marangoni spreading mechanism, Racz et al. [3] have argued that antifoam action requires that an oil film spreads rapidly over foam film surfaces. They argue that this oil film will drag oil lenses into those films to cause film rupture, which in turn results in emulsification of the antifoam. The emulsified oil is assumed to be totally inactive. The greater the extent of spreading before foam film rupture, the greater the amount of oil emulsified and the greater the deactivation. This mechanism is therefore inconsistent with the well-known practice of preparing emulsions of antifoams so that they may be readily applied to practical situations where the presence of undispersed neat antifoam is not desirable (e.g., in paint formulations). Such emulsions are active but, as we shall see, are less so than the neat antifoam. The mechanism of Racz et al. [3] is also, as we show below, inconsistent with different rates of deactivation associated with differences in oil viscosity over the range of viscosities where spreading rates are constant.

It has been suggested that deactivation of hydrophobed silica–polydimethylsiloxane antifoams concerns decreasing antifoam drop sizes during both foam generation and preparation of antifoam emulsions [2, 4]. As we have seen (Section 6.2), decreasing drop sizes represents the probable explanation for the deactivation of the weak antifoam effects of polydimethylsiloxane oils alone where these effects probably concern antifoam action in Plateau borders. Koczo et al. [2] extend this argument to hydrophobed silica–polydimethylsiloxane antifoams, claiming that small drops are readily excluded from foam films and that film rupture due to bridging drops also occurs primarily in Plateau borders. This means, for example, that single drop diameters would have to be of the order of 10–50 μm in diameter if they are to bridge Plateau borders in dry foam of <10 cm height [2]. Much larger drops are presumably implied if wet foams with larger Plateau borders are considered. However, it is well known that hydrophobed silica–polydimethylsiloxane antifoam effects have been reported for drop sizes ≪10 microns [6, 21, 22]. Koczo et al. [2] avoid this difficulty by claiming that any dimensional discrepancy could be accounted for by the presence of several drops in the Plateau borders. However, if this occurred easily, it would tend to undermine the proposition [2] that decreasing drop sizes necessarily leads to antifoam deactivation.

Another problem with this argument of Koczo et al. [2] is that there is strong evidence that that hydrophobed silica–polydimethylsiloxane antifoams actually function by bridging foam films (see, e.g., reference [5] and Section 4.8). It is therefore possible that the relatively small dimensions of foam films could mean that diminution of drop sizes in the case of silica–polydimethylsiloxane antifoams could lead to enhanced antifoam activity simply because the concomitant increase in number concentration of drops should lead to an increase in the probability of their presence in foam films [23].

Accepting that hydrophobed silica–polydimethylsiloxane antifoams function in foam films, Bergeron et al. [4] argue that smaller drops mean longer times for those films to drain to form bridging configurations. This would therefore lead to diminished frequencies of foam film rupture by individual antifoam drops. However, Dippenaar [24] has modeled the effect of size on the relative effectiveness of inert particles as antifoams (see Section 4.7.6), taking into account both the times for film drainage and the increase in concentration of antifoam entities accompanying reduction in size for a given weight concentration. He shows that the effect of the latter dominates so that the overall frequency of foam film rupture increases with decrease in size [24]. However, these arguments of Dippenaar [24] and Bergeron et al. [4] both consider only the time for film drainage to the dimensions of the antifoam entities so that a bridging drop may be formed. They therefore ignore the time for antifoam action by the bridging drop, which will be expected to decrease with decreasing drop size (as capillary forces driving antifoam action increase). As we have seen, Frye and Berg [25] in considering foam collapse by bridging inert particles, argue that the two processes of foam film drainage and antifoam action will conspire to produce an optimum particle size where foam film rupture frequencies are at a maximum (see Section 4.7.6). The model of Dippenaar [24] then only applies to sizes less than this putative optimum where film drainage times dominate in determining overall antifoam effect. The frequency of foam film rupture by oil drops of a size greater than the optimum will be dominated by the time for antifoam action. However, again, we may deduce that decreasing drop sizes for such drops will lead to increased frequencies of foam film rupture because of both increased frequencies of rupture by individual drops and increased number concentrations of drops. Using these arguments, decreasing antifoam drop size is therefore expected to lead to increased frequencies of foam film rupture irrespective of whether sizes are greater than or less than the optimum.

Bergeron et al. [4, 26] also argue that another possible cause of diminished antifoam effectiveness with decreasing size could concern the likely critical rupture thickness of pseudoemulsion films. Thus, if the theory of Vrij and Overbeek [27] (see Section 1.3.3) can be applied to such films, then the smaller the radius of the films, and therefore drops, the smaller the critical rupture thickness, the longer the drainage time, and therefore the lower the frequency of rupture. However, in applying this theory in this context, we should remember that the stability of the pseudoemulsion film is already severely compromised by the presence of the silica particles. It may therefore be irrelevant.

Evidence that hydrophobed silica–polydimethylsiloxane antifoam drop diameters actually decrease during deactivation is described by both Bergeron et al. [4] and Koczo et al. [2]. Both these groups of authors used similar techniques—neat antifoam was added directly to solutions that were aerated by shaking. Bergeron et al. [4] demonstrate, using light scattering, that deactivation is accompanied by decrease in drop diameters from >50 to <8 μm with an approximate mean diameter of ~3 μm. Similarly, Koczo et al. [2] claim, without describing their size measurement technique, that deactivation is accompanied by decrease in drop sizes from 2 to 30 μm down to <5 μm ("typically" 1–3 μm). Both groups of authors then state that decrease in antifoam drop diameter is the cause of deactivation. The only justification for this inference is apparently that hypotheses can be advanced, which suggest why decrease in antifoam drop diameter could lead to diminished antifoam effectiveness. However, direct evidence that the relevant hypotheses described in the foregoing paragraphs are applicable in the present context is lacking. Moreover, we have shown that it is possible to make objection to each one.

By contrast, Denkov et al. [6] have presented evidence that decreases in drop size per se may not be responsible for antifoam deactivation. In this work the drop diameters of a hydrophobed silica–polydimethylsiloxane antifoam emulsified in 0.01 M AOT solution were measured before and after deactivation. Using dynamic light scattering, mean drop diameters were shown to remain essentially constant within a range of 1.4–1.8 μm. Microscopic examination of the emulsion before and after deactivation also revealed little change in drop sizes. Relevant photomicrographs are reproduced in Figure 6.4. In contrast to dynamic light scattering, the contribution of suboptical drops to the overall size distribution is of course absent with photomicrography, which leads to larger estimates of the apparent mean drop size in the range of 2–6 microns. Careful examination of Figure 6.4 reveals that large drops before deactivation are slightly deformed, which suggests the presence of significant

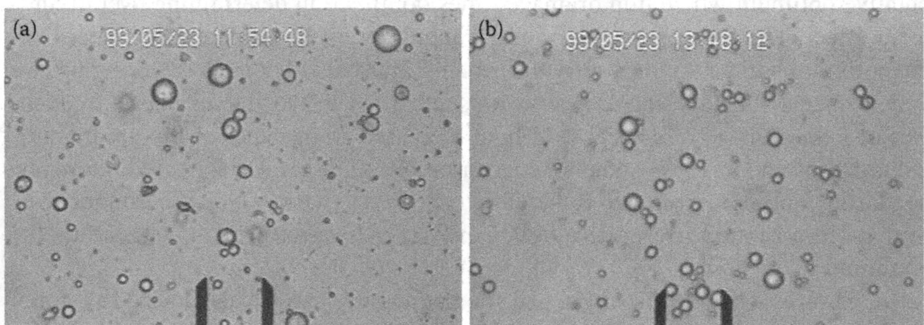

FIGURE 6.4 Photomicrographs of pre-emulsified hydrophobed silica–polydimethylsiloxane antifoam drops dispersed in 0.01 M AOT solution: (a) fresh emulsion; (b) deactivated emulsion. Bar, 31 microns. (Reprinted with permission from Denkov, N., Marinova, K.G., Christova, C., Hadjiiska, A., Cooper, P. *Langmuir*, 16, 2515. Copyright 2000 American Chemical Society.)

numbers of particles at the oil–water interface. However, large drops after deactivation reveal no such deformation, implying absence of such particles [6].

In a later study of the same system, Marinova et al. [7] showed that, rather than decreases in size of antifoam drops, large silica-rich agglomerates form during antifoam deactivation. Others have reported similar observations [1]. In yet another study, using 0.0045 M of a non-ionic surfactant (a commercial C_{12-14} alkyl glucopyranoside), initial emulsified drop diameters were similar to those in 0.01 M AOT [21]. Again no evidence was found for a significant decrease in drop diameters, which could correlate with the drastic reduction in antifoam effect associated with deactivation. In the case of these examples, deactivation of the antifoam must therefore be attributed to some process that occurs during agitation and aeration other than drop size diminution.

Hydrophobed silica–polydimethylsiloxane emulsions are less weight effective, both in initial effectiveness and durability of performance, than the neat antifoams from which they are prepared [2, 3, 6]. It seems likely that this may be attributed to partial deactivation by a similar process to that which occurs during foam generation. The experiments of both Koczo et al. [2] and Bergeron et al. [4] concerned addition of neat antifoam to solutions followed by foam generation. The decrease in drop size reported by these workers therefore presumably concerns a process of emulsification of the antifoam to form drops of small size during foam generation, which is analogous to that involved in the separate preparation of antifoam emulsions. Consistency of the findings of Koczo et al. [2] and Bergeron et al. [4] with those of Denkov et al. [6] therefore requires that addition of neat antifoam involves simultaneous drop breakup and antifoam deactivation during aeration where some other cause, apart from the decrease in drop size per se, must be sought as the reason for that deactivation.

6.4 DEACTIVATION OF HYDROPHOBED SILICA–POLYDIMETHYLSILOXANE ANTIFOAM BY DISPROPORTIONATION

6.4.1 Experimental Evidence

The polydimethylsiloxane oils used for antifoams usually spread on the air–water surfaces of surfactant solutions (see Section 3.6.2). At equilibrium, this process produces either complete wetting and duplex films for which $S^e = 0$ or pseudo-partial wetting and oil films in contact with lenses of bulk oil for which $S^e \leq 0$ (see Section 3.6.2.1). It has been shown by Racz et al. [3], and later confirmed by Denkov et al. [6, 7, 21], that deactivation of hydrophobed silica–polydimethylsiloxane antifoams correlates with the disappearance of this spread oil film. These studies used solutions of both anionic (SDS [3] and AOT [6, 7]) and non-ionic surfactants (alkyl glucopyranoside) [21]. Loss of the spread layer during deactivation is accompanied by an increase in surface tension to that of the pure surfactant solution [6]. It has also been directly observed using ellipsometry [21]. This finding is key to the understanding of deactivation because the presence of a spread layer of polydimethylsiloxane at the air–water surface is a clear indicator that oil has emerged into that surface.

Emergence into the air–water surface is of course a necessary prerequisite for antifoam action. The presence of the spread layer, once formed, also actually facilitates both the emergence of oil drops into air–water surfaces [28] and the formation of the unstable oil bridges, which lead to foam film rupture [29, 30].

Elimination of the spread layer, despite the continued presence of hydrophobed silica–polydimethylsiloxane in the aerated solution, clearly implies some change in the efficiency with which polydimethylsiloxane can transport to the air–water surface. As we have seen, this does not concern any irreversible change in the state of the hydrophobed silica because the antifoam effect can be restored by addition of polydimethylsiloxane alone. A convincing explanation has, however, been proposed by Denkov et al. [6, 7] who suggest that the cause of both the antifoam deactivation and the elimination of the spread layer derives from disproportionation of antifoam entities during emulsification rather than simply from any change in drop size accompanying emulsification per se. This emulsification may occur either in the bulk phase by applying shear with a mixing device, as a consequence of the formation of a spread layer or as a result of antifoam action by the bridging–stretching mechanism. It is proposed that these processes ultimately result in disproportionation of the drop population into two—some drops containing no particles and some drops enriched with respect to particles [6, 7]. Drops containing no particles will be ineffective in both supplying polydimethylsiloxane to the air–water surface and in antifoam action because of the stability of the relevant pseudoemulsion films. Drops containing excess silica are also ineffective but for reasons that are less obvious. These latter drops appear to coalesce during foam generation to form an increasing proportion of large, white non-deformable, gel-like agglomerates [1, 6]. It is possible that these agglomerates grow so large that they become ineffective as sites for spreading PDMS oil onto air–water surfaces, thereby contributing to the depletion of the spread oil films on those surfaces summarizing all of these processes is shown in Figure 6.5.

Marinova et al. [7] have made a detailed study of the state of the polydimethylsiloxane–hydrophobed silica antifoam entities during foam generation in solutions of 0.01 M AOT by shake–quiescent cycles. The study involved addition of neat antifoam to the solution. This implies that dispersal of the antifoam was superimposed on the deactivation processes. A micrograph obtained by these workers illustrating the nature of the irregular non-deformable silica-rich agglomerates is shown in Figure 6.6. Formation of such agglomerates has been previously observed by Pouchelon and Araud [1] who showed that they can grow to sizes in excess of 1000 microns during foam generation. Analysis revealed that the silica content of these agglomerates was 17% by weight, which compared with only 2.5% for the original antifoam [1]. Denkov et al. [6] report similar findings with agglomerates containing ~14–16% by weight silica.

The results of this study of Marinova et al. [7] are reproduced in Table 6.1. Here the change in time for foam collapse due to antifoam deactivation is compared with microscope observations of the state of the antifoam dispersion. It is obvious from Table 6.1 that the size and proportion of non-deformable gel-like agglomerates increases as the antifoam deactivates. Similarly, the proportion of (antifoam) oil drops decreases. The range of diameters of these oil drops appears to decline

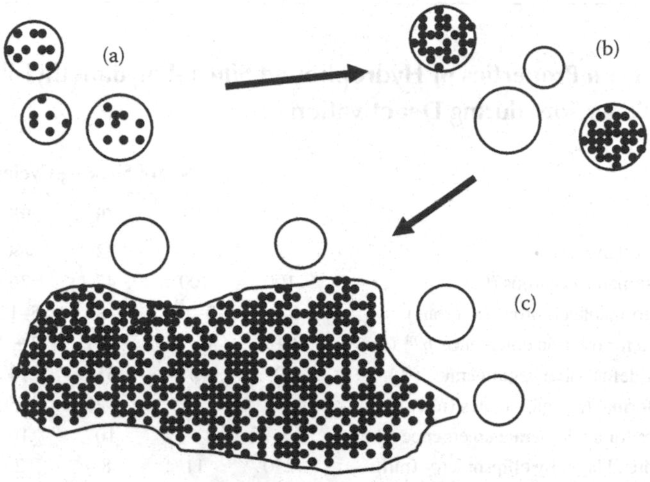

FIGURE 6.5 Schematic of deactivation by disproportionation. (a) Initial state of antifoam dispersion. (b) Deactivated antifoam dispersion formed after emulsification consisting of silica-free and silica-rich drops. (c) Silica-rich drops tend to coalesce to form an increasing proportion of large non-deformable agglomerates after further emulsification.

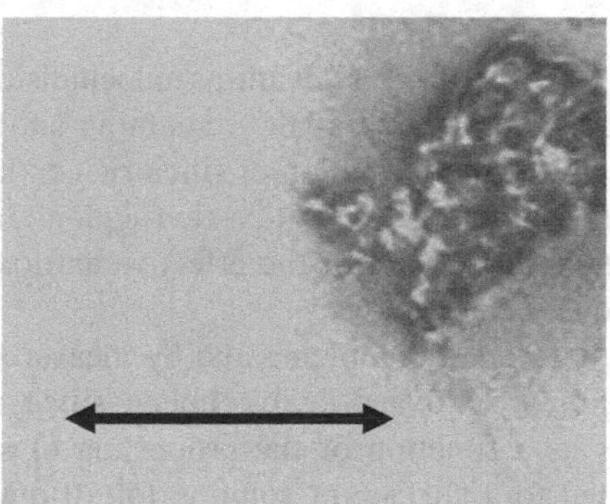

FIGURE 6.6 Photomicrograph of large non-deformable silica-rich agglomerate formed from hydrophobed silica–polydimethylsiloxane after 95 shake–quiescent cycles in 0.01 M AOT solution. Bar, 100 microns. (Adapted with permission from Marinova, K.G., Tcholakova, S., Denkov, N., Roussev, S., Deruelle, M. *Langmuir*, 19, 3084. Copyright 2003 American Chemical Society.)

slightly during deactivation. This is presumably due to superposition of antifoam dispersal from a neat state onto the deactivation processes. Also shown in the table is the critical pressure, p_c^{crit}, for emergence into the air–water surface, measured by the film trapping technique (see Sections 2.4.3 and 3.4.1), of these deformable oil drops. As deactivation proceeds, this pressure is seen to increase. This presumably

TABLE 6.1
Observations on Properties of Hydrophobed Silica–Polydimethylsiloxane Antifoam Dispersion[a] during Deactivation

	No. of Shaking Cycles				
	3	30	70	95	300
Time for foam collapse (s)	2	4	22	>60	>600
Fraction of deformable oil drops (%)	100	100	45	26	26
Diameter of deformable oil drops (microns)	6–24	7–25	6–25	5–12	3–7
Critical pressure for oil drop emergence p_c^{crit} (Pa)	3 ± 2	4 ± 2	9 ± 1	7 ± 2	18 ± 3
Fraction of non-deformable agglomerates (%)	0	0	55	74	74
Size of non-deformable agglomerates (microns)	–	–	6–26	4–50	5–100
Critical pressure for agglomerate emergence (Pa)	–	–	10	10	–
Thickness of spread layer by ellipsometry (nm)	250	11	8	2	<0.5

Source: Adapted with permission from Marinova, K.G., Tcholakova, S., Denkov, N., Roussev, S., Deruelle, M., *Langmuir*, 19, 3084. Copyright 2003 American Chemical Society.

[a] Antifoam added neat to 0.01 M AOT solution. Dispersion subjected to repeated shaking–quiescent foam generation cycles.

reflects the increasing stability of the relevant pseudoemulsion films as the spread layer disappears and the silica content of the remaining antifoam drops decreases due to disproportionation and the formation of silica-rich entities. The critical pressure for emergence of the non-deformable silica-rich agglomerates, where measured, is, however, seen to be similar to that of the effective antifoam oil drops after the same umber of shake–quiescent cycles.

These observations have been complemented by measurements of the durability of the same polydimethylsiloxane–hydrophobed silica antifoam in repeated shake–quiescent cycles as a function of the proportion of silica. Here again the neat antifoam was added to a surfactant solution (of 10 mM AOT) without predispersal so that a process of drop breakup must continuously occur as foam is generated. Results are shown in Figure 6.7 where durability is defined simply as the number of cycles required to produce foam that takes more than 60 s to collapse. A maximum in durability is found at an optimal silica content. Marinova et al. [7] suggest that increases in durability at low silica content are due to decreasing critical pressure for drop emergence. It is, however, easy to see how the initial antifoam effectiveness could be attributed to such a cause but not the durability. At high silica contents, antifoam deactivation becomes increasingly rapid due to the relatively rapid rate of formation of ineffective silica-rich agglomerates. Here we should emphasize that the effect of silica content on durability depicted in Figure 6.7 does not necessarily represent the effect of silica content on initial antifoam performance. Ross and Nishioka [31] claim, for example, that the initial effectiveness

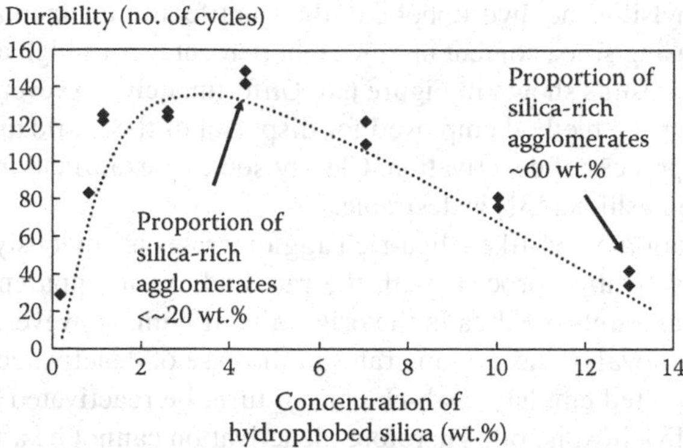

FIGURE 6.7 Antifoam durability as function of concentration of silica. Durability is defined as number of shake–quiescent foam generation cycles required to deactivate antifoam so that the foam takes more than 60 s to collapse. Antifoam was a mixture of polydimethylsiloxane of viscosity 1000 mPa s and hydrophobed pyrogenic silica particles. Antifoam concentration was 0.1 g dm^{-3} added neat to 0.01 M AOT solution. (Adapted with permission from Marinova, K.G., Tcholakova, S., Denkov, N., Roussev, S., Deruelle, M. *Langmuir*, 19, 3084. Copyright 2003 American Chemical Society.)

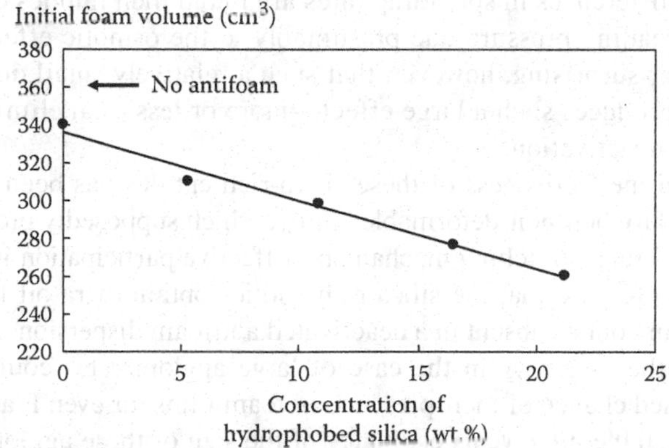

FIGURE 6.8 Effect on initial foam volume of increasing the proportion of hydrophobed silica in hydrophobed silica–polydimethylsiloxane antifoam. Foam generation by Ross–Miles method; ~6 × 10^{-4} g dm^{-3} antifoam dispersed in 0.014 M SDS solution. Antifoam prepared from mixture of polydimethylsiloxane of ~1000 mPa s viscosity and *in situ* hydrophobed silica Aerosil 200. (From Ross, S., Nishioka, G. Experimental researches on silicone antifoams, in *Emulsions, Lattices, and Dispersions*, Becher, P., Yudenfreund, M.N., eds., Marcel Dekker, New York, p 237, 1978.)

of a polydimethylsiloxane–hydrophobed silica antifoam can increase monotonically with increasing silica content up to weight percentages as high as 30%. This is illustrated by the results shown in Figure 6.8. Unfortunately, however, these authors do not make clear the method employed for dispersal of the antifoam and make no reference to the process of deactivation. Clearly some re-examination of these findings of Ross and Nishioka [31] is desirable.

Ease of formation of gel-like silica-rich agglomerates is obviously an important factor in the deactivation process with the rate and extent apparently increasing with increasing amounts of silica in the original antifoam. However, Denkov et al. [6] report that removal of the agglomerates in the case of deactivated antifoam can produce a deactivated emulsion, which can, in turn, be reactivated by addition of more polydimethylsiloxane oil. Therefore, deactivation cannot be attributed solely to removal of silica in the form of large agglomerates. It seems that high silica content per se is at least part of the reason for the inactivity of silica-rich entities regardless of size. However, this still leaves unanswered the reason for the ineffectiveness of these silica-rich entities in supplying oil to the air–water surface. It clearly does not concern the ease with which they emerge into the air–water surface because the critical pressure for emergence of these entities is similar to that of effective oil drops containing hydrophobed silica (see Table 6.1) [7]. On the other hand, it has been shown [6] that the presence of concentrations of hydrophobed silica typical of those found in the gel-like agglomerates reduces the rate of spreading of polydimethylsiloxane over the air–water surface by about a factor of 4. Denkov et al. [6] attribute the observed elimination of the spread layer on complete deactivation to this slow rate of spreading. The perimeter of spreading material present in these experiments appears to have been approximately constant. That differences in spreading rates are found then implies differences in equilibrium spreading pressure due presumably to the osmotic effect of the particles. It is rather surprising, however, that such a relatively small decrease in the spreading rate produces such a large effect—more or less total elimination of the spread layer on deactivation.

The antifoam ineffectiveness of these silica-rich entities has been attributed by Denkov et al. [6] to their non-deformable nature, which supposedly prevents participation in the bridging–stretching mechanism. Effective participation in that mechanism would then require that the silica-rich entities obtain extra oil from a spread layer, which is, of course, absent in a deactivated antifoam dispersion. An additional contribution to the inactivity in the case of large agglomerates could also derive from a diminished chance of their presence in foam films (or even Plateau borders). Examination of Table 6.1 reveals estimates of the size of these agglomerates, after significant deactivation, to be more than an order of magnitude greater than the initial sizes of effective antifoam drops, and where we remember Pouchelon and Araud [1] even report agglomerate diameters that must be up to three orders of magnitude greater than those of effective drops. These sizes imply number concentration differences between agglomerates and effective antifoam drops of from three to nine orders of magnitude. Such low number concentrations could clearly contribute to a significant degree of deactivation.

6.4.2 Theoretical Considerations Concerning Deactivation by Disproportionation

A significant amount of evidence for disproportionation as a cause of hydrophobed silica–polydimethylsiloxane antifoam deactivation now exists [6, 7, 21]. This process would appear to occur both during dispersal of antifoam to form an emulsion and during foam generation. Since disproportionation ultimately leads to total loss of antifoam activity, any amelioration could therefore lead to enhanced overall antifoam effectiveness, diminished antifoam requirements, and economic benefit. It is therefore important to examine the nature of this process to establish whether any amelioration is likely to be possible.

Arguably, the first challenge of theory is to establish how a process of continuous antifoam drop breakup and coalescence, occurring during either emulsification or foam generation, can lead to the formation of particle-free drops. The proportion of such drops in a fully deactivated antifoam dispersion can be estimated from the experimental finding that the dispersion consists only of particle-free drops and particle-rich drops or agglomerates if we suppose that the compositions of those particle-rich drops and agglomerates are identical. In effect, we assume that the agglomerates are formed by coalescence of the particle-rich drops. We can therefore deduce from mass balance that the volume fraction of particle-free drops $\tilde{V}_0$ is given by

$$\tilde{V}_0 = \frac{1-(w_s/\tilde{w}_s)}{1+w_s((\rho_o/\rho_p)-1)} \tag{6.1}$$

where w_s is the weight fraction of silica in the antifoam before deactivation, $\tilde{w}_s$ is the weight fraction of silica in the particle-rich drops and agglomerates after disproportionation, and ρ_o and ρ_p are the densities of polydimethylsiloxane oil and silica particles, respectively. The agglomerates have been found to contain 14–17 wt.% silica [1, 6]. From Equation 6.1, we therefore calculate that the corresponding volume fraction of particle-free drops would have to lie in the range of 0.63–0.69 in the case of a typical silicone antifoam with about 5 wt.% of hydrophobed silica if we assume realistically that the density of silica is double that of the polydimethylsiloxane oil.

One possibility is that deactivation is a consequence of the distribution of particles in a random manner across the drops formed upon dispersal of an antifoam to achieve a steady state by processes of drop splitting and coalescence. If this process is to explain deactivation, it should obviously result in a significant probability of formation of particle-free drops. Following a suggestion by Grassia [32], let us suppose then that the particles present in antifoam drops are classic distinguishable particles. Since we are dealing with colloidal particles, this is necessarily true. If there are M drops present, then the probability of a particle being present in a drop is $1/M$ and the probability of a particle being elsewhere is $(1 - 1/M)$. If there are $\tilde{N}$ particles present, then the probability of i particles being present in a drop is $(1/M)^i$ and the probability of the remaining $\tilde{N} - i$ particles being elsewhere is $(1 - 1/M)^{\tilde{N}-i}$. The overall probability, Q_i, of finding drops with i particles is therefore given

$$Q_i = \frac{\tilde{N}!}{(\tilde{N}-i)!\,i!}\left(\frac{1}{M}\right)^i\left(1-\frac{1}{M}\right)^{\tilde{N}-i} \tag{6.2}$$

where the factor $\tilde{N}!/(\tilde{N} - 1)!i!$ is the number of ways of achieving the arrangement given that we do not distinguish between the different particles present in a given drop or elsewhere. A similar approach has been adopted in Section 5.2.4 for calculating the probability of finding bubbles associated with a given number of antifoam entities.

The overall probability of finding drops without particles, Q_0, is given by Equation 6.2 if we set $i = 0$ (remembering that $0! = 1$) so that

$$Q_0 = \left(1-\frac{1}{M}\right)^{\tilde{N}} \tag{6.3}$$

We usually have $M \gg 1$ and the probability, Q_0, therefore reduces to [32]

$$Q_0 \sim \exp(-\tilde{N}/M) \tag{6.4}$$

It is easily shown that this result is independent of the path by which a given distribution is obtained, provided that at every stage particles are distributed randomly. Thus, if drops break up by forming arbitrary numbers of debris drops at each splitting, where the particles are randomly distributed over those debris drops, the resulting probability of particle-free drops will always be determined by the overall values of $\tilde{N}$ and M.

Let us assume that the particles are randomly distributed across drops irrespective of drop volume, so that even in antifoam emulsion the volume size distribution of drops is the same irrespective of the presence or absence of particles. The probability of obtaining particle-free drops is then equal to the total volume fraction of such drops (so that $Q_0 = \tilde{V}_0$). In which case, we calculate from Equation 6.4 that $\tilde{N}/M \leq 0.46$ if volume fractions of particle-free drops are to match the values of ≥ 0.63 indicated by experimental observation of the silica content of the agglomerates in deactivated antifoam dispersions. Since $\tilde{N}/M < 1$, this means that such high volume fractions are predicted to occur only in the case of dilute systems where the number of drops exceeds the number of particles. Obviously, this situation could be achieved as antifoam is dispersed where the number of drops increases, as a result of splitting, because the number of particles then remains constant so the ratio $\tilde{N}/M$ must decline. Here we note that if both particles and drops are spherical and monodisperse and the overall volume fraction of the particles is small, then

$$\tilde{N}/M \approx \frac{\tilde{V}_p r_{AF}^3}{(1-v_p)R_p^3} \tag{6.5}$$

where $\tilde{V}_p$ ($\ll$1) is the intrinsic total volume fraction of particles in the antifoam, R_p is the particle radius, and r_{AF} is the radius of antifoam drops after dispersal. The overall weight fraction of silica in those drops that contain that material, $\tilde{w}_s$, is given by Equation 6.1 if we set $Q_0 = \tilde{V}_0$ so that

$$\tilde{w}_s = \frac{w_s}{1 - Q_0 - Q_0 w_s \left(\frac{\rho_0}{\rho_p} - 1 \right)} \tag{6.6}$$

from which it is obvious that if $Q_0 = 0$, then we must have $\tilde{w}_s = w_s$. The overall weight fraction defined in this way must of course include all drops with $i > 0$ so that

$$\tilde{w}_s = \sum_{i>0}^{\tilde{N}} Q_i \tilde{w}_i \tag{6.7}$$

where $\tilde{w}_i$ is the weight fraction of silica in a drop containing i particles.

We can now use actual experimental results and Equations 6.4 and 6.5 to estimate the probability of finding particle-free drops in the case of a putative steady state where a random distribution of particles over drops is formed as a result of successive splitting and coalescence events. In the study of polydimethylsiloxane–hydrophobed silica antifoam deactivation given by Denkov et al. [6], drop radii, both before and after deactivation, lie in the range of 1–5 microns. The silica concentration is about 5% by weight and the silica density about 2 g cm^{-3} and that of the oil about 1 g cm^{-3}. Silica particle "sizes" lie in the range of 0.1–5 microns. Both drops and particles are clearly rather polydisperse. The ratio of antifoam drop to particle size is therefore in the range of $1 < r_{AF}/R_p \leq 100$ with an overall particle volume fraction of ~0.026.

Calculations of $\tilde{N}/M$, Q_0, and $\tilde{w}_s$ for a range of possible values of the ratio r_{AF}/R_p are presented in Table 6.2. Calculations were made using Equations 6.4 through 6.6, which means that we simplify by supposing that both the oil drops and particles are monodisperse spheres. As expected, it is clear that if $\tilde{N}/M < 1$, large values of Q_0 are found. That particle-free drops are found in such circumstances is, of course, a necessity regardless of the nature of the distribution! If, on the other hand, $\tilde{N}/M$ exceeds unity, Q_0 rapidly tends to zero and deactivation is not predicted. These conclusions are, however, derived for the case where monodisperse particles are dispersed in monodisperse drops. In reality, polydispersity is to be expected, which complicates any interpretation of the calculations shown in Table 6.2. Also, if some particles are present that are larger than some drops in such a polydisperse system, the relevance of Equation 6.2 is undermined. However, it is clear from Table 6.2 that if the number distribution of silica aggregate particle sizes in the polydisperse system described by Denkov et al. [6] is completely dominated by the submicron range (so that mostly $R_p \leq$ ~0.3 micron and therefore $\tilde{N}/M > 1$ for all drop sizes), then

TABLE 6.2
Deactivation of Polydimethylsiloxane–Hydrophobed Silica Antifoams; Random Distribution of Particles Over Oil Drops and Resultant Disproportionation

Ratio Radius of Oil Drops to Radius of Particles r_{AF}/R_p	$\tilde{N}/M$	Probability of Particle-Free Drops, Q_0	Overall Weight Fraction of Silica in Drops Not Particle-Free, $\tilde{w}_s$	Comments
2.0	0.21	0.87	0.24	
2.2	0.28	0.75	0.19	
2.3	**0.33**	**0.72**	**0.17**	**$\tilde{w}_s$ has same value as weight fractions of silica found in large agglomerates in deactivated antifoam**
2.5	**0.41**	**0.66**	**0.14**	
3.0	0.70	0.50	0.10	
4.0	1.66	0.19	0.06	Negligible proportion of particle-free drops so $\tilde{w}_s = w_s$
5.0	3.33	0.04	0.05	
10.0	26.7	5×10^{-12}	0.05	
50.0	3337	0	0.05	
100.0	2.67×10^4	0	0.05	

Note: Calculations assume monodisperse particles and drops using Equations 6.4 through 6.6 with $w_s = 0.05$ and $\tilde{V}_p = 0.026$ and $\rho_o/\rho_p = 0.5$.

$Q_0 \to 0$ and deactivation is not predicted. It is, moreover, clear that random distribution of particles between drops in the manner described by Equations 6.2 through 6.4 produces particle-free drops only in the trivial case where the antifoam is dispersed to the point where the number of drops exceeds the number of particles. That deactivation involving the production of a large proportion of particle-free drops is found in the system considered suggests that either the number distribution of particle sizes is dominated by large particles of similar size to the drops so that $\tilde{N}/M < 1$ and $Q_0 \gg 0$ or that a steady state where all particles are randomly distributed over drops is not attained. The absence of detailed knowledge of the silica particle size distribution in the work of Denkov et al. [6] means that the relevance to experiment of either of these two options cannot be rigorously distinguished from this analysis alone. All of this does suggest that a study of antifoam deactivation using monodisperse hydrophobic particles of various sizes could produce some helpful insights. However, any such study should ensure that particles do not aggregate inside drops because $\tilde{N}$ would then represent the number of aggregates rather than the number of individual particles. Clearly any such aggregation would favor deactivation and would probably mean some measure of polydispersity as the aggregation number would presumably vary.

From Table 6.2, it is clear then that we can calculate from Equations 6.4 and 6.5 probabilities of finding particle-free drops that can be representative of those found in deactivated antifoam dispersions provided $\tilde{N}/M < 1$. However, Equation 6.2 implies that, in the corresponding distribution of particles over drops, values of $Q_{i\geq 1}$ are all finite. Drops containing upward of a single particle would all be present despite mean weight fractions, $\tilde{w}_s$, equal to those present in an actual deactivated antifoam where the particle concentration in deactivated drops and agglomerates is uniformly high. The antifoam dispersion predicted from Equation 6.2 would therefore of necessity contain some drops with particle concentrations too low to be inactive—it could not therefore be totally deactivated. Moreover, the gradual formation, during deactivation, of large silica-rich agglomerates [1, 6, 7] suggests that a steady state with a random distribution of particles over a constant size distribution of drops is not in fact obtained. We therefore must conclude that a steady-state random distribution of particles over drops cannot represent a complete picture of antifoam deactivation by disproportionation.

Formation of large deactivated silica-rich agglomerates implies that the presence of a large concentration of silica particles may increase the viscosity of some drops so that the prevailing shear forces during emulsification or foam generation cannot cause splitting. Such drops will therefore gradually increase in size with successive coalescence events with other silica-rich drops. It is easy to see in fact that splitting and coalescence can result in drops with much higher concentrations of particles than the original drops. Consider, for example, drops containing three particles—one product of splitting could be two equal volume drops, one of which contains three particles. Coalescence of that drop with another drop containing three particles produces a drop of the original volume but containing six particles. Subsequent splitting could then produce a drop with six particles at double the original concentration, etc. Whether this process could account for the observed deactivation behavior could best be explored by computer simulation. This should include rules for coalescence and splitting superimposed on the relevant statistics. Here we note that coalescence is facilitated by the presence of particles adhering to the oil–water surface of drops. Thus particles, which promote the antifoam effect of oil drops by facilitating the rupture of pseudoemulsion films, will also facilitate the rupture of oil–water–oil films (see Chapter 4). Therefore, one rule should be that coalescence of particle-free drops with each other should not be permitted in the simulation. A more sophisticated rule would relate the probability of coalescence to the number of particles present in a drop so that the probability of coalescence is actually higher the larger the number of particles present. Another rule should concern splitting—this should not be permitted for drops with a certain threshold particle concentration because that would imply a viscosity too high to allow the relevant shear forces to prevail in causing drop splitting. Fractal silica particles, of the type often used in antifoams, would be expected to be especially effective at increasing viscosity because they occupy a large volume for a given weight.

Hitherto we have supposed that the distribution of particles into debris drops after splitting of antifoam drops is a random process. It is however possible that the nature of the hydrodynamics of splitting of oil drops containing particles during bulk phase emulsification could lead to a non-random distribution. Apparently the only experimental studies of such processes [33, 34] concern the breakup in silicone oil of single aqueous drops under controlled hydrodynamic conditions. Smith and van de Ven [34] have observed breakup of such drops containing low volume fractions (~0.03) of silanized resin particles under pure shear. In these circumstances, debris drops containing high volume fractions of particles are continuously ejected from the tips of deformed drops. This process leaves the original aqueous drops essentially intact but completely denuded of particles. The process is depicted in Figure 6.9. Unfortunately, the particles are more than an order of magnitude larger than those present in typical hydrophobed silica–polydimethylsiloxane antifoams and the viscosity ratio is the inverse of that found when such antifoams are dispersed in aqueous surfactant solutions. Nevertheless, there would appear to be merit in extending this type of study to the breakup, in aqueous surfactant solution, of antifoam oil drops containing particles under appropriate hydrodynamic conditions.

These large silica-rich agglomerates are reported to be white in overall appearance [1], which contrasts with the usually rather clear or slightly turbid appearance of the neat antifoam. This appearance could concern a tendency for emulsified aqueous phase to be incorporated into the entities during coalescence. However, it is difficult to envisage this in the case of coalescence between two drops—perhaps it may happen if several drops coalesce. Any aqueous emulsion drops incorporated in the silica-rich agglomerates would cause a significant amount of light scattering and could therefore account for their white appearance. If the volume fraction of such putative aqueous drops was sufficiently high, it could also contribute to an enhanced viscosity and a diminished tendency to split.

Not only must coalescence be involved in the antifoam deactivation process but it must obviously also be involved in the reactivation of deactivated antifoam dispersion by addition of fresh polydimethylsiloxane oil. Here the effectiveness of the added oil has been shown to decrease with decreasing drop size, being most effective if neat oil is simply added to the deactivated dispersion. It is obvious that this effect must concern the efficiency of the coalescence of neat oil drops with silica-rich, but

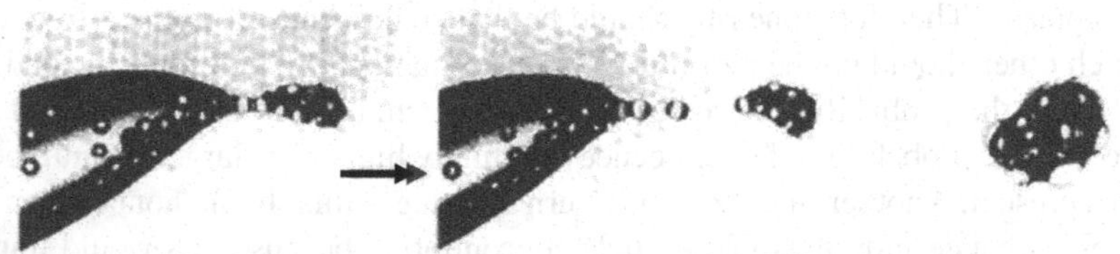

FIGURE 6.9 Photomicrograph of rupture of water drop containing silylated polystyrene resin particles dispersed in polydimethylsiloxane in a Couette device. Shear rate, 1.32 s^{-1}; viscosity of polydimethylsiloxane, ~10 Pa s; resin particle radius, 86 ± 10 microns; and initial radius of drop, ~0.11 cm. Particle volume fraction in debris formed by expulsion from the main drop is seen to be much higher than that of the latter. Particles are also seen to adhere to oil–water interface. (Reprinted from *Colloids Surf.*, 15, Smith, P.G., van de Ven, T.G.M., 191. Copyright 1985, with permission from Elsevier.)

inactive, drops and large agglomerates. Yan and Masliyah [35] report similar coalescence of large oil drops with particle-stabilized emulsions. Presumably the coalescence is promoted by the presence of the particles if they are preferentially wetted by the oil (Chapter 4). This coalescence means that the ratio of silica to polydimethylsiloxane in the agglomerates declines so that the viscosity also declines to permit breakup, redispersal, and restoration of antifoam activity. This in turn will, however, ultimately lead to a repeat of the original deactivation process through further disproportionation as drops continue to split and coalesce during foam generation as shown in Figure 6.3.

Addition of polydimethylsiloxane in the form of an emulsion of fine drops produces antifoam reactivation, which subsequently deactivates relatively rapidly compared with that achieved with neat antifoam. This in turn suggests that coalescence of silica-rich agglomerates with small drops is less efficient than coalescence with neat undispersed oil so that the concentration of silica in the drops formed by breakup of the agglomerates thereby facilitated is still high. As Marinova et al. [7] have shown, high concentrations of silica lead to low antifoam durability and rapid increase in the formation of large inactive agglomerates (see Figure 6.7).

It might seem surprising that similar problems of disproportionation do not bedevil the preparation of the particle-stabilized, so-called Pickering emulsions [36, 37]. We should, however, stress that in Pickering emulsions, the particles stabilize emulsion drops against coalescence—that is, they violate the condition for oil–water–oil film and pseudoemulsion film rupture, which characterizes the particles present in antifoams for aqueous foams. Emulsion drop splitting will nevertheless occur when the oil phase is dispersed in water during the preparation of an oil-in-water Pickering emulsion. However, in the case of such emulsions, the particles, are "partially" hydrophobic—they have a contact angle <90° when measured through the aqueous phase. Therefore, in preparing oil-in-water Pickering emulsions, the particles are first dispersed in the aqueous phase and the oil is subsequently emulsified in this particle dispersion (see, e.g., references [38–40]). Any tendency for drops to be produced without particles during the resultant splitting will therefore be opposed by the subsequent capture of particles from the aqueous phase. In this respect, the procedure differs radically from that involved in emulsification of hydrophobic particle–oil antifoams where the particles are preferentially wetted by the oil and are therefore usually first dispersed in the oil phase. In contrast to antifoam emulsification then, neither splitting nor coalescence can therefore result in ready disproportionation during preparation of Pickering emulsions.

6.5 EFFECT OF OIL VISCOSITY ON DEACTIVATION OF HYDROPHOBED SILICA–POLYDIMETHYLSILOXANE ANTIFOAMS

Viscosity of the oil has a marked effect on the rate of deactivation of hydrophobed silica–polydimethylsiloxane antifoams. Indeed it is stated by Denkov et al. [6] that there is an optimum viscosity. Antifoams with viscosities lower than the optimum exhibit high initial activity, which rapidly deactivates. This is presumed to occur because of easy drop breakup and disproportionation. Antifoams with viscosities higher than the optimum, on the other hand, have low activity, which does not derive from disproportionation but rather from slow rates of dispersal leading to low probabilities of inclusion in foam films. Too high viscosity is also presumed to diminish antifoam effectiveness because of slow rates of formation of spread layers and slow rates of deformation in the bridging–stretching mechanism [5]. The former leads to low antifoam activity because the absence of a spread layer is supposed to both stabilize the relevant pseudoemulsion films and inhibit the bridging–stretching mechanism [5]. Curiously, Koczo et al. [6] claim that increase in oil viscosity actually decreases pseudoemulsion film stability in the presence of hydrophobed silica particles. However, this observation concerns only oils with low viscosity where the spreading rate is in any case high and is essentially independent of viscosity (see Section 3.6.2.2).

There is little experimental evidence in the scientific literature concerning the existence of this supposed optimum. Indeed there have apparently been few published studies of the effect of oil viscosity on both the effectiveness and deactivation rates of hydrophobed silica–polydimethylsiloxane antifoams. Evidence that increase in polydimethylsiloxane oil viscosity can reduce the rate of deactivation of hydrophobed silica–oil antifoams is, for example, presented by Racz et al. [3]. These authors repeatedly pulled films, using a film frame, from surfactant solution upon which antifoam had been spread. The time for film rupture was measured. After several hundred films had been pulled, the film rupture time increased dramatically indicating partial antifoam deactivation. Increase in the viscosity of the oil in a hydrophobed silica–oil antifoam, from 200 to 1000 cSt, increased the number of films that could be pulled by more than 50%, implying a decrease in rate of deactivation [3].

Koczo et al. [2] have also reported a study of the effect of polydimethylsiloxane oil viscosity on hydrophobed silica–oil antifoam deactivation. Here the half-life of foam generated by shaking cylinders was measured in the presence of polydimethylsiloxane oils of various viscosities containing hydrophobed silica. Results are presented in Figure 6.10 where the foam half-life is plotted against the number of shaking tests with a solution of 0.06 M SDS. Unfortunately, it is not entirely clear how the antifoam was added to these solutions—whether neat or in some manner predispersed. Marked deactivation is seen to accompany repeated shaking tests with low-viscosity oils as indicated by increasing foam half-lives. By contrast, an antifoam prepared from an oil with a viscosity of ~60,000 mPa s is seen to exhibit negligible deactivation under these conditions. The low activity of the low-viscosity antifoams presumably reflects an extremely rapid rate of deactivation due to rapid drop breakup and disproportionation. It is not, however, obvious from these observations that an optimum viscosity for antifoam effectiveness exists within the range of viscosities studied (from 5 to 60,000 mPa s).

It is, however, obvious from Figure 6.10 that increasing the oil viscosity from 5 to 1000 mPa s produces a decrease in rate of deactivation. It has been shown (see Section 3.6.2.2) [4, 41] that the spreading rate of polydimethylsiloxane oils on water

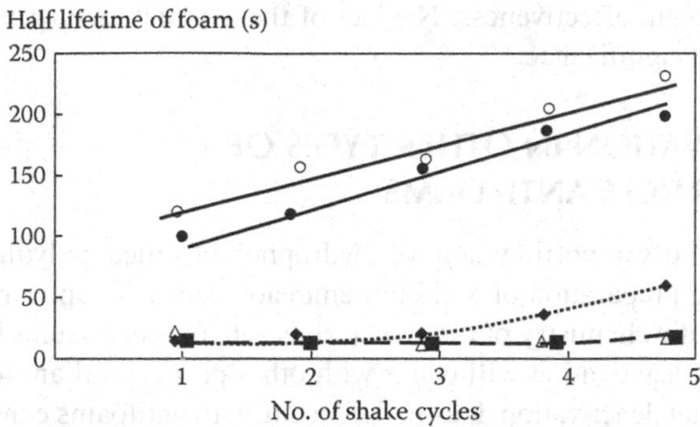

FIGURE 6.10 Effect of polydimethylsiloxane viscosity on rate of deactivation of hydrophobed silica–polydimethylsiloxane antifoams. Foam half-lifetimes are measured after shake cycles made with graduated cylinder. Surfactant solution 0.06 M SDS, total antifoam concentration 0.2 g dm^{-3}. Polydimethylsiloxane viscosities: ○, ~5 mPa s; ●, ~20 mPa s; ♦, ~1000 mPa s; ■, ~12,500 mPa s; △, ~60,000 mPa s. (Adapted from *J. Colloid Interface Sci.*, 166, Koczo, K., Koczone, J., Wasan, D., 225. Copyright 1994, with permission from Elsevier.)

is essentially independent of viscosity of the oil over this range of viscosities (see Section 3.6.2.2). The film pulling experiments of Racz et al. [3] described here also involved oils with viscosities lying within this range. These findings of both Kocz et al. [2] and Racz et al. [3] must therefore represent evidence that differences in antifoam deactivation rates do not necessarily correlate with differences in spreading rates. The hypothesis of Racz et al. [3] that differences in antifoam deactivation are attributable to differences in the extent of the spread layer at foam film rupture is also clearly not consistent with these findings.

Increasing the volume fraction of hydrophobed silica in a mixture with polydimethylsiloxane will increase the viscosity. It is therefore to be expected that the volume fraction of silica required to produce such high viscosities that antifoam effectiveness in inhibited would be less the higher the viscosity of the oil. This would imply that formation of inactive silica-rich drops and agglomerates would therefore be facilitated. However, as we have discussed in Section 6.4.2, formation of drops containing concentrations of silica higher than the initial concentration requires not only coalescence but a combination of drop coalescence and splitting. It seems likely that increasing viscosity of the polydimethylsiloxane will inhibit the rate of the latter process producing a lower rate of formation of inactive silica-rich drops and agglomerates. We therefore find a lower rate of disproportionation and deactivation despite the putative requirement for lower concentrations of silica before drops become inactive.

The latter argument is of course speculative and we are therefore forced to conclude from these limited studies that the effect of oil viscosity on hydrophobed silica–polydimethylsiloxane antifoam effectiveness and rate of deactivation is not fully understood. Clear evidence of an optimum viscosity, as suggested by Denkov et al. [6], is lacking as is the magnitude of the viscosity before the supposed onset of diminished antifoam effectiveness. Neglect of these issues is surprising in view of their commercial significance.

6.6 DEACTIVATION IN OTHER TYPES OF OIL–PARTICLE ANTIFOAMS

Deactivation by disproportionation of hydrophobed silica–polydimethylsiloxane antifoams during preparation of antifoam emulsions does not appear to necessarily concern the specific chemistry of these materials. Therefore, it seems likely that similar processes of deactivation will occur with other particle–oil antifoams. There is some evidence that deactivation does in fact occur with antifoams consisting of other oils in combination with hydrophobed silica. Reported examples are hydrophobed silica in combination with liquid paraffin [8, 10] and in combination with mixtures of liquid paraffin and perfluoroalkyl alkanes [9]. There are, however, no reports in the scientific literature of deactivation occurring in the case of oil–particle antifoams where the particle is not hydrophobed silica.

Garrett et al. [8] report studies with hydrophobed silica in liquid paraffin, which utilized Aerosil 200 silica particles hydrophobed to different extents using a trimethylchlorosilane. The surfactant was a commercial blend of ~1.4×10^{-3} M sodium alkylbenzene sulfonate. Foam generation was by shaking cylinders. The weak antifoam effect of the particles alone, if present, deactivated with agitation during foam generation. It seems probable that this effect may be attributed to the instability of the contact angle, rendering the particles more hydrophilic as they are exposed to the aqueous phase (as described by Blake and Kitchener [15]). Mixtures of the least hydrophobic of these hydrophobed silicas with liquid paraffin showed deactivation after repeated shake–quiescent cycles, which could also be due to contact angle instability. Mixtures of the more hydrophobic of these silicas with liquid paraffin did not, however, deactivate under these conditions despite deactivation of the relevant silica when used alone. It is unclear why deactivation due to either contact angle instability or disproportionation was apparently not observed with this antifoam under these conditions. Moreover, the study did not reveal significant differences in antifoam activity between emulsified antifoam and neat antifoam added without predispersal, which would again imply absence of deactivation due to disproportionation. Absence of any such deactivation may, however, be simply due to the limited agitation time and number of shake–quiescent cycles involved. Garrett et al. [8] do in fact report the formation of irregular agglomerates of size 1 mm or larger when mixtures of these silanized silica and liquid paraffin are dispersed in anionic surfactant solution using a high-speed emulsifier. Formation of similar entities in hydrophobed silica–polydimethylsiloxane antifoam is, as we have seen, associated with deactivation by disproportionation (see Table 6.1).

Preliminary observations of foam generation from a solution of SDS in the presence of a non-spreading hydrophobed silica–mineral oil antifoam, using a continuous circulation Ross–Miles apparatus, do in fact indicate that the antifoam deactivates [10]. This obviously implies that deactivation does not necessarily concern formation of spread layers. It is possible also that non-spreading oils do not function by the bridging–stretching mechanism but rather function by the bridging–dewetting mechanism (see Chapter 4). The latter mechanism may imply retention of the integrity of the antifoam drop as it induces foam film collapse, which could mean less disproportionation during foam generation and slower rates of deactivation for given rates of foam destruction.

Curtis et al. [9] report markedly different rates of deactivation during foam generation, of antifoams prepared from hydrocarbon and perfluoroalkane mixed oils of different air–oil surface tensions and viscosities. These differences do not appear to arise from instability of the contact angle because the same hydrophobed silica was used throughout. Increasing the viscosity of the antifoam, by addition of isobutene, reduced rates of deactivation. It seems likely that this latter effect may be attributed to diminished rates of drop breakup and therefore disproportionation. It is therefore likely to be similar in origin to the effect on deactivation of increasing the viscosity of the oil in hydrophobed silica–polydimethylsiloxane antifoams.

The fractal nature of hydrophobed silica particles suggests a possible reason for the tendency of oil–hydrophobed silica antifoams to be especially susceptible to deactivation. The effective volume fraction of fractal entities is of course much larger than the actual volume fraction, which leads to exaggerated consequences for the rheology of the relevant particle–oil mixtures. If non-fractal particle structures are selected, the volume (or weight) fraction of particles, which could be tolerated before inactive gel-like agglomerates form, could be higher than for silica particles. In turn, this could mean tolerance of greater extents of disproportionation before these inactive agglomerates acquire most of the particles present in the original antifoam.

We should note that deactivation can occur with polydimethylsiloxane-based antifoams mixed with particles other than hydrophobed silica. It is known, for example, that antifoam mixtures of ethylene distearamide particles and polydimethylsiloxane oils deactivate during the long periods of agitation and foam generation involved in the machine washing of textiles [42]. Dispersion of ethylene distearamide particles in polydimethylsiloxane is difficult—large particle sizes relative to the drop sizes for effective antifoam action may mean large values of the ratio $\tilde{N}/M$ favoring ready deactivation.

Mixtures of hydrocarbons with calcium alkyl phosphate particles are known to be effective antifoams in the context of domestic textile machine washing where wash cycles can last for up to 1 h and involve temperature ranges from 30°C to 95°C (see Section 8.2.4.1) [43, 44]. Under these circumstances, little or no deactivation of antifoam effectiveness is apparent. Mixtures of hydrocarbons with alkyl phosphoric acid esters also function as antifoams in this context provided the aqueous phase has a high enough pH and calcium ions are present [43, 44] so that the calcium salts can precipitate as particles *in situ* at the relevant hydrocarbon–water interface. This behavior is of course analogous to that shown by mixtures of oleic acid and hexadecane when dispersed in an aqueous phase under similar conditions [45]. As with the preformed calcium alkyl phosphate particles, no deactivation of antifoam effectiveness is observed in the case of *in situ* formation of the precipitates. Indeed, it has been observed that continuous aeration for several hours, using a circulating Ross–Miles apparatus at 90°C (see Section 2.2.3), of an aqueous solution of a blend of a commercial sodium dodecylbenzene sulfonate and an ethoxylated alcohol in the presence of mixtures of a hydrocarbon and an alkyl phosphoric acid ester (dispersed ultrasonically) produce effective antifoam action with no indication of deactivation over that time [46]. By contrast, mixtures of low-viscosity polydimethylsiloxane and hydrophobed silica (of viscosity ~3000 mPa s) deactivate significantly under the conditions prevailing during a textile machine wash (see Section 8.2.5.2). We should also note that these phosphate derivatives are effective in the relevant hydrocarbons even at 20 wt.% at which concentration polydimethylsiloxane–hydrophobed silica mixtures are relatively inactive. The densities of the relevant calcium alkyl phosphates are at least half those of hydrophobed silicas, which means that the volume fractions for a given weight fraction of the former are significantly higher. We are left with the tentative conclusion that the deactivation of polydimethylsiloxane–hydrophobed silica antifoams may largely concern a property of the hydrophobed silica.

None of the studies summarized above were concerned with deactivation per se of these hydrocarbon-based antifoams. They are, therefore, limited in scope and do not fully address the issues concerning deactivation in general and deactivation by disproportionation in particular. There is, therefore, a need for systematic studies of this issue with antifoams prepared from different hydrophobic particles and oils where prolonged foam generation is combined with observations of the state of dispersion of the antifoam. Obviously, the role of particle sizes and shapes, state of aggregation in the oil, and resultant rheology, etc., should also be included. The oil types should be selected to include consideration not only of the effect of viscosity but also of their spreading behavior at air–water surfaces.

6.7 THEORIES OF FOAM VOLUME GROWTH IN PRESENCE OF DEACTIVATING ANTIFOAM

6.7.1 Antifoam–Bubble Heterocoalescence–Kinetic Model of Pelton and Goddard for Foam Generation by Sparging [47]

The earliest attempt to produce a theoretical model of the growth of foam volumes in the presence of deactivating hydrophobed silica–polydimethylsiloxane antifoams is due to Pelton and Goddard [47]. Here the model attempts to describe the growth of foam volume in a sparging experiment where gas is passed through a porous frit into a vessel containing surfactant solution and emulsified antifoam. It implicitly assumes that the action of antifoam entities in rupturing several foam films eventually produces total deactivation of those entities so that the resulting debris cannot participate in any further antifoam action. This assumption apparently owes much to the theory proposed earlier by Kulkarni et al. [18, 19] where antifoam action is supposed to involve separation of the hydrophobed silica particles from the oil, which would inevitably lead to deactivation. The implicit assumption that antifoam drops are deactivated after rupturing several foam films is, however, also not entirely inconsistent with the more recent experimental observations of Denkov et al. [6, 7]. As we have seen (see Section 6.4), it seems probable that disproportionation of hydrophobed silica–polydimethylsiloxane antifoams accompanies foam rupture by the bridging–stretching mechanism. In turn, this process eventually leads to replacement of active antifoam by inactive antifoam entities containing either too much or too little silica.

Use of an antifoam emulsion is supposed to imply that antifoam drop sizes and size distributions are essentially unaffected by the process of foam generation. Again, this agrees with the observations of Denkov et al. [6, 7]. Pelton and Goddard [47] implicitly assume that any consequences of antifoam polydispersity can be subsumed within an average drop size, which remains constant during foam generation.

In this model, the gas–liquid mixture formed by sparging is, somewhat arbitrarily, divided into two regions—a liquid phase with a low gas volume fraction through which bubbles are freely passing and a foam column containing a high gas volume fraction. Some of the supposedly monodisperse "primary" bubbles formed at the frit are assumed to heterocoalesce with antifoam drops in the liquid phase. Again, somewhat arbitrarily, these primary bubbles are divided into groups of equal numbers of bubbles. Coalescence of all members of a group of primary bubbles to form a secondary bubble is then supposed to occur in the foam column if at least one of those bubbles has acquired an antifoam drop. All secondary bubbles therefore have the same volume. When these secondary bubbles reach the top surface of the foam column, they rupture, releasing the antifoam drops to the aqueous phase. It is implicitly assumed that these antifoam drops are totally deactivated. Why the antifoam drops should be deactivated at this stage rather than earlier, as they rupture the films between primary bubbles to form secondary bubbles, is not made clear. Any remaining primary bubbles in groups without antifoam are by contrast assumed to be indefinitely stable upon reaching the top of the foam. A schematic diagram of this model of foam generation is given in Figure 6.11.

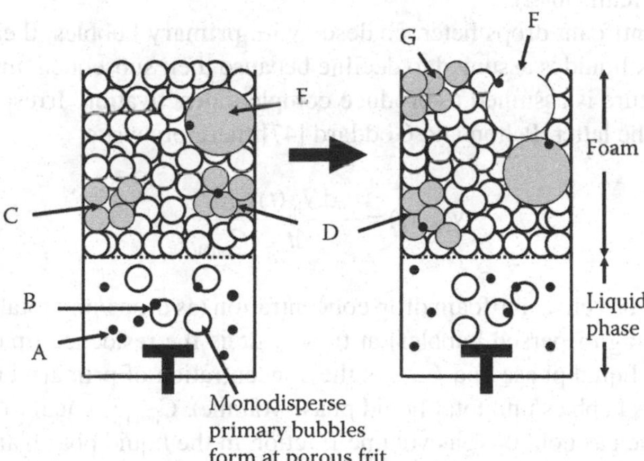

FIGURE 6.11 Schematic illustration of model of foam generation in presence of antifoam assumed by Pelton and Goddard [47]. A, antifoam drops (hydrophobed silica–polydimethylsiloxane) dispersed in surfactant solution; B, primary bubbles heterocoalesce with antifoam, reducing dispersed antifoam concentration; C, bubble group containing G primary bubbles is stable if no antifoam entities are present on any of the bubbles; D, bubble group with at least one antifoam entity on at least one bubble will coalesce to form one secondary bubble; E, secondary bubble; F, secondary bubble rises to top of foam column where the antifoam entity present causes it to rupture and is simultaneously deactivated; G, stable primary bubble groups without antifoams also move up to top of foam column but remain stable.

This basic model seems contrived, in that no direct experimental justification is given for it by Pelton and Goddard [47]. Indeed these workers argue [47] that "groups in this treatment are a mathematical convenience to facilitate the derivation and not a physically observable cluster of bubbles." However, in a later paper, which uses a similar model, Pelton [48] claims to observe formation of secondary bubbles in the foam column. It is claimed that they expand in size by coalescence until the buoyancy force exceeds the "yield stress" in the foam whereupon they rise rapidly to the top of the foam column and rupture.

Pelton and Goddard [47] define a measure of antifoam effectiveness, $f(t)$, as the rate of foam volume increase/volumetric gas flow rate. If the antifoam is totally ineffective or absent, $f(t) = 1$. On the other hand, if all groups of primary bubbles have at least one antifoam drop, then the foam mass is converted entirely into secondary bubbles, which rupture so that $f(t) = 0$. If $N_{\text{prim}}(t)$ is the average number of antifoam drops attached to each primary bubble at time t, then the average number $N_{\text{G}}(t)$ of antifoam drops for each nascent secondary bubble group is $N_{\text{prim}}(t)G$, where G is the number of primary bubbles in such a group. Provided $N_{\text{prim}}(t)G < 1$, then $N_{\text{G}}(t)$ becomes effectively the probability of a bubble group coalescing to form an unstable secondary bubble. Therefore

$$f(t) = 1 - N_{\text{sec}}(t) = 1 - N_{\text{prim}}(t)G \tag{6.8}$$

Clearly, in the limit when $N_{\text{prim}}(t)$ is large, we may have $N_{\text{sec}}(t) > 1$. Pelton and Goddard [47] then set $f(t) = 0$ (negative rates of foam volume increase are of course not physically realizable!).

As active antifoam drops heterocoalesce with primary bubbles, their concentration in the bulk liquid is assumed to decline because their subsequent involvement in foam film rupture is assumed to produce complete deactivation. Irrespective of the likelihood of the latter, Pelton and Goddard [47] therefore write

$$N_{\text{prim}}(t) = -\frac{\mathrm{d}N_{\text{lp}}(t)}{\mathrm{d}t}\frac{\Delta\tilde{t}}{C_{\text{prim}}} \tag{6.9}$$

where $N_{\text{lp}}(t)$ is the active antifoam drop concentration (as drops/unit total liquid phase volume including dispersed bubbles) at time t, $\Delta\tilde{t}$ is the residence time of primary bubbles in the liquid phase and C_{prim} is the concentration of primary bubbles in the liquid phase (as bubbles/unit total liquid phase volume). C_{prim} is equal to $3\Phi_{\text{lp}}/4\pi r_{\text{prim}}^3$, where Φ_{lp} is the gas hold up (gas volume fraction in the liquid phase) and r_{prim} is the radius of the primary bubbles. Φ_{lp} should be calculated from the increase in volume of the liquid phase due to the presence of the gas, making due allowance for the liquid held in the foam column. Equation 6.9 is clearly only valid either for small residence times, $\Delta\tilde{t}$, or if $\mathrm{d}N_{\text{lp}}(t)/\mathrm{d}t$ is constant.

The process of heterocoalescence is assumed to be rate determining so that the rate of antifoam drop concentration decline is given by a simple second-order rate equation

$$\frac{\mathrm{d}N_{\text{lp}}(t)}{\mathrm{d}t} = -k_{\text{p}}C_{\text{prim}}N_{\text{lp}}(t) \tag{6.10}$$

where k_p is a rate constant. Here we note that $N_{lp}(t)$ is proportional to $\tilde{c}(t)$, the weight concentration of antifoam at time t, because $N_{lp}(t) = 3\tilde{c}(t)/4\pi\rho_{AF}r_{AF}^3$, where r_{AF} is the average radius of antifoam drops and ρ_{AF} is the density of the antifoam. Therefore, if we combine Equations 6.8 through 6.10, we can write

$$f(t) = 1 - k_p G\Delta\tilde{t} \cdot N_{lp}(t) = 1 - \frac{3k_p G\Delta\tilde{t} \cdot \tilde{c}(t)}{4\pi\rho_{AF}r_{AF}^3} \tag{6.11}$$

and Equation 6.10 can be rewritten as

$$\frac{d\tilde{c}(t)}{dt} = -k_p C_{prim}\tilde{c}(t) \tag{6.12}$$

By integrating Equation 6.12, we obtain $\tilde{c}(t)$ as a function of time and $\tilde{c}(t = 0)$

$$\tilde{c}(t) = \tilde{c}(t = 0)\exp(-k_p C_{prim} t) \tag{6.13}$$

Combining Equations 6.11 and 6.13 gives the basic equation of the theory relating $f(t)$ to the initial antifoam concentration $\tilde{c}(t = 0)$ and time

$$f(t) = 1 - \frac{3k_p G\Delta\tilde{t} \cdot \tilde{c}(t = 0) \cdot \exp(-k_p C_{prim} t)}{4\pi\rho_{AF}r_{AF}^3} \tag{6.14}$$

A limitation with this equation is immediately apparent in that as $\tilde{c}(t = 0)$ becomes very large, $f(t)$ becomes negative. As we have seen, this means we must set $f(t) = 0$ [47].

For k_p in Equations 6.10 through 6.14, Pelton and Goddard [47] write

$$k_p = k_e K_s \tag{6.15}$$

where k_e is an "adjustable coagulation efficiency factor" and K_s is the Smoluschowski coagulation constant (see, e.g., reference [49]) given by

$$K_s = 4(r_{AF} + r_{prim})(1/r_{AF} + 1/r_{prim})(\bar{k}T/6\eta_W) \tag{6.16}$$

where η_W is the viscosity of the foaming solution, $\bar{k}$ is the Boltzmann constant, and T is the temperature.

Use of K_s implies that the supposed process of heterocoalescence between antifoam drops and bubbles is diffusion controlled. In practice, it seems likely, however, that convection will be important under the conditions assumed for this model. The adjustable coagulation efficiency factor presumably accounts for the probability of antifoam drops overcoming colloidal repulsion forces to adhere to bubbles to which they have diffused. It seems likely, in the case of antifoam drops and primary bubbles, that $r_{AF} \ll r_{prim}$ so that Equation 6.16 can be simplified and combined with Equation 6.15 to give

$$k_p \approx 4k_e \left(\frac{r_{prim}}{r_{AF}} \right) (\bar{k}T / 6\eta_W) \tag{6.17}$$

Absence of knowledge of k_e means that the rate constant k_p is essentially unknown. G is at best ambiguous. Other parameters in Equation 6.14 are at least in principle amenable to experimental observation. However, in deriving this equation, Pelton and Goddard [47] used an incorrect expression for $\Delta\tilde{t}$, equating it to the ratio of liquid height/linear flow rate of gas. As pointed out by Pelton in a later paper [48], the correct expression should be

$$\Delta\tilde{t} = \frac{\Phi_{lp} V_l}{V_G} \tag{6.18}$$

where V_l is the liquid phase volume including dispersed bubbles and V_G is the volumetric gas flow rate.

Despite these limitations, Pelton and Goddard [47] present a test of Equation 6.14 against experiment that would appear to be valid, at least in principle. However, they argue that agreement with experiment should be best for $f(t)$ at $t = 0$ where the exponent in Equation 6.14 becomes unity. This means that the predictive value of the theory is best under conditions where antifoam deactivation is essentially absent. Here reasonable values of G, r_{prim}, $\Delta\tilde{t}$, and k_e are used in fitting a combination of Equations 6.14 and 6.17 to experimental results for $f(t)$ at $t = 0$ as a function of the initial antifoam drop concentration, $c(t = 0)$, for an antifoam with a known average value of the drop radius r_{AF} (= 4 μm). The authors then calculate the corresponding behavior, measured at the same gas volumetric gas flow rate (and therefore the same r_{prim} and $\Delta\tilde{t}$) and assuming the same value of G for another antifoam with a different value of r_{AF} (= 8 μm). This procedure also assumes the same value of k_e for the second antifoam. Results are presented in Figure 6.12 as a plot of $f(t)$ against $\log(\tilde{c}(t = 0))$ despite the implication of Equation 6.14 that $f(t)$ should be proportional to $\tilde{c}(t = 0)$ if $t = 0$. The agreement is not striking. Discrepancy between theory and experiment is marked for the log(antifoam concentration) dependence of $f(t)$ for the antifoam emulsion with r_{AF} = 8 μm. The reason for this discrepancy is not made clear. However, it could in part concern the value of r_{AF} because the authors give no experimental evidence for the respective drop sizes and size distributions in the two chosen commercial polydimethylsiloxane–hydrophobed silica antifoam emulsions.

Clearly then, the theory has limited predictive value even when restricted to circumstances where antifoam deactivation is not apparent (i.e., when $t \to 0$). The assumptions concerning the rate, and therefore mode, of antifoam deactivation are therefore untested and may be open to question.

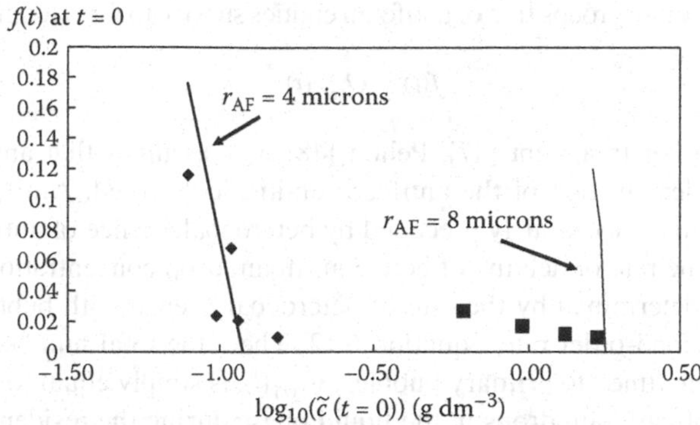

FIGURE 6.12 Comparison of theory of Pelton and Goddard [47] with experimental results for ratio $f(t)$ (rate of foam volume increase/volumetric gas flow rate) at $t = 0$ as function of logarithm of initial antifoam concentration, $\log(\tilde{c}(t = 0))$, for emulsions of hydrophobed silica–polydimethylsiloxane antifoam in 0.017 M SDS solution. Continuous lines represent prediction of theory for $f(t)$ against $\log(\tilde{c}(t = 0))$ for an emulsion with average antifoam drop radius, r_{AF}, of 8 microns using parameters obtained to fit experimental results for an emulsion with average drop radius of 4 microns. Foam generation by passing nitrogen through porous frit with gas flow rate 8.5 cm^{-3} s^{-1}. (Adapted from *Colloids Surf.*, 21, Pelton, R.H., Goddard, E.D., 167. Copyright 1986, with permission from Elsevier.)

6.7.2 Modified Antifoam–Bubble Heterocoalescence-Kinetic Model Using a Statistical Distribution of Antifoam Drops over Bubbles [48]

An obvious limitation of Equation 6.14 is that as $\tilde{c}(t = 0) \rightarrow \infty$, then $f(t) < 0$ and negative foam heights are predicted. This deficiency is circumvented inconveniently by setting $f(t) = 0$ if $f(t)$ is predicted to be <0. In an attempt to remove this limitation, Pelton [48] has presented a modified version of the theory. Again the theory concerns foam generation by sparging in the presence of a deactivating dispersion of a hydrophobed silica–polydimethylsiloxane antifoam. However, a statistical approach is used to assess the distribution of antifoam drops over all secondary bubble groups rather than simply assuming some average as used in Equation 6.8. Pelton [48] assumes that the total number of antifoam drops $N_{TOT}(t)$, adhering at time t to primary bubbles, is randomly distributed over those groups. All bubbles are assumed to be members of a group and therefore the total number of bubble groups is S $(= C_{prim}V_l/G)$. The probability of one antifoam entity being present in a given group is therefore $1/S$ and the probability of the entity being absent is $1 - 1/S$. Pelton [48] then deduces that the probability $Q_{free}(t)$ of a group being free of antifoam drops is

$$Q_{free}(t) = \left(1 - \frac{1}{S}\right)^{N_{TOT}(t)} \tag{6.19}$$

which is of course the equivalent of Equation 6.3.

Because only those groups free of antifoam entities survive to form foam we must have

$$f(t) = Q_{\text{free}}(t) \tag{6.20}$$

As in the earlier treatment [47], Pelton [48] now assumes that antifoam action produces total deactivation of the antifoam entities concerned. Antifoam action is again assumed to be necessarily preceded by heterocoalescence of antifoam entities with bubbles. The rate of decline of active antifoam drop concentration is therefore assumed to be determined by the rate of heterocoalescence with bubbles, which is given by the second-order rate Equation 6.12. Then, the total number of antifoam drops adhering at time t to primary bubbles, $n_{\text{TOT}}(t)$, is simply equal to the change in total numbers of antifoam drops in the liquid phase during the residence time $\Delta\tilde{t}$ of the primary bubbles. Therefore

$$n_{\text{TOT}}(t) = \frac{3(\tilde{c}(t+\Delta\tilde{t}) - \tilde{c}(t))V_1}{4\pi\rho_{\text{AF}}r_{\text{AF}}^3} = \frac{3V_1}{4\pi\rho_{\text{AF}}r_{\text{AF}}^3}\cdot\frac{\mathrm{d}\tilde{c}(t)}{\mathrm{d}t}\cdot\Delta\tilde{t} \tag{6.21}$$

if the residence time $\Delta\tilde{t}$ is short. On combining Equation 6.21 with Equations 6.12 and 6.13, we therefore have

$$n_{\text{TOT}}(t) = \frac{3\Delta\tilde{t}V_1k_pC_{\text{prim}}\cdot\tilde{c}(t=0)}{4\pi\rho_{\text{AF}}r_{\text{AF}}^3}\exp(-k_pC_{\text{prim}}t) \tag{6.22}$$

By combining Equations 6.19, 6.20, and 6.22, it is possible to derive an expression analogous to Equation 6.14 relating $f(t)$ to t and the initial antifoam weight concentration $c(t = 0)$. Thus, we obtain

$$f(t) = \left[1 - \frac{1}{S}\right]^{A\cdot\tilde{c}(t=0).\exp-k_pC_{\text{prim}}t} = \exp\left[m_1\tilde{c}(t=0)\cdot\exp(-m_2t)\right] \tag{6.23}$$

where $A = \dfrac{3\Delta\tilde{t}V_1k_pC_{\text{prim}}}{4\pi\rho_{\text{AF}}r_{\text{AF}}^3}$, $m_1 = A\cdot\ln\left(1 - \dfrac{1}{S}\right)$, and $m_2 = k_pC_{\text{prim}}$. Both m_1 and m_2 are constants for a given antifoam and given set of foam generation conditions. We must obviously have $m_1 < 0$ and $m_2 > 0$.

Equation 6.23 can be compared with experimental measurements of foam volumes against time for various concentrations of a given antifoam. To make such comparison, it is necessary to integrate Equation 6.23 to obtain the foam volume $V(t)$ as a function of time. Since we have $\dfrac{\mathrm{d}V(t)}{\mathrm{d}t} = V_{\text{G}}f(t)$, where V_{G} is the volumetric gas flow rate, then we must have

$$V(t) = V_{\text{G}}\int_0^t f(t)\mathrm{d}t \tag{6.24}$$

which corrects the error in the expression for $V(t)$ given by Pelton [48]. Substitution of Equation 6.23 in 6.24 means that we can write

$$V(t) = V_G \int_0^t \exp\left[m_1\tilde{c}(t=0)\cdot\exp(-m_2 t)\right]\mathrm{d}t \tag{6.25}$$

Integration of Equation 6.25 can be done numerically. It is clear from Equations 6.23 and 6.25 that, because $m_1 < 0$, $f(t) \to 0$ and $V(t) \to 0$ as $\tilde{c}(t=0) \to \infty$ and $t \to 0$. Also $f(t) \to 1$ and $V(t) \to V_G t$ as $\tilde{c}(t=0) \to 0$. This of course represents the expected behavior of an antifoam at the extremes of concentration. It is also clear that as $t \to 0$, then $f(t) \to \exp[m_1\tilde{c}(t=0)]$ so that $f(t)$ becomes time independent and plots of $V(t)$ against time are linear with a gradient of $V_G \exp[m_1\tilde{c}(t=0)]$. Therefore, Equations 6.23 through 6.25 predict linear plots of foam volume against time in the initial stages of foam generation. Finally, another feature of Equations 6.23 through 6.25 is that $f(t) \to 1$ and the gradient $\mathrm{d}V(t)/\mathrm{d}t \to V_G$ as $t \to \infty$. This of course must happen when the antifoam is completely deactivated—the rate of foam volume increase must then equal the volumetric gas flow rate provided the foam is intrinsically stable.

Pelton [48] has presented experimental plots of foam volume against time for solutions of a commercial dishwashing liquid containing polydimethylsiloxane–hydrophobed silica antifoam emulsion. Foam was generated by passing nitrogen through a porous frit. Examples of the results, which should afford a test of Equation 6.25, are presented in Figure 6.13. A linear plot of $V(t)$ against t from $t = 0$ is obtained

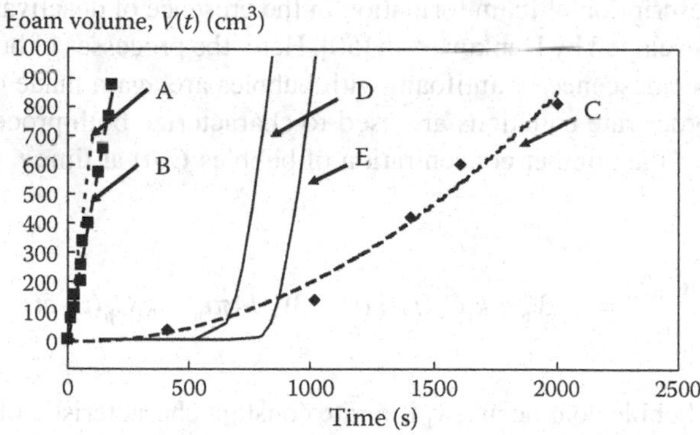

FIGURE 6.13 Comparison of theory of Pelton [48] (Equation 6.25) with experimental results for $V_{foam}(t)$ against time for various concentrations of a hydrophobed silica–polydimethylsiloxane antifoam emulsion dispersed in 5 g dm^{-3} solution of commercial dishwashing liquid. Foam generation was by passing nitrogen through porous frit at flow rate of 7 cm^3 s^{-1}. Here m_1 and m_2 were calculated using measured values for r_{pb}, V_l, and V_G together with values of G, k_e, and Φ_{lp} adjusted to give a curve "in the same time scale" as the experimental curve for $\tilde{c}(t=0) = 0.5$ g dm^{-3}. A, experiment in absence of antifoam, $\tilde{c}(t=0) = 0$. B, experiment for antifoam concentration $\tilde{c}(t=0) = 0.1$ g dm^{-3}. C, experiment for antifoam concentration $\tilde{c}(t=0) = 0.5$ g dm^{-3}. D, theory for antifoam concentration $\tilde{c}(t=0) = 0.025$ g dm^{-3}. E, theory for antifoam concentration $\tilde{c}(t=0) = 0.5$ g dm^{-3}.

in the absence of antifoam. Experimental measurements at a low concentration of antifoam (0.1 g dm^{-3}) also give a linear plot from $t = 0$. The plot at an antifoam concentration of 0.5 g dm^{-3} deviates markedly from linearity presumably as a result of antifoam deactivation.

Pelton [48] has attempted to reproduce the experimental results shown in Figure 6.13 using Equation 6.25 and explicit expressions for m_1 and m_2. Here, values for r_{prim}, V_l, and V_G were measured, a value of $r_{AF} = 3$ μm was assumed and values of G, k_e, and Φ_{lp} (giving $\Delta\tilde{t} = \Phi_{lp}V_l/V_G$, $C_{\text{prim}} = 3\Phi_{lp}/4\pi r_{\text{prim}}^3$, and $S = C_{\text{prim}}V_l/G$) were adjusted to give a foam volume against time curve in "the same time scale" as that found by experiment for an antifoam concentration, $\tilde{c}(t = 0)$, of 0.5 g dm^{-3}. The results of this exercise are also reproduced in Figure 6.21. Clearly, the agreement is at best poor for an initial antifoam concentration of 0.5 g dm^{-3} where, despite some qualitative resemblance to the experimental results, quantitative agreement is lacking. That the agreement is poor is especially apparent when low antifoam concentrations are considered. The linear foam volume against time dependence from $t = 0$ for those concentrations of antifoam is not reproduced and initial antifoam effectiveness is grossly overestimated. Clearly, the explicit expressions used by Pelton [48] for m_1 and m_2 have little predictive value. This presumably arises in part because some parameters are ill defined, such as the size of secondary bubbles, and other parameters are not defined, such as the "adjustable coagulation efficiency factor," k_e.

6.7.3 Combination of Kinetic Model of Antifoam Deactivation with Kinetic Model of Antifoam Action [50]

A theoretical description of foam formation in the presence of deactivating antifoam has also been developed by Denkov et al. [50]. Here, the processes of bubble destruction and heterocoalescence of antifoam with bubbles are again made explicitly distinct. Second-order rate equations are used to characterize both processes. For the rate of change of the number concentration of bubbles $C_b(t)$ at time t, Denkov et al. [50] write

$$\frac{dC_b(t)}{dt} = V_G/V_b - \tilde{k}_1 C_b(t)\cdot\tilde{c}(t) = 3V_G/4\pi r_b^3 - \tilde{k}_1 C_b(t)\cdot\tilde{c}(t) \qquad (6.26)$$

where V_b is the bubble volume and $\tilde{k}_1$ is a rate constant characteristic of the antifoam effect. The term V_G/V_b in the equation describes the rate of increase in the number of bubbles in the foam in the absence of antifoam, which is constant. The second term in this equation describes the effect of antifoam on the rate of bubble buildup, which is assumed proportional to the product of the number concentration of bubbles and the antifoam concentration.

A problem with Equation 6.26 concerns the exact definition of $C_b(t)$. If the concentration of bubbles is simply that which is present in the liquid phase, then $C_b(t)$ is constant and depends only on the gas flow rate. This is the definition used by Pelton and Goddard [47] and again later by Pelton [48]. If, on the other hand, $C_b(t)$ is the concentration of bubbles in a foam, then that will be determined by the gas phase volume fraction and will be only weakly dependent on the volume of foam generated (because gas phase volume fraction will vary only from ~75% to >99% in a foam, depending on the circumstances of foam generation). It seems then Denkov et al. [50] really mean that $C_b(t)$ is the total number of bubbles per unit volume of the vessel that contains both gas bubbles dispersed in the liquid phase and foam. In which case, absence of foam does not imply $C_b(t) = 0$. However, absence of increase in foam volume with time of aeration does imply

$$\frac{dC_b(t)}{dt} = 0 \tag{6.27}$$

The rate of volume increase of foam is then simply $V_b \cdot dC_b(t)/dt$ regardless of the presence or otherwise of antifoam.

In this treatment of Denkov et al. [50], the rate of antifoam deactivation is assumed equal to the rate of heterocoalescence between antifoam drops and bubbles, which is supposed given by another second-order rate equation so that

$$\frac{d\tilde{c}(t)}{dt} = -\tilde{k}_2 C_b(t) \cdot \tilde{c}(t) \tag{6.28}$$

where $\tilde{k}_2$ is a rate constant characteristic of antifoam deactivation. Equation 6.28 superficially resembles the second-order rate Equations 6.10 and 6.12 used by Pelton and Goddard [47] and later by Pelton [48] to describe the rate of antifoam deactivation. However, as we have seen, an important difference concerns the definition of bubble concentration where Pelton and Goddard [47] use the number concentration of bubbles in unit volume of the liquid phase, which is constant for given aeration conditions.

Denkov et al. [50] now use Equations 6.26 through 6.28 to develop arguments that permit comparison with the foam volume against time plots typified by Figure 6.14 where two stages are defined. Here foam was generated by the so-called foam rise method, where monodisperse foam was continuously generated by passing nitrogen through a collection of capillaries into a dispersion of silicone antifoam in 0.01 M AOT solution. For stage 1 where $t < t^*$, little foam is formed and the rate of change in bubble concentration is given by Equation 6.27. For stage 2 where $t > t^*$, then the foam volume is seen to rapidly increase approximately linearly with time but at a rate less than that in the absence of antifoam. Clearly then some active antifoam is still present at $t > t^*$.

Equation 6.27 implies that we can deduce from Equation 6.26 that

$$C_b(t) \cdot \tilde{c}(t) = V_G / \tilde{k}_1 V_b \tag{6.29}$$

where $V_G / \tilde{k}_1 V_b$ is a constant for given foam-generating conditions. Substituting this equation into Equation 6.28 and integrating yields the following expression for the antifoam concentration as a function of time, $\tilde{c}(t)$

$$\tilde{c}(t) = \tilde{c}(t = 0) - \tilde{K}_3 t \tag{6.30}$$

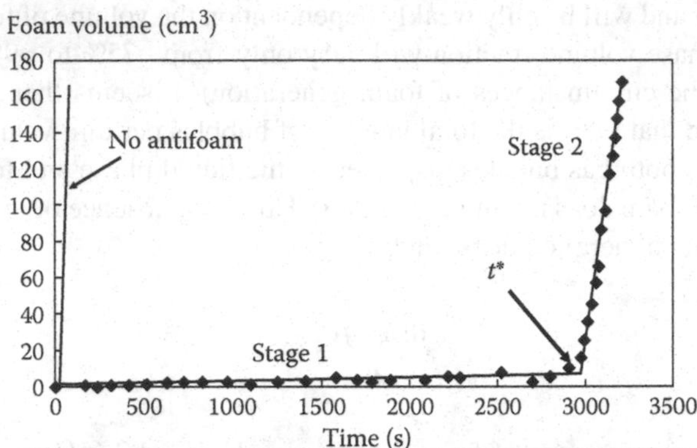

FIGURE 6.14 Plot of foam volume against time by the foam rise method illustrating the definition of stage 1, stage 2, and t^*. 0.2 g dm^{-3} of a polydimethylsiloxane–hydrophobed silica antifoam emulsion (polydimethylsiloxane oil, 1000 mPa s; 4.2 wt.% hydrophobed pyrogenic silica) stabilized by a mixture of sorbitan monostearate and Span 60 and dispersed in 0.01 M AOT solution. $V_G = 3.75$ cm^3 s^{-1}; $V_b = 0.015$ cm^3. (From Denkov, N.D., Marinova, K., Tcholakova, S., Deruelle, M. Mechanism of foam destruction by emulsions of PDMS–silica mixtures, in *Proceedings of 3rd World Congress on Emulsions*, 24–27 September, 2002, Lyon, France, paper 1-D-199.)

where $\tilde{K}_3$ is a constant given by

$$\tilde{K}_3 = \frac{\tilde{k}_2}{\tilde{k}_1} \cdot \frac{V_G}{V_b} \tag{6.31}$$

Substitution of Equation 6.30 into Equation 6.29 then yields an expression for the total concentration of bubbles, $C_b(t)$, in the foam during "stage 1"

$$C_b(t) = \frac{V_G}{\tilde{k}_1 V_b (\tilde{c}(t=0) - \tilde{K}_3 t)} \tag{6.32}$$

Clearly then the reasoning implied by Equations 6.26 through 6.29 has led to an expression for $C_b(t)$ under "stage 1" behavior, which is a function of time. In turn, this must mean that $dC_b(t)/dt \neq 0$, which violates one of the assumptions (Equation 6.27) upon which this analysis is based. The origin of the inconsistency is made obvious if we consider Equation 6.29 where the product of the antifoam and bubble concentrations is seen to be a constant. However, if there is essentially no foam, then $C_b(t)$ will be essentially constant under "stage 1" conditions. In which case there must also be no change in antifoam concentration as well if Equation 6.29 is to hold true. Obviously Equations 6.27 and 6.29 are only intended to be approximate but nevertheless they would still imply virtually no antifoam deactivation over "stage 1" if the basic rate Equations 6.26 and 6.28 have relevance for "stage 1" behavior.

Despite these limitations, Denkov et al. [50] have made some comparison with experiment. Here, for example, $\tilde{k}_1$ and $\tilde{k}_2$ were determined from foam volume against time plots for one method of foam generation. Assuming these constants are independent of foam generation method, they were then combined with an experimental value of t^* for another method of foam generation in an application of deductions from Equations 6.30 through 6.31. A reasonable estimate of V_G/V_b for that method of foam generation was obtained.

Finally it should be emphasized that this approach would appear to have no direct relevance for "stage 2" behavior where foam volume, and therefore $C_b(t)$, grows dramatically with time. Indeed "stage 2" behavior has been observed by other workers [48, 51] from $t = 0$ (so that $t^* = 0$) for low polydimethylsiloxane–silica antifoam concentrations. A more general treatment for the phenomenon of foam volume increase during continuous aeration by gas bubbling in the presence of a deactivating antifoam is therefore clearly desirable, which we describe in Section 6.7.4.

6.7.4 Combination of Kinetic Model of Antifoam Deactivation by Disproportionation with Empirical Expression for Antifoam Action

Here we attempt to develop a simple theoretical model of the effect of antifoam deactivation on the rate of formation of foam, which accommodates recent findings concerning the nature of that deactivation. It will be applied to continuous foam generation by sparging. As with all models of antifoam effect, it is supposed that the antifoam is dispersed in the form of an emulsion. Deactivation is represented as a decrease in active antifoam concentration. The link between that concentration and foam behavior is provided by an empirical expression for which there is at least some limited theoretical basis provided by the statistical model described earlier (see Section 5.2).

In the case of an emulsified polydimethylsiloxane–hydrophobed silica antifoam, the work of Denkov et al. [6, 7, 21] has revealed that deactivation arises because of disproportionation. The antifoam forms two distinct populations—one of drops devoid of particles and the other of silica-rich entities ranging in size from drops to large agglomerates. Neither of these populations are effective antifoams. As we have shown (see Section 6.4), disproportionation is probably an inevitable consequence of the effect of drop splitting and coalescence, processes that are likely to occur even under steady-state agitation conditions. Clearly the number concentration of active antifoam drops must be decreasing with time because of both the formation of inactive silica-rich drops and agglomerates (see Section 6.4) by a net coalescence process and the formation of essentially inactive particle-free drops by a net drop splitting process.

We may therefore represent the rate of decline of active antifoam concentration $\tilde{c}(t)$ as the sum of a first-order drop splitting and a second-order coalescence so that

$$-\frac{d\tilde{c}(t)}{dt} = k_1^0\tilde{c}(t) + k_2^0\tilde{c}(t)^2 \tag{6.33}$$

where k_1^0 and k_2^0 are first- and second-order rate constants, respectively. Obviously this approach represents a gross oversimplification of the actual processes occurring during antifoam deactivation as outlined in Section 6.4. In particular, the complexities involved in the repeated coalescence–splitting processes required to produce particle-rich agglomerates are probably poorly represented by a simple second-order rate process.

Equation 6.33 may be integrated to give

$$\tilde{c}(t) = \frac{k_1^0}{\left[\left(\dfrac{k_1^0 + k_2^0 \tilde{c}(t=0)}{\tilde{c}(t=0)}\right)\exp k_1^0 t - k_2^0\right]} \tag{6.34}$$

We are now required to relate the antifoam concentration to the foam volume or F (= volume of air in foam in presence of antifoam/volume of air in foam in absence of antifoam) in order to make comparison between the implications of Equation 6.34 and observations of foam generation. The experimental concentration dependence of a commercial polydimethylsiloxane antifoam on foam volume, and therefore F, in the case of sparging solutions of sodium oleate, has been reported by Kulkarni et al. [51]. Results for short-duration sparging (≤200 s), where negligible deactivation of the antifoam is to be expected, have been shown by Garrett [23] to be represented by a simple exponential so that

$$F = \exp(\tilde{m}\tilde{c}) \tag{6.35}$$

where $\tilde{m}$ is a constant, independent of antifoam concentration and the process of deactivation. Since increasing F must accompany decreasing $\tilde{c}$, we must have $\tilde{m} < 0$. Also Equation 6.35 implies that we must also have $F = 1$ as $\tilde{c} = 0$ and $F = 0$ as $\tilde{c} \to \infty$, which are desirable characteristics of any useful empirical expression in this context. This equation can in fact be derived in the form of Equation 5.32 from the statistical theory of antifoam action for the case of a monodisperse foam. We will therefore use this simple exponential in exploring the implications of Equation 6.34 for foam generation by sparging in the presence of deactivating silicone antifoams.

Equation 6.35 may now be combined with Equation 6.34 to derive an expression for $\tilde{F}(t)$, the instantaneous value of F at time t, where the antifoam concentration has declined to $c(t)$ as a result of disproportionation (so substituting $\tilde{c}(t)$ for $\tilde{c}$). We therefore find that

$$\tilde{F}(t) = \exp[\tilde{m}\tilde{c}(t)] = \exp\left[\frac{\tilde{m}k_1^0 \tilde{c}(t=0)}{\{k_1^0 + k_2^0\tilde{c}(t=0)\}\exp k_1^0 t - k_2^0 \tilde{c}(t=0)}\right] \tag{6.36}$$

Here $\tilde{F}(t)$ is the value of F for a hypothetical situation where $\tilde{c}(t)$ has prevailed throughout the foam generation process. However, in a prolonged foam-generating situation, producing significant antifoam deactivation, both $\tilde{c}(t)$ and therefore $\tilde{F}(t)$

change with time throughout that process. We can allow for that by considering increments in the buildup of the foam in the presence of deactivating antifoam. Thus, we can write for the volume of foam $V(t)$ at time t

$$V(t) = V_G \mathrm{d}t \cdot \tilde{F}(t = \mathrm{d}t) + V_G \mathrm{d}t \cdot \tilde{F}(t = 2\mathrm{d}t) + V_G \mathrm{d}t \cdot \tilde{F}(t = 3\mathrm{d}t) \ldots\ldots + V_G \mathrm{d}t \cdot \tilde{F}(t) \quad (6.37)$$

where in the limit $\mathrm{d}t \to 0$ so that $\tilde{F}(t)$ is essentially constant over that time interval. Here we remember that V_G is the volumetric gas flow rate in the case of sparging. We can now write for $F(t)$

$$F(t) = \frac{V(t)}{V_G t} = \frac{(\tilde{F}(t = \mathrm{d}t) + \tilde{F}(t = 2\mathrm{d}t) + \tilde{F}(t = 3\mathrm{d}t) \ldots\ldots + \tilde{F}(t))\mathrm{d}t}{t}$$

$$= \frac{1}{t}\int_0^t \tilde{F}(t)\,\mathrm{d}t \quad (6.38)$$

and

$$V(t) = V_G t F(t) = V_G \int_0^t \tilde{F}(t)\,\mathrm{d}t \quad (6.39)$$

where it is obvious from Equation 6.24 that $\tilde{F}(t)$ is identical to the ratio $f(t)$ defined by Pelton and Goddard [47]. Substituting Equation 6.36 in Equation 6.39 means that we can write for the foam volume as a function of time

$$V(t) = V_G \int_0^t \exp\left[\frac{\tilde{m}k_1^0 \tilde{c}(t=0)}{\{k_1^0 + k_2^0 \tilde{c}(t=0)\}\exp k_1^0 t - k_2^0 \tilde{c}(t=0)}\right]\mathrm{d}t \quad (6.40)$$

This equation may be integrated numerically to afford comparison with experiment. It predicts that as $t \to 0$

$$V(t) = V_G \int_0^t \exp(\tilde{m}\tilde{c}(t=0))\,\mathrm{d}t = V_G \exp(\tilde{m}\tilde{c}(t=0))t \quad (6.41)$$

which means that plots of $V(t)$ against time become linear with a gradient of $V_G \exp(\tilde{m}\tilde{c}(t=0))$ where $V(t) = 0$ at $t = 0$. In the limit, where $\tilde{c}(t=0) = 0$, Equation 6.40 reduces to $V(t) = V_G t$. However, when $\tilde{c}(t=0) \to \infty$, Equation 6.40 becomes

$$V(t) = V_G \int_0^t \exp\left[\frac{\tilde{m}k_1^0}{k_2^0(\exp k_1^0 - 1)}\right]\mathrm{d}t \quad (6.42)$$

which clearly reduces to $V(t) = 0$ (because $\tilde{m} < 0$, $k_1^0 > 0$, and $k_2^0 > 0$) only when $t \to 0$. Also it is easy to show that Equations 6.40 and 6.42 both imply that $dV(t)/dt = V_G$ as $t \to \infty$. This means that regardless of the concentration and effectiveness of an antifoam, deactivation will eventually triumph so that the antifoam becomes totally ineffective given a sufficiently long duration of aeration. The foam volume will then increase at a rate equal to the volumetric gas flow rate, provided the foam is intrinsically stable. As we have seen, similar limiting behavior is also predicted by Equation 6.25, which is based on the theory of Pelton [48]. This is not surprising because any model that has any claim to consistency with experimental observation of foam generation by sparging in the presence of a deactivating antifoam should exhibit these characteristics.

A comparison of Equation 6.40 with experiment can be made using the plots of foam volume against time for solutions of a commercial dishwashing liquid containing various concentrations of polydimethylsiloxane–hydrophobed silica antifoam emulsion given by Pelton [48]. Here Equation 6.40 was fitted to the plot for an antifoam concentration high enough (at $\tilde{c}(t = 0) = 0.5$ g dm^{-3}) to clearly show deactivation and therefore marked deviation from linearity. This yielded values of $\tilde{m}$, k_1^0, and k_2^0. The values of these parameters may then be used with Equation 6.40 to calculate foam volume against time plots at different antifoam concentrations in a procedure that involves no fitting. Results are presented in Figure 6.15. It is clear that Equation 6.40 is in, at least, semiquantitative agreement with experiment and compares favorably with attempts to relate earlier theory [48] to the same results despite the crudity of the model implied by Equation 6.33.

We may also fit Equation 6.40 to the foam rise results of Denkov et al. [50] shown in Figure 6.14. Comparison is difficult because the foam volume is close to zero as

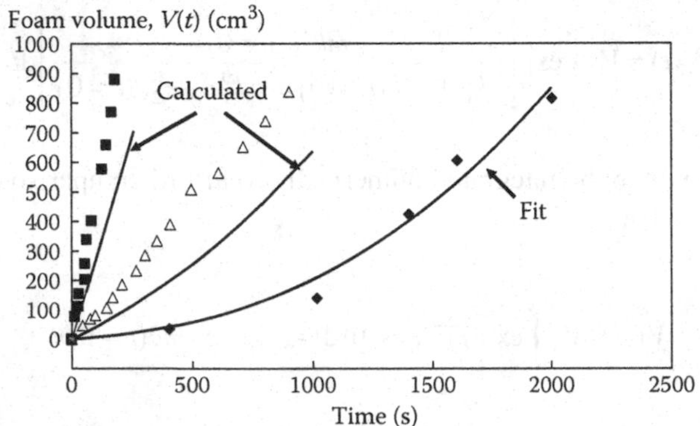

FIGURE 6.15 Comparison of Equation 6.40 with experimental results for foam volume, $V(t)$, against time obtained by Pelton [48] for various concentrations of a hydrophobed silica–polydimethylsiloxane antifoam emulsion dispersed in a 5 g dm^{-3} solution of commercial dishwashing liquid. Foam was generated by passing nitrogen through porous frit at gas flow rate of 7 cm^3 s^{-1}. Antifoam concentrations: ■, 0.1 g dm^{-3}; △, 0.3 g dm^{-3}; ◆, 0.5 g dm^{-3}. Fit to results at 0.5 g dm^{-3} yielded $\tilde{m} = -9.65\ \text{g}^{-1}\ \text{dm}^3$, $k_1^0 = 1.3\times10^{-6}\ \text{s}^{-1}$, $k_2^0 = 1.36\times10^{-3}\ \text{g}^{-1}\ \text{dm}^3\ \text{s}^{-1}$. These values were used to calculate the plots shown for 0.1 and 0.3 g dm^{-3}.

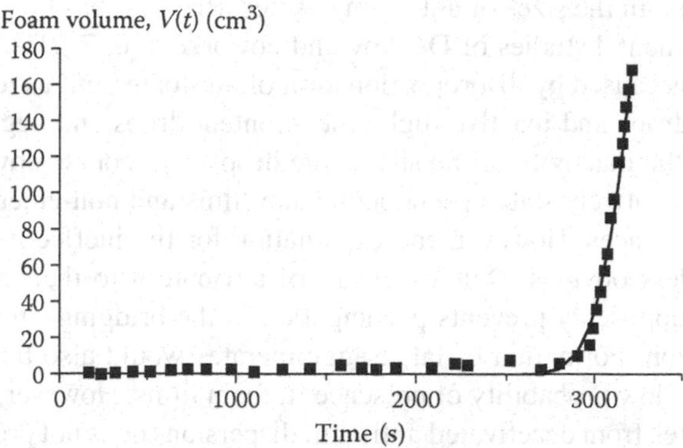

FIGURE 6.16 Comparison of Equation 6.40 with experimental plots by Denkov et al. [50] of foam volume, $V(t)$, against time by foam rise method with 0.2 g dm^{-3} of polydimethylsiloxane–hydrophobed silica antifoam emulsion (polydimethylsiloxane oil 1000 mPa s; 4.2 wt.% hydrophobed pyrogenic silica) stabilized by a mixture of sorbitan monostearate and Span 60 and dispersed in 0.01 M AOT solution. Monodisperse foam generated by blowing nitrogen through 27 glass capillaries at gas flow rate of 3.75 cm^3 s^{-1}. Experimental results: ■, Solid line is fit to results with $\tilde{m} = -1.6\times10^5$ g^{-1} dm^3, $k_1^0 = 3\times10^{-3}$ s^{-1}, $k_2^0 = 1.1\times10^{-2}$ g^{-1} dm^3 s^{-1}.

$t \rightarrow 0$, which means that accurate estimates of $\tilde{m}$ cannot be made. It is therefore limited to establishing whether the equation can at least fit the results using suitable values of $\tilde{m}$, k_1^0, and k_2^0. It is clear from Figure 6.15 that provided large enough values of $|\tilde{m}|$ are selected, the basic features of the experimental results are reproduced. That the fit does not represent accurately low values of foam volume at low times could simply be due to the near-zero capillary pressures prevailing in the foam at low foam heights. Antifoam effectiveness under those conditions could then be low irrespective of the extent of deactivation of the antifoam and Equation 6.35 would not be expected to hold. A small residual volume of foam would then be apparent even under conditions where Equation 6.35 would predict that the antifoam concentration would be high enough to destroy all foam despite some deactivation.

The fact that one equation can represent, albeit qualitatively, the apparently very different foam volume against time plots shown in both Figures 6.15 and 6.16 could indicate that the same phenomenon underlies both sets of results. The huge difference in the values of $\tilde{m}$ for these two plots in the main derives from a combination of enormous differences in antifoam effectiveness and a probable large difference in bubble sizes (where, as we have described in Section 5.2, a statistical theory of antifoam action predicts that larger bubbles are likely to require a smaller concentration of antifoam to effect rupture).

6.8 SUMMARIZING REMARKS

There is now a significant body of published data that reveal that antifoam mixtures of polydimethylsiloxane and hydrophobed silica deactivate as a result of both emulsification and antifoam action [1–7]. The process does not appear to simply concern reduction in the sizes of antifoam entities. Rather it has been shown, by the excellent experimental studies of Denkov and coworkers [6, 7, 21], that the process of deactivation is caused by disproportionation of antifoam entities to produce inactive silica-free drops and inactive high-silica-content drops and agglomerates. The explanation for the inactivity of the silica-free drops is of course obvious—absence of silica means relatively stable pseudoemulsion films and non-emergence of drops into air–water surfaces. However, the explanation for the ineffectiveness of silica-rich entities is less obvious. Denkov et al. [6] attribute it to their non-deformable nature, which supposedly prevents participation in the bridging–stretching process of antifoam action. Formation of large agglomerates would also be a contributory cause because of low probability of presence in foam films. However, the removal of such agglomerates from deactivated antifoam dispersions does not preclude reactivation by addition of the oil, which tends to suggest that much particulate material is still present in the remaining drops [6].

There has been little study of the deactivation of other hydrophobic oil–hydrophobic particle antifoams where the oil is not a polydimethylsiloxane and the particle is not a hydrophobed silica. Apart from practical relevance, such studies would help explore the mechanism of deactivation in more detail. A matter of some interest here is elimination of any putative role for oil film spreading in determining the process of drop splitting by study of deactivation in the case of an antifoam based on a non-spreading oil exhibiting partial wetting. The possibility that the fractal nature of hydrophobic silica particles (or any other property unique to that material) renders antifoams relatively susceptible to deactivation requires attention. A detailed study of deactivation involving consideration of particle properties such as size, shape, and aggregation behavior in the relevant oil would seem to be timely.

We have explored the possibility that deactivation of polydimethylsiloxane–hydrophobed silica antifoams by disproportionation is a consequence of the random distribution of particles across the drops formed when the antifoam is dispersed to achieve a steady state by processes of drop splitting and coalescence. Simple mass balance considerations permit the estimation of the composition of deactivated antifoam dispersions if we assume they are made up of particle-free drops together with particle-rich drops and agglomerates of known silica content. From that composition, it is possible to calculate the ratio of the number of particles to the number of drops, $\tilde{N}/M$, in a deactivated antifoam if the distribution of particles across drops is assumed to be random. This analysis reveals that for monodisperse particles and drops where $\tilde{N}/M < 1$ this necessarily implies a proportion of particle-free drops irrespective of the nature of any distribution! If, on the other hand, $\tilde{N}/M > 1$, then the probability of finding particle-free drops is always vanishingly small.

There are two obvious limitations with this approach. Although it is possible to account for a realistically large proportion of particle-free drops, it is not possible to account for inactive particle-rich drops and agglomerates where particle concentrations are uniformly high. The gradual accumulation of large agglomerates, as an antifoam dispersion deactivates, implies some irreversible coalescence which possibility is obviously ignored if deactivation is considered to simply involve a random distribution of particles over oil drops in a steady state. Another limitation concerns the nature of the distribution of particles over drops—randomness implies that the uniform high particle concentrations will not be found but rather a distribution of particles with drops containing particle concentrations low enough to be effective. Total deactivation of an antifoam into inactive particle-free drops and inactive particle-rich drops and agglomerates cannot therefore be accounted for by a steady-state random distribution of particles over drops. Perhaps a better approach to understanding the process is computer-based simulation where rules for drop coalescence and splitting are superimposed upon the basic statistics. For example, coalescence is, as we have seen, dependent on the presence of particles at the oil–water interface of drops. One rule could therefore be that drops without particles should not coalesce. On the other hand, drop splitting is likely to be inhibited by increases in viscosity and therefore increases in particle volume fraction. Particle-rich drops could therefore continue to coalesce but without splitting so that increasing amounts of material are contained in agglomerates of increasing size as the antifoam deactivates. Another rule could therefore simply be that drops cannot split if the volume fraction of particles reaches a certain threshold value.

Irrespective of these considerations, disproportionation obviously involves a redistribution of particles across oil drops. Increasing the number of particles at a fixed total particle volume fraction would therefore seem to decrease the rate of disproportionation to produce a large volume fraction of particle-free drops without enhancing the formation of high particle volume fraction agglomerates with high viscosities. Conversely, any factors that can lead to formation of such large particle-rich agglomerates should be minimized. In the main, this would appear to concern the effect of particles on the viscosity of oil drops. Particles that form fractal aggregates, as exemplified by the hydrophobed silicas used in silicone antifoam formulation, would therefore not appear to be ideal. Such fractal particles occupy a large volume for their weight and would therefore be expected to be effective in increasing drop viscosities at relatively low weight concentrations with an adverse effect on drop splitting.

Apart from these considerations, there is a clear need for improved understanding of both the role of oil viscosity in antifoam deactivation and the extent to which the deactivation behavior of hydrophobed silica–polydimethylsiloxane antifoams may be generalized to other hydrophobic particle–oil antifoams.

Overall theories concerning the effect of deactivating antifoam on the volume of foam generated by continuous aeration have been of limited success. Perhaps it is worth emphasizing that if such theories are to be at least relevant, they should satisfy certain obvious criteria concerning the built-in relation between antifoam concentration and foam volume $f(t)$ or $F(t)$. Thus, they should always satisfy the limit that foam volume is never <0 even as antifoam concentration tends to high values; that is, $V(t), f(t)$, or $F(t) \rightarrow 0$ as $\tilde{c}(t=0) \rightarrow \infty$! The early theory of Pelton and Goddard [47] (Section 6.7.1) fails in that respect (Equation 6.11). It would appear to have little or no predictive value. We should also expect to have $V(t) = 0$ as $t = 0$. It is important that as the initial antifoam concentration, $\tilde{c}(t=0)$, tends to zero, the predicted foam volume or rate of foam generation tends to the intrinsic value; that is, $f(t) \rightarrow 1$ and $V(t) \rightarrow V_G t$ as $\tilde{c}(t=0) \rightarrow 0$. If, as appears to be the case, antifoam concentration falls to zero with time as a result of deactivation, then we would also expect $f(t) \rightarrow 1$ and $dV(t)/dt \rightarrow V_G$ (the rate of aeration) as $t \rightarrow \infty$. Finally, if the antifoam does not deactivate, then we would expect $f(t)$ to be a constant, independent of time. The later theory of Pelton [48] (Section 6.7.2) and the theory presented here (Section 6.7.4) both satisfy all these criteria. However, only the latter appears to have predictive value, and even then only of a semiquantitative nature.

Such overall theories of foam generation in the presence of deactivating antifoam must of course take account of the kinetics of antifoam deactivation. Pelton and Goddard [47] and later Pelton [48] both attribute deactivation to heterocoalescence of antifoam entities with bubbles in the liquid phase below the foam during foam generation by sparging. They assume deactivation is described by a second-order rate equation where the rate is proportional to the bubble and antifoam concentrations. Denkov et al. [50] (Section 6.7.3) also use a second-order rate equation involving a product of antifoam concentration and bubble concentration. Neither of these rate equations take account of the fact that deactivation does not in fact apparently require coalescence with bubbles because deactivation can apparently occur during emulsification. By contrast, the theory presented here (Section 6.7.4) takes account of the evidence that deactivation is accompanied by disproportionation in that a rate equation is used with a first-order term for drop splitting and a second-order term used for drop coalescence. This, combined with a semiempirical expression for the concentration dependence of foam generation, appears to give the closest approximation to experimental observation.

ACKNOWLEDGMENT

The author is particularly grateful to Dr. Paul Grassia for many discussions about the statistics of disproportionation in mixed particle–oil drops. The author is also grateful to Dr. Robin Curtis for fitting Equation 6.40 to experimental data.

REFERENCES

1. Pouchelon, A., Araud, C. *J. Disper. Sci. Technol.*, 14(4), 447, 1993.
2. Koczo, K., Koczone, J. Wasan, D. *J. Colloid Interface Sci.*, 166, 225, 1994.
3. Racz, G., Koczo, K., Wasan, D. *J. Colloid Interface Sci.*, 181, 124, 1996.
4. Bergeron, V., Cooper, P., Fischer, C., Giermanska-Kahn, J., Langevin, D., Pouchelon, A. *Colloids Surf. A*, 122, 103, 1997.
5. Denkov, N., Cooper, P., Martin, J. *Langmuir*, 15, 8514, 1999.
6. Denkov, N., Marinova, K.G., Christova, C., Hadjiiska, A., Cooper, P. *Langmuir*, 16(6), 2515, 2000.
7. Marinova, K.G., Tcholakova, S., Denkov, N., Roussev, S., Deruelle, M. *Langmuir*, 19(7), 3084, 2003.
8. Garrett, P.R., Davis, J., Rendall, H.M. *Colloids Surf. A*, 85, 159, 1994.
9. Curtis, J.C., Garrett, P.R., Nicholls, M., Yorke, J.W.H. Preliminary observations concerning the use of perfluoroalkyl alkanes in hydrocarbon-based antifoams, in *Emulsions, Foams, and Thin Films* (Mittal, K.L., Kumar, P., eds.), Marcel Dekker, New York, 2000, p 177.
10. Wicks, S.P. The effect of high volume fractions of latex particles on foaming and antifoam action in surfactant solutions, PhD Thesis, University of Manchester, 2006, p 183.
11. Ross, S., *J. Phys. Colloid Chem.*, 54(3), 429, 1950.
12. Basheva, E.S., Ganchev, D., Denkov, N.D., Kasuga, K., Satoh, N., Tsujii, K. *Langmuir*, 16(3), 1000, 2000.

13. Yorke, J.W.H. unpublished work.
14. Niibori, Y., Kunita, M., Tochiyama, O., Chida, T.J. *Nucl. Sci. Technol.*, 37(4), 349, 2000.
15. Blake, T., Kitchener, J. *J.C.S. Faraday I*, 68, 1435, 1972.
16. Vigil, G., Xu, Z., Steinberg, S., Israelachvili, J. *J. Colloid Interface Sci.*, 165, 367, 1994.
17. Ross, S., Nishioka, G. *J. Colloid Interface Sci.*, 65(2), 216, 1978.
18. Kulkarni, R., Goddard, E. *Croatica Chem. Acta*, 50(1–4), 163, 1977.
19. Kulkarni R., Goddard, E., Kanner, B. *Ind. Eng. Fundam.*, 16(4), 472, 1977.
20. Bergeron, V. PDMS based antifoams: mechanisms and performance, in *Les Mousses: Moussage et Démoussage* (Lagerge, S., ed.), EDP Sciences, Cahiers de Formulation, Paris, 2000, Vol 9, p 116.
21. Marinova, K., Denkov, N.D. *Langmuir*, 17(8), 2426, 2001.
22. Veber, V., Pauchek, M. *Acta Fac. Pharm. Univ. Comenianae*, 26, 221, 1974.
23. Garrett, P.R. *Langmuir*, 11(9), 3576, 1995.
24. Dippenaar, A. *Int. J. Miner. Process.*, 9, 1, 1982.
25. Frye, G.C., Berg, J.C. *J. Colloid Interface Sci.*, 127(1), 222, 1989.
26. Bergeron, V. Forces and structure in surfactant-laden thin liquid films, PhD Thesis, University of California, 1993.
27. Vrij, A., Overbeek, J. *J. Am. Chem. Soc.*, 90(12), 153, 1968.
28. Denkov, N.D., Tcholakova, S., Marinova, K. G., Hadjiiski, A. *Langmuir*, 18(15), 5810, 2002.
29. Denkov, N.D., Cooper, P., Martin, J. *Langmuir*, 15(24), 8514, 1999.
30. Denkov, N.D. *Langmuir*, 15(24), 8530, 1999.
31. Ross, S., Nishioka, G. Experimental researches on silicone antifoams, in *Emulsions, Lattices, and Dispersions* (Becher, P., Yudenfreund, M.N., eds.), Marcel Dekker, New York, 1978, p 237.
32. Grassia, P. personal communication.
33. Powell, R.L., Mason, S.G. *AIChE J.*, 28(2), 286, 1982.
34. Smith, P.G., van de Ven, T.G.M. *Colloids Surf.*, 15, 191, 1985.
35. Yan, Y., Masliyah, J.H. *Colloids Surf. A*, 75, 123, 1993.
36. Pickering, S.U. *J. Chem. Soc.*, 91, 2001, 1907.
37. Ramsden, W. *Proc. Royal. Soc.*, 72, 156, 1903.
38. Binks, B., Lumsdon, S. *Langmuir*, 16, 3748, 2000.
39. Binks, B., Lumsdon, S. *Langmuir*, 16, 2539, 2000.
40. Binks, B., Lumsdon, S. *Langmuir*, 17, 4540, 2001.
41. Huh, C., Inoue, M., Mason, S. *Can. J. Chem. Eng.*, 53, 367, 1975.
42. Garrett, P.R. (assigned to Unilever PLC), EP 0075433, 30 March 1983, filed 14 September 1982.
43. Carter, M.N.A. (assigned to Unilever Ltd.), GB 1,571, 501, 16 July 1980, filed 23 Jan 1976.
44. Carter, M.N.A., Garrett, P.R. (assigned to Unilever Ltd.), GB 1,571,502, 16 Jul 1980, filed 23 Jan 1976.
45. Zhang, H., Miller, C.A., Garrett, P.R., Raney, K.H. *J. Colloid Interface Sci.*, 263, 633, 2003.
46. Yorke, J.W.H., Garrett, P.R., Giles, D. unpublished work.
47. Pelton, R.H., Goddard, E.D. *Colloids Surf.*, 21, 167, 1986.
48. Pelton, R.H. *Chem. Eng. Sci.*, 51(19), 4437, 1996.
49. Evans, D.F., Wennerström, H. *The Colloid Domain*, Second Edition, Wiley-VCH, New York, 1999, p 420.
50. Denkov, N.D., Marinova, K., Tcholakova, S., Deruelle, M. Mechanism of foam destruction by emulsions of PDMS–silica mixtures, in *Proceedings of 3rd World Congress on Emulsions*, 24–27 September, 2002, Lyon, France, paper 1-D-199.
51. Kulkarni, R., Goddard, E., Kanner, B. *J. Colloid Interface Sci.*, 59(3), 468, 1977.

7 Mechanical Methods for Defoaming

7.1 INTRODUCTION

Use of antifoams in controlling unwanted foam is in some cases considered to involve serious disadvantages. An obvious example concerns aerated fermentation processes where gas is sparged into reaction vessels by various means. Foaming should be minimized in such processes to maximize the working volumes of containing vessels and minimize any escape of foam leading to loss of reactants or products accumulated therein. Minimal foam requires low rates of aeration and agitation or, preferably, the use of some means of foam control [1–4]. Although foam in this context can be eliminated by antifoams, their presence can be undesirable [3–7]. Antifoams are, for example, known to reduce mass transfer rates and inhibit fermentation reaction rates (presumably as a result of contamination of gas–liquid surfaces). In some cases, they may also exhibit toxicity problems, although use of food-grade or pharmaceutical-grade polydimethylsiloxane (PDMS) antifoams should minimize that factor. Antifoams may also contaminate products so that effective separation and purification is adversely affected. In consequence, foam control by use of mechanical devices is often used to avoid these problems. Perhaps an extreme illustration of the need for such approaches is revealed by a patent application concerned with defoaming, using a centrifuge, of blood circulating externally from an artificial lung [8] (see also Chapter 11). However, in severe cases of overfoaming, the limitations of mechanical defoaming devices mean that they can often only be used successfully in combination with antifoams.

Of the various types of mechanical defoaming devices, the most commonly used employ a rotary action. Such devices include rotating discs, various impeller arrangements, centrifugal baskets, spinning cones, and cyclones. However, other types of devices have also been investigated. These latter include ultrasonic devices (see, e.g., reference [7] and references cited therein), application of vacuum through a suitable membrane [1,9–11], use of packing materials of appropriate wettability (see Section 7.4), and even application of liquid sprays [1,10,12]. Here we review the use of most of these mechanical defoaming devices and include speculation concerning the mechanisms by which they function.

7.2 DEFOAMING USING ROTARY DEVICES

7.2.1 Designs of Rotary Devices Described in Scientific Literature

Rotary devices are generally used to apply centrifugal and shear forces to foam to produce film rupture. As we discuss in Section 7.2.2, understanding of precisely how a given device functions is generally absent—the designs are therefore essentially empirical. Many different rotary defoaming devices are in fact described in the patent literature. Viesturs et al. [1], for example, listed 70 patents in a review published 30 years ago, concluding that "clearly theoretical works on the efficiency of rotor-type mechanical foam breakers are only few in number." A number of experimental studies that are intended to at least systematically examine the relative performance of different defoamer designs have been published since then. However, little direct attention is paid in these studies to the defoaming mechanisms of such devices. They note that the nature of both the foam generation technique and the foaming liquid can influence the practicality or otherwise of mechanical defoaming. In the case of the foam generation technique, Pandit [13], for example, makes a distinction between sparged bubble columns (where agitation is provided by the bubbly flow from the sparger) and mechanically stirred sparged vessels where smaller bubble sizes and more stable foam usually characterize the latter. The foam generated from stirred vessels is therefore supposed to be harder to defoam using rotary devices.

Deshpande and Barigou [14] have tabulated a useful summary of the many published empirical studies of defoaming using various rotary devices. In Appendix 7.1, Table 7.A1, we produce an updated version of that table where later studies are included and an error in a citation is rectified. Whether performance has been studied in either bubble columns or stirred vessels is indicated. Most of the reported studies are at laboratory scale, concerned with controlling the foam of dilute aqueous surfactant solutions in sparged vessels of <0.5 m diameter. Only two reported studies concern pilot scale with practical systems—Kraft (paper) mill effluent [15] and recombinant *Bacillus* fermentation [16].

The configuration of a typical laboratory-scale stirred vessel [17], which employs rotary devices for foam control, is shown schematically in Figure 7.1. A similar arrangement is used for bubble columns but with no agitator. The simplest rotary devices include various turbines, vaned discs, and paddles that are applied directly to a rising foam column. Takesono, Ohkawa, and coworkers describe the use of these devices for controlling the foam in both bubble columns [18] and stirred vessels [17]. The designs of these rotary devices are illustrated by way of example in Figure 7.2. Ng et al. [15] have also used various vaned disc turbines for controlling Kraft Mill effluent foam in a stirred vessel. Defoaming effectiveness and power consumption were shown to be relatively insensitive to the number of blades on the disc. The angular rotational velocity, ω, required for elimination of foam, declined linearly with increasing diameter of the disc.

Deshpande and Barigou [6,14] have compared the performance of novel "split" and "needle" two-blade paddles with more conventional turbines and paddles in both bubble column and stirred vessels. The designs of devices used in this study are illustrated in Figure 7.3. A plot of ω for foam control against gas flow rate (and therefore rate of foam generation) for each of these rotary devices is given in Figure 7.4 for a 6×10^{-5} M submicellar solution of sodium dodecylbenzene sulfonate. It is clear that increasing the rotor velocity is necessary for foam control if the gas flow is increased. The performance of the various devices is strikingly similar despite the radical difference in design—only an ~12% increase in angular velocity separates the most effective from the least effective and only ~5% difference in the case

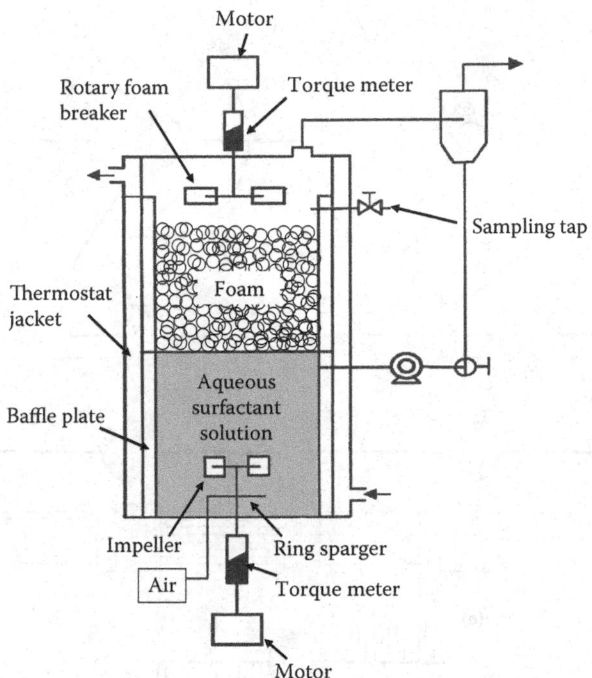

FIGURE 7.1 Schematic diagram of typical laboratory-scale stirred vessel using rotary devices for foam breaking. (From Takesono, S., Onodera, M., Ito, A., Yoshida, M., Yamagiwa, K., Ohkawa, A., *J. Chem. Technol. Biotechnol.* 2003. 78. 48. Copyright Wiley-VCH Verlag GmbH & Co. KGaA. Reproduced with permission.)

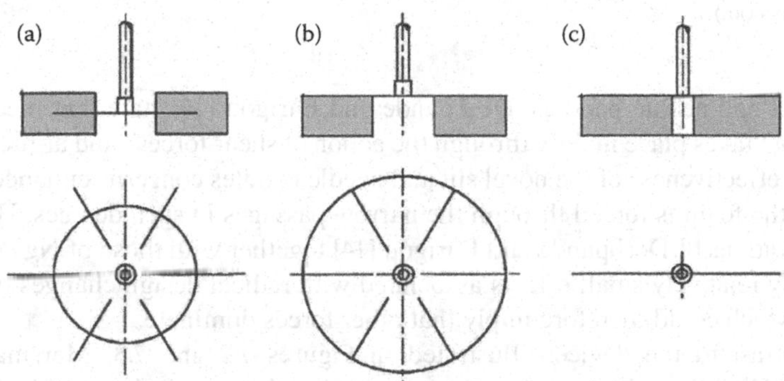

FIGURE 7.2 Designs of rotary foam breakers. (a) Six-blade turbine. (b) Six-blade vaned disc. (c) Two-blade paddle. (From Takesono, S., Onodera, M., Ito, A., Yoshida, M., Yamagiwa, K., Ohkawa, A., *J. Chem. Technol. Biotechnol.* 2003. 78. 48. Copyright Wiley-VCH Verlag GmbH & Co. KGaA. Reproduced with permission.)

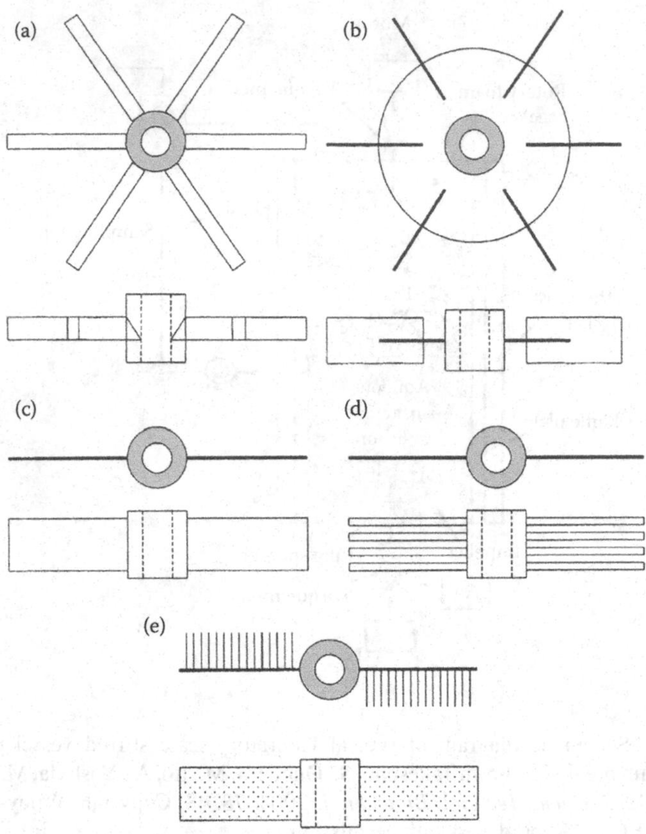

FIGURE 7.3 Designs of rotary foam breakers after Deshpande and Barigou [6]. (a) Pitched-blade turbine. (b) Disc turbine. (c) Two-blade paddle. (d) Two-blade paddle with three slits. (e) Two-blade paddle with 168 needles. (After Deshpande, N.S., Barigou, B., *J. Chem. Technol. Biotechnol.* 1999. 74. 979. Copyright Wiley-VCH Verlag GmbH & Co. KGaA. Reproduced with permission.)

of the split and needle paddles. Deshpande and Barigou [14] state that mechanical defoaming "takes place mainly through the action of shear forces" and argue that the increased effectiveness of the novel slit and needle paddles concerns enhanced shear forces as the foam is forced through the narrow passages in such devices. However, the observations of Deshpande and Barigou [14] together with those of Ng et al. [15] reveal only relatively small effects associated with radical design changes of rotary devices, which could therefore imply that other forces dominate.

In contrast to the devices illustrated in Figures 7.2 and 7.3, Mersmann and coworkers [19,20] used a rotor–stator system rather than just a simple rotor for controlling the foam in a stirred vessel. They considered two types of such devices, a so-called blade foam breaker and a radial accelerator foam breaker, both of which are illustrated in Figure 7.5. The former is a 6- or 12-blade rotor that rotates above a stator plate, which covers the whole surface of the sparged stirred vessel. A small

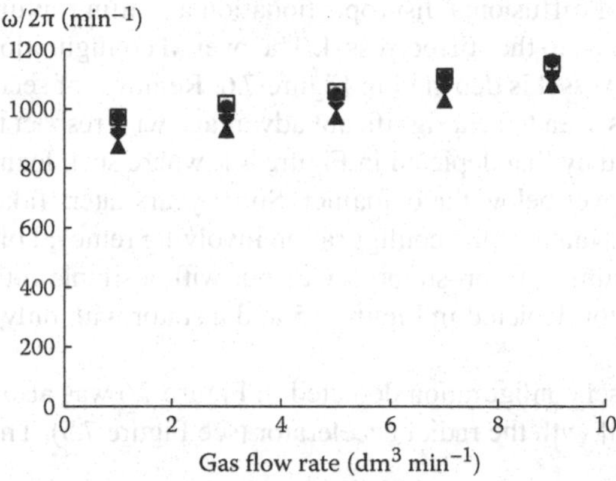

FIGURE 7.4 Plot of critical angular rotational velocity, ω, for foam control against sparged gas flow rate in a stirred vessel with various rotary foam breakers depicted in Figure 7.3 for 6×10^{-5} M submicellar aqueous solution of sodium dodecylbenzene sulfonate. ●, Pitched-blade turbine; +, disc turbine; □, two-blade paddle; ◆, two-blade paddle with three slits; ▲, two-blade paddle with 168 needles. (From Deshpande, N.S., Barigou, B., *J. Chem. Technol. Biotechnol.* 1999. 74. 979. Copyright Wiley-VCH Verlag GmbH & Co. KGaA. Reproduced with permission.)

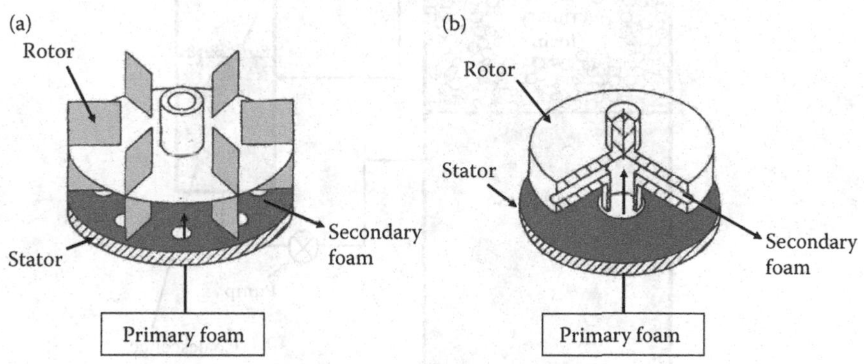

FIGURE 7.5 Rotor–stator foam breakers. (a) Blade foam breaker. (b) Radial accelerator. (From Gutwald, S., Mersmann, A., *Chem. Eng. Technol.* 1997. 20. 76. Copyright Wiley-VCH Verlag GmbH & Co. KGaA. Reproduced with permission.)

gap separates the rotor from the stator. Foam is forced into that gap through holes in the stator, foam film rupture supposedly occurring at that point. As is common with use of rotary devices, complete destruction of foam does not occur and the so-called secondary foam is formed, which cannot be broken by these devices. The latter is characterized by smaller bubble diameters than the primary foam. In the later design of Gutwald and Mersmann [20], further defoaming occurs by direct impact on the blades and surrounding "defoamer box" where the secondary foam accumulates. The latter is drained away and allowed to slowly collapse in a coalescence column by a combination of diffusional disproportionation and film rupture, whereupon the liquid can be returned to the stirred vessel. The overall configuration of the defoamer device and stirred vessel is depicted in Figure 7.6. Removal of secondary foam with this arrangement is seen to be a significant advantage with respect to earlier configurations, exemplified by that depicted in Figure 7.1, where such foam simply accumulates in the foam layer below the defoamer. Some years later, Takesono et al. [4,21] also realized the advantage of a configuration involving removal of secondary foam. They proposed a similar rotor–stator device but with a simple rotating disc instead of the six-blade rotor depicted in Figure 7.5 and a stator with only one large central hole [4,21].

The stirred vessel configuration depicted in Figure 7.6 was also used by Gutwald and Mersmann [20] with the radial accelerator (see Figure 7.5). This device consists

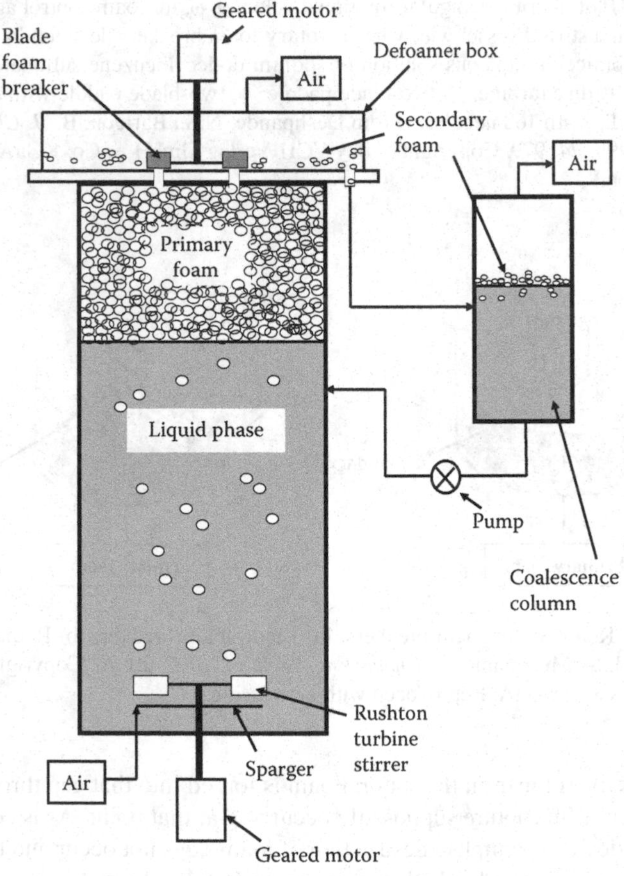

FIGURE 7.6 Schematic diagram of stirred vessel configuration showing use of rotor–stator blade foam breaker and secondary foam coalescence column. (From Gutwald, S., Mersmann, A., *Chem. Eng. Technol.* 1997. 20. 76. Copyright Wiley-VCH Verlag GmbH & Co. KGaA. Reproduced with permission.)

of a disc rotor with radial holes. The drive shaft is located in the stator. Foam is drawn through the drive shaft into the disc, whereupon it is "accelerated" through the radial holes and thrown against the wall of the retaining "defoamer box." Foam film rupture is then supposed to occur as a result of the impact force, any secondary foam being separately collected as shown in Figure 7.6. In this case, the role of the stator is simply to separate any secondary foam from primary foam. This device requires higher angular velocities for equivalent foam control to that obtained with the blade foam breaker of the same authors.

For over 30 years Takesono, Okhawa, and coworkers [17,22–33] have studied a rotating disc defoamer, which functions by spraying liquid drops over the foam generated in both bubble columns and stirred vessels (see Appendix 7.1, Table 7.A1). The design of this type of device is schematically illustrated in Figure 7.7. Inherent liquid (i.e., the liquid from which the foam is generated) is pumped into the gap between a rotating disc and an "impact plate." Centrifugal force accelerates the liquid out of that gap to form a spray of drops, which defoams the foam flowing upward into the space between the impact plate and the baffles attached to the foam-generating vessel [22]. No provision is made by Takesono, Okhawa, and coworkers [17,22–33] for dealing with secondary foam in the sparged vessel configurations used with this device. Their device should also not be confused with the simple rotating disc used by Goldberg and Rubin [10]. The latter involves pouring foam onto a spinning disc upon the surface of which foam collapse occurs. The latter is attributed [10] to direct application of shear by the rotating disc to the foam. However, no clear indication is given of the mechanism of foam film rupture by that shear.

A characteristic of the device of Takesono, Okhawa, and coworkers is that the critical angular rotation velocity for defoaming increases as either the inherent liquid flow rate from the device declines or the gas flow rate increases [22,26]. However, increasing the viscosity of the liquid decreases the critical angular rotation velocity. This surprising observation is attributed [26] to the effect of increase in viscosity

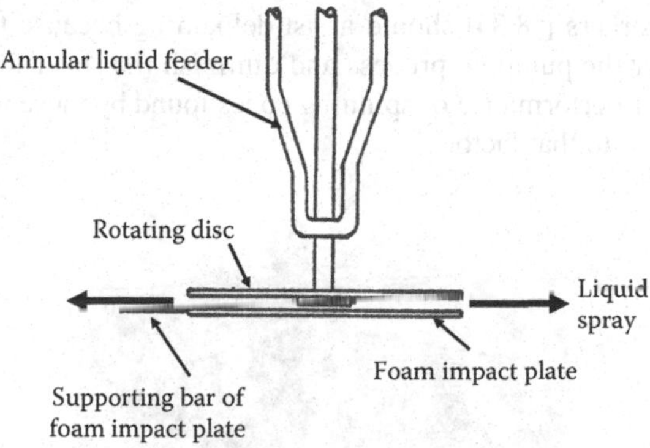

FIGURE 7.7 Inherent liquid spray foam breaker. (After Ohkawa, A. et al., *J. Ferment. Technol.*, 56, 428, 1978. Takesono, S., Onodera, M., Ito, A., Yoshida, M., Yamagiwa, K., Ohkawa, A., *J. Chem. Technol. Biotechnol.* 2003. 78. 48. Copyright Wiley-VCH Verlag GmbH & Co. KGaA. Reproduced with permission.)

upon the intrinsic properties of the foam rather than to enhanced defoaming efficiency. A similar observation concerning the effect of increasing liquid viscosity is reported by Deshpande and Barigou [14], which was attributed to the same cause. The latter were of course using a completely different type of defoamer to that used by Takesono, Okhawa, and coworkers [26].

Spinning cones represent another type of rotary device. Both Cooke et al. [16,34] and Takesono, Ohkawa, and coworkers [17,18] have reported measurements with such devices. An example of the type of spinning cone used by Cooke et al. [34] is shown in Figure 7.8 where the cone is open at both an inlet at the bottom and an outlet at the top. The design used by Takesono, Okhawa, and coworkers [17,18] differs from that depicted in Figure 7.8 in that blades are inserted in the cone. Cooke et al. [16, 34] used a spinning cone for control of foam in a stirred vessel where the cone and stirrer are both mounted on the same drive shaft. The configuration is shown in Figure 7.9. An image of a spinning cone actually controlling the foam in such a vessel is shown in Figure 7.10.

In the device used by Cooke et al. [16,34], centrifugal force pulls liquid from the Plateau borders in the foam toward the surface of the cone, which would increase the capillary pressure in the Plateau borders of the foam. That centrifugal force increases as the radius of the cone increases, which means that the liquid is pumped up both the inside and outside of the cone whereupon it is thrown out from the edge tip. This pumping action leads to shearing of the foam, which may also be subject to impact force due to drops of liquid ejected from the upper edge tip of the cone. Foam film rupture may therefore be caused by a combination of high Plateau border capillary pressure due to the action of centrifugal force, deformation due to shear and the force of drop impacts. These issues are discussed in detail in Section 7.2.3. Examination of Figure 7.10, however, reveals that the foam level inside the spinning cone is lower than in the surrounding vessel. This implies that defoaming is occurring inside the cone and that therefore drop impact cannot be the exclusive cause of defoaming.

It is not clear why inserting blades in the cone as described by Takesono and Ohkawa and coworkers [18,34] should assist defoaming because the blades would presumably hinder the pumping process and diminish the shear force on the foam. The relatively poor performance of spinning cones found by these workers may possibly be attributable to that factor.

FIGURE 7.8 Spinning cone foam breaker. (Reprinted from *Chem. Eng. Sci.*, 60, Stocks, S.M., Cooke, M., Heggs, P.J., 2231. Copyright 2005, with permission from Elsevier.)

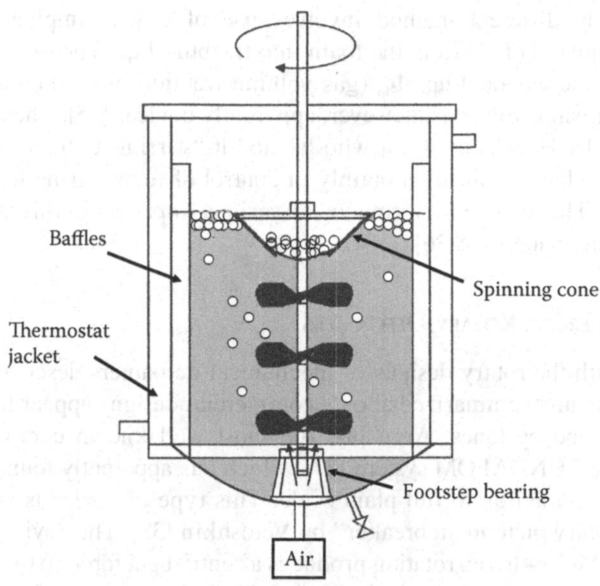

FIGURE 7.9 Schematic diagram of configuration of stirred vessel with spinning cone foam breaker. (After Cooke, M., Heggs, P.J., Eaglesham A., Housley, D., *Trans. Inst. Chem. Eng. Part A, Chem. Eng. Res. Design*, 82, 719, 2004; Stocks, S.M., Cooke, M., Heggs, P.J., *Chem. Eng. Sci.*, *60*, 2231, 2005. Reprinted from *Trans. Inst. Chem. Eng. Part A, Chem. Eng. Res. Design*, 82, Cooke, M., Heggs, P.J., Eaglesham A., Housley, D., 719. Copyright 2004, with permission from Elsevier.)

FIGURE 7.10 Foam control by spinning cone foam breaker in stirred vessel (at 16 Hz, aqueous solution of 4×10^{-6} M polypropylene glycol + 10 ppm dishwashing liquid). (Reprinted from *Trans. Inst. Chem. Eng. Part A, Chem. Eng. Res. Design*, 82, Cooke, M., Heggs, P.J., Eaglesham A., Housley, D., 719. Copyright 2004, with permission from Elsevier.)

A completely different method involves use of down-pumping impellers to enhance entrainment of air from the foam into the bulk liquid phase. This produces an increase in the gas hold-up, Φ_{lp} (gas volume fraction in the liquid phase). The precise mechanism involved is, however, apparently unclear [35]. The approach was first described by Hoecks et al. [3] who labeled it "stirring as foam disruption." It would appear to have application mainly in control of foam during fermentation in stirred vessels. The relative performance of various impellers in this context is also described in later papers [35,36].

7.2.2 Commercial Rotary Defoamers

By contrast with the rotary designs of mechanical defoamers described in the scientific literature and summarized above, commercial designs appear to be confined to centrifuges and cyclones. Arguably, the most well known commercial rotary defoamer is the FUNDAFOM system [37], which has apparently found some application in fermentation up to full plant scale. This type of device is described as a "centrifugal rotary plate foam breaker" by Vetoshkin [38]. The device is illustrated in Figure 7.11. Motor-driven rotation produces a centrifugal force so that liquid phase is forced back into the foam generation vessel. The gas phase passes through a "gas-bleeding fitting" as shown in Figure 7.11. Vetoshkin [38] presents a theoretical model

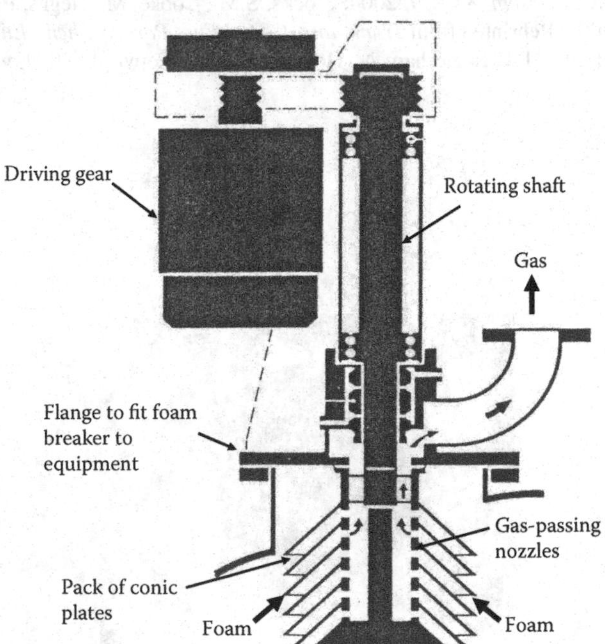

FIGURE 7.11 Commercial centrifugal rotary plate foam breaker with conic plates and peripheral foam feed. (From FUNDAFOM, Chemap AG, 1977. With kind permission from Springer Science+Business Media: *Theor. Found. Chem. Eng.*, 37, 2001, 372, Vetoshkin, A.G.)

of the process that permits the conditions for defoaming to be calculated together with the power requirement. The results are claimed to agree with experiment to within 5% but unfortunately no detailed comparison with the relevant experimental results is given.

Another approach to full plant-scale mechanical defoaming using rotary devices concerns the use of cyclones, especially in the gas–oil separation tasks of crude oil production and refining. Such devices have the advantage that, unlike centrifuges, no moving parts are involved. Use of cyclones in defoaming has recently been reviewed by Hoffmann and Stein [39]. These authors point out that "relatively little fundamental information on the design and performance of foam-breaking cyclones appears in the literature." They note that the design of defoaming cyclones is "based largely on tests and experience" and it is therefore difficult to provide any firm design guidelines.

A simplified schematic of a defoaming cyclone system is depicted in Figure 7.12. The foam enters tangentially into the cyclone to generate a swirling motion,

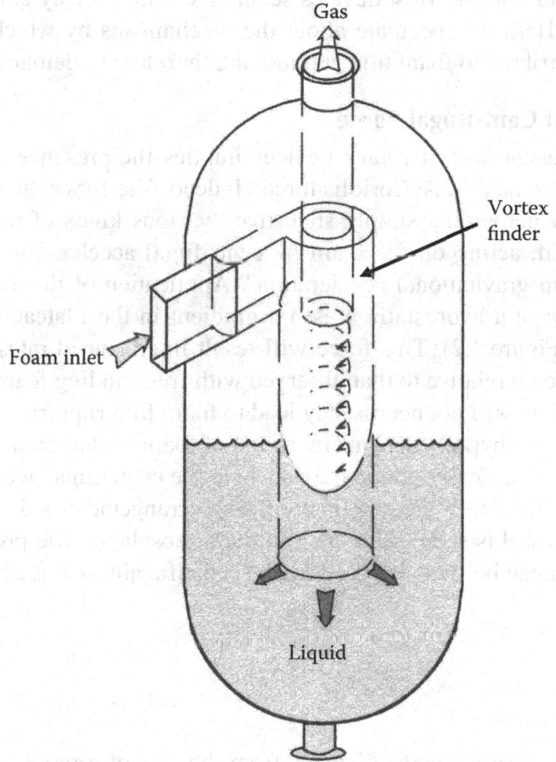

FIGURE 7.12 Simplified schematic diagram of cyclone foam breaker. (With kind permission from Springer Science+Business Media: *Gas Cyclones and Swirl Tubes; Principles, Design and Operation*, Foam-breaking cyclones, 2008, p 327, Hoffmann, A.C., Stein, L.E., Chapter 14.)

generated by the pressure drop across the device. This creates a centrifugal field. The liquid phase is then driven to the walls of the cyclone. A combination of centrifugal and shear force causes foam film rupture, leading to the separation of a continuous gas phase along the cyclone axis, presumably in the form of an inner vortex. That gas phase escapes through a so-called vortex finder. The liquid phase drops out of the base of the vessel, which is immersed in liquid. A particular design feature is ensuring that the base of the vessel is immersed to a sufficient depth so that the hydrostatic head prevents escape of gas. Such cyclones can handle superficial gas velocities in excess of 1200 cm s^{-1} [39].

7.2.3 Defoaming Mechanisms of Rotary Devices

The large number of different rotary devices described in both the scientific and patent literature attests to an absence of consensus about optimum design. It is tempting to conclude that this reflects, at least in part, a lack of clear understanding concerning the relevant mechanisms of defoaming action. However, despite this limitation, it is possible to at least identify the involvement of three types of force—centrifugal, shear/inertial, and impact. Most devices seem to simultaneously employ more than one such force. Here we speculate about the mechanisms by which each of these forces could contribute to foam film rupture and therefore to defoaming.

7.2.3.1 Role of Centrifugal Force

Regardless of design, use of rotary devices implies the presence of a centrifugal force (or even in some cases a Coriolis force). Indeed, Viesturs et al. [1] in reviewing rotor designs for defoaming simply state that "various kinds of mechanical foam breakers are used, acting on the foam by centrifugal acceleration which is many times higher than gravitational acceleration." Application of this centrifugal force to foam will induce a hydrostatic pressure gradient in the Plateau border network (see Chapter 1, Figure 1.2). This force will result in enhanced rates of drainage of liquid from the foam relative to that observed with freestanding foam if it exceeds g. However, that alone will not necessarily lead to foam film rupture.

Consider then the hypothetical arrangement of foam contained in a vessel subject to rotation about a vertically oriented axis where the centrifugal acceleration is $\gg g$ so that gravitational effects can be ignored. The arrangement is depicted in Figure 7.13 where the vessel is seen to be open to the atmosphere. The pressure gradient, $dP_r/d\tilde{r}$, in the Plateau borders, induced by the centrifugal force, is given [38,40]

$$\frac{dP_r}{d\tilde{r}} = \rho_W \omega^2 \tilde{r} \tag{7.1}$$

where P_r is the pressure, $\tilde{r}$ is the distance from the axis of rotation, ω is the angular rotational velocity (usually in rad s^{-1}), and ρ_W is the liquid density. Now suppose that this foam is in a state of mechanical equilibrium analogous to that described in Section 1.4.1 for a foam in the gravity field where the capillary pressure balances the hydrostatic head and drainage has ceased. Equation 7.1 can then be integrated to give

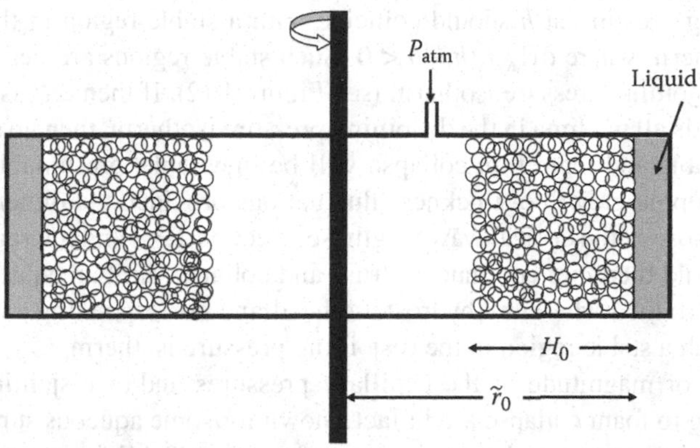

FIGURE 7.13 Foam in cylindrical vessel subject to centrifugal field and in state of mechanical equilibrium.

the contribution to the total pressure drop across the Plateau border network due to the centrifugal force, ΔP_r, as

$$\Delta P_r = 0.5\rho_W \omega^2 H_0 (2\tilde{r}_0 - H_0^e) \tag{7.2}$$

where $\tilde{r}_0$ is the distance from the axis of rotation to the bottom of the foam layer (see Figure 7.13) and H_0^e is the height of the equilibrium foam. At equilibrium, this pressure drop will be balanced by the capillary pressure at the upper continuous surface of the Plateau borders as described in Section 1.4.1 for the case of an equilibrium foam in the gravity field. That capillary pressure is negative and is given as $-\sigma_{AW}|\kappa_t|$ where σ_{AW} is the air–liquid surface tension and κ_t is the curvature of the Plateau border surfaces at the top of the foam column. We can therefore write, by analogy with Equation 1.10 (see Section 1.4.1), for the total equilibrium Plateau border pressure in the liquid at the bottom of the foam column at $\tilde{r}_0$ that

$$P_{PB}(\tilde{r} = \tilde{r}_0) = P_{atm} + \Delta P_r - \sigma_{AW}|\kappa_t| = P_{atm} \tag{7.3}$$

where we neglect the contribution due to the mass of liquid in the foam by supposing that $\Phi_G^{foam} \rightarrow 1$. Therefore, at equilibrium we have to a first approximation

$$\Delta P_r \approx \sigma_{AW}|\kappa_t| = 0.5\rho_W \omega^2 H_0 (2\tilde{r}_0 - H_0^e) \tag{7.4}$$

However, at equilibrium, the foam should be stable so that the capillary pressure in the Plateau border equals the disjoining pressure $\Pi_{AWA}(h)$ in the adjacent air–liquid–air (AWA) foam films at the top of the foam column (where h is the film thickness and where we neglect any net curvature of the film). As described in Section 1.3.3, that disjoining pressure at h should coincide with a stable region in the disjoining pressure isotherm where $d\Pi_{AWA}(h)/dh < 0$. Such stable regions are defined by maxima in the disjoining pressure isotherm (see Figure 1.12). If then ΔP_r is so high that $\sigma_{AW} |\kappa_t|$ exceeds all maxima in the disjoining pressure isotherm, then no stable region will be available and foam film collapse will be inevitable, presumably as a result of the catastrophic growth of thickness fluctuations provided the dimensions of the film exceed those of the critical wavelength (see Section 1.3.3 and references [41,42]). The foam would therefore continue to drain and collapse until an equilibrium foam height H_0 is attained where the hydrostatic head and corresponding capillary pressure can match a stable region in the disjoining pressure isotherm.

The order of magnitude of the capillary pressures and/or disjoining pressures corresponding to foam collapse are in fact known for some aqueous surfactant solutions. Kruglyakov and coworkers (see, e.g., references [43,44] for summaries of this work) have reported many studies of the effect of an applied pressure drop on the drainage and stability of foam columns under both centrifuged and static conditions. These studies establish "critical" capillary pressures at which the foam collapses as foam films rupture in an "avalanche-like" manner. The critical capillary pressures measured using static methods are in fact somewhat higher than those measured using centrifuges where the differences are attributed to "mechanical effects" such as vibrations, which are unavoidable with the latter [44].

Vilkova and Kruglyakov [43] report values of centrifugal critical capillary pressures for foams of 0.5 and 1.5 cm height prepared from solutions of Triton X-100 (TX-100) and SDS of the order of 10 kPa. The concentration of TX-100 was unfortunately not revealed in this work; however, solutions of 10^{-3} M SDS in 0.5 M NaCl were used. Similarly, Aronson et al. [45] report "critical" foam film rupture capillary pressures for aqueous SDS solutions measured using a Scheludko cell with a porous frit (see Section 2.3.1.2). In the case of submicellar solutions of 10^{-3} M SDS, rupture pressures range from 0.3 to 1.5 kPa, regardless of the presence or absence of added NaCl. However, for micellar SDS solutions at 10^{-3} M SDS in 0.18 M NaCl, measured rupture pressures range from 4 to 11 kPa, which is of the same order as those measured by Vilkova and Kruglyakov [43] using a centrifuge with the same concentration of SDS but at a higher salt concentration. At 0.017 M, SDS Aronson et al. [45] find that films are stable up to 30 kPa, the maximum measurable with the apparatus used. Bergeron [46] reports foam film rupture pressures for a series of aqueous solutions of alkyl trimethylammonium bromides where films of aqueous C_{14}TAB and C_{16}TAB at the CMC were both stable up to 30 kPa, again the maximum measureable.

Curiously, many accounts of the defoaming mechanisms of the devices listed in Table 7.A1 ignore a possible direct contribution from the centrifugal force, which is undoubtedly present in all cases. Clearly, such forces are important in commercial systems employing cyclones and "centrifugal rotary plate foam breakers" (the FUNDAFOM device) [37,38]. However, Goldberg and Rubin [10] argue that even in the case of centrifugal foam breakers, the defoaming action is "primarily due to shear of the bubbles in the spinning bowl." In the case of the spinning cone described by Cooke et al. [16,34], it has, however, been suggested, as we describe in Section 7.2.1, that defoaming could be due, at least in part, to the enhanced capillary pressure in Plateau borders due to centrifugal force. We can use Equation 7.4 to roughly estimate the magnitude of the capillary pressures on a layer of foam only 1 cm deep (i.e., $H_0 = 0.01$ m) at the tip of the cone. In the case of a cone of 0.14 m diameter with an angular rotational velocity of 17 s^{-1}, we find a capillary pressure of ~0.2 kPa. Unfortunately, Cooke et al. [16] have not used a solution for which the relevant critical capillary pressures are known. They in fact used aqueous solutions of 4×10^{-6} M polypropylene glycol together with 10 ppm of a dishwashing liquid. Assuming the latter consists entirely of typical surfactants for such liquids, then the concentration must be of order 2×10^{-5} M. It is tempting therefore to suggest that the solutions were submicellar. The measurement of Aronson et al. [45], however, suggest that the critical capillary pressures for submicellar solutions could be <1 kPa, which is of the correct order for the likely magnitude of the relevant centrifugally induced pressures. This of course hardly represents unequivocal proof that enhanced Plateau border capillary pressures due to centrifugal force are responsible for the defoaming effect with the spinning cone!

By contrast, Gutwald and Mersmann [20] have used the stator–rotor devices shown in Figure 7.5 and 7.6 in defoaming of surfactant solutions for which critical capillary pressures for foam and foam film rupture are available. The solutions included C_{16}TAB and TX-100 at concentrations of from 0.5 × CMC to 2 × CMC. The nature of the forces involved during defoaming is, however, rather complex. The use of a 6- or 12-blade rotor is obviously not ideal for application of centrifugal defoaming. Indeed, Gutwald and Mersmann [20] argue that defoaming is only by shear and impact. Nevertheless, it is instructive to make order of magnitude calculations of the relevant capillary pressures using Equation 7.4 with the angular rotation velocities and rotor dimensions given in Appendix 7.1, Table 7.A1, again setting $H_0 = 1$ cm. The calculated capillary pressures lie in the range of 7–400 kPa, which would appear to be at least consistent with the measured critical capillary pressures quoted here for these surfactants [43, 46]. If the approach is valid, this should mean effective defoaming. The formation of secondary foam represents a separate issue. The process of foam film rupture described by Vrij [41,42], where $d\Pi_{ALA}(h)/dh < 0$, involves the growth of a hydrodynamic instability of a critical wavelength. Bubbles with foam films of dimensions less than that wavelength will be stable. Such a phenomenon could therefore offer an explanation for the formation of secondary foam consisting of bubbles of smaller size than the original so-called primary foam. In a polydisperse foam, bubbles of a size smaller than average are always present but are not readily destroyed.

According to Exerowa and Kruglyakov [44], the critical capillary pressure in centrifuged foams is insensitive to the rate of rotation—however, increasing ω so that ΔP_r exceeds the critical capillary pressure increases the rate of foam drainage and therefore the rate of buildup of capillary pressure to the critical level so that the overall foam lifetime is diminished. This factor is, of course, important when considering a centrifugal device coping with a continuous supply of foam in a vertically oriented vessel such as a bubble column. If the rate of supply of fresh foam is too high, then the centrifuge device may become "flooded" with foam so that the critical capillary pressure in the Plateau borders is never attained even though ΔP_r is greater than the critical value. In these circumstances, we would therefore expect the angular rotational velocity of the device ω (and therefore ΔP_r) required for foam breaking to increase with increasing rate of supply of foam as measured by the gas flow rate. This would ensure that drainage from the Plateau borders to the critical capillary pressure occurs rapidly before more foam arrives at the device. As shown in Figure 7.4, Deshpande and Barigou [6], for example, observe that the critical angular rotation velocity for foam breaking does in fact increase with the gas flow rate in the case of a sparged bubble column. Another factor affecting the rate of attainment of the critical Plateau border capillary pressures concerns the extent to which the foam generated in, for example, a bubble column is allowed to drain under gravity before it reaches any centrifugal device. The higher the position of the device, the longer the time for foam drainage and the less liquid need be pumped out of the Plateau borders by the centrifuge before the critical capillary pressures are attained. In turn, that means the higher the position of the centrifugal device, the lower the required rotational velocities as found experimentally by Deshpande and Barigou [6,14].

7.2.3.2 Role of Shear and Impact Forces on Bubbles in Mechanical Defoaming

It is often argued that defoaming by rotary devices is largely due to the shear forces induced by the rotation. However, these arguments are also often speculative—little direct experimental or theoretical evidence is usually involved. Study of the defoaming of aqueous solutions by direct presentation of the foam to rotating discs represents an example of this type of approach. Versions of this technique have been described by both Goldberg and Rubin [10] and Takesono et al. [4,21]. Both studies used aqueous solutions of TX-100, micellar (1 × CMC and 4 × CMC) by the former and submicellar (0.4 × CMC) by the latter. The respective arrangements used in the application of these devices by these two groups of workers are schematically depicted in Figure 7.14. In the case of Goldberg and Rubin [10], the foam, exhibiting plug flow, was simply poured onto the disc. Foam collapse seemed to involve only the layer of bubbles adjacent to the disc and, according to these authors, was a result of the applied shear. No evidence of the magnitude of that shear is given nor is there any explanation of the resulting putative foam film rupture mechanism.

A more complicated arrangement was used by Takesono et al. [4,21] where foam from a vertical vessel was presented to the underside of a spinning disc through a relatively large orifice in a stator plate. Foam collapse is supposed to occur by the action of shear in the gap between the stator and the rotating disc as shown in Figure 7.14b. The collapse is not, however, complete; small bubble secondary foam is ejected from that gap. However, it is difficult to understand how the deformation under shear of bubbles in that gap could produce foam film rupture—rather it seems more likely that it will simply produce bubble fission and an increase in the number of foam films. According to Walstra and Smulders [47], if, as seems likely in this case, the gap width is comparable to the diameter of the bubbles, then the flow regimen is laminar and bubble fission will occur if the capillary number, C_a is exceeded where

$$C_a = \frac{K_a \eta_W \dot{\gamma}}{2\sigma_{AW}/r_b} \tag{7.5}$$

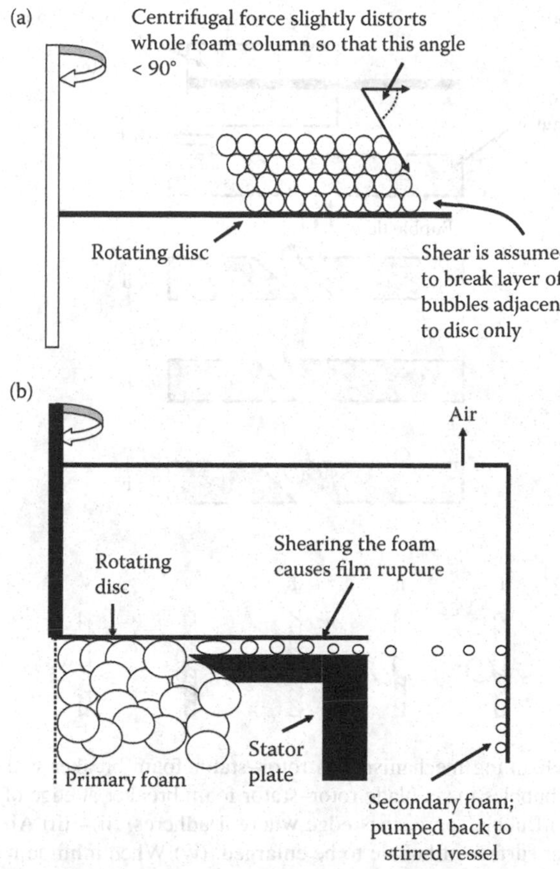

FIGURE 7.14 Speculation about role of shear in causing foam collapse by shear in case of rotating disc foam breakers. (a) Pouring foam directly onto rotating disc. (After Goldberg, M., Rubin, E., *Ind. Eng. Chem. Process Design Dev.*, 6, 195, 1967.) (b) Effect of shearing a foam between rotating disc and stator. (After Takesono, S. et al., *J. Chem. Eng. Jpn.*, 37, 1488, 2004; Takesono, S. et al., *J. Chem. Eng. Jpn.*, 40, 565, 2007.)

and where η_W is the viscosity of the aqueous phase, $\dot{\gamma}$ is the shear rate (determined by the tangential velocity gradient in a rotating system), σ_{AW} is the air–water surface tension, r_b is the bubble radius, and K_a is a constant of value ≥ 1.

Gutwald and Mersmann [20] also use arguments based on bubble fission in explanation of the defoaming behavior of their bladed rotor–stator device shown in Figure 7.5. We have of course already discussed the possibility that centrifugal force could drive the defoaming of that device. By contrast, Gutwald and Mersmann [20] argue that defoaming occurs as a result of three processes. The first involves inertial interaction of bubbles, emerging from stator holes, with a turbulent gas flow supposedly caused by the moving blade. The direction and magnitude of that gas flow has been inferred by measurements with air rather than foam, making the assumption that "the pressure field in the box (containing the rotor–stator) in the presence of foam

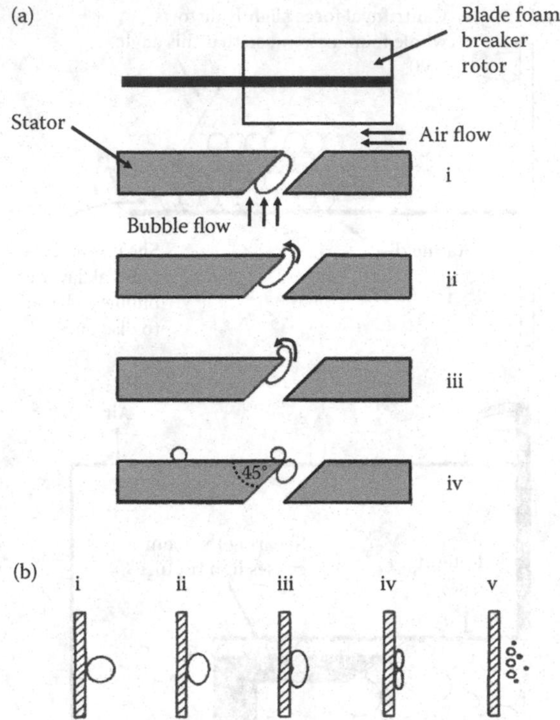

FIGURE 7.15 Defoaming mechanisms in rotor–stator foam breaker with blade rotor. (a) "Gas shearing" of bubbles in six-blade rotor–stator foam breaker at edge of stator hole with acute angle. (i)–(ii) Bubble approaches edge where it adheres. (ii)–(iii) Air flow above stator causes air–water surface of bubble to be enlarged. (iv) When minimum radius of curvature is reached bubble fission occurs. (b) Collision between foam bubble and blade of rotor. (i) First contact. (ii)–(iii) Deformation of bubble to an ellipsoid. (iv) Rupture of lamella at positions of most deformation. (v) Formation of secondary bubbles and drops. (Gutwald, S., Mersmann, A., *Chem. Eng. Technol.* 1997. 20. 76. Copyright Wiley-VCH Verlag GmbH & Co. KGaA. Reproduced with permission.)

is the same as for air because the volume fraction of the liquid in the foam is very small." This assumption would appear to deny the existence of a distinct foam rheology! The diagram of the resulting "gas shearing" of bubbles emerging from a stator hole given by Gutwald and Mersmann [20] is reproduced in Figure 7.15a. It is seen to be a process of bubble fission, which these authors interpret to predict the effect of rotor velocity on maximum bubble sizes using Kolmogorov turbulence theory.* In the second and subsequent process with their blade rotor–stator defoamer, Gutwald and Mersmann [20] argue that bubbles are accelerated to the rotors where the impact

* It is worth noting here that Gutwald and Mersmann [20] state that bubbles tend to be "fixed" at the edge of the stator hole. They are in fact depicted as having a contact angle at the stator surface in Figure 7.15a. That, combined with the acute angle characterizing the stator hole, could suggest foam film rupture by the antifoam mechanism described in Section 4.7.

supposedly produces further bubble fission together with foam film rupture as a result of a stretching deformation to the point where surfactant adsorption is essentially zero. The diagram of this process given by the authors is also reproduced in Figure 7.15b. A further impact event occurs as the bubbles are subsequently transported to the walls of the box containing the rotor–stator. Each of these three processes is supposed to produce defoaming and bubble fission where the latter gives rise to the secondary foam. That the impact events could give rise to stretching of foam films to the point where the air–water surfaces are essentially free of surfactant seems improbable. In reality, it would be a bank of foam rather than individual bubbles actually interacting directly with the blade. Unless, that is, most of the defoaming occurs at the edge of the stator holes (see Figure 7.15a) where apparently only bubble fission is proposed.

Clearly all these interpretations emphasize bubble *fission*, whereas defoaming requires bubble *coalescence*. However, in all cases with rotary devices, there must be centrifugally induced pressure drops in Plateau borders, which can, in principle, give rise to foam film rupture and therefore coalescence (especially with the ambient gas phase) by mechanisms that have been revealed by both well-established theory and experiment. Nevertheless, we should also note that dynamic stretching of foam films can, in principle, lead to depletion of adsorbed surfactant, inadequate surface tension gradients, accelerated foam film drainage, and ultimately plug flow and foam film rupture (see Section 1.3.1). Moreover, Lucassen [48] has shown that differences in Gibbs elasticities across foam films can also give rise to catastrophic thinning of the thinnest part of a foam film when the film is stretched by some external force. These considerations are still speculative but may nevertheless suggest alternative approaches to mechanical defoaming.

7.2.3.3 Defoaming by Inherent Liquid Spray

The rotary device of Takesono, Okhawa, and coworkers [17,22–33], shown in Figure 7.7, involves spraying drops of inherent liquid over foam (remembering that "inherent" liquid is the same as that in the continuous phase of the foam). That liquid is accelerated by injection into the gap between a disc rotating at 21–380 s^{-1} and a stator to form a spray of drops that impact the foam and cause foam film rupture. Pahl and Meinecke [12] also describe the use of inherent liquid spray for defoaming a bubble column. In this case, the spray is generated by pumping the inherent liquid through a sprinkler device. The latter rotates slowly (0.1–2 s^{-1}) to ensure that successive drops do not impact the same part of the foam. It is claimed by Pahl and Meinecke [12] that such an inherent liquid spray can give more efficient foam control and less secondary foam than can be obtained with commercial rotary devices. This comparison did not, however, include the rotary device described by Takesono, Okhawa, and coworkers [17,22–33], which of course, although rotary, also uses an inherent liquid spray. We should note that the tangential velocity of the rotor used in various versions of the latter is up to 50 m s^{-1}, which implies drop velocities of a similar order. This compares with drop velocities of 2–4 m s^{-1} in the case of the device described by Pahl and Meinecke [12].

Franke and Pahl [49] have made a study of the effect of inherent liquid drop impact on the stability of horizontal liquid films. Their experimental observations are summarized in Figure 7.16 where the probability of foam film rupture is plotted against film thickness. All observations concerned submicellar solutions of a technical grade alkyl sulfonate. Increasing the surfactant concentration to 5×10^{-4} M (0.25 × CMC) is seen to produce films where rupture is confined to submicron thicknesses. However, films of lower concentrations appear to exhibit rather complex probability plots with regions of stability interspersed with regions of instability where the latter extends to thicknesses in excess of 10 microns. High-speed photography suggest that under the conditions investigated, drops can pass straight through films, accompanied by some of the film liquid [12,49]. The films then succumb to oscillations, which may lead to rupture. Modeling of the process [49] suggests that the oscillations can give rise to thin regions, which stretch differentially, perhaps giving rise to foam film rupture as suggested by Lucassen [48] (see Section 1.3.1).

It is well known that impact of inherent liquid drops can also cause *air entrainment* (see, e.g., reference [50]), a process that is in fact the basis of the well-known Ross–Miles pour test [51] (see Section 2.2.3). That drop impact can cause both the formation and destruction of foam would seem to be contradictory. It is, however, clearly a real effect that probably concerns the relative magnitude of variables such as the gas phase volume fraction in the foam, foam film thickness, and drop properties such as size, velocity, and frequency. Such a vague statement suggests the need for improved understanding of the phenomena associated with inherent liquid drop impact on foam and foam films.

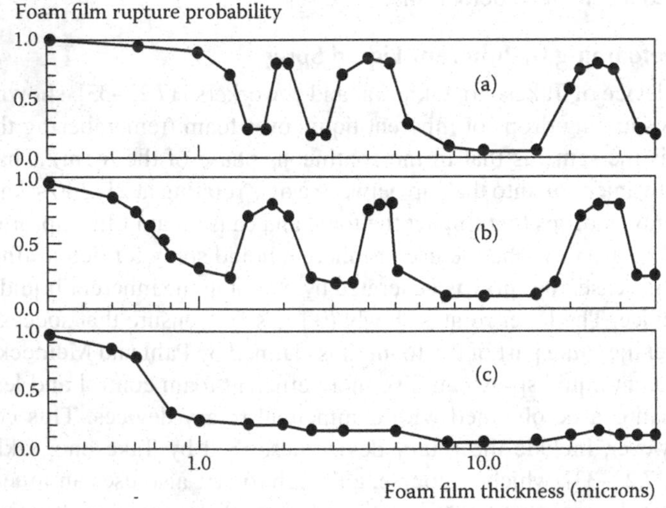

FIGURE 7.16 Probability of foam film rupture after impact by drops of inherent liquid as function of film thickness for various surfactant concentrations. Drop velocity: 2.74 m s^{-1}. Surfactant: technical grade alkyl sulfonate. (a) 5×10^{-5} M. (b) 10^{-4} M. (c) 5×10^{-4} M. (Reprinted from *Chem. Eng. Process.*, 36, Franke, D., Pahl, M.H., 175. Copyright 1997, with permission from Elsevier.)

7.3 DEFOAMING USING ULTRASOUND

7.3.1 Brief History of Defoaming by Ultrasound

Use of ultrasound for defoaming was first proposed by Ross and McBain [52] more than 60 years ago. However, perhaps the first demonstration of the potential effectiveness of that method of defoaming was provided by Shiou-Chuan Sun [53] a few years later. Sun used ultrasound in the frequency range of 3–34 kHz, generated by use of a siren, to demonstrate a defoaming rate of a concentrated flotation froth decreasing almost linearly with increasing frequency over the range 6–20 kHz.* According to Sun [53], this effect may be due to the observed decrease in the power of the siren with increasing frequency. The method, however, proved impractical—the volumetric air flow required by the siren exceeded that supplied for aeration (of a mineralized flotation froth) by four orders of magnitude! In a later description of the use of an air-driven ultrasonic generator by Dorsey [54], again the volumetric air flow required by the generator exceeded that required for aeration in a fermentation bubble column. A frequency range of 26–34 kHz was used in this work, where again defoaming effectiveness decreased with increasing frequency.

The potential practical use of ultrasound had to await the development of piezoelectric ultrasonic generators in the late 1970s. This development is exemplified by the work of Gallego-Juarez and coworkers [55,56] who described the application of a novel piezoelectric ultrasonic transducer. Laboratory-scale measurements with such a device are described by Rodriguez-Corral et al. [56]. These measurements involved application of ultrasound at power levels of 50–100 W, with an ultrasonic frequency of 20 kHz, to foam generated by sparging in a stirred vessel. Results with persistent foams of extremely high viscosity and bubble diameter <0.1 cm (such as the molasses formed in the sugar industry) were unpromising. However, the foam formed from a concentrated solution of polyvinyl alcohol of viscosity ~300 mPa s with bubble diameters of 0.5–3 mm was destroyed by ultrasound at a rate five times faster than it decayed without use of any agency. Even more rapid foam destruction was observed with foam formed from a dilute aqueous solution of a low foam detergent of unspecified composition. Application of ultrasound reduced the lifetime of the foam from a few minutes to <1 s. Rodriguez-Corral et al. [56] also described several pilot-scale devices using the same type of transducer together with a configuration where the transducer is positioned on a bottle filling line in order to defoam the liquid in each individual bottle [57]. Riera et al. [58] have recently produced a brief review of the potential practical defoaming applications of the Gallego-Juarez type of ultrasonic generator [55].

Other studies followed the early work of Gallego-Juarez and coworkers [55,56], again using ultrasound of a fixed frequency of 20 kHz. Sandor and Stein [59], for example, studied the effect of the ultrasonic generator tip size, power level, and distance from the top of a bubble column. A sparged submicellar solution of 2.5×10^{-3} M SDS was used throughout where the generator was either applied to the foam generated after a fixed time or was applied during the whole time of sparging. The

* Here we note that the frequency, $\tilde{\nu}$, is quoted in Hz, which in this context is equivalent to cycles s^{-1}.

experimental arrangement is shown schematically in Figure 7.17. Results for a given tip size and a given (but unspecified) power level are shown in Figure 7.18. In this experiment, the gas flow into the sparger is maintained for 180 s whereupon the flow is stopped and the foam is allowed to decay. The foam profile in the absence of ultrasound is given by the line ABCF in the figure. If the ultrasonic generator is switched on at C, at the point where the gas flow is stopped, then the foam decays along the line CE. Clearly, ultrasound has accelerated the rate of foam decay (compare CE with CF). The foam profile ABD is obtained if the generator is switched on throughout. It is clear that the ultrasound is without effect until the foam reaches the point B, 10–15 mm below the location of the generator tip, which presumably implies attenuation of the ultrasound intensity and thicker and more resilient foam films at lower foam heights. However, that D is at a lower foam height than B implies ultrasound induced foam decay even though the distance is attenuating the signal relative to B, presumably as a consequence of foam film drainage in the region BD.

Other observations by Sandor and Stein [59] reveal that defoaming is significantly enhanced for a given power level when using a generator tip of a larger diameter. However, increasing the power level for a given tip size also enhances defoaming significantly.

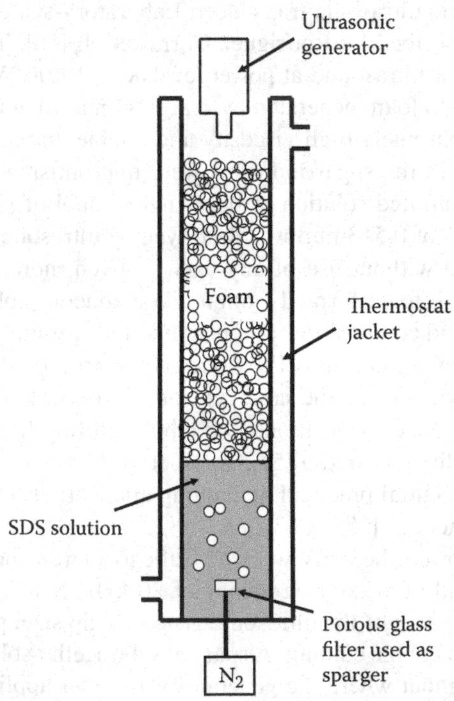

FIGURE 7.17 Schematic diagram of arrangement used by Sandor and Stein to study effect of ultrasound on foam generated in bubble column from a submicellar aqueous solution of SDS. (From Sandor, N., Stein, H.N., *J. Colloid Interface Sci.*, 161, 265, 1993.)

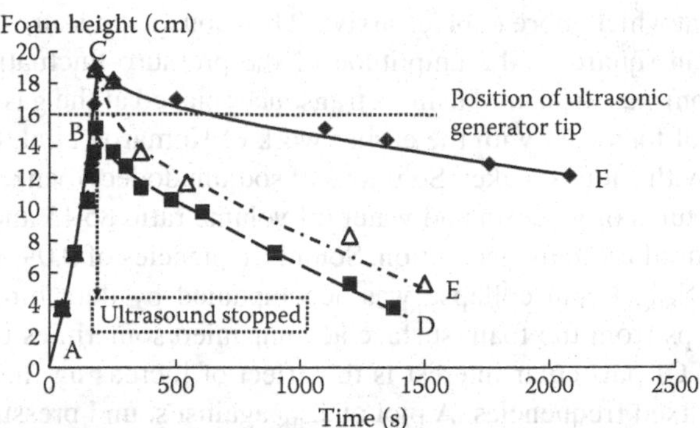

FIGURE 7.18 Effect of ultrasound on foam height in a bubble column as function of time. Gas supply switched of at 180 s in all cases. ■, Foam behavior in absence of ultrasound (ACF). ◆, Foam behavior with ultrasound maintained constantly (ABD). △, Foam behavior where ultrasound started at 180 s (ACE). (After Sandor, N., Stein, H.N., *J. Colloid Interface Sci.*, 161, 265, 1993.)

Morey et al. [7] have used a similar experimental approach, also using an ultrasonic generator at 20 kHz. Aqueous solutions of micellar (1.25 × CMC) sodium dodecylbenzene sulfonate were sparged continuously, the foam height being maintained at a constant value by varying the amplitude of the ultrasound vibrator (and therefore presumably the acoustic pressure amplitude). In general, the greater the amplitude, the more effective the defoaming and the lower the resulting steady-state foam height at a constant gas flow rate.

The role of ultrasound frequency, in the range 0.2–15 kHz, on the foaming and defoaming of solutions of a poorly characterized commercial detergent of has been explored by Komarov et al. [60]. Most of this frequency range lies in the acoustic range and therefore sound was generated using a loudspeaker mounted on a sparged bubble column. Solutions were prepared with either water or 50:50 volume ratios of water to glycerin of respective viscosities of 1 and 7.7 mPa s. At the chosen levels of sound intensity and gas flow rate, the effect of ultrasound on the rate of foam generation was essentially negligible. However, plots of the rate of defoaming against frequency revealed a maximum, which occurred at lower frequencies for the more viscous solution. The defoaming effect decreased at frequencies >2–3 kHz with increasing frequency and decreased at frequencies <2–3 kHz with decreasing frequencies. However, for much of the chosen frequency range, rates of defoaming in the case of the less viscous solution were below those observed in the absence of (ultra)sound.

Komarov et al. [61] have also studied the effect of acoustic pressure on the residence time of gas in foam generated by nitrogen sparging a bubble column containing highly viscous glycerin–water solutions (of viscosity >100 mPa s). The residence time of gas in the foam in a bubble column is of course the "foaminess," Σ_{BIK}, of Bikerman [62]. It is measured in the case where continuous injection of gas results in the formation of a foam of a steady height at which the rate of bubble rupture equals the rate at which more bubbles arrive. The acoustic pressure (usually quoted as the root mean square of the amplitude of the pressure fluctuations caused by sound generation) was measured using a transducer placed at the gas–liquid surface in the absence of foam. As with the earlier work of Komarov et al. [60], the sound was generated with a loudspeaker. Solutions of sodium dodecylbenzene sulfonate in water or in mixtures of glycerin and water (of volume ratio 85:15 and 90:10, rich in glycerin) were used for foam generation. Sound frequencies of 0.09–10 kHz in most cases reduced Σ_{BIK}. Foam collapse was accompanied by detachment of a "huge number" of drops from the foam surface accompanied sometimes by cavitation in the foam layer. Of particular interest is the effect of increasing the acoustic pressure on Σ_{BIK} at fixed frequencies. A plot of Σ_{BIK} against sound pressure obtained by Komarov et al. [61] for the viscous glycerin–water mixtures is reproduced in Figure 7.19. Defoaming at a given frequency is seen to occur at a threshold value of the acoustic pressure. That threshold increases with the frequency but decreases with the increase in viscosity accompanying an increase in the proportion of glycerin in the liquid mixture.

One of the advantages of the use of ultrasound for defoaming is that physical contact between the generator and the foam is not in principle necessary. This possibility has suggested application in the context of metallurgical process where foaming of an oxide melt—slag—can be a problem. Since the temperature of this type of system is usually >1000°C, use of mechanical devices that contact the foam is therefore clearly undesirable! Fine carbonaceous particles are not readily wetted by the slag and therefore can function as antifoams by the mechanism described in Section 4.7 but can also have an adverse effect on the chemistry of the slag [63].

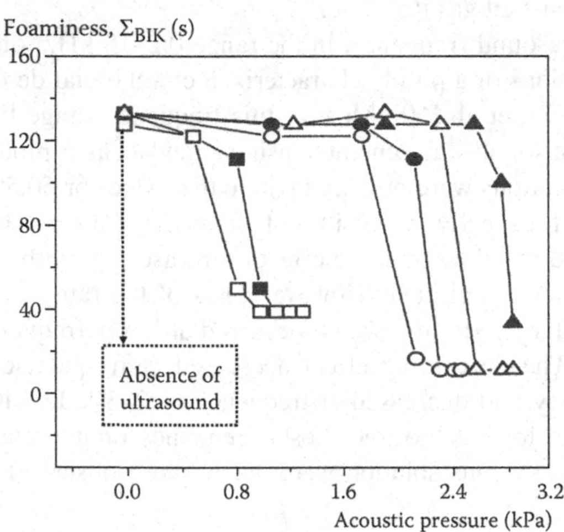

FIGURE 7.19 Residence time of gas in foam (foaminess, Σ_{BIK}) of a bubble column as function of acoustic pressure and frequency. Solutions of 85:15 volume ratio glycerin–water: ■, $\tilde{\nu} = 90$ Hz; ●, $\tilde{\nu} = 289$ Hz; ▲, $\tilde{\nu} = 465$ Hz. Solutions of 90:10 volume ratio glycerin–water: □, $\tilde{\nu} = 90$ Hz; ○, $\tilde{\nu} = 289$ Hz; △, $\tilde{\nu} = 465$ Hz. (Reprinted from *Ultrason. Sonochem.*, 7, Komarov, S.V., Kuwabara, M., Sano, M., 193. Copyright 2000, with permission from Elsevier.)

The possibility of applying ultrasound in order to suppress the foaming tendency of slag has therefore been explored [61,64]. Komarov et al. [61,64] have shown, using laboratory-scale experiments, that the foaming rate in a sparged (with argon gas) bubble column of model oxide slags at temperatures >1200°C can be suppressed by sound at relatively low frequencies in the range 0.1–10 kHz. This range of course overlaps the acoustic range and again these authors used a loudspeaker to generate the sound. A curious feature of plots of the ratio of foaming rate in the presence of (ultra)sound to that in the absence of sound as a function of frequency is the presence of marked peaks and troughs superimposed upon an overall decline in defoaming effectiveness with increase in frequency. Komarov et al. [61,64] argue that this is essentially a resonance phenomenon where the sound pressure drastically increases at the various resonance frequencies where defoaming is enhanced. That defoaming can increase markedly at threshold sound pressures is of course revealed by Figure 7.19.

Yet another investigation of the effect of ultrasound on the stability of foam after generation in a sparged bubble column is described by Dedhia et al. [65]. This study used an ultrasonic horn, at a frequency of 20 kHz, positioned at the top of a bubble column. Submicellar solutions of SDS were used throughout. This study established that the foam collapse rate increased with increase in the ratio of the horn diameter to bubble column diameter—a finding consistent with that of Sandor and Stein [59] concerning the effect of increases in sonic generator tip diameters. Dedhia et al. [65] identify the high cost of defoaming by ultrasound as a cause of limited commercial application. As a consequence, they have explored the use of periodic pulses of ultrasound, which they show could produce significant energy savings, which in turn could reduce costs and render the approach more economically viable.

Most designs of apparatus for study of the effect of ultrasound on the generation or stability of foam in a bubble column place the sonic generator in the air above the foam. However, the attenuation coefficient of sound in air is almost three orders of magnitude greater than that in liquids like water. Therefore, air is not as efficient a medium for the transport of sound energy as the liquid used to generate the foam. Therefore, Winterburn and Martin [66] have explored the use of a sparged bubble column where the ultrasonic transducer was placed in direct contact with the liquid phase at the bottom of the column. There are, however, two problems with this approach. First, the acoustic intensity will be almost totally reflected at the water–foam boundary if, as seems likely, the acoustic impedances of water and foam differ markedly (see Section 7.3.2). Transmission into the foam will therefore be low. Second, the acoustic attenuation coefficient of sound in foam is extremely high so that any sound that is transmitted at the water–foam boundary and that arrives at the top of a foam column will be significantly attenuated. It is of course foam collapse at the top of a foam column that contributes to most measures of defoaming effectiveness.

Winterburn and Martin [66] used transducers of frequencies 28 and 40 kHz with aqueous (5 vol.%) "Teepol" detergent solution. The ultrasound was applied to a fixed volume of foam formed by sparging at known rates. Foam generated in this manner was, however, intrinsically unstable in the absence of ultrasound, declining almost completely after a few minutes. Unsurprisingly, in view of the potential problems with the chosen experimental arrangement, application of ultrasound at either frequency had little or no defoaming effect. Moreover, application of ultrasound at 40 kHz even appeared to reduce the rate of decline in liquid volume fraction in the foam.

Rodriguez et al. [67] argue that ultrasonic defoaming is not simply a function of the total energy delivered to the foam but is also dependent on the acoustic intensity at a given frequency. The acoustic intensity, I_{ac}, is the rate of delivery of acoustic energy per unit area (or power/unit area) and is given by

$$I_{ac} = \frac{P_{ac}^2}{\rho v_s} \tag{7.6}$$

where P_{ac} is the root mean square (RMS) of the acoustic pressure and v_s is the velocity of sound in the relevant medium. The total energy is therefore the product of acoustic intensity, time, and area of foam exposed to sound. Rodriguez et al. [67] therefore measured the extent of defoaming as a function of total acoustic energy at different acoustic intensities. This required application of ultrasound of a given intensity to a given volume of foam for various lengths of time. Foam was generated by sparging a small cell containing an aqueous solution of an unspecified "soap." The source of the ultrasound was a focusing stepped-plate ultrasonic radiator with a frequency of 25.8 kHz. A system of two rotors permitted exposure times of a foam cell to ultrasound in the range of 8–100 ms. Plots of the volume of foam destroyed as a function of acoustic intensity for various exposure times and as a function of total energy at various acoustic intensities are given in Figure 7.20. Similar results are published by Rodriguez et al. [67] for a lower concentration of the unspecified soap; indeed the total energy plots for the different concentrations are identical despite

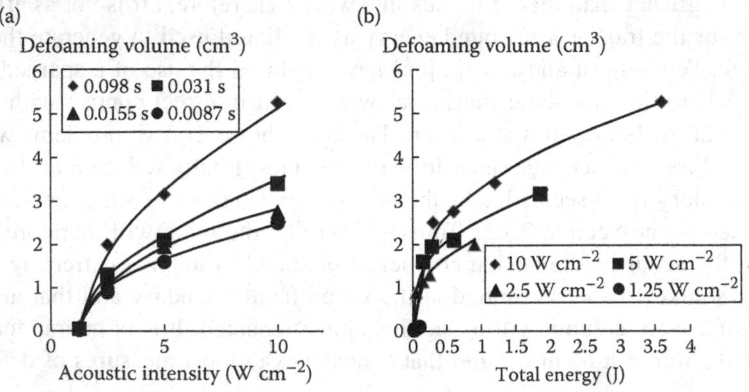

FIGURE 7.20 Plots of extent of defoaming by ultrasound where foam was generated from an aqueous solution containing "soap" of concentration 2.85 kg m^{-3}. (a) Defoaming volume as a function of acoustic intensity for various times of exposure to ultrasound. (b) Defoaming volume as function of total acoustic energy for various acoustic intensities. (Reprinted from *Phys. Procedia*, 3, Rodriguez, G., Riera, E., Gallego-Juarez, J.A., Acosta, V.M., Pinto, A., Martinez, I., Blanco, A., 135. Copyright 2010, with permission from Elsevier.)

significant experimental scatter, which suggests a reproduction error! Defoaming clearly increases with both the acoustic intensity and time of exposure to ultrasound. Since the acoustic intensity is proportional to the square of the acoustic pressure for a given area of foam and a given sound-transmitting medium, then this result is not inconsistent with the findings of Komarov et al. [61] shown in Figure 7.18 where the foaminess decreases with increasing acoustic pressure. The apparent difference concerns the nature of the experiments. Those of Komarov et al. [61] concerned a steady state (independent of time) during continuous sparging at various values of acoustic pressure (and therefore acoustic intensity) and those of Rodriguez et al. [67] concerned various values of acoustic intensity (and therefore acoustic pressure) for various fixed time intervals on a fixed amount of already generated foam.

7.3.2 Defoaming Mechanism of Ultrasound

As we have seen it has been known for more than 60 years that ultrasound can produce defoaming effects [62]. Despite this fact, authors are still forced to write in recent publications, for example, that "the mechanisms of ultrasonic defoaming are not well known but it can be assured that they include the effects of the acoustic pressure, the radiation pressure, bubble resonance, streaming and liquid film cavitation" [67]. This presumably reflects the difficult nature of the subject, although it may also reflect the limited extent of practical application of ultrasound in this context. It is even possible that these two factors mutually reinforce one another.

The propagation of sound in foam is already a difficult topic. The velocity of sound in aqueous foam of gas volume fraction, $\Phi_g > 0.99$ is significantly less than that in either air or water (remembering that the velocity of sound in water is more than three times faster than in air). This is illustrated by a plot of experimental values of the velocity of sound in aqueous foam as a function of Φ_g for $\Phi_g > 0.99$ reproduced by Kann [68], which is shown in Figure 7.21. Sound is also attenuated as it propagates through foam due to absorption resulting from the induced oscillations of foam

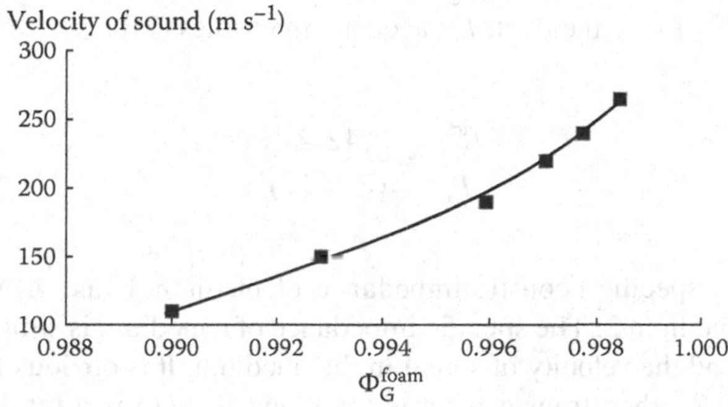

FIGURE 7.21 Velocity of sound in foam as function of gas volume fraction, Φ_G^{foam}. (From Kann, K.B., *Colloids Surf. A*, 263, 315, 2005.) For comparison, the velocity of sound in humid air at 20 kHz is 344.7 m s^{-1}. (From Lide, D.R. ed., *Handbook of Chemistry and Physics*, 85th Edition, CRC Press, pp 14–43, 2004.)

films as they interact with the pressure wave. This attenuation apparently exceeds that in liquids and gas by between 7 and 10 orders of magnitude [68]! Komarov et al. [60] reproduce an expression for the attenuation of the acoustic intensity (defined by Equation 7.6) in a foam due to Tolstov [69], which states

$$\frac{I_{ac}}{I_{ac}^{i}} = \exp(-2\xi_{att}H) \tag{7.7}$$

where I_{ac}^{i} is the incident acoustic intensity arriving at the surface of a foam layer of thickness H, and where ξ_{att} is the attenuation coefficient, which is dependent on the sound frequency and liquid volume fraction in the foam. Tolstov [69] gives an empirical expression for the attenuation coefficient for sound in foam of the form

$$\xi_{att} = 0.036(1-\Phi_{G}^{foam})^{0.625}\,\tilde{\nu}^{1.25} \tag{7.8}$$

where $(1 - \Phi_{G}^{foam})$ is the liquid volume fraction in the foam and $\tilde{\nu}$ is the frequency (in Hz). This expression is, however, only valid for frequencies in the range of 0.3–2 kHz, which lie outside the range of almost all studies of defoaming by sound. Nevertheless, since it appears that generally the attenuation coefficient of sound increases with increasing frequency for both liquids and gases [70], it is also likely to be the case for all foams.

Following Komarov et al. [60], we present a schematic illustration in Figure 7.22 of the acoustic pressure wave propagation in a bubble column containing an aqueous foam at the top of which is placed a sonic generator. Progressive waves are indicated by solid lines and reflected waves by dotted lines. The progressive wave, after passing through air, is seen to be attenuated as it passes through foam. Whether it is transmitted at the air–foam surface or at the foam–water surface is determined by the respective specific acoustic impedances of the relevant media. Thus, in the case of transmission between two homogeneous media, we can write for the ratios of the transmitted, I_{ac}^{trans}, to the incident, I_{ac}^{i}, acoustic intensities [71]

$$\frac{I_{ac}^{trans}}{I_{ac}^{i}} = \frac{4Z_1Z_2}{(Z_1+Z_2)^2} \tag{7.9}$$

where Z_1 is the specific acoustic impedance of medium 1 and Z_2 is the specific impedance of medium 2. The specific impedance of a medium is simply the product of the density and the velocity of sound in that medium. It is obvious from Equation 7.9 that if $Z_1 = Z_2$, then transmission is total, and if there is a marked imbalance between acoustic impedances so that $Z_1 \gg Z_2$ or $Z_2 \gg Z_1$, then transmission is minimal in either case.

There would appear to be no knowledge of the specific acoustic impedance of foam. It seems probable, however, that the impedance at, for example, the top of

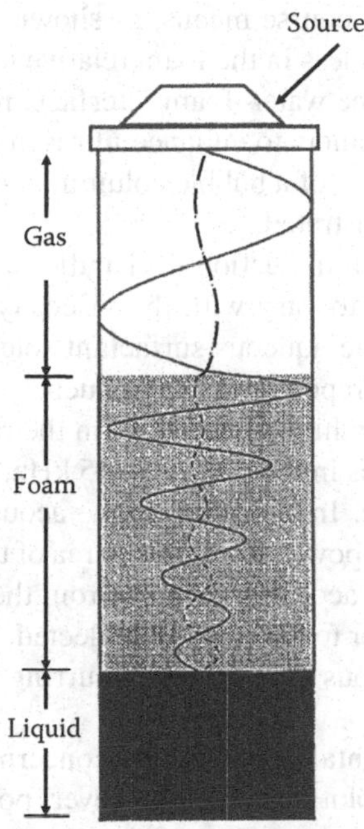

FIGURE 7.22 Schematic diagram illustrating propagation of sound waves through foam in bubble column. Continuous lines represent progressive waves and broken lines represent reflected waves. (After Komarov, S.V., Kuwabara, M., Sano, M. *ISIJ Int.*, 39, 1207, 1999. Originally published in *ISIJ International.*)

a foam column will be much closer to that of air than water. Thus, if we suppose, realistically, that the gas volume fraction at the top of a foam column is 0.99, then the velocity of sound in the foam is ~100 m s^{-1} (see Figure 7.21) and the density is ~11.2 kg m^{-3} giving a specific acoustic impedance of 1.12 kPa s m^{-1}, which compares with that of humid air of 0.414 kPa s m^{-1} and water of 1.5×10^3 kPa s m^{-1}. This estimate of the specific acoustic impedance of foam together with Equation 7.9 then implies an acoustic transmission of almost 80% at the air–foam interface.

Bubbles at the foam–water interface in a bubble column are subject to minimal capillary pressure and will therefore be essentially spherical. In which case, the gas volume fraction of foam at the foam–water interface in a bubble column in the case of a polydisperse foam is ~0.72 [72]. The velocity of sound at $\Phi_G^{foam} = 0.72$ in foam is ~50 m s^{-1} according to estimates quoted by Kann and Kislitsyn [73] and the density is ~281 kg m^{-3}, giving an apparent specific acoustic impedance of ~14 kPa s m^{-1}, which is two orders of magnitude less than that of water. The relatively high value for water means that $Z_1 \gg Z_2$ in Equation 7.9 and sound is almost totally reflected from the foam–water interface, regardless of whether it is arriving at the interface from water or foam. As indicated in Figure 7.22, a sound wave, attenuated after passing through the foam, is totally reflected from the foam–water interface whereupon it is further attenuated as it again passes through the foam. That the velocity of sound is less in foam than in air of course means, as shown in the figure, that the wavelength of the pressure wave is less in the foam relative to that in air. It should also be clear that total reflection at the water–foam interface, regardless of the direction of propagation, implies that attempts to enhance ultrasonic defoaming by locating the ultrasonic generator at the base of a bubble column, in direct contact with the liquid phase [66], will tend to be frustrated.

The observations described in Section 7.3.1 indicate that the frequency range for effective defoaming appears to vary with the viscosity of the continuous phase in the foam. In the case of dilute aqueous surfactant solutions of viscosity ~1 mPa s, defoaming effects have been reported in the frequency range of 20–25.8 kHz. In the case of liquids of significantly higher viscosity (in the range 8–200 mPa s), defoaming effects involve frequencies in the range 0.1–15 kHz, which puts most such observations in the acoustic range. In some cases, the acoustic pressure on the foam is quoted and in other cases the power output/unit area of the sonic generator is quoted. It is possible to calculate the acoustic pressure from the latter using Equation 7.6 if attenuation from the generator to the foam is neglected. We therefore find that observations of defoaming of various systems are occurring with acoustic pressures over the range of 0.5–6 kPa.

Although these experimental observations concern different techniques, often with poorly characterized solutions, it is, however, possible to make some generalizations. We find then that, regardless of the viscosity of the continuous phase, defoaming at a given frequency is enhanced by increasing the acoustic pressure and prolonging the time of exposure to that sound. A steady state is possible in a bubble column where the rate of defoaming with continuous exposure to sound of a given amplitude and frequency is matched by the rate of foaming as gas is continuously sparged into the column. However, if the frequency is increased at a given amplitude, then defoaming effectiveness decreases. This is particularly clearly revealed by the observations of Komarov et al. [61], concerning foam prepared from liquids of viscosities ≥100 mPa s, shown in Figure 7.19. The only reported exception to this effect concerns the results, also obtained by Komarov et al. [60], showing diminished defoaming with decreasing frequencies in the frequency range of 0.2–1.0 kHz. These authors also report evidence of resonance in defoaming at particular frequencies, at least in the difficult systems represented by slag defoaming at extremely high temperatures [61,64]. Finally, it seems that increase in viscosity of the continuous phase increases the susceptibility of foam to sound of a given amplitude (see Figure 7.19).

Clearly then, the acoustic pressure is a key factor in determining defoaming effectiveness. However, the amplitude of the pressure fluctuations in the foam determines not only the magnitude of the volume fluctuations of the bubbles but also the rate of overall acoustic energy delivery to the foam as revealed by Equation 7.6. The latter will obviously produce temperature changes in foam films, which could lead to evaporative destabilization. Fluctuations in bubble volumes will of course be accompanied by fluctuations in foam film areas and thicknesses as films are alternately stretched and compressed. If, for example, a foam element of initial gas volume fraction 0.99 at atmospheric pressure (~10^2 kPa) is compressed by pressure changes over the range apparently used for defoaming of 0.5–6 kPa, then bubble volumes will be reduced by only of order 0.5%–6% and the gas–liquid surface areas by only of order 0.3%–4%. Clearly not a dramatic effect—not sufficient to cause cavitation or the bubble explosions of the speculations of Boucher and Weiner [74]! Such an effect is, however, occurring at frequencies > 0.1 kHz where it is likely that surfactants adsorbed at air–water surface exhibit insoluble monolayer behavior [75]. In other words, the surface expansion and contraction rates are so fast that the adsorbed molecules do not have time to adsorb or desorb.

As we have seen, defoaming of foams prepared from dilute, low-viscosity, aqueous surfactant solutions has been reported mainly with ultrasound in the frequency range of 20–25.8 kHz. The only reported exception to this concerns the extremely weak effect described by Komarov et al. [60] in the frequency range of 2–7 kHz. If we reasonably ignore the latter, then we can use the plot of the sound velocities for $0.99 < \Phi_G^{foam} < 0.999$ shown in Figure 7.21 to estimate the range of wavelengths implied by the effective observed frequency range of 20–25.8 kHz as 4–15 mm. The latter is close to the bubble diameter range expected for aqueous foams prepared by sparging. Foam bubble sizes are rarely in fact measured in the study of ultrasound defoaming. However, the work of Rodriguez et al. [67] represents an exception where both the gas volume fraction and range of bubble sizes have been measured. Using the frequency of 25.8 kHz reported by Rodriguez et al. [67], together with the velocity of sound inferred from both their experimental values of $1 - \Phi_G^{foam}$ and Figure 7.21, yields acoustic wavelengths of ~4 mm in the case of a foam with measured bubble diameters in the range 0.2–2 mm and ~6 mm in the case of a foam with measured bubble sizes in the range of 0.2–10 mm.

We therefore have the possibility that the acoustic wavelength is of the same order as the dimensions of bubbles and therefore the lateral dimensions of foam films. This possibility is ignored in the theoretical treatment of acoustic defoaming by Nevolin [76]. That acoustic waves could cause symmetrical thickness oscillations in those films of a similar wavelength has in fact been suggested by Sandor and Stein [59]. This implies some kind of resonance where the lateral dimension of the film is some integer of the acoustic wavelength. It is perhaps easier to see how this could happen in the case of a film vertically oriented parallel to the sound pressure wave. Clearly, however, the actual orientations of films in a typical polydisperse foam are essentially mostly random, with, however, the exception of the dome-shaped films at the top of a foam column.

The reported acoustic pressures of waves (i.e., the RMS average pressure) that are used to achieve defoaming effects lie in the range of 0.5–6 kPa, which compares with film rupture capillary pressures for foam films formed from submicellar aqueous solutions of the same order (see Section 7.2.3.1). A foam film experiencing such an external pressure wave may react by movement of fluid in the film in response to pressure differences provided that the frequency of the wave is not so high that such movement cannot take place in time—the film would then tend to behave as an elastic solid of extremely low compressibility. The movement of fluid at any point in the film would be oscillatory to coincide with the pressure oscillations of the acoustic wave. Presumably, the greater the acoustic pressure, the faster the induced fluid flow in the film and the greater the amplitude of thickness oscillations induced by the passage of one wavelength of acoustic pressure. However, the higher the frequency for a given acoustic pressure, the shorter the time interval between the reversal of fluid flow at any point in the film and the lower the amplitude of the film thickness oscillations.

Growth of thickness fluctuations will also be resisted by the resulting differences in capillary pressure and will be either enhanced or resisted by disjoining pressure in the film provided it is thin enough. The former is likely to be augmented by the high surface dilatational moduli associated with any high-frequency longitudinal surface waves [77] caused by the thickness fluctuations, which are necessarily accompanied by surface area fluctuations. The situation is therefore not too dissimilar to that analyzed by Vrij [41,42] when considering the stability of a foam film subject to random thermally induced capillary waves which is briefly described in Chapter 1 (see Section 1.3.3). By analogy with that analysis, it is obvious that if the wavelength of the sound wave is too short, it will be damped by capillarity. Conversely, if it is too long, it will grow too slowly because the movement of fluid over relatively long distances will be resisted by viscous forces—the physical size of the film clearly in any case represents a constraint. As with the analysis of Vrij [41,42], an optimum wavelength for rapid foam film rupture, longer than the critical wavelength, is therefore predicted.

We tentatively propose then that a critical sound wavelength will exist above which thickness fluctuations will rapidly grow catastrophically leading to film rupture. That would require the minimum relevant lateral film dimension and therefore the bubble size to be of the order of that wavelength. This optimum wavelength would be determined by the relative magnitude of capillary pressure, acoustic pressure, and disjoining pressure (provided $d\Pi_{AWA}(h)/dh > 0$; see Section 1.3.3). If the acoustic frequency is too high, then the wavelength will be too short so that capillary force will suppress the fluctuations, which will be in any case of limited amplitude because of the short time between flow reversals due to the change in sign of the pressure wave. Film rupture will not then occur if the frequency is too high. However, as the film thins, disjoining forces will reinforce the growth of the fluctuations so that the critical wavelength declines. It is therefore possible that the effective frequency for foam collapse will increase as the foam films drain. This implies that it may be desirable to use acoustic waves with a range of frequencies for defoaming a typical polydisperse foam containing films of different sizes and thicknesses.

In the case of defoaming with a given acoustic frequency, it seems then that foam film rupture may often only occur after the film has drained to lower thicknesses. This factor could explain the lack of instantaneous foam collapse after exposure to ultrasound of a frequency such that up to 20,000 oscillations have occurred in 1 s and where total foam collapse times upon exposure to ultrasound take several minutes (see, e.g., Figure 7.18). Sandor and Stein [59] argue that interaction between sound waves and foam films may not only facilitate film rupture but may also in fact lead to increased film drainage rates by enhancing the fluctuations in film thickness, which characterize asymmetric film drainage (see Section 1.3.2). However, it has also been argued that the oscillations of bubbles induced by ultrasound passing through the whole body of the foam can lead to squeezing motions in Plateau borders and increased rates of overall drainage [60,78]. Greater attenuation of the sound at increased frequencies may, however, diminish such effects by minimizing the intensity of the sound in the regions of the foam furthest from the source.

All of these arguments are of course purely speculative—there is, however, little direct experimental evidence available concerning the actual mode of action of defoaming by ultrasound against which to test them. Clearly, a major weakness concerns the nature of the interaction between a sound wave and inducement of thickness fluctuations in foam films. We should also note that the results of Komarov et al. [61] with foam generated from viscous liquids shown in Figure 7.19 involve some defoaming effects with sound of only 0.09 kHz. This means that the velocity of sound with foam prepared from such liquids would have to be of order of 0.09–0.9 m s^{-1} if wavelengths are to be of the same order as the expected diameters of bubbles (probably of order 1–10 mm)! More realistically, it implies wavelengths of sound far in excess of bubble diameters. Of the arguments given here, that leaves only the effect of sound on foam drainage in the case of foam prepared from these viscous liquids. Presumably, the greater the amplitude and the lower the frequency, the greater any putative enhancement of foam drainage through induced motion in Plateau borders.

Finally, we note that Komarov et al. [60] also present an argument for a contribution to foam collapse by the so-called acoustic radiation pressure, which is present even at the interface between materials with matched acoustic impedances. The potential relative contribution of this phenomenon is, however, unclear.

7.4 DEFOAMING USING PACKED BEDS OF APPROPRIATE WETTABILITY

There are foam control problems where neither rotary nor ultrasonic devices nor antifoams are acceptable. Such situations are characterized by systems that preclude the use of rotary or ultrasonic devices because of either unacceptable power consumption or effectiveness or both. Similarly, use of antifoams implies contamination of the foaming medium with dispersed hydrophobic material, the presence of which may be unacceptable. Even non-toxic PDMS–hydrophobed silica antifoams can present unacceptable problems in some contexts (see Chapter 11). Attempts to immobilize such antifoams using porous media have, not surprisingly, failed to avoid contamination. However, there exists the possibility of turning the porous medium itself into an inert defoamer if the contact angle formed by the gas–foaming liquid surface against the relevant substrate is high enough.

As we have described at length in Chapter 4, inert hydrophobic particles can be effective in destroying the foam of aqueous surfactant solutions provided the air–water contact angle exceeds a critical value determined by the particle geometry. That combination of contact angle and geometry can also of course be replicated in a fixed porous bed through which the foam is passed. Since defoaming is then provided by such a bed, no contamination of the foaming liquid with antifoam is implied. A porous bed containing sharp edges could mean that relatively low contact angles would produce effective defoaming. On the other hand, observations of foam breaking by hydrophobic cylinders suggest that porous beds made from cylindrical fibers would probably require contact angles in excess of 90°. Suitable hydrophobic substrates are likely to possess perfluorinated surfaces.

An example of an application of this concept is described by Ward et al. [79]. This application concerns fuel cell development. The relevant fuel cell cathode employs an aqueous solution of a redox catalyst, which must be regenerated by aerobic oxygenation (see, e.g., reference [80] for a description of this type of fuel cell). Sparging the solution with air produces an unacceptable foam in a situation where contamination with conventional antifoam would damage the integrity of the process. Use of rotary foam devices, on the other hand, consumes an unacceptable proportion of the output of the fuel cell. The resultant foam is, however, readily destroyed by a packed porous bed consisting of a polytetrafluoroethylene (PTFE) mesh.

The mesh structure consists of knitted cylindrical fibers. The surface tension of the aqueous solution of the oxygenated redox catalyst is ~75 mN m^{-1} at 20°C, which implies a negative surface excess of the redox electrolyte. It seems likely therefore that the air–solution contact angle with this system is >90° since that at the air–water surface is >90° and a negative surface excess is to be expected also at the PTFE–water surface. Such high contact angles for the redox catalyst solution are consistent with the destruction of foam by the cylindrical fibers in the mesh.

7.5 SUMMARIZING REMARKS

Arguably the most striking feature of the literature concerning mechanical defoaming is the dearth of fundamental studies concerning their mode of action. Comments about lack of understanding have been a feature of that literature for at least the past quarter of a century. Some speculations concerning their mode of action have, however, been included here.

The justifications given for the use of mechanical defoaming devices rather than antifoams mostly concerns cost and/or the undesirability of the chemical contamination represented by the latter. Deshpande and Barigou [6,7] also add that the use of antifoams is empirical and based upon a "hit-and-miss" approach. However, developments in the understanding of the mode of action of antifoams over the past quarter of a century (reviewed earlier in reference [81] and here in Chapters 3–6) suggest that such a comment is unjustified. It would appear to be better directed at the use of the mechanical defoaming methods, which were the subject of the relevant publications [6,7].

In the case of mechanical defoaming, systematic empirical studies concerning the effect of surfactant variables appear to be rare. The use of reasonably well-characterized surfactant solutions, where existing knowledge of relevant physical properties such as CMC, equilibrium and dynamic surface behavior, thin film behavior, etc., would contribute to mechanistic understanding, is largely absent. Indeed there would appear to be no comparative study of the effect on defoaming effectiveness, with a given device, of increasing surfactant concentration from submicellar to micellar as a function of the CMC of the surfactant. A recent and otherwise useful paper concerning defoaming using ultrasound gives only the surfactant concentration used with no mention of chemical nature or solution properties at all save an obscure reference to the use of "soap" [67]! Such accounts do not, of course, readily permit assessment of the relative effectiveness of different devices! The concentrations used in measuring the effectiveness of perhaps all such studies should include measurements with standard surfactant solutions so that such comparisons can be readily made (e.g., using micellar SDS of 2×10^{-2} M with and without 0.1 M NaCl). Of particular concern in this context is the use of dilute submicellar solutions of surfactants with low CMCs, the foam of which can be readily controlled with intrinsically inefficient devices.

In the case of the mode of action of rotary devices, the main controversy concerns the relative importance of centrifugal action and shear. Much could presumably be learnt by high-speed video imaging of representative devices built of transparent materials. This could, in principle, produce evidence for actual foam film rupture events as rotor blades intercept foam bubbles. Centrifugal pressure gradients could be compared with static measures of film collapse pressures as suggested here. The combination of continuous centrifugal action and foam supply could be modeled, perhaps revisiting some of the arguments of Vetoshkin [38].

There would appear to be no thorough systematic study of the defoaming effect of ultrasound (at frequencies > 20 Hz) as a function of acoustic pressure and frequency in the case of foams prepared from dilute aqueous surfactant solutions. This could be made with foam of different bubble sizes and low polydispersity to establish, for example, whether there exists a relation between frequency, bubble size, and defoaming. It should include consideration of the effect of foam age and therefore drainage. The application of the relevant method to additionally study the effects of the continuous phase viscosity could also be made. The mechanism of defoaming could be probed further if such studies could be combined with study of the effects of changes in the surface dynamic and rheological behavior of the surfactant solutions.

Direct measurement of the intensity of reflected sound from the upper surface of foam as a function of frequency may also give some indication of the processes giving rise to foam film rupture. The effect of thickness of the bubble layer (from a monolayer upward) and drainage on such measurements could also be instructive. Perhaps direct observation, using high-speed imaging with reflected light, of the effect of ultrasound on the drainage behavior of single foam films oriented both horizontally and vertically may also help to elucidate mechanism. Modern developments in ultrasound imaging may have application in this context.

In conclusion, it should be stated that the present primitive level of understanding of rotary and ultrasonic defoaming may prevent design of significantly improved devices and realization of the true potential of these methods. Finally, we note that the use of fixed porous packed beds of appropriate wettability may have potential in either replacing or supplementing such devices in certain circumstances where the presence of antifoams is unacceptable.

APPENDIX 7.1

TABLE 7.A1
Summary of Empirical Studies of Mechanical Defoaming by Rotary Devices

Type and Ang. Velocity of Rotary Defoamer	Mode of Operation	Comments	D_{FV} (m)	D_{RD} (m)	Superficial Gas Velocity (cm s^{-1})	Surfactant (All as Aqueous Solutions)	Ref.
Rotating disc, centrifugal basket ($\omega \geq 261\ s^{-1}$)	Foam fed to disc from above		–	0.1	–	Triton X-100	[10]
Three to eight vaned disc turbines ($\omega = 82 - 189\ s^{-1}$)	Foam fed from outside source	Pilot scale	0.9 × 0.9 square	0.15–0.38	0.78	Kraft Mill effluent	[15]
Glass paddle stirrer ($\omega = 0.05\ s^{-1}$)	Sparging/bubble column		0.055	0.052	–	Bovine sebum albumin + salts	[82]
Rotating disc/inherent liquid spray ($\omega = 21–210\ s^{-1}$)	Sparging/stirred vessel		0.23	0.18	0.19–0.76	Diluted "detergent" solution	[22, 23]
Rotating disc/inherent liquid spray ($\omega \leq 200\ s^{-1}$)	Sparging/stirred vessel		0.23	0.18	0.19–0.76	Commercial "soft-type detergent"	[24]
Three-blade marine propeller ($\omega = 48–188\ s^{-1}$)	Sparging/bubble column + sparging stirred vessel	Comparison stirred tank and unstirred bubble column	0.088 bubble column; 0.15 stirred tank	0.06	0.2–8.0	Teepol	[13]
Six-blade turbine in rotor/stator assembly ($\omega = 47–440\ s^{-1}$)	Sparging/stirred vessel	Recommendation for scale-up. Effect of viscosity studied using glycerol	0.45	0.15	3.5	Technical-grade sodium alkyl sulfonate	[19]
Rotating disc/inherent liquid spray ($\omega = 125–314\ s^{-1}$)	Sparging/stirred vessel	Power consumption for mechanical foam breakdown is smaller in stirred vessels containing antifoams due to decreased hold-up	0.23	0.17–0.18	0.38–0.76	Tween 40, Tween 60, egg albumin, "commercial detergent"	[25]

Rotating disc/inherent liquid spray ($\omega = 125–314\ s^{-1}$)	Sparging/bubble column	Mechanical foam control better than chemical control for mass transfer. Also used silicone antifoam for comparison	0.19	0.15	1–5	Commercial (anionic + non-ionic detergent) with or without either corn syrup or baker's yeast	[26]
Rotating disc/inherent liquid spray ($\omega = 125–380\ s^{-1}$)	Sparging/bubble column	Liquid hold-up, critical speed, and power consumption correlations. Foam breaking difficulty quantitatively related to intrinsic foamability of liquid. Correlations given for critical disc speed for foam breaking with foamability	0.19–0.31	0.13–0.25	1–5	Commercial detergent, Triton X-100, Tween 40, Tween 60, saponin, egg albumin	[27–29]
Rotating disc/inherent liquid spray ($\omega = 125–380\ s^{-1}$)	Sparging/bubble column	Correlations for critical disc speed extended to complex biological media	0.19–0.31	0.15–0.24	1–5	Commercial detergent, Triton X-100, Tween 40, Tween 60, saponin, egg albumin, various biological media	[30]
Rotating disc/inherent liquid spray	Sparging/stirred vessel	Foam passed through perforated plate placed below foam breaker	0.19	0.15	0.5–2.4	Commercial anionic detergent	[31]
Rotating disc/inherent liquid spray ($\omega = 188–314\ s^{-1}$)	Sparging bubble column	Effect of gas sparger type studied	0.19	0.15	2.1–5.8	Commercial detergent, Triton X-100, Tween 60, saponin all in aqueous sodium sulfite solution	[32]

(continued)

TABLE 7.A1 (Continued)
Summary of Empirical Studies of Mechanical Defoaming by Rotary Devices

Type and Ang. Velocity of Rotary Defoamer	Mode of Operation	Comments	D_{FV} (m)	D_{RD} (m)	Superficial Gas Velocity (cm s^{-1})	Surfactant (All as Aqueous Solutions)	Ref.
Six-blade turbine, six-blade vaned disc, two-blade paddle, spinning cone (ω = 100–220 s^{-1})	Sparging/bubble column	Required rotational speed highest for conical rotor. Significant amounts of liquids entrained with exhaust air	0.19	Cone 0.1; others 0.115	2.1–6.4	Commercial detergent, Triton X-100, Tween 60, saponin, egg albumin	[18]
Comparison of 6–12-blade rotor/stator with orifices combination and radial accelerator/stator combination (ω = 94–523 s^{-1})	Sparging/stirred vessel	Blade–rotor/stator combination relatively effective Provision made for removal of the secondary foam, which is always formed regardless of methodology	0.45	0.15–0.3	0.87	C_{16} TAB, Triton X-100, Technical grade sodium alkyl sulfonate	[20]
Use of novel devices –Two-blade paddles with three slits –Two-blade paddle with 168 thin needles (ω = 10–120 s^{-1})	Both sparging/stirred vessel and sparging/bubble column	Novel devices gave reduced power consumption and speeds compared with two-blade paddle, six-pitched blade turbine, and six-blade disc turbine	0.16	0.11	0.083–0.83	Sodium dodecyl benzene sulfonate and chicken egg albumin with viscosity modified by glycerol	[6, 14]
Rotating disc/inherent liquid spray (ω = 138–281 s^{-1})	Sparging/stirred vessel	Foam-breaking effectiveness quantitatively related to intrinsic foamability of liquid	0.19–0.27	0.15–0.21	0.38–0.76	Commercial detergent, Triton X-100, Tween 40, saponin, egg albumin	[33]

Rotating disc/inherent liquid spray compared with six-blade turbine, six-blade vaned disc, two-blade paddle, spinning cone, and fluid impact dispersion apparatus ($\omega = 150–250\ s^{-1}$)	Sparging/stirred vessel	Only rotating disc, six-blade turbine, and six-blade vaned disc effective for all aerating conditions. However, power consumption for rotating disc/inherent liquid spray lowest	0.23	Cone 0.1; other rotary devices 0.115	0.38–0.76 for rotary devices	Commercial detergent, egg albumin, soybean meal	[17]
Rotating disc/stator with orifice combination ($\omega = 94–314\ s^{-1}$)	Sparging/stirred vessel	Provision made for removal of secondary foam	0.19	0.20	0.65–0.31	Triton X-100 with or without sodium sulfite	[4, 21]
Spinning cones ($\omega = 17–31\ s^{-1}$)	Sparging/stirred vessel	Effective defoaming but no comparisons with other rotary devices made	0.61–0.91	Cone outlet diameter 0.14–0.30	0.11	Commercial dish washing liquid + polypropyl glycol (with or without glycerol to control viscosity)	[16]
Spinning cones ($\omega \approx 50\ s^{-1}$)	Pilot-scale sparging/stirred fermentation vessel	Pilot-scale foam control of commercially exploited recombinant *Bacillus* fermentation achieved without need of added antifoam	0.688	Cone outlet diameter 0.344		*Bacillus* fermentations of an unspecified nature	[34]

Source: Adapted and updated from Deshpande, N.S., Barigou, M., *Chem. Eng. Process.*, 39, 207, 2000.

Note: D_{FV} = diameter of foaming vessel; D_{RD} = diameter of rotary device; ω = angular rotation velocity (in rad. s^{-1}) required to cause defoaming.

REFERENCES

1. Viesturs, U.E., Kistapsons, M.Z., Levitans, E.S. *Adv. Biochem. Eng.*, 21, 169, 1982.
2. Barigou M. *Chem. Eng. Technol.*, 24(6), 659, 2001.
3. Hoeks, F.W.J.M.M., van Wees-Tangerman, C., Lubyen, K.Ch.A.M., Gasser, K., Schmind, S., Mommers, H.M. *Can. J. Chem. Eng.*, 75, 1018, 1997.
4. Takesono, S., Onodera, M., Yoshida, M., Yamagiwa, K., Ohkawa, A. *J. Chem. Eng. Jpn.*, 37(12), 1488, 2004.
5. Yagi, H., Yoshida, F. *J. Ferment. Technol.*, 52, 905, 1974.
6. Deshpande, N.S., Barigou, B. *J. Chem. Technol. Biotechnol.*, 74, 979, 1999.
7. Morey, M.D., Deshpande, N.S., Barigou, M. *J. Colloid Interface Sci.*, 219, 90, 1999.
8. Ito, A. (assigned to Terumo Kabushiki Kaisha, Japan), US 2007/0110612 A1, 17 May 2007, filed 9 November 2006.
9. Haas, P.A., Johnson, H.F. *Am. Inst. Chem. Eng. J.*, 11, 319, 1965.
10. Goldberg, M., Rubin, E. *Ind. Eng. Chem. Process Design Dev.*, 6(2), 195, 1967.
11. Kruglyakov P.M. et al. USSR 524557, 1976.
12. Pahl, M., Meinecke, H. *Dechema-Monogr.*, 114, 433, 1989.
13. Pandit, A.B. *Adv. Biochem. Eng., VCH Inc.*, 45, 1989.
14. Deshpande, N.S., Barigou, M. *Chem. Eng. Process.*, 39, 207, 2000.
15. Ng K.S., Mueller, J.C., Walden, C.C. *Can. J. Chem. Eng.*, 55, 439, 1977.
16. Cooke, M., Heggs, P.J., Eaglesham A., Housley, D. *Trans. Inst. Chem. Eng. Part A, Chem. Eng. Res. Design*, 82(A6), 719, 2004.
17. Takesono, S., Onodera, M., Ito, A., Yoshida, M., Yamagiwa, K., Ohkawa, A. *J. Chem. Technol. Biotechnol.*, 78, 48, 2003.
18. Andou, S., Yamagiwa, K., Ohkawa, A. *J. Chem. Technol. Biotechnol.*, 68, 94, 1997.
19. Furchner, B., Mersmann, A. *Chem. Eng. Technol.*, 13, 86, 1990.
20. Gutwald, S., Mersmann, A. *Chem. Eng. Technol.*, 20, 76, 1997.
21. Takesono, S., Onodera, M., Ito, A., Yoshida, M., Yamagiwa, K., Ohkawa, A. *J. Chem. Eng. Jpn.*, 40(7), 565, 2007.
22. Ohkawa, A., Sakagami, M., Sakai, N., Futai, N., Takahara Y. *J. Ferment. Technol.*, 56(4), 428, 1978.
23. Ohkawa, A., Sakai, N., Imai, H., Endoh, K. *J. Ferment. Technol.*, 62(2), 179, 1984.
24. Ohkawa, A., Sakai, N., Imai, H., Endoh, K. *J. Chem. Technol. Biotechnol.,* 34B, 87, 1984.
25. Yasukawa, M., Onodera, M., Yamagiwa, K., Ohkawa, A. *J. Chem. Eng. Jpn.*, 24(2), 188, 1991.
26. Takesono, S., Yasukawa, M., Onodera, M., Izawa, K., Yamagiwa, K., Ohkawa, A. *J. Chem. Technol. Biotechnol.*, 56, 97, 1993.
27. Takesono, S., Onodera, M., Yamagiwa, K., Ohkawa, A. *J. Chem. Technol. Biotechnol.*, 57, 237, 1993.
28. Takesono, S., Onodera, M., Nagai, J., Yamagiwa, K., Mori, A., Ohkawa, A. *J. Ferment. Bioeng.*, 75(4), 314, 1993.
29. Takesono, S., Onodera, M., Yamagiwa, K., Mori, A., Ohkawa, A. *J. Ferment. Bioeng.*, 77(2), 221, 1994.
30. Takesono, S., Onodera, M., Yamagiwa, K., Ohkawa, A. *J. Chem. Technol. Biotechnol.*, 60, 125, 1994.
31. Kunii, M., Hosokai, K., Onodera, M., Yamagiwa, K., Ohkawa, A. *Can. J. Chem. Eng.*, 72, 212, 1994.
32. Andou, S., Yamagiwa, K., Ohkawa, A. *J. Chem. Technol. Biotechnol.*, 66, 65, 1996.
33. Takesono, S., Onodera, M., Ito, A., Yoshida, M., Yamagiwa, K., Ohkawa, A. *J. Chem. Technol. Biotechnol.*, 76, 355, 2001.
34. Stocks, S.M., Cooke, M., Heggs, P.J. *Chem. Eng. Sci.*, 60, 2231, 2005.

35. Boon, L.A., Hoeks, F.W.J.M.M., van der Lans, R.G.J.M., Bujalski, W., Wolff, M.O., Nienow, A.W. *Biochem. Eng. J.*, 10, 183, 2002.
36. Boon, L.A., Hoecks, F.W.J.M.M., van der Lans, R.G.J.M., Bujalski, W., Nienow, A.W. *Can. J. Chem. Eng.*, 78, 884, 2000.
37. FUNDAFOM, Chemap AG, 1977.
38. Vetoshkin, A.G. *Theor. Found. Chem. Eng.*, 37(4), 372, 2001.
39. Hoffmann, A.C., Stein, L.E. Foam-breaking cyclones, in *Gas Cyclones and Swirl Tubes; Principles, Design and Operation*, Springer, Berlin, 2008, Chapter 14, p 327.
40. Vetoshkin, A.G., Kutepov, A.M. *J. Appl. Chem. (of USSR)*, 56(3), 584, 1984.
41. Vrij, A. *Disc. Faraday Soc.*, 42, 23, 1996.
42. Vrij, A., Overbeek, J.Th.G. *J. Am. Chem. Soc.*, 90, 3074, 1968.
43. Vilkova, N.G., Kruglyakov, P.M. *Kollodn. Zh.*, 58, 159, 1996.
44. Exerowa, D., Kruglyakov, P.M. *Foam and Foam Films; Theory, Experiment and Application*, Elsevier, Amsterdam, 1998, Chapter 6, p 486.
45. Aronson, A.S., Bergeron, V., Fagan, M.E., Radke, C.J. *Colloids Surf. A*, 83, 109, 1994.
46. Bergeron, V. *Langmuir*, 13(13), 3474, 1997.
47. Walstra, P., Smulders, P.E.A. Emulsion formation, in *Modern Aspects of Emulsion Science* (Binks, B.P., ed.), Royal Soc. Chem., Cambridge, UK, 1998, Chapter 2, p 56.
48. Lucassen, J. Dynamic properties of free liquid films and foams, in *Anionic Surfactants, Physical Chemistry of Surfactant Action* (Lucassen-Reynders, E.H., ed.), Marcel Dekker, New York, Surfactant Sci. Series, 1988, Vol 29, Chapter 1, p 1.
49. Franke, D., Pahl, M.H. *Chem. Eng. Process.*, 36, 175, 1997.
50. Engel, O., *J. Appl. Phys.*, 37(4), 1798, 1966; ibid, 38(10), 3935, 1967.
51. Ross, J., Miles, G.D. *Oil Soap, May*, 99, 1941.
52. Ross, S., McBain, J.W. *Ind. Eng. Chem.*, 36(6), 570, 1944.
53. Sun, S.-C. *Mining Eng.*, October, 865, 1951.
54. Dorsey, A.E. *J. Biochem. Microbiol. Technol. Eng.*, 1(3), 289, 1959.
55. Gallego-Juarez, J.A., Rodriguez-Corral, G., Gaete-Garrreton, L. *Ultrasonics*, 16(6), 267, 1978.
56. Rodriguez-Corral, G., Gallego-Juarez, J.A., Gallegos, E.A. *Quim. Ind. (Madrid)*, 32(6), 469, 1986.
57. Gallego-Juarez, J.A. Some applications of air-borne power ultrasound to food processing, in *Ultrasonics in Food Processing* (Povey, M.J.W., Mason, T.J., eds.), Thomson Science, London, 1998, p 127.
58. Riera E., Gallego-Juarez, J.A., Mason, T.J. *Ultrason. Sonochem.*, 13, 107, 2006.
59. Sandor, N., Stein, H.N. *J. Colloid Interface Sci.*, 161, 265, 1993.
60. Komarov, S.V., Kuwabara, M., Sano, M. *ISIJ Int.*, 39(12), 1207, 1999.
61. Komarov, S.V., Kuwabara, M., Sano, M. *Ultrason. Sonochem.*, 7, 193, 2000.
62. Bikerman, J.J. *Foams*, Springer, New York, 1973, p 80.
63. Ogawa Y., Katayama, H., Hirata, H., Tokumitsu, N., Yamauchi M. *ISIJ Int.*, 32, 87, 1997.
64. Komarov, S.V., Kuwabara, M., Sano, M., *ISIJ Int.*, 40(5), 431, 2000.
65. Dedhia, A.C., Ambulgekar, P.V., Pandit, A.B. *Ultrason. Sonochem.*, 11, 67, 2004.
66. Winterburn, J.B., Martin, P.J. *Asia-Pac. J. Chem. Eng.*, 4, 184, 2009.
67. Rodriguez, G., Riera, E., Gallego-Juarez, J.A., Acosta, V.M., Pinto, A., Martinez, I., Blanco, A. *Phys. Procedia*, 3, 135, 2010.
68. Kann, K.B. *Colloids Surf. A*, 263, 315, 2005.
69. Tolstov, G.S. *Sov. Phys. Acoust.*, 38, 596, 1992.
70. Lide, D.R., ed., *Handbook of Chemistry and Physics*, 85th Edition, CRC Press, 2004, pp 14–43.
71. Rayleigh, J.W.S. *The Theory of Sound*, Vol. 2, Dover Publications, 1945, Chapter 13, p 69.
72. Princen, H.M. *Langmuir*, 2(4), 519, 1986.
73. Kann, K.B., Kislitsyn, A.A. *Kolloid Zh.*, 65(1), 31, 2003.

74. Boucher, R.M.G., Weiner, A.L. *Br. Chem. Eng.*, 8(12), 808, 1963.
75. Lucassen, J., van den Tempel, M. *Chem. Eng. Sci.*, 27, 1283, 1972.
76. Nevolin, V.G. *J. Eng. Phys.*, 61(3), 1070, 1991.
77. Lucassen-Reynders, E.H., Lucassen, J. *Adv. Colloid Interface Sci.*, 2, 347, 1969.
78. Vafina, F.I., Goldfarb, I.I., Shreider, I.R. *Sov. Phys. Acoust.*, 38, 1, 1992.
79. Ward, D.B., Hine, M.J., Longman, R.J., Clarkson, B.G. (Applicant ACAL Energy Ltd.), Application GB 1203565.5, filed 29 February 2012.
80. Creeth, A.M., Ward, D. (Applicant ACAL Energy Ltd.), WO 2010/128333 A1, 11 November 2010, Filed 7 May 2010.
81. Garrett, P.R. The mode of action of antifoams, in *Defoaming, Theory and Industrial Applications* (Garrett, P.R., ed.), Marcel Dekker, New York, Surfactant Sci. Series, Vol. 45, Chapter 1, p 1.
82. Bumbullis, W., Schurgel, K. *Eur. J. Appl. Microbiol. Biotechnol.*, 11, 106, 1981.

8 Antifoams for Detergent Products

8.1 INTRODUCTION

Foam is often regarded by consumers as a beneficial aspect of detergent product performance. In the case, for example, of washing textiles or dishes by hand, the presence of foam is considered to be both an indication of cleaning effectiveness and a contribution to any aesthetic appeal associated with those activities. It would, however, seem that the former only has a firm basis in physical reality when phase separation of the surfactant as a result of interaction with polyvalent water hardness ions (to form crystalline or liquid crystalline entities) means that much of that ingredient is essentially not available for either cleaning action or foam generation. Use of certain surfactant types and surfactant mixtures together with various chelating agents ("builders") in modern laundry products minimizes such effects and thereby minimizes any clear correlation between high foam and efficient detergency. This is fortunate because a large proportion of the market for detergent products for washing textiles or dishes consists of consumers who use foam-intolerant washing machines.

Failure to control the foam in certain types of textile washing machines can produce dramatic effects as illustrated in Figure 8.1. Apart from the obvious difficulty with intrusion of foam into spaces where it is unwelcome, overfoaming inhibits the flexing of textiles in the machine, which will adversely affect detergency. The machine shown in Figure 8.1 is an example of the type used for domestic washing of laundry in Europe where water can be expensive and kitchen space is at a premium. Avoidance of overfoaming is usually achieved by integrating antifoams in the formulations designed for washing by such machines. Apart from obvious considerations of effectiveness, selection of such antifoams also involves issues of cost and ease of incorporation in products. The antifoam should also be compatible with the product because otherwise there may be loss of antifoam activity after prolonged periods of storage or even deterioration in the ease of delivery of the product to the wash. Such issues will be considered in this chapter in sections devoted to textile machine washing with products in both powder and liquid form, which are characterized by distinct differences in the practicalities of antifoam incorporation.

Use of foam-intolerant machines for washing dishes also represents an important global market. There are, however, a number of important differences with respect to textile washing. The latter involves movement of the textile, which is subject to flexing. By contrast, machine dishwashing involves the use of sprays directed at the relevant surfaces, which are not subject to any movement. Soil levels on dishes are often high relative to those on textiles. Moreover, in contrast to textile washing, acceptable dishwashing requires essentially zero levels of residual cleaning product

FIGURE 8.1 Overfoaming in front-loading drum-type washing machine due to absence of antifoam in detergent formulation.

and soil for both aesthetic and hygienic reasons. To achieve this, washing is done at high temperatures (≥~60°C) using highly alkaline formulations that hydrolyze food particles to produce surface-active materials, which in turn stabilize unacceptable levels of foam. To ensure minimal product residues, foam control is often achieved using cloud-point antifoams (see Section 4.6.3.2). These latter are effective only at temperatures greater than the cloud point. A corollary of the use of this approach is that surfactants other than the cloud point antifoams are usually omitted from such formulations because they can contribute to unacceptable increases in the cloud points of the antifoam. Unlike the combination of typical hydrophobic particle–oil antifoams and surfactants used for machine textile washing, these cloud point antifoams leave little deposit on ceramic or glass surfaces; indeed, they appear to have a dual function as rinse aids in this context.

There are situations where excessive foam can be produced even if machines are not involved in the cleaning process. Although, for example, copious amounts of foam are preferred by consumers during hand washing of dishes, there is a lack of enthusiasm for the concomitant effort associated with rinsing the foam away after the wash. Similarly, general-purpose cleaners for domestic hard surfaces such as floors and work surfaces should not form copious amounts of foam, again presumably because of the effort of removing the foam after the cleaning operation is complete. Some novel approaches to formulation design to provide control of foam in these general cleaning contexts are described in the patent literature, which we also review here.

Unfortunately, there is little published scientific literature concerned with the application of antifoams in the context of detergent products. We are therefore obliged to consult the relevant patent literature to infer the empirical generalizations and principles involved. This is not, however, intended to be a thorough review of that literature—we select patents only to illustrate what appear to be important generic principles or interesting effects. Patents of selection or "me-too" patents that describe minor variations on a revealed theme are in the main not considered.

8.2 POWDERS FOR MACHINE WASHING OF LAUNDRY

8.2.1 Front-Loading Drum-Type Textile Washing Machines

There are three basic designs of domestic textile washing machines—the top-loading agitator type, the top-loading impeller type, and the front-loading drum type. The latter is the predominant design in European markets, whereas the agitator type is predominant in the United States and Canada, and the impellor type in Japan [1]. Of these, the front-loading drum-type machine is characterized by low water usage and therefore low wash liquor-to-textile ratios. In turn, this permits the economic use of high concentrations of detergent. Concentrations of ~5–13 kg m^{-3} are usually used [1]. This compares with levels some 4–5 times lower in top-loading machines, which must function with much higher wash liquor-to-textile ratios [1]. Typical detergent powders for use with front-loading drum machines contain 10–15 wt.% of surfactant, which implies surfactant concentrations of 0.5–2.0 kg m^{-3} and ~$(1.1–4.4) \times 10^{-3}$ M (assuming average molecular masses of 450), well above the CMC at the relevant ionic strengths of the type of surfactants normally included in such powders.

A typical front-loading drum-type washing machine is depicted in Figure 8.1. A schematic diagram showing the essential features of the action of this type of machine is shown in Figure 8.2. It consists essentially of a perforated inner drum, which contains the textile wash load. The drum revolves around a horizontal axis, reversing the direction of rotation occasionally. The angular rotational velocity, ω, is usually in the range of 6–12 s^{-1} (i.e., 1–2 rps). Baffles attached to the drum lift the wash load until gravity overcomes friction and the weak centrifugal force, causing the load to fall back into the wash liquor reservoir. The latter is contained by a fixed outer drum of radius slightly larger than the rotating drum as shown in Figure 8.2. The detergent powder can be automatically introduced into the wash liquor using the dispensing arrangement shown in the figure. The powder is held in a dispensing tray from which it is removed by a water spray. The temperature of the wash liquor is controlled by heaters. During a normal wash cycle, cold water is continuously heated until the desired temperature is reached. Wash cycles with such machines are readily automated.

An obvious problem with the drum type of machine design concerns the process where a porous textile is continuously lifted out of the wash liquor and subsequently reinserted. Significant air entrainment is an inevitable consequence. High surfactant concentrations ensure that the entrained air builds up as foam, leading in the extreme to overfoaming as depicted in Figure 8.1. The prospect of such large amounts of foam oozing out of the machine over kitchen floors during a wash cycle is not of

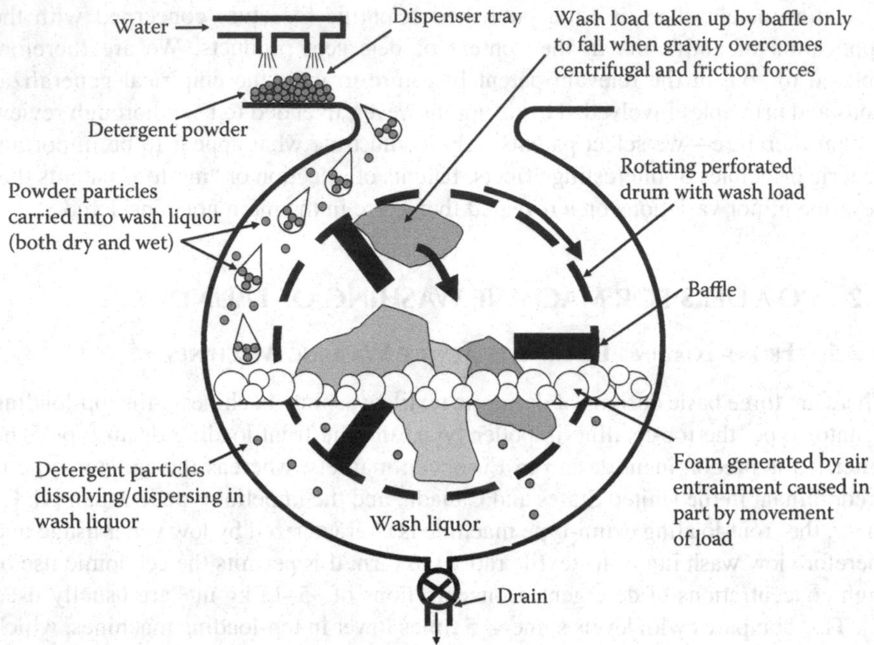

FIGURE 8.2 Schematic illustrating basic mode of action of front-loading drum-type washing machines.

course likely to be warmly appreciated by consumers. Moreover, it is known that if the air space above the wash liquor is filled with foam, the movement of the textile load is impeded and the required convection within the textile yarn pore space is inhibited so that detergency become less efficient.

As a consequence of this obvious problem with foam generation, it is necessary to either use low concentrations of low foaming surfactants or include antifoams in the relevant detergent formulations suitable for use with front-loading drum-type textile washing machines. The former approach is often adopted in the case of machine dishwashing but risks seriously inefficient detergency in the case of textile machine washing. Therefore, antifoams are usually used. Formulations for textile machine washing usually contain enough surfactant to produce overfoaming in minutes in the absence of antifoams. As a consequence, foam profiles in the absence of antifoam are not usually measured in this context and are therefore absent from those presented here.

Control of foam in front-loading drum-type textile washing machines probably represents the most important application of antifoam technology by the detergent industry. Suitable antifoams and methods of incorporation in detergent powders and liquids for this application are therefore described in this section and in Section 8.4, respectively.

8.2.2 Use of Fatty Acids and Soaps

It is well known that precipitation of calcium soaps (i.e., calcium alkyl carboxylates) is often accompanied by reductions in foamability and foam stability of aqueous solutions of other surfactants. The available evidence, as we have discussed in Section 4.7.6, suggests that these effects concern the bridging mechanism of hydrophobic particles.

That significant antifoam effects in front-loading drum-type washing machines due to the presence of sodium soaps can be attributed to calcium soap formation has been described by Ferch and Leonard [2]. The effect of water hardness on the antifoam behaviorof sodium behenate for a detergent of formulation containing only 3 wt.% sodium alkylbenzene sulfonate as a function of temperature is shown by way of illustration in Figure 8.3. Dependence of the effect on calcium ion concentration strongly implicates calcium soap precipitation (where, e.g., the solubility product of calcium behenate is $>10^{-18}$ at an ionic strength of 0.1M and 50°C [3]) as the antifoam.

Practical laundry detergent formulations usually include builders for control of polyvalent metal ion activities so that the relevant salts of the anionic surfactant, which is usually present, are not precipitated out of solution. The presence of builders also has implications for the effectiveness of antifoam action because they may prevent the precipitation of calcium soaps. We can illustrate the relevant issues if we consider, for example, the concentration of calcium hardness necessary to just initiate calcium soap precipitation in the case of a solution built by sodium tripolyphosphate at a pH ≥10.5. If we ignore the effect of solubilization of soaps and calcium binding in anionic surfactant micelles, then, according to Irani and Callis [4],

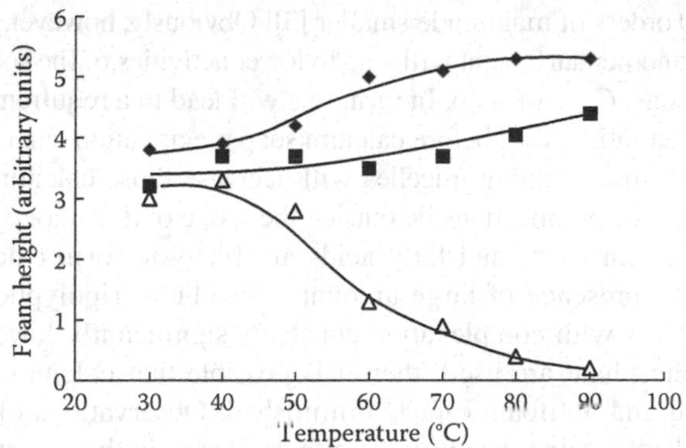

FIGURE 8.3 Antifoam behavior of sodium behenate as function of Ca^{2+} concentration in front-loading drum-type washing machine with 100 g prewash and 100 g main wash of detergent (clean wash load) with 3.1 wt.% linear alkylbenzene sulfonate. ◆, Deionized water; ■, 0.54 mM Ca^{2+}; △, 2.9 mM Ca^{2+}. Foam height at full porthole = 4. (After Ferch, H., Leonhardt, W. Foam control in detergent products, in *Defoaming, Theory and Applications*, First Edition, Garrett, P.R. ed., Marcel Dekker, Surfactant Science Series Vol 45, Chpt 6, p 221.)

the free calcium ion concentration is determined by two equilibria:the solubility product, K_1, for the calcium soap

$$K_1 = [Ca^{2+}][CH_3\,(CH_2)n\,COO^-]^2 \tag{8.1}$$

and the complexation constant, K_2, for the builder

$$K_2 = \frac{[Ca^{2+}]\left[P_3O_{10}^{3-}\right]}{\left[CaP_3O_{10}^{3-}\right]} \tag{8.2}$$

Applying a simple mass balance, ignoring activity coefficients,and setting the calcium soap concentration at a vanishing small level (effectively zero) means that we can easily show that

$$C_{Ca} = \frac{K_1}{C_{soap}^2} + \frac{C_{STP}}{\left(1 + K_2\,C_{soap}^2/K_1\right)} \tag{8.3}$$

where C_{Ca} is the total calcium concentration required to just initiate calcium soap precipitation. C_{STP} and C_{soap} are the sodium tripolyphosphate and total soap concentrations, respectively. Clearly the higher the soap concentration and the higher the ratio K_2/K_1, the lower the calcium hardness needed to cause calcium soap formation and the onset of antifoam effects despite the presence of the builder. According to Irani and Callis [4], we have for the complexation constant of sodium tripolyphosphate $4 \times 10^{-7} < K_2 < 5 \times 10^{-7}$ at 0.1M ionic strength in the temperature range of 25–50°C. By contrast, for example, the solubility products over the same temperature range for calcium stearate are more than 10 orders of magnitude smaller [3]! Obviously, however, comicellization of the soap with another surfactant will lead to lower activities of the soap and therefore lower concentrations, C_{soap}, of soap. In turn, this will lead to a requirement for a higher calcium ion concentration, C_{Ca}, before calcium soap precipitation can occur. Moreover, calcium binding to the resulting micelles will decrease those calcium ion concentrations. However, these complications lie outside the scope of this book.

Clearly, long-chain soaps and fatty acids are likely to form calcium soap antifoams even in the presence of large amounts of sodium tripolyphosphate builder. However, if builders with complexation constants significantly lower than those of sodium tripolyphosphate are used, then it is possible that calcium soap formation may be inhibited and antifoam effects diminished. Observations of Schmadel and Kurzendorfer [5] comparing the antifoam effect of soaps in the presence of nitrilotriacetic acid and sodium tripolyphosphate are consistent with that expectation.

Ferch and Leonhardt [2] have made a comparison of the antifoam effects of blends of sodium tallow (C_{16}–C_{18}) and behenic (C_{20}–C_{22}) soaps on a wash cycle at a water hardness of 2.85 mM. A detergent powder formulated with a mixture of sodium alkylbenzene sulfonate and non-ionic surfactants was used together with a mixture of zeolite and sodium tripolyphosphate as builder. The resulting foam profiles are shown in Figure 8.4. An initial foam peak at the low temperature start of the wash

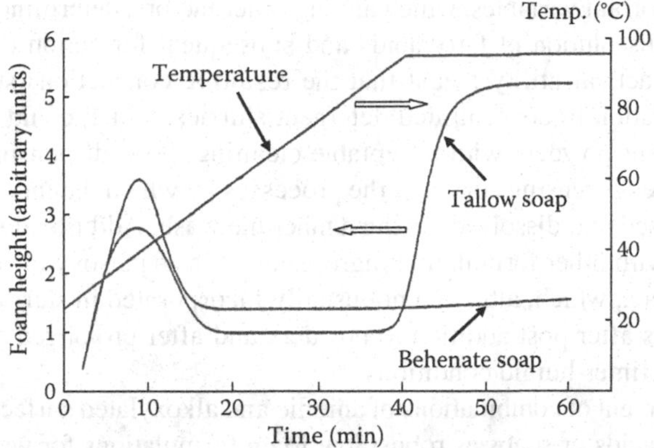

FIGURE 8.4 Comparison of antifoam behavior of tallow (C_{16-18}) and behenate (C_{20-22}) soaps in front-loading drum-type washing machine with clean load and wash cycle heating up to 95°C. Detergent concentration: 8.5 kg m^{-3} in 2.85 mM Ca^{2+}. Detergent formulation: 8 wt.% linear alkylbenzene sulfonate, 4 wt.% non-ionic surfactant, 16 wt.% zeolites, 22 wt.% sodium tripolyphosphate, 3.5 wt.% soap. Foam height at full porthole = 4. (After Ferch, H., Leonhardt, W. Foam control in detergent products, in *Defoaming, Theory and Applications*, First Edition, Garrett, P.R. ed., Marcel Dekker, Surfactant Science Series Vol 45, Chpt 6, p 221.)

cycle is apparent. It is probable that formation of effective calcium soap antifoam requires the formation of finely dispersed calcium soap precipitates. This will be facilitated by dissolution of the sodium soap before reaction with calcium. The low solubilities of these soaps at low temperature will inhibit such a process leading to the observed foam peak. However, the solubility product of calcium soaps increases strongly with increasing temperature (from ~1.3×10^{-16} at 50°C to ~2×10^{-14} at 80°C for calcium stearate and ~6×10^{-20} at 50°C to ~6×10^{-18} at 80°C for calcium eicosoate at an ionic strength of 0.1M [3]). Ferch and Leonhardt [2] suggest that in consequence, the calcium tallow soaps begin to dissolve as the temperature reaches 90°C and the antifoam effect is lost, whereas the precipitated calcium behenate soaps do not dissolve and continue to function. It is, however, difficult to reconcile that view with Equation 8.3 and the relative magnitudes of the complexation constant for sodium tripolyphosphate given in reference [4] together with the solubility products of the relevant calcium soaps (unless the former becomes increasingly temperature dependent as the temperature increases).

As in the example of the detergent formulation used to obtain the foam profiles shown in Figure 8.4, early attempts to develop practical washing powder formulations suitable for foam-intolerant, front-loading drum-type laundry washing machines utilized the antifoam effect of calcium soaps in mixed anionic and ethoxylated alcohol surfactant systems. The approach appears to have been entirely empirical. The foam behavior of ternary mixtures of anionic surfactants, such as sodium alkylbenzene sulfonates, various ethoxylated alcohols, and soaps, was optimized for low foamability in relevant machines by systematic trial. Obviously, polyvalent metal ion (especially calcium) activity as determined by temperature, water hardness, and builder type and concentration represented an additional variable. Wash temperature and pH were other important variables, which among other factors, determine the conditions necessary for dissolution of fatty acids and subsequent formation of calcium soap precipitates. Practical utility meant that the resulting concoctions should be capable of incorporation in concentrated detergent slurries, which could be spray-dried to form detergent powders with acceptable cleaning, flow, dispensing, and storage properties. Here dispensing concerns the processes by which the ingredients in powders are dispersed and dissolved during a machine wash. Additional factors concern compatibility with other formulation ingredients such as per-oxy bleach compounds and enzymes, etc., which, although not usually incorporated in slurries, can interact with surfactants after post-addition to powders and after prolonged periods of storage under sometimes humid conditions.

The development of combinations of anionic and alkoxylated surfactants in admixture with fatty acids or soaps as robust low foam formulations for washing machine application is exemplified by an early patent, filed more than half a century ago [6]. The basic technical proposition in this patent is illustrated by an example that is reproduced in Figure 8.5. Essentially, the foamability in a front-loading washing machine of a detergent composition, based on a sodium dodecylbenzene sulfonate as primary surfactant, is synergistically diminished by the addition of ethoxylated non-ionic surfactant and saturated fatty acids. In Figure 8.5, the amount of product that can be dissolved before a given foam height is realized is plotted against that foam height. Obviously, the greater the foaming tendency of the product, the less of it can be tolerated before a given foam height is realized. The non-ionic in the example was a polyethoxylated–polyoxypropylene (EO_nPO_m) of molecular weight 3000 and a cloud point of 58°C; however, the claims include ethoxylated nonyl phenol surfactants as

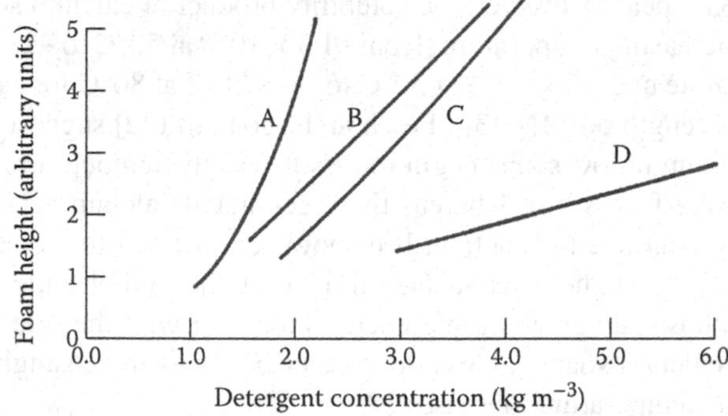

FIGURE 8.5 Effect of surfactant composition on concentration of detergent, which can be dissolved before a given foam height is realized in front-loading drum-type washing machine with "naturally soiled load" at 60°C with water hardness of 1.2 mM Ca^{2+}. Surfactant compositions: A, 16 wt.% linear alkylbenzene sulfonate (LAS); B, 16 wt.% LAS + 5 wt.% fatty acid (C_{16-18}); C, 12 wt.% LAS + 4 wt.% EO_nPO_m block copolymer; D, 12 wt.% LAS + 4 wt.% EO_nPO_m block copolymer + 5 wt.% fatty acid (C_{16-18}). Rest of each formulation made up to 100 wt.% with same proportions of silicates, sodium carbonate, and sodium sulfate. Foam height at full porthole = 5. (After Schwoeppe, E.A. (assigned to Procter & Gamble Co), US 2,954,348; 27 September 1960, filed 28 May 1956; and GB 808,945; 11 February 1959, filed 16 May 1957.)

having equivalent effectiveness. The wash temperature was 60°C, which is just above the cloud point of the non-ionic. However, it seems likely that the presence of the anionic surfactant will increase the cloud point as a result of comicellization and, if so, no cloud phase will have been formed. The fatty acid blend ranged from palmitic to behenic acid. The formulation also contained sodium tripolyphosphate as builder and significant proportions of alkaline inorganics. Water hardness was ~1.2 mM. It seems possible therefore that a combination of temperature, water hardness, and co-micellization would mean that at least some of the fatty acid blend would be solubilized as soap only to be subsequently precipitated as calcium soap despite the presence of the sodium tripolyphosphate builder. That calcium soap would be expected to act as an antifoam. This of course leaves the role of the non-ionic surfactant in contributing to the synergistic antifoam effect of fatty acid and ethoxylated non-ionic surfactant without explanation. It would, however, seem to be a manifestation of a wider generality. As we will describe in Section 8.2.5, the presence of ethoxylated non-ionic surfactants also significantly enhances the effectiveness in front-loading washing machines even of polydimethylsiloxane (PDMS)–hydrophobed silica antifoams [7]. This implies that the effect is unlikely to concern a unique aspect of the fatty acid (or calcium soap)–ethoxylated surfactant combination.

8.2.3 Use of Non-Soap Particulate Antifoams

Monoalkyl and dialkyl phosphoric acid esters also form precipitates at high pH with the ions of polyvalent metals, such as calcium, which are present in hard water. These precipitates exhibit antifoam effects that have been claimed as effective for the control of foam in textile washing machines [8–13]. Patent claims suggest that the alkyl phosphoric acid esters can be incorporated into detergent powders by spraying a solution of the esters in an ethoxylated alcohol onto a suitable carrier to form, for example, "noodles," which can be subsequently added to the detergent powder [8]. They may also be incorporated by introducing such solutions into the high-pressure line carrying detergent slurry to spray-drying nozzles [9]. The latter procedure would appear to be necessary to minimize exposure to highly alkaline detergent slurries at high temperatures. It is possible that these esters may hydrolyze and become ineffective if exposed to the conditions prevalent during the preparation of those slurries. In this respect, at least it would seem that soaps offer a more robust option.

The monoalkyl phosphoric acid esters have been described as offering superior foam control to the secondary alkyl esters in washing machines [8]. This presumably concerns the relative solubilities of the calcium salts and possibly the contact angles of the precipitates. It has also been suggested that suitable calcium monoalkyl phosphate precipitates exhibit superior antifoam effectiveness to that of calcium soap precipitates [2]. It is unlikely that such a conclusion reflects a thorough comparison under scientifically determined conditions where all relevant chain lengths of each type of compound, other formulation ingredients, water hardness, temperature, etc., are considered! Confidence is not inspired by the occasional reported use of soiled clothes in foam testing where the presence of variable amounts of soil antifoam may confuse the assessment of relative effectiveness [8]. Such practices have now apparently ceased after washing machine manufacturers started to test detergent powders in their machines with clean rather than soiled clothes. The recommendations of such manufacturers are, of course, of some relevance for the marketing of detergent products!

Reliance on the reaction of hardness ions with soap or alkyl phosphate ions to form calcium soap or calcium alkyl phosphate precipitates for foam control has an obvious limitation in the case of soft water. This factor has encouraged a search for alternative materials that can also form finely divided hydrophobic particles. Hydrocarbon waxes represent an obvious choice in this context.

Microcrystalline hydrocarbon waxes have been shown to control the foam in washing machines [14, 15]. By comparison, waxes of polar materials, such as carnauba wax and beeswax, were totally ineffective. Results for a detergent powder formulation containing a mixture of an ethoxylated alcohol and sodium dodecylbenzene sulfonate together with other typical ingredients such as a builder and a bleach additive are shown in Figure 8.6. The results concern washes with soiled loads and therefore probably include more than usual variability as a consequence of the presence of the antifoam effects of variable amounts of soil. The waxes were first dissolved in the ethoxylated alcohol and then sprayed onto carrier granules to ensure their delivery into the wash liquor in a finely divided form. As we have described in Section 4.7.7, the waxes would be expected to lose antifoam effectiveness at temperatures where they completely liquefy because of the relative ineffectiveness of liquid drops due to the expected stability of the relevant pseudoemulsion films. Clearly, their effectiveness is significantly diminished at temperatures above the quoted melting "points." However, complete elimination of antifoam effectiveness is not apparent. This could concern either the width of the melting range with these commercial waxes or even the lower stability of the pseudoemulsion films in the case of these

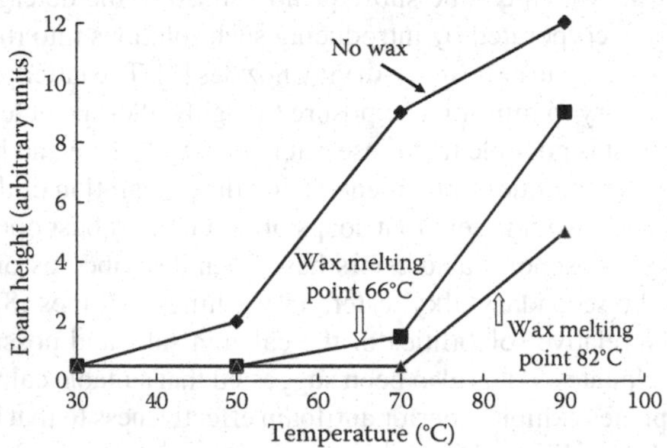

FIGURE 8.6 Effect of commercial microcrystalline hydrocarbon waxes on foam height during wash cycle with "realistically soiled load" in a "miniature drum" washing machine heating up to 90°C in 45 min. Formulation contained 12 wt.% commercial ethoxylated alcohol ($C_{14-15}.EO_7$) and 1 wt.% LAS as surfactants together with 1 wt.% wax if present, and was used at total concentration of 5 kg m^{-3}. (After Tate, J.R., McRitchie A.C. (assigned to Procter & Gamble CO.), GB 1,492,938; 23 November 1977, filed 11 January 1974.)

formulations overwhelmingly rich in an ethoxylated alcohol. We should note that the inventors report that replacing the ethoxylated alcohol with sodium dodecylbenzene sulfonate essentially eliminates the antifoam effect found with these waxes in this context. Again, we have a manifestation of the enhancement of antifoam effectiveness in the presence of ethoxylated surfactants.

Other particulate antifoams, which should function effectively in soft water, have been suggested as alternatives to calcium soaps. These have included monalkylamides [16, 17] and melamine derivatives [18–26]. The former can apparently be introduced directly into the detergent slurry before it is spray-dried. The latter is first emulsified using suitable surfactant whereupon it can either be directly sprayed onto the detergent powder or be sprayed onto a carrier, which can be subsequently incorporated.

8.2.4 Use of Hydrocarbon–Hydrophobic Particle Mixtures

8.2.4.1 Hydrocarbon Mixtures with Alkyl Phosphoric Acid Derivatives

The levels of particulate antifoam, exemplified by the soaps or fatty acid precursors used in formulations for effective foam control of detergents in front-loading automatic textile washing machines, are usually of the order of several percent by weight (typically about 4–5%; see reference [6]) even when significant proportions of ethoxylated surfactant are present. Such approaches have largely been replaced by recourse to antifoam mixtures of hydrophobic particles and oils as described in Section 4.8 and exemplified by numerous patents for many different applications. As we have seen, such mixtures exhibit synergistic antifoam behavior. They, therefore, permit effective foam control at much lower levels than is usual with soaps—typically <1% by weight of a detergent powder formulation. This provides space in the formulation for other ingredients and can lead to significant cost savings. As we will describe here, these advantages are partially offset by difficulties with incorporation deriving from deactivation during through-slurry processing, storage deactivation, and in some cases deactivation during wash cycles. The latter is probably a manifestation of the general problem with deactivation often found with some types of particle–oil antifoams after dispersal in a solution subject to continuous aeration as discussed in Chapter 6. Difficulties with dispensing and dispersal of detergent powders can also be induced by this type of antifoam.

Here, we consider mixtures of hydrocarbon oils and hydrophobic particles. We also include consideration of precursors of such mixtures. This includes mixtures where the hydrocarbon is a wax, which must first melt before it disperses. It also includes mixtures where a precursor of the hydrophobic particle is present, such as a fatty acid, a water-soluble soap, or an alkyl phosphate acid ester, which under certain circumstances can react at the oil–water interface to form hydrophobic particles of calcium soap [27] or calciumalkyl phosphate.

That mixtures of alkyl phosphoric acid esters and mineral oils produce synergistic antifoam effects when incorporated into detergent formulations was apparently first described by Hathaway and Heile [28]. Mixtures of mineral oil and stearyl acid phosphate were particularly effective. Carter [29] subsequently demonstrated that the effect required the presence of calcium in the aqueous phase. Conversion of the alkyl phosphoric acid ester to a precipitate of calcium alkyl phosphate and subsequent admixture with a mineral oil reproduced the synergistic effect without the need for calcium ions and high pH in the aqueous phase [30–31]. Such mixtures were shown to be effective antifoams for detergent powders in the context of front-loading textile washing machines [30–31].

The antifoam behavior of mixtures of liquid paraffin and calcium stearyl acid phosphate for an aqueous solution of a commercial linear alkylbenzene sulfonate by hand shaking of cylinders is shown in Figure 8.7 where marked synergy is clearly apparent. That the calcium alkyl phosphate particles adhere to the oil–water interface in this case is shown in Figure 4.77. Clearly, therefore, any attempt to incorporate this type of antifoam in a detergent powder must ideally ensure that the two components of the antifoam remain in intimate admixture. The foam profile in a washing machine achieved by simply adding the antifoam components separately to a detergent powder is compared with that achieved by adding an intimate mixture of a commercial calcium alkyl phosphate and liquid paraffin in Figure 8.8. Foam control is clearly inadequate if the antifoam components are added separately. However, the decline in the foam profile toward the end of the wash cycle in the case of separate addition of antifoam components suggests that some heterocoalescence has occurred. Use of the alkyl phosphoric acid ester precursor in similar experiments, not surprisingly, yields the same conclusion provided calcium ions are present in the wash liquor. That heterocoalescence can occur with other types of particle is also revealed in Table 4.9, where it was observed in the case of hydrophobed silica and liquid paraffin.

We have discussed the issue of deactivation by disproportionation of PDMS–hydrophobed silica antifoams during foam generation in some detail in Chapter 6. By contrast, there seems little evidence that mixtures of hydrocarbons and either calcium alkyl phosphates or alkyl phosphoric acid esters deactivate during washing

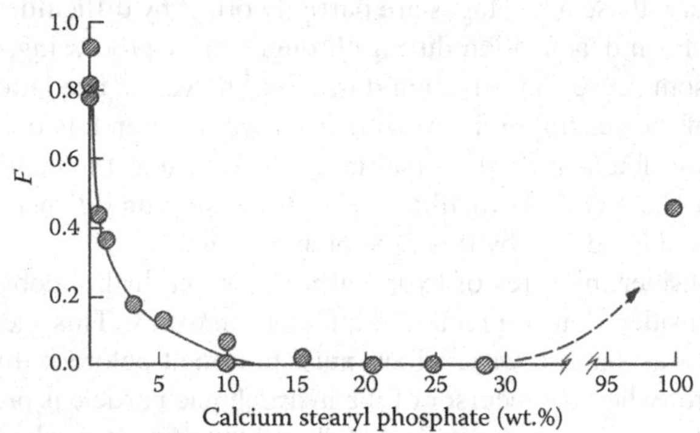

FIGURE 8.7 Antifoam effectiveness F = (volume of air in foam with antifoam/volume of air in absence of antifoam) of calcium mono-stearyl phosphate–liquid paraffin mixtures as function of proportion of calcium mono-stearyl phosphate. Aqueous surfactant solution: 0.5 kg m^{-3} commercial sodium alkyl (C_{10-14}) benzene sulfonate. Concentration antifoam: 1.2 kg m^{-3}. Foam generated by cylinder shaking at ambient temperature (22 ± 2°C). (Reprinted from *Colloids Surf. A*, 85, Garrett, P.R., Davis, J., Rendall, H.M., 159. Copyright 1994, with permission from Elsevier.)

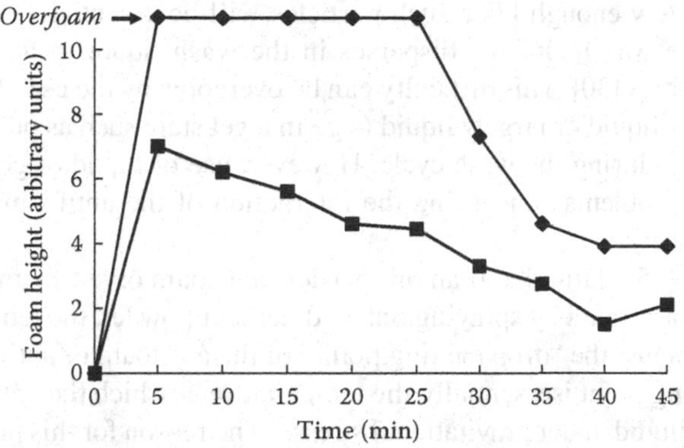

FIGURE 8.8 Effect of separate addition of calcium mono-alkyl (C_{16-18}) phosphate antifoam particles and liquid paraffin oil on foam profile in machine wash. ■, Intimate mixture of calcium mono-alkyl stearate–liquid paraffin (weight ratio 1:4) added directly to detergent powder, mixed, and immediately dispersed in wash liquor. ◆, Calcium mono-alkyl stearate and liquid paraffin added separately to detergent powder and immediately dispersed in wash liquor. Detergent powder composition: commercial LAS, 7.5 wt.%; ethoxylated alcohol, 2.5 wt.%; antifoam, 1.2 wt.%; various inorganics, etc., up 100 wt.%. Concentration of detergent powder in wash liquor: 6.3 kg m^{-3}. Wash conditions: Miele 429 machine, 95°C main wash, clean load, foam height at full port hole = 10. (After Yorke, J.W.H., Garrett, P.R., Giles, D. unpublished work.)

machine cycles. Indeed, observation of the antifoam effect of mixtures of the latter dispersed in an aqueous solution of a mixture of sodium dodecylbenzene sulfonate and an ethoxylated alcohol revealed no deterioration after several hours of continuous foam generation at 90°C using a circulating Ross–Miles apparatus [33].

As we have seen, the available evidence suggests that incorporation of alkyl phosphoric acid esters alone into the concentrated detergent slurries prepared for spray drying deactivates whatever antifoam action they may possess. It seems likely therefore that a similar fate awaits direct incorporation of mixtures of hydrocarbons and such esters in detergent slurries. Experience is consistent with that view. Therefore, most of the relevant patents describe a preferred option of incorporation of such mixtures directly by spraying onto powders or carriers and subsequently adding the carrier to the powder [30, 31].

It is claimed that the physical state of suitable hydrocarbons for alkyl phosphoric acid ester–hydrocarbon antifoam precursor mixtures can range from liquids to waxes and gel-forming hydrocarbons such a petroleum jelly [30, 31]. It is obvious from the description of the mode of action of mixed hydrophobic particle–oil antifoams described in Chapters 4 through 6 that the antifoam entities should be effectively dispersed as liquid drops of a few microns diameter with the particles adhering to the oil–water interface. Incorporation of, say, a wax–alkyl phosphate acid ester mixture by, for example, spraying a molten mixture directly onto a detergent powder (or onto a carrier material such as soda ash) produces macro-drops, which will rapidly freeze to form particles of order of a few hundred microns diameter if the relevant temperature is low enough [34]. Such particles will be essentially completely ineffective until the wax melts and disperses in the wash liquor to form the required oil–particle entities [30]. This difficulty can be overcome by the use of hydrocarbons, which are either liquid or largely liquid (e.g., in a gel state such as petroleum jelly) at all temperatures during the wash cycle. However, use of liquid or gel hydrocarbons presents other problems concerning the interaction of the antifoam with detergent powder.

Curtis et al. [35] claim that if an oil–particle antifoam or antifoam precursor is to be directly incorporated by spraying onto a detergent powder, the temperature of the latter must be below the "drop melting point" of the antifoam or antifoam precursor. The drop melting point isessentially the temperature at which the antifoam becomes a free-flowing liquid under gravitational stress.* The reason for this provision derives from a significant insight. Consider, for example, an experiment where a detergent powder (or even a typical inorganic ingredient such as sodium tripolyphosphate) is contained in a vessel similar to that shown in Figure 8.9 where the powder is held at a temperature significantly above the drop melting point of analkyl phosphoric acid ester–hydrocarbon antifoam precursor. It has been shown that if the molten antifoam precursor is poured onto such a column of detergent powder, a "chromatographic" process occurs where the hydrocarbon drains through the column at a much faster rate than the ester. The eluate is essentially free of the alkyl phosphoric acid ester for columns containing both detergent powders and sodium tripolyphosphate. The process effectively separates the hydrocarbon from the alkyl phosphate, which will necessarily lead to loss of antifoam effectiveness as shown in Figure 8.8. This example of course concerns the specific case of an antifoam precursor that is soluble in the hydrocarbon. In the case of a hydrocarbon–hydrophobic particle antifoam where the latter is insoluble in the former, then it is possible that separation may also occur as a result of interaction with detergent powder if the hydrocarbon is liquid. However, the mechanism is likely to be different so that it resembles filtration where, for example, excluded volume effects complicate the flow of a particle–oil mixture through a porous bed.

Obviously then, any situation where an alkyl phosphoric acid ester–hydrocarbon mixture is liquid or even partially liquid will lead to some measure of selective migration into the powder and therefore antifoam deactivation. Such situations could occur during periods of storage of the powder at elevated temperatures or when the antifoam is sprayed directly onto hot powder immediately after production from a spray-drying tower, as described by Curtis et al. [35]. The antifoam exemplified in this patent consisted of a mixture of a commercial (C_{16}–C_{18}) alkyl phosphoric acid monoester and a petroleum jelly, which is a hydrocarbon gel at ambient temperatures with a carbon number >25. Mixing petroleum jelly with the phosphoric acid monoester in the preferred ratio of 1:3 by weight produces an effective antifoam precursor with a drop melting point of ~58°C. Curtis et al. [35] describe both full-plant scale

* The drop melting point is the temperature at which a material (usually a wax) becomes sufficiently fluid to drop from the thermometer used for the determination under controlled conditions (specified in ASTM D127-08 "Standard Test Method for Drop Melting Point of Petroleum Wax, Including Petrolatum ("petroleum jelly")).

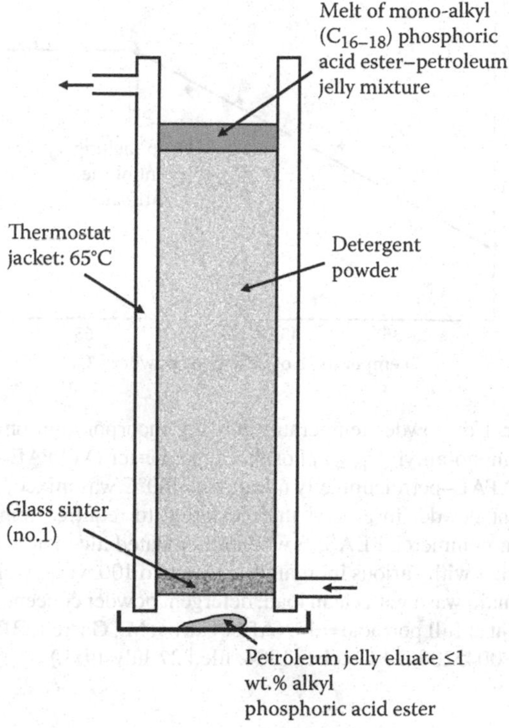

FIGURE 8.9 Schematic illustrating column for demonstrating separation of mono-alkyl (C_{16-18}) phosphoric acid ester (MAPAE) from molten mixture with petroleum jelly (of nominal composition 80 wt.% petroleum jelly). Detergent powder composition is same as described in caption to Figure 8.8. Height of column is ~30 cm and was thermostatted at 65°C. ^{31}P-NMR analysis indicated initial MAPAE concentration of 17.3 wt.% against nominal 20 wt.% and eluate concentration of only 0.8 wt.%. Similar experiment using only sodium tripolyphosphate powder produced an eluate containing only ~0.3 wt.% MAPAE. (After Giles, D., Garrett, P.R., unpublished work.)

and laboratory-scale experiments, which clearly reveal the advantage of adding the antifoam to detergent powder at temperatures significantly below thattemperature. We illustrate their findings in Figure 8.10, where the end-of-wash foam height measured in a front-loading textile washing machine is plotted against the temperature of the detergent powder to which the antifoam was added. Continuous improvement in antifoam performance is seen as the temperature of the powder is decreased. Such a continuous improvement appears to correlate with increases in solid content of the antifoam precursor with decreasing temperatures as revealed by nuclear magnetic resonance (NMR) measurements where, however, even at 22°C the mixture is still ~65% liquid [36]. That the end-of-wash foam control for powders at temperatures above the drop melting point is constant suggests total deactivation due to complete separation of the phosphoric acid ester from the hydrocarbon. However, some residual foam control must be present because otherwise, overfoaming would occur

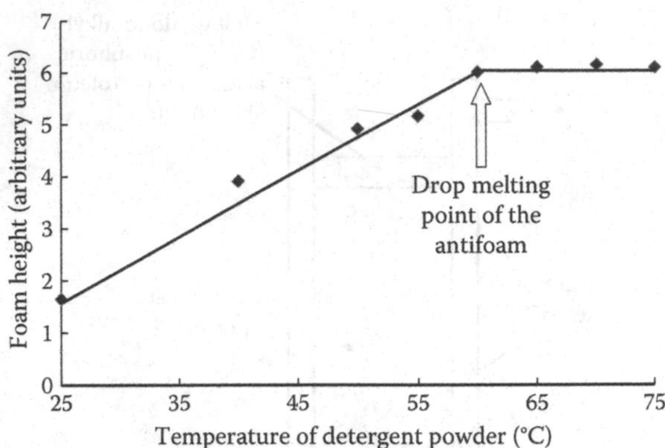

FIGURE 8.10 Effect of powder temperature during incorporation on antifoam effectiveness of mixtures of mono-alkyl (C_{16-18}) phosphoric acid ester (MAPAE)–petroleum jelly (of weight ratio 1:3). MAPAE–petroleum jelly mixture at ~80°C was mixed, using domestic food mixer, with detergent powder in a bowl thermostatted to required temperature. Detergent powder composition: commercial LAS, 6 wt.%; ethoxylated alcohol, 4 wt.%; antifoam precursor, 1 wt.%, together with various inorganics, etc., up to 100 wt.%. Wash conditions: Miele 429 machine, 95°C main wash with clean load, detergent powder concentration 9.4 kg m^{-3}, 2.4 mM Ca^{2+}, foam height at full porthole = 10. (After Curtis, M., Garrett, P.R., Mead J. (assigned to Unilever Ltd.), EP0045208; 24 October 1984, filed 27 July 1981.)

under end-of-wash conditions. This residual control could result from the intrinsic antifoam effect of calcium alkyl phosphate particles combined with the possibility of heterocoalescence. A corollary of the plot shown in Figure 8.10 is of course that the higher the storage temperature (up to the drop melting point) of detergent powders containing directly incorporated alkyl phosphoric acid ester–hydrocarbon antifoam precursor, the greater the expected deterioration in foam control. Finally, we should note that mixtures of calcium alkyl phosphate particles and hydrocarbons exhibit essentially the same behavior as the alkyl phosphoric acid ester–hydrocarbon mixtures with respect to deactivation upon incorporation in detergent powders as a function of powder temperature [36]. This result implies that separation of particles from oil can also occur if these mixtures are directly incorporated into detergent powders. Direct evidence, using, for example, the experiment shown in Figure 8.9, is, however, lacking.

8.2.4.2 Hydrocarbon Mixtures with Non-Phosphorous-Containing Organic Compounds

It is known that fatty acids and soaps act as effective promoters for hydrocarbon oil-based antifoams [27]. Such mixtures exhibit the usual synergistic effect and have, not surprisingly, been claimed as antifoams and equivalent precursors for detergent powders. Mixtures of hydrocarbons and either fatty acids or soaps are, for example, claimed by Schweigl and Best [37] to be suitable for front-loading drum-type washing machines. However, the claims are exemplified by use of combinations of soaps and hydrocarbons where these materials are incorporated into a detergent powder by separate addition into a detergent slurry prepared for spray drying. The proportion of soap in the combination varies from 25 to 67 wt.%. This composition implies absence of knowledge of the nature of the synergy expected with oil–particle mixtures where the proportion of soap would be expected to be ≤25 wt.% (see Tables 4.A1 through 4.A3). It is, however, possible that the high proportions of soap or fatty acid claimed favor heterocoalescence of the separately added components in the wash liquor. Nevertheless, it seems unlikely that optimal antifoam effectiveness would be realized by separate through-slurry incorporation of the hydrocarbon and the soap. It is also known that mixtures of petroleum jelly and a commercial long-chain fatty acid blend (C_{16}, 10%; C_{18}, 35%; C_{20} + C_{22}, 50%) should be sprayed onto detergent powder at temperatures significantly lower than the drop melting point of the mixture in a manner exactly analogous to that found with mixtures of alkyl phosphoric acid ester–petroleum jelly [36]. Again, this finding suggests, by analogy, that separation of the fatty acid and hydrocarbon diminishes effectiveness. However, the effectiveness of this fatty acid–petroleum jelly mixture is significantly inferior, in a 95°C machine wash, to that of the (C_{16} –C_{18}) alkyl phosphoric acid monoester–petroleum jelly mixture claimed by Curtis et al. [35] where both antifoam precursors are incorporated at ambient temperature [36].

Mixtures of ethylene and methylene distearamide with hydrocarbons have also been claimed to be effective particle–oil antifoams for control of detergent foam in front-loading textile washing machines [38]. Such mixtures represent well-known antifoams for general application (see, e.g., Table 4.A3). They may be prepared by milling the mixtures or by cooling a melt of the mixtures (since the solubility of distearamides at ambient temperatures is low) to form particulate dispersions in hydrocarbons. The particles of these alkylene distearamides are intrinsically hydrophobic [39] and have the properties necessary for rupture of the relevant pseudo-emulsion films (see Chapter 4). Unlike the soaps and alkyl phosphoric acid esters, there is no requirement for interaction with any ingredient, such as water hardness, present in the wash solution.

Preferred embodiments of this patent [38] include both use of petroleum jelly as hydrocarbon and incorporation, by direct spraying, of a melt of the intimate mixture of the hydrocarbon and the distearamide directly onto detergent powder. Alternatively, a melt may be sprayed onto a suitable porous inorganic carrier material to form an adjunct, which can be pose-dosed onto the powder. Evidence is presented of deactivation of the antifoam if it is added directly to the detergent slurry prepared for spray drying of a powder. The presence of finely divided ethylene distearamide particles appears to significantly structure hydrocarbon liquids. The drop melting point of petroleum jelly is, for example, increased by almost 20°C if ethylene distearamide is present at a proportion of 25 wt.%. This suggests enhanced stability of the antifoam when present in detergent powders at elevated temperatures.

Mixtures of hydrophobed silica and hydrocarbons represent another common class of synergistic oil–particle antifoams (see Table 4.A2), which have been the subject of a number of claims in the context of control of detergent foam in front-loading drum-type textile washing machines [40–44]. The silicas considered in this context may be hydrophobed using a variety of methods, including reaction with alkyl chlorosilanes and even long-chain alcohols [45]. Use of the so-called *in situ* hydrophobing by treating silica with PDMS oils [2, 46] tends, however, to be confined to PDMS-based antifoams where hydrocarbon is absent.

Preferred incorporation methods involve spraying the antifoam either directly onto a detergent powder or onto a suitable carrier. They do not include direct addition to the detergent slurry prepared for spray drying of a powder. This implies deactivation of antifoam effectiveness as a result of exposure of the hydrophobed silica–hydrocarbon mixture to an environment that involves the presence of concentrated surfactant, alkaline salts, and high temperatures of ≥70°C. Since it is known that the aqueous solubility of silica increases markedly with temperature and pH [47], it is possible that the silica may actually dissolve in the slurry!

Again, it has been shown that antifoam effectiveness is significantly diminished if the hydrophobed silica and hydrocarbon are separately dispersed in solution [36]. Moreover, mixtures of hydrophobed silica particles and hydrocarbon when sprayed onto detergent powder at temperatures significantly above the drop melting point are partially deactivated much as observed with, for example, mixtures of calcium alkyl phosphate particles and hydrocarbons [36]. Obviously then, the same problems of deactivation during incorporation and storage may occur with this type of hydrocarbon–particle antifoam as found with other antifoams. In consequence, Atkinson and Ross [41] claim the use of hydrocarbon waxes together with hydrophobed silica and an ethoxylated alcohol dispersant as antifoams for detergent powders. Use of a wax means that such antifoams are essentially in a solid state when interacting with the detergent powder during storage (or even when directly sprayed onto the powder provided the latter is "cool"). It does, however, also mean that the antifoam will be ineffective until the wash temperature reaches the melting temperature of the wax. Atkinson and Ross [41] argue that this provides an advantage for low temperature hand washing where high foam levels are appreciated by consumers. A simultaneous publication [42], however, reveals an approach to incorporation of a hydrophobed silica–hydrocarbon mixture where the hydrocarbon is a blend that is >70 wt.% liquid paraffin. This predominantly liquid hydrocarbon means that the antifoam is effective at ambient temperatures. Blending the antifoam as a melt with a mixture of a "compatibilizing agent" such as urea and a highly ethoxylated alcohol serves to prevent selective migration of the liquid hydrocarbon into the detergent powder and therefore partial deactivation. It is claimed that this rather complex concoction could be prepared in the form of noodles and directly added to the detergent powder or alternatively sprayed onto a suitable carrier, which could also be directly added to the detergent powder.

Descriptions of other approaches to the problem of providing both adequate low-temperature foam control and stability of the effectiveness of hydrophobed silica–hydrocarbon antifoams after storage in detergent powders are to be found in the patent literature. Ho Tan Tai [43], for example, teaches that use of hydrocarbon blends consisting of relatively high proportions of wax can be used provided the antifoam is sprayed onto suitable carriers such as gelatinized starch and sodium perborate monohydrate. Typical foam profiles in a front-loading drum-type textile washing machine are reproduced in Figure 8.11 for the case where the liquid hydrocarbon component is a white mineral oil. It is clear that a peak in foam height appears at the low-temperature part of the wash cycle. Presumably, this reflects the predominance

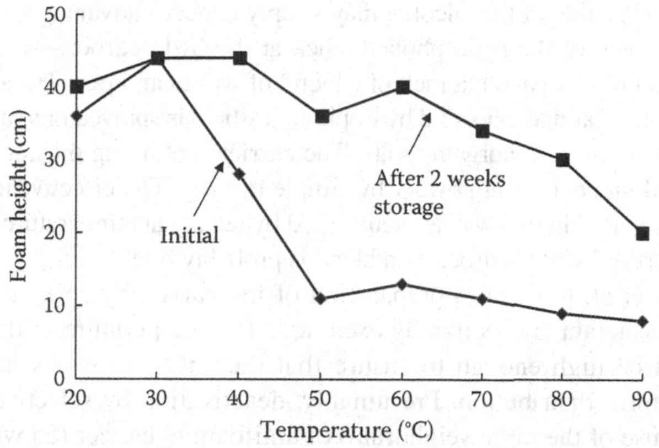

FIGURE 8.11 Foam profiles illustrating effect of storage on effectiveness of hydrocarbon wax + mineral oil–hydrophobed silica antifoam sprayed onto granular sodium perborate monohydrate carrier. Detergent powder composition: commercial LAS, 9 wt.%; ethoxylated alcohol, 4 wt.%; antifoam granules, 1 wt.%, with various inorganics, etc., up to 100 wt.%. Antifoam granules: 50 wt.% carrier + 50 wt.% antifoam. Antifoam composition: white mineral oil, 11.3 wt.%; paraffin wax (melting point 40°C), 11.3 wt.%; paraffin wax (melting point 50–52°C), 21.4 wt.%. Wash conditions: Brandt 433 machine, 90°C main wash with clean load, 3.0 mM Ca^{2+}. Detergent powder storage conditions: 37°C, 60% relative humidity for 2 weeks. (After Ho Tan Tai, L. (assigned to Unilever PLC), EP 0109247; 23 May 1984, filed 8 November 1983.)

of solid wax in the hydrocarbon blend. The latter is presumably a solid–liquid mixture, the solid component of which is only effective at temperatures where it begins to melt and disperse into the wash liquor. There is, however, clear evidence of deactivation after storage at 37°C. At that temperature, it seems likely that a significant part of the hydrocarbon blend will be liquid to reflect a composition of ~25 wt.% white mineral oil, ~25 wt.% wax (melting point 40°C), and ~50 wt.% wax (melting point 50°C). Selective migration of the liquid hydrocarbon component would therefore seem to be the cause of such deactivation.

A feature of this patent of Ho Tan Tai [43] is the absence of a control experiment where the antifoam is directly sprayed onto cold detergent powder. This means that definite evidence for the effectiveness of the chosen carrier materials in facilitating dispersal of the antifoam, despite a high proportion of solid wax, is not revealed. It also means that there is no evidence either that the chosen carrier materials inhibit storage deactivation. Finally, we should note that the claimed hydrocarbon–hydrophobed silica antifoams are exemplified by the use of hydrocarbon blends containing either white mineral oil or spindle oil "Velocite 6" (manufactured by Mobile). According to the relevant product data sheet [48], the latter material contains a "defoamant"!

Yet another approach to incorporation of hydrocarbon–hydrophobed silica mixtures in detergent powders is described by Wuhrmann et al. [49]. A branched-chain alcohol is added to the antifoam to "enhance effectiveness and improve processability." The role of this alcohol may simply concern advantageous modification of the contact angle of the hydrophobed silica at the hydrocarbon–water interface. In one embodiment of this patent, a melt of a blend of waxes and petroleum jelly (together with the branched-chain alcohol and hydrophobed silica) is sprayed onto a porous carrier composed of a mixture of inorganic salts. The carrier containing the antifoam can then be incorporated in a detergent powder by simple mixing. The effectiveness of the antifoam system prepared in this way is exemplified by testing at temperatures ≥40°C where at least a proportion of the hydrocarbon blend is probably liquid.

Wuhrmann et al. [49] claim preparation of the carrier by spray drying a slurry from which surfactant is specifically excluded. The temperature of this spray-dried carrier should be high enough to ensure that the antifoam wicks into the carrier to form a uniform distribution. Presumably, deactivation by selective migration is minimal because of the high weight ratio of antifoam to carrier (≥5 wt.% antifoam). Formation of a uniform distribution implies that the antifoam wicks into the smallest pores in the carrier. This could help stabilize the antifoam against deactivation under high-temperature storage conditions where partial melting of the hydrocarbon occurs. Capillary action will mean that selective migration of the liquid component of the hydrocarbon into the detergent powder will be inhibited if the pore sizes in the carrier are smaller than those in the detergent powder.

Another embodiment of the patent of Wuhrmann et al. [49] is emulsification of the molten hydrocarbon–hydrophobed silica antifoam in the inorganic carrier slurry before spray drying. This slurry is claimed to contain high concentrations of sodium silicate. The rate of any potential dissolution of the hydrophobed silica may possibly be inhibited by the presence of the latter despite high temperature and pH.

Finally we note that comparison in Figure 5.5 of the antifoam effectiveness of the calcium salt of a commercial alkyl acid phosphate and a commercial hydrophobed silica (D17, ex-Degussa), both dispersed in liquid paraffin, suggests that the former may in general be potentially superior to the latter. Systematic comparison of hydrophobed silicas and the polyvalent metal salts of alkyl acid phosphates as antifoam ingredients in this context, where the chemistry, particle size, shape, and contact angles are all considered, has not been made and it is therefore not possible to form firm conclusions about the potential relative merits of either material. Indeed the difficulty of making such systematic comparisons means that the same conclusion applies equally to any of the particulate materials considered here.

8.2.5 Use of Polydimethylsiloxane-Based Antifoams

8.2.5.1 General Properties

Mixtures of polydimethylsiloxanes oils and hydrophobed silicas have formed the basis of many patents concerned with the control of detergent foam in front-loading drum-type textile washing machines. It is likely that this type of antifoam is actually included in many commercial detergent powders designed for that application.

These antifoams may be prepared by mixing preformed hydrophobed silica particles with PDMS oils. They may also be prepared by mixing untreated silica particles with these oils in the presence of a catalyst that facilitates an *in situ* reaction between PDMSs and the surface hydroxyl groups of silica. A useful brief summary of the relevant chemistry is to be found elsewhere in the review of Ferch and Leonhardt [2]. The use of alternative particles such as alkylene dialkamides, the polyvalent metal salts of alkyl phosphoric acid esters and polyvalent metal soaps, have also been claimed [50].

PDMS–hydrophobed silica antifoams are generally more effective than hydrocarbon–hydrophobed particle antifoams, giving adequate foam control at incorporation levels in powders often at significantly less than 1 wt.%. They are, however, not necessarily always more cost-effective. Successful application of these PDMS-based antifoams in the context of textile machine washing must, however, take account of certain basic properties. The first concerns continuous and relatively rapid deactivation in applications involving continuous air entrainment. We have considered this property in some detail in Chapter 6. Deactivation has been shown to be an inevitable process of disproportionation where splitting and coalescence of antifoam drops accompanying agitation and antifoam action result in the formation of a combination of inactive oil drops without particles and inactive drops and large agglomerates containing excessive concentrations of particles. This process has been studied in depth in the case of PDMS–hydrophobed silica antifoams; however, little attention has been paid to the possibility that it may occur in hydrocarbon-based antifoams. There is some limited evidence that it may, at least in the case of hydrocarbon–hydrophobed silica mixtures. However, as we have seen (see Sections 6.6 and 8.2.4.1), there is also some evidence that such deactivation is absent in the case of hydrocarbon–alkyl phosphoric acid esters. The antifoam effect, for example, of an intimate mixture of calcium mono-alkyl phosphate–liquid paraffin on the foam profile depicted in Figure 8.8 reveals little evidence of deactivation after a 45-min wash cycle. This finding suggests a tentative conclusion that it is a property of the hydrophobed silica that determines the deactivation, although mixtures of ethylene distearamide and PDMS also show deactivation [50].

Increasing the viscosity of the PDMS oil reduces the rate of deactivation during air entrainment (see Section 6.5). However, the rate of dispersal of the antifoam is also diminished. A comparison of the foam control achieved with a PDMS–hydrophobed silica of viscosity ~2.5 × 10^3 mPa s with one of ~55 × 10^3 mPa s (using the same hydrophobed silica) in a textile washing machine is shown in Figure 8.12 by way of example. Here the antifoams were simply directly inserted into the textile load so that issues concerning interaction with detergent powder were avoided. The low-viscosity antifoam is seen to be relatively effective at the beginning of the wash cycle but rapidly deactivates. By contrast, the high-viscosity antifoam is less effective at the start of the cycle but shows little evidence of deactivation during the course of the cycle. It is likely, however, that the advantage of the high-viscosity antifoam with respect to deactivation is diminished if it is predispersed as finely divided drops. For example, Sawicki [7] reports foam profiles with granules containing “very high viscosity” PDMS-based antifoams predispersed in an organic “binder” that clearly exhibit significant deactivation during a typical wash cycle as illustrated in Figure 8.13. However, a detailed characterization of the antifoam (including measurements of the viscosity) is absent in this work, which significantly diminishes its contribution to our understanding of this issue.

A second basic property concerns incorporation in detergent powders. As with hydrocarbon–hydrophobic silica antifoams, incorporation in the detergent slurry

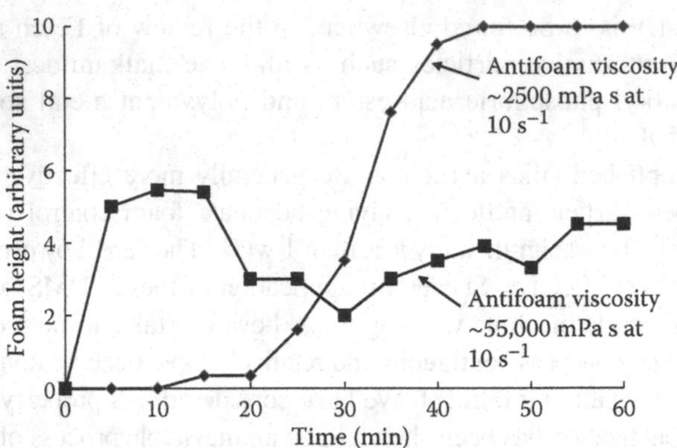

FIGURE 8.12 Foam profiles illustrating effect of increasing PDMS viscosity (and therefore molecular weight) on rate of deactivation of PDMS–hydrophobed silica antifoams. Detergent powder: LAS, 9 wt.%; ethoxylated alcohol, 4 wt.%; antifoam, 0.5 wt.%, with various inorganics, etc., up to 100 wt.%. Antifoams: 87.5 wt.% PDMS blends to yield relevant viscosity +10 wt.% hydrophobed precipitated silica (D10, ex Degussa) + 2.5 wt.% polyoxyalkylene–polyorganosiloxane copolymer as dispersion aid (DC 190, ex Dow Corning), antifoams added directly to wash load. Wash conditions: Miele 756 machine, 40°C main wash with clean load, detergent powder concentration 12.5 kg m^{-3}, antifoam concentration 0.0625 kg m^{-3}, 2.4 mM Ca^{2+}, foam height at full porthole = 10. (After Yorke, J.W.H., unpublished work.)

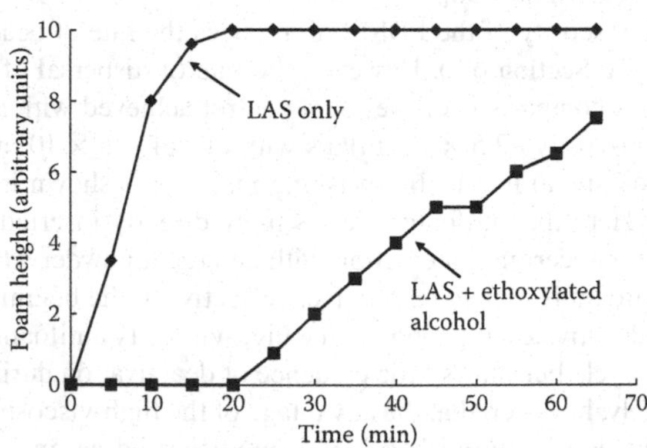

FIGURE 8.13 Effect of addition of ethoxylated alcohol to LAS-based detergent solution on foam profile with "very high viscosity" PDMS-based antifoam granule. Solutions: ◆, LAS 1.5 kg m^{-3}, PDMS-based antifoam 0.035 kg m^{-3}. ■, LAS 1.5 kg m^{-3}, C_{12-13}. EO_5 (branched oxo-alcohol, ex Sasol) 0.6 kg m^{-3}, PDMS–hydrophobed silica antifoam 0.035 kg m^{-3}; all made up with zeolites and other inorganics. Antifoam granule: PDMS–hydrophobed silica dispersed in organic "binder" and granulated with inorganic carrier. Wash conditions: Miele W377 machine, 40°C main wash with clean load, 1.3 mM Ca^{2+}–Mg^{2+}, foam height at full porthole = 10. (After Sawicki, G.C., *Colloids Surf. A*, 263, 226, 2005.)

before spray drying results in rapid and total deactivation of the antifoam [52] for reasons that may even concern the chemistry possible in that environment. Addition of a PDMS–hydrophobed silica antifoam directly to a detergent powder also results in a gradual deactivation of the antifoam [52, 53]. Sawicki [53] has suggested that this may be a consequence of migration of the PDMS oil into the detergent powder so that smaller non-optimal drop sizes of antifoam are formed on dissolution of the powder in the wash liquor. No direct evidence has, however, been presented in favor of this hypothesis. Denkov et al. [54] and Marinova et al. [55] have, on the contrary, shown convincingly that deactivation during air entrainment does not concern decreases in drop size but rather concerns disproportionation. It seems likely then that deactivation in detergent powders indeed involves migration of the oil but that it also leads to segregation of the oil from the particles in a process essentially analogous in outcome to that which occurs with dispersed antifoam drops during air entrainment. A similar argument has been used by Berg et al. [56]. We have of course made that argument for the case of hydrocarbon-based antifoams and have presented direct evidence in the case of mixtures of petroleum jelly and alkyl phosphoric acid esters (see Figure 8.9). Unfortunately, it is not possible in the case of PDMS-based antifoams to resort to the equivalent of waxes that are immobile in the powder but melt in the wash liquor. It is therefore necessary to encapsulate PDMS antifoams if they are to be successfully incorporated in detergent powders.

As we have seen, the presence of ethoxylated non-ionic surface-active compounds can enhance the susceptibility of the foam of solutions of anionic surfactants to antifoam. This appears to be a general phenomenon that is also manifest with PDMS–hydrophobed silica antifoams in wash cycles with drum-type, front-loading, textile washing machines. This well-known effect is exemplified in Figure 8.13 where the addition of ethoxylated alcohols is seen to diminish the foam profile of solutions of sodium alkylbenzene sulfonate (LAS) in the presence of PDMS–hydrophobed silica antifoam. Sawicki [7] has shown that the effect of these ethoxylated compounds does not concern either the precipitation of cloud phase drops (see Section 4.6.3.2) or marked changes in dynamic or equilibrium air–water solution surface tensions. One possible explanation could concern a putative inhibiting effect of ethoxylated compounds upon the rate of PDMS–hydrophobed silica antifoam deactivation. However, this would afford no explanation for the effect of those compounds on the antifoam action of hydrophobic precipitates where no oil is present (see Section 8.2.2).

Sawicki [7] notes that these ethoxylated compounds appear to coadsorb with anionic surfactants at the air–water surface and speculates that this could produce a reduction in the stability of the relevant oil–water–air pseudoemulsion films. There have, however, apparently been no relevant experimental studies of this possibility. Moreover, since the antifoam enhancement effect is also manifest with particulate precipitates in the absence of oil, the generality of both phenomenon and explanation would have to imply an equivalent effect of the ethoxylated compounds on the stability of particle–water–air films. Clearly we have an issue that demands attention!

8.2.5.2 Storage Deactivation and Incorporation in Detergent Powders

A list of early patents concerning the incorporation of PDMSs in detergent powders designed for the washing of textiles in drum-type front-loading machines is given in Appendix 8.1, Table 8.A1. The list includes brief descriptions of the nature of the relevant inventions. It is, however, not comprehensive but rather illustrative of the approaches adopted to overcome the difficulties that are intrinsic to the use of these materials. Although many of the patents make claims for antifoams prepared from mixtures of hydrophobed silica and polyorganosiloxanes in general, it is usually clear from the quoted examples that any practical application is likely to concern specifically PDMSs. Not surprisingly, no patents claim addition of the antifoam directly to detergent slurries before they are spray dried because it is generally known that this procedure totally deactivates the antifoam.

By far the majority of the patents summarized in Appendix 8.1, Table 8.A1, are concerned with preserving the integrity of PDMS–hydrophobed silica antifoams after incorporation in detergent powders. Such powders often have to be stored for significant periods before use, sometimes under adverse conditions of high temperature and humidity. Of particular concern then is the integrity of the antifoam under those storage conditions. All approaches involve some method of post-dosing the antifoam onto detergent powder already prepared by, for example, spray drying. If migrational segregation is to be avoided, this means immobilizing the antifoam when under storage conditions in a detergent powder without at the same time preventing release and dispersal in the wash liquor to produce the desired foam control. A popular version of this approach is to immobilize the PDMS–hydrophobed silica mixture as an emulsion in a molten matrix of a water-soluble material that has a melting temperature above the temperatures likely to be experienced during detergent powder storage [52, 57, 65, 67–69]. Examples of suitable matrix-forming materials include ethoxy compounds such as polyethylene oxides [52, 57, 65] and highly ethoxylated alcohols [52, 57]. The preparation of such emulsions can be facilitated by the use of appropriate alkoxysiloxane compounds that are surface active at the PDMS interface with the ethoxy matrix-forming compound [61]. Inclusion of the latter in the overall formulation may actually enhance the effectiveness of the antifoam as shown by Sawicki [7]. Simply spraying the emulsion in a molten form onto a detergent powder risks deactivating the antifoam if the emulsion breaks during such a process. Solidification of the matrix before addition to the powder by, for example, spray drying the emulsion to form granules presents other problems. The effectiveness of PDMS–hydrophobed silica antifoams is such that the amount of such granules is likely to be so low that homogeneous mixing with the detergent powder will be difficult. This difficulty will probably be accentuated by the likely tendency of such granules to aggregate. That tendency will also lead to handling difficulties—the granules will not readily form a free-flowing powder. All these difficulties are usually overcome by spraying the emulsion of PDMS–hydrophobed silica in the molten ethoxy compound matrix onto a suitable inorganic particulate material using a fluidized bed. Examples of suitable inorganic materials include sodium tripolyphosphate, soda ash, or a bleach component such as sodium perborate [52, 57, 67, 68]. The resulting free-flowing powders can be readily incorporated into detergent powders by direct mixing.

Other versions of this basic concept include the use of matrices of materials that simply melt and disperse in the wash liquor rather than dissolve. Examples include fatty acids [67], fatty alcohols [67], glycerol monoesters [68], etc. The difficulty with this approach is that antifoam action only occurs at the melting point of the matrix, the temperature of which should exceed the detergent powder storage temperature. Since storage tests often involve temperatures near 40°C, such an approach may prejudice foam control in the case of the increasingly popular 40°C machine wash.

Another variation of this theme is the use of matrix-forming materials that do not aggregate and do therefore form free-flowing powders that can be easily handled and directly post-dosed into detergent powders. The need for preparing granules based on an inorganic support is thereby obviated [65, 69]. For example, it is claimed by Appel et al. [69] that PDMS–hydrophobed silica antifoams can be emulsified in a molten matrix prepared from fatty acid–alkali metal soap mixtures [69]. The resulting emulsion can be spray cooled to produce "prills" suitable for incorporation into detergent powder.

It is claimed also that stabilization of a PDMS–hydrophobed silica antifoam against migrational segregation can also be achieved if the antifoam is directly sprayed onto porous inorganic supports of defined porosity and pore size range [66, 71]. The stability of the resulting granules under storage conditions is supposed achieved by a capillary pressure imbalance between the pores of the granule and those in the detergent powder. If the latter are larger than the former, the liquid PDMS cannot migrate away from the granule. This approach requires that the inorganic support does not exhibit any adverse chemical reaction with the antifoam.

Preparation of an aqueous slurry of suitable inorganic materials in which the antifoam is emulsified [58, 63], using polymers such as sodium carboxymethylcellulose and methylcellulose to stabilize the emulsion, represents another version of this approach. The inorganic materials include sulfates, carbonates, silicates, phosphates, etc. This slurry should avoid the presence of the usual suspects, which supposedly cause antifoam deactivation in through slurry processing—anionic surfactant and high alkalinity [57]. However, the inorganic materials claimed in the relevant patents include carbonates and silicates, which will obviously form alkaline slurries. Presumably, the spray-dried granules produced by this method will contain PDMS–hydrophobed silica absorbed in pores such that a capillary pressure imbalance will stabilize the antifoam against migrational separation after their incorporation in detergent powders. Reuter et al. [60] even claim that cospraying through separate nozzles of a detergent slurry and this PDMS–hydrophobed silica-rich slurry can produce detergent powders where the benefits of the latter are retained.

In yet another variant [57] of this concept, a trace of the antifoam is added to the inorganic slurry before spraying. The remainder of the antifoam is then added to the porous particles prepared in this way. Presumably, the addition of traces of PDMS–hydrophobed silica to the slurry represents an attempt to render the surfaces of the pores in the resulting powder hydrophobic in order to increase the capillary pressure imbalance and thereby increase stability when the granules are stored in detergent powders. The likely effect of hydrophobing the surfaces of high-energy salts such sodium sulfate on the contact angle of PDMS against such salts in air is, however, apparently not known.

Finally, it is worth noting that granules containing PDMS–hydrophobed silica mixtures should ideally also be of a similar particle size to that of the detergent powder into which they are incorporated. If the particle size of the granules is significantly larger than that of the detergent powder, any vibration of the complete product could result in segregation of the granules and inconsistent foam control. Such segregation is of course a manifestation of the well-known "Brazil nut" effect [76].

8.2.5.3 Dispensing

An occasional problem with the incorporation of PDMS–hydrophobed silica antifoams in detergent powders concerns dispensing. The delivery of detergent powders into the wash with front-loading washing machines is usually through a dispenser tray. Powder is poured into that tray and flushed out into the wash by means of a water spray. If this process is to be successful, it is necessary for the water to penetrate the detergent powder to effect rapid dispersal. That process is assisted by capillary action. It is, however, inhibited by the growth of surfactant mesophases into the interparticle pore space. Formation of such mesophases as a result of the mutual interdiffusion of water and surfactant is a feature of most surfactants [77]. If the dispersal of the powder is too slow, then the outer edge of the interparticle pore space can fill up with highly viscous mesophase. The presence of this layer prevents further penetration of water into the powder. High viscosity of the mesophase means that considerable force is necessary to break up this cohering layer, which can mean powder dispensing fails completely. Three properties of the detergent powder influence the possibility of such an outcome. Mesophase blocking of the interparticle pore space is clearly influenced by the interparticle porosity and the total surfactant loading. Attempting to both produce high-density detergent powders by decreasing the pore space and simultaneously increasing the surfactant content will produce a higher risk of dispensing failure. Lowering the ionic strength of solutions by substituting insoluble zeolites for sodium tripolyphosphate will also enhance mesophase formation and therefore the risk of such failure. However, another factor concerns the wettability of the powder. A simple experiment serves to reveal the role of wettability in this context. It has been shown by Garrett and Gratton [78] that the dispensing of silica gel particles from a washing machine dispensing tray is reduced by hydrophobing the silica with trimethylchlorosilane. Indeed, at an air–water–silica contact angle of 90°, water penetration and dispensing essentially cease, whereas with untreated hydrophilic silica, gel dispensing is complete.

It is of course well known that PDMS can be used to hydrophobe surfaces. It seems likely then that contamination of a detergent powder with PDMS will tend to diminish wettability by water. Iley [72] reveals that as little as 0.002 wt.% of PDMS of viscosity ~1000 mPa s is sufficient to cause unacceptable residues in dispenser trays in the case of detergent powders with high surfactant levels and an absence of sodium tripolyphosphate as builder. This finding quotes the work of White [81], which involved contaminating an inorganic salt with low levels of PDMS and mixing the salt with detergent powder, followed by dispensing tests using washing machines. The extreme sensitivity to low levels of PDMS revealed by this work suggests that PDMS contamination is essentially a surface phenomenon. If the analysis given here is correct, we would expect such sensitivity to be specific to detergent powders where small changes in wettability induced by PDMS are enough to slow down imbibition of water into the powder so that mesophase formation can block the interparticle pore space.

Approaches to solve this problem in the probably rare situations where it occurs include incorporation of effervescent materials in the detergent powder [80]. Inclusion of organic acids such as citric acid can react with any carbonate salts present to produce effervescence when the powder is wetted. This disrupts mesophase blocking of pores by producing extra pore space. Iley [72], on the other hand, has risen to the challenge of encapsulating a PDMS–hydrophobed silica antifoam so completely that PDMS contamination of a detergent powder does not occur even at the required extremely low levels. The granules prepared by absorbing PDMS–hydrophobed silica antifoams into an inorganic carrier as revealed by Garrett et al. [71] are used as the basis of this approach. After compaction, the resulting spheroidal granules are coated,using a fluidized bed, with a latex that is soluble or dispersible in aqueous alkaline solution. Dispensing residues were reduced by ≥75 wt.% relative to those provided by the untreated granules when using a detergent powder with high surfactant levels, containing no sodium tripolyphosphate and using zeolites as builder.

8.2.5.4 Enhancement of Antifoam Effectiveness

Surprisingly, somewhat less attention has been paid to enhancing the effectiveness of PDMS–hydrophobed silica antifoams in this context. Clearly any such enhancement should concern the marked deactivation depicted in Figure 8.12 for the "low"-viscosity PDMS–hydrophobed silica antifoams. Not only does that deactivation pose a risk of overfoaming at the end of a wash cycle but it also probably means, as discussed in Chapter 6, the formation of large silica-rich agglomerates that may deposit on machine parts and even textile loads.

Clearly the risk of overfoaming at the end of wash cycles will be reduced by increasing the viscosity of the antifoam. This appears to reduce the rate of deactivation of PDMS–hydrophobed silica antifoams at the expense of higher initial foam heights as depicted in Figure 8.12. Unfortunately, there would appear to be no reports concerning the effect of systematically increasing the viscosity of these antifoams on the foam profile throughout a wash cycle. However, Akay et al. [75] report a problem with incorporation of PDMS–hydrophobed silica antifoams with a "high" viscosity of ~3×10^4 mPa s in detergent powders using inorganic carriers. This antifoam actually deactivates rapidly in a matter of days after absorption into the intra- and interparticle pores of porous inorganic carriers such as soda ash and sodium monoperborate even if stored separately from detergent powders. This deactivation is also apparent with carriers prepared from finely divided sodium bicarbonate and sodium chloride, which have no intraparticle porosity. It is also apparent with PDMS–hydrophobed silica antifoams of "low" viscosity of ~3×10^3 mPa s but is significantly less marked [51]—otherwise, incorporation in detergent powders of such antifoams by use of inorganic carriers as described in the patent literature would be futile.

Similar observations of PDMS–hydrophobed silica antifoam deactivation on inorganic carrier materials have been reported in a pharmaceutical context. The antifoam performance of antiflatulent preparations with PDMS–hydrophobed silica mixed with basic antacid carriers, such as aluminum hydroxide and magnesium carbonate, has been shown to deactivate after the ingredients were either granulated or compressed into tablets [81, 82]. Analysis of the molecular weight distribution of

PDMS–hydrophobed silica antifoams, after storage in soda ash, revealed no change in molecular weight. This suggests that molecular cleavage at the Si–O bond in PDMS to produce lower molecular weight fragments of lower viscosity cannot provide an explanation for the phenomenon [79].

By contrast, absorption of PDMS–hydrophobed silica antifoam of viscosity ~3 × 10^4 mPa s onto organic carriers produced no deactivation of the antifoam when the resultant granules were stored separately from detergent powders [75]. A comparison of the antifoam effectiveness obtained with various inorganic and organic carriers after separate storage is given in Table 8.1. The relative stability of

TABLE 8.1
Effect of Carrier Type on Stability of High-Viscosity PDMS–Hydrophobed Silica Antifoam before Incorporation in Detergent Powder (Antifoam Viscosity ~30,000 mPa s)

Carrier[a]	Mean Particle Size (microns)	Interparticle Porosity	Intraparticle Porosity	Total Porosity	Storage Period (weeks)	Foam Height After 45 Min (arbitrary units)[b]
None, antifoam simply added to wash load directly	–	–	–	–	–	15[c]
Sodium carbonate	120	0.84	0.28	1.14	1	100
Sodium bicarbonate	80	0.36		0.36	1	100
Hydrated alumina		3.23		3.23	4	100
Microcrystalline cellulose	50	1.97	0.18	2.15	1	33[c]
Polyvinyl alcohol (mol. wt. 2000)	350	–	–	–	7	10
Gelatin	330	0.77	–	0.77	8	2
Urea	540	–	–		5	55
Wood flour	340	4.23	–	4.23	2	9

Source: Akay, G., Garrett, P.R., Yorke, J.W.H. (assigned to Unilever PLC), WO93/01269; 21 January 1993, filed 1 July 1992.

[a] With 25 wt.% PDMS–hydrophobed silica antifoam.

[b] Miele washing machine with clean load; 0.2 kg of detergent powder containing 0.5 wt.% PDMS–hydrophobed silica. PDMS–hydrophobed silica carrier particles added directly to wash load before starting the machine. Surfactant content: 9 wt.% anionic, 5 wt.% ethoxylated alcohol. For foam height, 100 = full porthole (i.e., overfoaming).

[c] After Yorke [51].

the antifoam after absorption onto the organic carriers is obvious. However, addition of the resultant granules to a detergent powder produces storage deactivation by migrational segregation. In consequence, practical application requires that the granules must be coated with another material to prevent that process. Suitable coatings are exemplified by organic polymers, polyethoxylated compounds, and mixtures of soap with fatty acid [75]. Gowland et al. [64], for example, claim that emulsions of high-viscosity (18×10^3 mPa s) PDMS–hydrophobed silica antifoams in highly ethoxylated tallow alcohol can be sprayed onto sodium tripolyphosphate particles in a fluidized bed without apparently resulting in deactivation either before or after the addition of the resulting granules to a detergent powder.

Addition of a secondary antifoam may represent another approach to mitigating the adverse effect of the deactivation of "low"-viscosity PDMS–hydrophobed silica antifoams by disproportionation during a wash cycle. Since wash cycles can finish at any temperatures from 30°C to 95°C, a secondary antifoam would have to cut in at temperatures ≤30°C if it is to be universally applicable. However, release of a secondary antifoam need not be controlled only by temperature. It could, for example, be controlled by the delayed release associated with the dissolution of a coating. The time of release would have to be less than the shortest wash time at which the risk of overfoaming becomes significant. That will depend on the concentration of PDMS–hydrophobed silica antifoam present and the agitation conditions in a given machine. The concentration of course depends on the consumer dose.

The difficulties associated with controlled or delayed release suggest that the secondary antifoam should preferably function during most of the wash cycle, but without deactivating as rapidly as the relevant PDMS-based antifoam. Possible candidates are soaps or fatty acids, hydrocarbon waxes [83], and mixtures of hydrocarbons and hydrophobic particles [84] or particle precursors [71]. In the case of hydrocarbon–hydrophobic particle precursor antifoams, such as hydrocarbon–alkyl phosphoric acid esters, there is a probability that antifoam deactivation during the wash cycle is minimal (see Section 6.6). Use of such secondary antifoams also implies, where relevant, application of incorporation method that prevents storage deactivation.

Deactivation by disproportionation may also in principle be mitigated by delayed release of a portion of the antifoam. Examination of the foam profile shown in Figure 8.12 suggests that the concentration of "low"-viscosity antifoam is more than enough to provide adequate foam control in the early part of the wash. Reduction of the concentration is then possible without overfoaming. If the antifoam aliquot saved by that is subsequently released at a later time in the wash cycle, it will have less time for deactivation before the end of that cycle. The overall foam height at the end of the cycle will therefore be lower than would otherwise have been the case. If the original end of cycle foam height was acceptable, then the concentration released after delay could be reduced relative to the amount saved by this design of introduction of antifoam into the wash. A cost saving is therefore in principle possible albeit at the expense of some additional cost associated with incorporation of part of the antifoam into a delayed release granule. A schematic diagram illustrating this concept is shown in Figure 8.14. Using arguments similar to those given in Section 6.7.4 together with experimental knowledge of the antifoam concentration dependence

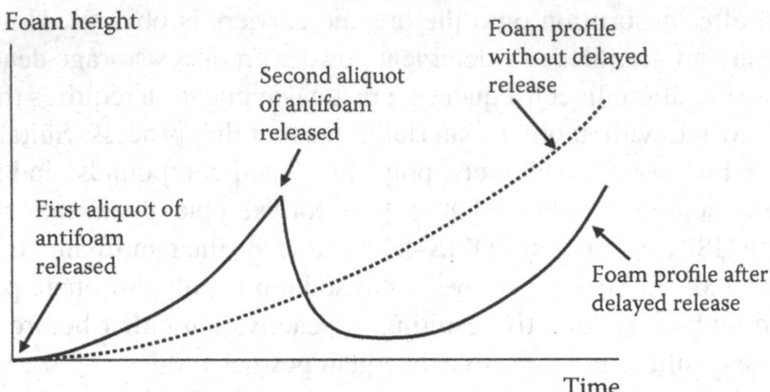

FIGURE 8.14 Schematic illustration of potential effect of delayed release on washing machine foam profile of deactivating PDMS–hydrophobed silica antifoam.

of foam heights as a function of time in a given washing machine, Garrett [85] has shown that savings in antifoam requirement of up to 30 wt.% are in principle possible if a criterion of a fixed maximum foam height allowable during the cycle is used. Powell et al. [74] appear to describe a practical embodiment of this concept. They describe a two-stage antifoam system for detergent powders containing high-foaming surfactants. The first involves spraying an emulsion of PDMS–hydrophobed silica in a carrier fluid directly onto the detergent powder. The carrier fluid is exemplified by an ethoxylated alcohol containing an alkoxysiloxane emulsion stabilizer. This stage is designed to release the antifoam immediately into the wash liquor. The second stage is exemplified by an emulsion of PDMS–hydrophobed silica in tallow alcohol ethoxylate (TA.EO$_{80}$) and fatty acid incorporated onto a starch carrier, which would appear to be likely to only slowly release the antifoam into the wash liquor.

8.2.6 Hydrocarbon-Based Simulation of Dimethylsiloxane-Based Antifoams

Polydimethylsiloxane-based antifoams are effective in most practical situations involving front-loading drum-type machine washing of textiles. They are, however, relatively expensive and their application can present problems. For example, susceptibility to deactivation upon interaction with detergent powders or even certain ingredients means that successful incorporation often requires relative complex processes involving encapsulation and the use of carriers. Also PDMS–hydrophobed silica antifoams of "low" viscosity (i.e., <~5000 mPa s) disproportionate and deactivate relatively rapidly during typical machine wash cycles. The corollary of such disproportionation is of course that large agglomerates of silica-rich PDMS fluid are formed, which could deposit on textiles or machine parts. Moreover, the presence of even extremely low levels of PDMS contamination can, in some circumstances, lead to enhanced problems with the dispensing of detergent powders. As a consequence, there has been some interest in developing antifoams with at least the same effectiveness as PDMS–hydrophobed silica mixtures but with some mitigation of the disadvantages of the latter.

Attempts to find alternative antifoam ingredients to PDMS have been confined to reproducing the viscosities and the air–liquid surface tensions typical of such materials (typically ~21 mN m^{-1} at 25°C, as we have described in Chapter 3) using hydrocarbons. Aronson et al. [88] describe the use of alkylaminosiloxanes for reduction of the air–oil surface tension of mineral oils from typical values of ~30 to ~22 mN m^{-1}, which is close to that found with PDMSs. These alkylaminosiloxanes have an additional advantage in that they permit *in situ* hydrophobing of otherwise hydrophilic silica dispersed in the mineral oil. Aronson et al. [86] have tested the efficiency of antifoams prepared in this manner with respect to controlling the foam of a typical detergent powder but using sparging for an interval of only 10 min as a foam-generating method. The effectiveness under the continuous agitation for >30 min of a machine wash cycle is therefore not established. A key aspect of the performance of any oil–particle antifoam in that context is of course the rate of deactivation by disproportionation. Other issues involving incorporation in detergent powders have not been considered.

Curtis et al. [87] describe the simulation of PDMS antifoams using perfluoroalkyl alkanes to reduce the air–oil surface tensions of blends of mineral oil and polyisobutene. This combination permits simulation of both the air–oil surface tensions and the viscosities of the typical PDMS oils used in antifoams. Preliminary tests with machine wash cycles suggested that at least comparable foam control would be achievable, using a proprietary hydrophobed silica, to that found with typical commercial PDMS–hydrophobed silica antifoams [88]. As with the latter, some deactivation is apparent through the wash cycle, presumably as a result of disproportionation. However, unlike PDMSs, addition of neat perfluoroalkylalkanes to detergent powders with high surfactant levels and an absence of sodium tripolyphosphate did not cause dispensing failure [88]. An obvious problem with this approach concerns the use of perfluorocarbon derivatives, which could cause pollution problems. However, those compounds would represent <10 wt.% of the antifoam, which in turn typically represents <1 wt.% of a detergent powder.

8.3 LIQUIDS FOR MACHINE WASHING OF LAUNDRY

8.3.1 General Properties

Incorporation of antifoams in liquid detergents presents problems that are at least of the same order of difficulty as those involved in detergent powders. Perhaps the most readily adopted approach concerns incorporation of fatty acids and soaps in formulations containing mixtures of anionic and ethoxylated non-ionic surfactants. These compounds simply form calcium soap precipitates upon interaction with hard water at sufficiently high pH. The precipitates can be effective antifoams [89] if present at high enough concentrations in formulations that contain significant amounts of ethoxylated alcohols [6]. Beers et al. [90] have, for example, recently described the application of this approach for controlling the foam of an aqueous liquid.

In contrast to the simplicity of using fatty acids or soaps, hydrophobic oil–hydrophobic particle antifoams or their precursors (e.g., substances which transform into hydrophobic particles *in situ*) must be incorporated as stable emulsions where coalescence, creaming, or sedimentation upon storage in the detergent liquid is unacceptable. We should remember here that emulsion coalescence is in fact likely to be facilitated by the presence of the particles in such antifoams (see Section 4.8) if the relevant wettability conditions are satisfied. An additional problem concerns incorporation in clear liquids where any marked differences in refractive index between the antifoam oil and the detergent liquid will cause turbidity. Deactivation of the antifoam by selective solubilization of the antifoam oil in the concentrated surfactant mixture present in liquid detergents designed for textile machine washing represents another potential problem. However, there is some evidence that subsequent dilution of the liquid detergent during use could lead, in certain circumstances, to the nucleation of precipitating solubilized oil onto the hydrophobic particles present. Some restoration of the synergistic oil–particle antifoam action could thereby occur (see Section 8.5). Intimate mixing with detergent ingredients means, however, that other possible causes of deactivation may exist.

8.3.2 Incorporation of Polyorganosiloxane–Hydrophobic Silica Antifoams in Detergent Liquids

Recent patent literature suggests there is little evidence of any interest in incorporation of hydrocarbon–hydrophobic particle antifoams in detergent liquids for machine washing of textiles (although there may be for liquid detergents for other applications; see Section 8.5). Here, we are therefore concerned entirely with incorporation of polyorganosiloxane–hydrophobed silica antifoams in liquid detergents designed for application in front-loading drum-type textile washing machines. However, it is not intended to make a complete review of the relevant patents but merely to select certain patents, filed over the past 20 years or so, in order to illustrate the issues involved in preparing and incorporating suitable antifoams for liquid detergents.

It is apparent from the patent literature that the emulsion stability of polyorganosiloxane–hydrophobed silica antifoam drops in detergent liquids represents the main challenge to successful incorporation of these materials. The emulsions should be stable under the stagnant conditions prevailing during product storage for periods of up to 3 months. A major cause of instability may concern a tendency of antifoam drops to rapidly coalesce should they colloid. Such coalescence will be facilitated by the presence of hydrophobed silica at the oil-water surfaces of drops dispersed in aqueous liquids (as described in Section 4.8.5.10).

The rate of collision of antifoam drops will be determined by a combination of diffusion and buoyancy. Here we simplify by considering these two processes separately in order to illustrate their relative significance. Under stagnant and neutral buoyancy conditions the rate of collision will be determined by diffusion and can be described by so-called Smoluchowski kinetics [91]. If we assume that the emulsion is monodisperse we can therefore write for the initial rate of colloision dN_{AF}/dt, where N_{AF} is the concentration of antifoam drops per unit volume of detergent liquid, that

$$\frac{dN_{AF}}{dt} = -16\pi .\ r_{AF} D_{AF} N_{AF}^2 \tag{8.4}$$

where r_{AF} and D_{AF} are the radii and diffusion coefficients of the antifoam drops, respectively. The concentration of antifoam drops at $t = 0$ is given

$$N_{AF}(t=0) = \frac{v_{AF}}{4/3\pi r_{AF}^3} \tag{8.5}$$

where v_{AF} is the volume of the antifoam per unit volume of detergent liquid. The diffusion coefficient (assuming that Einstein's law of diffusion is relevant) is given

$$D_{AF} = \frac{\bar{k}T}{6\pi\eta_L r_{AF}} \tag{8.6}$$

where T is the temperature, $\bar{k}$ is the Boltzmann constant, and η is the detergent liquid viscosity. Integrating Equation 8.4 and combining the result with Equations 8.5 and 8.6 yields for the half-life, $t_{1/2}$, of the concentration of unflocculated drops as

$$t_{1/2} = \frac{\pi.\eta_L r_{AF}^3}{2v_{AF}\bar{k}T} \tag{8.7}$$

We therefore find that under stagnant and neutral buoyancy conditions, the relative rate of diffusion controlled collision is dominated by the radius of the antifoam drops—the larger the drops, the slower the rate of collision because both the number concentration and the diffusion coefficients of drops are lower. Increasing viscosity of the detergent liquid and decreasing the volume fraction of the antifoam also both decrease the rate of flocculation.

In deriving Equation 8.7, we have ignored the complex rheology of detergent liquids, a subject that lies outside the scope of this book. Those liquids are not usually Newtonian fluids but rather are often viscoelastic structured fluids that exhibit shear thinning behavior (see, e.g., reference [92]). Nevertheless, it is of interest to use Equation 8.7 to make order of magnitude estimates of the required viscosities needed to ensure that the half-life of coalesced drops exceeds the storage period of up to 3 months to which liquid detergent products may be subject. It is then instructive to compare those viscosities with the range of viscosities reported for such detergent liquids. In making such estimates, it is necessary to assume a value for r_{AF}. Unfortunately, the patent literature only occasionally makes reference to the emulsion drop sizes of the polyorganosiloxane antifoams incorporated in detergent liquids. Exceptionally, Jones et al. [93] quote mean volume average radii of 15–30 microns, which we will therefore assume here. Setting the half-life at 3 months, the volume fraction at 0.01 and T at 25°C yields a detergent liquid viscosity of 3.5–28.0 mPa s required to prevent any collision and possible coalescence of 50% of the drop population. Actual viscosities of liquid detergents tend to be much lower—usually $\leq 5 \times 10^3$ mPa s but measured at relatively high shear rates of up to 20 s^{-1}, which may not accurately reflect the conditions during storage (see, e.g., references [94, 95]). Clearly then diffusion controlled collisions at the initial antifoam concentrations typically present in liquid detergents will not lead to significant coalescence. However, buoyancy driven processes will produce enhanced local concentrations of antifoam which means a corresponding increase in local collision rates and coalescence.

If either creaming or sedimentation occurs during storage, then accumulation of emulsified drops at either the top or bottom of the containing vessel will locally increase v_{AF} to decease the time for collision by diffusion to occur. The velocity, u_{AF}, of movement of a drop in the gravity field is easily shown (assuming Stokes relationship for the viscous resistance of a rigid sphere) to be given by

$$u_{AF} = 0.22\, r_{AF}^2\, \Delta\rho g/\eta_L \quad (8.8)$$

where $\Delta\rho$ is the density difference between the antifoam and the detergent liquid. The velocity increases markedly with increase in drop size. Stability with respect to creaming (or sedimentation) is clearly favored by decreasing drop sizes, which is in marked contrast to stability with respect to diffusion, which is favored by increasing drop sizes. However, stability is again enhanced by increasing the detergent liquid viscosity. The process of creaming (or sedimentation) can, in principle, be eliminated by matching the density of the antifoam to that of the continuous phase. In the case of polyorganosiloxane–hydrophobed silica antifoam, an excess of silica is sometimes added over that needed for antifoam action in an attempt to avoid creaming in liquid detergents (see, e.g., reference [93]). Such an approach suffers from the obvious difficulty that the coefficient of thermal expansion is rarely identical for two different fluids. Neutral buoyancy at one storage temperature does not guarantee neutral buoyancy at another! It would appear then that stability against creaming is best secured by relying on the rheological properties of the detergent liquid.

We have thus far considered only those factors which can minimise any tendency of antifoam drops to cream, collide and coalesce. Collision may however result in the formation of only loosely bound flocculated drops. This does not of course represent a desirable product attribute even though such drops may be readily redispersed by simply shaking the containing vessel. Moreover if those drops should subsequently coalesce, then redispersal by such means may be extremely difficult. The concomitant loss of antifoam action may be permanent. This situation is obviously to be avoided if possible by selection of appropriate stabilizing surfactants or polymers. However, little is apparently known at all about the coalescence of polyorganosiloxane–hydrophobed silica drops in the types of environment provided by detergent liquids.

Actual dispersal of polyorganosiloxane–hydrophobed silica antifoam in liquid detergents to produce homogenous emulsions can also present problems. However, despite all these difficulties, the patent literature is replete with many different approaches to obtaining storage stable polyorganosiloxane-based antifoams for detergent liquids. Surutzidis and Jones [96], for example, teach the preparation of a PDMS–hydrophobed silica antifoam where an ethoxylated alcohol is absorbed into the silica before mixing with the PDMS oil. Homogeneous dispersal in a detergent liquid containing 29 wt.% of an anionic–ethoxylated alcohol mixture, 10 wt.% fatty acid, and 30 wt.% water produced a stable emulsion with no "flocculation" after 6 weeks storage in contrast to a control. The presence of the ethoxylated alcohol in the antifoam mix presumably facilitated dispersal with reduced drop sizes so that the rate of creaming was reduced.

Fleuren et al. [97] note that the foam control after storage of unbuilt unstructured detergent liquids containing polyorganosiloxane–hydrophobed silica antifoams often deteriorates. They attribute this problem not to changes in the intrinsic effectiveness of the antifoam but to creaming and coalescence, which lead to an inhomogeneous distribution of antifoam in the liquid. Fleuren et al. [97] claim that dispersal in an unstructured detergent liquid (with a water content of ~48 wt.%) of a mixture of a branched chain PDMS–hydrophobed silica antifoam with an equal proportion of a polyalkylsiloxane–polyoxyalkylene surfactant formed an emulsion of the antifoam with no coalescence after several weeks of storage. However, some creaming and flocculation was still apparent, but the resulting flocculated antifoam drops could be readily redispersed. By contrast, a control using a linear PDMS–hydrophobed silica antifoam produced an unstable emulsion that exhibited significant coalescence after a storage, in some cases for only a few minutes. It is possible that marked differences in the rheology of the branched and linear PDMS oils may be responsible for the differences in stability to coalescence.

Jones et al. [93] describe a complex approach to the avoidance of the accumulation of antifoam flocculates at the top of the detergent liquid after storage. In this approach, a polyorganosiloxane–hydrophobed silica is prepared *in situ*. This mixture is subsequently dispersed in a non-aqueous liquid to form a concentrated emulsion that is stabilized by finely divided "moderately" hydrophobed silica particles. The non-aqueous liquid can be any polyoxyalkylene derivative, which should be either soluble or at least dispersible in the detergent liquid. References to the wettability of the moderately hydrophobed silica particles suggest that a Pickering emulsion is formed where optionally a polyalkylsiloxane–polyoxyalkylene emulsion stabilizer may also be present (such compounds should be surface active at the antifoam–non-aqueous liquid interface). Dispersal of this emulsion in a detergent liquid produced a uniform dispersion of the antifoam, which formed no visible flocculates in the body of the liquid, and minimal thickness of a layer of antifoam accumulated at the top of the body of the liquid around the wall of the retaining vessel after a few days relative to that of various controls. It is tempting to conclude that more prolonged periods of storage would have produced rather more such accumulation! Other variations of the basic concept described by Jones et al. [93] have also been described where the "moderately" hydrophobed silica is replaced by either untreated hydrophilic silica [98] or a polymer-based latex such as a polystyrene-based latex [99].

It is obvious from Equations 8.7 and 8.8 that the rheology of the liquid detergent represents an important factor in determining the storage stability of antifoam emulsions. It will not only determine the rate of transport of antifoam drops in the body of the liquid but will also represent an important factor in determining the stability against flocculation and coalescence. Not surprising then, we find a patent that concerns a method of antifoam incorporation specific to liquid detergents containing ingredients specifically designed to enhance that stability. Huber and Panandiker [100] teach an incorporation method specifically for liquid detergents containing polyhydroxy fatty acid amides. Essentially it involves dispersing the antifoam in a polyoxyalkylene liquid using a polyalkylsiloxane–polyalkoxylene surfactant as an emulsion stabilizer and subsequently dispersing the emulsion in the detergent liquid. If a polypropylene glycol is used as the polyoxyalkylene, then the antifoam emulsion in the detergent liquid is unstable—the antifoam separates out within a few days. This leaves the detergent liquid depleted of antifoam, leading to overfoaming. Surprisingly, replacement of the polypropylene glycol with either polyethylene glycol or a polyethylene glycol–polypropylene glycol copolymer appears to solve the problem.

A non-ionic detergent liquid is described by Garrett et al. [101] in which a PDMS–hydrophobed silica antifoam can be incorporated by simple mixing. The non-ionic liquid consists essentially of an ethoxylated alcohol continuous phase containing a dispersion of inorganic particulate materials, including hydrophobed silica, where there is a tendency for the inorganic particles to sediment after storage for several weeks. That tendency is increased significantly by the presence of PDMS–hydrophobed silica antifoam of viscosity 3000 mPa s. However, if the viscosity of the antifoam is increased up to 30,000 mPa s, the effect is diminished especially if the antifoam is mixed with a polyalkylsiloxane–polyalkoxylene surfactant. It is tempting to suggest that the problem occurs because antifoam drops flocculate and coalesce with the inorganic particles to produce relatively large aggregates that sediment more rapidly as suggested by Equation 8.8. Under the stagnant conditions prevailing during storage, it seems likely that this flocculation is diffusion controlled—the larger the antifoam drops, the slower the process as indicated by Equation 8.7. Increasing the viscosity of the antifoam is likely to produce larger mean drop sizes and therefore slower rates of diffusion and flocculation, which will lead to a diminished tendency for large agglomerates to form over a given time period so that adverse effects on the rate of sedimentation are diminished.

As with the example described by Garrett et al. [101], detergent liquids are often formulated with high proportions of ethoxylated alcohol as surfactant. Jager [102] has shown that the foam of such liquids is readily controlled in the main machine wash but that they will exhibit a tendency to form excessive amounts of foam during the rinse cycle. Such problems do not usually occur in the case of detergents formulated with high proportions of anionic surfactant. Absence of a rinse cycle foam problem with the latter arises because dilution with water during that cycle causes a decline in both concentration of surfactant and antifoam together with an increase in the polyvalent metal ion activity as the builder concentration declines. As a consequence, precipitation of much of the excess surfactant as the calcium salt occurs, leading to little tendency for formation of excessive amounts of foam (see Section 4.3). This factor is largely absent in the case of detergents with high concentrations of ethoxylated alcohol, which means relatively high concentrations of residual surfactant in the rinse and excessive rinse cycle foam. The problem is accentuated if "low"-viscosity (i.e., ~3000 mPa s) PDMS–hydrophobed silica antifoams are used. As shown in Figure 8.12, such antifoams are known to deactivate significantly by disproportionation by the end of the main wash cycle. A combination of relatively high residual surfactant concentrations and low levels of partially deactivated PDMS-based antifoam accentuates the problem of rinse cycle foam. Use of "high"-viscosity PDMS-based antifoams would be expected to mean less deactivation toward the end of the wash cycle, as shown in Figure 8.12, and therefore may help alleviate the problem.

There have clearly been many different attempts at incorporation of polyorganosiloxane–hydrophobed silica antifoams as stable emulsions in detergent liquids. It would appear that no method is entirely satisfactory in eliminating creaming, flocculation, and coalescence. It is not then altogether surprising then that a very recent patent of Delbrassinne et al. [103] concerning the use of polyorganosiloxane–hydrophobed silica antifoams in this context states that, "the stabilization of a silicone foam control agents in a liquid detergent system is a challenge and to date no robust solutions have been identified which provide good physical and chemical stability across the wide range of heavy duty liquid detergents." Delbrassinne et al. [103] in fact claim that a dispersion of a particular complex blend of organopolysiloxanes as antifoam in a detergent liquid is effective in minimizing accumulation of a ring of coalesced antifoam at the top of the liquid when contained in a suitable vessel. The quoted examples, however, make clear that the process nevertheless eventually occurs, albeit after periods ranging from a few weeks to 2 months—it is only the rate of creaming and coalescence that is reduced by selection of the claimed composition as antifoam. It seems much hangs on the exact meaning of the term "robust solution."

8.4 MACHINE DISHWASHING

Both domestic batch-scale and industrial continuous dishwashing machines are employed. A schematic diagram of the former is depicted in Figure 8.15. The device includes racks to support dishes, which are subject to aqueous sprays directed from below. The impact of the spray onto already wet surfaces results in aeration by a process that may, in some measure, resemble that occurring during a continuous Ross–Miles foam test (see Section 2.2.3). The presence of surface-active material therefore leads to foam formation.

Domestic batch machines usually take in cold water, which is used in a cold prewash. Subsequently, the water is heated to 65–70°C for the main wash, using heaters located in a reservoir at the bottom of the machine. Detergent is added from a dispenser and the hot solution is sprayed over the dishes. It then falls back into the reservoir whereupon it is recirculated until the cycle is complete. This solution, containing deterged debris, is then pumped out of the machine to be replaced by clean water for rinsing. The rinse cycle is a repetition of the main wash, again at

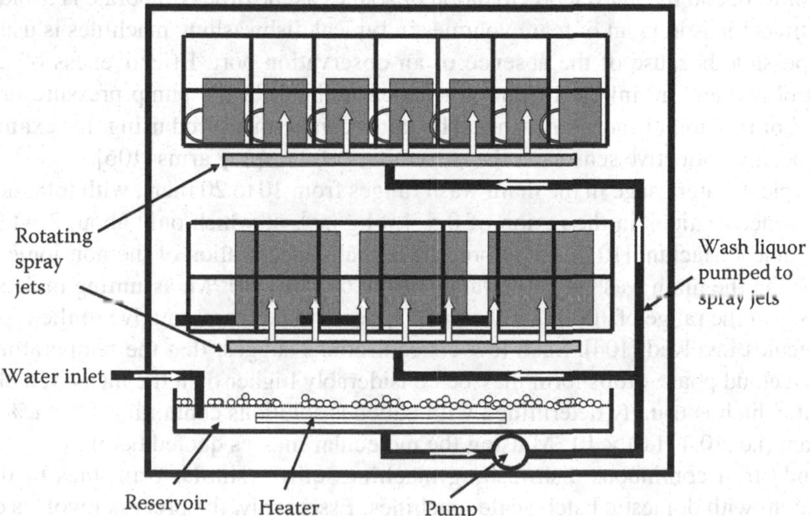

FIGURE 8.15 Schematic illustration of mode of action of domestic dishwashing machine.

similar temperatures, but without detergent, although the so-called rinse aids can be added at this stage from the dispenser. The wash cycle is completed by a drying step where hot air heats the dishes up to temperatures of about 90°C. Drying by condensation onto parts of the machine cooled by various means is alternatively used. As with machine washing of textiles, the whole process is readily automated.

Dishwash detergent formulations contain significant amounts of inorganic salts, which in some cases can produce an extremely alkaline environment (pH ≥11). High pHs and high temperatures conspire to produce foam film–stabilizing surfactants such as soaps from the hydrolysis of fats present in food debris. The presence of proteinaceous material can also contribute to foam film stability. High rates of aeration together with these factors mean foam can rapidly fill machines so that, for example, the circulation pumps fail and the sprays become ineffective in removing soils. It is necessary therefore to include some antifoam function in the relevant formulations. In consequence, dishwash formulations usually contain low levels of ethoxylated (or propoxylated) surfactants with cloud temperatures below the wash temperature. Examples of suitable surfactants are listed by Blease et al. [104] and Gabriel et al. [105]. At temperatures above the cloud temperatures, partial miscibility exists in the case of such surfactants where a surfactant-rich conjugate phase separates out as the so-called cloud phase drops, which can function as antifoams as described in Section 4.6.3. The dilute conjugate is, however, also a micellar solution of the surfactant that can adsorb on relevant surfaces and thereby assist detergency in both the removal of food debris and in ensuring minimal redeposition of that debris. Deposition of the cloud phase on the surfaces of ceramics and glass is not apparently a problem—all the surfactant is apparently readily removed in the rinse step. Indeed, rinse aids are often essentially relatively concentrated (~25 wt.%) solutions of this type of surfactant where the dilute micellar conjugate can assist in stabilizing thin films of solution on glassware and ceramics during drying. Formation of drops of liquid on drying surfaces is thereby minimized, and the uneven precipitation of solutes as the drops evaporate is avoided.

Direct measurement of foam volumes in typical dishwashing machines is usually not possible because of the absence of an observation port. Effectiveness of foam control is therefore inferred from either measurement of the pump pressure or the speed of rotation of the spray arms. The latter can be measured using, for example, magnet and inductive sensors at the extremities of the spray arms [106].

Typical water usage in the main wash ranges from 10 to 20 liters, with total detergent concentrations in the region of 0.5–1.0 kg m^{-3}, of which only about 3 wt.% is non-ionic surfactant [104]. Therefore, the actual concentration of the non-ionic surfactant in the main wash is in the range of ~$(0.6–3.0) \times 10^{-5}$M, assuming molecular masses in the range of 1000–2500, which appears to be representative of the type of molecules involved [104]. Such low concentrations suggest that the temperature at which cloud phase drops form may be considerably higher than the measured cloud point, which is usually determined with aqueous solutions containing 1.0 wt.% surfactant (i.e., $(0.4–1.0) \times 10^{-2}$M using the molecular masses quoted here).

Industrial continuous dishwashing machines utilize similar principles to those apparent with domestic batch-scale machines. Essentially, the process involves conveying continuous processions of dishes through various chambers representing prewash, main wash, rinse, and drying stages. Again foam generation presents potential problems resulting in the necessary inclusion of antifoam function in the relevant formulations. Cloud point ethoxylated or propoxylated compounds often again represent the antifoam of choice.

The usual approach to formulation design for machine dishwashing, involving use of high concentrations of inorganic salts, high pH, and low concentrations of ethoxylated, propoxylated, or EO_nPO_m non-ionic surfactants as cloud point antifoams, has some obvious limitations. The first concerns the corrosive nature of these formulations, which can, for example, lead to roughening the surface of certain types of glass. In turn, this can produce both lead crystal glassware with a cloudy appearance and a measure of consumer discontent. Other difficulties concern the biodegradation and aquatic toxicity of these non-ionic compounds, which have led to regulations limiting the selection of possible molecular structures [104].

It has been argued that these limitations could be avoided by formulation to produce less corrosive conditions where effective detergency is achieved using instead, relatively high concentrations of suitable surfactants. However, as with machine washing of textiles, this would produce a requirement for inclusion of effective antifoams. This, in turn, would imply the presence of relatively large amounts of weak antifoams such as soaps or the presence of from 0.1 to 1.0 kg m^{-3} of effective oil–hydrophobed particle mixtures. In either case, there is a significant risk of deposition of antifoam on glass or ceramic surfaces. As we have seen, incorporation of oil–hydrophobed particle mixtures in detergent products is also not without problems. It is therefore perhaps not so surprising that Chao et al. [106] write in a recent paper that, "Surprisingly, in the household area, the automatic dishwashing application has remained without a suitable antifoam solution" and that "due to the absence of a suitable antifoam the inclusion of organic surfactants has been significantly limited, and this represents a constraint." Chao et al. [106] do in fact present results that imply that it is possible to use polyorganosiloxane-based antifoam technology for effective foam control in machine dishwashing tablets with none of the erstwhile problems of deposits of antifoam on glass or ceramic surfaces. Unfortunately, the absence of any details of the nature of the polyorganosiloxane-based antifoam diminishes the scientific merit of this paper. In another approach to provision of suitable antifoams in this context, Angevaare et al. [107] claim the use of mixtures of hydrophobic particles (particularly long-chain ketones) with high-viscosity (>500 mPa s) hydrocarbon polymers such as polyisobutene. Both effective foam control and effective detergency of glass in dishwashing machines are claimed for combinations of surfactant with this antifoam type.

8.5 GENERAL HARD-SURFACE CLEANING PRODUCTS

Cleaning of floors and work surfaces in kitchens without machines represents examples of the so-called general hard-surface cleaning tasks. Products designed to assist in this context are invariably aqueous liquids, which may be applied neat or diluted before use. However, formation of excessive amounts of foam produces a concomitant rinsing task. Application of some means of foam control especially during rinsing would appear to be desirable.

These products tend to be used at ambient or near-ambient temperatures. Foam control by use of cloud phase antifoams under those conditions would require the use of non-ionic surfactants with low cloud points. Such an approach would probably compromise detergency and would also probably risk phase separation during storage because these products would often be stored and used at the same temperatures. Low viscosities and the absence of structuring would mean that use of conventional antifoams such as polyorganosiloxane–hydrophobed silica would present even more problems due to creaming and coalescence than are found with laundry liquids. However, incorporation of soaps in these general cleaning products so that calcium soap antifoam formation can occur upon dilution with hard water is often successful in providing some foam control.

Garrett et al. [108] and later Ashcroft et al. [109] have claimed that a rather novel application of the antifoam synergy found with hydrocarbon–calcium soap mixtures can be applied in this context. Consider then a cleaning liquid where both the hydrocarbon and a sodium soap (or fatty acid) are solubilized in a concentrated aqueous micellar solution of another surfactant. If use of the product involves dilution in hard water, then both the solubilized hydrocarbon and soap will be precipitated as the concentration of micelles decreases and the water hardness increases. It is claimed [108, 109] that synergistic foam control is then observed as exemplified by the results shown in Table 8.2. Possibly precipitation of bulk phase hydrocarbon is nucleated on the calcium soap particles so that a hydrocarbon–calcium soap antifoam entity could be formed *in situ* giving rise to that synergy.

TABLE 8.2
Effect of Dilution with Hard Water on Foam Stability of Cleaning Liquid Containing Synergistic Mixture of Soap and Solubilized Hydrocarbon

Composition of Cleaning Liquid (wt.%)					Foam Collapse Time[b] (s) after Dilution with Water of (1.2–1.5) × 10^{-3} M Hardness to Yield a Cleaning Liquid Concentration of 6 kg m^{-3}
SAS[a]	AEO_m^a	Soap[a]	Branched Hydrocarbon[a]	Water	
4	8			88	>300
4	8	1		87	150
4	8		1	87	>300
4	**8**	**1**	**1**	**86**	**20**
	16			84	>300
	16	1.2		82.8	120
	16		2	82	>300
	16	**1.2**	**2**	**80.8**	**5**

Source: Garrett, P.R., Instone, T., Puerari, F.M., Roscoe, D., Sams, P.J. (assigned to Unilever PLC), EP 0559,472; 8 September 1993, filed 4 March 1993.

Note: Bold numbers indicate presence of both soap and hydrocarbon to yield synergistic foam control.

[a] SAS, commercial secondary alkyl sulfate; AEO_m, commercial ethoxylated alcohol; soap is sodium soap of coconut fatty acids; branched hydrocarbon is odor-free commercial hydrocarbon with a boiling point range of 171–191°C.

[b] Foam generated by pouring 5 dm^3 hard water onto 30 cm^3 of cleaning liquid in a suitable vessel. Foam collapse time taken as time taken for foam to collapse so that about half of area of air–water surface was clear of foam.

8.6 SUMMARIZING REMARKS

The patent literature is replete with descriptions of methods for incorporation of polyorganosiloxane–hydrophobed silica in all manner of detergent products, ranging from laundry powders and liquids to machine dishwashing tablets. There are clearly problems of incorporation of these antifoams that are specific to product types. However, common to all is the deactivation of the antifoam during long wash cycles if PDMSs of relatively low viscosity are used. Obviously, this effect can be ameliorated by use of extremely high-viscosity PDMSs (see Figure 8.12), but at the expense of both enhanced difficulties with incorporation and inferior initial foam control. As described in Chapter 6, the deactivation of PDMS–hydrophobed silica antifoams appears to concern disproportionation of the antifoam during prolonged aeration or emulsification to produce inactive particle-free drops and inactive particle-rich drops or large (≤1 mm) gel-like agglomerates. However, understanding of the nature of this process of disproportionation appears to be incomplete. It does not correlate with the spreading rates of PDMS on air–water surfaces (see Section 6.5). The available evidence tentatively suggests that it concerns all oil–hydrophobed silica antifoams, even hydrocarbon–hydrophobed silica mixtures. However, as we have seen here, it seems to be absent in the specific case of hydrocarbon–calcium alkyl phosphates and hydrocarbon–alkyl phosphoric acid esters. Clearly there is a need for increased understanding of this phenomenon where studies are generalized to include different oil and particle types.

A common feature of the use of all antifoams (or precursors) in most types of detergent products concerns the effect of ethoxylated non-ionic surfactants. Replacement of anionic surfactant with ethoxylated alcohols or even addition of such compounds produces an enhanced susceptibility to foam control by all types of antifoams ranging from calcium soaps to PDMS–hydrophobed silica mixtures. It has been shown, unsurprisingly, that the phenomenon does not concern formation of cloud phase drops of the non-ionic surfactant. It is also not a simple consequence of a reduction in the intrinsic foamability of a solution of an anionic surfactant by the addition of alkoxylated surfactants—overfoaming within minutes in the absence of antifoam usually occurs regardless of the presence or absence of the latter. Theoretical speculation suggests that it may concern the effect of adsorbed ethoxylated alcohol on the stability of the asymmetrical air–water–antifoam pseudoemulsion films, which must first rupture if antifoam action is to occur. However, no evidence is available to substantiate such speculation. Commercial significance suggests that the phenomenon deserves some attention.

We have seen that incorporation of polyorganosiloxane–hydrophobed silica antifoams in detergent liquids is fraught with difficulties. There is clearly a need for improved understanding of the factors determining the stability to coalescence of such antifoams in detergent liquids exhibiting various rheologies and of compositions ranging from concentrated aqueous surfactant mixtures to completely nonaqueous surfactant mixtures. The role of the hydrophobed silica in determining the stability to coalescence of such antifoams would be of particular interest.

APPENDIX 8.1

TABLE 8.A1
Examples of Patents Claiming Incorporation of Polysiloxane–Hydrophobed Silica Antifoams in Detergent Powders for Machine Washing of Textiles

Claimed Polysiloxanes	Claimed Hydrophobed Silica	Method of Incorporation	Comments	Year	Ref.
Any polydialkylsilane or mixture of PDMS + siloxane resin	Any hydrophobic silica including silanated silica	Dispersed in non-surface-active carrier, e.g., polyethylene glycol or TA.EO_{25}. Spray the result onto particulate ingredient (e.g., sodium tripolyphosphate) to produce free-flowing granule, which is post-dosed to detergent powder	Avoids storage deactivation	1975	[52]
Preferably PDMS, viscosity ~10^4 mPa s	Silanated silica	Antifoam granules essentially similar to reference [53]. Claims a mix of spray-dried detergent powder + antifoam granule + ethoxylated alcohol granule	Avoids dispensing problems; gives even distribution of antifoam in detergent	1979	[57]
PDMS	Silanated silica	Antifoam granules prepared by spray drying an aqueous slurry of antifoam + ethoxylated alcohol and Na.cmc as emulsion stabilizers + inorganic structurants	Reduces adverse detergency effect of PDMS antifoam	1982	[58]
Almost any polydialkylsilane	Silanated silica	Antifoam granules prepared by (1) spray drying aqueous slurry of inorganic structurant containing part of polydilakylsiloxane + silica; (2) add rest of antifoam to this spray-dried powder	States storage deactivation due to migrational segregation where pore structure of granules reduces this by capillary action	1981	[57]
Any silicone oil	Any hydrophobic particle	Spherical core material of sucrose, enzyme, etc. Granulated with an absorbent such as starch, TiO_2, etc. Solution of antifoam in organic solvent then absorbed and solvent evaporated. Solution of paraffin wax in organic solvent then added leaving a protective coating after evaporation of this solvent	Avoids storage deactivation	1983	[59]

Polysiloxanes	Hydrophobed silica or alumina	An aqueous slurry containing emulsified antifoam + film forming polymer (e.g., carboxymethyl cellulose, Na.cmc) sprayed together with detergent slurry through separate nozzles	Avoids storage deactivation	1985	[60]
PDMS	Silica of unspecified hydrophobicity	Emulsion of antifoam in ethoxylated alcohol stabilized by a siloxane-alkoxyalkylene directly sprayed onto detergent powder	Avoids storage deactivation and dispensing problems	1985	[61]
Any silicone oil	Hydrophobed silica	Antifoam sprayed onto gelatinized starch. Solution of paraffin wax dissolved in chloroform subsequently sprayed onto granules, which were dusted with silica to reduce adherence. Granules post-dosed onto detergent powder	Avoids storage deactivation	1983	[62]
Organopolysiloxanes	Finely divided silica of unspecified hydrophobicity	An aqueous slurry containing emulsified antifoam, carboxymethyl cellulose + Na.cmc, and inorganic structurants sprayed to form granules for post-dosing onto detergent powder	Avoids storage deactivation	1986	[63]
High-viscosity blend of two PDMS polymers of ~1.8×10^4 mPa s	Silica hydrophobed by various means	Antifoam dispersed in TA EO_{25-80} using siloxane–alkoxyalkylene as emulsion stabilizer and melt sprayed onto STP in fluidized bed to form granules for post-dosing onto detergent powder	Improved foam control—especially suitable for powders with high surfactant content	1985	[64]
Any alkylated polysiloxanes	Silica hydrophobed by any means	Antifoam incorporated within water-soluble, non-hygroscopic non-surface-active carrier impermeable to detergents and alkalinity, exemplified by polyethylene glycols Carrier post-dosed	Post-dosed prills of a sufficiently large dimension yield storage stable foam control	1987	[65]
Any organopolysiloxane and polysiloxane resins	Silica hydrophobed by any means	Incorporation of antifoam in neutral or mildly alkaline water-soluble salts (e.g., anhydrous sodium sulfate) and subsequent compaction yields granules for post-dosing into detergent powder	Storage deactivation avoided perhaps because of "capillary fixing" of the oil component in the granule	1988	[66]

(*continued*)

TABLE 8.A1 (Continued)
Examples of Patents Claiming Incorporation of Polysiloxane–Hydrophobed Silica Antifoams in Detergent Powders for Machine Washing of Textiles

Claimed Polysiloxanes	Claimed Hydrophobed Silica	Method of Incorporation	Comments	Year	Ref.
Polydimethylsiloxane with trimethyl end-blocking units. Preferred viscosities in range of ~(2.0–4.5) × 10^4 mPa s	Silica hydrophobed with dimethyl or trimethyl silyl groups	Antifoam incorporated in a molten matrix of a fatty acid or fatty alcohol, which is insoluble in water and of melting points ranging from 45°C to 80°C. The resulting emulsion is sprayed onto a fluidized bed of carrier particles (e.g., STP) for post-dosing into detergent powder	The antifoam is only released when the melting point of the matrix is exceeded. Some deterioration in foam control observed after storage	1987	[67]
Polydiorganosiloxane	Silica of unspecified hydrophobicity	As for reference [69] but using monoesters of glycerol of melting point range 50–85°C	The antifoam is only released when the melting point of the matrix is exceeded. Storage stability relatively better than reference [69]	1987	[68]
Polysiloxanes, preferably PDMS (other liquid oils such as hydrocarbons also claimed)	Any type of hydrophobic silica (other types of particles also claimed)	Antifoam incorporated in molten matrix of a mixture of a fatty acid and an alkali metal soap with a liquid phase content at 40°C of preferably ≤50 wt.%. Prills for post-dosing into a detergent powder were formed by spraying into a tower	Avoids storage deactivation	1988	[69]
Polysiloxane preferably of high viscosity, inferred from claimed molecular structures of ~(0.8–6.4) × 10^4 mPa s	Hydrophobed precipitated or pyrogenic silica	Antifoam incorporated in granules of cellulose-based water-soluble/dispersible powder + water-soluble inorganic salts + anionic surfactant dispersing agent Resulting granules coated with aqueous alkali-soluble/dispersible film-forming synthetic polymers. Whole can be post-dosed into detergent powders	Particularly effective foam control	1989	[70]

Any polysiloxane	Any hydrophobed silica	Antifoam incorporated onto inorganic carrier (e.g., soda ash or spray-dried Burkeite) of defined porosity and pore size range. Can be post-dosed into detergent powder	Avoids storage deactivation. Can be used in combination with hydrocarbon-based antifoams	1988	[71]
Any polysiloxane	Any hydrophobed silica or other particulates as claimed in reference [51]	Antifoam incorporated onto inorganic carrier as claimed in reference [73]. Then, porosity reduced by any convenient means. Resulting granule coated with alkaline solution soluble/dispersible latex in a fluidized bed at a temperature above the glass transition temperature of the latex. Can be post-dosed into detergent powder	Avoids dispensing problems that can occur when even minor amounts of PDMS are present in certain types of detergent powders	1992	[72]
Any polysiloxanes yielding antifoams with preferred viscosities in the range of (2–4.5) × 10^4 mPa s	Any silica type hydrophobed preferably with dimethyl or trimethyl silyl groups	A mixture of glycerol and antifoam is added to a carrier of starch, kieselguhr or Fuller's earth. Subsequently add a water-soluble/dispersible organic coating material, e.g., polyethylene glycols. Resulting granules can be post-dosed into detergent powders	Avoids storage deactivation and produces free-flowing granules, which can be easily incorporated into detergent powders. Glycerol is supposed to improve penetration of antifoam into the carrier particles	1992	[73]
As for references [63, 69, 70]	As for references [63, 69, 70]	Two-part incorporation of PDMS–silica antifoam: 1. Spray-onto detergent powder as for reference [63] 2. Post-dose particulate as described in references [69, 70]	Effective for high foaming powders with high surfactant levels and high foaming surfactants	1995	[74]
PDMS of viscosity ≥10^4 mPa s	Silica of unspecified hydrophobicity	Antifoam incorporated on particulate porous organic carrier (e.g., cellulosic materials). Carrier then coated with organic material exemplified by polyethylene glycol. Effervescent material may also be incorporated in the coating Resulting granules can be post-dosed into detergent powders	Avoids deactivation of high-viscosity PDMS-based antifoam both during incorporation in granule and upon incorporation in detergent powders Adverse effects on dispensing are also avoided	1993	[75]

Note: PDMS, polydimethylsiloxane; $TAEO_m$, tallow alcohol ethoxylated with *m* ethoxy groups; Na.cmc, sodium carboxymethyl cellulose; STP, sodium tripolyphosphate (builder).

REFERENCES

1. Jakobi, G., Lohr, A. In *Detergents and Textile Washing, Principles and Practice*, VCH, Weinheim, 1987, p 206.
2. Ferch, H., Leonhardt, W. Foam control in detergent products, in *Defoaming, Theory and Applications* First Edition (Garrett, P.R., ed.), Marcel Dekker, New York, 1993, Surfactant Science Series Vol 45. Chpt 6, p 221.
3. Brouwer, H. unpublished work.
4. Irani, R.R., Callis, C.F. *J. Phys. Chem.*, 64, 1741, 1960.
5. Schmadel, E., Kurzendorfer, C.P. *Waschmittelchemie* (Henkel KGaA, ed.), Huthig, Heidelberg, 1976, p 122.
6. Schwoeppe, E.A. (assigned to Procter & Gamble Co), US 2,954,348; 27 September 1960, filed 28 May 1956; and GB 808,945; 11 February 1959, filed 16 May 1957.
7. Sawicki, G.C. *Colloids Surf. A*, 263, 226, 2005.
8. Farren, D.W., Stuttard, L.W. (assigned to Unilever Ltd.) GB 1,526,932; 4 October 1978, filed 28 August 1974.
9. Key, M.D., McNee, W.G. (assigned to Unilever Ltd.), GB 1,570,603, 2 July 1980, filed 5 March 1976.
10. Fethke, W.P. (assigned to Procter & Gamble Co.), GB 1,148,997; 16 April 1969, filed 29 March 1968.
11. Schlecht, H., Distler, H., Stockigzt, D. (assigned to BASF AG), DE-OS 1,362,570; June 26 1975, filed December 17, 1973.
12. Distler, H., Diessel, P. (assigned to BASF AG), DE 2,727,382; 4 January 1979, filed 18 June 1977.
13. Henning K., Kandler J. (assigned to Hoechst AG), DE-OS 2,532, 804;10 February 1977, filed 27 July 1975.
14. Tate, J.R., McRitchie A.C. (assigned to Procter & Gamble Co.), GB 1,492,938; 23 November 1977, filed 11 January 1974.
15. Scheffler, D.G. (assigned to Procter & Gamble Co.), GB 1,533,118; 22 November 1978, filed 17 March 1975.
16. Lew, H.Y. (assigned to Chevron Research Ltd.), US 3,231,508; 25 January 1966, filed 18 March 1964.
17. Lew, H.Y. (assigned to Chevron Research Ltd.), US 3,285, 856; 15 November 1966, filed 18 March 1964.
18. Schmadel, E. *Fette Seifen Anstrichmittel*, 70, 491, 1968.
19. Gotte, E., Schmadel, E. (assigned to Henkel & Cie GmbH), DE 1,257,338; July 25, 1968; filed February 11, 1965.
20. Schmadel, E. (assigned to Henkel & Cie GmbH), DE 1,467,620; 6 August 1970; filed 15 December 1965.
21. Lehmann, H.J., Schmadel E. (assigned to Henkel & Cie GmbH) DE 1,617,116; 16 October 1975; filed 25 June 1966.
22. Berg, M., Schmadel, E. (assigned to Henkel & Cie GmbH), DE 1,617,127; 18 February 1971, filed 1 April 1967.
23. Amberg, G., Saran, H. (assigned to Henkel & Cie GmbH), DE-OS 2,333,568; 30 January 1975, filed July 2, 1973.
24. Glasl, J., Saran, H., Hoffmeister, J., Berg, M. (assigned to Henkel & Cie GmbH), DE-OS 2,431, 581; 22 January 1976; filed 1 July 1974.
25. Amberg, G. (assigned to Henkel & Cie GmbH), DE-OS 2,544, 034; 7 April 1977. filed 10 February 1977.
26. Perner, J., Frey, G., Helfert, H. (assigned to BASF AG), DE 2,710,355;10 August 1978; filed 10 March 1977.
27. Zhang, H., Miller, C.A., Garrett, P.R., Raney, K.H. *J. Colloid Interface Sci.*, 263, 633, 2003.

28. Hathaway, H.D., Heile, B.J. (assigned to Procter & Gamble Co.), GB 1,113,712; 15 May 1968, filed 4 January 1967.
29. Carter, M.N.A. unpublished work.
30. Carter, M.N.A., Garrett, P.R. (assigned to Unilever Ltd.) GB1,571,502; 16 July 1980, filed 23 Jan 1976.
31. Carter, M.N.A. (assigned to Unilever Ltd.), GB1,571,501; 16 July 1980, filed 23 January 1976.
32. Garrett, P.R., Davis, J. Rendall, H.M. *Colloids Surf. A*, 85, 159, 1994.
33. Yorke, J.W.H., Garrett, P.R., Giles, D. unpublished work.
34. Garrett, P.R., Carter, M.N.A. unpublished work.
35. Curtis, M., Garrett, P.R., Mead J. (assigned to Unilever Ltd.), EP0045208; 24 October 1984, filed 27 July 1981.
36. Giles, D., Garrett, P.R. unpublished work.
37. Schweigel, O.F., Best, G.P. (assigned to Unilever Ltd.), GB 1,099,502; 17 January 1964, filed July 8 1965.
38. Carter, M.N.A., Garrett, P.R., Giles, D., Naik, A.R. (assigned Unilever NV), EP 0087233; 31 August 1983, filed 3 Feb 1983.
39. Aveyard, R., Cooper, P., Fletcher, P.D.I., Rutherford, C.E. *Langmuir*, 9(2), 604, 1993.
40. Peltre, P., Lafleur, A. (assigned to Procter and Gamble Co.), EP 0000216; 10 January 1979, filed 12 June 1978.
41. Atkinson, R.E., Ross, D.A. (assigned to Procter & Gamble Co.), EP 0008829; 19 March 1980, filed 28 August 1979.
42. Gandolfo, D., Cooper, D.J. (assigned to Procter & Gamble Co.), EP0008830; 19 March 1980, filed 29 August 1979.
43. Ho Tan Tai, L. (assigned to Unilever PLC), EP 0109247; 23 May 1984, filed 8 November 1983.
44. Briand, J.P., Storer, C.C. (assigned Unilever NV), FR 2,559,400; 16 August 1985, filed 14 February 1984.
45. Iler, R.K. (assigned to E.I. du Pont de Nemours), US 2,657,149; 27 October 1953, filed 21 October 1952.
46. Ross, S., Nishioka, G. *J. Colloid Interface Sci.*, 65(2), 216, 1978.
47. Alexander, G.B., Heston, W.M., Iler, R.K. *J. Phys. Chem.*, 58(6), 453, 1954.
48. Mobile Oil Company, Product Data Sheet, Mobile Velocite Oil Numbered Series.
49. Wuhrmann, J.C., Seiter, W., Giesen, B., Schmadel, E. (assigned to Henkel KGaA), US 4,590,237; 20 May 1986, filed 2 January 1985.
50. Garrett, P.R. (assigned to Unilever PLC), EP0075433; 30 Mar 1983, filed 14 Sept 1982.
51. Yorke, J.W.H. unpublished work.
52. Bartolotta G., de Oude, N.T., Gunkel, A.A. (assigned to Procter & Gamble Co.), GB 1,407, 997; 1 October1975, filed 1 August 1972.
53. Sawicki, G.C. *J. Am. Oil Chem. Soc.*, 65(6), 1013, 1988.
54. Denkov, N., Marinova, K.G., Christova, C., Hadjiiska, A., Cooper, P. *Langmuir*, 16(6), 2515, 2000.
55. Marinova, K.G., Tcholakova, S., Denkov, N., Roussev, S., Deruelle, M. *Langmuir*, 19(7), 3084, 2003.
56. Berg, M., Vogt, G., Smolka, G., Reuter, H. (assigned to Henkel KCaA), EP 36,162; 23 September 1981, filed 9 March 1981.
57. Boeck, A., Krings, P., Smulders, E. (assigned to Henkel KGaA), GB 2009223; 13 June 1979, filed 2 December 1977.
58. Hachmann, K., Jung, D., Boeck, A. (assigned to Henkel KGaA), EP 13028; 10 March 1982, filed 24 December 1979.
59. Ho Tan Tai, L. (assigned to Unilever PLC), EP40,091; 14 September 1983, filed 11 May 1981.

60. Reuter, H., Saran, H., Witthaus, M. (assigned to Henkel KGaA), EP 70491; 30 January 1985, filed 12 July 1982.
61. Dhanani, S., Mac Donald, J.S., Clunie, J.S., Brooks, M.C. (assigned to Procter & Gamble Co.), EP 46342; 13 February 1985, filed 23 July 1981.
62. Ho Tan Tai, L. (assigned to Unilever PLC), EP 71,481; 9 February 1983, filed 27 July 1982.
63. Reuter, H., Seiter, W. (assigned to Henkel KGaA), DE-OS 3,436,194; 10 April 1986, filed 3 October 1984.
64. Gowland, M.S., Johnson, S.A., Pell, R. (assigned to Procter & Gamble Co.), EP 142,910; 29 May 1985, filed 21 August 1984.
65. Baginski, R.M., Dems, B.C., Ross, L.A., Soule, R.H. (assigned to Procter & Gamble Co.), US 4,652,392; 24 March 1987, filed 30 June 1985.
66. Schulz, P., Waldemann, J., Carduck, F., Witthaus, M., Schmadel, E. (assigned Henkel KGaA), US 4,832,866; 23 May 1989, filed 1 October 1987.
67. Burrill, P.M. (assigned to Dow Corning Ltd.), EP 210,721; 2 April 1987, filed 6 May 1986.
68. Burrill, P.M. (assigned to Dow Corning Ltd.), EP 210,731; 2 April 1987, filed 9 June 1986.
69. Appel, P.W., Bartolotti, F., Delwel, F., Tomlinson, A.D., Willemse, S., Hornung, F. (assigned to Unilever PLC), EP 256,833; 24 February 1988, filed 10 August 1987.
70. Asbeck, A., Meffert, A., Rombey, G., Schmid, K.H. (assigned to Henkel KGaA), EP301,412; 1 February 1989, filed 21 July 1988.
71. Garrett, P.R., Hewitt, M., Iley, J., Knight, P.C., Pilidis, A.P., Ho Tan Tai, L., Taylor, T., Yorke, J.W.H. (assigned to Unilever PLC), EP 266,863; 11 May 1988, filed 10 August 1987.
72. Iley, J. (assigned to Unilever PLC), EP 0484081; 6 May 1992, filed 28 October 1991.
73. De Cupere, M.J.J. (assigned to Procter & Gamble Co.), EP 0495345; 22 July 1992, filed 16 January 1991.
74. Powell, S., Thoen, C.A.J.K. (assigned to Procter & Gamble Co.), WO 95/02665; 26 January 1995, filed 7 July 1994.
75. Akay, G., Garrett, P.R., Yorke, J.W.H. (assigned to Unilever PLC), WO93/01269; 21 January 1993, filed 1 July 1992.
76. Rosato, A., Strandburg, K., Prinz, F., Swendsen, R. *Phys. Rev. Lett.*, 58(10), 1038, 1987.
77. Laughlin, R.G. *The Aqueous Phase Behaviour of Surfactants*, Academic Press, London, 1994.
78. Garrett, P.R., Gratton, P. unpublished work.
79. White, M. unpublished work.
80. Akay, G., Garrett, P.R., Knight, P.C., Yorke, J.W.H. (assigned to Unilever PLC), EP 0534,525; 31 March 1993, filed 11 Sept 1992.
81. Rezak, M. *J. Pharm. Sci.*, 55, 538, 1966.
82. Stead, J.A., Wilkins, R.A., Ashford, J.J. *J. Pharm. Pharmacol.*, 30, 350, 1978.
83. McRitchie, A.C. (assigned to Procter and Gamble Ltd.), GB 1,492 939; 23 November 1975, filed 10 March 1975.
84. Foret, R., Ho Tan Tai, L. (assigned to Unilever PLC), EP 0206522; 30 December 1986, filed 21 May 1986.
85. Garrett, P.R. unpublished work.
86. Aronson, M.P., Lin, S.O., Policello, G.A. (assigned to Unilever PLC), EP 0397 297; 14 November 1990, filed 2 January 1990.
87. Curtis, R.J., Garrett, P.R., Nicholls, M., Yorke, J.W.H. Preliminary observations concerning the use of perfluoroalkyl alkanes in hydrocarbon-based antifoams, in *Emulsions, Foams and Thin Films* (Mittal, K.L., Kumar, P., eds.), Marcel Dekker, New York, 2000, Chpt 10, p 177.

88. Curtis, R.J., Garrett, P.R., Nicholls, M., Yorke, J.W.H. unpublished work.
89. Peper, H. *J. Colloid Sci.*, 13, 199, 1958.
90. Beers, C.P., Delroisse, M.G.J., van Schadewijk, M., Storey, J.L., Veerman, S.M. (assigned Unilever N.V.), EP 1,616,936; 18 January 2006, filed 9 June 2005.
91. von Smoluchowski, M. *Physik. Z.*, 17, 557, 1916; ibid 17, 92, 1917.
92. Ho Tan Tai, L. (assigned to Unilever PLC), EP 0081908; 22 June 1983, filed 11 November 1982.
93. Jones, R.J., Fisk, A.A., Suritzidis, A. (assigned to Procter & Gamble Co.), EP 0635 564; 25 January 1995, filed 22 July 1993.
94. Fagg, A.J., Campens, R.T.G. (assigned to Procter & Gamble Co.), US 6,095,380; August 1 2000, filed 27 October 1998.
95. Boutique, J.P., Braeckman, K.G. (assigned to Procter & Gamble Co.), US 2008/0234169; 25 September 2008, filed 19 March 2008.
96. Surutzidis, A., Jones, R.J. (assigned to Procter & Gamble Co.), EP 0573,699; 15 December 1993, filed 6 June 1992.
97. Fleuren, R.H.M., Mallen, E.F., L'Hostis, J., Renauld, F.A.D. (assigned to Dow Corning S.A.), EP 0549,232; 30 June 1993, filed 14 December 1992.
98. Jones, R.J., Meyer, A., Surutzidis, A., Buyaert, H.M.R. (assigned to Procter & Gamble Co.), EP 0666 301; 9 August 1995, filed 4 February 1994.
99. Jeuniaux, E.M.B. (assigned to Procter & Gamble Co.), WO 93/18126, 16 September 1993, filed 1 March 1993.
100. Huber, A.C., Panandiker, R.K. (assigned to Procter & Gamble Co.), WO 93/25647; 23 December 1993, filed 1 June 1993.
101. Garrett, P.R., Kowalski, A.J., van der Hoeven, P.C., Prescott, A.J., Yorke, J.W.H. (assigned to Unilever PLC), WO 94/29427; 22 December 1994, filed 1 June 1994.
102. Jager, H.U. *Comunicaciones Presentadas a la Jornadas del Comite Espanol de la Detergencia*, 20, 45, 1989.
103. Delbrassinne, P., L'Hostis, J., Zeng, J. (assigned to Dow Corning Corp.), WO 2010/091044; 12 August 2010, filed 3 February 2010.
104. Blease, T.G., Evans, J.G., Hughes, L. Surfactant antifoams, in *Defoaming, Theory and Applications*, First Edition (Garrett, P.R., ed.), Marcel Dekker, New York, 1993, Surfactant Science Series Vol 45. Chpt 8, p 299.
105. Gabriel, R., Aronson, M.P., Steyn, P.L. (assigned to Unilever NV), EP 0337 760; 18 October 1989, filed 12 April, 1989.
106. Chao, S., Wipret, A., Henault, B., Roidl, J.T., Hilberer, A., Ugazio, S. *SOFW J.*, 135(11), 40, 2009.
107. Angevaare, P.A., Beers, O., Yorke, J.W.H., Garrett, P.R., Tartakovsky A. (assigned to Unilever N.V.), WO 97/13832; 17 April 1997, filed 20 August 1996.
108. Garrett, P.R., Instone, T., Puerari, F.M., Roscoe, D., Sams, P.J. (assigned to Unilever PLC), EP 0559,472; 8 September 1993, filed 4 March 1993.
109. Ashcroft, A.T., Carvell, M., Crowley, G.J. (assigned to Unilever PLC), EP 1607472; 21 December 2005, filed 17 June 2004.

9 Control of Foam in Waterborne Latex Paints and Varnishes

9.1 INTRODUCTION

Paints make an important contribution to both the preservation and aesthetics of many different types of artifacts. They are applied to an enormous range of substrates ranging from various metals, woods, and plaster to paper. Traditionally many types of paints have been formulated to consist of solid pigment particles and "binders," dispersed in volatile non-aqueous solvents. After application, the solvent evaporates leaving the binder to coat the substrate with a coherent film containing the dispersed pigment. However, such solvents are usually classified as "volatile organic compounds" (VOCs), which are known to represent environmental hazards. Some solvents are known carcinogens, the vapors of others are known greenhouse gases, and most contribute to the formation of excessive tropospheric ozone levels (through an ultraviolet-induced chain of reactions with oxides of nitrogen). In consequence, legislation in many countries has been introduced that seeks to minimize the use of VOCs [1].

The industrial reaction to this situation is one of compliance, with an increasing use of waterborne paint formulations based on the use of "synthetic latices" as binders especially in paints for domestic application. Such latices are usually polymeric colloids, of volume fraction from 0.2 to 0.5, dispersed in aqueous surfactant solution, prepared by a process of emulsion polymerization. These dispersions have a slightly turbid appearance, often with a low viscosity of order 1 mPa s. The latices can be readily prepared as near-monodisperse colloids.

The preparation of waterborne latices by emulsion polymerization usually employs polymerization in aqueous micellar surfactant solutions. The simplest manifestation of the process involves the presence in an aqueous medium of emulsified monomer drops, micellar surfactant, and a water-soluble polymerization (free-radical) initiator. A combination of monomers is often used, exemplified by combinations of compounds such as methyl methacrylate, butyl acrylate, and styrene. Typically monomers have only slight water solubility. The classic qualitative picture of this process was described by Harkins [2] more than 60 years ago. A schematic of that process is shown in Figure 9.1. The key step is solubilization of the monomers in the micelles where polymerization is initiated. Both the low solubility of the monomer and the relatively low surface area of the monomer emulsion drops means that initiation of polymerization is essentially confined to the micelles. Polymerization now proceeds as more monomer is transported to the swelling micelles from the

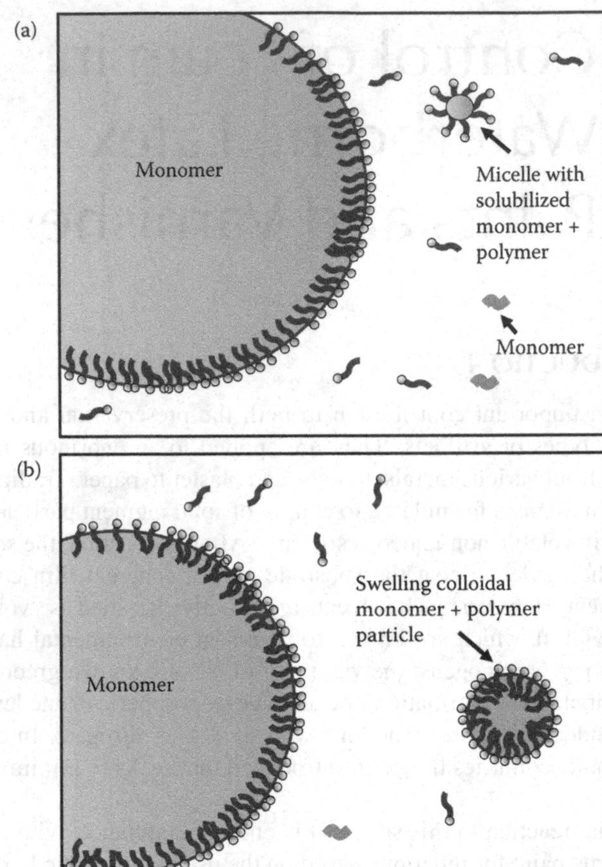

FIGURE 9.1 Schematic illustration of model of emulsion polymerization by Harkins. (a) Emulsified monomer supplies monomer to micelles where polymerization is initiated. (b) Process continues as micelles transformed into colloidal particles, which continue to swell until emulsified monomer entirely consumed. (From Harkins, W.D., *J. Am. Chem. Soc.*, 69, 1428, 1947.)

monomer emulsion drops. Eventually the solubilization capacity of the micelles is exceeded whereupon they are transformed into colloidal polymer–monomer particulate nuclei stabilized by adsorbed surfactant. No further nuclei form. However, transport of monomer to the existing nuclei continues from the monomer emulsion drops. The nuclei continue to grow into particles of colloidal dimensions (~100–500 nm) until the monomer drops disappear. A scanning electromicrograph of a typical latex that results from this process is shown in Figure 9.2. At this stage any further polymerization concerns only the monomer present in the colloidal particles. Presumably the low polydispersity of the resulting latex derives at least in part from the low polydispersity of the micellar incubator of the colloid. References [3–6] can be consulted for more detailed accounts of emulsion polymerization.

FIGURE 9.2 Scanning electron micrograph of acidic styrene–acrylate block copolymer latex dispersed in aqueous 0.03 M aerosol OT solution. (Reprinted from *Colloids Surf. A*, 283, Garrett, P.R., Wicks, S.P., Fowles, E., 307. Copyright 2006, with permission from Elsevier.)

The presence of surfactant is therefore intrinsic to latex prepared in this manner. Separation is possible only using, for example, dialysis, centrifugation, etc. (see reference [6]) and is not usually made in the case of latex paints. Moreover, the presence of surfactant is desirable both for stabilizing other ingredients, such as certain pigments, and as a wetting aid. Surfactant types typically include anionic surfactants such as sodium alkyl sulfate, sodium alkyl sulfosuccinates, or sodium alkylbenzene sulfonates often in admixture with ethoxylated alcohols [7]. However, as a consequence of the inevitable presence of surfactants, waterborne latex paints and varnishes tend to form stable foams when subject to aeration during, for example, the filling of cans and the application to substrates. In the case of the former, the presence of excessive foam can lead to difficulties in supplying gravimetrically quantitative measures. Aeration to form bubbles during application can leave paint defects as the water evaporates and bubbles rupture. In consequence, it is usually necessary to include antifoams in the formulation of waterborne latex paints and varnishes. In fact, problems due to unwanted foam are also present even in the case of mixed water–organic solvent latex coating products also necessitating the inclusion of antifoams in those formulations.

Ready formation of films on various substrates is of course the main practical application of latices. A schematic illustration of the process of latex film formation is shown in Figure 9.3. The first step in forming such films requires that the latex effectively wets the substrate, preferably with a zero contact angle. According to Fitch [6] a multilayer of close packed latex particles, often as a regular array, is then formed as the water evaporates. Fusion of the particles at the upper surface of the film then occurs, induced by capillary action due to a residual aqueous meniscus and the proximity of the minimum film forming temperature. The latter is the temperature at which the modulus of the film is exceeded by the forces promoting coalescence—practically it is within a few degrees of the glass transition temperature [8].

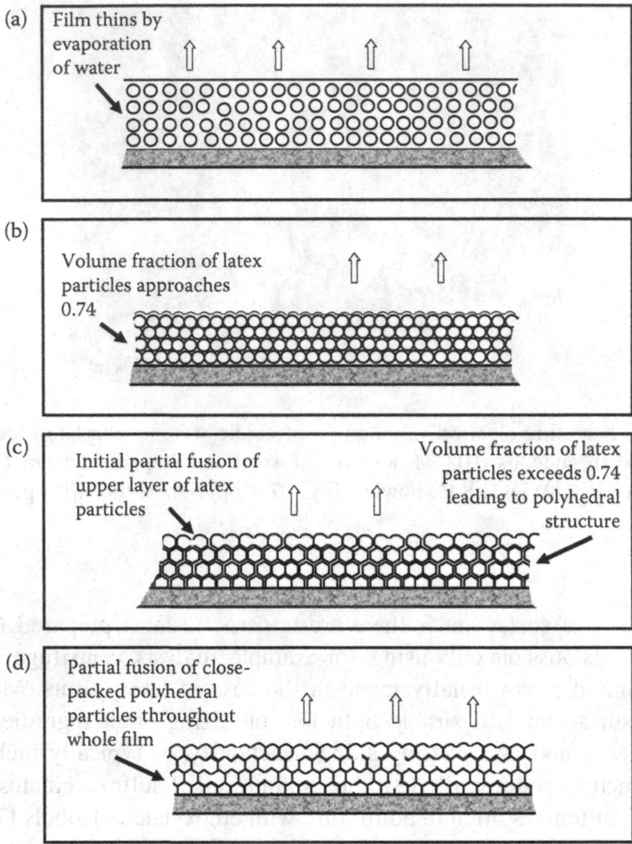

FIGURE 9.3 Simplified schematic illustration of processes accompanying thinning of latex film, comprising spherical polymer particles, on solid substrate by evaporation of water. (a) Initially the volume fraction of particles could be less than that for close packing of spheres even allowing for double layer repulsion. (b) Volume fraction approaches close packing limit of 0.74. (c) Volume fraction exceeds close packing limit so that particles deform to polyhedral structure. Some fusion of polyhedral occurs at top of film, inhibiting the rate of transport of water from the remainder of the film. (d) Whole film dries to form close-packed polyhedral structure where, however, extensive fusion of particles occurs.

After fusion of the particles at the upper surface of the film, the remaining water must diffuse through the resulting continuous film. As the water diffuses and evaporates away, the volume fraction throughout the film eventually exceeds ~0.74 and the monodisperse particles deform into polyhedra—in fact, rhombic dodecahedra [9, 10]. At this stage, the film acquires enhanced strength as polymer segmental interdiffusion occurs across the original particle boundaries. However, all of this leaves the ultimate location of non-volatile solutes such as surfactants and insoluble paint components such as pigments and antifoams open to question.

There have in fact been studies of the ultimate location of soluble anionic surfactants in a drying latex film. These studies present evidence that the surfactant desorbs from the coalescing polymer particles and precipitates as crystals in the film as the water evaporates [11, 12]. Lam et al. [12], for example, report an atomic force microscopy study of the drying of an SDS stabilized butyl acrylate–styrene–acrylic acid latex film cast on a glass substrate. They present evidence that the surfactant moves with the evaporating water as it migrates to the film–air surface where it crystallizes to form a continuous separate phase covering the whole film surface. Kientz and Holl [11], on the other hand, argue that the ultimate surfactant location could be either at the film–air surface or the film–substrate surface depending on the flux of evaporating water. If that is too slow, then the fusion of the latex particles near the film–air surface supposedly prevents migration of surfactant with the water so that the surfactant crystallizes close to the film–substrate surface. This suggests that the effect of water evaporation on the ultimate location of antifoam drops could also be of significance—if, for example, the drops accumulate close to the film–air border, then they may spread to form an oil film, which could have adverse effects on the adherence of a second coat of latex.

Waterborne latices obviously represent the basic ingredient of many waterborne paints and varnishes. However, formulation of paint requires the presence of other ingredients—mostly pigments such as TiO_2, additional surfactants to ensure effective wetting of relevant substrates, antifoams, and polymeric thickeners to assist achievement of the required rheology. Formulation of varnishes is simpler—pigments are often absent. In all cases, however, the foam behavior is likely to be dominated by the aqueous phase containing both surfactant and a high volume fraction (from 0.2 to 0.5) of latex polymer particles. It is known that presence of such particles can stabilize foam films even in the absence of surfactant, leading to significant foam formation [13–16]. Film thinning occurs in a stepwise manner, stabilization resulting from a long-range oscillatory structural force. Similar behavior is found in the presence of surfactant [17].

In this chapter, we consider the foam and antifoam behavior of waterborne latex paints and varnishes, particularly with respect to the role of the latex polymer in this context. Oil–hydrophobed particle–based antifoams are often used in paints and varnishes. We briefly consider the difficulties associated with their incorporation in these products. However, even though they are introduced to eliminate the coating defects associated with foam bubbles, they are often the cause of other defects. The latter represents a key aspect of the application of antifoams in this context and therefore we consider the causes of such defects in some detail.

9.2 FOAM AND ANTIFOAM BEHAVIOR

9.2.1 General Considerations

According to Schulte and Hofer [18], typical waterborne latex paints for domestic exterior application usually contain about 30 wt.% water and about 15 wt.% of latex polymer together with about 0.3 wt.% "dispersant" (which presumably means "surfactant" although it is not clear whether that is in addition to the surfactant intrinsic to the latex), about 0.5 wt.% viscosity modifying polymer ("thickener"), and about 0.25 wt.% antifoam. In addition, more than 50 wt.% consists of various inorganics, including $CaCO_3$ and TiO_2 pigment in the case of white paint. Similar components are present in waterborne latex paints for domestic interior application but with a higher water content of about 40 wt.% and only about 10 wt.% of latex polymer. Waterborne latex varnishes, on the other hand, are characterized by the absence of these inorganic particles except in the case of matt varnishes where particles of matching refractive index are usually present. We find then that the latex component of waterborne paints (i.e., excluding inorganic pigment) and varnishes contains from about 20 to 35 wt.% polymer and from 65 to 80 wt.% aqueous phase. The latter is in fact a solution containing about 5–10 kg m^{-3} of "dispersant" (i.e., surfactant in addition to that intrinsic to the latex), and a similar concentration of thickener, in which is also dispersed from about 5 to 8 kg m^{-3} of antifoam.

One obvious property likely to affect foam film stability concerns the rheology of the paint. In general, the higher the viscosity, the more slowly draining both foam films and pseudoemulsion films. Therefore, the lifetimes of even inherently unstable films will be extended by increases in viscosity. The frequency of action of antifoams will also be diminished. That waterborne latex paints and varnishes are usually shear thinning non-Newtonian fluids is unlikely to alter the relevance of those generalizations. Obviously then the presence of any components in the paint, including finely divided inorganic and latex polymer particles and thickeners, which increases the viscosity at any relevant shear rate, will increase the stability of foam films and render antifoams less effective.

The foam properties of such paints are unlikely to be significantly determined by the presence of the inorganics such as finely divided TiO_2 pigment particles except in so far as their presence influences the rheology of the paint. Such particles are known to be hydrophilic so that they do not adhere to air–water surfaces and neither stabilize nor destabilize foam films (see Section 4.7). Extender pigments such as $CaCO_3$ are also often added to paints "for decorative purposes" [19]. Anionic surfactant can adsorb onto the surfaces of divalent metal salts such as $CaCO_3$ to render the relevant surfaces hydrophobic at least in the case of submicellar solutions [20]. The adsorption in this context essentially occurs at the solid-vapor surface as first described by Smolders [21] and emphasized in the classic paper on wettability and contact angles by Johnson and Dettre [22]. Adsorption at the solid liquid surface as claimed by Cui et al. [20] would appear to be in violation of the Gibbs adsorption equation because hydrophobing the $CaCO_3$ surface will actually increase the solid–water surface tension! However, as the surfactant concentration approaches the CMC, then bilayer adsorption can occur at the surfactant solid–liquid surface to reduce the solid–water surface tension. Cui et al. [20] show that adsorption of anionic surfactants such as SDS and aerosol OT adsorption onto $CaCO_3$ at submicellar concentrations increases the air–water contact angle significantly so that the particles can actually stick to air–water surfaces and enhance foam stability. This effect is of course lost as surfactant concentrations approach the CMC. It is, however, worth noting that $CaCO_3$ is absent from varnishes and also from the typical waterborne latex paint formulation quoted by Porter [23].

The antifoams used in this context are usually hydrocarbon-based according to both Schulte and Hofer [18] and Porter [23]. It is clear from the examples given by Porter [23] that the relevant hydrophobic particles include fatty acids, alkyl monoglycerides, ethylene distearamides, hydrophobed silica, etc. However, Shah et al. [24] teach the use of a propylene glycol monoester of a fatty acid as oil mixed with hydrophobed silica to form an antifoam for waterborne paints. They show that these antifoams are as effective in controlling foam as hydrocarbon mineral oil–hydrophobed silica mixtures but have the advantage of facilitating fusion of latex particles in a drying film. Thus, these esters actually decrease the minimum film-forming temperature in latex paints in contrast to mineral oils, which increase that temperature. Unfortunately, Shah et al. [24] describe the antifoam oils as mere "carriers" and the hydrophobed silica as a "defoamer-active" component as was common in the patent literature several decades ago (see Section 4.8.2) and in total denial of the experimental evidence concerning the role of the oil accumulated over the past quarter of a century.

Other components are often added to the basic oil–hydrophobic particle mixture. The use, for example, of the reaction product of certain alkoxylated compounds with epichlorohydrin as a supplementary component of hydrocarbon–hydrophobic particle antifoams for waterborne latex paints has been claimed to give improved foam control [25]. Unfortunately, the claim is made without so much as a hint concerning the mode of action of these reaction products. One possibility is that these compounds reduce the contact angle of the hydrophobic particles at the oil–water interface (see Section 4.8.5.2 for indications of other possibilities). The use of this approach with, for example, hydrocarbon lubricating oils (of API Group II Base Oil specification), using ethylene distearamide as an hydrophobic particulate, is claimed to produce an antifoam for waterborne paints with a VOC significantly below that using conventional mineral oils [26]. Kahn et al. [27] claim that these reaction products of alkoxylated compounds and epichlorohydrin can also be admixed with polyalkoxylated–polyorganosiloxane block copolymers to produce effective antifoams for waterborne paints even though hydrophobic particles may be absent. Such antifoams are also claimed to produce none of the paint film defects that so often occur if polyorganosiloxane-based antifoams are incorporated in waterborne paints (see Section 9.3.2). These block copolymers are cloud-point antifoams. Their effectiveness must rely on a low tendency to comicellize with any other surfactant present and, in the absence of hydrophobic particles, for the pseudoemulsion air–water–cloud phase films to exhibit low stability. These matters are discussed in some detail in Section 4.6.3.2.

Despite a reputation for causing paint film defects, even polydimethylsiloxane–hydrophobed silica antifoams can also apparently find use in waterborne latex paints if certain preventive measures are applied (see, e.g., reference [28] and Section 9.3.2). It is tempting then to infer that, despite the complexity of waterborne latex paint formulations, the conventional oil–hydrophobed particle antifoams listed in Chapter 4 (Tables 4.A1 through 4.A3) find ready application in this context, albeit with additional components such as emulsifiers designed to ease dispersal or structurants to ensure stability of the stored antifoam against sedimentation. Practical experience then implies that the rupture of foam films containing high volume fractions of latex polymer particles by conventional antifoams is readily achieved.

9.2.2 Effect of Stratified Layers of Polymer Latex Particles on Foam and Pseudoemulsion Film Stability

As we have seen, the process of emulsion polymerization is complete when micelles are absent from the aqueous dispersion. The aqueous phase surfactant concentration present in latices must therefore be less than or equal to the CMC. However, significant amounts of surfactant must be present but adsorbed at the polymer surface—indeed, estimates of (1.3–1.5) × 10^{-6} M m^{-2} on a styrene–acrylate copolymer have been made for SDS and aerosol OT at bulk phase concentrations close to the CMC [17]. In preparing complete formulations of paints and varnishes, further amounts of surfactant are usually added as wetting agents. This may well mean that surfactant solutions in the aqueous phase of latex paints are micellar.

A study [17] has been made of the effect of a dialyzed styrene–acrylate copolymer latex on the foam and the resistance to antifoam of three different surfactants—SDS, aerosol OT (sodium bis-diethylhexyl sulfosuccinate), and Triton X-100 (OP.EO_{10})—all at a nominal concentration of 0.03 M. The polymer particles were dispersed in the surfactant solutions at a proportion of 25.5 wt.%. Adsorption of the surfactant onto the polymer particles significantly reduced the concentration of free surfactant in solution. A comparison was therefore made between the foam and resistance to antifoam behavior of the latex polymer–containing surfactant solution and a surfactant solution at the same depleted surfactant concentration, but containing no polymer. These depleted solutions were all submicellar—from about 80% to 99.9% of the surfactant (depending on the surfactant) was lost by adsorption onto the polymer–water surface.

Comparison of the drainage behavior of vertical foam films (see Section 2.3.3) drawn from these solutions in the absence of antifoam revealed that the films containing latex particles drained more slowly and exhibited stratification involving ordered layers of polymer particles in the presence of the polymer latex. The typical configuration of such a film is shown schematically in Figure 9.4. Similar observations of stratification together with enhanced film lifetimes were observed in the presence of the polymer particles in the case of air–water–hexadecane pseudoemulsion films (made using the apparatus shown in Figure 2.12). These observations of pseudoemulsion film lifetimes are compared with the antifoam effect of hexadecane in Table 9.1. In all cases, the bridging coefficient, B, is positive, which implies that the oil has the potential to function as an antifoam in this context depending, however, on the stability of the relevant pseudoemulsion film. The stabilizing effect of the polymer on pseudoemulsion film lifetimes is obvious and correlates with the absence of an effect of hexadecane on the stability of the foam as indicated by the relative values of F (= volume of air in foam in presence of antifoam/volume of air in foam in absence of antifoam) at various foam ages. The effect probably concerns the rheology of the stratified films leading to diminished rates of drainage rather than to any effect on, for example, the disjoining pressure.

A similar comparison of values of F with pseudoemulsion film stability is shown in Table 9.2, but using mixtures of hexadecane and hydrophobed silica. The usual antifoam synergy is observed regardless of the presence or absence of the latex polymer. No evidence of enhanced pseudoemulsion stability, due to the presence of the

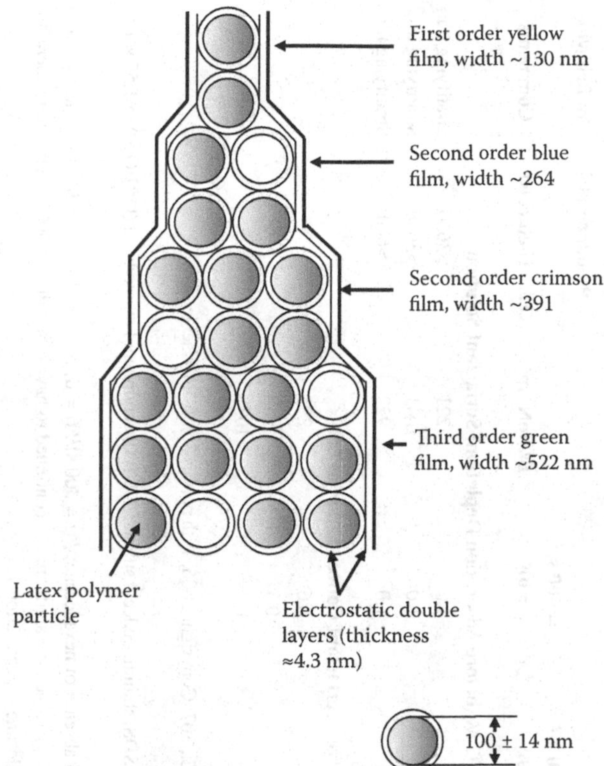

FIGURE 9.4 Schematic representation of probable structure of stratified foam film prepared from 25.5 wt.% latex in aqueous 0.03 M AOT solution (actual free AOT concentration after depletion by adsorption on latex: 2.2 mM). Film age: 120 s; generated using apparatus depicted in Figure 2.11 and observed in white light. (Reprinted from *Colloids Surf. A*, 283, Garrett, P.R., Wicks, S.P., Fowles, E., 307. Copyright 2006, with permission from Elsevier.)

latex polymer, is apparent. It is possible that the hydrophobed silica ruptures the pseudoemulsion film before a stratified film can form. The presence of polydisperse silica particles could even disrupt the packing of polymer particles in the film, so that stratified films are not formed [16]. It is nevertheless clear that concentrated latex polymer particles do not necessarily inhibit antifoam action in the case of oil–hydrophobed particle antifoams, a finding that would appear to correlate with practical experience.

9.3 SPECIFIC ISSUES CONCERNING OIL-BASED ANTIFOAMS

9.3.1 Incorporation in Paints and Varnishes

As we have seen, suitable antifoams for latex paints and varnishes consist of the familiar mixtures of hydrophobic oils and hydrophobic particles described in Chapter 4. Preparation of such mixtures can be difficult; hydrocarbon–ethylene distearamide

TABLE 9.1
Effect of Latex Polymer[a] on Antifoam Effect of Hexadecane

Surfactant[b]	Actual Free Surfactant Concn. Allowing for Depletion (M)	$F(t = 0)$[c]	$F(t = 300\ s)/F(t = 0)$[c]	$F(t = 600\ s)/F(t = 0)$[c]	B[d] ($mN^2\ m^2$)	Pseudoemulsion Film Behavior: Film Lifetime(s)	Pseudoemulsion Film Behavior: Comments
Latex with 25.5 wt.% Polymer and 74.5 wt.% 0.03 M (before Adsorption Depletion) Surfactant Solution							
AOT	2.2×10^{-3}	1.0	1.0	1.0	221	>3600	Stratifying film
SDS	6.7×10^{-3}	1.0	1.0	1.0	1041	>3600	Stratifying film
TX-100	3.7×10^{-5}	0.9	1.0	1.0	927	>600	Stratifying film
Surfactant Solutions with Same Free Surfactant Concentration as Latex							
AOT	2.2×10^{-3}	1.0	1.0	0.6	295	~70	–
SDS	6.7×10^{-3}	1.0	0.9	0.3	936	~30	–
TX-100	3.7×10^{-5}	0.1	0.0	0.0	1038	~7	–

Source: Adapted from *Colloids Surf. A*, 283, Garrett, P.R., Wicks, S.P., Fowles, E., 307. Copyright 2006, with permission from Elsevier.

[a] Acidic styrene–acrylate block copolymer latex as depicted in Figure 9.2.

[b] AOT, aerosol OT, sodium bis-diethylhexyl sulfosuccinate in 3×10^{-4} M EDTA; SDS, sodium dodecyl sulfate; TX-100, octylphenol ethoxylated (OP.EO10) in 1.55×10^{-3} M NaCl.

[c] F = volume of air in foam in presence of hexadecane/volume of air in foam in absence of hexadecane; $F(t = 300\ s)/F(t = 0)$ is ratio of F at $t = 300$ s to F at $t = 0$ as a measure of antifoam effects as the foam ages. If, for example, $F(300\ s)/F(t = 0) < 1$, then the antifoam has continued to have effect during 300 s after foam generation ceased. Foam measurements made by hand shaking measuring cylinders. Hexadecane concentration 1.0 kg m^{-3}.

[d] B is the bridging coefficient. If $B > 1$, then unstable liquid bridges should potentially form, leading to foam film collapse provided the pseudoemulsion films are unstable, too. For interpretation, see Section 4.5.1.

TABLE 9.2
Effect of Latex Polymer[a] on Antifoam Effect of Hexadecane–Hydrophobed Silica[b]

Surfactant[c]	Actual Free Surfactant Concn. Allowing for Depletion (M)	$F\ (t = 0)$[d]	$F\ (t = 300\ s)/F\ (t = 0)$[d]	$F\ (t = 600\ s)/F\ (t = 0)$[d]	B ($mN^2\ m^2$)[e]	Pseudoemulsion Film Behavior: Film Lifetime (s)	Pseudoemulsion Film Behavior: Comments
Latex with 25.5 wt.% Polymer and 74.5 wt.% 0.03 M (before Adsorption Depletion) Surfactant Solution							
AOT	2.2×10^{-3}	0.21	0.19	0.0	221	<1	Instantaneous film rupture
SDS	6.7×10^{-3}	0.04	0.0	0.0	1014	<1	Instantaneous film rupture
TX-100	3.7×10^{-5}	0.0	0.0	0.0	927	<1	Instantaneous film rupture
Surfactant Solutions with Same Free Surfactant Concentration as Latex							
AOT	2.2×10^{-3}	0.19	0.0	0.0	295	<1	Instantaneous film rupture
SDS	6.7×10^{-3}	0.0	0.0	0.0	936	<1	Instantaneous film rupture
TX-100	3.7×10^{-5}	0.0	0.0	0.0	1038	<1	Instantaneous film rupture

Source: Adapted from *Colloids Surf. A*, 283, Garrett, P.R., Wicks, S.P., Fowles, E., 307. Copyright 2006, with permission from Elsevier.

[a] Acidic styrene–acrylate block copolymer latex as depicted in Figure 9.2.

[b] AOT, aerosol OT, sodium bis-diethylhexyl sulfosuccinate in 3×10^{-4} M EDTA; SDS, sodium dodecyl sulfate; TX-100, octylphenol ethoxylated ($OPEO_{10}$) in 1.55×10^{-3} M NaCl.

[c] Antifoam contained 10 wt.% hydrophobed silica.

[d] F = volume of air in foam in presence of hexadecane/volume of air in foam in absence of hexadecane; $F\ (t = 300\ s)/F\ (t = 0)$ is ratio of F at $t = 300$ s to F at $t = 0$ as a measure of antifoam effects as the foam ages. If, for example, $F\ (300\ s)/F\ (t = 0) < 1$, then the antifoam has continued to have effect during 300 s after foam generation ceased. Foam measurements made by hand shaking measuring cylinders. Antifoam concentration: 1.0 kg m^{-3}.

[e] B is the bridging coefficient. If $B > 1$, then unstable liquid bridges should potentially form leading to foam film collapse provided the pseudoemulsion films are also unstable. For interpretation, see Section 4.5.1.

mixtures are, for example, extremely viscous, presumably as a result of relatively strong structuring by the particles.

There are essentially two options for incorporation of antifoams in latex paints and varnishes. The first involves simple addition of the neat antifoam to the paint or varnish. This may preclude effective monitoring of the dispersion state of the antifoam—potentially a serious issue if coating defects such as cratering by antifoam drops are to be minimized as we discuss in Section 9.3.2. The second involves preparation of an aqueous dispersion of the antifoam, which can be subsequently added to the relevant formulation at an appropriate stage in the mixing process. As we have discussed in Chapter 6, this process can produce some deactivation of antifoam effectiveness as a result of partial disproportionation, involving the risk of formation of large inactive agglomerates, which could cause large defects in coatings. Storage stability of the emulsion is also potentially a problem—creaming, flocculation, and coalescence can all have adverse effects as described in the context of detergent liquids in Section 8.3.2. Use of emulsion stabilizers will of course be necessary and add to costs without necessarily offering any improvement in the properties of the paint or varnish. A possible advantage of this approach is, however, ease of control and monitoring of the state of dispersal of the antifoam before addition to the paint. As we will discuss below, the problem of cratering due to the presence of antifoam drops can be minimized in a given system by reducing antifoam drop sizes.

9.3.2 Defect Formation in Drying Paint Films

9.3.2.1 General Considerations

There would appear to be several types of paint film defect associated with foams and antifoams. Most are described as "cratering," where crater-shaped depressions are present in films, which are attributed to a variety of causes by Kornum and Nielsen [29]. One cause concerns bubbles that burst only after the drying latex film has at least partially fused so that the shape of the bubble is largely preserved in the paint film as depicted in the sketch in Figure 9.5. Antifoams are of course included in paint formulations in order to rupture such foam films before fusion of the latex film. They may, however, themselves cause defects. Thus, Kornum and Nielsen [29] attribute some defects caused by antifoams to Marangoni flows induced by, for example, spreading from silicone oil drops. Such flows produce only transient depressions in

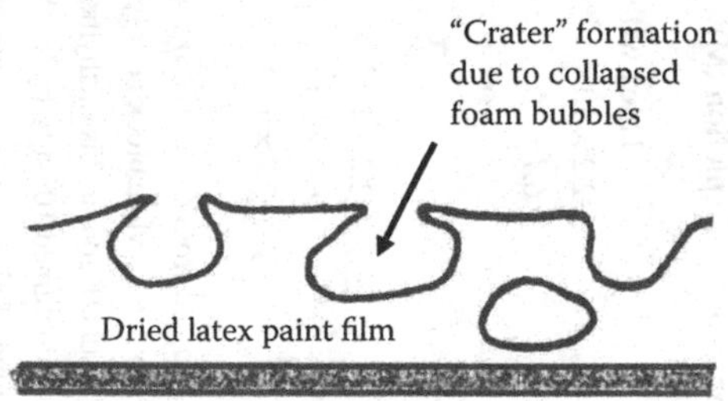

FIGURE 9.5 Schematic sketch of "crater" formation in dried latex paint film due to presence of foam bubbles in film.

films if evaporative drying and latex fusion is absent. However, observation [30] and mathematical simulation [31] reveals that the combined effect of the Marangoni flow and drying produces surprisingly large crater-like depressions in paint films with high rims and central peaks as shown in Figure 9.6. As indicated in the figure, the central peak can even be surrounded by dewetted regions. An image of a paint film "decorated" with this type of defect is shown in Figure 9.7.

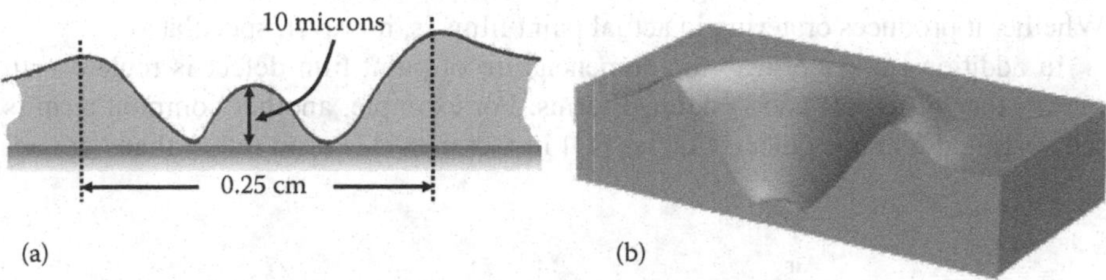

FIGURE 9.6 Crater formation in paint film due to Marangoni effect induced by surface tension gradient. (a) Experimental observation of profile of crater formed by spreading of silicone oil drop on a film of alkyd-amino latex thinning by evaporation. (Modified from Weh, L., Linde, H., *Plaste Kautsch*, 20, 849, 1973.) (b) Computer-generated cut-away view of a crater created by spreading from a bolus in paint film after thinning by evaporation until dry—based on a mathematical model of process. (Reprinted from *J. Colloid Interface Sci.*, 227, Evans, P.L., Schwartz, L.W., Roy, R.V., 191. Copyright 2000, with permission from Elsevier.)

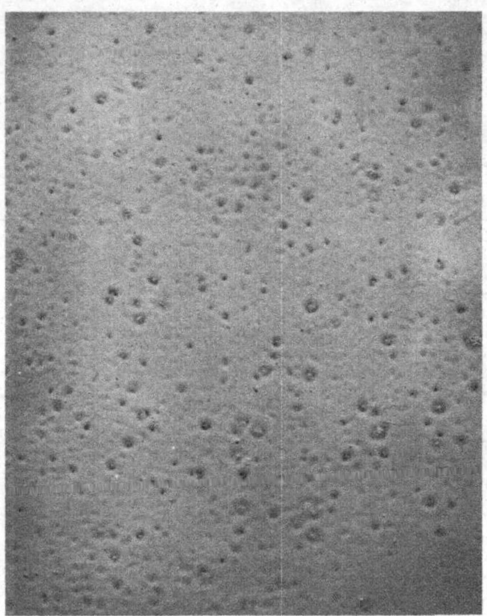

FIGURE 9.7 Effect of crater formation on appearance of a paint film. Craters were supposed caused by spreading from drops of low air–oil surface tension insoluble fluid over the surface of freshly applied paint coating. (Reprinted from *Prog. Org. Coat.*, 3, Bierwagen, G.P., 101. Copyright 1975, with permission from Elsevier.)

Another possible cause of crater formation by antifoams concerns the effect of non-spreading antifoam oils on the stability of films that imperfectly wet the relevant substrates. If the oil drop bridges the film, the resulting configuration can lead to a capillary instability as the film thins by evaporation whereupon a circular dry patch forms on the substrate and the oil drop is left as a separate central peak, superficially resembling that formed by a spreading oil. The phenomenon is depicted in Figures 9.8 and 9.9. Unlike Marangoni effects, it is not dependent on surface tension gradient driven flows. The phenomenon is well known [33–42] and bears more than a passing resemblance to the bridging–stretching antifoam mechanism (see Section 4.5.1). Whether it produces cratering in actual paint films is, however, speculative.

In addition to "cratering," the nomenclature of paint film defect is replete with many other different, poorly defined terms. For example, another common term is the so-called fish-eye defect. Cooper [43] in fact uses this term rather than "crater"

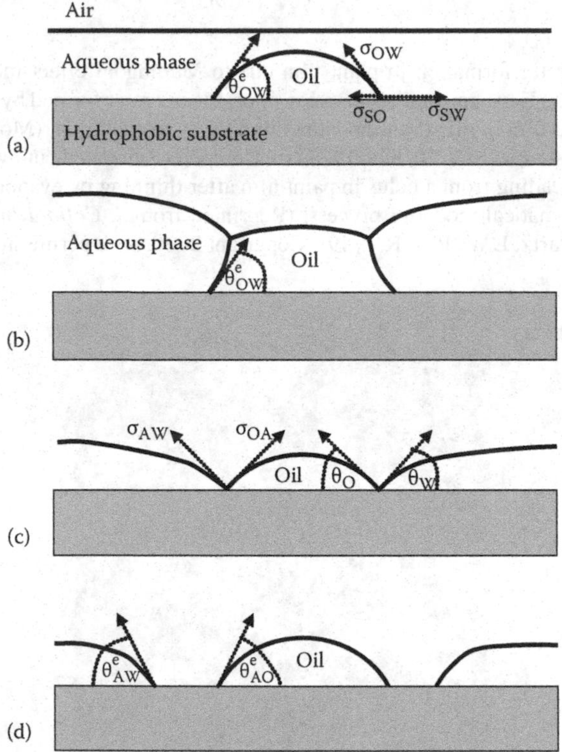

FIGURE 9.8 Possible configurations of bridging oil drop in aqueous film coating a hydrophobic surface in case of ideal wetting (i.e., no contact angle hysteresis). (a) Oil drop totally submerged in aqueous phase. (b) Partially submerged drop bridging aqueous film. (c) Configuration where the drop terminates at a four-phase contact line so that the oil–water surface area is eliminated. (d) If conditions 9.3 and 9.4 are satisfied as described in the text, then the four-phase contact line shown in c is unstable and the aqueous film peels away from the oil drop leaving dry patch.

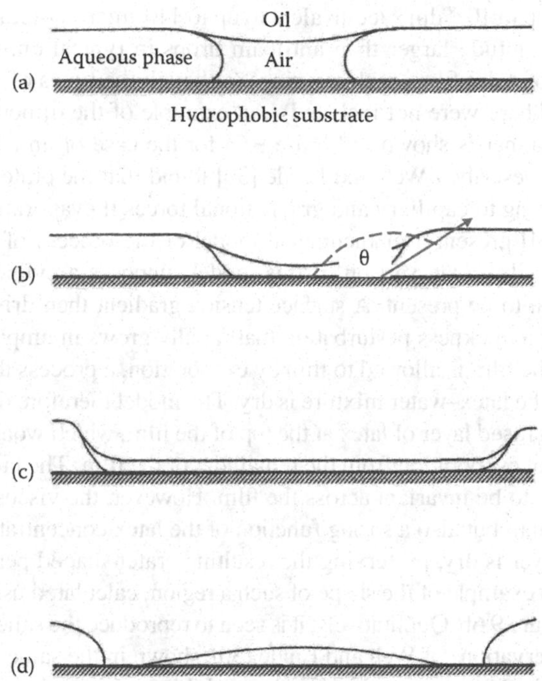

FIGURE 9.9 Schematic illustration of experimental observations of process of formation of dry patch induced by a bridging oil drop in case of aqueous film coating a hydrophobic surface where oil–water contact angle exhibits hysteresis. (a) Initial configuration. (b) As aqueous film thins, oil–water contact angle adopts advancing angle $\theta^a_{OW} > 90°$. (c) Further thinning forces the oil to spread at that angle, displacing the aqueous phase. (d) Capillary instability leads to formation of dry patch on substrate. (Reprinted from *J. Colloid Interface Sci.*, 68, Wilkinson, M.C., Zettlemoyer, A.C., Aronson, M.P., Vanderhoff, J.W., 508. Copyright 1979, with permission from Elsevier.)

to describe the formation of a "dry" patch in the substrate by an oil drop as shown in Figures 9.6, 9.8 and 9.9. However, here, we somewhat arbitrarily use the term "fish-eye" to define another potential type of defect. Thus, bridging oil drops in water-borne films may be caught in place by a drying paint film to form a defect without forming a dry patch. Such a defect could represent a serious weakness in the film if the oil is at all volatile. Gradual evaporation of the oil would eventually leave a hole in the film, forming a corrosion site if, for example, the substrate is steel.

Clearly much confusion could be eliminated if a generally agreed terminology for types of paint film defect could be devised. Ideally that should be based on the relevant physicochemical processes involved.

9.3.2.2 Experimental and Theoretical Studies of Cratering Caused by Marangoni Effect Induced by Spreading Oil Drops

Experimental studies of cratering by spreading antifoam oils such as silicone oils (i.e., polyorganosiloxanes) in paint films appear to be limited. Perhaps the most quoted is that of Weh and Linde [30] who observed cratering caused by the addition of silicone drops of volume up to 10^{-8} dm^{-3} (equivalent to up to 140-micron radius, which is about two orders of magnitude larger than antifoam drops in typical antifoam emulsions) to various types of paint films with a syringe (although the types of waterborne latex paints considered here were not included). An example of the dimensions of a crater created in this manner is shown in Figure 9.6a for the case of an alkyd-amine resin film. As we have described Weh and Linde [30] found that the craters were transient features, succumbing to capillary and gravitational forces if evaporation is prevented.

Evans et al. [31] present a mathematical model of the process of crater formation in a drying paint film. One version of this model supposes a cylindrical "bolus" of spreading material to be present. A surface tension gradient then drives the film fluid outward, creating a thickness perturbation that rapidly grows in amplitude and radius. Simultaneously, the film is allowed to thin by evaporation, a process that is assumed to be constant until the latex–water mixture is dry. The model therefore does not allow for the formation of a fused layer of latex at the top of the film, which would be expected to decrease the rate of evaporation from the remainder of the film. The latex concentration is in fact assumed to be invariant across the film. However, the viscosity of the film is supposed Newtonian but also a strong function of the latex concentration. Flow ceases when the paint layer is dry, preserving the resulting crater-shaped perturbed region of the paint film. An example of the shape of such a region, calculated using this model, is reproduced in Figure 9.6b. Qualitatively, it is seen to reproduce the salient features of the experimental observations of Weh and Linde [30], shown in the same figure.

Clearly then, inclusion of antifoams that exhibit positive spreading coefficients on aqueous latex surfaces, such as those based on polyorganosiloxanes, is likely to cause crater defects in paint and varnish films. Indeed, according to Kornum and Nielsen [29], Saitoh and Takano [44] write, after an investigation of cratering tendency, that ".... it is concluded that the silicone oil added to the paint should be completely soluble, otherwise it may become the origin of the craters in coating films." Cooper [43] has reported a comparative study of the foaming and film defect forming characteristics of silicone–hydrophobed silica and hydrocarbon–hydrophobed silica (where the hydrocarbon was kerosene or mineral oil) antifoams in an acrylic-based aqueous gloss emulsion applied to a clean glass surface. This study distinguished between craters caused by foam bubbles present due to ineffective antifoams and loss of film "quality" due to craters formed by local dewetting around drops of antifoam. After drying, the hydrocarbon–hydrophobed silica–containing films were characterized by the former defects and the silicone–hydrophobed silica–containing films were characterized by the latter defects. The hydrocarbon-based antifoam was therefore less effective in breaking foam films than the silicone antifoam. However, the silicone antifoam produced craters containing oil drops as the central peak surrounded by a region of dewetted paint film. It is tempting therefore to conclude that the hydrocarbon-based antifoam exhibits partial wetting and therefore does not spread over the paint surface so that craters do not form, whereas the silicone-based antifoam does spread (as is usually the case with aqueous systems; see Section 3.6.2) so that craters are inevitably formed as suggested by Saitoh and Takano [44].

Antifoams based on mixtures of polyorganosiloxanes and hydrophobed silica are clearly extremely effective in eliminating foamability, even in the present context. As a consequence, attempts have been made to overcome the concomitant problems of paint film defect formation. Arguably, these in the main concern reduction of the air–water surface tension of the paint, and therefore the spreading coefficient of the antifoam, by addition of a suitable surfactant to the paint [31, 32, 45, 46]. This approach has the added advantage of improving the wetting behavior of the paint on hydrophobic surfaces. According to Schmidt [28], use of alkoxylated polydimethylsiloxanes would appear to be particularly useful in this context. Indeed, Binks and Dong [47] report partial wetting of polydimethylsiloxane on solutions of an ethoxylated silicone surfactant. Such behavior implies lens formation at the air–water surface with no spreading at all (see Section 3.6.2.1) and therefore no Marangoni effect. However, the absence of a spread layer of PDMS may adversely affect antifoam performance as implied by Denkov et al. [48] (see Section 4.8.5). Nevertheless, Schmidt [28] reports that PDMS–hydrophobed silica antifoams can be effective for waterborne coatings despite the presence of an alkoxylated polydimethylsiloxane wetting agent dissolved in the paint. Unfortunately, this study does not include any observations of the incidence of paint defects due to the antifoam.

Other possible approaches to minimizing the formation of craters by Marangoni spreading in drying paint films are suggested by the mathematical model of the process given by Evans et al. [31]. They argue that both increasing the viscosity of the paint and increasing the drying rate of the film reduce the rate of crater development. Increasing the viscosity of the paint would, however, increase the stability of foam and decrease the defoaming efficiency of antifoams. Koerner et al. [49] suggest that such paint film defects may also be minimized by reducing antifoam drop sizes. This would of course increase the number of spreading sites that could in turn help reduce the surface tension gradients driving the Marangoni effect.

9.3.2.3 Putative Craters Caused by Non-Spreading Oil Drops Bridging Paint Films

The obvious requirement for crater formation by Marangoni spreading in drying paint films due to antifoam oil drops is that the oil must have a positive initial (or semi-initial) spreading coefficient (see Section 3.2). However, it is well known [33–42] that a non-spreading oil drop, exhibiting partial wetting, which bridges across an aqueous film covering a solid substrate, may lead to the partial dewetting of the film and the separation of the oil drop as the film thins if certain wetting conditions are satisfied. The process is schematically illustrated in Figures 9.8 and 9.9. If such a process occurred in a paint film due to the presence of antifoam drops, then it would obviously also mean the formation of unacceptable crater defects.

In Figure 9.8, we illustrate the various stages in the formation of a bridging oil drop in a thinning aqueous film, covering the surface of a substrate that is preferentially wetted by the oil. All contact angles are assumed to be ideal so that hysteresis is absent. Initially it is supposed that the drop is completely submerged in the film where the equilibrium contact angle, θ^e_{OW}, is $< 90°$ and is given by the Young equation so that

$$\sigma_{SW} = \sigma_{SO} + \sigma_{OW} \cos\theta^e_{OW} \tag{9.1}$$

where σ_{SW} is the substrate–water surface tension, σ_{SO} is the substrate–oil surface tension, and σ_{OW} is the oil–water interfacial tension. As the film thins, the oil drop emerges into the air–water surface of the film to form a bridging configuration with three surfaces—oil–air, oil–substrate, and oil–water. Further thinning gradually eliminates the oil–water surface until the configuration shown as Figure 9.8c is formed. According to Aronson et al. [36], equilibrium then requires that

$$\sigma_{AO}\cos\theta_O = \sigma_{AW}\cos\theta_W + \sigma_{OW}\cos\theta^e_{OW} \tag{9.2}$$

where θ_O and θ_W are defined by Figure 9.8c and where σ_{AO} and σ_{AW} are the air–oil and air–water surface tensions, respectively. Aronson et al. [36] show that the configuration is unstable with respect to the formation of substrate–air surface as depicted in Figure 9.8d if the inequalities

$$\theta^e_{AO} > \theta_O \tag{9.3}$$

and

$$\theta^e_{AW} > \theta_W \tag{9.4}$$

are satisfied, where θ^e_{AO} and θ^e_{AW} are the equilibrium air–oil and air–water contact angles, respectively.

This analysis assumes ideal surface behavior. In practice, contact angle hysteresis is to be expected with most real systems. Wilkinson et al. [39] in fact report an experimental study of the stability of these bridging drop configurations where hysteresis is a common feature of the systems studied. Typical behavior is shown in Figure 9.9, which represents actual observations. Here an oil drop is forced to spread with an advancing contact angle, $\theta^a_{OW} > 90°$ against the hydrophobic substrate as the aqueous film thins. Formation of an isolated oil drop with a dewetted region of substrate–air surface is again observed. In contrast to Figure 9.8d, this is occurring even though the oil–water surface area is still finite.

We therefore have a mechanism that could, at least in principle, exist in drying paint films containing non-spreading antifoam drops. It would clearly mainly concern aqueous paint films on hydrophobic surfaces. However, it would not be confined to oils of low viscosity—the phenomenon occurs in aqueous films even with oils of viscosities as high as 10^4 mPa s [39], although the effect of the high viscosities of the aqueous phase likely to prevail in drying paint films is not apparently known.

There would therefore appear to be some merit in establishing whether this phenomenon can occur in drying waterborne latex paint films. Farinha et al. [50] have in fact reported a study of the effect of a drying aqueous latex film on the wetting behavior of a hydrophobic oil (triolein) contaminating both hydrophilic and hydrophobic substrates. Unfortunately, the oil drops present in these films were too small to form bridging configurations. Clearly the phenomena depicted in Figures 9.8 and 9.9 require the diameters of the antifoam drops to be of the same order as that of the relevant paint films. Avoidance of the type of putative paint film defect described here therefore simply involves control of antifoam drop sizes.

9.4 SUMMARIZING REMARKS

It is clear that control of foam in the case of waterborne latex paints and varnishes is easily achieved using the well-known concepts realized in mixtures of hydrophobic oils and hydrophobic particles. Occasional references in the relevant patent and scientific literature to the oil component as a mere "carrier" does, however, suggest a failure to embrace modern understanding about the mode of action of such antifoams [18, 24, 25, 51]. This is surprising since that understanding is based on experimental evidence, carefully garnered over more than a quarter of a century and published in respected, peer reviewed, scientific journals.

A key issue in the application of oil-based antifoams for waterborne paints concerns avoidance of defects in paint films. The spreading behavior of the oil on emergence into the air–film surface of a drying paint film would appear to determine whether defects occur. Such oils may, however, exhibit a variety of spreading behaviors. These may in principle range from complete wetting (i.e., duplex film formation) to pseudo-partial wetting where the oil forms a lens in equilibrium with an oil-contaminated surface to partial wetting where the oil does not spread but just forms a lens (see Chapter 3). Which of these possibilities is realized depends not only on the intrinsic properties of the oil but also on the air–surface properties of the paint film as determined largely by the surfactant present.

It seems that a Marangoni effect due to spreading from antifoam oils could be the cause of crater-like defects in paint films. Clearly that effect is only possible if the oil exhibits either complete wetting or pseudo-partial wetting of the paint film surface because it is only with those behaviors that surface tension gradients can be formed. That a Marangoni effect due to spreading of the oil could be the main cause of paint defects may, however, at first seem surprising since such effects are not now considered to be central to the defoaming mode of action of the relevant antifoam entities. However, we should note that defoaming effects are induced by both spreading and non-spreading oils. By contrast, practical experience suggests that non-spreading oils do not apparently give rise to crater defects, whereas spreading oils do. Moreover, Marangoni effects would appear to be transient in both foam films and paint films in the absence of evaporation. Crater formation in the case of the latter only occurs because those otherwise transient structures are "frozen" in place by the fusion of the latex particles as water evaporates.

Although plausible, this view of crater defect formation due to antifoam oils has not apparently been subject to a detailed rigorous study of the role of the effect of oil spreading behavior in this context. Obviously the stability of the pseudoemulsion film as influenced by the presence of hydrophobic particles is also an important factor. If that film is stable, then the oil will not emerge into the air–paint film surface so that spreading will not occur regardless of the intrinsic properties of the oil. Therefore, the absence of defect formation due to that cause should correlate with ineffective antifoam behavior, which is also dependent on the stability of the pseudoemulsion film. There is also the possibility that oils that exhibit non-spreading, partial wetting, behavior may cause defects by forming bridging configurations as shown in Figures 9.8 and 9.9. Superimposed on all these effects is of course the role of the rate of evaporation and latex fusion. Systematic exploration of all these issues might help produce understanding that could reduce the amount of testing necessary in selection of antifoams for given waterborne paints.

REFERENCES

1. Wigglesworth D.J. Volatile organics—Legislation and the drive to compliance, in *The Chemistry and Physics of Coatings* (Marrion, A.R., ed.), The Royal Society of Chemistry, Cambridge, UK, 1994.
2. Harkins, W.D. *J. Am. Chem. Soc.*, 69, 1428, 1947.
3. El-Asser, M.S., Sudol, E.D. Features of emulsion polymerisation, in *Emulsion Polymerisation and Emulsion Polymers* (Lovell, P.A., El-Asser, M.S., eds.), Wiley, New York, 1997, Chpt 2, p 37.
4. Dunn, A.S. Harkins, Smith-Ewart and related theories, in *Emulsion Polymerisation and Emulsion Polymers* (Lovell, P.A., El-Asser, M.S., eds.), Wiley, New York, 1997, Chpt 4, p 126.
5. Dunk, W.A.E. Disperse phase polymers, in *The Chemistry and Physics of Coatings* (Marrion, A.R., ed.), The Royal Soc. of Chem, Cambridge, UK, 1994, Chpt 8, p 142.
6. Fitch, R.M. Emulsion polymerisation, in *Polymer Colloids—A Comprehensive Introduction*, Academic Press, San Diego, California, 1997.
7. Hellgren, A., Weissenborn, P., Holmberg, K. *Prog. Org. Coat.*, 35, 79, 1999.
8. Chainey, M., Reynolds, P.A. Mechanical properties of composite polymer latex films, in *Colloidal Polymer Particles* (Goodwin, J.W., Buscall, R., eds.), Academic Press, London, 1995, p 181.
9. Chevalier Y., Pichot, C., Graillat, C., Joanicot, M., Wong, K., Maquet, J., Lindner, P., Cabane B. *Colloid Polym. Sci.*, 270, 806, 1992.
10. Joanicot, M., Wong, K., Maquet, J., Chevalier Y., Pichot, C., Graillat, C., Lindner, P., Rios, L., Cabane B. *Prog. Colloid Polym. Sci.*, 81, 175, 1992.
11. Kientz, E., Holl, Y. *Colloids Surf. A*, 78, 255, 1993.
12. Lam, S., Hellgren, A.C., Sjoberg, M., Holmberg, K., Schoonbrood, H.A.S., Unzue, M.J., Asua, J.M., Tauer, K., Sherrington, D.C., Goni, A.M. *J. Appl. Polym. Sci.*, 66, 187, 1997.
13. Basheva, E.S., Nikolov, A.D., Kralchevsky P.A., Ivanov I.B., Wasan, D.T. *Surfactants in Solution* (Mittal, K.L., Shah, D.O., eds.), Plenum Press, New York, Vol 11, 1991, p 152.
14. Nikolov A.D., Wasan, D.T. *Langmuir*, 8, 2985, 1992.
15. Sethumadhaven, G.N., Nikolov, A.D., Wasan, D.T. *J. Colloid Int. Sci.*, 240, 105, 2001.
16. Sethumadhaven, G.N., Bindal, S., Nikolov, A.D., Wasan, D.T. *Colloids Surf. A*, 204, 51, 2002.
17. Garrett, P.R., Wicks, S.P., Fowles, E. *Colloids Surf. A*, 283, 307, 2006.
18. Schulte, H.G., Hofer, R. Uses of anti-foaming agents in paints and surface coatings, in *Surfactants in Polymers, Coatings, Inks and Adhesives* (Karsa, D.R., ed.), Applied Surfactant Series, Vol 1, Chpt 4, p 93.
19. Baxter, K.F. Formulations of coating compositions, in *The Chemistry and Physics of Coatings* (Marrion, A.R., ed.), The Royal Society of Chemistry, Cambridge, UK, 1994, Chpt 9, p 165.
20. Cui, Z.-G., Cui, Y.-Z., Cui, C.-F., Chen, Z., Binks, B.P. *Langmuir*, 26(15), 12567, 2010.
21. Smolders, C.A. *Rec. Trav. Chim.*, 80, 650, 1961.
22. Johnson, R.E., Dettre, R.H. Wettability and contact angles, in *Surface and Colloid Science* (Matijevic, E., ed.), Wiley, New York, 1969, Vol 2, p 85.
23. Porter, M.R. Antifoams for paints, in *Defoaming, Theory and Industrial Applications*, Edn. 1 (Garrett, P.R., ed.), Marcel Dekker, New York, Surfactant Science Series, Vol 45, 1993, Chpt 7, p 269.
24. Shah, S.C., Herzog, D., Bene, P., Wiggins, M.S. (assigned to Cognis IP Management GmbH), WO 2009/106252; 3 September 2009, filed 18 February 2008.
25. Wiggins, M.S., Broadbent, R.W. (assigned to Cognis Corp.), WO 02/00319; 3 January 2002, filed 26 June 2001.

26. Mangano, J.B., Balasubramanian, V., Herzog, D. (assigned To Cognis IP Management GmbH), EP 2,374,517; 12 October, 2011, filed 27 January 2011.
27. Kahn, A., Firman, S.A., Brown, D.W., Nowicki, J.E. (assigned to Cognis Corp.), WO 02/13940;21 February 2002, filed 1 August 2001.
28. Schmidt, O. *Eur. Coat. J.*, 4, 40, 2010.
29. Kornum, L.O., Nielsen, H.K.R. *Prog. Org. Coat.*, 8, 275, 1980.
30. Weh, L., Linde, H. *Plaste Kautsch*, 20, 849, 1973.
31. Evans, P.L., Schwartz, L.W., Roy, R.V. *J. Colloid Interface Sci.*, 227, 191, 2000.
32. Bierwagen, G.P. *Prog. Org. Coat.*, 3, 101, 1975.
33. Clayfield, E.J., Dear, J.A., Matthews, J.B., Whittam, T.V. *Proc. Int. Congress Surface Activity, 2nd*, 3, 165, 1957.
34. Zettlemoyer, A.C., Aronson, M.P., Lavelle, J.A. *J. Colloid Interface Sci.*, 34(4), 545, 1970.
35. Wilkinson, M.C., Aronson, M.P., Zettlemoyer, A.C. *J. Colloid Interface Sci.*, 37(2), 498, 1971.
36. Aronson, M.P., Zettlemoyer, A.C., Wilkinson, M.C. *J. Phys. Chem.*, 77(3), 318, 1973.
37. Aronson, M.P., Zettlemoyer, A.C., Codell, R., Wilkinson, M.C. *J. Colloid Interface Sci.*, 52(1), 1, 1975.
38. Aronson, M.P., Zettlemoyer, A.C., Codell, R., Wilkinson, M.C. *J. Colloid Interface Sci.*, 54(1), 134, 1976.
39. Wilkinson, M.C., Zettlemoyer, A.C., Aronson, M.P., Vanderhoff, J.W. *J. Colloid Interface Sci.*, 68(3), 508, 1979.
40. Wilkinson, M.C., Ellis, R., Aronson, M.P., Vanderhoff, J.W., Zettlemoyer, A.C. *J. Colloid Interface Sci.*, 68(3), 545, 1979.
41. Wilkinson, M.C., Mattison, I.C., Zettlemoyer, A.C., Vanderhoff, J.W., Aronson, M.P. *J. Colloid Interface Sci.*, 68(3), 560, 1979.
42. Wilkinson, M.C., Aronson, M.P., Zettlemoyer, A.C., Vanderhoff, J.W. *J. Colloid Interface Sci.*, 68(3), 575, 1979.
43. Cooper, P. *Surf. Coat. Aust.* 38(3), 8, 2001.
44. Saitoh, T., Takano, N. *Colour Mater. (in Japanese)*, 45(7), 349, 356, 1972; ibid., 46(8), 419, 1973; ibid., 47(9), 402, 1974.
45. Hahn, F.J. *J. Paint Technol.*, 43, 58, 1971.
46. Hahn, F.J., Steinhauer, S. *J. Paint Technol.*, 47, 54, 1975.
47. Binks, B., Dong, J. *JCS FaradayTrans.*, 94(3), 401,1998.
48. Denkov, N.D., Tcholakova, S., Marinova, K.G., Hadjiiski, A. *Langmuir*, 18(15), 5810, 2002.
49. Koerner, G., Fink, F., Berger, R., Heilen, W. *World Surfactant Congress, IV*, 211, 1984.
50. Farinha, J.P.S., Winnik, M.A., Hahn, K.G. *Langmuir*, 16(7), 3391, 2000.
51. Karsa, D.R. Process aids and additives for latices and thermoplastics, in *Surfactants in Polymers, Coatings, Inks and Adhesives* (Karsa, D.R., ed.), Applied Surfactant Series, Blackwell, Oxford, Vol 1, Chpt 9, p 245.

10 Antifoams for Gas–Oil Separation in Crude Oil Production

10.1 INTRODUCTION

The "live" crude oil produced at well heads contains dissolved gas and dispersed saline water. That oil is often at pressures as high as 10 MPa and at temperatures in excess of 100°C. After production, it must be reduced to ambient temperature and pressure for storage, either immediately or after pipeline transportation. However, reduction of pressure releases dissolved "natural" gas, potentially forming foam. The process of pressure reduction occurs in a series of so-called gas–oil separators, each representing various stages in pressure reduction down to ambient. These separators are essentially large tanks, in which foam is eliminated and out of which separate streams of oil and gas flow and water is collected. Foaming is potentially most severe in the first stage high-pressure separator (see, e.g., reference [1]). Elimination of that foam is necessary to achieve efficient separation of oil from gas. Failure to complete this process can mean, for example, that foam is carried downstream to contaminate gas lines with liquid, leading to flooding of both vessels and gas compressors and therefore plant shutdown [1–4].

A schematic diagram illustrating the main features of a gas–oil separator is shown in Figure 10.1. The foam generated is usually intrinsically unstable and therefore increasing the residence time in these vessels potentially assists separation of the gas phase. Indeed, baffles to hold back foam flow are usually incorporated to increase those times [5]. In practice, residence times are typically of the order of minutes. However, simply relying on residence times long enough for total foam collapse implies ever larger tanks, depending on the stability of the foam. This approach does not therefore usually represent an acceptable solution to any foam problem. This is especially true of a combination of both highly viscous crudes, which can form exceptionally stable foam [6], and expensive offshore installations where space is at a premium. It is therefore common practice to add antifoams to the crude oil stream as shown in Figure 10.1.

The use of antifoams in gas–oil separation arguably represents the "largest single application" of such additives in the petroleum industry according to Pape [2]. Polyorganosiloxane derivatives are often employed as antifoams in this context. However, as so often with antifoams, there can be undesirable consequences associated with their use. One such potential problem concerns the deposition of polydimethylsiloxanes (PDMSs) on downstream cracking catalyst surfaces, leading to diminished effectiveness [2, 7]. As a consequence, mechanical methods of

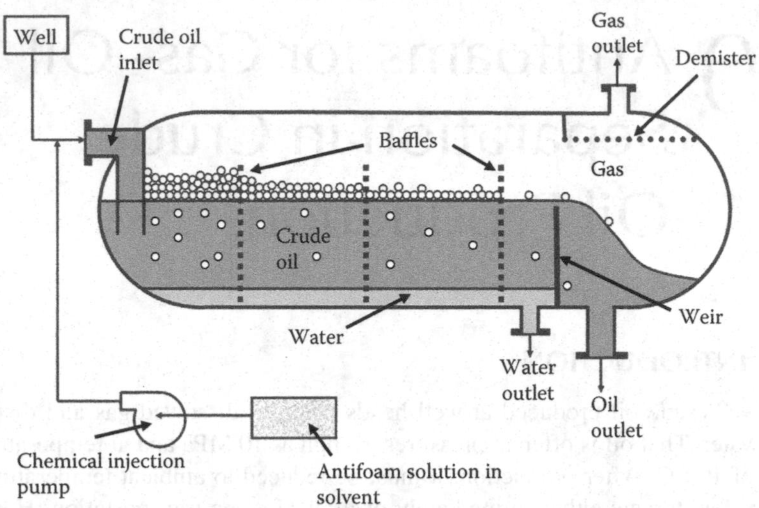

FIGURE 10.1 Schematic diagram showing some typical features of gas–oil separator, including manner of antifoam injection. Baffles act to increase residence times of fluids and to moderate surges. Demisters (of stainless-steel mesh) remove any crude oil drops from outlet gas stream. Water originates in the oil reservoir where it usually coexists with crude oil.

defoaming, such as cyclones, centrifuges [8], and even ultrasound [9], have sometimes been advocated. However, according to Pape [2], such devices usually only produce a reduction of foaming tendency and antifoams are still generally necessary [10]. Indeed, Callaghan [5] claims that cyclonic devices may in fact worsen foam control in "some instances." A complete review of the mode of action of mechanical defoaming devices is to be found in Chapter 7.

In this chapter, we first consider the nature of surface activity in gas–hydrocarbon interfaces in general and in gas–crude oil systems in particular. That represents an issue fundamental to understanding the causes of foam formation in gas–oil separators and has relevance for the mode of action of antifoams in that context. In a separate section, we consider the possible causes of foam formation in gas–crude oil systems. There we review the observations of foam behavior in gas–crude oil systems and make the limited comparison with theory which that permits. Finally we consider foam control in gas–oil separators, which usually involves the use of antifoams. Therefore, we describe the design criteria for suitable antifoams and the evidence available concerning their mode of action in a non-aqueous medium such as crude oil.

10.2 SURFACE ACTIVITY AT GAS–HYDROCARBON AND GAS–CRUDE OIL INTERFACES

In the case of gas–crude oil systems, surface activity is defined by the Gibbs adsorption equation, which relates the adsorption (strictly the surface excess) of a solute, Γ_s^{GC}, at the gas–crude oil interface to the derivative of the gas–crude oil surface tension, σ_{GC}, with respect to the chemical potential, μ, of that solute [11]. Since the chemical potential is proportional to the logarithm of the activity, $\bar{a}$, of the solute, we can write the Gibbs equation in the form

$$\Gamma_s^{GC} = -\frac{d\sigma_{GC}}{d\mu} = -n\frac{d\sigma_{GC}}{RTd\ln\bar{a}} \tag{10.1}$$

where n is a numerical constant dependent on the degree of ionization of the species. Clearly then adsorption will not occur if the solute cannot lower the surface tension of the liquid.

Adsorption is determined not only by the free energy of the interactions of the solute in the surface but also by those with the solvent in the bulk phase. The latter can be maximized by increasing the activity in the bulk phase until the solubility (or in some cases the CMC) is reached. In some circumstances, this will result in a close-packed adsorbed monolayer, which must have a surface tension lower than that of the pure solvent.

A rough indication of the surface tension of a liquid surface is given by the empirical critical wetting tension concept of Zisman [12] if we accept that there is some equivalence between the state of a solid surface covered with the same chemical groups as exposed at the outer molecular surface of the liquid. We present a table of critical wetting tensions for some of the groups relevant in the present context at ambient temperatures in Table 10.1. They indicate that in the case of a mainly hydrocarbon mixture like crude oil, with a relatively high ratio of CH_2 to CH_3 molecular groups, surface tensions will be close to 31 mN m^{-1}. This compares with measured values at ambient temperatures of the air–oil surface tensions for degassed, so-called dead, crude oils in the range of, for example, 26.3–28.7 mN m^{-1} [4] and 31.0 mN m^{-1} [13]. Differences in the surface tensions between these crude oils could simply reflect differences in the CH_2 to CH_3 ratios. By contrast, the long chain polydimethylsiloxanes, often used as antifoams for crude oils, all have air–oil surface tensions

TABLE 10.1
Critical Wetting Tensions of Some Chemical Groups at 20°C

Surface Groups	Critical Wetting Tension (mN m^{-1})
CF_3	6
CF_2	18
CH_3 (crystal)	22
CH_3 (monolayer)	20–24
CH_2	31

Source: Zisman, W.A., *Adv. Chem. Ser.*, 43, 1, 1964.

in the region of 20 mN m^{-1} (see, e.g., reference [14]), largely as a result of a surface dominated by a combination of low surface energy CH_3 groups pendant upon an extremely flexible siloxane backbone [15]. Alkoxylated PDMS oils, also used as antifoams in this context, have air–oil surface tensions in the region of 22 mN m^{-1} [13] where the surface would again appear to be largely dominated by CH_3 groups. Perfluoroalkyl-substituted PDMSs represent yet another class of suitable antifoams, this time with the surface dominated by CF_2 groups and having air–oil surface tensions also in the region of 20 mN m^{-1} [16].

In principle, adsorption of these PDMSs and their derivatives at air–crude oil surfaces should therefore be possible, potentially leading to close-packed adsorption layers and surface tensions reduced to about 20 mN m^{-1}. However, the solubilities of these compounds in crude oil are relatively low, which could mean that bulk phase activities are never sufficiently high to realize close-packed monolayers—reductions in air–crude oil surface tensions could therefore be correspondingly modest. Mannheimer [17] has, for example, shown that the surface tension reduction of a variety of hydrocarbon oils by a given PDMS oil is greater the greater the apparent solubility of the PDMS—in no case, however, was the surface tension reduced to the expected surface tension of a close-packed CH_3-rich layer.

Schaefer [18] has shown that the solubilities of PDMS in hydrocarbons can be increased by partial substitution of long-chain hydrocarbons for some methyl groups in the PDMS to produce so-called silicone waxes. In turn increased solubility of silicone waxes means that the surface tensions of their solutions in hydrocarbon can be reduced to more closely reach the potential of close-packed methyl groups, with air–hydrocarbon surface tensions as low as 22 mN m^{-1}. We would expect therefore that silicone waxes would show an enhanced tendency to adsorb at gas–oil surfaces relative to underivatized PDMSs. This would presumably, in turn, lead to an enhanced tendency to act as pro-foamers. By analogy, we would also expect that partial substitution of more polar groups for methyl groups in PDMSs would decrease solubilities in hydrocarbons and therefore lead to less effective reduction of air–hydrocarbon surface tensions and therefore less adsorption. Fraga et al. [13] in fact report reductions of the surface tension of a degassed crude oil by ethoxylated PDMSs, at concentrations close to the solubility limit, of only 1–3 mN m^{-1} from an intrinsic value for the oil of 31.0 mN m^{-1}.

The adsorption at air–water surfaces of conventional surfactants with hydrocarbon, PDMS, or perfluorocarbon chains from aqueous solutions is mainly driven by the hydrophobic effect. That, in turn, is driven by the entropic consequences of diminished disorder imposed on H-bonded solvent molecules in the vicinity of these chains. The latter exhibit only weak attractive interactions with water molecules. This diminishes the number of configurations that the water molecules can adopt in the vicinity of a hydrocarbon chain, leading to a decrease in entropy. Adsorption of surfactants with hydrocarbon chains dissolved in hydrocarbon environments such as crude oil cannot therefore be driven by the same mechanism. The intermolecular affinity between the like chains in solute and solvent would be too great. On the other hand, adsorption cannot involve the exposure of a surfactant polar group at the gas–oil surface because this would increase the surface energy, which would in turn produce negative surface excesses according to the Gibbs equation (Equation 10.1).

Based on this reasoning, we would not therefore expect simple surfactants with hydrocarbon chains, if present in crude oils, to adsorb and modify gas–oil surface properties. It would seem unlikely then that presence of such surfactants will significantly enhance the lifetime of foam films unless that can occur even with negative surface excesses. Rather surprisingly then, we find that Callaghan et al. [19] report that alkaline extraction of several different crude oils removes short chain (i.e., $<C_{11}$) alkanoic and phenolalkanoic acids, thereby eliminating both any foaming tendency as measured by foaminess, Σ_{BIK} (defined in Section 2.2.5) and any measurable surface dilatational relaxation spectrum. This finding deserves further investigation, perhaps involving addition of the supposed foam-enhancing alkanoic and phenolalkanioic acid surfactants to oil that had been subjected to the same alkaline extraction in order to confirm the specific chemical nature of the effect.

By contrast, the specific components of crude oil that have been most closely associated with foam behavior reveal radically different chemistry to these simple hydrocarbon chain surfactants. As exemplified by the work of Poindexter et al. [4, 20], those components can be listed as asphaltenes, resins, and waxes. Of these, arguably asphaltenes are the most important. These components are derivatives of polycyclic aromatics, which are distinguished from other crude oil components by insolubility in short-chain n-alkanes such as n-heptane. They are, however, soluble in toluene. Resins are soluble in short-chain alkanes and are therefore usually extracted from crude oil by adsorption onto silica from solution. Both asphaltenes and resins can even each be present in crude oil at concentrations in excess of 15 wt.%. Such extremely high concentrations usually lead to crude oils of high density and high viscosity—so-called heavy crudes (see, e.g., reference [4]).

Asphaltenes are believed to form a hierarchy of structures in crude oil and other solvents, such as toluene. These structures are described by Mullins [21] after an original version by Yen [22]. They are depicted in Figure 10.2. The details vary with

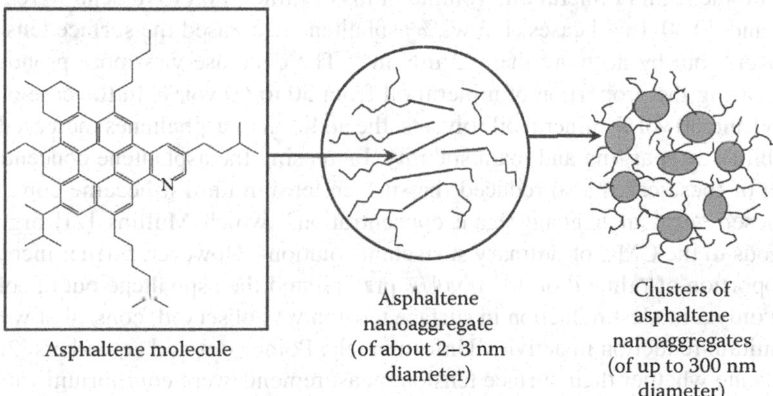

FIGURE 10.2 The modified Yen model [22] of asphaltene structure showing likely hierarchical aggregation behavior in crude oil. The structure of the asphaltene molecule depicted is typical but will vary for different crude oils. (Reprinted with permission from Mullins, O.C. *Energy Fuels*, 24, 2179. Copyright 2010 American Chemical Society.)

the origin of various crude oils but the usual primary molecular structure is of a single polycyclic aromatic core with about seven benzene rings decorated by linear and cyclic saturated hydrocarbon chains. Hetero atoms may be present as shown in the figure. Ionizable groups are absent. In suitable solvents such as toluene and crude oil, asphaltene molecules aggregate by molecular stacking of the polycyclic regions through π–π interactions to form "nanoaggregates" of aggregation number 4–10. The hydrocarbon chains surround the polycyclic core to give a structure of about 2–3 nm diameter having a superficial resemblance to a surfactant micelle and described by Mullins [21] as a "hairy tennis ball!" Depending on the nature of the medium in which the aggregates are dispersed, they may further associate to form clusters of size up to 300 nm diameter.

The structure of resins is less well understood than that of asphaltenes. Speight [23] argues that they may be regarded as similar molecules to asphaltenes but with lower molecular weights and a greater proportion of saturated hydrocarbon attached to the polycyclic aromatic core of the molecules. This of course correlates with a higher solubility of resins in n-alkanes.

Bauget et al. [24] have studied the surface activity of both an asphaltene and a resin at the air–toluene surface. The asphaltene is clearly surface active at that surface. However, the rate of reduction in air–toluene surface tension is extremely slow—even after 1 day equilibrium was not achieved at 20 wt.% asphaltene concentration. After that time, there was clear evidence of formation of a solid skin on the surface from concentrated solutions of the asphaltene. Bauget et al. [24] infer that the rate of adsorption is not diffusion controlled but rather is controlled by the formation of coalesced clusters in the surface. Addition of resin significantly increased the surface tension of asphaltene solutions and ameliorated the formation of the solid skin, implying solubilization of the asphaltene.

The effect of asphaltene and resin on the surface tension of solvents has also been described by Poindexter and coworkers [20]. Here crude oil was simulated by mixtures of toluene and mineral oil. Volume ratios of mineral oil to toluene were 50:50, 60:40, and 70:30. In all cases, 1–3 wt.% asphaltene decreased the surface tension of the solvent, but by no more than ~2 mN m^{-1}. The decrease was more pronounced on increasing the proportion of mineral oil from 50 to 60 vol.%. In the case of both these 50 and 60 vol.% mineral oil solvents, the addition of asphaltenes increased both foamability by sparging and foam stability. Increasing the asphaltene concentration in both of these cases also reduced the surface tension until it became constant at a supposed "critical nanoaggregate concentration," which Mullins [21] argues is analogous to the CMC of ordinary surfactant solutions. However, further increasing the proportion of mineral oil to 70 vol.% precipitated the asphaltene out of solution so that only a modest reduction in surface tension was observed, consistent with the concomitant reduction in activity. Unfortunately, Poindexter and coworkers [20] did not indicate whether their surface tension measurements were equilibrium values.

A comparison of the surface activity of two different resins was also made by Poindexter and coworkers [20] in the same mineral oil–toluene mixtures. Spectroscopy revealed that one of these resins was "highly polar," with carboxyl and carbonyl groups present, and the other not so. Both reduced the air–solvent surface tensions of these solvent mixtures. However, the more polar resin was more effective.

Just why a more polar compound can adsorb at a non-polar hydrocarbon surface to reduce the surface tension to a greater extent is rather difficult to explain!

We therefore conclude that there is some evidence that certain components of crude oils, including asphaltenes and resins, can adsorb at the air–liquid surface of non-polar hydrocarbon liquids. However, the molecular origin of such surface activity is far from clear, especially in those cases where significant polarity is present. Moreover, there appears to be little direct evidence that, for example, adsorption of asphaltenes and resins occurs at gas–crude oil surfaces. The main challenge therefore remains the difficult task of verifying that such surface activity is actually present in crude oils.

10.3 CAUSES OF FOAM FORMATION IN GAS–CRUDE OIL SYSTEMS

10.3.1 Disjoining Pressures

The dielectric constant (i.e., static relative permittivity) of crude oil is typically of a similar order to that of other hydrocarbons at ~2 at 20°C (see, e.g., reference [25]), whereas that of water is ~80. The low dielectric constant of the oil means that electrostatic attraction between charged entities is higher and dissociation of any dissolved ionic species present is therefore likely to be low. In turn, this will lead to decreased solubilities of electrolytes, low ionic strengths, increased Debye lengths, and low electrostatic potentials at surfaces. In turn, this will mean low positive contributions to the disjoining pressure isotherm from adsorbed ionic species, which, in the absence of any other positive contribution, will necessarily mean that films will drain to a critical thickness and then succumb to rupture. This process occurs as random thickness fluctuations, amplified by negative van der Waals contributions to the disjoining pressure, grow catastrophically. A simplified account of this mechanism, based on the arguments of Vrij [26, 27], is to be found in Section 1.3.3. In this respect, a major contribution to foam film stability present in aqueous solutions is absent from crude oil systems. Rapidly draining foams formed from low-viscosity crude oils therefore tend to be transient, a fact utilized in gas–oil separators, which are sometimes designed to produce residence times consistent with achieving significant foam decay even in the absence of antifoams.

Adsorbed non-ionic species of surfactants and polymers present in oil may, on the other hand, in principle, make positive contributions to disjoining pressure isotherms as a result of steric effects. As we have seen, adsorption at crude oil–gas surfaces generally implies molecules with low surface energy groups such as methyl siloxane derivatives, perfluoromethyl, and perfluoromethylene derivatives. Evidence of significant positive contributions to the disjoining pressure isotherms of gas–hydrocarbon–gas foam films from the adsorption of such molecules is, however, rare. Perhaps the only clear example is the effect of the adsorption of a perfluoroalkyl polymeric ester at air–hydrocarbon surfaces. Callaghan et al. [28] claim that solutions of this compound in crude oil, at concentrations as low as 10^{-4} wt.%, are sufficient to increase the foaminess, Σ_{BIK} by ~15%. Bergeron et al. [29] claim that this perfluoroalkyl polymeric ester also produces extremely stable foam in the air–dodecane system.

Measurements of the disjoining forces in films prepared from 1 wt.% solutions of this perfluoroester in dodecane reveal no repulsion until the film thickness becomes ≤~17.5 nm. Bergeron et al. [29] suggest that this behavior represents strong evidence of positive contributions to the disjoining pressure isotherm from steric interactions. They find that such air–dodecane–air films do not rupture even with an external capillary suction pressure of >30 kPa (which corresponds to >~0.3 bar).

Perfluoroalkyl polymeric esters obviously do not have an intrinsic presence in crude oil. However, we should stress that this pro-foaming behavior illustrates the danger of adding siloxane or perfluoroalkyl compounds to gas–oil separators at concentrations below their solubility limit in the relevant live crude system in an attempt to achieve defoaming! That PDMS antifoams can actually promote foam formation if dissolved in hydrocarbons is of course well known (see, e.g., reference [30]). A corollary of this observation is that if a crude oil sample has been subject to foam control using, for example, a PDMS oil, and subsequently filtered or centrifuged to remove undissolved material then the foamability could be higher than intrinsic because of the presence of dissolved PDMS. All of this represents yet another reason for care in sampling crude oil for foam and antifoam studies. Clearly live crude oil sampled before any addition of antifoam is best!

10.3.2 Origin of Surface Tension Gradients at Gas–Crude Oil Interfaces

As we have seen, crude oil foams are usually transient (see, e.g., reference [31]). The absence of significant positive contributions to disjoining pressures means that foam collapse is inevitable when the constituent films drain to their so-called critical thickness. This means that the process of film drainage largely determines foam persistence. In turn, that process is usually dominated by the tangential stress boundary condition, Equation 1.1, which equates the viscous shear stress to the surface tension gradient in the draining foam film.

It is well known that pure liquids do not usually form foam. For example, the lifetimes of bubbles formed on shaking a vessel containing extremely pure water last ≤1 s. Surface tension gradients due to variations in surface compositions cannot form in such liquids,* which means in turn that the response to any external force (gravity, Plateau border suction) cannot include viscous shear resistance because velocity gradients will not then exist (see Section 1.3.1). Such films will exhibit plug flow and will therefore usually succumb rapidly to rupture as film elements are accelerated toward the direction of the perturbing force. Since shear forces are not involved, the shear viscosity of the fluid will be irrelevant, only high extensional viscosities will influence the outcome (where the extensional viscosity is usually three times the

* An exception to this concerns the formation of monomolecular crystalline layers on the surfaces of molten alkane waxes at temperatures close to the melting point as reported by, for example, Gang et al. [32]. This layer causes increases in the lifetimes of the transient froths formed by such materials. Presumably, the stress at the surface due to such a layer will result in the formation of velocity gradients and therefore shear stress to resist drainage. Gang et al. [32] state, however, that "the stabilization of a foam by surface crystallization or surface-induced crystallization of the interior, by a non-surfactant component, is clearly not common but may be an additional mechanism at work in complex systems where a monolayer crystallizes at a higher temperature than the bulk."

shear viscosity). The effect of the latter can *in extremis* cause films to form even in the absence of the shearing caused by surface tension gradients.

In the case of all liquid mixtures, such as crude oil, surface compositional changes are possible as a consequence of the presence of many components having different surface activities. Surface tension gradients can occur in response to fluid flow in the bulk phase because dynamic surface compositional changes, which cause the gradients, cannot relax sufficiently rapidly. In turn, such surface tension gradients can play a key role in determining the growth of the hydrodynamic and capillary instabilities, which dominate film drainage [33] and rupture [26, 27] as briefly outlined in Section 1.3.

10.3.3 Experimental Observations of Foam Behavior

Callaghan and Neustadter [31] have made a study of the foam stabilities of air–crude oil and natural gas–crude oil systems using a variety of light crude oils of viscosities ≤14 mPa s. This study, at ambient temperature using a sparging method, concerned so-called dead oils from which natural gas had been separated. It also involved a comparison of the foam behavior with critical film rupture thicknesses, bulk phase, and surface shear viscosities together with dilatational surface properties.

As shown in Figure 10.3 for the natural gas–crude oil system, this study revealed a linear correlation (albeit with a correlation coefficient of only 0.84) between increasing bulk phase shear viscosity and increasing foam stability as indicated by the average foam lifetime, L_F, which is defined as

$$L_F = \frac{1}{H_0} \int_0^{t^*} H_t \mathrm{d}t \qquad (10.2)$$

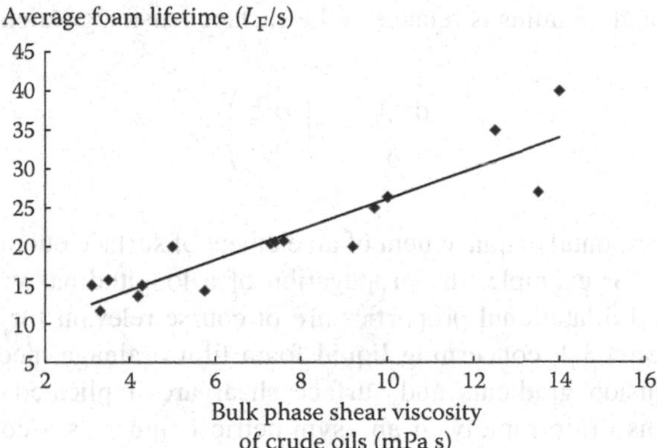

FIGURE 10.3 Average foam lifetime, L_F (defined by Equation 10.2) against bulk phase shear viscosity of various degassed crude oils. Foam generated by sparging with natural gas. (Callaghan, I.C., Neustadter, E.L.: *Chem. Ind.* 1981. January. 55. Copyright Wiley-VCH Verlag GmbH & Co. KGaA. Reproduced with permission.)

Here H_0 is the initial foam height, H_t is the foam height at time t, and t^* is the time for total foam collapse, where all quantities are relative to those at the situation immediately after sparging has ceased. A similar correlation was observed for the air–crude oil system—the foam stability of which was lower than for the natural gas–crude oil system. The foam is seen to be extremely unstable—collapsing with average lifetimes < 45 s even with oils of viscosities up to an order of magnitude higher than that of water.

Critical rupture thicknesses, h_{crit}, are defined in Section 1.3.3. They were measured by Callaghan and Neustadter [31] using microscopic films generated in a Scheludko cell (see Section 2.3.1). Measured critical film rupture thicknesses were all <20 nm but with apparently no clear correlation with foam stabilities. No dimensions of the films were, however, revealed by these workers. Since the critical rupture thickness increases with the film radius [26, 27, 34], it is therefore difficult to make rigorous comparison of these measurements of h_{crit} with others. However, reported measurements of h_{crit} for aqueous films are apparently usually somewhat higher even for films of diameter as low as 200 microns [34]. This is consistent with the theory of Vrij [26, 27], which indicates higher values of the critical rupture thickness of aqueous films are to be expected relative to hydrocarbon films because of stronger van der Waals attractive forces in the former as indicated by the relative Hamaker constants (hydrocarbons ~0.5×10^{-19} J; water ~1.5×10^{-19} J [35]). The absence of a clear correlation between h_{crit} and foam stabilities may suggest that film and overall drainage rates dominate in determining relative foam lifetimes in these systems.

In general, Callaghan and Neustadter [31] find that increasing foam stability of these crude oils also correlates with increasing values of surface shear viscosities and surface dilatational elasticities and viscosities. Here the respective dilatational properties are the real and imaginary parts of the surface dilatational modulus, ε, usually measured using an oscillating longitudinal wave at different frequencies [36], although Callaghan and Neustadter [31] use a different approach involving a Fourier transform from a surface relaxation time to a frequency measurement. The surface dilatational modulus is related to the surface tension gradient, $d\sigma_{GC}/dy$, by

$$\frac{d\sigma_{GC}}{dy} = \varepsilon\left(\frac{\partial^2\xi}{\partial y^2}\right) \tag{10.3}$$

where ξ is the horizontal displacement of an element of surface out of its equilibrium position during, for example, the propagation of a longitudinal wave [37]. These surface shear and dilatational properties are of course relevant for the theory, discussed in Section 1.3.2, concerning liquid foam film drainage mechanisms where both surface tension gradients and surface shear are implicated in determining whether the films drain rapidly in an asymmetric manner as a consequence of a hydrodynamic instability. The maximum surface shear viscosity measured in this study was, however, 0.9 kg s^{-1}, which compares with values of 0.1–0.2 kg s^{-1} required to suppress rapid asymmetric drainage in films formed from aqueous solutions of SDS and dodecanol [33]. At sufficiently high surface shear viscosities, the latter forms extremely slow draining axisymmetric films and stable foams in contrast to the foams of these crude oil systems, which have average lifetimes of <45 s. It is difficult to understand why the foam of these crude oils is so unstable if the suppression of hydrodynamic instability in thin films is dominated by the magnitude of surface shear viscosity as argued convincingly by Joye et al. [33]. Clearly there is a need for more comparisons of both the surface rheology and foam film behavior of non-aqueous and aqueous systems.

Extremely viscous so-called heavy oils are often produced from wells in Canada, Venezuela, and China. These oils often have reported viscosities in the range of (3–30) × 10^3 mPa s [38–40] and are often produced at the well head as a gas-in-oil emulsion with a gas volume fractions of from 0.05 to 0.40 [41], which has the appearance of "chocolate mousse" [38]. The foams formed from such gas-in-oil emulsions upon standing can be extremely stable, persisting for several hours in open vessels [38].

Production of this type of heavy oil, or "foamy-oil," is much higher than would be expected by conventional processes based on reservoir characteristics such as pressures, viscosities, and permeabilities. Production is through a so-called solution gas drive. As the pressure declines in a reservoir, it will eventually drop below the point at which gas bubbles nucleate in the oil. With such viscous oils, neither coalescence nor migration of the bubbles occurs and they remain finely distributed throughout the reservoir where the relevant pressure prevails. Removal of oil instead of reducing the reservoir pressure, and thereby reducing the production rate, supposedly simply causes more gas to nucleate at constant pressure. The production rate is therefore preserved until either the pressure drops to the point where so much gas has been produced that it becomes mobile as a separated phase [38, 39] or the vapour pressure of the gas declines significantly as the concentration of dissolved gas in the oil declines. Other explanations for the anomalous high production of heavy oil, concerning the effect of bubbles on the rheology of the oil, are given by Abivin et al. [40].

The high viscosity of heavy oils has been ascribed to high concentrations of asphaltenes [42, 43]. That high concentrations of asphaltenes and resins can lead to high crude oil viscosities is exemplified by the data presented in Figure 10.4 for a series of degassed oils at ambient temperature. It has also been argued that asphaltenes represent nucleation sites for gas bubble formation [39] and can adsorb

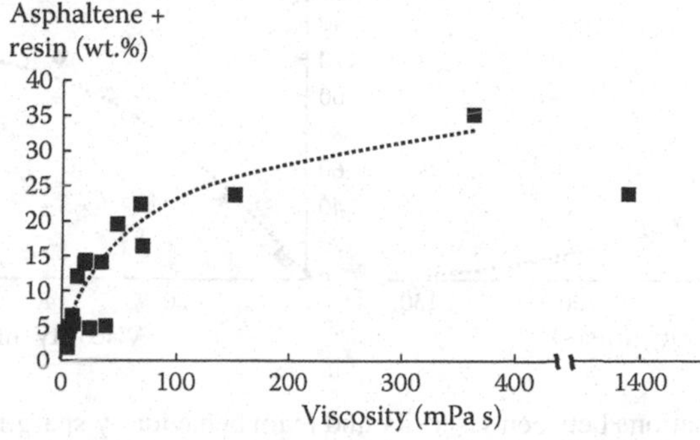

FIGURE 10.4 Plot of total asphaltene + resin content against bulk shear viscosity at 38°C for a range of degassed crude oils from various fields. (After Poindexter, M.K. et al., *Energy Fuels*, 16, 700, 2002.)

[39, 44] at gas bubble surfaces to contribute to the extreme stability of gas-in-oil emulsions. However, as we have seen, the evidence for such adsorption is weak and even if it occurs, it is difficult to see how bubble–bubble coalescence at volume fractions <0.4 typical of gas-in-oil emulsions in a medium of viscosity >> 3 Pa s could be affected by such adsorption. Foam and therefore foam films only form at gas volume fractions ≥ 0.72 [45]. Formation of transient foam films by colliding bubbles in a medium of such high viscosity at gas volume fractions < 0.4 would seem improbable.

All of this may imply that a strong correlation exists between high foam stability, high bulk phase viscosity, and the presence of high concentrations of asphaltenes in crude oil. However, a study by Poindexter et al. [4] of foam behavior using 20 different crude oils reveals a more complex picture. The study involved "dead" oils, from which natural gas had been removed, and foam generation by sparging with nitrogen. It involved two different selections of crude oils. The first, group A (changing the designation used by Poindexter et al. [4] in the interests of clarity), consisted of a wide range of oils from different sources with different viscosities, and asphaltene and resin contents. The second, group B, consisted of Gulf of Mexico crude oils with known foaming problems and asphaltene/resin ratio < 1. In turn, this divided into two subgroups, one with relatively high asphaltene content (subgroup B1) and one with relatively low asphaltene content (subgroup B2).

In most cases, Poindexter et al. [4] find that sparging results in the attainment of a steady-state foam height where bubble collapse rates equal the gas flow rate. It is of interest to establish whether the correlation between oil viscosity and average foam lifetime, L_F, reported by Callaghan and Neustadter [31] is replicated using the data of Poindexter et al. [4]. However, plots of foam volume and L_F against viscosity for group A crude oils reveal no linear correlation as shown in Figure 10.5. The most that can be inferred from the figure is that, in general, increasing viscosity tends to produce less foam by this sparging method and tends to increase average foam

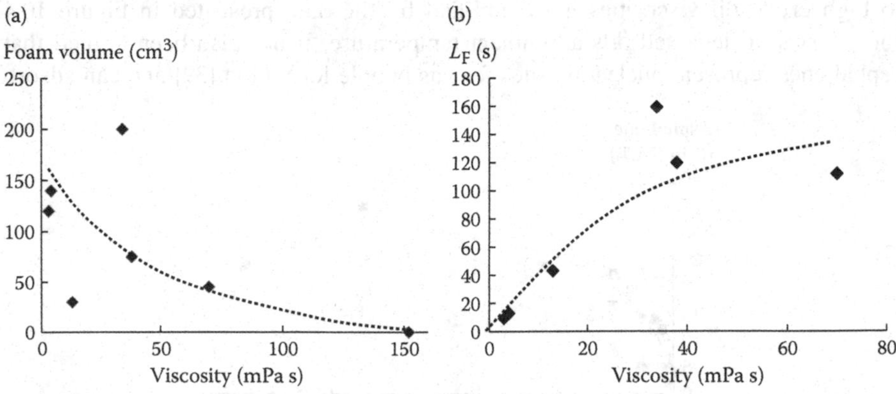

FIGURE 10.5 Relations between viscosity and foam behavior by sparging of a selection of degassed crude oils from many different fields (i.e., of group A; see text Section 10.2.2.2). (a) Foamability (as indicated by steady-state foam volume) against viscosity. (b) Foam stability (as indicated by average foam lifetime, L_F, defined by Equation 10.2) against viscosity. (After Poindexter, M.K. et al., *Energy Fuels*, 16, 700, 2002.)

lifetimes. The latter suggests a role for film drainage in determining foam stability. The absence of a clear linear correlation does, however, suggest, unsurprisingly, that in all probability the gas–oil surface properties also have a role in this context. The range of viscosities considered by Poindexter et al. extend to ~70 mPa s, whereas the linear correlation found by Callaghan and Neustadter [31] shown in Figure 10.3 concerns a range of viscosities of <20 mPa s, which could mean that the role of viscosity is dominant at low viscosities where drainage is therefore extremely rapid.

Poindexter et al. [4] also fail to find any linear correlation between foam properties and any other property with group B crude oils save two apparent exceptions. The first concerned the four group B1 oils only, where a linear correlation between foam stability as (measured by the rate of foam collapse) with each of oil viscosity, asphaltene content, density, and surface tensions is claimed. These apparent correlations are, however, unproven by the selected data because it concerns two pairs of oils, each pair having similar properties. A high correlation coefficient is therefore inevitably calculated as would be the case if a straight line is drawn through two points! The second correlation concerned the group B2 oils where a linear correlation is found between foam volume and gas–oil surface tension. Somewhat surprisingly this indicates that the higher the surface tension the higher the foam volume, for which Poindexter et al. [4] offer no convincing explanation.

We find then that direct understanding of the causes of foam formation in crude oils is largely limited to practical generalizations from field experience and rather uncertain correlations between foam properties and basic physical and compositional variables. Moreover, much of that understanding has been derived from sparging degassed oils at ambient temperatures and pressures, whereas gas–oil separation involves nucleation of natural gas bubbles upon reduction of the pressure, all at high temperatures and pressures. Gas–oil separation is also physically radically different from sparging, not least because the release of dissolved hydrocarbon gas can change the solubilities of asphaltenes in the oil.

It is more than 30 years since Callaghan and Neustadter [4] made a study of oil foaming, which combined surface rheological and thin film observations. Perhaps it is time this approach is revisited with some emphasis on the difficult issue of measurements on live crude oils at elevated temperatures. Such measurements could be combined with application of modern theories about foam drainage, foam film drainage, and rupture (see Chapter 1 for a brief introduction to these topics).

10.4 USE OF ANTIFOAMS

10.4.1 General Considerations

All of the concepts concerning the mode of action of liquid oils as antifoams for aqueous systems can be transformed into their equivalents for non-aqueous foaming liquids in general and to crude oils in particular. Such concepts include entry coefficients, spreading coefficients, and bridging coefficients, which describe the necessary properties of an antifoam liquid to function in crude oil. Moreover, the types of wetting behavior by drops of the putative antifoam on crude oil can be described in principle by the same terms—partial wetting (lens formation with no tendency to form films at the air–crude oil surface), pseudo-partial wetting (lens formation in equilibrium with a film formed at the air–crude oil surface), and complete wetting (where the antifoam oil forms a duplex film over the crude oil surface and where the antifoam film has the same surface tensions as the corresponding bulk phase). Unfortunately, however, there would appear to have been little systematic experimental observation of the wetting behavior of any antifoam oil on any crude oil.

The definitions of the classic entry and spreading coefficients, given by Equations 3.1 through 3.12 in Section 3.2, and for the bridging coefficient, given by Equations 4.29 through 4.31 in Section 4.5.1, can be transformed to the present context if we make the substitutions

$$\sigma_{GC} \text{ for } \sigma_{AW}\,;\ \sigma^{i}_{GC} \text{ for } \sigma^{i}_{AW} \text{ and } \sigma^{e}_{GC} \text{ for } \sigma^{e}_{AW}$$

where σ_{GC}, σ^{i}_{GC}, and σ^{e}_{GC} are the relevant gas–crude oil surface tensions;

$$\sigma_{FC} \text{ for } \sigma_{OW}\,;\ \sigma^{i}_{FC} \text{ for } \sigma^{i}_{OW} \text{ and } \sigma^{e}_{FC} \text{ for } \sigma^{e}_{OW}$$

where σ_{FC}, σ^{i}_{FC}, and σ^{e}_{FC} are the relevant antifoam–crude oil surface tensions;

$$\sigma_{GF} \text{ for } \sigma_{AO}\,;\ \sigma^{i}_{GF} \text{ for } \sigma^{i}_{AO} \text{ and } \sigma^{e}_{GF} \text{ for } \sigma^{e}_{AO}$$

where σ_{GF}, σ^{i}_{GF}, and σ^{e}_{GF} are the relevant gas–antifoam surface tensions. Here we remember that the superscripts i and e refer to initial (i.e., non-equilibrium conditions) and equilibrium conditions, respectively, and where the absence of a superscript refers to any equilibrium or non-equilibrium condition. We can therefore write, for example, for the bridging coefficient B

$$B = \sigma^{2}_{GC} + \sigma^{2}_{FC} - \sigma^{2}_{GF} \tag{10.4}$$

where we note that it is possible to satisfy the requirement that $B > 0$ if σ_{FC} is large enough even if the air–oil surface tension of the crude oil is less than that of the antifoam, $\sigma_{GF} > \sigma_{GC}$. This possibility is probably more readily realized in the present context than is likely in the equivalent aqueous context where high levels of surfactant are usually present so that oil–water interfacial tensions are often extremely low and it is therefore usually necessary that the air–oil surface tension of the antifoam be less than the air–water surface tension of the foaming liquid.

We have shown in Chapter 4 that the stability of pseudoemulsion air–water–antifoam oil films often dominates in determining the effectiveness of oils as antifoams for aqueous solutions regardless of the magnitude of the bridging coefficient. As a consequence, hydrophobic particles are usually admixed with the antifoam oils in order to rupture those aqueous pseudoemulsion films. That those films are otherwise stable must derive from positive contributions to the disjoining pressure isotherm originating with the adsorption of surfactant at both air–water and oil–water surfaces. Those positive contributions would appear to be in the main electrostatic, resulting from overlapping electrostatic double layers. Such interactions are, however, as we have discussed here, likely to be essentially absent in a low dielectric medium such as crude oil. In the absence of any direct measurements on pseudoemulsion air–crude oil–antifoam films, it is tempting to predict that their stabilities will be extremely low. In consequence, addition of suitable particles to the relevant antifoam fluids in order to destabilize pseudoemulsion films could be unnecessary. This is consistent with practice where, for example, PDMS and perfluorosilicone oils are always used neat as fluid antifoams for crude oil, albeit often dissolved in suitable crude oil soluble solvents to assist dispersal and handling [2]. It does not, however, preclude the possibility that there exist differences between pseudoemulsion film stabilities with different antifoam oils to yield differences in overall effectiveness. Such differences could even have their origin in adsorption of components such as asphaltenes at the crude oil–antifoam interface. All this could even suggest a role for particles in this context, although achieving the necessary contact angles at the relevant surfaces in pseudoemulsion films may be more difficult than in an aqueous context.

10.4.2 Polydimethylsiloxanes and Substituted Polydimethylsiloxanes

Antifoams for gas–oil separation usually consist of either PDMSs or their derivatives. Concentrations of antifoam for this application typically lie in the range of $(1–50) \times 10^{-3}$ g dm^{-3}—significantly lower than is usual for aqueous systems. However, solubility in crude oils of these materials is often significant, unlike in aqueous systems where PDMS solubilities at relevant molecular weights are $<10^{-6}$ g dm^{-3} [46]. Therefore, relative antifoam effectiveness of PDMS and derivatives in gas–oil separators can often be determined by solubility rather than intrinsic effectiveness.

Unfortunately, a complete picture of the relevant physical properties of dispersions and solutions of PDMSs and PDMS derivatives in various crude oils is not apparently available. We can, however, infer the likely properties if we combine consideration of actual available observations with extrapolations based on the behavior of such polymeric compounds in other non-aqueous media, including other hydrocarbons.

10.4.2.1 Effect of Solubility of Antifoam Oils

In Chapter 4, we have discussed at length the requirement that antifoams should be present at concentrations that exceed their solubility limit in the solution to be defoamed. As shown by Shearer and Akers [30], more than half a century ago, the presence of dissolved PDMS acts as a so-called pro-foamer for hydrocarbon lube oils. Similar observations have been reported by Centers [47] concerning the effect of PDMS oils on the foam behavior of a synthetic ester turbine lubricant. Callaghan et al. [3] have also shown that addition of short-chain alkanes such as hexane to a degassed "dead" crude oil not surprisingly increases the solubility of PDMS in the oil because PDMS is completely miscible with such alkanes at ambient temperatures. The antifoam effect of the PDMS can therefore be changed to a pro-foaming effect if sufficient hexane is added to the crude oil to solubilize the antifoam, as shown in Figure 10.6 (where $F < 1$ represents antifoaming and $F > 1$ represents pro-foaming).

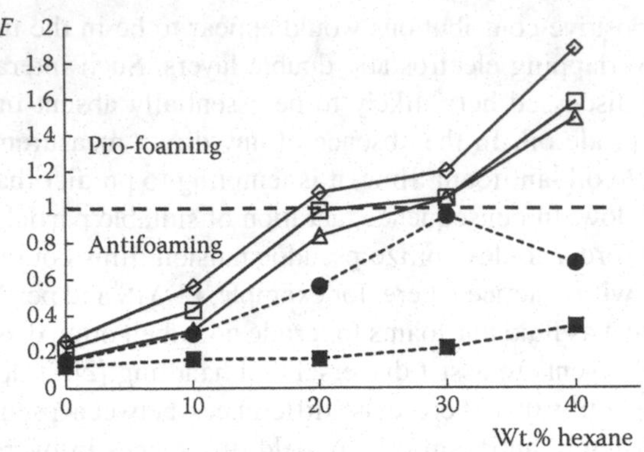

FIGURE 10.6 Effect of various antifoam liquids on foamability, measured by sparging, of mixtures of degassed crude oil with hexane. F = foam height with antifoam/foam height without antifoam, where $F < 1$ is antifoaming behavior and $F > 1$ is pro-foaming behavior due to dissolution of antifoam. Concentration of antifoams, 5×10^{-3} g dm^{-3}. ◇, PDMS of viscosity ~12.5 Pa s; □, PDMS of viscosity ~60 Pa s; △ PDMS of viscosity ~10^3 Pa s; ●, fluorosiloxane; ■, ethoxylated siloxane (silicone glycol). (After Callaghan, I.C., Hickman, S.A., Lawrence, F.T., Melton, P.M. Antifoams in gas–oil separation, in *Industrial Applications of Surfactants*, Karsa, D. ed., 59, 48, 1987. Reproduced by permission of The Royal Society of Chemistry.)

It is clear from the figure that increasing the viscosity, and therefore the molecular weight, of polydimethylsiloxane oils decreases the solubility and slightly increases the persistence of antifoam effects up to higher concentrations of hexane.

Improved antifoam performance of PDMSs with increasing molecular weight is also described by L'Hostis and Renauld [48]. They teach that increasing the molecular weight of PDMSs from 240,000 (of viscosity 600 Pa s) to >1,900,000, with polydispersity <4, produces enhanced foam control in sparging of crude oils. Pape [2] additionally reports that increasing PDMS viscosity can increase antifoam effectiveness in practical gas–oil separation, citing experience on the Ekofisk Platform in the North Sea. Thus relative efficiencies were reported to increase fourfold on increasing the PDMS viscosity from 12.5 to 60 Pa s. This effect is attributed by Pape [2] to decreasing solubility of the PDMS with increasing viscosity, and therefore molecular weight, rather than increasing intrinsic antifoam effectiveness. It contrasts with the use of neat PDMS in aqueous media where the solubility is so low that it is not an issue. In that context, increasing the viscosity of the PDMS is expected to reduce the rate at which drop sizes decline to suboptimal dimensions so that effective antifoam action persists for longer periods of foam generation (see Section 6.2).

The study of Callaghan et al. [3] involving crude oil–hexane mixtures also included a comparison of the effectiveness of various polydimethylsiloxanes with both an ethoxylated and a perfluoroalkyl-substituted PDMS. In contrast with the polydimethylsiloxanes, antifoam effects with those derivatives prevailed, as shown in Figure 10.6, even up to crude oil-to-hexane ratios as high as 60:40 vol.%. It seems likely that the enhanced antifoam effects at these high crude oil-to hexane ratios of the derivatives concern relatively low solubility. Thus, both derivatives are shown by Callaghan et al. [3] to be less soluble than polydimethylsiloxanes in hexane. Ethoxylated PDMS graft copolymers, including that used in this work, are in fact the subject of patent claims concerning their superiority as antifoams to PDMS for gas–crude oil separators because of their lower solubility in the oil [49].

These observations of Callaghan et al. [3] imply that crude oil compositional differences can significantly influence antifoam effectiveness in the case of PDMS oils. It is possible, for example, that degassing a crude oil will remove the lower hydrocarbon components to decrease the solubility of the antifoam and enhance effectiveness. Antifoams that are effective in degassed "dead" crude oils may therefore be ineffective in live crude oils. Thus, Callaghan et al. [3] cite an example where both a PDMS oil and an ethoxylated PDMS were effective in degassed crude oil but only the ethoxylated PDMS was effective as an antifoam in the "live" crude oil, presumably as a consequence of lower solubility.

The temperature represents another factor. Shearer and Akers [30] have shown that the solubility of PDMS in a hydrocarbon lube oil increases with increase in temperature—antifoam effectiveness at any concentration of PDMS therefore progressively declines as the temperature is increased eventually changing to pro-foaming behavior. It is probable therefore that similar behavior will be exhibited by dispersions of PDMS in crude oil. We should also note an additional complication—the rate of solubilization of PDMS in hydrocarbons has been shown by Mannheimer [17] to decrease with increasing viscosity of that hydrocarbon. Moreover, the rate of solubilization of PDMS in hydrocarbons appears to decrease with increase in molecular weight, and therefore of the viscosity, of the polymer [3]. This combined with decreased solubilities suggests an advantage for the use of the highest viscosity PDMS available. However, it is also necessary that the antifoam be present as finely divided emulsion drops to ensure a high probability of presence in foam films. Dispersal as a finely divided emulsion may be difficult with extremely high viscosity PDMS—often, therefore, such materials are added as solutions in a suitable crude oil soluble solvent, which has the additional advantage of facilitating processing. The antifoam concentrations in gas–oil separators are usually ≤ 0.1 g dm^{-3}, which is often applied as ~5–30 wt.% solutions in either diesel oil for PDMS [2] and toluene for ethoxylated PDMS derivatives [13]. The implied low levels of solvents such as diesel oil or toluene are unlikely to have any effect on the foam properties of the relevant crude oils [13].

Use of perfluoroalkyl-substituted polydimethylsiloxanes apparently finds a particular advantage with respect to polydimethylsiloxanes in foam control of "sour" crude oils containing appreciable quantities of hydrogen sulfide and mercaptans. It is argued that polydimethylsiloxanes are either deactivated or solubilized by hydrogen sulfide, whereas perfluoroalkyl-substituted PDMS is unaffected [50, 51]. However, perfluoroalkyl derivatives, such as those exemplified by the structure shown in Figure 10.7 and described by Fink et al. [52], are highly viscous with viscosities of order 1000 Pa s [50, 51], which indicates that some vehicle must be used to facilitate dispersal in crude oil. Callaghan and Macleod [50] claim that dissolution of the perfluoroalkyl-substituted PDMS in a solvent consisting mixtures of an alkane, toluene, and a chlorofluorocarbon is effective for this purpose. Taylor and Macleod [51], on the other hand, teach the preparation of a concentrated oil-in-water emulsion

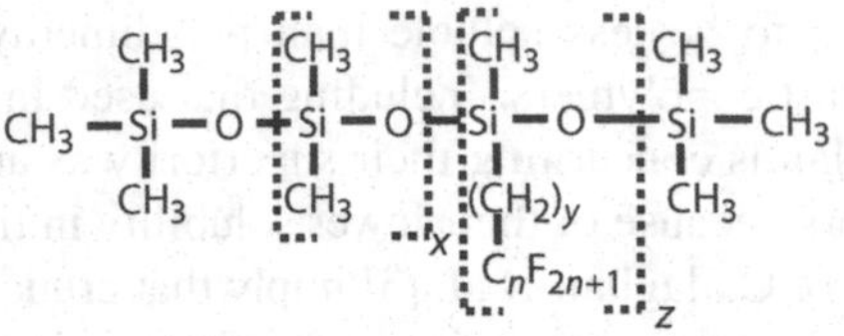

$$CH_3-\underset{CH_3}{\overset{CH_3}{Si}}-O-\left[\underset{CH_3}{\overset{CH_3}{Si}}-O\right]_x-\left[\underset{(CH_2)_y C_nF_{2n+1}}{\overset{CH_3}{Si}}-O\right]_z-\underset{CH_3}{\overset{CH_3}{Si}}-CH_3$$

FIGURE 10.7 Chemical structures of typical examples of perfluoroalkyl-substituted PDMSs claimed by Fink et al. where $4 \leq n \leq 16$, and, for example, $0 \leq x \leq 25$ with $5 \leq z \leq 58$ and $y = 2$. (From Fink, H. et al. (assigned to TH. Goldschmidt AG), GB 2196976; 11 May 1988, filed 15 October 1986.)

as a suitable vehicle for dispersing these perfluoroalkyl derivatives into crude oil. By contrast, Pape [2] argues the use of tributyl acetate as a suitable solvent for such compounds in this context.

10.4.2.2 Mode of Antifoam Action in Crude Oils

As we have described in Chapter 4, the mode of action of antifoam oils appears in general to be best described by a bridging mechanism [53]. This involves formation of unstable bridging configurations in foam films, which lead to film collapse either by the bridging–dewetting mechanism, or more commonly, by the bridging–stretching mechanism [54, 55]. These bridging mechanisms are essentially characterized by inequality 4.31, which in the present context can be rewritten as

$$0 < B \leq 2\sigma_{FC}\sigma_{GC} \tag{10.5}$$

which embraces bridge formation of oils showing either partial, pseudo-partial, or complete wetting behavior by the antifoam oil on the surface of the crude oil as outlined in general in Section 4.5.1. It is the only mechanism that can explain the antifoam effects of oils showing any of those possible wetting behaviors—all other proposed mechanisms involving the putative actions of oils at surfaces (e.g., Marangoni spreading) impose more restrictive conditions that cannot therefore explain all relevant experimental observations.

In establishing the mode of action of any antifoam oil, it is first necessary to observe the wetting behavior on the relevant substrate. As we have seen, this requirement is, however, rarely observed with anything like rigor or even at all in the present context. For example, Callaghan et al. [28] report a study of the effect of PDMS oils on the surface dilatational rheological behavior of crude oils, seeking correlations with antifoam action. They present no observations concerning the wetting behavior of antifoam drops at all. They do find, however, that antifoam action appears to correlate with decreases of, for example, the dilatational elasticity at 1.0 mHz, of a degassed crude oil of from ~1.3 mN m^{-1} to 0.51–0.99 mN m^{-1}, depending on the nature of the PDMS. This represents a small effect compared with the dramatic antifoam effects associated with the same polymers. It would be interesting to compare these trends in surface dilatational elasticities with those at concentrations just at the solubility limit where pro-foaming prevails. Moreover, measurements on other systems suggest that it is unlikely that such dilatational properties are of fundamental significance for antifoam action in this context. Thus, McKendrick et al. [56] find, for example, that antifoam action in the case of a random copolymer of PDMS + methyltrifluoropropylsiloxane on a "medium gas oil" hydrocarbon appears to correlate with increases in dilatational elasticities upon addition of the antifoam from ~2 mN m^{-1} to about 50–60 mN m^{-1} irrespective of frequency. However, that both Callaghan et al. [28] and McKendrick et al. [56] find changes to the dilatational properties at all of crude oil and "medium gas oil," respectively, upon addition of antifoam implies contamination of the air–oil surface and therefore either complete wetting or, more likely, pseudo-partial wetting by antifoam drops in both cases.

Mansur and coworkers have made two studies [13, 57] of the antifoam effects of several alkoxylated PDMSs on the foam behavior of crude oil combined with measurements of the relevant surface tensions, which should, in principle, permit the wetting behavior to be established. The general structures of these alkoxylated polydimethylsiloxanes are shown in Figure 10.8. Foam was generated by first pressurizing the crude oil with air until equilibrium was attained and subsequently releasing the pressure, discharging the foaming crude to a suitable vessel for observation of foam stability.

The first such study concerned two alkoxylated polydimethylsiloxanes of structure shown in Figure 10.8a, having molecular weights of 19,600 and an EO-to-PO

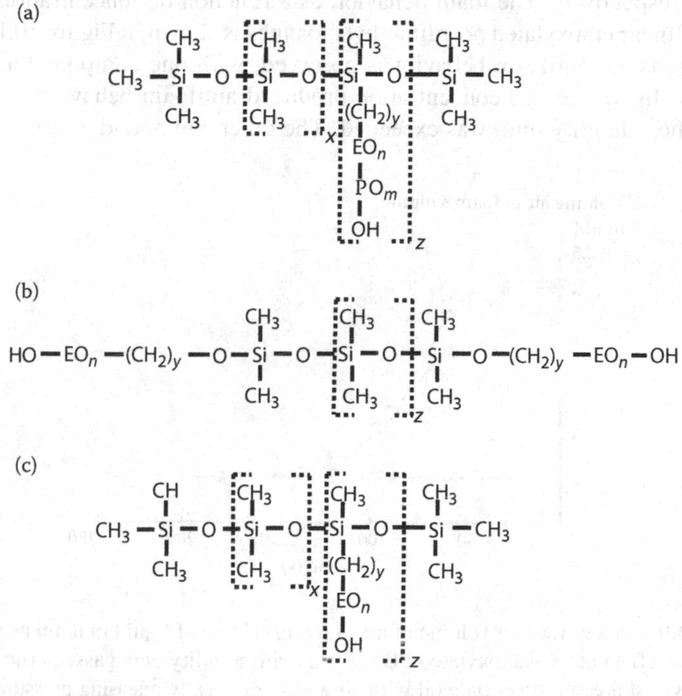

(a)

$$CH_3-Si(CH_3)_2-O-[Si(CH_3)_2-O]_x-[Si(CH_3)((CH_2)_y-EO_n-PO_m-OH)-O]_z-Si(CH_3)_2-CH_3$$

(b)

$$HO-EO_n-(CH_2)_y-O-Si(CH_3)_2-O-[Si(CH_3)_2-O]_z-Si(CH_3)_2-O-(CH_2)_y-EO_n-OH$$

(c)

$$CH_3-Si(CH)(CH_3)-O-[Si(CH_3)_2-O]_x-[Si(CH_3)((CH_2)_y-EO_n-OH)-O]_z-Si(CH_3)_2-CH_3$$

FIGURE 10.8 Chemical structures of various alkoxylated PDMSs. (a) Branched-chain structure with ethoxylated + propoxylated chains where $n/m \geq 1$. (b) Linear structure with ethoxylated chains. (c) Branched-chain structure with ethoxylated chains. (After Fraga, A.K. et al., *J. Appl. Polym. Sci.*, 124, 4149, 2012; Rezende, D.A. et al., *J. Petrol. Sci. Eng.*, 78, 172, 2011.)

ratio of 1.1 and 3300 with an EO-to-PO ratio of 5.0, respectively [57]. Plots of foam collapse against time are presented in Figure 10.9. Of these two polymers, only the low molecular weight example proved to be an effective antifoam, presumably because of lower solubility in the crude oil due to higher polarity. However, there was little evidence of overall pro-foaming activity by the high molecular weight polymer, possibly because of a combination of pro-foaming and antifoaming in a case of partial solubilization of the polymer. Rezende and coworkers [57] calculated positive spreading coefficients for the effective lower molecular weight polymer on crude oil from the relevant surface tensions. That the quoted spreading coefficient is positive means that it must be an initial or semi-initial value because an equilibrium value must be ≤0 (see Section 3.2). They also calculate a positive bridging coefficient. However, a positive initial spreading coefficient means that any calculated bridging coefficient using the same surface tensions is meaningless (see Section 4.5.1). A positive initial spreading coefficient indicates that the polymer may exhibit either complete or pseudo-partial wetting behavior on the relevant crude oil. Observations of the consequences of direct addition of drops of the antifoam to the crude oil surface would be necessary in order to distinguish between these two options.

Fraga and coworkers [13] have made a second study of alkoxylated polydimethylsiloxanes, but of linear and branched ethoxylated structures shown as Figures 10.8b and 10.8c, respectively. The foam behavior, as a function of concentration in crude oil, of two linear ethoxylated polydimethylsiloxanes is shown in Figure 10.10. At low concentrations, no antifoam behavior is observed, with one compound acting as a pro-foamer. Increasing the concentration produced antifoam behavior in one compound as the solubility limit was exceeded. The other compound, presumably more

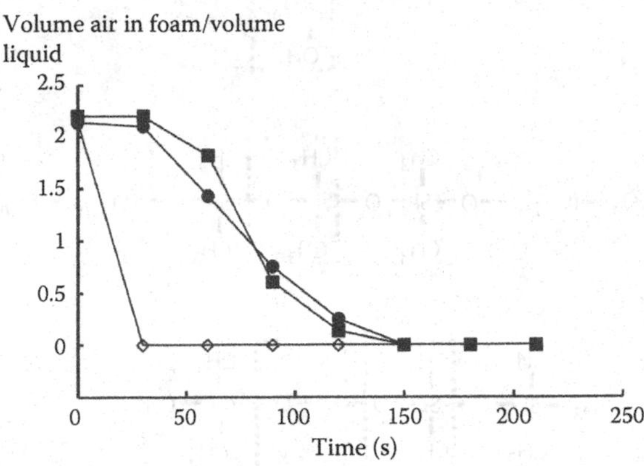

FIGURE 10.9 Plot of ratio of volume of air in foam/volume of liquid in foam as function of time showing effect of two alkoxylated PDMSs on foam stability of degassed crude oil. Foam generated by first pressurizing crude oil with air and subsequently releasing pressure. ■, crude oil; ●, after addition of alkoxylated PDMS of molecular weight 19,600 and EO-to-PO ratio of 1.1; ◇, after addition of alkoxylated PDMS of molecular weight 3300 and EO-to-PO ratio of 5.0. Concentration of alkoxylated PDMSs: 40×10^{-3} g dm^{-3}; structures shown in Figure 10.8a. (Adapted minimally from *J. Petrol. Sci. Eng.*, 78, Rezende, D.A., Bittencourt, R.R., Mansur, C.R.E., 172. Copyright 2011, with permission from Elsevier.)

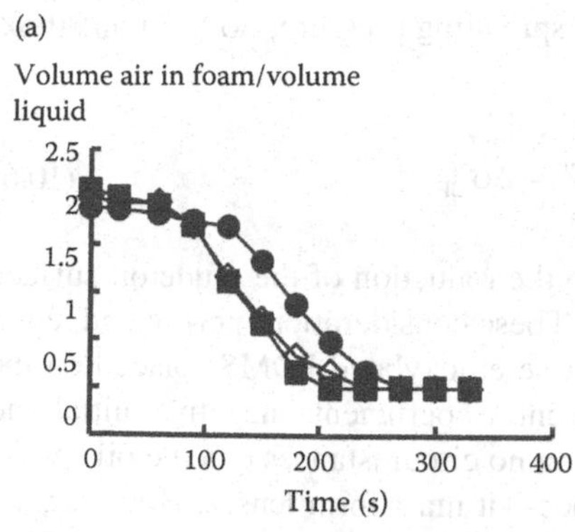

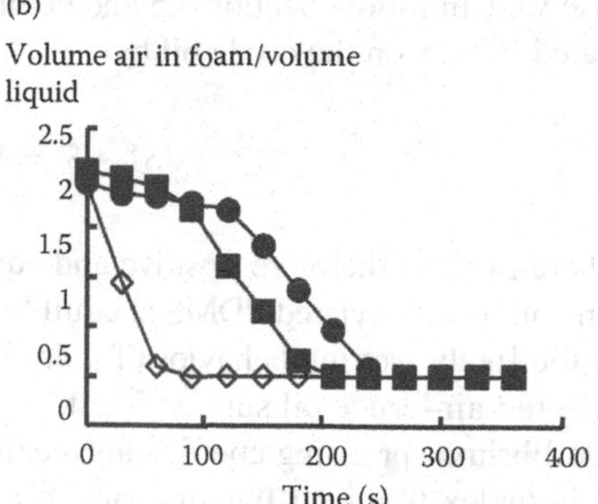

FIGURE 10.10 Plot of ratio of volume of air in foam/volume of liquid in foam as function of time, comparing effects of two linear ethoxylated polydimethylsiloxanes (structure shown in Figure 10.8b) on foam of degassed crude oil at different concentrations. Foam generated by first pressurizing crude oil with air and subsequently releasing the pressure. ■, crude oil; ●, after addition of ethoxylated PDMS with number of EO units $n = 15$ ($z = 8, y = 3$); ◇, after addition of ethoxylated polydimethylsiloxanes with number of EO units $n = 19$ ($z = 7, y = 3$). (a) Concentration of ethoxylated PDMSs 20×10^{-3} g dm^{-3}. (b) Concentration ethoxylated polydimethylsiloxanes 50×10^{-3} g dm^{-3}. (Reproduced from Fraga, A.K. et al., *J. Appl. Polym. Sci.*, 124, 4149. Copyright 2012 Wiley Periodicals Inc., with the kind permission of John Wiley and Sons.)

soluble, still showed pro-foaming behavior. Not shown in the figure are the results of the foam behavior of the branched and more soluble ethoxylated polydimethylsiloxanes, all of which acted as pro-foamers.

All of the ethoxylated polydimethylsiloxanes used in this second study by Fraga and coworkers [13] were relatively polar compounds that exhibited high surface tensions at the interface with crude oil and low air–ethoxylated PDMS surface tensions due to the PDMS chain integral to their molecular structure. Entry coefficients at the air–crude oil surface were therefore all strongly positive and initial spreading coefficients were all markedly negative. As a consequence of the latter, meaningful bridging coefficients could be calculated. In all cases, these bridging coefficients were strongly positive and consistent with condition 10.5. This implies that all these ethoxylated polydimethylsiloxanes have the potential to function as antifoams with the given crude oil provided they are present at concentrations in excess of their solubility limit, which would appear to be relatively high in most cases. We now examine the evidence for this expectation by considering the surface tension results given by Fraga and coworkers [13].

Dispersal of these linear and branched ethoxylated polydimethylsiloxanes in the selected crude oil lowered the air–crude oil surface tension slightly by ~1–3 mN m^{-1}, depending on the compound, at an apparent equilibrium. The equilibrium spreading coefficient, S^e, will therefore be slightly more negative than the initial spreading coefficient, S^i, where the latter is calculated using the surface tension of the crude oil before contamination with ethoxylated PDMS. If we suppose reasonably that components of the crude oil do not partition into ethoxylated PDMS, then there should be no distinction between the initial and semi-initial spreading coefficients. In which case we can follow Section 3.5 and define the spreading pressure, $\Delta\sigma_{GF}^{sp}$, of an ethoxylated PDMS on the crude oil by

$$S^{si} - S^{e} = S^{i} - S^{e} = \Delta\sigma_{GF}^{sp} \tag{10.6}$$

where $\Delta\sigma_{GF}^{sp}$ is therefore positive and equal to the reduction of the crude oil surface tension by ethoxylated PDMS at equilibrium. These considerations permit deduction of the likely wetting behavior of a drop of these ethoxylated PDMSs placed on the selected air–crude oil surface. First, positive entry coefficients, negative initial and equilibrium spreading coefficients mean that in no circumstances can the oil spread as a duplex film over the air–crude oil surface—it must form lenses. However, we must instead have direct spreading of a pseudo-partial, probably monomeric film, from the lens unless ethoxylated PDMS is already present at the crude oil surface by, for example, adsorption from the bulk crude oil phase already contaminated by partial dissolution of the antifoam. Since the latter seems likely and bridging coefficients are positive, it is difficult to avoid the conclusion of Fraga and coworkers [13] that any antifoam action in these systems must involve the formation of unstable configurations as antifoam oil lenses bridge crude oil foam films.

10.4.3 Other Materials

Use of siloxane chemistry in the context of degassing of crude oils is not entirely ubiquitous. There have been reported (and indeed patented) accounts of the use of essentially polar materials in this context. In the main, the motivation for seeking alternatives to siloxane derivatives has concerned cost and possible downstream problems such as contamination of refining catalysts and even adverse effect on the properties of refined products. For example, Wylde [7] describes an example of the latter, claiming that PDMS contamination both adversely affects the efficacy of refining of high-foaming heavy oil and adversely affects the rheology of the asphalt product. The process involves first heating the oil to a relatively high temperature of 50–85°C, which despite low gas–oil ratios leads to unacceptable degassing and the formation of a viscous, persistent foam. Surprisingly, both a sodium sulfosuccinate and a particular long-chain fatty alcohol were found to be effective antifoams in this context. The fatty alcohol was apparently a solid at room temperature and it is known that some sodium alkyl sulfosuccinates exhibit significant solubility in hydrocarbons. No physical measurements of relevant properties such as the solubility in the oil or the effect of the additive on oil viscosity or surface tension at the relevant temperature were, however, reported so it is not possible to infer anything about the mode of action of these materials. Possibly, these additives simply solubilize the asphaltenes to reduce the oil viscosity and therefore reduce foam stability.

It is worth noting here that Hart [58] also teaches the use of sodium di-octyl sulfosuccinate (together with a wide range of alkyl sulfonates and alkyl phosphonates) as an antifoam for use in the distillation and coking of crude oils where temperatures range from 150°C to 500°C. Again no physical data are available concerning the state of these materials after dispersal in the crude oil at those temperatures.

10.5 SUMMARIZING REMARKS

The control of foam during the production and refining of crude oil arguably represents the most important non-aqueous use of defoaming technology. The oil industry is also one of the largest and most profitable of industries. Despite all this, the level of understanding of the relevant processes as revealed by published literature appears to be at a primitive level. This is particularly striking if we consider the origins of foam problems. The available evidence suggests that crude oil foam has a transient stability in cases of low oil viscosity where gravitational (or centrifugal if cyclones or centrifuges are used) forces induce rapid drainage in the Plateau borders and foam films. The probable absence of significant positive contributions to the disjoining pressures in such films means that those films rupture at thicknesses where they become susceptible to negative van der Waals contributions. Drainage of films to such thicknesses is likely to be determined by a, probably subtle, combination of viscous shear and surface tension gradient. Increasing the crude oil viscosity is believed to be a major factor in reducing the rate of such drainage and prolonging foam lifetimes despite the low intrinsic stability of the foam films once they thin to a critical thickness. This view, while consistent with observation, is, however, largely speculative. It clearly implicates bulk phase viscosity, surface tension gradients, foam film drainage mechanisms, critical foam film rupture thicknesses, and disjoining forces as important determining factors. Surprisingly, only one published study of the origins of foam in crude oil addresses more than one of these properties [31]! It is also noteworthy that such studies resort to simple correlations with individual measured properties rather than comparison with a relevant theory [4, 31]. These correlations tend to ignore the real complexities of the phenomena and therefore suffer increasing numbers of exceptions as more crude oils are included. Seeking correlations within subcategories defined by dubious justification is sadly often the response to such exceptions.

Another speculation concerns the putative role of certain crude oil components in determining foam behavior. It has been claimed that components such as alkanoic acids [19], asphaltenes, and resins [20, 39, 44] can adsorb at gas–crude oil surfaces. This putative adsorption would of course influence properties such as surface tension gradients and disjoining forces in foam films. These components are, however, more polar than crude oil of a largely aliphatic nature and it is therefore difficult to understand the molecular origin of such adsorption. We should also note that Poindexter et al. [4] have compared the foam behavior of pure alkanes and a heavy mineral oil with various crude oils. Not surprisingly, the pure alkanes do not form any foam. However, the foamability and foam stability of this mineral oil, of bulk phase shear viscosity of 68 mPa s, are both significantly greater than those of a crude oil of 70 mPa s containing a combination of 7.5 wt.% asphaltene and 8.9 wt.% resin. Clearly the foam behavior of the mineral oil must be determined only by a combination of viscous shear and surface tension gradients where the latter have an origin in the fact that the oil is a mixture of many components exhibiting different surface activities. All this would tend to support the view that convincing evidence that adsorption of asphaltenes and resins at gas–crude oil surfaces to promote enhanced foamabilities and foam stabilities is still lacking.

By contrast, understanding of the role of antifoams in gas–oil separation would appear to have a slightly firmer basis than understanding of the causes of foam formation in crude oil. Basically antifoam effectiveness requires that the antifoam be present as a separate phase, be capable of emerging into the gas–oil surface to form unstable bridging configurations in foam films, and be in a sufficiently finely divided state to ensure reasonably high probability of presence in foam films. This leads inexorably to selection of PDMSs with their low air–oil surface tensions and relatively low solubilities in crude oil. Substitution of alkoxy and perfluoroalkyl groups in the PDMS molecule appears to concern only decreases in the solubility of the antifoam in crude oils. Dissolution in crude oils not only removes the antifoam but the dissolved antifoam is an effective pro-foamer. Despite the importance of solubility of the antifoam in this context, there would appear to be no published accounts of actual measurements of antifoam solubility in crude oil as, for example, a function of temperature and molecular structure. Little systematic knowledge of the relation between crude oil composition and measured solubilities of PDMSs and their derivatives appears to exist.

The absence of positive contributions to the disjoining pressures of gas–crude oil–antifoam pseudoemulsion films of an electrostatic origin suggests that these films will be unstable. The need for the presence of the particles required for effective rupture of air–water–antifoam pseudoemulsion films in the context of aqueous surfactant solution foams is apparently therefore obviated. However, there is apparently no knowledge at all of the behavior of, for example, gas–crude oil–PDMS pseudoemulsion films.

That the likely action of antifoam oils in aqueous surfactant solutions concerns formation of unstable bridging oil drops in foam films has been developing over the past three decades (see Section 4.5.1). In principle, antifoam oils in crude oils should also function by the same mechanism. No direct studies of the putative phenomenon in that context have been made. Indeed, the wetting behavior of antifoam oils on any crude oil has not been reported in the published literature at all. Studies of such behavior represent one of the first steps in establishing the relevance of the bridging mechanism. Measurements of all the relevant surface tensions are also essential, as exemplified by the recent studies of Mansur and coworkers [13, 57] from which we can deduce indirect evidence for a bridging mechanism in the case of an ethoxylated PDMS antifoam oil dispersed in a crude oil.

Finally, we should draw attention to the prevalent use of air–degassed crude oil systems and foam generation by sparging at ambient temperatures and pressures. It is known that solubilities of asphaltenes, resins, PDMS, and PDMS derivatives are likely to be influenced by temperature and dissolution of natural gas. Moreover, sparging represents a poor model for foam generation in gas–oil separators, which involves depressurization and nucleation of bubbles. Use of apparatus designed to replicate the conditions in actual gas–oil separators for basic studies should therefore be encouraged.

REFERENCES

1. Callaghan, I.C., Gould, C.M., Reid, A.J., Seaton, D.H. *J. Petrol. Technol.*, 37, 2211, 1985.
2. Pape, P.G. *J. Petrol. Technol.*, 35, 1197, 1983.

3. Callaghan, I.C., Hickman, S.A., Lawrence, F.T., Melton, P.M. Antifoams in gas–oil separation, in *Industrial Applications of Surfactants* (Karsa, D., ed.), Royal Society of Chemistry Special Publications no. 59, 1987, p 48.
4. Poindexter, M.K., Zaki, N.N., Kilpatrick, P.K., Marsh, S.C., Emmons, D.H. *Energy Fuels*, 16(3), 700, 2002.
5. Callaghan, I.C. Antifoams in the oil industry, in *Defoaming, Theory and Industrial Applications* (Garrett, P.R., ed.), Marcel Dekker, New York, 1993, Surfactant Science Series, Vol 45, Chpt 2, p 119.
6. Maini, B.B., Sarma, H.K., George, A.E. *J. Can. Petrol. Technol.*, 32(9), 50, 1993.
7. Wylde, J. *SPE Prod. Oper.*, 25(1), 25, 2010.
8. Birmingham, D.P. (assigned to Hudson Products Corp.), US 6,004,385; 21 December 1999, filed 4 May 1998.
9. Varadaraj, R. (assigned to Exxon-Mobil Research and Engineering Co.), US 2002/0128328; 12 September 2002, filed 9 March 2001.
10. Chin, R., Inlow, H., Keja, T., Hebert, P., Bennett, J., Yin, T., Chemical defoamer reduction with new internals in the MARS TLP separators; SPE paper 56705, presented at SPE Annual Technical Conference and Exhibition, Houston, TX, October 1999.
11. Guggenheim E.A. *Thermodynamics: An Advanced Treatment for Chemists and Physicists*, North-Holland Personal Library, Amsterdam, 1967, p 48.
12. Zisman, W.A. *Adv. Chem. Ser.*, 43, 1, 1964.
13. Fraga, A.K., Santos, R.F., Mansur, C.R.E. *J. Appl. Polym. Sci.*, 124(5), 4149, 2012.
14. Binks, B., Dong, J. *JCS Faraday Trans.*, 94(3), 401, 1998.
15. Kobayashi, H., Owen, M.J. *Macromolecules*, 23, 4929, 1990.
16. Owen, M.J., Groh, J.L. *J. Appl. Polym. Sci.*, 40, 789, 1990.
17. Mannheimer, J.J. *Chem. Eng. Commun.*, 113, 183, 1992.
18. Schaefer, D. *Tenside Surf. Deterg.*, 27, 154, 1990.
19. Callaghan, I.C., McKechnie, A.L., Ray, J.E., Wainwright, J.C. *Soc. Petrol. Eng. J.*, 171, April 1985.
20. Zaki, N.N., Poindexter, M.K., Kilpatrick, P.K. *Energy Fuels*, 16(3), 711, 2002.
21. Mullins, O.C. *Energy Fuels*, 24, 2179, 2010.
22. Yen, T.F. Asphaltic materials, in *Encyclopedia of Polymer Science and Engineering*, Second Edition (Mark, H.S., Bikales, N.M., Overberger, C.G., Menges, G., eds.), Wiley, New York, 1989, Supplementary Volume, p 1.
23. Speight, J.G. *The Chemistry and Technology of Petroleum*, Third Edition, Marcel Dekker, New York, 1998, p 471.
24. Bauget, F., Langevin, D., Lenormand, R. *J. Colloid Interface Sci.*, 239, 501, 2001.
25. Xiuwen, G., Yinzu, Y., Da, L., Shufang, Z. *J. Petrol. Technol.* 35(10), 1797, 1983.
26. Vrij, A. *Diss. Faraday Soc.*, 42, 23, 1966.
27. Vrij, A., Overbeek, J.Th.G. *J. Am. Chem. Soc.*, 90, 3074, 1968.
28. Callaghan, I.C., Gould, C.M., Hamilton, R.J., Neustadter, E.L. *Colloid Surf.*, 8, 17, 1983.
29. Bergeron, V., Hanssen, J.E., Shoghl, F.N. *Colloid Surf. A.*, 123–124, 609, 1997.
30. Shearer, L.T., Akers, W.W. *J. Phys. Chem.*, 62, 1264 and 1269, 1958.
31. Callaghan, I.C., Neustadter, E.L. *Chem. Ind.*, January, 55, 1981.
32. Gang, H., Patel, J., Wu, X.Z., Deutsch, M., Gang, O., Ocko, B.M., Sirota, E.B. *Europhys. Lett.*, 43(3), 314, 1998.
33. Joye, J., Hirasaki, G.J., Miller, C.A. *Langmuir*, 10(9), 3174, 1994.
34. Exerowa, D., Kruglyakov, P.M. *Foam and Foam Films*, Elsevier, Amsterdam, 1998, p 118.
35. Israelachvili, J.N. *Intermolecular and Surface Forces, with Applications to Colloidal and Biological Systems*, Academic Press, London, 1985, p 139.
36. Lucassen, J., van den Tempel, M. *Chem. Eng. Sci.*, 27, 1281, 1972.

37. Lucassen, J., van den Tempel, M. *J. Colloid Interface Sci.*, 41(3), 491, 1972.
38. Maini, B.B., Sarma, H.K., George, A.E. *J. Can. Petrol. Technol.*, 32(9), 50, 1993.
39. Adil, I., Maini, B.B. *J. Can. Petrol. Technol.*, 46(4), 18, 2007.
40. Abivin, P., Hénaut, I., Argillier, J.-F., Moan, M. *Petrol. Sci. Technol.*, 26, 1545, 2008.
41. Bauget, F. Production d'huiles lourdes par dépressurisation: Etudes des interfaces huiles-air et modélisation du procédé. PhD Thesis, University of Paris-Sud, Orsay, France, 2002.
42. Hénaut, I., Barré, L., Argillier, J.-F., Brucy, F., Bouchard, R. Rheological and structural properties of heavy crude oils in relation with their asphaltene content. SPE 65020, Oil Field Chem. Houston, TX, February 13–16, 2001.
43. Pierre, C., Barré, L., Pina, A., Moan, M. *Oil Gas Sci. Technol.*, 59, 489, 2004.
44. Claridge, E.L., Prats, M. A proposed model and mechanism for anomalous foamy heavy oil behaviour, SPE 29243, paper presented at Heavy Oil Symposium, Calgary, Alberta, Canada, June 1985.
45. Princen, H.M. *Langmuir*, 2(4), 519, 1986.
46. Varaprath, S., Frye, C.L., Hamelink, J. *Environ. Toxicology Chem.*, 15(8), 1263, 1996.
47. Centers, P.W. *Tribology Trans.*, 36(3), 381, 1993.
48. L'Hostis, J., Renauld, F. (assigned to Dow Corning Ltd.), EP 0916 377; 19 May 1999, filed 13 November 1998.
49. Callaghan, I.C., Gould, C.M., Grabowski, W. (assigned to British Petroleum Ltd.), EP 0167 361; 8 January 1986, filed 27 June 1985.
50. Callaghan, I.C., Macleod, M. (assigned to The British Petroleum Company Plc), GB 2234978; 20 February 1991, filed 17 May 1990.
51. Taylor, A.S., Macleod, M. (assigned to The British Petroleum Company Plc.), GB 2244279; 27 November 1991, filed 24 May 1990.
52. Fink, H., Koerner, G., Berger, R., Weltmeyer, C. (assigned to TH. Goldschmidt AG), GB 2196976; 11 May 1988, filed 15 October 1986.
53. Garrett, P.R. *J. Colloid Interface Sci.*, 76(2), 587, 1980.
54. Denkov, N.D. *Langmuir*, 20(22), 463, 2004.
55. Denkov, N.D. *Langmuir*, 15(24), 8530, 1999.
56. McKendrick, C.B., Smith, S.J., Stevenson, P.A. *Colloids Surf.*, 52, 47, 1991.
57. Rezende, D.A., Bittencourt, R.R., Mansur, C.R.E. *J. Petrol. Sci. Eng.*, 78, 172, 2011.
58. Hart, P.R. (assigned to Betz Laboratories, Inc.), US 5,169,560; 8 December 1992, filed 17 September 1990.

11 Medical Applications of Defoaming

11.1 INTRODUCTION

There are two main medical applications of defoaming. The first concerns the treatment of symptoms associated with excessive gas in the gastrointestinal tract. The second concerns the treatment of blood during cardiopulmonary bypass surgery.

Formation of excessive gas in the gastrointestinal tract is often associated with hyperacidity and can produce chronic pain. The prevalence of this problem derives from the formation of extremely stable mucous foam where gas is introduced either through swallowing during food ingestion or through gas generation during digestion. Not only does this foam produce unpleasant symptoms but it also hinders diagnosis of various conditions using techniques such as endoscopy. These problems of excessive foam usually succumb to either oral or enteric administration of PDMS–hydrophobed silica antifoam, which is often given the generic name "simethicone" in this context. Here, we review the issues associated with the application of simethicone.

Procedures such as open heart surgery and heart transplants have been practiced for several decades with progressively increasing survival rates. Death rates after such cardiac surgery in the United States, for example, have recently been reported as now less than 1% [1]. This has been made possible by the development of ever more efficient cardiopulmonary bypass (heart–lung) machines. A key function of such machines concerns the oxygenator where venous blood is supplied with oxygen and carbon dioxide is removed. Early designs of oxygenators involved gas bubblers where excess gas is removed using defoamer/filters [2]. These defoamers usually consisted of porous matrices impregnated with PDMS–hydrophobed silica antifoam. Although use of such devices reduces the risk of gas emboli, the antifoam is slowly dispersed in the blood supply. This can also result in a serious incidence of emboli due to the entrapment of antifoam drops in blood capillaries [3–5]. Such problems are alleviated by use of membrane oxygenators where gas exchange with venous blood occurs by molecular diffusion across permeable membranes. Defoamers are still used with such devices, however, to minimize the risk of gas emboli due to inadvertent introduction of air bubbles. This is a particular problem when "cardiotomy" blood from the "surgical field," containing both detritus from surgery and a risk of the presence of air bubbles, is reunited with venous blood for recirculation.

Difficulty of decontamination of either type of oxygenator after use means that they have to be affordably disposable. Even 15 years ago, it was estimated that more than 1 million disposable oxygenators were used globally where about 85–90% were of the membrane type with a clear trend toward elimination of the use of bubble oxygenators [2].

The designs of various defoamers for both types of oxygenators are reviewed here together with the evidence concerning emboli formation by polydimethylsiloxane (PDMS)-based antifoam. Although PDMS–hydrophobed silica is the preferred antifoam, other materials and even defoamers that do not use antifoams have been suggested, the practicality of which we also consider.

11.2 USE OF SIMETHICONE ANTIFOAM IN TREATMENT OF GASTROINTESTINAL GAS

Here we briefly review the issues involved in the use of simethicone in the treatment of gastrointestinal gas following the earlier reviews of Burton [6] and Berger [7].

11.2.1 Therapeutic Application

It has been known for more than half a century that antifoams consisting of mixtures of PDMS and hydrophobic silica can be used to dispel excessive gas in the gastrointestinal tract. However, the first use of antifoams in this context was reported by Quin et al. [8] more than 60 years ago and concerned treatment of bloat in ruminants. That the hydrophobed silica is a necessary component of an effective PDMS-based antifoam *in vivo* was first demonstrated by Birtley et al. [9] using X-ray observation of the stomachs of rats in which foam had been artificially produced.

The pharmaceutical grades of PDMS–hydrophobed silicas are without toxic effect and have been formulated into tablets or capsules but with various degrees of success. The excessive gas, for which the use of this simethicone is usually an effective remedy, is often accompanied by hyperacidity. Antacids such as various alkaline inorganic salts represent suitable remedies for such a condition but are rendered less potent if stable mucous foam is present. Therefore, simethicone is often combined with antacids to form tablets for convenient application. Rezak [10] has, however, shown that the antifoam can lose effectiveness after incorporation in such tablets. This study used tablets containing only 3.4 wt.% antifoam, 54 wt.% antacid, with the remainder consisting of various organic materials. Particularly striking was the almost total loss of antifoam activity (as measured by a cylinder shaking test) in the case of antacids such as aluminum hydroxide and magnesium carbonate after storage at 45°C for <24 h. These effects were apparent even in the case where tablet ingredients were simply granulated without compression. Other inorganic antacids, such as magnesium hydroxide and sodium bicarbonate, also caused deactivation but at a significantly slower rate, taking several weeks to lose antifoam activity. Deactivation of simethicone did not, however, occur in tablets prepared with organic antacids such as glycine and sodium citrate. Stead et al. [11] have studied the stability of simethicone–antacid tablets that contained significantly higher proportions (33 wt.%) of simethicone. This study involved both *in vitro* and *in vivo* observations. Again tablets containing aluminum hydroxide as antacid showed immediate deactivation of the antifoam effect of simethicone. Not only was the antifoam deactivated but the effectiveness of the antacid was also diminished. It has therefore been suggested by

Stead et al. [11] that deactivation on inorganic salts like aluminum hydroxide may simply concern adsorption of the antifoam onto the salt, rendering both materials less available. By contrast, tablets based on hydrotalcite as antacid showed stable antifoam and antacid behavior even after storage for 1.8 years at 37°C. If adsorption of simethicone onto the antacid is the cause of deactivation, then it is not, however, obvious why it should not occur on a carbonate clay like hydrotalcite.

Similar observations by Akay et al. [12] concerning the stability of PDMS–hydrophobed silica antifoams have been made in the case of addition of these antifoams to inorganic carriers before incorporation in detergent powders (see Section 8.2.5.4). The adverse interaction between PDMS–hydrophobed silica antifoam and certain inorganic salts does not, however, appear to concern molecular cleavage of the siloxane chain induced by these alkaline materials [13]. Again it was observed by Akay et al. [12] that admixture of the antifoam with organic materials did not produce deactivation. However, these observations concerning deactivation on inorganic carriers were largely confined to combinations of "high"-viscosity (~3×10^4 mPa s) PDMS–hydrophobed silica antifoams and salts like sodium carbonate and sodium monoperborate, the effect being significantly less apparent with antifoams of "low" viscosity (~3×10^3 mPa s). Unfortunately, neither Rezac [10] nor Stead et al. [11] reported the viscosity the simethicone used in their work so direct comparison is not possible.

A comparison of antifoam effectiveness with a Fourier transform infrared spectroscopy assay of tablets containing simethicone, including those based on magnesium hydroxide, has been made by Torrado et al. [14]. The experimental procedure for evaluation of antifoam effectiveness involved first grinding the tablets to a "fine powder" followed by ultrasonic dispersal for 5 min in a surfactant solution before assessing foam stability. Despite relatively high PDMS levels in the tablets of ≥17 wt.%, the antifoam effectiveness of those containing magnesium hydroxide was negligible. This effect did not simply concern slow release of the antifoam because ultrasonication for 5 min dispersed 60 wt.% of the PDMS—further periods of ultrasonication increased the release to ~90 wt.% but without increasing antifoam effectiveness. Torrado et al. [14] have, however, noticed changes in the spectrum of PDMS in the magnesium hydroxide tablets relative to nominal where the intensity of the symmetrical deformation vibration of the methyl peak is diminished relative to the stretching vibrations of the Si–O–Si group. An analysis of simethicone-containing aqueous suspensions and tablets, based on mixtures of magnesium hydroxide and aluminum hydroxide, has also been made by Moore et al. [15] who used a reversed phase liquid chromatographic technique together with evaporative light scattering detection for assay of the PDMS. Curiously, the analysis revealed nominal levels of PDMS in the aqueous suspensions but only ~87 wt.% of nominal in the case of tablets that had been stored for 4 months. Moore et al. [15] offer no explanation for the discrepancy, which lies well outside the margins of error reported by these workers. It may, of course, be possible that it has the same cause as the antifoam deactivation reported for simethicone tablets by Torrado et al. [14] and others [10, 11].

Thus far then there is no convincing explanation for the deactivation of PDMS–hydrophobed silica antifoam when admixed with certain inorganic salts. A possible explanation concerns separation of the hydrophobed silica from the PDMS oil either during admixture with the inorganic salt or after storage. This would depend on the pore structure and even wettability for PDMS of the tablet material. It is even possible that deactivation concerns the stability of the hydrophobic surface of the silica rather than the PDMS. This deactivation could, in principle, be mitigated in the case of antacid tablets by following the approaches used in applying PDMS–hydrophobed silica antifoams in detergent powders. For example, emulsification of the simethicone in a suitable non-toxic water-soluble organic matrix (such as polyethylene glycol or non-toxic EO–PO alkoxylated block copolymers) followed by incorporation in a tablet might represent a successful approach (see Section 8.2.5.2).

In the event that simethicone is to be administered in situations without an accompanying antacid, use of an oil–water emulsion stabilized by a suitable aqueous non-toxic surfactant solution would be preferable. However, as we describe in Chapter 6, formation of an emulsion inevitably also causes some deactivation of the antifoam. Such deactivation is, however, usually modest compared with that reported after incorporation in tablets based on certain inorganic salts.

There have been numerous controlled trials, using placebos, of the therapeutic effectiveness of simethicone for conditions associated with gastrointestinal gas such as flatulence, dyspepsia, and postoperative abdominal discomfort, etc. (see, e.g., references [16–21]). In most cases, significant benefit has been found for the use of simethicone [16–19]. However, there have been exceptions where the expected benefit is small or absent altogether (see, e.g., references [20, 21]). A feature of most of these reported trials has been a lack of sufficient detail concerning both the chemical nature of the simethicone and the composition and method of preparation of the delivery vehicle (usually a tablet). For example, the effectiveness of PDMS–hydrophobed silica antifoams is known to be a strong function of the viscosity of the antifoam [22, 23]—that property is never quoted in this literature. Also the deactivation of the antifoam, as a result of adverse interactions with inorganic antacids, increases with storage time and can be total—again the composition and age (and storage conditions if relevant) of tablets are often not quoted [18, 19, 21]. The absence of such details makes interpretation of the findings of controlled trials rather difficult despite the care involved in double-blind comparisons with placebos.

11.2.2 Use of Simethicone in Endoscopy

Use of simethicone as an aid to endoscopy is well established but not without controversy. A number of controlled trials involving placebos have established the general utility of the antifoam in preparing organs for endoscopy [24–26]. However, some trials have produced more equivocal results. For example, Bertoni et al. [27], after a placebo controlled trial, claim to find that simethicone is useful for endoscopy of the upper gastrointestinal tract only in cases where patient has been subject to a gastric resection. Wu et al. [28] suggest that, although the use of simethicone removes air bubbles from the colonic lumen, the effect on overall diagnosis "remains controversial."

Parikh and Khanduja [29] note that colonoscopy can actually introduce air to form a foam that obscures visualization of the relevant organ. They have argued that application as an aqueous "solution" via the biopsy channel of the colonoscope has particular merit. Here we should note that polydimethylsiloxanes of the molecular weight characteristic of PDMS–hydrophobed silica antifoams are essentially totally

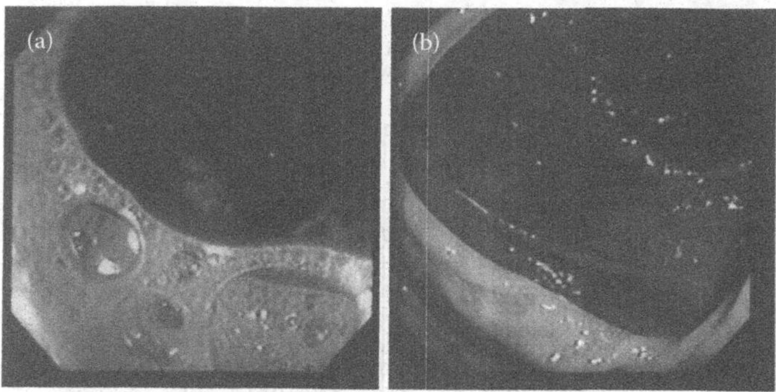

FIGURE 11.1 Elimination of foam during colonoscopy by use of simethicone (PDMS–hydrophobed silica antifoam. (a) Endoscopic view showing foam. (b) View after introduction of simethicone using the biopsy channel of colonoscope. (With kind permission of Springer Science+Business Media: *Dis. Colon Rectum*, 38, 1995, 1007, Parikh, V.A., Khanduja, K.S.)

insoluble in water [30]. It seems likely then that Parikh and Khanduja [29] had used an oil–water emulsion. Figure 11.1 illustrates the effectiveness of their procedure.

Conventional endoscopy is unable to satisfactorily access the small bowel. The relatively new technique of capsule endoscopy is, however, finding application in that context. In this technique, the patient swallows a small wireless camera that transmits images to sensors secured to the abdominal wall. There appears to be little agreement about preparatory treatment of the small bowel with simethicone before the use of capsule endoscopy. A search through relevant databases revealed six controlled trials favoring such treatment with respect to alternatives [31–36] and only two indicating a preference for alternative treatments [37, 38].

11.3 DEFOAMING OF BLOOD DURING CARDIOPULMONARY BYPASS SURGERY

11.3.1 Gas Bubble Oxygenators and Use of Antifoams

Cardiopulmonary bypass makes possible routine open heart surgery and heart transplants. This bypass essentially replaces the functions of the heart and lung with a pump and an oxygenator. Early approaches to oxygenation of blood in such devices involved direct sparging with oxygen. This produces foam, which must be removed before returning blood to the patient lest gas emboli form leading to organ damage. Emboli that form in the brain are particularly serious, potentially leading, for example, to partial paralysis and loss of cognitive function.

The approach used for the design of bubble oxygenator–defoamer combinations is exemplified by that of Gremel and Grant [39] who claim a combination of a sparging oxygenator and the use of a PDMS–hydrophobed silica antifoam. A simplified illustration of their device is shown in Figure 11.2. In this device, oxygenated blood leaves the sparger containing an unspecified gas volume fraction of dispersed bubbles. It then passes into a cylindrical defoamer vessel where it must in turn pass through three porous annular elements before passing through a reservoir en route

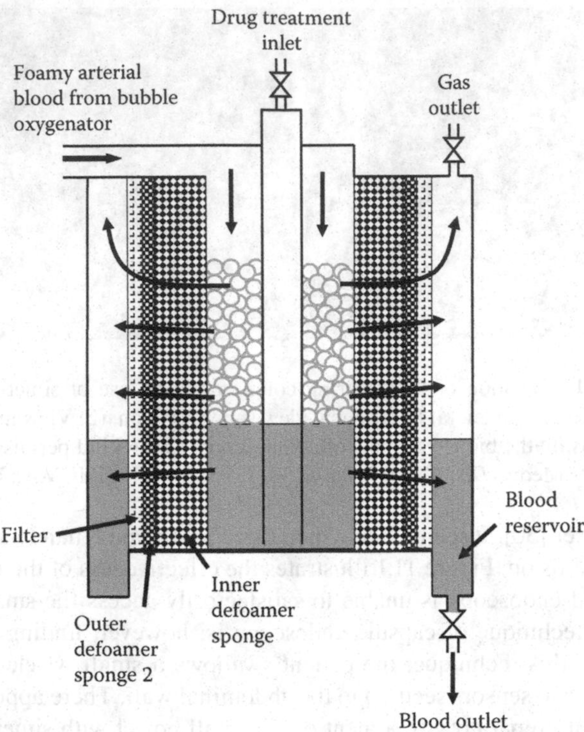

FIGURE 11.2 Simplified schematic of vessel used for defoaming foamy blood after bubble oxygenation. Inner defoamer sponge is a porous open-celled polyurethane foam coated with PDMS–hydrophobed silica antifoam. Outer defoamer sponge is similar to inner sponge but with smaller pores. Filter is polyester fabric with pore sizes of ~100 microns. (After Gremel, R.F., Galt, K.M. (assigned to Shiley Inc., California), US 4,568,367; 4 February 1986, filed 15 November 1982.)

to the patient. The first two such elements are defoamers, preferably fabricated from a sponge-like porous open-celled polyurethane foam. The pore sizes of the first of these elements (see Figure 11.2) are defined, somewhat eccentrically, as 7–10 pores per centimeter (possibly meaning pores of about 1 mm diameter) and those of the second, as about 30 pores per centimeter (possibly of 300 microns diameter). These two polyurethane sponge-like elements are coated with a PDMS–hydrophobed silica antifoam. Gemel and Grant [39] do not, however, explain how that coating is accomplished. The third element is a fine fabric filter with smaller pores sizes than either of the defoaming elements. A preferred filter element consists of a plain square weave nylon or polyester fabric with a mesh opening of about 100 microns. Such a filter would be considered too course by modern standards. It would risk allowing gaseous or particulate microemboli to pass into the arterial blood supply of a size sufficiently large to potentially cause patient morbidity. In this we follow Nussmeier [40] and distinguish somewhat arbitrarily between macroemboli of size > 200 microns and microemboli of size < 40 microns.

Variations of this arrangement of defoamer and filter elements are typical of those claimed for de-aerating blood regardless of source or oxygenator type (see, e.g., references [39, 41–50]). Often the PDMS–hydrophobed silica antifoam, which should be of pharmaceutical grade, is specified as "Antifoam Type A" of Dow Corning, which trade literature suggests implies an antifoam of viscosity 1500 mPa s [51]. Homogeneous coating of the relevant porous element to give concentrations of antifoam typically at 0.5 mg cm^{-3} [48] may be difficult with such high-viscosity fluids. However, Tsai and Haynes [52] state that coating the substrate usually simply involves dipping it in a "solution" of the antifoam in a halogenated hydrocarbon followed by evaporation of the solvent. It is easy to see how this process could separate the hydrophobed silica from the antifoam, possibly leading to partial deactivation.

11.3.2 Mechanism of Polydimethylsiloxane–Hydrophobed Silica–Coated Porous Defoamers

The mode of action of PDMS–hydrophobed silica antifoams in aqueous surfactant solutions has been extensively studied by Denkov et al. [53] and reviewed in detail in Chapter 4. Essentially the hydrophobed silica particles rupture the so-called air–water–oil pseudoemulsion film, thereby enabling the oil to emerge into the air–water surface. It is known that once they emerge into the air–water surface, drops of PDMS oils usually initially spread over that surface, exhibiting either complete wetting or pseudo-partial wetting behavior (see Section 3.6.2). This means that the oil spreads as either a thick duplex layer or spreads and breaks up into lenses in equilibrium with a thin oil layer. Since such behavior is ubiquitous with aqueous surfactant solutions, it is reasonable to expect similar behavior when PDMS oil drops are introduced into the gas–blood surface. It is not, however, known whether complete or pseudo-partial wetting behavior is to be expected.

If a PDMS oil bridge of sufficient size forms across both surfaces of a foam film, then, regardless of the wetting behavior, the resulting configuration is unstable—the bridge is drawn apart by capillarity so that it ruptures by the so-called bridging–stretching mechanism [54, 55]. In turn this causes the foam film to collapse. Another role for the hydrophobed silica is facilitating the formation of such bridges (see Section 4.8.5.2).

The usual approach to delivery of these antifoam phenomena is to disperse the antifoam as finely divided drops in the solution to be defoamed to enhance the probability of its presence in foam films. However, delivery of the antifoam in a blood circuit defoamer does not follow that practice. The antifoam is "immobilized" in the pores of a hydrophobic structure such as a sponge. It is of course subject to shear as blood flows through the defoamer; however, we remember that the often preferred PDMS–hydrophobed silica antifoam in this context is "Antifoam Type A," which has a viscosity of about 1500 mPa s [51], where the PDMS oil alone is a Newtonian fluid of viscosity ~1000 mPa s. Blood is a viscoelastic shear thinning fluid at shear rates < ~10 s^{-1} and a Newtonian fluid at higher shear rates [56, 57]. There is apparently no knowledge of the shear rates prevailing in a defoamer sponge; however, if, as seems likely, they exceed the threshold for Newtonian behavior, then the viscosity of blood is ~4 mPa s [56, 57], almost three orders of magnitude lower than that of the antifoam. Continuity of shear stress at the antifoam–blood interface means that the induced shear rates in the antifoam will also be nearly three orders of magnitude lower than those prevailing in the blood flow. Significant direct emulsification of the antifoam and dispersal into the blood flow as a result of the shear stress applied by the latter is therefore unlikely to occur.

The actual antifoam delivery method probably concerns direct contact between the foam bubbles and the antifoam. The presence of hydrophobed silica particles should ensure that any gas–blood–oil pseudoemulsion film separating a bubble from direct contact with the antifoam should readily rupture. In consequence, we would expect configurations similar to that depicted in Figure 11.3 to form as gas bubbles interact with the "immobilized" antifoam, which is held in the pores of the defoamer sponge by capillarity. Probable positive initial spreading coefficients will mean that the PDMS oil component of the antifoam will spread over the remaining gas–blood surfaces of adjacent bubbles to produce a continuous thick duplex layer (or a thinner layer in equilibrium with oil lenses) on both sides of the foam films separating the bubbles. The rate of spreading of PDMS oils over aqueous surfaces is independent of the viscosity of the oil for oils of viscosity < 1000 mPa s [58, 59] (see Section 3.6.2.2). This behavior arises because the oil spreads like a rigid solid with virtually zero difference in velocity at the air–oil and oil-water interfaces [58]. It means also that the silica particles will therefore be dragged along with the oil. Measurements by Bergeron et al. [59] of the rate of radial spreading of PDMS oils suggests that such a layer would spread over a bubble of 1 mm radius in about 10 ms. As we have discussed in Section 4.8.5, the presence of such layers on the surface of foam films will lead to the formation of unstable bridging configurations and foam film collapse. The antifoam debris formed after that collapse would presumably be emulsified and could transport through the Plateau borders to other parts of the foam and eventually to the bulk liquid phase. It is unlikely to return to the "immobilized" state and it therefore inevitably contaminates the blood flow. It is therefore possible that such emulsified antifoam debris could participate in further foam collapse even if the

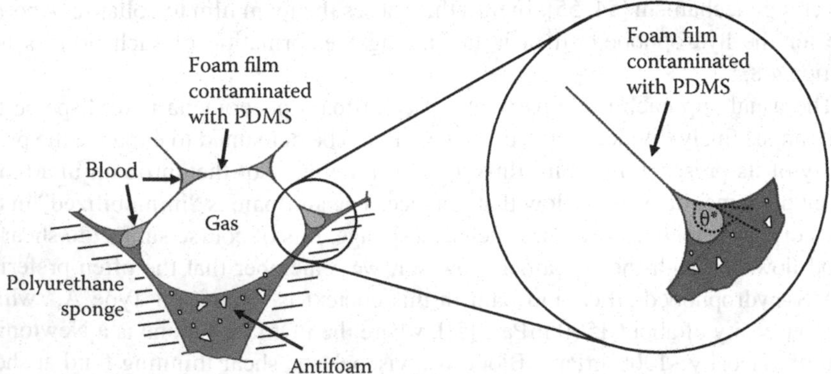

FIGURE 11.3 Schematic illustration of probable configuration formed by foam bubbles in direct contact with PDMS–hydrophobed silica antifoam trapped in pores of a hydrophobic polyurethane sponge typical of those used in cardiotomy reservoirs.

defoamer sponge becomes depleted. However, continuous involvement of such emulsified material in foam generation is known to lead to disproportionation into inactive particle-free drops of PDMS and large inactive silica-rich agglomerates [60, 61]. This emulsified antifoam material will therefore gradually deactivate—a detailed account of the process is given in Chapter 6.

Examination of the configuration shown in Figure 11.3 suggests another possible mechanism by which PDMS oil bridges could form in foam films—one that does not require either that the antifoam initially spread over the bubble surfaces. As indicated in the figure, some foam films must be contiguous to Plateau borders located partially on the surface of the antifoam. This means that an oil–blood meniscus should exist in the Plateau border where the curvature of the meniscus is determined by the angle between the gas–blood surface and the oil–blood surface, θ^*. A necessary condition for antifoam action requires that $\theta^* > 90°$ (see Section 4.5.1), which means that the curvature of the meniscus is positive in the direction of the contiguous foam film. The presence of a meniscus in the Plateau border could then mean that the capillary pressure in the Plateau border becomes greater than both the pressure in the foam film and that in any connected Plateau border not in contact with such a meniscus. This situation would probably lead to a displacement of blood toward those other Plateau borders and a movement of the meniscus toward with film. In turn, this would lead to an increasing capillary pressure driving the process. Eventually, the oil would be sucked into the film to form a bridging configuration and foam film collapse. Such a mechanism is of course largely speculative in the absence of observational evidence. It is in fact rather similar to that proposed in Section 4.5.3 for antifoam action by drops bridging Plateau borders.

Finally, it is worth noting that, before application, the defoamer consists of a hydrophobic porous medium containing a hydrophobic fluid trapped in the smaller pores with the remainder of the pore space occupied by air. Before exposure to blood flow, it is usual to "prime" the defoamer with a suitable fluid to remove air. However, we should remember that use of a fluid to displace another fluid (air in this case) with which it is not miscible in a porous medium is not usually complete as those familiar with crude oil production will attest. Perhaps the best approach to this problem is to first flush out the defoamer with CO_2 followed by an aqueous priming fluid.

11.3.3 Polydimethylsiloxane–Hydrophobed Silica Antifoam as Source of Emboli

Pharmaceutical grade PDMS–hydrophobed silica antifoam is non-toxic. However, this does not exclude the possibility that such material, dispersed in circulating blood, could cause blockages in capillaries, leading to organ damage analogous to that formed by gas bubbles. The need to use significant amounts of antifoam in combination with bubble oxygenators together with the likelihood that the antifoam will in fact contaminate circulating blood, despite the presence of filters, has implicated this material as a possible source of such emboli and occasional postoperative deaths.

An early warning of possible problems arising from the use of PDMS–hydrophobed silica as an antifoam in this context was made by Cassie et al. [5] who subjected dogs to cardiopulmonary bypass using a bubble oxygenator that did not

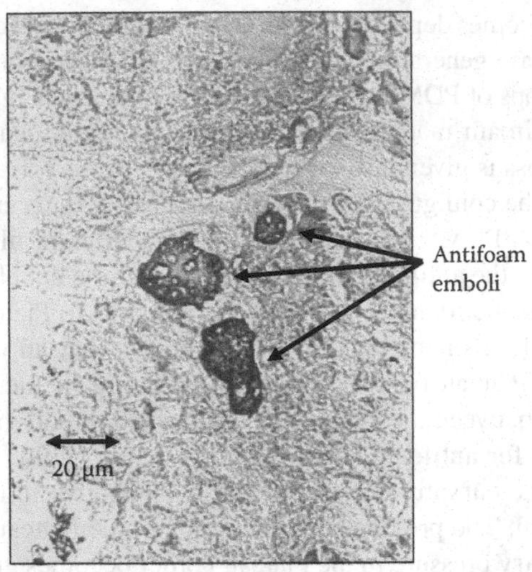

FIGURE 11.4 Bright-field photomicrograph showing PDMS–hydrophobed silica antifoam drops trapped in glomerular capillaries of a kidney after using bubble oxygenator for cardiopulmonary bypass. (Reprinted from *Hum. Pathol.*, 13, Orenstein, J.M., Sato, N., Aaron, B., Buchholz, B., Bloom, S., 1082. Copyright 1982, with permission from Elsevier.)

include filters in the defoamer. The mortality rate was about 74%, which implies use of high levels of antifoam. Emboli due to drops of antifoam blocking blood capillaries were apparent in all the dogs regardless of survival. That those emboli had caused strokes through interruption of the blood supply was also clear. Cassie et al. [5] concluded that "the bubble oxygenator carries an inherent risk of cerebral emboli from the antifoam." Many other observations of this phenomenon followed [62–64].

Some 20 years after Cassie et al. [5] published their paper, Orenstein et al. [3] made a study of autopsy material arising from a series of deaths of patients who had undergone cardiopulmonary bypass involving the use of bubble oxygenators. The study included light microscopy, scanning and transmission electron microscopy, and scanning electron microscopy X-ray microprobe spectroscopy. Light microscopy revealed the presence of antifoam drops in the capillaries of many organs, including the brain, heart, and liver. Microprobe spectroscopy confirmed that the observed entities contained silicon and must therefore have been antifoam. They were invariably present in the kidneys where they were also present in greatest abundance. A bright-field photomicrograph showing the presence of drops of PDMS–hydrophobed silica trapped in the glomerular capillaries* of a kidney is shown in Figure 11.4. The drops are seen to be of irregular shape but are clearly containing silica particles.

Orenstein et al. [3] showed that there was evidence of antifoam drops on both the venous and arterial sides of the circulatory system, which implied continuous circulation. They also state that the clinical significance of microemboli formed by the PDMS–hydrophobed silica used in their study had not been "fully established."

* The glomerular capillaries constitute a major part of the basic filtration unit of the kidney.

Nevertheless, they conclude that "the results of this study suggest that the continued presence of such microemboli has the potential to produce delayed or long-term post-operative complications following cardiopulmonary bypass surgery." They express the view that retention of antifoam materials by organs is potentially harmful and suggest that this problem may be solved by use of alternative types of oxygenator such as the membrane type, which rely on gas diffusion and do not therefore involve bubble formation and the consequent need for antifoams. Within 10 years of the expression of this opinion, use of membrane oxygenators had more or less totally displaced use of bubble oxygenators in the United States [2].

11.3.4 Cardiotomy Defoaming

Another source of gas bubbles in blood during cardiopulmonary bypass, apart from bubble oxygenators, concerns aspiration of blood from the "surgical field" (i.e., the surgery operations on the patient). Cardiopulmonary bypass surgery of extended duration implies the release of significant amounts of blood into that surgical field, which sometimes, of necessity, must be returned to the overall blood flow. However, cardiotomy blood contains detritus from the surgery, such as lipids, together with significant quantities of air bubbles, including microbubbles of < 40 microns diameter [65]. The potential of the latter to cause emboli is even more serious than is the case for oxygen bubbles introduced from a bubble oxygenator because they are air bubbles containing mostly nitrogen, which is less soluble in, for example, venous blood than oxygen [65, 66]. Removal of such potential sources of emboli is therefore desirable as argued by Pearson et al. [65].

Several devices have been patented with the specific function of cardiotomy defoaming and filtering [41–44, 46]. These patented "cardiotomy reservoir" devices usually embrace the same approach to defoaming to that described for bubble oxygenator defoaming involving the use of PDMS–hydrophobed silica antifoam. They are exemplified by the device claimed by Servas et al. [44], which is similar in concept to that shown in Figure 11.2. The relevant defoamer/filter system is depicted in plan view in Figure 11.5. It consists of alternate annular elements of defoamer and filter as shown in the figure. Cardiotomy blood first meets a relatively course defoaming element preferably an open-celled polyurethane foam with only 8 pores per centimeter and coated with a PDMS–hydrophobed silica antifoam. This element is also intended to filter out course surgical detritus. Next is a particle filter for removing small-size surgery detritus. This filter is made from non-woven polyester or polypropylene fabric with one "fluffy" surface (to enhance filtration efficiency) and pore sizes in the range of 50–90 microns. This is followed by another defoamer layer, which differs from the first only in the use of a material with larger pores. The assembly is completed by yet another fabric filter of pore size ~40 microns. After passing through that filter, the blood passes into a temporary storage "cardiotomy reservoir."

The whole assembly should preferably remove from blood any bubbles or detritus of size > 20 microns. We should, however, note here that, in an ultrasonic study of the relative efficiency of blood defoamer/filters, Pearson et al. [65] argue that 20-micron filters are desirable if microbubbles of size ≥ 20 microns are to be removed. The device of Servas et al. [44] could clearly be improved by use of a finer filter.

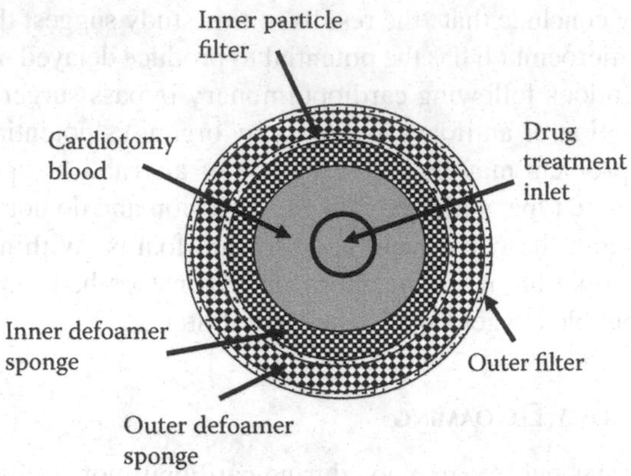

FIGURE 11.5 Simplified plan view of defoamer and filter elements in typical cardiotomy reservoir. Both defoamer elements are sponge-like open-celled polyurethane foams coated with PDMS–hydrophobed silica antifoam. Inner particle filter has pore sizes in range of 50–90 microns and outer filter has pore size of ~40 microns. (After Servas, F.M., Gremel, R.F., Ryan, T.C. (assigned to Shiley Inc.), US 4,743,371; 10 May 1988, filed 28 February, 1986.)

Despite use of these defoamer/filters, the presence of lipids, PDMS antifoam particles, and microbubbles of air conspire to make cardiotomy blood the major cause of emboli in modern cardiopulmonary bypass surgery when using membrane oxygenators [67]. Superior outcomes can often be achieved by simply discarding that blood [67], although this is not usually a practical option.

11.3.5 Defoaming in Cardiopulmonary Bypass Blood Circuits, Which Include Membrane Oxygenators and Cardiotomy/Venous Reservoirs

Membrane oxygenators have now mostly replaced bubble oxygenators in cardiopulmonary bypass surgery. These devices involve oxygen and carbon dioxide exchange by direct molecular diffusion through permeable materials (see, e.g., a review by Gaylor [68]). No gas bubbles are involved and therefore the need for a defoamer is apparently obviated. However, there always exists the possibility of inadvertent aeration in the extracorporeal blood circulatory system as a result of, for example, the action of pumps or as blood is trickled over heat exchanger coils. The imperative of avoiding the possibility of gas emboli in patients means that some defoaming capacity is apparently preferred.

Cardiopulmonary bypass surgery also involves return of cardiotomy blood to the patient. Unlike the venous blood supplied to the oxygenator, cardiotomy blood is, as we have discussed, likely to be both aerated and contain surgical detritus. It is also likely to be flowing at a significantly slower rate than venous blood.

It is argued that combining venous and cardiotomy in a single compact vessel has some advantages, including minimizing required priming volumes and possible sources of aeration. Thus, for example, mounting a combined venous and cardiotomy defoamer/filter above a membrane oxygenator rather than having separate parallel facilities minimizes the presence of air traps during priming [45]. As a consequence, many such combined devices have been described in the patent literature (see, e.g., references [45, 47–50, 69].

An example of a combined defoamer/filter and oxygenator/heat exchanger system is described by Raneri et al. [49]. A simplified schematic illustration of their device is shown in Figure 11.6. The venous and cardiotomy blood inlets are separate and lead to separate chambers in the defoamer/filter vessel. As shown in the figure, the relatively contaminated nature of cardiotomy blood requires a sequence of defoamer–filter–defoamer where the defoamers are presumably based on a PDMS–hydrophobed silica–coated sponge-like open-cell polyurethane foam. By contrast, the venous blood requires only one defoamer, which is in fact formed by extending the outer cardiotomy defoamer as shown in Figure 11.6. After passing through that outer defoamer, both venous and cardiotomy blood pass through a fabric "sock," are united in a single chamber, and then pass onto the membrane oxygenator/heat

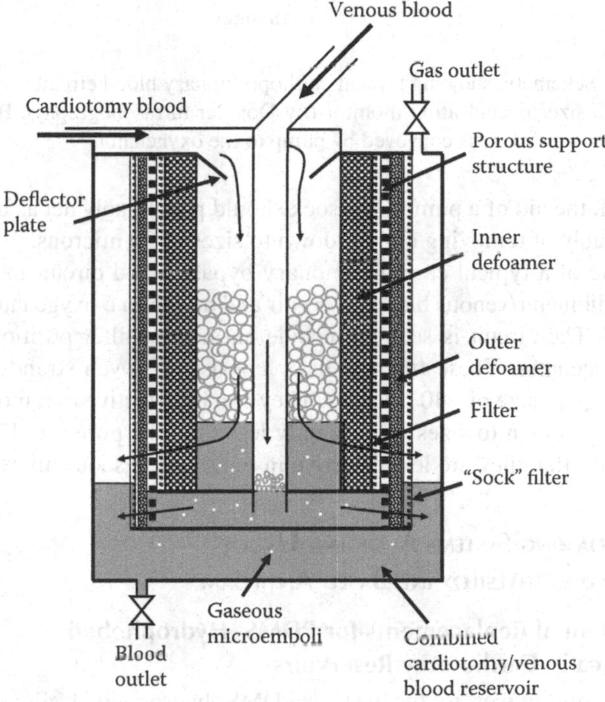

FIGURE 11.6 Simplified schematic of typical combined cardiotomy and venous blood defoamer/reservoir assembly. Cardiotomy blood is both foamy and contaminated with surgical detritus and therefore requires more extensive defoaming/filtration. Deflector plates direct the flow of cardiotomy blood to minimize aeration. (After Raneri, J.J., Buckler, K.E., Stanley, C.L., Intoccia, P. (assigned to C.R.Bard Inc.), US 5,849,186; 15 December 1998, filed 15 November 1996.)

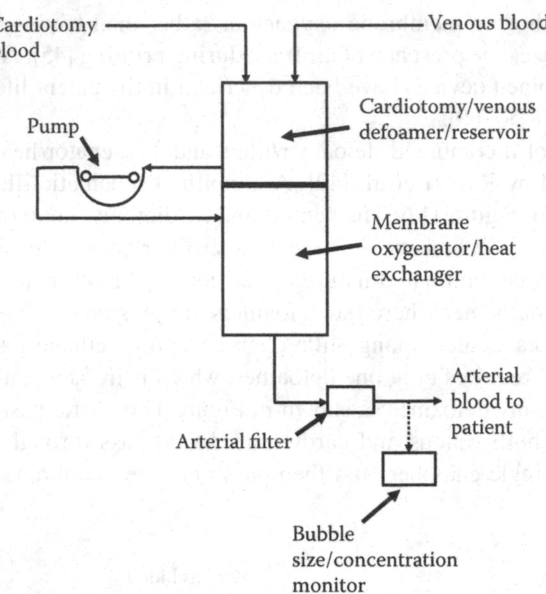

FIGURE 11.7 Schematic showing typical cardiopulmonary blood circuit including arterial filter and bubble size/concentration monitor (by Doppler ultrasonography). Blood from the cardiotomy/venous reservoir is conveyed by pump to the oxygenator.

exchanger with the aid of a pump. The sock should presumably act as a filter ideally capable preferably of removing emboli down to sizes of 20 microns.

A schematic of a typical cardiopulmonary bypass blood circuit incorporating a combined cardiometer/venous blood reservoir and membrane oxygenator is depicted in Figure 11.7. The circuit is seen to include an arterial filter positioned after the membrane oxygenator. These filters usually consist of woven strands of polyester with a uniform pore size of ~40 microns. They can be effective in removing particulate microemboli down to sizes significantly less than the pore size [70]. However, there is evidence that they are less effective in removing gaseous microemboli [71].

11.3.6 Defoaming Systems Avoiding Use of Polydimethylsiloxane-Based Antifoam

11.3.6.1 Potential Replacements for PDMS–Hydrophobed Silica in Cardiotomy Reservoirs

A number of alternatives to the use of PDMS–hydrophobed silica antifoam in cardiotomy defoamers have been proposed [52, 72, 73]. The reasons for seeking alternatives vary from concern over the supposed effects of PDMS on the immune system [72], use of polluting chlorofluorocarbons as solvents for the coating process employed in preparing defoamer elements [52, 73], and hydrophobing the relevant surfaces by PDMS leading to slow transport of fluids through those elements [52]. In addition, there are of course concerns about the evidence that PDMS–hydrophobed silica antifoam can cause emboli potentially leading to patient morbidity (see Section 11.3.3) [52].

In this context, Sacco et al. [72] claim the use of mixtures of triglycerides and hydrophobic particles as antifoams for cardiotomy defoaming. Mixtures of castor oil and hydrophobed silica appear to be a preferred provision. This antifoam can be used to coat a sponge-like polyurethane defoamer element using isopropanol as solvent. A cardiotomy reservoir prepared with this defoamer functioned as well as one prepared using a defoamer containing a proprietary PDMS–hydrophobed silica antifoam. Not surprisingly, Sacco et al. [72] demonstrate that the antifoam is gradually released into the blood flow during use, which, as we have discussed in Section 11.3.2, is an inevitable consequence of the mode of action of such materials. They claim that, unlike PDMS, castor oil thereby released will be metabolized *in vivo*. This of course ignores the fate of the hydrophobed silica and the known potential hazards of fat emboli. The latter include hydrolysis to form free fatty acids, which can produce toxic effects (see, e.g., references [74, 75]).

The viscosity of castor oil alone is relatively high, about 500 mPa s at 29°C [76], and the presence of hydrophobed silica is likely to increase it. As with PDMS–hydrophobed silica antifoams, we would therefore not expect mixtures of castor oil and hydrophobed silica to readily disperse as a result of the shear force applied by the flowing blood in a defoamer. However, it seems probable that castor oil will initially spread over the surface of water and probably blood. The air–oil surface tension of castor oil at 20°C is about 35 mN m^{-1} [77] with an oil–water interfacial tension of about 15 mN m^{-1} at room temperature [78]. This gives a positive initial spreading coefficient on water at 20°C despite the relatively high air–oil surface tension. It must also mean a positive initial spreading coefficient on blood because the air–water surface tension of that fluid is about 60 mN m^{-1} at 20°C [79] and the protein present can only reduce the castor oil–water interfacial tension. Pulido-Mayoral and Galindo [76] have in fact shown also that drops of castor oil dispersed in both water and aqueous protein solutions will capture air bubbles, which subsequently adhere to the oil–water interface with an angle $\theta^* \gg 90°$ (and therefore a bridging coefficient $B > 0$). This implies pseudo-partial wetting behavior on both water and protein solutions. In turn, the latter implies similar behavior on the surface of blood. It suggests that castor oil–hydrophobed silica will show qualitatively similar behavior to that of PDMS–hydrophobed silica in this context with a similar mode of action if incorporated in a porous medium.

Friedman [73] makes similar claims to those of Sacco et al. [72] but for an antifoam composed of lecithin and inorganic particles, including hydrophobed silica. Again it is used to coat a sponge-like defoamer element in a cardiotomy reservoir using a solution in ethanol or isopropanol or an emulsion in water. In practice, the reported antifoam performance in a cardiotomy defoamer appeared inconsistent unless admixed with PDMS–hydrophobed silica antifoam.

Tsai and Haynes [52] by contrast claim the use of cloud point antifoams for cardiotomy and general extracorporeal blood processing. These compounds are exemplified by polyoxyethylene-polyoxypropylene block copolymers of formula $OH.EO_nPO_mEO_nH$ with cloud points preferably < 20°C. They are non-toxic and compatible with blood. At temperatures > 20°C in aqueous media, these polymers form dispersions of polymer-rich cloud phase drops in equilibrium with dissolved monomer. The existing evidence described in Section 4.6.3.2 suggests that the cloud phase functions by the same bridging mechanism as that of the oil–particle systems described here. It would seem, however, that the pseudoemulsion films between drops of these polymers and the air–water surface are intrinsically unstable, obviating the necessity for the presence of hydrophobic particles. In the case of blood, it seems likely that the dissolved monomers adsorb at the air–blood surface, displacing the adsorbed protein and destabilizing the relevant pseudoemulsion film.

Sponge-like defoamer elements can be coated with these polymers by dip-coating in an isopropanol solution. The resulting defoamers are slightly more effective than a proprietary PDMS–hydrophobed silica antifoam in destroying foam. Neither are, however, effective in removing gaseous microemboli. Such microbubbles more closely resemble a dilute gas-in-blood emulsion where the gas volume fraction is much less than the 0.72 required if foam films are to be formed (see Section 1.2), except in a transient sense in low probability bubble–bubble collisions. Since the antifoams function by bridging such films, their lack of effectiveness in this context is hardly surprising.

11.3.6.2 Potential Use of Defoamer Elements with High Air–Blood Contact Angles

There does not appear to have been any fundamental studies concerning either the nature of foam formation in blood or the stability of the resulting foam. The surface tension of whole blood is about 52 mN m^{-1} at normal body temperature, increasing to about 60 mN m^{-1} at 20°C [79]. Such high values for an aqueous system suggest that the main foam stabilizing protein in blood, human albumin, is not highly surface active at the gas–blood interface. In particular, the surface tension is much higher than the critical wetting tension of hydrophobic materials such as polytetrafluoroethylene [80]. If we suppose that the surface tension of blood lies on the empirical plot of cos θ_{AW}, against air–liquid surface tension given by Bernett and Zisman [81] for a variety of aqueous surfactant solutions on polytetrafluoroethylene, then we can deduce that the advancing contact angle, θ_{AW}, of blood on that surface at 25°C should be >90°. If, in addition, we consider the antifoam mechanism of polytetrafluoroethylene particles discussed in Section 4.7.2, it seems likely that a PTFE membrane of suitable geometry would be effective in destroying blood foam. Indeed, a PTFE membrane, consisting of woven cylindrical fibers, has been claimed to be effective in another context where again the air–water surface tension of the foaming liquid is significantly higher than the critical wetting tension of PTFE so that $\theta_{AW} > 90°$ (see Section 7.4). It is, however, unlikely that such a membrane could be designed to eliminate microbubbles of size < 40 microns, although it could be effective as a cardiotomy defoamer, thereby eliminating the risk of emboli due to antifoam drops.

Sevastianov [82] reports the preparation of an open-celled polyurethane foam chemically hydrophobed with a perfluorosiloxane as a suitable defoaming material for defoaming blood. This material was initially as effective as the same polyurethane sponge coated with a proprietary PDMS–hydrophobed silica antifoam for defoaming a solution of human albumin (supposedly the main foam stabilizing protein in blood). However, the effectiveness of the latter diminished relatively rapidly (presumably as a result of dispersal of the antifoam in the solution) in contrast to that of the perfluorosiloxane treated foam. This is of course the behavior to be expected of a porous material sufficiently hydrophobic to give blood–air stable contact angles in excess of 90°.

Ito [83] describes a centrifugal device for de-aerating blood. Here any foam that accumulates despite the action of the centrifuge impacts a filter. This filter is intended to collapse the foam so that the blood is retained and gas is drawn through it. We note that the filter is either prepared from a material treated to have a hydrophobic surface or is a film prepared from an intrinsically hydrophobic solid. Examples of the latter given by Ito [83] exclusively concern fluorinated polymers such as polytetrafluoroethylene. These films are rendered porous by a variety of methods, including, for example, electron beam etching. Again it seems possible that foam collapse is induced by a mechanism analogous to that attributed to solid materials with high contact angles and described in Section 4.7.2. It is noteworthy that the centrifugal device of Ito [83] does not obviate the need for the supplementary defoaming usually required when using centrifugal devices (see Section 7.2.3)!

11.3.6.3 Removal of Gaseous Microemboli

Schonburg et al. [84] note that many studies have shown that "serious postoperative psycho-neurological dysfunction occurs in 2–8% of all patients undergoing cardiopulmonary bypass." Less serious adverse outcomes apparently occur in up to 70% of cases [84, 85]. The latter include cognitive dysfunctions such as comprehension, attention and perception impairment, and memory loss [84, 85]. Although these outcomes are sometimes reversible, it can take patients several months to return to normal [86]. They have been attributed to microemboli, which are either gaseous or particulate, where the latter can presumably even include antifoam drops.

Gaseous microemboli, if large enough, can become lodged in cerebral blood vessels so that blood flow is blocked resulting in cell death for lack of oxygen [87]. Other problems can result from thrombo-inflammatory responses caused by either air damage to the cells lining blood vessels or interactions between platelets and denatured proteins adsorbed at the gas–blood surface [88]. It could be argued that if this damage is to be largely avoided, diameters of any bubbles passing through the filters in blood circuits should not be allowed to exceed the diameters of cerebral capillaries, which appear to lie in the region of 6–7 microns [89]. However, air bubbles containing mostly nitrogen can partially dissolve as they pass from the arterial filter (see Figure 11.7) to the patient. We can estimate the order of magnitude of such shrinkage if we use the simple model of the process presented by Hlastala and Farhi [90]. Their model uses boundary layer theory to describe the rate of dissolution of a spherical air bubble in a free-flowing blood stream. Simplifying, by considering the bubble to be of pure nitrogen, we obtain for the rate of decrease of the bubble radius

$$\frac{\mathrm{d}r_{\mathrm{b}}}{\mathrm{d}t} = -K_{\mathrm{m}} H_{\mathrm{E}} \Delta P \tag{11.1}$$

where r_b is the radius of the bubble, K_m is a mass transfer coefficient, H_E is Henry's law constant and ΔP is the difference in partial pressure between the nitrogen in the bubble and that due to the dissolved nitrogen in the blood. If we also set the partial pressure of nitrogen in the blood as zero (i.e., the blood is oxygenated with pure oxygen), we can therefore write

$$\Delta P = P_{atm} + \frac{2\sigma_{GB}}{r_b} \tag{11.2}$$

where P_{atm} is the atmospheric pressure, $2\sigma_{GB}/r_b$ is the capillary pressure, and σ_{GB} is the gas–blood surface tension. Combining Equations 11.1 and 11.2 and integrating yields

$$r_b(t=0) - r_b(t=\Delta t) = K_M H_E P_{atm} \Delta t - \frac{2\sigma_{GB}}{P_{atm}} . \log_e \left\{ \frac{P_{atm} r_b(t=\Delta t) + 2\sigma_{GB}}{P_{atm} r_b(t=0) + 2\sigma_{GB}} \right\} \tag{11.3}$$

where $r_b(t = 0)$ is the initial bubble radius and $r_b(t = \Delta t)$ is the bubble radius after a time interval Δt. We can solve the equation iteratively using the data quoted by Dexter and Hindman [91] where H_E for nitrogen in blood at 37°C is 1.25×10^{-7} Pa^{-1} and the mass transfer coefficient, K_m, is 24×10^{-6} m s^{-1}. The transit time for blood to reach the brain capillaries is about 4 s [92], which we set equal to Δt. Using a value of 52 mN m^{-1} [79] for the surface tension of human blood, we can now use Equation 11.3 to estimate the maximum initial diameter of a nitrogen bubble as ~9 microns if it is to shrink during transit to a diameter equal to or less than that of the cerebral capillaries (of diameter ~ 6 microns). Bubbles of greater diameters will potentially cause transient blockage of cerebral blood capillaries with the risk of injury. Bubbles of significantly smaller diameters should, however, move through those capillaries with a faster velocity than the blood flow velocity because of an excluded volume effect (see Section 5.2.2 and Appendix 5.1). They should therefore present no embolic hazard where we note, for example, that microbubbles of sizes ≤ 6 microns are even used as contrast agents in ultrasonography (see, e.g., reference [93]). Here we should, however, stress that if the activity of nitrogen in the blood is non-zero as a consequence, for example, of oxygenating with a gas mix containing nitrogen, then the rate of dissolution of a nitrogen microbubble will be diminished and the critical initial diameter before a significant risk of capillary blockage will be reduced. We should also note that the capillary pressure contribution to the overall pressure difference driving dissolution of bubbles may be minimized by the effects of adsorbed surface-active agents, such as human albumin, on the air–blood surface dilatational rheology. Thus if, for example, albumin adsorbs irreversibly, then the adsorption will increase when the bubble shrinks as the gas dissolves. In turn, the air–blood surface tension must then decrease, which decreases the capillary pressure and the overall pressure driving dissolution.

Filters on the combined venous and cardiotomy/venous blood reservoirs usually remove particulate microemboli of size > 40 microns. Similarly, arterial filters placed after membrane oxygenators are designed to remove any further microemboli introduced after the reservoirs in the blood circuit. Such filters also remove microbubbles; however, they operate in this respect by the maximum bubble pressure concept. They are probably most effective in that context if they are perfectly wetted by the blood—that is, the air–blood contact angle at the surface of the filter is zero measured through blood. If then the pressure drop ΔP across the filter is less than $2\sigma_{GB}/R_{fp}$, where R_{fp} is the radius of the filter pores, then all bubbles of radius greater than the radius of the pores of the filter will be retained by the filter; smaller bubbles would of course pass through it. However, if the material of the filter is hydrophobic, the air–blood surface will then have a finite contact angle at the surface of the filter. Depending on the size of that contact angle, it is possible that bubbles of larger radius than R_{fp} could be permitted to pass at the same pressure drop. If the contact angle is high enough, it could even cause coalescence of bubbles. Arterial filters are usually prepared from woven polyester or nylon and are therefore hydrophobic. An example is described by Kim et al. [94]. Use of such hydrophobic filters may therefore explain the reason for reports described in references [84, 95] that filters with a specification of, for example, 40 microns can still apparently allow microbubbles of > 40 microns diameter to pass into the arterial blood line and on to the patient. As a result of all these limitations, it has been claimed [71, 84, 96] that significant quantities of microbubbles of air even of > 40 microns diameter can be transported in arterial blood to patients during cardiopulmonary bypass surgery.

These considerations imply that passage into cerebral arterioles and capillaries of microbubbles of diameters ≥ 40 microns is possible in some cardiopulmonary bypass surgery. However, such bubbles, even if trapped in those arterioles and capillaries, will continue to dissolve. Assessment of potential injury therefore concerns the time taken for that dissolution into cerebral tissue. As we have argued here, some dissolution will in any case occur during transit from the arterial filter to the cerebral capillaries. Using Equation 11.3, we can calculate that, for example, a 40-micron nitrogen bubble would shrink to a diameter of 37.5 microns during that time. Such bubbles are of course too large to be admitted into blood capillaries. Estimating the time taken for such a microbubble trapped in a cerebral arteriole to dissolve into the surrounding tissue can, however, be made using, for example, the diffusion model of Branger and Eckmann [97]. The model assumes that the bubble adopts a cylindrical shape in blood vessels where the cylinder is closed by hemispherical end caps and where the aspect ratio is defined as the length of the cylinder divided by the radius of the hemispherical end-caps. Experimental observation of the dissolution of such bubbles in blood arterioles and capillaries reveals that the aspect ratio declines with time as the cylinder shrinks and the end caps approach one another until a sphere is formed.

This model of Branger and Eckmann [97] also assumes that nitrogen diffusing into the surrounding tissue is removed by the general tissue perfusion. It predicts that the maximum dissolution time occurs for bubbles of an initial aspect ratio of 2.6 regardless of the initial volume of the bubble. In this circumstance, the dissolution time, Δt_{diss}^{max}, is given (in minutes) by

$$\Delta t_{diss}^{max} \approx 709\ V_b^{2/3} \tag{11.4}$$

where V_b is the bubble volume (in mm³). The minimum dissolution time, Δt_{diss}^{min}, is predicted to occur for spherical bubbles for which the aspect ratio is zero and

$$\Delta t_{diss}^{min} \approx 472\ V_b^{2/3} \tag{11.5}$$

In using the theoretical model of Branger and Eckmann [97] to calculate the dissolution time of this bubble of initially 37.5 microns diameter, we again assume that the blood is oxygenated with pure oxygen and that therefore the partial pressure of nitrogen in the blood is zero. We calculate then, using Equations 11.4 and 11.5, that this bubble will dissolve in a minimum time of about 26 s and a maximum time of about 38 s. Such times are clearly significant when compared with total arterial transit times for cerebral blood flow of <1 s [98]. This might indicate that localized interruption of that flow for periods even as brief as the few tens of seconds implied by the analysis using the theory of Branger and Eckmann [97] could cause injury and account for the reported incidents of cognitive dysfunction after cardiopulmonary bypass [84, 85]. However, we should also note that the view of de Somer [66], for example, is that bubble dissolution times even as long as 120 s should be too short to cause primary ischemic injury. Unfortunately, de Somer [66] gives no clear evidence for that view.

Clearly, removal of microbubbles of sizes > 10 microns or so would appear to be desirable if transient blocks to blood flow into cerebral capillaries are to be completely avoided. Removal of a dilute dispersion of microbubbles is not, however, possible using conventional antifoams. It is even possible that the latter could facilitate coalescence of bubbles during low-probability collisions, involving transient formation of foam films, to make larger bubbles that take longer times to dissolve! The performance of arterial filters in removing microbubbles may, however, apparently be improved by use of a "dynamic bubble trap" [70, 84, 95, 99, 100]. This device is depicted in Figure 11.8a. It consists essentially of a fixed three-channel helix, which rotates the blood flow. In turn, this applies a centrifugal force to any microbubbles present, concentrating them into the center of the bubble flow in a diffuser chamber as shown in the figure. Those bubbles may then be collected by a suitably located tube and returned to the cardiotomy reservoir. A computer-generated simulation assisted the design of the apparatus. The movement of a bubble is complex—being a function of the design of the apparatus, initial position of a given bubble in the blood flow, blood viscosity, and mainly blood flow rate and bubble size. A simulation of the trajectory of bubbles of two different sizes is shown in Figure 11.8b. The flow of the large bubble is seen to be stable so that it may be readily picked up by the collection tube. However, the flow of the smaller bubble is less stable, representing a lower probability of collection.

A clinical study reported by Schonburg et al. [84] employed Doppler ultrasonography to measure and count bubbles before and after the location of a dynamic bubble trap in the arterial lines of patients. The efficiency of removal of bubbles of size > 40 microns was >90% and that for bubbles of size range 11–20 microns was ~75%. Direct measurement of the number of microbubbles entering cerebral arteries using trans-cranial Doppler ultrasonography revealed a significant reduction

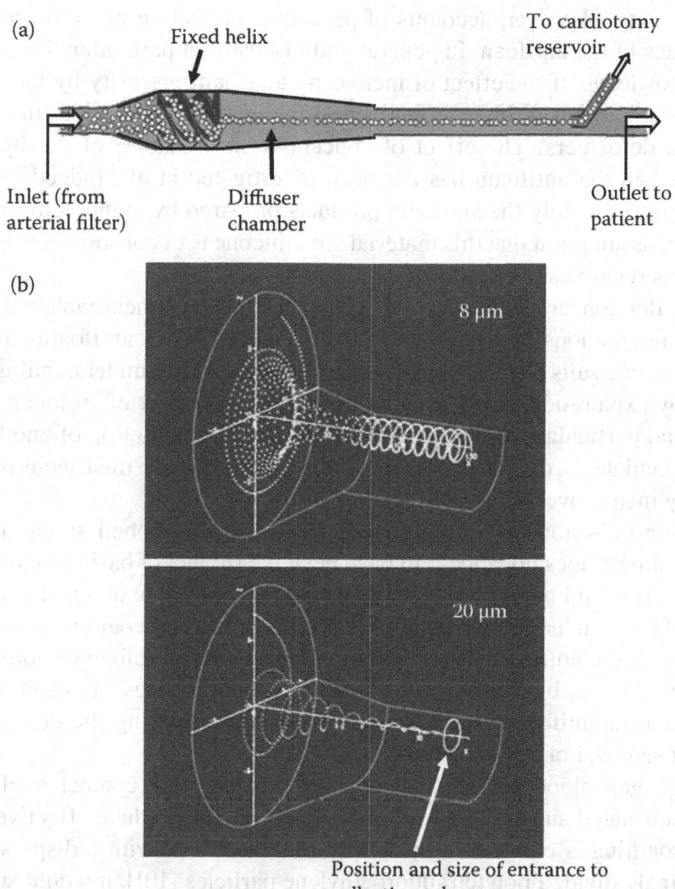

FIGURE 11.8 Dynamic bubble trap. (a) Cross section of apparatus. (From Perthel, M., Kseibl S., Bendisch, A., Laas, J., *Perfusion*, 18, 325, 2003. Copyright by SAGE Publications, reprinted by permission of SAGE Publications.) (b) Simulation of flow of bubbles through apparatus showing effect of bubble size where the trajectory of smaller bubble of diameter 8 microns is seen to be unstable so that the probability of capture by the collection tube is lower than is the case of 20-micron bubble. (From Schonburg, M., Erhardt, G., Kraus, B., Taborski, U., Muhling, A., Hein, S., Roth, M., Klovekorn, W.P., *Perfusion*, 16, 19, 2001. Copyright by SAGE Publications, reprinted by permission of SAGE Publications.)

accompanying the use of the dynamic bubble trap [84]. Several German studies have now therefore concluded that this technique could be of advantage to patients in protecting their neurocognitive function [70, 84, 99, 100].

11.4 SUMMARIZING REMARKS

Use of PDMS–hydrophobed silica as an antifoam is ubiquitous in all the clinical contexts considered here. In the main, this choice derives from both its effectiveness and non-toxicity. However, accounts of practical application make no reference to the properties of the antifoam in general and viscosity in particular. There is apparently no knowledge of the effect of increasing antifoam viscosity by increasing the PDMS oil viscosity on antifoam effectiveness in tablets for gastrointestinal treatment or in blood defoamers. The effect of concentration or nature of the hydrophobic particle used in the antifoam has not been investigated at all. Indeed all reported studies seem to use only the antifoam products prepared by manufacturers, making the apparent assumption that this material simethicone is of constant consistency and effectiveness regardless of source.

A particular concern in the case of gastrointestinal treatment tablets is the cause of adverse interactions between PDMS–hydrophobed silica antifoams and certain antacid inorganic salts [10–12, 14]. A start could be made in understanding this phenomenon by exploration, using *in vitro* studies, of the effect of systematic changes in the oil and particulate component of the antifoam. Substitution of another hydrophobic particulate of different chemistry for hydrophobed silica would clearly be particularly instructive.

The mode of action of "immobilized" PDMS–hydrophobed silica antifoam in porous defoamers does not appear to have been the subject of basic research. Of particular interest would be studies of the effectiveness of the use of a non-spreading oil in order to better understand the role of the initial spreading coefficient in determining delivery of the antifoam to foamed blood. The role of defoamer pore structure, antifoam properties, blood flow rates, and foam bubble sizes in determining drop sizes of released antifoam would be of interest in considering the design of filters suitable for removal of that material.

The high gas–blood surface tension [79] suggests that contact angles against solid perfluorinated substrates could be high enough to produce effective antifoam action. If foaming is eliminated by simply shaking blood with a dispersion of, for example, finely divided polytetrafluoroethylene particles [101], it would suggest that a fixed porous matrix of that material of suitable geometry could represent a practical defoamer element for cardiotomy blood. Such a defoamer element would not contaminate the blood flow and would therefore possess that advantage over the use of PDMS–hydrophobed silica–coated porous defoamer elements.

It is clear that microbubbles of a size ≤ ~10 microns diameter after leaving the blood circuit in cardiopulmonary bypass surgery are unlikely to cause cerebral injury. However, it appears to be possible that larger microbubbles can cause problems [84–87, 97]. The question of just how much larger before such problems arise does not at, present, appear to have an unequivocal answer. Calculations using the diffusion model of Branger and Eckmann [97] indicate that even nitrogen microbubbles with diameters as low as 40 microns could cause injury. There are, however, significant uncertainties about that, including the issue of just how long the blood flow in a cerebral capillary has to be interrupted by transient gaseous microemboli before permanent localized injury can occur.

Experimental *in vivo* animal testing of the model of Branger and Eckmann [97], upon which the estimates of the time for dissolution of microbubbles in cerebral tissues are based, appear to validate the model. The model could be applied to various situations ranging from the dissolution of oxygen microbubbles in blood saturated with oxygen at atmospheric pressure to the dissolution of air microbubbles in blood saturated with oxygen–nitrogen mixtures at different partial pressures. Whether the resulting estimates accurately describe dissolution times in human cerebral tissue is, however, uncertain. Moreover, application of such models requires knowledge of the actual sizes of microbubbles to be expected given blood circuit properties such as filter pore sizes. As shown by Goritz et al. [95], the relation between the specification of the filters in the blood circuit and the actual diameters of bubbles that pass those filters is uncertain. Thus, they demonstrate that an arterial filter with a nominal pore size of 40 microns permits the passage about 26% of bubbles in the size range of 45–130 microns. The novel concept of the dynamic bubble trap [70, 84, 95, 99, 100] produces an improved performance under the same conditions and would therefore appear to have merit in this context. Perhaps arterial filter design could also be improved—the role of contact angle and geometry of the material of those filters could be explored in seeking improved performance.

REFERENCES

1. Stoney, W.S. *Circulation*, 119, 2844, 2009.
2. High, K.M., Bashein, G., Kurusz, M. Principles of oxygenator function: gas exchange, heat transfer, and operation, in *Cardiopulmonary Bypass: Principles and Practice*, Second Edition (Gravlee, G.P., Davis, R.F., Kurusz, M., Utley, J.R., eds.), Lippincott, Williams, Wilkins, Philadelphia, 2000, Chpt 4, p 49.
3. Orenstein, J.M., Sato, N., Aaron, B., Buchholz, B., Bloom, S. *Hum. Pathol.*, 13(12), 1082, 1982.
4. Harrington, J.S. *Thorax*, 16, 120, 1961.
5. Cassie, A.B., Riddell, A.G., Yates, P.O. *Thorax*, 15, 22, 1960.
6. Burton, J.S. Foams and their clinical implications, in *Foams* (Akers, R.J., ed.), Academic Press, London, 1976, from original conference proceedings, p 115.
7. Berger, R. Application of antifoams in pharmaceuticals, in *Defoaming, Theory and Industrial Applications* (Garrett, P.R., ed.), Marcel Dekker, New York, 1993, Chpt 4, p 177.
8. Quin, A.H., Austin, J.A., Ratcliff, K. *J. Am. Vet. Med. Assoc.*, 114, 313, 1949.
9. Birtley, R.D.N., Burton, J.S., Kellett, D.N., Oswald, B.J., Pennington, J.C. *J. Pharm. Pharmacol.*, 25, 859, 1973.
10. Rezac, M. *J. Pharmacol. Sci.*, 5, 538, 1966.
11. Stead, J.A., Wilkins, R.A., Ashford, J.J. *J. Pharm. Pharmacol.*, 30, 350, 1978.
12. Akay, G., Garrett, P.R., Yorke, J.W.H. (assigned to Unilever PLC),WO93/01269; 21 January 1993, filed 1 July 1992.
13. White, M. unpublished work.
14. Torrado, G., Garcia-Arieta, A., de los Rios, F., Menendaez, J.C., Torrado, S. *J. Pharm. Biomed. Anal.*, 19, 285, 1999.
15. Moore, D.E., Liu, T.X., Miao, W.G., Edwards, A., Elliss, R. *J. Pharm. Biomed. Anal.*, 30, 273, 2003.
16. Coffin, B., Bortolloti, C., Bourgeois, O., Denicourt, L. *Clin. Res. Hepatol. Gastroenterol.*, 35(6–7), 494, 2011.
17. Holtmann, G., Gschossmann, J., Mayr, P., Talley, N.J. *Aliment. Pharmacol. Ther.*, 16(9), 1641, 2002.
18. Voepel-Lewis, T.D., Malviya, S., Burke, C., D'Agostino, R., Hadden, S.M., Siewart, M. *J. Clin. Anesth.*, 10(2), 91, 1998.
19. Lifschitz, C.H., Irving, C.S., Smith, E.O. *Digest. Dis. Sci.*, 30(5), 426, 1985.

20. Friis, H., Bode, S., Rumessen, J.J., Gudmand-Hoyer, E. *Digestion*, 49(4), 227, 1991.
21. Danhof, I.E., Stavola, J.J. *Obstet. Gynecol.*, 44(1), 148, 1974.
22. Koczo, K., Koczone, J., Wasan, D.J. *J. Colloid Interface Sci.*, 166, 225, 1994.
23. Racz, G., Koczo, K., Wasan, D.J. *J. Colloid Interface Sci.*, 181, 124, 1996.
24. Sudduth, R.H., DeAngelis, S., Sherman, K.E., McNally, P.R. *Gastrointest. Endosc.*, 42(5), 413, 1995.
25. Hosseini Asi, S.M.K., Sivandzadeh, G.R. *World J. Gastroenterol.*, 17(37), 4213, 2011.
26. Tongprasert, S., Sobhonslidsuk, A., Rattanasiri, S. *World J. Gastroenterol.*, 15(24), 3032, 2009.
27. Bertoni, G., Gumina, C., Conigliaro, R., Ricci, E., Staffetti, J., Mortilla, M.G., Pacchione, D. *Endoscopy*, 24(4), 268, 1992.
28. Wu, L., Cao, Y., Liao, C., Huang, J., Gao, F. *Scand. J. Gastroenterol.*, 46(2), 227, 2011.
29. Parikh, V.A., Khanduja, K.S. *Dis. Colon Rectum*, 38(9), 1007, 1995.
30. Varaprath, S., Frye, C.L., Hamelink, J. *Environ. Toxicol. Chem.*, 15(8), 1263, 1996.
31. Chen, H., Huang, Y., Chen, S., Song, H., Li, X., Dai, D., Xie, J., He, S., Zhao, Y., Huang, C., Zhang, S., Yang, L. *J. Clin. Gastroenterol.*, 45(4), 337, 2011.
32. Wei, W., Ge, Z.Z., Lu, H., Gao, Y.J., Hu, Y.B., Xiao, S.D. *Am. J. Gastroenterol.*, 103(1), 77, 2008.
33. Shiotani, A., Opekun, A.R., Graham, D.Y. *Digest. Dis. Sci.*, 52(4), 1019, 2007.
34. Ge, Z.Z., Chen, H.Y., Gao, Y.J., Hu, Y.B., Xiao, S.D. *Endoscopy*, 38(8), 836, 2006.
35. Albert, J., Gobel, C.M., Lesske, J., Lotterer, E., Nietsch, H., Fleig, W.E. *Gastrointest. Endosc.*, 59(4), 487, 2004.
36. Fang, Y., Chen, C., Zhang, B. *J. Zhejiang Univ. Sci. B*, 10(1), 46, 2009.
37. Esaki, M., Matsumoto, T., Kudo, T., Yanaru-Fujisawa, R., Nakamura, S., Iida, M. *Gastrointest. Endosc.*, 69(1), 94, 2009.
38. Spada, C., Riccioi, M.E., Familiari, P., Spera, G., Pirozzi, G.A., Marchese, M., Bizzotto, A., Ingrosso, M., Costamagna, G. *Digest. Liver Dis.*, 42(5), 365, 2010.
39. Gremel, R.F., Galt, K.M. (assigned to Shiley Inc., California), US 4,568,367; 4 February 1986, filed 15 November 1982.
40. Nussmeier, N.A. *J. Extra Corpor. Technol.*, 34, 4, 2002.
41. Bentley, D.J. (assigned to Bentley Laboratories), US 3,507,395; 21 April, 1970, filed 1 December 1967.
42. Tipler, D.R., Ghadiali, P.E. (assigned to National Heart and Chest Hospitals), GB 1,407,982; 1 October 1975, filed 15 October 1971.
43. Raible, D.A. (assigned to Bentley Laboratories Inc.), US 4,253,967; 3 March 1981, filed 9 April 1979.
44. Servas, F.M., Gremel, R.F., Ryan, T.C. (assigned to Shiley Inc.), US 4,743,371; 10 May 1988, filed 28 February, 1986.
45. Bringham, R.L., Gordon, L.S., Mosch, K.E. (assigned to Baxter Int. Inc.), US 4,876,066, 24 October 1989, filed 14 July 1987.
46. Steg, R.F., Peterson, D.M., Cusi, D.S. (assigned to Galen Medical Inc.), WO 94/03098; 17 February 1994, filed 30 July 1992.
47. Thor, E.J., McIntosh, K.D., Jones, B.R., Dando, J.D. (assigned to Avecor Cardiovascular Inc.), US 5,411,705; 2 May 1995, filed 14 January 1994.
48. van Driel, M.R. (assigned to Medtronic Inc.), US 5,759,396; 2 June 1998, filed 10 October 1996.
49. Raneri, J.J., Buckler, K.E., Stanley, C.L., Intoccia, P. (assigned to C.R.Bard Inc.), US 5,849,186; 15 December 1998, filed 15 November 1996.
50. Lindsay, E.J. (assigned to Terumo Cardiovascular Systems Corp.), US 6,180,058; 30 January 2001, filed 17 July 1998.
51. Dow Corning Product Literature, 2001.

52. Tsai, C., Haynes, R.E. (assigned to Cobe Cardiovascular Inc.), US 6,506,340; 14 January 2003, filed 1 June 1993.
53. Denkov, N.D., Cooper, P., Martin, J. *Langmuir*, 15(24), 8514, 1999.
54. Denkov, N.D. *Langmuir*, 20(22), 9463, 2004.
55. Denkov, N.D. *Langmuir*, 15(24), 8530, 1999.
56. Somer, T., Meiselman, H.J. *Ann. Med.*, 25, 31, 1993.
57. Chien, S., Dormandy, J., Ernst, E., Matral, A. *Clinical Hemorheology*, Martinus Nijhoff, Dordrecht, 1987.
58. Hu, C., Inoue, M., Mason, S. *Can. J. Chem. Eng.*, 53, 367, 1975.
59. Bergeron, V., Cooper, P., Fischer, C., Giermanska-Kahn, J., Langevin, D., Pouchelon, A. *Colloids Surf. A*, 122, 103, 1997.
60. Denkov, N.D., Marinova, K.G., Christova, C., Hadjiiska, A., Cooper, P. *Langmuir*, 16(6), 2515, 2000.
61. Marinova, K.G., Tcholakova, S., Denkov, N.D., Roussev, S., Deruelle, M. *Langmuir*, 19(7), 3084, 2003.
62. Lindberg, D.A.B., Lucas, F.V., Sheagren, J., Malm, J.R. *Am. J. Pathol.*, 39(2), 129, 1961.
63. Williams, I.M., Stephens, J.F., Brunckhorst, L.F., Brodie, G.N. *J. Histochem. Cytochem.*, 23, 149, 1975.
64. Valentin, N., Vilhelmsen, R. *J. Cardiovasc. Surg.*, 17, 20, 1976.
65. Pearson, D.T., Watson, B.G., Waterhouse, P.S. *Thorax*, 33, 352, 1978.
66. de Somer, F. *J. Extra Corpor. Technol.*, 39(4), 271, 2007.
67. Brown, W.R., Moody, D.M., Challa, V.R., Stump, D.A., Hammon, J.W. *Stroke*, 31, 707, 2000.
68. Gaylor, J.D.S. *J. Biomed. Eng.*, 10, 541, 1988.
69. Lindsay, E.J. (assigned to Lindsay, E.J.), US2002/0009386; 24 January 2002, filed 28 September 2001.
70. Perthel, M., Kseibi, S., Bendisch, A., Laas, J. *Perfusion*, 20, 151, 2005.
71. Urbanek, S., Tiedtke, H.J. *Perfusion* 6, 429, 2002.
72. Sacco, S., Porro, G., Rinaldi, S., Della Ciana, L. (assigned to Dideco S.p.A), EP 0774 285; 21 May 1997, filed 14 November 1996.
73. Friedman, R.S. (assigned to Medtronic Inc.), US 6,254,825; 3 July 2001, filed 1 April 1999.
74. Batra, P. *J. Thorac. Imaging*, 2(3), 12, 1987.
75. Szabo, G., Magyar, Z., Reffy, A. *Injury*, 8(4), 278, 1977.
76. Pulido-Mayoral, N., Galindo, E. *Biotechnol. Prog.*, 20, 1608, 2004.
77. Stadler, T., Fornes, A., Buteler, M. *Bull. Insectol.*, 58(1), 57, 2005.
78. Than, P., Preziosi, L., Joseph, D.D., Arney, M. *J. Colloid Interface Sci.*, 124(2), 552, 1988.
79. Rosina, J., Kvasnak, E., Suta, D., Kolarova, H., Malek, J., Krajci, L. *Physiol. Res.*, 56 (Suppl.1), S93, 2007.
80. Zisman, W.A. *Adv. Chem. Ser.*, 43, 1, 1964.
81. Bernett, M.K., Zisman, W.A. *J. Phys. Chem.*, 63, 1241, 1959.
82. Sevastianov, V.I. (assigned to Sevastianov, V.I.), US 2003/0026730; 6 February 2003, filed 9 July 2002.
83. Ito, A. (assigned to Terumo Kabushiki Kaisha), US 2007/0110612; 17 May 2007, filed 9 November 2006.
84. Schonburg, M., Erhardt, G., Kraus, B., Taborski, U., Muhling, A., Hein, S., Roth, M., Klovekorn, W.P. *Perfusion*, 16, 19, 2001.
85. Pugsley, W., Kinger, L., Paschalis, C., Treasure, T., Harrison, M., Newman, S. *Stroke*, 25, 1393, 1994.
86. Clark, R.E., Brillman, J., Davis, D.A., Lovell, M.R. *J. Thorac. Cardiovasc. Surg.*, 109, 249, 1995.

87. Kort, A., Kronzon, I. *J. Clin. Ultrasound*, 10, 117, 1982.
88. Ryu, K.H., Hindman, B.J., Reasoner, D.K., Dexter, F. *Stroke*, 27, 303, 1996.
89. Karbowski, J. *PLoS One*, 6(10), e26709. Doi:10.1371/journal.pone.0026709.
90. Hlastala, M.P., Farhi, L.E. *J. Appl. Physiol.*, 33(3), 311, 1973.
91. Dexter, F., Hindman, B.J. *Perfusion*, 11, 445, 1996.
92. Kilian, W., Seifert, F., Rinneberg, H. *Magn. Reson. Med.*, 51, 843, 2004.
93. Quinones, M.A. *Circulation (Supplement III)*, 83(5), 104, 1991.
94. Kim, W.G., Kin, K., Yoon, C.J. *Artif. Organs*, 24(11), 874, 2000.
95. Goritz, S., Schelkle, H., Rein, J., Urbanek, S. *Perfusion*, 21, 367, 2006.
96. Taylor, R.L., Borger, M.A., Weisel, R.D., Fedorko, L., Feindel, C.M. *Ann. Thorac. Surg.*, 68, 89, 1999.
97. Branger, A.B., Eckmann, D.M. *J. Appl. Physiol.*, 87(4), 1287, 1999.
98. Wang, J., Alsop, D.C., Song, H.K., Maldjian, J.A., Tang, K., Salvucci, A.E., Detre, J.A. *Magn. Reson. Med.*, 50, 599, 2003.
99. Perthel, M., Kseibl S., Bendisch, A., Laas, J. *Perfusion*, 18, 325, 2003.
100. Taborski, U., Urbanek, P., Erhardt, G., Sconburg, M., Basser, S., Wohlgemuth, L., Heidinger, K., Klovekorn, W. *Artif. Organs*, 27(8), 736, 2003.
101. Garrett, P.R. *J. Colloid Interface Sci.*, 69(1), 107, 1979.

Frequently Used Symbols and Abbreviations

SYMBOLS

B, B^e: Unspecified and equilibrium bridging coefficients
Ca: Capillary number
$\tilde{c}_i$: Concentration of antifoam i
$\tilde{c}_i(t)$: Time-dependent concentration of antifoam
D: Diffusion coefficient
F: Ratio of volume of air in foam in presence of antifoam/volume of air in foam in absence of antifoam.
E, E^i, E^{si}, E^e: Unspecified, initial, semi-initial, and equilibrium classical entry coefficients
E_g: Generalized entry coefficient
g: Gravitational constant
h: Thickness of liquid film
h_i: Thickness of film i
h_{crit}: Critical rupture thickness of liquid film
H_o: Initial foam height without antifoam
H_o^e : Equilibrium foam height after foam drainage has ceased
H_t: Foam height at time t
H_E: Henry's law constant
$\bar{k}$: Boltzmann constant
$N_i(r_b)$: Number of antifoam entities in film of a bubble of radius r_b at the top of a foam
P_{atm}: Atmospheric pressure
$P(r_b,t)$: Bubble size distribution as a function of r_b and t
p_c^i: Capillary pressure at i (with respect to some reference pressure)
p_c^{crit}: Critical applied capillary pressure for rupture of pseudoemulsion film
p_c^{PB}: Plateau border capillary pressure
Δp_c^{ij}: Capillary pressure jump across surface ij where i and j are different fluids
r_{AF}: Radius of antifoam entity (oil drop or oil/particle mixture)
r_f: Radius of cylindrical foam film
r_b: Radius of spherical bubble or radius of equivalent spherical bubble
S, S^i, S^{si}, S^e: Unspecified, initial, semi-initial, and equilibrium classical spreading coefficients
S_g, S_g^i, S_g^{si}, S_g^e: Unspecified, initial, semi-initial, and equilibrium generalized spreading coefficients
T: Temperature

t: Time
u_y: Fluid velocity in the y direction
$\bar{u}_{AF}$: Mean velocity of antifoam entity in draining film
u_{AF}: Velocity of antifoam in gravity field
V_b: Bubble volume
V_G: Volumetric gas flow rate
$V(t)$: Volume of foam generated after time t
V_T: Total volume of air in a foam

GREEK SYMBOLS

Γ_k^{ij}: Surface excess of species k at the surface ij where i and j are different fluids
$\dot{\gamma}_i$: Shear rate in fluid i
ε: Surface dilatational modulus (of elasticity)
ε_G: Gibbs elasticity of foam film
η_i: Viscosity of fluid i
$\theta_{ij}^*, \theta_{ki}^*$: Angles determining shape of lens of liquid i at the surface of fluids j and k by the surfaces ij and ki according to Neumann's triangle
θ^*: Angle formed at a lens of liquid i at the surfaces of fluids j and k by the interfaces ij and jk
θ_{AW}: Contact angle of a solid against the air–water surface measured through the aqueous phase
θ_{OW}: Contact angle of a solid against the oil–water surface measured through the aqueous phase
κ_t: Curvature of the plateau borders at the top of a draining foam
μ_i: Chemical potential of species i
$\tilde{\nu}$: Frequency of sound
$\Pi_{ALA}(h)$: Disjoining pressure in an air–liquid–air foam film of thickness h
$\Pi_{AWA}(h)$: Disjoining pressure in an air–water–air foam film of thickness h
$\Pi_{ALO}(h)$: Disjoining pressure in an air–liquid–oil pseudoemulsion film of thickness h
$\Pi_{AWO}(h)$: Disjoining pressure in an air–water–oil pseudoemulsion film of thickness h
ρ_i: Density of fluid i (where $i = W$ = water)
$\sigma_{ij}, \sigma_{ij}^i, \sigma_{ij}^e$: Unspecified, initial, and equilibrium surface tensions between fluids i and j
$\sigma_{AWO}(h)$: Air–water oil pseudoemulsion film tension
$\sigma_{AOW}(h)$: Air–oil–water spread film tension
$\Delta\sigma_{ij}^{sp}$: Classical spreading pressure of liquid at ij surface
Σ_{BIK}: Foaminess as defined by Bikerman
$\Delta\sigma_{ij}^{sp}(h = h^e)$: Generalized spreading pressure at ij surface to form an equilibrium film of thickness h^e
τ: Line tension
ω: Angular rotational velocity (radians unit time^{-1})
Φ_G^{foam}: Gas volume fraction in foam
Φ_L^{foam}: Liquid volume fraction in foam
Φ_{lp}: Volume fraction of gas in liquid phase below foam in bubbling method of foam generation (the gas hold up)

ABBREVIATIONS

CAC: Critical aggregation concentration
CMC: Critical micelle concentration
FTT: Film trapping technique

CHEMICALS

AOT: Sodium bis-dioctyl sulfosuccinate
C_{10}TAB: Decyl tetra-ammonium bromide
C_{12}TAB: Dodecyl tetra-ammonium bromide
C_{14}TAB: Tetradecyl tetra-ammonium bromide
C_{16}TAB: Hexadecyl tetra-ammonium bromide
C_nEO_m: $C_nH_{2n+1}(CH_2CH_2O)_mOH$
$C_{12}6\Phi SO_3Na$: Sodium 6-phenyl dodecyl benzene sulfonate
$C_{12}\Phi SO_3Na$: Commercial sodium dodecyl benzene sulfonate (branched chain blend)
C_{12}APB: Commercial dodecyl aminopropyl betaine
$C_{12}EO_{2.5}SO_4Na$: Commercial sodium dodecyl ether sulfate
PDMS: Polydimethylsiloxane
SAS: Commercial secondary alkyl sulfate
SDS: Sodium dodecyl sulfate
Triton X-100: Octylphenyl $(CH_2CH_2O)_{10}OH$

Index

Page numbers followed by f and t indicate figures and tables, respectively.

A

Abbreviations, frequently used, 557
Acoustic impedance, 416
Acoustic pressure, 411, 412, 412f, 414, 418
  wave propagation, 416
Acoustic radiation pressure, 421
Additives, 116
Adsorption, 62, 74, 95, 96, 124, 347, 486, 488, 546
  of antifoam, 531
  of hydrocarbon groups, 233
  oil-soluble surface-active impurities, 264
  of polydimethylsiloxane, 289
  of solute, 504, 505, 506
  surfactant, 169
Aeration, 36, 319, 377
Aerosil 200 silica, 355f, 366
Agglomerates, 352, 353f, 356, 357, 367
Air (in foam), volume of, 330–334. *See also* Antifoam action, statistical theory of
Air–blood contact angles, 544–545. *See also* Polydimethylsiloxane-based antifoams
Air bubbles, 545
Air entrainment, 330, 408, 433
Air–liquid–air disjoining pressure, 9, 16f
Air–liquid–air foam films, 401
Air–water–air foam film rupture, 222t
Air–water–oil pseudoemulsion film
  disjoining pressure isotherms measurements, 48, 48f
  pseudoemulsion films, observation of, 46–47, 47f
  rupture, 273f, 280t
  rupture pressure measurements, 49–52, 49f, 50f, 51f
Air–water–solid films, stability of, 233–238, 234f, 235f, 236f, 238f
Air–water surface, 141, 216, 220f, 225
  of axially symmetrical particle, 230t
Air–water surface tension, 72, 124
Alcohol ethoxylate surfactant, 76
Alcohols
  ethoxylated, 120, 158f
  on foamability and foam stability, 174f
  long-chain, 171
  short-chain, 167–175, 169f, 172t, 173f, 174f. *See also* Neat oils
Alkanediols, 170
Alkanes, 118
  short-chain, 517
Alkoxysiloxane emulsion stabilizer, 460
Alkylbenzene chain, 91
Alkylbenzene sulfonate, 91, 249
Alkyl chlorosilanes, 447
Alkyl ethoxylates, 93
Alkyl glucopyranoside, 291f
Alkyl phosphoric acid derivatives, 441–446, 442f, 443f, 445f, 446f. *See also* Hydrocarbon–hydrophobic particle mixtures
Alkyl phosphoric acid esters, 367, 441, 442
Alkyl sulfonate, 408f
Alkyl trimethylammonium bromide, 93f, 97, 102, 258, 259
Aluminum hydroxide, 530
Anionic surfactant, 486
Antacids, 530, 531
Antifoam, 34, 38, 314f, 389. *See also* Polydimethylsiloxane-based antifoams
  for aqueous systems, 115
  compound, 346
  concentration, 324, 325, 336
    excluded volume on, in draining film, 339–340
    Reynolds drainage and, 338–340
  deactivation. *See also* Deactivation of mixed oil–particle antifoams
    with antifoam action, 376–379, 378f
    by disproportionation, 379–383, 382f, 383f
  dimensions and kinetics, 290–292, 291f. *See also* Antifoam mechanism
  dispersion, 310, 311f, 312f, 361
  drop, 41–43, 42f. *See also* Single foam films
  drop size, 42, 348–351
  durability, 289, 291f, 354, 355f
  effectiveness
    enhancement, 457–460, 458t, 460f. *See also* Polydimethylsiloxane-based antifoams
    measure of, 370
  for gas–oil separation. *See* Gas–oil separation
  hydrophobed silica, 255t
  hydrophobic oil–hydrophobic particle mixed, 180

materials, liquid, 166
mixed dispersion of, 313–318, 314f, 317f
oils, spreading behavior of, 52–53
PDMS–hydrophobed silica, 451
polydimethylsiloxane-based, 327, 460
synergy, 249–251, 250f, 346
triglyceride-based, 125
Antifoam action, statistical theory of
assumptions, 320–324, 322f
in foam film (number of entities), 324–330, 325t, 328f
limitations, 334–336
volume of air, calculation of, 330–334
Antifoam–bubble heterocoalescence–kinetic model, 368–372, 369f, 373f. *See also* Deactivation of mixed oil–particle antifoams
Antifoam entities
in foam film, 324–330, 325t, 328f
in foam structures, 318–320
mean flow velocity of, 338–339, 338f
Antifoam mechanism
emulsified liquids
neat oils in aqueous foaming systems, 165–183, 169f, 172t, 173f, 174f, 176t, 177f, 178f, 179f, 181t
neat oils in non-aqueous foaming systems, 183–185
partial miscibility, 185–201, 185f, 186f, 188f, 189f, 191f, 195f, 199f, 200t
on foamability of mesophase precipitation, 121–128, 122f, 123f, 124f, 125f, 126f, 127f
hydrophobic particles and oils, mixtures of
antifoam dimensions and kinetics, 290–292, 291f
antifoam synergy, 249–251, 250f
oil in synergistic oil–particle antifoams, 252–263, 253f, 255t, 257f, 258f, 260f, 261f, 262f
particles in synergistic oil–particle antifoams, 263–266, 265f, 266t, 267–290, 268f, 269f, 270t, 271f, 273f, 275f, 277f, 278f, 280t, 282f, 283t, 284f, 286f, 288t, 290f
patent literature, early, 251–252
inert hydrophobic particles and capillary theories of, 201–203, 203f
calcium soaps, 243–247, 244f, 246f
contact angles, 203–216, 204f–211f, 213f–216f
hydrophobic materials, melting of, 247–249, 248f
particle geometry, 216–228, 217f, 218f, 220f, 222t, 226f, 227f
particle size/kinetics of foam film rupture, 239–243, 241f
rugosities, effect of, 228–233, 229f, 230t, 231f
rugosities and stability of air–water–solid films, 233–238, 234f, 235f, 236f, 238f
oil bridges
in foam films, 141–153, 142f, 143f, 147f, 148f, 149f, 150f, 152f
line tensions, 153–157, 154f, 156t
in plateau borders, 157–165, 158f, 159f, 160f, 163f, 164f
pseudoemulsion films, stability of, 157–165
and solubilized oils, 116–121, 119f, 120f
surface tension gradients, 128–129
elimination of, 138–140
induced by spreading antifoam, 129–137, 130f, 132f, 136f
Antifoam oils. *See also* Gas–oil separation
and cratering, 495–498
insolubility of, 183
mode of action of, 520–524, 521f, 522f, 523f
solubility of, effect of, 517–520, 518f, 520f
Antonow's rule, 60, 146
Aqueous surfactant solutions surfaces. *See also* Hydrocarbon oils
complete/pseudo-partial wetting behavior of hydrocarbons, 87–94, 88f, 89f, 90f, 92f, 93f
non-spreading (partial wetting) by hydrocarbons, 94–96
Argon gas, 413
Arterial filters, 546, 547, 548
Asphaltenes, 507, 507f, 508, 513, 513f
Asymmetric drainage, 11
Attenuation, 415, 416
Automated shake tests, 35. *See also* Foam
Axisymmetric drainage, 11

B

Bartsch method, 33–34, 34f. *See also* Foam
Betaine foam booster, 182
Bikerman method, 314f
Binary liquid mixtures, partial miscibility in, 185–188, 185f, 186f, 188f. *See also* Partial miscibility
Binomial distribution, 335
Biopsy channel of colonoscope, 532, 533f
Blade foam breaker, 392, 393f
Blood treatment during bypass surgery. *See* Defoaming, medical applications of
Boltzmann constant, 223, 371, 463
Boylan's product, 252
Boyle's law, 25
Brazil nut effect, 1, 456
Brewster angle microscopy, 181f, 190
Bridging coefficients, 145, 146, 150, 172t, 180, 181t, 197, 252, 253, 254, 263, 516, 523

Bridging–dewetting mechanism, 151f, 152, 218, 520
Bridging drops, 149f, 150, 151f
Bridging–stretching mechanism, 152, 152f, 153, 185, 356, 364, 366, 535
Bright-field photomicrograph, 538, 538f
Bubble(s), 234, 403
  coarsening, 24–28, 25f, 27f. *See also* Foam
  concentration, 377
  in film, 492f
  at foam–water interface, 17
  movement, 1
  shear/impact forces on, 404, 405f, 406f
Bubble size distributions, 321, 330, 332, 333
  measurement of, 39–41, 40f, 41f. *See also* Foam
Buoyancy, neutral, 464
Butylcyclohexane, 89
2-butyloctanol, 116, 117, 171, 172, 173

C

Calcium alkyl carboxylates, 128
Calcium alkyl phosphate particles, 367
Calcium laurate, 245
Calcium mono-alkyl stearate, 443f
Calcium mono-stearyl phosphate, 442f, 443f
Calcium oleate, 245, 246f, 272
Calcium palmitate, 245
Calcium soap, 439
  antifoam effects of, 243–247, 244f, 246f. *See also* Capillary theories (for aqueous systems)
  precipitation, 435, 461
Calcium stearyl acid phosphate, 442
Calcium stearyl phosphate, 250, 254
Calcium stearyl phosphate–liquid paraffin, 313, 317, 326
Capillary pressure, 43, 73f, 74, 79, 147f, 148, 401, 402
  critical applied, 73, 73f, 78, 106, 107t
  gradients, 18–21, 19f. *See also* Foam
  in Plateau border, 240
  rugosities on, 236
Capillary theories (for aqueous systems). *See also* Antifoam mechanism
  calcium soaps, 243–247, 244t, 246t
  contact angles and particle bridging mechanism, 203–216, 204f, 205f, 206f, 207f, 208f, 209f, 210f, 211f, 213f, 214f, 215f, 216f
  hydrophobic materials, melting of, 247–249, 248f
  particle geometry, 216–228, 217f, 218f, 220f, 222t, 226f, 227f
  particle size and kinetics of foam film rupture, 239–243, 241f
  rugosities
    effect of, 228–233, 229f, 230t, 231f
    and stability of air–water–solid films, 233–238, 234f, 235f, 236f, 238f
Carbon dioxide, 26
Carcinogens, 481
Cardiopulmonary bypass surgery, blood treatment and. *See* Defoaming, medical applications of
Cardiotomy, 529. *See also* Defoaming, medical applications of
  defoaming, 539–540, 540f
  reservoirs. *See also* Polydimethylsiloxane-based antifoams
    blood treatment and, 540–542, 541f
    replacements for PDMS–hydrophobed silica in, 542–544
Castor oil, 543
Centrifugal device for de-aerating blood, 545
Centrifugal force, 395, 396, 400–404, 401f. *See also* Defoaming, mechanical methods for
Centrifugal rotary plate foam breaker, 398, 398f, 402
Cetyltrimethylammonium bromide, 208
Chemical potential, 26, 95
Chlorofluorocarbons, 542
Cinematographic film frames, high-speed, 214f
Cloud point, 186f, 187, 194–198, 195, 195f, 432, 469. *See also* Partial miscibility
Coagulation efficiency factor, 371
Coalescence, 321, 333, 357, 359, 361, 363, 369, 379, 407, 462, 514
Coalescence–splitting processes, 380, 383f
Coarsening, 24–28, 25f, 27, 27f. *See also* Foam
Coking of crude oils, 524
Collection efficiency of rough quartz particles, 235, 235f, 238
Colonoscopy, 533f
Comicellization, 436, 439
Commercial rotary defoamers, 398–400, 398f, 399f. *See also* Defoaming, mechanical methods for
Common white film, 62, 65, 68
Compatibilizing agent, 448
Complete wetting, 58, 93f, 141, 274, 286
Compound antifoams, 346
Compression, 135
Contact angles
  of antifoam particles, 288t
  measurement, 95, 208, 209, 287
  and particle bridging mechanism, 203–216, 204f, 205f, 206f, 207f, 208f, 209f, 210f, 211f, 213f, 214f, 215f, 216f. *See also* Capillary theories (for aqueous systems)
Coriolis force, 400
Cospraying, 455

Cost savings, 459
Couette device, 362f
Crater formation. *See also* Drying paint films, defect formation in
by antifoams, 493, 493f, 494
by non-spreading antifoam oils, 497–498
by spreading antifoam oils, 495–497
Critical aggregation concentration (CAC), 121, 123
Critical applied capillary pressure, 73, 73f, 78, 106, 107t
Critical capillary pressures, 160, 172t, 181t, 182, 227, 270t, 271t, 402
Critical micelle concentration (CMC), 7, 18, 118, 488
Critical nanoaggregate concentration, 508
Critical thickness, 510
Critical wetting, 84, 505, 505t
Crude oil, 519
coking of, 524
degassed, 505, 506, 511f, 513, 514f, 517, 518f, 522f
viscosity, 525
Crystal-like geometry, 280
Crystalline particle geometry, 212
Crystalline particles, 128
Cyclone foam breaker, 399f
Cyclones in defoaming, 399
Cylinder shaking, 175, 231, 232, 313, 320
test, 530

D

Darcy's law, 23
Deactivation, 343, 379, 451
by disproportionation, 459
of simethicone, 530
storage, 453–456. *See also* Polydimethylsiloxane-based antifoams
Deactivation of mixed oil–particle antifoams
foam volume growth, theories of
antifoam–bubble heterocoalescence–kinetic model, 368–372, 369f, 373f
kinetic models of antifoam deactivation/antifoam action, 376–379, 378f, 379–383, 382f, 383f
statistical distribution, use of, 373–376, 375f
of hydrophobed silica–polydimethylsiloxane antifoams
deactivation, emulsification, and drop sizes, 348–351, 350f
by disproportionation, 351–363, 353f, 354t, 355f, 360t, 362f
equilibration and deactivation, 347–348
oil viscosity on, 363–366, 365f
silica separation from oil, 346–347, 347f
of polydimethylsiloxane oils without particles, 344–345, 345f
Dead oils, 505, 517. *See also* Degassed crude oils
Debye screening, 291f
Decane, 91, 92f, 118, 119, 119f, 120, 120f, 176
Decontamination, 529
Decyltrimethylammonium bromide, 93f
Defoamer box, 395
Defoamers, 529, 544–545. *See also* Polydimethylsiloxane-based antifoams
Defoaming, mechanical methods for
packed beds of appropriate wettability, 421–422
rotary devices, use of
centrifugal force, role of, 400–404, 401f
commercial rotary defoamers, 398–400, 398f, 399f
designs in scientific literature, 389–398, 391f, 392f, 393f, 394f, 395f, 396f, 397f
inherent liquid spray, 407–408, 408f
shear/impact forces on bubbles, 404–407, 405f, 406f
ultrasound, use of
background, 409–415, 410f, 411f, 412f
defoaming mechanism of ultrasound, 415–421, 415f, 417f
Defoaming, medical applications of
blood treatment during bypass surgery
cardiotomy defoaming, 539–540, 540f
cardiotomy/venous reservoirs, 540–542, 541f
gas bubble oxygenators, 533–535, 534f
membrane oxygenators, 540–542, 541f, 542f
PDMS–hydrophobed silica antifoam, 537–539, 538f
PDMS–hydrophobed silica-coated porous defoamers, 535–537, 536f
polydimethylsiloxane-based antifoams, 542–549, 549f
gastrointestinal gas treatment, simethicone antifoam in
simethicone in endoscopy, 532–533, 533f
therapeutic application, 530–532
overview of, 529–530
Defoaming cyclones, 399
Degassed crude oils, 505, 506, 511f, 513, 514f, 517, 518f, 522f
Demisters, 504f
Derjaguin–Landau–Verwey–Overbeek (DLVO) theory, 17
Detergent powder
composition, 449f, 452f
storage conditions, 449f
Detergent products, antifoams for
hard-surface cleaning products, 469–470, 470t

liquids for machine washing
polyorganosiloxane–hydrophobic silica antifoams in, 462–467
properties of, 461–462
machine dishwashing, 467–469, 467f
overview of, 431–433, 432f
powders for machine washing
dimethylsiloxane-based antifoams, 460–461
fatty acids and soaps, 435–439, 435f, 437f, 438f
front-loading drum-type textile washing machines, 433–434, 434f
hydrocarbon–hydrophobic particle mixtures, 441–450, 442f, 443f, 445f, 446f, 449f
non-soap particulate antifoams, 439–441, 440f
polydimethylsiloxane-based antifoams, 450–460, 452f, 458t, 460f
Dewetting, 215, 496
Diagonal orientation, 225
Dialkyl phosphoric acid esters, 439
Dielectric constant, 80, 509
Diffusional disproportionation, 24–28, 25f, 27, 27f, 28. *See also* Foam
Diffusion coefficient, 25
Dihedral angle, 89, 89f, 176, 218
Dilatational effects, 17
Dimethylsiloxane-based antifoams, hydrocarbon-based simulation of, 460–461. *See also* Detergent products, antifoams for
Dimple, 9, 10, 10f
Dippenaar cells, 44–45. *See also* Single foam films
Direct air injection, 38–39, 39f. *See also* Foam
Disc turbine, 392
Dishwash detergent formulations, 468
Dishwashing liquid, 382, 382f
Disjoining forces, 14–18, 15f, 16f
Disjoining pressure, 509–510. *See also* Gas–oil separation; Oils at interfaces
and stability of pseudoemulsion films, 72–78, 73f, 75f, 77f
Disjoining pressure isotherm, 64, 65f, 66f, 68, 70, 77, 77f, 82, 84, 121, 141, 193, 234, 402, 509
measurement of, 43–44, 48, 48f. *See also* Air–water–oil pseudoemulsion film; Single foam films
Dispensing (detergent powders), 456–457. *See also* Polydimethylsiloxane-based antifoams
Dispersion, 197
of ethylene distearamide particles, 367
of hydrophobic particles, 247
mixed, 313–318, 314f, 317f
single antifoam, 310, 311f, 312f
Disproportionation, 53, 459
deactivation and, 351–363, 353f, 354t, 355f, 360t, 362f
Dissolution, 548
Dodecane, 71, 73, 93, 102, 276, 510
lens, 156t
Dodecanol, 11, 116, 172
Dodecyl trimethylammonium bromide, 88, 90f
Doppler ultrasonography, 542f, 548
Double-walled lamellar surfactant vesicle, 123f
Down-pumping impellers, 398
Drainage, 9–14, 10f, 11f, 12f, 13f
foam, 21–24, 22f
Drive shaft, 395
Drop melting point, 444, 446
defined, 444
Drop size, antifoam, 349–351, 350f
Drop splitting, 379
Drying paint films, defect formation in. *See also* Oil-based antifoams
assumptions, 492–495, 493f, 494f, 495f
cratering by Marangoni effect, 495–497
putative craters by non-spreading oil drops, 497–498
Dry powder, 259
Duplex film, 58, 61, 82, 83, 93f, 97, 104, 138, 139
formation, 145, 146
spreading, 133, 134
Duplex film-forming oil, 275f
Durability, defined, 354, 355f
Dynamic bubble trap, 548, 549f
Dynamic wetting effects, 276

E

Edge angles, 218
Electron micrograph
calcium stearyl phosphate–liquid paraffin, 268f
of hydrophobed silica, 268f
Electrostatic forces, 64
Electrostatic repulsions, 65, 121, 233
Ellipsometry, 172
Emboli, 529, 533, 537–539
Emulsification, 353f, 536. *See also* Deactivation of mixed oil–particle antifoams
and deactivation/drop sizes, 348–351, 350f
Emulsified liquids. *See* Antifoam mechanism
Emulsion(s), 462
drop splitting, 363
stabilizer, 465
Endoscopy, simethicone in, 532–533, 533f. *See also* Defoaming, medical applications of
Energy minimization technique, 153, 163, 225

Entry coefficients, 141, 154, 157, 166, 172t, 523. *See* Oils at interfaces
  equilibrium, 59, 61
  PDMS, 97
Epichlorohydrin, 487
Equilibration and deactivation, 347–348. *See also* Deactivation of mixed oil–particle antifoams
Equilibrium spreading coefficients, 93, 104
Ethanol, 166
Ethoxylated alcohol, 453, 470t
Ethoxylated nonyl phenol surfactants, 438
Ethylene distearamide, 367, 447
Ethylene glycol, 198, 199f
Evaporation, 484, 484f, 493f

F

Fatty acids and soaps, 435–439, 435f, 437f, 438f, 446
Fick's first law, 24
Film rupture probability, 208f
  by octadecyltrichlorosilane-treated glass, 209f
Film tension, 62, 63, 79
Film trapping technique (FTT), 72, 73f, 159, 160, 291f
Filter, 545, 546
Filtration, 127, 196
Fish-eye defect, 494, 495
Fish eyes, 261, 262f
Fission, bubble, 404, 405, 406f, 407
Foam
  aging of
    bubble coarsening, 24–28, 25f, 27f
    capillary pressure gradients, 18–21, 19f
    foam drainage, 21–24, 22f
  and antifoam behavior, 485–487. *See also* Waterborne latex paints and varnishes, foam control in
    stratified layers of polymer latex particles, effect of, 488–489, 489f, 490t–491t
  behavior, 511–515, 511f, 513f, 514f. *See also* Gas–oil separation
  boosters, 180, 182
  collapse, 313
  control, 94, 389, 442
    in waterborne latex paints/varnishes. *See* Waterborne latex paints and varnishes, foam control in
  destabilization, 117
    by neat oil drops, 161
  dry polyhedral, 40f
  formation, 188
    in gas–crude oil systems. *See* Gas–oil separation
  generation, 322f
  lifetime, average, 511, 511f
  measurement of, 172
    automated shake tests, 35
    Bartsch method, 33–34, 34f
    bubble size distributions, measurement of, 39–41, 40f, 41f
    direct air injection, 38–39, 39f
    gas bubbling, 37–38
    by Ross–Miles apparatus, 35, 36f, 317, 317f
    tumbling cylinders, 36–37, 37f
  polydisperse, 1
  polyhedral, 1, 3f
  rise method, 377, 378f
  stability, 5, 120, 124f, 192, 195f
    hydrophobed spherical glass particles on, 211
  stability measurements, 173
  structure of, 1–5, 2f, 3f, 4f
  volume, 382, 382f
  volume growth, theories of. *See* Deactivation of mixed oil–particle antifoams
  wet, 40f
Foamability, 5, 117, 119, 313, 438, 514f
  control of, 309
  of mesophase precipitation, 121–128, 122f, 123f, 124f, 125f, 126f, 127f. *See also* Antifoam mechanism
  and turbidity, 122f
Foam-breaking cyclones, 399
Foam film. *See also* Single foam films
  antifoam drops in, 41, 42f
  collapse, 16, 21, 119, 135, 178
  destabilization, 137
  disjoining forces, 14–18, 15f, 16f
  drainage, 132, 132f, 140, 325, 349
    processes in, 9–14, 10f, 11f, 12f, 13f
    time, 11f
  oil bridges in, 141–153, 142f, 143f, 147f, 148f, 149f, 150f, 152f. *See also* Antifoam mechanism
  rupture, 130f, 151f, 165, 205f, 214, 239–243, 241f, 395, 396, 403, 408, 408f. *See also* Capillary theories (for aqueous systems)
    air–water–air, 222t
  stability, 5–9, 18, 121
    polymer latex particles on, 488–489, 489f, 490t–491t
  surface tension gradients, 5–9
Foam generation, 344, 366, 379f, 390
  by cylinder shaking, 232
  by hand shaking, 216f, 254
  by passing nitrogen through porous frit, 373f, 375, 375f
  pneumatic method of, 184
  by Ross–Miles method, 355f
  by sparging, 368

Foaming, 503
minimization of, 389
Foam-inhibiting agent, 167
Foamy-oil, 513
Fourier transform infrared spectroscopy, 531
Free drainage, 21, 23
Free energy, 185, 185f, 187, 188f
excess, 153
Front-loading drum-type textile washing machines, 433–434, 434f. *See also* Detergent products, antifoams for
Frothers, 212
Froth flotation, 234
Fuel cell cathode, 422
FUNDAFOM system. *See* Commercial rotary defoamers

G

Gas-bleeding fitting, 398
Gas–blood surface tension, 546
Gas bubble oxygenators, 533–535, 534f. *See also* Defoaming, medical applications of
Gas bubbling, 37–38. *See also* Foam
Gas diffusion, 20, 21
Gas–oil separation
antifoams, use of
polydimethylsiloxanes/substituted polydimethylsiloxanes, 517–524, 518f, 520f, 521f, 522f, 523f
siloxane chemistry, 524
foam formation, causes of
disjoining pressures, 509–510
foam behavior, 511–515, 511f, 513f, 514f
surface tension gradients, 510–511
overview, 503–504, 504f
surface activity, 504–509, 505t, 507f
Gas shearing, 406, 406f
Gastrointestinal gas treatment, simethicone antifoam in. *See* Defoaming, medical applications of
Gel hydrocarbons, 443, 444
Gibbs adsorption equation, 504
Gibbs–Duhem equation, 82
Gibbs elasticity, 7, 8f, 407
Gibbs equation, 95
Globular protein, 108
Glucopyranoside, 292
Gravitational force, 22
Gravity drainage, 132, 132f
Greenhouse gases, 481

H

Half-life ratios, 211, 211f
Hamaker constants, 64, 75, 80, 85, 106, 233
Hand shaking measuring cylinders, 33–34, 34f. *See also* Foam
Hard-surface cleaning products, 469–470, 470t. *See also* Detergent products, antifoams for
Hard water, 435, 439
on foam stability, 470t
Heat exchanger, 541
Heavy crudes, 507, 513
Helmholtz free energy, 215
Henry's law, 25, 26
Henry's law constant, 545
Heptanol, 172
Heterocoalescence, 370, 377, 442, 446
Hexadecane, 88, 89f, 179, 272, 367, 488
lens, 156t
Hexadecyl ammonium bromide, 175
Hexadecyl trimethylammonium bromide, 209f
Hexamethylhexacosane, 90
Hexane, 518f, 519
High-speed video-microscopy, 261
Hydration forces, 65
Hydrocarbon
as antifoams, 85
chemical potential of, 87
entry/spreading coefficients for, on pure water, 86t
gel, 443, 444
nonvolatile, 118
particles, destabilizing effect of, 202
Hydrocarbon–ethylene distearamide, 489, 492
Hydrocarbon–hydrophobic particle mixtures. *See also* Detergent products, antifoams for
with alkyl phosphoric acid derivatives, 441–446, 442f, 443f, 445f, 446f
with non-phosphorous-containing organic compounds, 446–450, 449f
Hydrocarbon–hydrophobic silica antifoams, 451, 453, 455
Hydrocarbon oils. *See also* Oils at interfaces
on aqueous surfactant solutions surfaces
complete/pseudo-partial wetting behavior, 87–94, 88f, 89f, 90f, 92f, 93f
non-spreading (partial wetting) by, 94–96
on pure water surface
spreading/wetting behavior of, 85–87, 86t
Hydrodynamic instabilities, 12, 12f, 239, 329, 403
Hydrogen sulfide, 519
Hydrophobed glass particles, antifoam effect of, 216f
Hydrophobed silica, 268f, 269f
antifoam durability, 35
in liquid paraffin, 366
on pseudoemulsion film stability, 270

Hydrophobed silica–polydimethylsiloxane antifoam. *See* Deactivation of mixed oil–particle antifoams
  dispersion, 354t
Hydrophobic forces, 64, 77, 80, 233
  short-range, 75
Hydrophobicity, 289, 290, 290f, 291f
Hydrophobic materials
  dispersion of, 247
  inert, and capillary theories. *See* Antifoam mechanism
  melting of, 247–249, 248f. *See also* Capillary theories (for aqueous systems)
  and oils, mixtures of. *See* Antifoam mechanism
Hydrophobic oils, 344
Hydrophobization of silica, 291f
Hysteresis, 287, 498

I

Impact forces, 404
Inherent liquid spray, 407–408, 408f. *See also* Defoaming, mechanical methods for
Initial entry coefficient, 95
Initial spreading coefficient, 59, 82, 95, 96
In situ hydrophobing, 448, 461
Interdigitation model, 93, 103
Intrusive methods, 39
Isobutene, 367
Isohexyl-neopentanoate, 117
Isopropanol, 543
Isotherm, 14, 15f, 16, 68

K

Kelvin cells, 2, 3f, 4f, 5
Kinetic energy, 223
Kinetic models. *See also* Deactivation of mixed oil-particle antifoams
  antifoam–bubble heterocoalescence model of Pelton and Goddard, 368–372, 369f, 373f
  of antifoam deactivation/antifoam action (combination of), 376–379, 378f
  antifoam deactivation by disproportionation/ antifoam action (combination of), 379–383, 382f, 383f
  statistical distribution, use of, 373–376, 375f
Kolmogorov turbulence theory, 406
Krafft temperature, 127, 128
Kugelschaum, 1

L

Laminar flow, 234f
Laplace equation, 37, 204
Laplace pressure jump, 143
Latex, polymer-based, 465
Lecithin, 543
Lenses of PDMS–hydrophobed silica antifoam, 261f
Lens formation, 252
Light microscopy, 538
Light scattering, 117, 350, 362
Line tensions, 153–157, 154f, 156t. *See also* Antifoam mechanism
Liquid paraffin, 249, 250, 266t, 366
Liquid phase, 400
Liquids for machine washing of laundry. *See* Detergent products, antifoams for

M

Machine dishwashing, 467–469, 467f. *See also* Detergent products, antifoams for
Magnesium hydroxide, 531
Marangoni spreading mechanism, 129, 130, 130f, 131, 137, 140, 166, 167, 182, 251, 254, 258, 263, 347, 348, 492, 493, 495–498, 520
Marginal regeneration, 12, 13, 13f, 14
Mass transfer coefficient, 545
Mechanical defoaming. *See also* Defoaming, mechanical methods for
  by rotary devices (studies of), 424t–427t
Medium gas oil, 521
Melting of hydrophobic materials, 247–249, 248f
Membrane oxygenators, 540–542. *See also* Defoaming, medical applications of
Mesophase
  precipitation, foamability of, 121–128, 122f, 123f, 124f, 125f, 126f, 127f. *See also* Antifoam mechanism
  viscosity of, 456
Metastable pseudoemulsion film, 67
Methanol, 166
Methyl acetate, 198, 199f
Methyl acetate–ethylene glycol
  entry, spreading, and bridging coefficients for, 200t
Methylene distearamide, 447
Methyl isobutyl carbinol, 117
Micellar solution, 197
Micellar surfactant solution, 93, 100t, 103, 104
Microbubbles, 544, 547, 548
Microcrystalline hydrocarbon waxes, 440
Microemboli, 534
  removal, gaseous, 545–549, 549f. *See also* Polydimethylsiloxane-based antifoams
Microlenses, 87
Microprobe spectroscopy, 538
Molecular packing, 121
Monoalkyl phosphoric acid esters (MAPAE), 439, 445f

Monodisperse foam generation, 377
Monomers, 481, 482
Motor-driven rotation, 398

N

n-alkanes, 107t, 118, 175–180, 177f, 178f, 179f, 249. *See also* Neat oils
  classic entry/spreading/bridging coefficients for, 176t
Navier–Stokes equation, 6
n-butanol, 168
n-docosane, 247
Neat oils. *See also* Antifoam mechanism
  in aqueous foaming systems, 165–167
    n-alkanes, 175–180, 176t, 177f, 178f, 179f
    neat polydimethylsiloxane oils, 180–183, 181t
    short-chain alcohols, 167–175, 169f, 172t, 173f, 174f
  in non-aqueous foaming systems, 183–185
Neat polydimethylsiloxane oils, 180–183, 181t. *See also* Neat oils
  entry/spreading/bridging coefficients for, 181t
Neumann's triangle, 63, 63f, 142, 142f, 146, 154, 154f, 164, 164f, 217
Neutron reflection measurements, 90, 90f
Newton black films, 17
Newton white films, 62, 65, 76
n-heptane, 507
n-heptanol, 168
Nitrilotriacetic acid, 36
Nitrogen, 118, 547
n-octane, 76, 85, 87
Nonintrusive methods, 39
Non-ionic ethoxylated alcohols, 92, 94
Non-phosphorous-containing organic compounds, 446–450, 449f. *See also* Hydrocarbon–hydrophobic particle mixtures
Non-soap particulate antifoams, 439–441, 440f. *See also* Detergent products, antifoams for
Non-spreading by hydrocarbons, 94–96. *See also* Hydrocarbon oils
Nonyl phenol ethoxylate, 191
Nuclear magnetic resonance (NMR) spectroscopy, 247, 445, 446f

O

Octadecyltrichlorosilane, 209
Octane lenses, 156t
Octanol, 170, 171
Octylphenyl ethoxylate, 189, 189f, 194
Oil-based antifoams. *See also* Waterborne latex paints and varnishes, foam control in
  drying paint films, defect formation in, 492–498, 492f, 493f, 494f, 495f
  incorporation in paints/varnishes, 489, 492
Oil bridges. *See* Antifoam mechanism
Oil drops in Plateau borders, 164f
Oil in synergistic oil–particle antifoams, 252–263, 253f, 255t, 257f, 258f, 260f, 261f, 262f
Oil lenses, antifoam behavior of, 153–157, 154f, 156t. *See also* Antifoam mechanism
Oil-on-water wetting, 177
Oil–particle antifoam mixture, 288t
Oils, antifoam, 52–53
Oils at interfaces
  definitions, 61–65, 63f, 65
  entry coefficients, 58–61
    greater than zero, 65–68, 66f, 67f
    less than zero, 68–70, 69f
    magnitude of, 70–72, 71f
  pseudoemulsion films
    disjoining pressures and stability of, 72–78, 73f, 75f, 77f
    surface tension gradients and stability of, 78–79
  pseudo-partial wetting, 58
  spreading behavior on aqueous surfaces
    effect on pseudoemulsion films, 106–107, 107t
    hydrocarbon oils, 85–96, 86t, 88f, 89f, 90f, 92f, 93f
    polydimethylsiloxane (PDMS) oils, 96–106, 97f, 98t, 100t–101t, 102f, 105f, 106f
  spreading coefficients, 58–61
    and thin film forces, 79–85, 81f, 83f
  surfactant transport, 108
Oil viscosity on deactivation, 363–366, 365f. *See also* Deactivation of mixed oil-particle antifoams
Oil–water surface, 283t
Oleic acid, 367
Optical microscopy, 118
Optical photomicrograph of hydrophobed silica, 268f
Organic acids, 457
Organopolysiloxanes, 467
Orthorhombic galena particles, 212, 213
Oscillating jet method, 170
Overfoaming, 432f
Oxygenators. *See also* Defoaming, medical applications of
  gas bubble, 533–535, 534f
  membrane, 540–542

P

Packed beds of wettability, 421–422
Paint films, 492f, 493
  crater formation in, 493f

Paints and varnishes, oil-based antifoams and, 489, 492
Palmitic acid, 244
Paraffin, liquid, 91, 442, 443f
Partial miscibility, 171. *See also* Antifoam mechanism
  with higher critical temperatures, 198–201, 199f, 200t
  with lower critical temperatures
    cloud point antifoams, 194–198, 195f
    phase separation, 188–194, 189f, 191f
  origins of, in binary liquid mixtures, 185–188, 185f, 186f, 188f
Partial wetting, 58, 82, 94, 141, 252, 496, 515
  by hydrocarbons, 94–96. *See also* Hydrocarbon oils
Particle bridging mechanism. *See* Contact angles
Particle collection efficiencies, 235, 238
Particle-free drops, 357
Particle geometry. *See also* Antifoam mechanism
  with curved surfaces/no edges, particles, 216
  with no edges, particles, 216–224, 217f, 218f, 220f, 222t
  surface energy minimization technique, 224–228, 226f, 227f
Particles, undissolved, 116
Particles in synergistic oil–particle antifoams, 263–266, 265f, 266t. *See also* Antifoam mechanism
  experimental observation, 267–272, 268f, 269f, 270t, 271f
  pseudoemulsion film, rupture of, 272–277, 273f, 277f
  rough particles with edges, 281–290, 282f, 283t, 284f, 286f, 288t
  smooth particles with edges
    in absence of spread oil layers, 277–281, 278f, 280t
    in presence of spread oil layers, 281
  spherical particles, 272, 275f
  spread oil layer, 272, 277
Particle size distribution, 360
Particle size/kinetics of foam film rupture, 239–243, 241f. *See also* Capillary theories (for aqueous systems)
Patents for polysiloxane-hydrophobed silica antifoams in detergents, 472t–475t
PDMS. *See* Polydimethylsiloxane
Perfluoroalkylalkanes, 461
Perfluoroalkyl polymeric ester, 509, 510
Permeability, 23
Phase separation, 188–194, 189f, 191f. *See also* Partial miscibility
Phenolalkanoic acids, 507
Photodiode, 196
Photomicrographs, 350, 350f, 353f, 362f
Photo-polymerizable hydrophobic oil, 216–217
Pickering emulsions, 363, 465
Piezoelectric ultrasonic transducer, 409
Pitched-blade turbine, 392f, 393f
Placebos, 532
Plastic-crystalline rotator phases, 249
Plateau borders, 1, 2, 3f, 4, 4f
  collapse, 345
  oil bridges in, 157–165, 158f, 159f, 160f, 163f, 164f
  pressure, 19
Poiseuille flow, 13, 21, 22, 239, 240
Polyalkoxylated-polyorganosiloxane block copolymers, 487
Polyalkylsiloxane-polyoxyalkylene emulsion stabilizer, 465
Polydimethylsiloxane (PDMS), 35, 116, 345f, 363
  alkoxylated, 497, 521f
  chemical structure, 520f, 521f
  ethoxylated, 519, 523, 523f
  perfluoroalkyl-substituted, 506, 519
  solubilization of, 519
  viscosity, 181t, 452f, 458t
Polydimethylsiloxane-based antifoams. *See also* Defoaming, medical applications of; Detergent products, antifoams for
  antifoam effectiveness enhancement, 457–460, 458t, 460f
  defoamer elements for high air-blood contact angles, 544–545
  dispensing, 456–457
  gaseous microemboli, removal of, 545–549, 549f
  properties of, 450–453, 452f
  replacements for PDMS–hydrophobed silica, 542–544
  storage deactivation, 453–456
Polydimethylsiloxane–hydrophobed silica, replacements for, 542–544. *See also* Polydimethylsiloxane-based antifoams
Polydimethylsiloxane–hydrophobed silica antifoam. *See also* Defoaming, medical applications of
  as source of emboli, 537–539, 538f
Polydimethylsiloxane–hydrophobed silica-coated porous defoamers, 535–537, 536f. *See also* Defoaming, medical applications of
Polydimethylsiloxane–lube oil combinations, 133
Polydimethylsiloxane (PDMS) oils, 132, 139. *See also* Oils at interfaces
  complete and pseudo-partial wetting behavior, 96
  deactivation of, 344–345, 345f
  entry and spreading coefficients
    on aqueous micellar surfactant solutions, 100t–101t
    on pure water, 98t

molecular structure, 97f
neat, 180–183, 181t. *See also* Neat oils
radial spreading rate of, 105, 106f
Polydimethylsiloxanes in gas–oil separation. *See also* Gas–oil separation
mode of action of antifoam oils, 520–524, 521f, 522f, 523f
solubility of antifoam oils, effect of, 517–520, 518f, 520f
Polydispersity, 1, 18, 24, 38, 326, 359, 360
Polyederschaum, 1
Polyethoxylated–polyoxypropylene, 438
Polyethoxy–polypropoxy block copolymers, 196
Polyethylene glycol–polypropylene glycol (EO–PO), 194
Polyhedral foam (polyederschaum), 2f, 3f
Polymerization, 481, 482f
Polymer latex particles on pseudoemulsion film stability, 488–489, 489f, 490t–491t
Polyorganosiloxane derivatives, 503
Polyorganosiloxane–hydrophobic silica antifoams in detergent liquids, 462–467. *See also* Detergent products, antifoams for
Polyorganosiloxanes, 454, 496
Polyoxyethylene, grafted, 184
Polyoxyethylene–polyoxypropylene block copolymers, 543
Polypropylene glycols, 192
Polysiloxane–hydrophobed silica antifoams
patents, examples of, 472t–475t
Polytetrafluoroethylene (PTFE), 203, 204f, 208f, 215, 321, 422
Polyurethane foam, open-celled, 534, 539, 540f, 544
Polyvinyl alcohol, 409
Powders for machine washing. *See* Detergent products, antifoams for
Precipitation
of lamellar phase, 124, 125, 126f, 128
phase diagram, 124, 125f
Prediction of theory, 373f
Primary alcohols, 171
Primary bubbles, 369, 369f, 370, 374
Primary foam, 403
Primary surfactant, 438
Probability theory, 335
Pro-foamers, 517, 523
Progressive waves, 416
Propylene glycol monoester, 487
Pseudoemulsion films, 18, 168, 516, 517. *See also* Air–water–oil pseudoemulsion film; Oils at interfaces
disjoining pressures and stability of, 72–78, 73f, 75f, 77f
metastability of, 145, 157
non-rupture of, 290f
rupture of, 272–277, 273f, 277f. *See also* Particles in synergistic oil-particle antifoams
spreading behavior on, 106–107, 107t
stability of, 115, 157–165, 344. *See also* Antifoam mechanism
polymer latex particles on, 488–489, 489f, 490t–491t
surface tension gradients and stability of, 78–79
Pseudo-partial wetting, 58, 61, 80, 81f, 82, 84, 87, 91, 93f, 99, 141, 176, 252, 257, 263, 274, 276, 286, 516, 535. *See also* Oils at interfaces
Putative craters by non-spreading oil drops, 497–498. *See also* Drying paint films, defect formation in

Q

Quartz particles, 235f

R

Radial accelerator foam breaker, 392, 393f
Rapid foam collapse, 313
Recycled paper system, 197
Redox electrolyte, 422
Refractive index, 462
Resins, 507, 508
Reynolds drainage, 329, 330
and antifoam concentration, 338
Reynolds equation, 9, 16, 240, 243, 327, 328, 328f
Reynolds number, 34
Rinse aids, 468
Ross–Miles apparatus, 162, 163f, 254, 312, 317, 317f, 319, 366, 367, 443
Ross–Miles foam generation, 125, 126f
Ross–Miles foam test, 467
Ross–Miles method, 35, 36f, 172, 173f, 179, 180, 191f. *See also* Foam
foam generation by, 355f
Rotary devices, use of. *See* Defoaming, mechanical methods for
Rotary foam breakers, designs of, 390, 391f, 392f
Rotating disc defoamer, 395
Rotating disc foam breakers, 405f
Rotational orientation, 225, 226f
Rotor–stator foam breaker, 392, 393f, 394f, 406f
Rough particles with edges, 281–290, 282f, 283t, 284f, 286f, 288t. *See also* Particles in synergistic oil-particle antifoams
Rugosities. *See also* Capillary theories (for aqueous systems)
effect of, 228–233, 229f, 230t, 231f
and stability of air-water-solid films, 233–238, 234f, 235f, 236f, 238f

Rupture. *See also* Pseudoemulsion films
pressure measurements, pseudoemulsion, 49–52, 49f, 50f, 51f. *See also* Air–water–oil pseudoemulsion
of pseudoemulsion films, 72, 272–277, 273f, 277f
times, 242

S

Scanning and transmission electron microscopy, 538
Scanning electron micrograph, 215f, 483f
Scheludko cells, 179, 246, 262. *See* Single foam films
SDS. *See* Sodium dodecyl sulfate
Secondary alkyl sulfate (SAS), 470t
Secondary antifoam, 459
Secondary bubbles, 369, 369f, 370, 373
Secondary foam, 393, 404
coalescence column, 394f
Sedimentation, 464, 466
Semi-initial entry coefficient, 59
Semi-initial spreading coefficient, 59
Senescence, 5
Shaking tests, 364, 366
Shear forces, 258, 404, 510
on bubbles, 404–407, 405f, 406f. *See also* Defoaming, mechanical methods for
Shear mixer, 312
Shear rates, 535, 536
Shear stress, 133
Short-chain alcohols, 167–175, 169f, 172t, 173f, 174f. *See also* Neat oils
Silane hydrophobed silica, 312f
Silica
hydrophobization, 291f
precipitated, 228
separation from oil, 346–347, 347f. *See also* Deactivation of mixed oil–particle antifoams
weight fraction of, 359
Silica-rich drops, 353f
Silicone waxes, 506
Siloxane chemistry, 524
Silylated polystyrene resin particles, 362f
Simethicone, 529, 530
in endoscopy, 532–533, 533f. *See also* Defoaming, medical applications of
Single antifoam, dispersion of, 310–313, 311f, 312f
Single foam films. *See also* Foam film
Dippenaar cells, 44–45, 45f
Scheludko cells
antifoam drops, 41–43, 42f
disjoining pressure isotherms, measurement of, 43–44
vertical films, large, 45–46, 46f
Six-blade turbine, 391f
Six-blade vaned disc, 391f
Smoluchowski kinetics, 462
Smoluschowski coagulation constant, 371
Smooth particles with edges. *See also* Particles in synergistic oil–particle antifoams
in absence of spread oil layers, 277–281, 278f, 280t
in presence of spread oil layers, 281
Sodium alkylbenzene sulfonate, 36, 123, 254, 255t, 266t, 312, 312f, 313, 314f, 435, 453
Sodium behenate, antifoam behavior of, 435f
Sodium bis-diethylhexyl sulfosuccinate, 91, 488
Sodium bis-ethylhexyl sulfosuccinate, 346
Sodium bis-octyl sulfosuccinate, 122
Sodium caseinate, 108, 135, 136f
Sodium chloride, 122, 123
Sodium dodecylbenzene sulfonate, 107, 116, 117, 120, 170, 173, 173f, 174, 174f, 244f, 393f, 411, 412
Sodium dodecyl 4-phenylsulfonate, 124, 125f, 126f, 127f
Sodium dodecyl polyoxyethylene sulfate, 163
Sodium dodecyl-polyoxyethylene-3-sulfate, 159f
Sodium dodecyl sulfate (SDS), 7, 11, 70, 117, 128, 346, 410f, 413
Sodium dodecyl sulfate (SDS)-dodecane system, 71
Sodium laurate, 244
Sodium lauryl sulfate, 267
Sodium 6-phenyltridecanesulfonate, 123
Sodium soap, 435, 470t
Sodium sulfosuccinate, 524
Sodium tridecylbenzene sulfonate, 231, 247
Sodium tripolyphosphate, 436, 439
Solid particles, 201
Solitary waves, 23
Solubilization, 91, 93, 117, 118, 121
Solubilized oils, 116–121, 119f, 120f. *See also* Antifoam mechanism
Solution gas drive, 513
Sorbitan monostearate, 291, 383f
Sound wave propagation, 415, 417f
Sparging, 38, 212, 232, 368, 380, 381, 409, 413, 422, 461, 511f, 518f
Spherical particles, 272, 275f. *See also* Particles in synergistic oil–particle antifoams
Spinning cone, 402, 403
Spinning cone foam breaker, 396, 396f, 397f
Splitting, 361
Spray drying, 441, 444, 447, 450, 454
Spreading coefficients, 58–61, 136, 137, 142, 146, 154, 155, 166, 172t, 181t, 190, 516, 522, 523. *See also* Oils at interfaces
near-zero, 280
semi-initial, 257, 258f
and thin film forces, 79–85, 81f, 83f

Spreading mechanism, 129–137, 130f, 132f, 136f, 139
on aqueous surfaces. *See* Oils at interfaces
Spreading pressure
comparison of, 93, 93f, 102f
defined, 82
Spread oil layer, 272, 277–281, 287, 290f. *See also* Particles in synergistic oil-particle antifoams
Sprinkler, 407
Squalane, 91, 95, 118
Statistical distribution of antifoam drops, 373–376, 375f. *See also* Deactivation of mixed oil–particle antifoams
Stearic acid, 231f, 313
Steric forces, 65, 92
Stirred vessel, laboratory-scale, 390, 391f
Stirring as foam disruption, 398
Stratification, 76
Superquadratic equation, 225
Surface activity at gas–crude oil interfaces, 504–509, 505t, 507f. *See also* Gas-oil separation
Surface chemistry, 313
Surface crystallization, 510
Surface dilatational modulus, 136
Surface elasticity, 184
Surface energy, 506
minimization technique, 224–228, 226f, 227f. *See also* Particle geometry
Surface Evolver software, 153, 163, 164f, 224, 225, 226f, 227f
Surface rheology, 327
Surface shear viscosity, 11f, 12f
Surface tension, 37, 59, 87, 88, 96, 323, 506
air–water, 72
of blood, 544
gradient, 5–9, 115, 139, 140, 200, 494, 496. *See* Antifoam mechanism; Foam film; Oils at interfaces
at gas–crude oil interfaces, 510–511. *See also* Gas–oil separation
and stability of pseudoemulsion films, 78–70
measurements, 252
Surfactant
alcohol ethoxylate, 76
concentration, 88
double-chain, 121
non-ionic ethoxylated alcohol, 92, 94
sodium alkylbenzene sulfonate, 312, 312f
transport, 108. *See also* Oils at interfaces
Symbols, frequently used, 555–556
Synergistic oil–particle antifoams. *See also* Antifoam mechanism
oil in, 252–263, 253f, 255t, 257f, 258f, 260f, 261f, 262f
particles in, 263–266, 265f, 266t
experimental observation, 267–272, 268f, 269f, 270t, 271f
pseudoemulsion film, rupture of, 272–277, 273f, 277f
rough particles with edges, 281–290, 282f, 283t, 284f, 286f, 288t
smooth particles with edges, 277–281, 278f, 280t
spherical particles, 272, 275f
spread oil layer, 272, 277
Synthetic latices, 481
Syringe, 42

T

Teepol solutions, 321
Ternary mixtures, 437
Tessellation of Kelvin cells, 4
Tetralin, 70
Textile washing machines, types of, 433
Thermostatting, 35
Thin film, 6, 7, 9, 15, 327, 328, 485, 495f, 513
forces. *See also* Oils at interfaces
and spreading coefficients, 79–85, 81f, 83f
Toluene, 519
Top-loading agitator type, 433
Top-loading impeller type, 433
Trans-cranial Doppler ultrasonography, 548
Tributyl acetate, 520
Tributyl phosphate, 117
Triglycerides, 256
Trimethylchlorosilane, 346
Triolein, 313, 498
Triton X-100, 196, 208f, 289, 402, 488
Tropospheric ozone levels, 481
Tumbling cylinders, 36–37, 37f. *See also* Foam
Two-blade paddle, 391f

U

Ultrasonication, 531
Ultrasonic defoaming, 414
Ultrasonic generator, 409, 410, 418
Ultrasound, use of. *See* Defoaming, mechanical methods for

V

Van der Waals forces, 14, 15f, 17, 64, 75, 80, 92, 146
Varnishes
formulation of, 485
and paints, oil-based antifoams and, 489, 492
Venous reservoirs, 540–542, 541f. *See also* Defoaming, medical applications of

Vertical films, 45–46, 46f. *See also* Single foam films
Video-enhanced micrograph, 125, 126f
Video-microscopy, 256
Viscosity, 5, 105, 463
of castor oil, 543
crude oil, 525
of heavy oils, 513
of mesophase, 456
oil, on deactivation, 363–366, 365f. *See also* Deactivation of mixed oil–particle antifoams
PDMS, 181t
Volatile organic compounds (VOC), 481
Vortex finder, 400
Vrij theory, 78

W

Washing machines, foam-intolerant, 431
Waterborne latex paints and varnishes, foam control in
foam and antifoam behavior, 485–487
stratified layers of polymer latex particles, 488–489, 489f, 490t–491t
oil-based antifoams
drying paint films, defect formation in, 492–498, 492f, 493f, 494f, 495f
in paints/varnishes, 489, 492
overview of, 481–485, 482f, 483f, 484f
Waterborne latices, 481, 485
Water drop, rupture of, 362f
Water hardness, 435
Wave propagation, 136
Waxes, 440
Wet foam, 40f
Wettability, 465
in antifoam effectiveness, 203, 204
Wetting, 141, 494f, 505, 505t
complete, 58, 93f
partial, 58, 82, 94–96. *See also* Hydrocarbon oils
pseudo-partial, 58, 61, 80, 81f, 82, 84, 87, 91, 93f, 99. *See also* Oils at interfaces
Winsor I microemulsion, 93
Winsor I state, 119

X

Xanthanated crystals, 212

Y

Young equation, 221

Z

Zeolites, 456
Zwitterionic fluorinated betaine surfactant, 259, 289
Zwitterionic surfactants, 128

Printed in the United States
by Baker & Taylor Publisher Services